EDWARD ELGAR

A CREATIVE LIFE

EDWARD ELGAR

(photograph by E. T. Holding)
National Portrait Gallery

EDWARD ELGAR

A CREATIVE LIFE

JERROLD NORTHROP MOORE

Oxford New York Melbourne
OXFORD UNIVERSITY PRESS
1984

Oxford University Press, Walton Street, Oxford OX2 6DP

London New York Toronto
Delhi Bombay Calcutta Madras Karachi
Kuala Lumpur Singapore Hong Kong Tokyo
Nairobi Dar es Salaam Cape Town
Melbourne Auckland

and associated companies in
Beirut Berlin Ibadan Mexico City Nicosia

Oxford is a trade mark of Oxford University Press

Published in the United States
by Oxford University Press, New York

British Library Cataloguing in Publication Data
Moore, Jerrold Northrop
Edward Elgar.
1. Elgar, Edward   2. Composers—England
—Biography
I. Title
780'.92'4   ML410. E41
ISBN 0-19-315447-1

Library of Congress Cataloging in Publication Data
Moore, Jerrold Northrop.
Edward Elgar: a creative life.
Bibliography: p.
Includes index.
1. Elgar, Edward, 1857–1934.   2. Composers—England—
Biography.   I. Title.
ML410.E41M65   1984   780'.92'4 [B]   83-23616
ISBN 0-19-315447-1

Set by Wyvern Typesetting Ltd, Bristol
Printed in Great Britain by
Butler & Tanner Ltd,
Frome and London

212564

The 'taste' of the poet is, at bottom and so far as the poet in him prevails over everything else, his active sense of life: in accordance with which truth to keep one's hand on it is to hold the silver clue to the whole labyrinth of his consciousness . . .

As the whole conduct of life consists of things done, which do other things in their turn, just so our behaviour and its fruits are one and continuous and persistent and unquenchable, so the act has its way of abiding and showing and testifying, and so, among our innumerable acts, are no arbitrary, no senseless separations.

Henry James
Preface to *The Golden Bowl*

Preface

IN the old days of biography, an artist's career was often recorded in two sections, life and works. The division puzzled me, for a creative life can have no general significance apart from its works, and works have no cumulative significance outside the worker's life. What I wanted was some means of combining the two modes, so that each new adventure in theme and form could be understood as a chapter in the spiritual autobiography of the whole man. Such a book would seek its goal neither in musicology nor in the sensual man: rather it would seek the influence of each on each.

Thus I have tried to understand what it is to look at the world through creative eyes, to listen through creative ears. The artist, like the rest of us, is torn by various desires competing within himself. But unlike the rest of us, he makes each of those desires into an element for use in his art. Then he seeks to synthesize his elements all together to form a style. The sign of a successful synthesis is a unified and unique style plain for all to recognize. So it is that a successful style can seem to its audience full of indefinably familiar things—and at the same time invested with godlike power of 'understanding' that is far indeed from the daily round. The process by which a man has forged such a unity is the most profound and most exalted of human stories.

My earliest interest in Elgar did not aim at any such discovery: it came simply from the grip of the music. Then growing familiarity brought home to me the remarkable richness of historical record which surrounds Elgar's music. It opens with his eldest sister's manuscript account of their childhood, beginning on the day of Edward's birth. It continues in books of poetry, prose, quotation, and observation kept by his mother. She also began a book of newspaper cuttings to document Edward's public appearances from the age of nine through that golden era of local journalism and reportage. Those books were carried on by his wife until she was filling a hundred-page folio volume every year at the climax of his career. She kept her own daily diaries throughout the thirty years of their marriage—which were precisely the thirty years of his creative achievement. Later his daughter kept her own diaries and the cutting-books. Together they saved so large a proportion of the letters they received that more than ten thousand survive to this day. The recipients of his replies kept and treasured those. And he himself saved musical sketches from all periods of his life—the record of a ruminative man, fascinated with the course of his own development.

From all this I gradually formed the view that Elgar's music offered, in addition to everything else, the finest spiritual record of a finely documented life. He himself, when asked the 'meaning' of his last large work—his music's culmination—in the Violoncello Concerto, answered: 'A man's attitude to life'.

No book will hinder Elgar's music from speaking and singing for itself. But if this book can tell the story—for those who share his problems, his joys, his fundamental dissatisfaction with the world—of how he scrutinized and explored those experiences until they generated their own harmony, then it may shed light on the generosity of spirit which is creative art.

Contents

⁊ⱷⱷⱷⱨ

Acknowledgements

ᛁᚾᚲᛟᛁᚲᛉᛁ

It is a happy but daunting task to marshal the names of those who have given significant help to the project for this book over more than twenty years. At the head of the list comes the composer's daughter, Carice Elgar Blake, who received all my questions with a welcome and intelligence that were a biographer's dream. Nothing was too much trouble to investigate, nothing I raised was declared off limits, and through the years I was privileged to know her she showed herself her father's daughter in delightful and astonishing ways.

To her introduction I owed my friendship with a galaxy of her father's relations, friends, and near-contemporaries still alive in the late 1950s and early 1960s, and eager to tell me what they vividly remembered of a great and much-loved man whose reputation among the fashionable critics was still largely in eclipse. The devotion of all these people to Sir Edward's memory made an unforgettable encouragement to the very young man I was then, to pursue and try to understand the character of the man they remembered.

The first of Carice Elgar Blake's introductions was to her cousins May and Madeline Grafton (the elder daughters of Pollie). Theirs had been a more normal upbringing than Carice's, and the well-known Grafton charm was turned on me as it had been turned, in days gone by, on their favourite uncle. (The charm and the intelligence were shared by their youngest brother Roland and inherited by his sons, of whom Paul is the present family historian.)

Graftons were cousins to Leicesters, and here I met the longest, deepest, and in many ways most valuable friendship shared with Sir Edward. This was with Hubert Leicester's daughter-in-law Nella. For more than twenty years she made me welcome whenever I could come to stay at The Homestead, where she had lived since 1926 and where Sir Edward came often: it was only round the corner from Marl Bank. Through hours and years of conversation Nella Leicester recovered for me family papers and documents, her own keen impressions, and a way of domestic life intimate to Sir Edward himself but encountered fully by me in no other house. It is impossible to put a value on the enrichment which this unique opportunity afforded me, and I wish to record my certainty that this book would have been very much less than it is without her. To her husband Philip Leicester (who died just as our friendship was ripening) and his brothers, Brig. B. W. Leicester and the Revd Fr. Edward Dering Leicester (whose friendship I was privileged to enjoy for many years) I owe a wide range of memories of turn-of-the-century Worcester which have filled and coloured the impressions I could gather for myself right up to the night of Nella Leicester's death in March 1980.

Specific aspects of Elgar's life were illuminated by other friends of Carice who became friends of mine to larger or smaller degrees. Here I acknowledge the kindness and great goodwill of Lady Bantock, who first welcomed me to her house when she was ninety, and her son Raymond; Sir Barry Jackson, who talked to me with verve and affection of 'Edward', and whose memories were later supplemented by Tom English; Agnes Nicholls (Lady Harty); Mr and Mrs Richard Powell ('Dorabella' of the 'Enigma' Variations) and their sons; Beatrice and Margaret Harrison; Harriet Cohen; Yehudi Menuhin; Astra Desmond (Lady Neame); Sir Steuart Wilson; Sir Percy and Lady Hull; Mr and Mrs E. Wulstan Atkins; Charles Brewer; Dr and Mrs Herbert

Sumsion; Herbert Howells; Rutland Boughton; Mrs Ernest Newman; Marjorie Ffrangçon-Davies; Lilian J. Dyke Griffith (Troyte's younger sister); Billy Reed's son Francis; Fred Gaisberg's niece Isabella Wallich. All these I owe to Carice's introduction.

In the years since her death I have been privileged to share the memories of Mrs Mathilde Loeb and Miss Sylvia Loeb (Hans Richter's daughter and granddaughter); L. A. Foster (Muriel Foster's son); Mrs Moore Ede; Mrs Davis and Mr Bowden, who gave me important pieces of the Broadheath background; Norah Masefield Parker, Norah Crowe, and Winifred Whitwell, three ladies whose memories of Elgar took me vividly back into the 1890s; Winifred Davidson Christie, Lady Rasch (Kitty Petre), and Gwladys Battiscombe Lee, all girlhood friends of Carice whose observations of the Elgar household were invaluable; Olive Gosden, on whose extensive memories of the Malvern scene and its inhabitants from the turn of the century onwards I have drawn again and again; Imogen Holst; Patricia Neal; and Sir Adrian and Lady Boult, who have given unfailing help and encouragement in many directions.

John L. Nevinson and Bettye Lucas untangled for me the confusing epicycles of the Nevinson family. Ruth Finlay gave me help over the Gedges, as did Rodney Baldwyn over his family, Mrs Brooksbank over her father-in-law W. M. Baker, Juliet Berkeley over the Berkeleys of Spetchley Park, Robert and Richard Conquest over the Acworths, Mrs Christopher Norbury and Patricia Egerton-Smith over the Norbury and Capel-Cure families. Penelope Spencer, Winifred Lawford, and Stella Gowers shared with me their recollections of Ina Pelly (Mrs Christopher Lowther).

My gratitude goes to other friends who have helped this book in many ways: the four Miss Elgars (granddaughters of Edward's brother Frank), Mr and Mrs Paul Grafton, Mrs Clifford Baylis, Mr and Mrs P. Hedley-Dent, R. G. Emblem, Ronald Sinnott, Norman Wand, Mr and Mrs John Savage, Felix Aprahamian, Richard Osborne, the Caulfeild family, the daughters of Laurence Binyon, Miss Geehl, Anthony Thorley, Timothy Hooke, Nicholas Priestley, Michael Dowty, Mrs Anthony Bernard, Dr and Mrs Vernon Butcher, James Tanis, Oliver Neighbour, J. H. Britton, Michael Pope, Dennis Mellor, Richard Brookes, Raymond Monk, Alec Hyatt-King, Ivor Newton.

At the Elgar Birthplace it has been my privilege to have the friendship of all the curators from the first, Mr and Mrs John Goodman (chosen by Carice Elgar Blake in the 1930s); Mr and Mrs Alan Webb; Jack McKenzie, and also Mrs McKenzie; James Bennett. All have given of their time and interest far beyond the limits of any duty, and their help has made possible a more complete study of the documents at Broadheath than would otherwise have been possible.

Elgar's music publishers and their families have extended many helping hands. At Novellos I have been able to draw on publishing-house memories of John Littleton, H. S. P. Brooke, Harry Fowle, and Arthur Holdaway. Mrs Mary Fraser (A. J. Jaeger's daughter) and Stanley West (John E. West's son) have taken me back to the turn-of-the-century days of their own childhoods. Of the present Novello firm George Rizza, Margaret Pace, Bernard Axcell, and all their staff have given unfailing help and generous understanding at every turn. Mr and Mrs Robert Elkin have shared their memories and letters. Mr and Mrs Leslie Boosey gave me much help, and at Boosey and Hawkes, Muriel James and Malcolm Smith have been as generous. Eric Ashdown at Edwin Ashdown, John Garrett at Chappell, Patrick Howgill at Keith Prowse, Geoffrey Brand at R. Smith, and Peter Makings at Schott opened their archives to me. George Neighbour at the Performing Rights Society shared many of his archives with me, as did Emmie Tillett for Ibbs and Tillett. At EMI J. K. R. Whittle, Peter Andry, John Milliner, David Bicknell, and Leonard Petts did the same, and I have benefited from the memories of William Laundon Streeton and Bernard Wratten.

To Elgar's previous biographers I owe a special debt. Basil Maine gave me a vivid account of his biographical meetings with his subject. Percy Young (who enjoyed the closest friendship with Carice Elgar Blake) has shown in book after Elgarian book a wonderful ability to see and show what is important in opening new horizons everywhere. Diana McVeagh brought a fine critical intelligence to give Elgar's music its first modern evaluation. Michael Kennedy opened the human side of Elgar in an unforgettable way. I wish to add my gratitude to Michael Holroyd for a conversation on the subject of biography as a whole. On the shoulders of all these I have stood: I hope my book may do as much for those who come after.

Over.musical aspects I wish to place on record a deep debt of gratitude to Anthony Mulgan at the Oxford University Press; to Watkins Shaw, Peter Dennison, Christopher Kent, Bernard Keeffe, Sir Michael Tippett, and Sir Clifford Curzon.

For discussion of medical matters I am indebted to John Barretto, Sir Arthur Thomson, Sir William Trethowan, J. A. Chalmers; and most especially to Anthony Storr, who read the entire manuscript: it has benefited much from his wisdom.

The religious aspect of the book has been guided by many discussions. In particular I wish to thank Fr. Francis Edwards, SJ for sharing insights, and for generous and patient guidance over matters which a non-Catholic could not unravel for himself. The Revd Harold Goodwin has given me much help over the *Apostles* chapter. My grateful thanks go also to Fr. Kavanagh and his successors at St. George's Church, Worcester; and to Fr. Stephen Dessain and Fr. Thomas Gornall, SJ at The Oratory, Birmingham.

Other librarians and curators have been very generous in wide-ranging co-operation. Here I acknowledge first and foremost Miss Margaret Henderson and her staff at the Hereford and Worcester County Record Office (now housed in St Helen's Church, where the teenage Elgar and Hubert Leicester used to ring the bells): every scholar who has worked there will understand my gratitude. Mrs Bridget Johnston at the Worcester Cathedral Library made the vital linkage of the Davison family with Castle House. Douglas Guest and Donald Hunt opened the treasures of the Cathedral Music Library founded by Sir Ivor Atkins. The Custodian of Worcester City Records made available the old Electoral Registers, thus illuminating the movements of the Elgars during the early 1860s.

At the British Library much help was given by Pamela Willetts and her staff; and at the Royal College of Music by Oliver Davies. The Librarians at the Royal Academy of Music, the Athenaeum, and the BBC Written Archives Centre have all helped in important matters. The Alice Ottley School (formerly the Worcester Girls High School) archives were opened for me by the Headmistress, Miss Millest. The President and Fellows of Magdalene College, Cambridge made it possible for me to read and quote from the diaries of A. C. Benson (an exercise which would have been practically impossible without the index compiled by David Newsome). I wish to add the names of three distinguished dealers in antiquarian music who have aided my searches and my acquisitions: Albi Rosenthal, Hermann Baron, and Richard Macnutt.

For support through many years of research and writing, it gives me very great pleasure to acknowledge the understanding help of the Leverhulme Trust Fund and of the Hinrichsen Foundation: they and their officers gave me peace of mind – without which what follows could not have emerged. For a year of the Leverhulme grant, the Dean and Students of Christ Church, Oxford offered me their hospitality during a particularly difficult phase of writing. Earlier grants, and encouragement when I most needed it, came from the University of Rochester through Professor Wilbur Dunkel and the American Philosophical Society through Dr George Corner.

Grateful thanks are due to the Trustees of the Elgar Will Trust for permission to quote from the music and writings of Sir Edward Elgar, and from the writings of Carice Elgar Blake; to Novello & Co. for correspondence between Elgar and members of the firm, and for extracts from the Elgar music they publish; to Boosey and Hawkes for correspondence and for quotations from the *Pomp and Circumstance* Marches, *Sea Pictures*, and the *Coronation Ode*, to Messrs Schott, Breitkopf & Härtel, Chappell, and Keith Prowse for brief quotations from works under their control; to Claud Powell for permission to quote from his mother's *Edward Elgar: memories of a variation*; to Mrs I. M. Fresson for quotations from *Edward Elgar: the record of a friendship* by Rosa Burley and Frank Carruthers; to Victor Gollancz Ltd. for quotations from W. H. Reed's *Elgar as I Knew Him*; to the BBC for letters from Owen Mase; and to the publishers and copyright holders of other quoted material as cited in footnotes. In three instances I have failed to trace copyright ownership: to these owners the publishers and I offer apology.

My final thanks are reserved for my closest friends who have given me their insights and patience over the years I have taken to develop and shape the ideas herein. It is the greatest pleasure to say over their names. First among them is my mother, who all through my life has shown me ways of drawing on the past to enrich the present; Andrew Neill, Edwin Buckhalter, John Knowles, Alan Sanders, Anthony Petley-Jones, Jonathan Cadman, Nigel Edwards, Robert Donington, Leonard Barkan, William Wimsatt, Evelyn Hutchinson, Janet Ilford, Barbara Gibbs, Isabella Wallich, Henry and Sybil Wohlfeld. It is their book, and Nella's, as it is my own.

JERROLD NORTHROP MOORE

Abbreviations

ᴵᴺᴼᴵᴼᴬᴴᴵ

Manuscript Sources

BL   British Library.

Blake MS   Manuscript notes by Elgar's daughter, Carice Elgar Blake, of
                     conversation with Hubert Leicester, c.1935, HWRO.

[Blake TS]   Typewritten notes by Carice Elgar Blake of conversation with
                     Hubert Leicester, c.1935, Elgar Birthplace Museum, Broadheath.
                     This TS, although certainly by Carice Elgar Blake, is not signed, and
                     for this reason her name is cited in square brackets.

Elgar Birthplace   Elgar Birthplace Museum, Broadheath.

HWRO   Hereford and Worcester County Record Office.

Printed Sources

Letters   Letters of Edward Elgar, ed. Percy M. Young, Bles, 1956.

Maine, *Life*   Basil Maine, *Elgar: his Life and Works*, vol. 1 (Bell, 1933).

Maine, *Works*   Basil Maine, *Elgar: his Life and Works*, vol. 2 (Bell, 1933).

Works frequently quoted and referred to in shortened form

Bax, Arnold, *Farewell, My Youth* (Longmans, Green, 1943).
Bliss, Arthur, *As I Remember* (Faber & Faber, 1970).
Buckley, R. J., *Sir Edward Elgar* (John Lane The Bodley Head, 1905).
Burley, Rosa, and Carruthers, Frank, *Edward Elgar: the Record of a Friendship* (Barrie
    & Jenkins, 1972).
Elgar, Edward, *'A Future for English Music' and other lectures*, ed. Percy M. Young
    (Dennis Dobson, 1966).
—— *My Friends Pictured Within* (Novello, 1949).
Letters to Nimrod, ed. Percy M. Young (Dennis Dobson, 1965).
Kennedy, M., *Portrait of Elgar* (OUP, 1968).
McVeagh, D., *Edward Elgar: his life and music* (Dent, 1955).
Moore, J. N., *Elgar on Record* (OUP, 1974).
Powell, D., *Edward Elgar: Memories of a Variation* (Methuen, 1949).
Reed, W. H., *Elgar* (Dent, 1946).
—— *Elgar as I Knew Him* (Gollancz, 1936).
Young, Percy M. *Alice Elgar: Enigma of a Victorian Lady* (Dennis Dobson, 1978).
—— *Elgar, O.M.* (2nd edn, White Lion Publishers, 1973).

Note: Many letters to and from Elgar contain double and triple underlinings. In this
    biography no attempt has been made to distinguish between levels of emphasis;
    all underlinings appear in italics.

BOOK I

THE WAND OF YOUTH

I

The Better Land

ᴵᴺᴖᴉ

MUSIC and biography share the expression of time remembered. For each finds its form through recapitulation. When Edward Elgar's career had attained its peak, in 1912, his eldest sister Lucy set down what she alone survived to remember of its beginning. Before the beginning, she sketched a portrait of their father, William Henry Elgar, born in September 1821:

My father was a native of Dover. His patronymic is of Saxon origin, which, being interpreted, means 'fairy-spear'. As a young man he entered the music-publishing house of Coventry & Hollier in London, but in the year 1841 came to Worcester and commenced his profession as a piano-forte tuner, teacher, and organist—and was a talented performer on the violin. He was much more of a musician than a business man, and devoted his whole thoughts to matters musical.[1]

The circumstances of that move were described many years later by Hubert Leicester, the nephew of a man who was to befriend William Elgar soon after his arrival in Worcester:

No music shops in those days. Went to Latimers for musical requisites . . . Uncle [John also] had these things [at the family printing shop at No. 6 High Street] as he was very keen on music.
    Stratfords—a stationer who'd a shop in The Cross – their musical side developed & applied to Broadwood's for a tuner & they sent W. H. E.[2]

He had not the temperament for steady working, and found that too much tuning made his head 'very queer indeed, though it soon goes off again'.[3] 'He was good-looking, with a great charm of manner—a gentleman in fact—never happier than when astride a horse.'[4] These traits were to be described later and more trenchantly by his son Edward: 'My father used to ride a thoroughbred

---

[1] Lucy Elgar Pipe, 'Reflections' (MS now in the Hereford and Worcester County Record Office). Subsequent quotations from Lucy Elgar Pipe have been taken from this source unless otherwise noted. Mrs Pipe seems to have drawn some background for her narrative from an article by F. G. Edwards in *The Musical Times*, 1 Oct. 1900, based on conversation with Edward Elgar. Some of the article's sentences are practically reproduced by Mrs Pipe; but in the main her narrative is unique.

[2] Carice Elgar Blake, MS notes of conversation with Hubert Leicester, c.1935 (HWRO).

[3] Quoted in Percy M. Young, *Elgar, O.M.* (2nd edn., White Lion Publishers, 1973), p. 23.

[4] [Blake], TS notes of conversation with Hubert Leicester, c.1935 (Elgar Birthplace).

mare when he went to tune a piano. He never did a stroke of work in his life.'[5]
Yet there was no doubt of his skill, and he knew how to be agreeable to the
County magnates whose custom he secured.

The most distinguished country house on his rounds was Witley Court,
twelve miles north-west of Worcester. It was the seat of Lord Dudley, and was
even let for three years to the Dowager Queen Adelaide, widow of William IV.
All this while W. H. Elgar attended to the pianos there. One day the august
connection brought a rare chance (as Hubert Leicester recalled hearing of it):

W. H. rather notorious as a tuner & producing wonderful effects on piano—he had
finest touch on a keyed instrument. At Witley Court Lord Dudley sent for him to tune
piano; in an adjoining room heard him trying it after finishing tuning. Lord D. came
in—W. H. stopped—after inducement played something. Lord Dudley wanted him to
do more than be an ordinary tuner—offered to pay for a course to make him finest
player in England. Refused—too nervous.[6]

But the nervous refusal did not prevent him from advertising his connection
with Witley Court: for years after it would be commemorated on his
bill-headings and business stationery.

The tuning connection expanded. William thought of his youngest brother
Henry, then a boy of twelve or thirteen, and wrote back to Dover to ask
whether Henry might come to Worcester as his apprentice: 'What I can
recollect of him he will make a good musician.'[7] Henry arrived, and served out
the years of his apprenticeship. But temperamental differences dogged their
relations, for William was a dreamer and Henry a realist. When he came of age
Henry took himself off to an independent piano-tuning career elsewhere, and
William once more worked alone.

For all his skill at the piano, William Elgar found his real happiness in the
amateur musical life which flourished in Worcester then. He was a good piano
accompanist, and knew his way through the second violin parts of the older
repertory from Corelli onwards. So he was welcomed in the Worcester Glee
Club, whose weekly meetings at the Crown Hotel often included an
instrumental group. There he made friends with John Leicester (whose family
kept a printing business in the High Street).

Leicester was a member of the Roman Catholic Church. When he
discovered that his Protestant friend Elgar could play the organ, he told him of
a difficulty facing the Catholic congregation in Worcester:

At St. George's Church there was wildly eccentric organist who'd made a fortune in
America—could not keep time—had to be superseded on this account.
    W. H. E. had met Uncle [John]—leading light of choir—who put him up for the job.
He applied & Uncle advised priest as to candidates. W. H. E. appointed 184[6].[8]

    [5] Conversation with Mr and Mrs Alan Webb, 13 Jan. 1931 (MS notes written by Alan Webb
immediately afterward).
    [6] Blake, MS notes.
    [7] Quoted in Stanley Godman, 'The Elgars of Dover', in *The Musical Times*, July 1946.
    [8] Blake, MS notes.

The organist's duties at St. George's were varied. He had to choose all the music, which sometimes involved composing, play voluntaries and accompany the singing, train the choir and rehearse them with any instrumentalists who might be assisting (for on many Sundays the Mass was sung in one of the big settings by Haydn, Mozart, Beethoven, Hummel, or Weber). William Elgar was an occasional and sometimes unwilling composer, as he himself confessed: the priest 'expects me to do these things—but I don't at present feel inspired, my mind wanders too much in my business'.[9]

Meanwhile there was one thing the young man did for himself (as Hubert Leicester would recall being told):

Had to be economical & lodged in Mealcheapen St—(The Shades Tavern)—not a public house but more a refreshment place (cafe)—centre of market—market held in Corn Market. Kept by man whose wife was a Miss Greening & she got her sister to help in the business. W. H. E. lodged with them.[10]

Soon the Greenings took him into their family home in the village of Claines, three miles north of Worcester.

The sister's name was Ann:

a tall, fine woman with big features and splendid carriage, with a great personality and presence. Although she had practically no opportunities, her love of reading was remarkable: she read every available book and her great longing was to know more and more of the world of literature.[11]

She seemed a good partner for another idealist, a self-made 'gentleman' (as their eldest daughter wrote):

In [January] 1848 my Mother and he were married. My Mother was the daughter of a Herefordshire farmer who came of a strong healthy long-lived stock—thus perhaps accounting for my Mother's splendid constitution and sincere love of the country and all things pastoral.

Her own young life had been peopled from noble books, and it was in their pages she had met her friends and companions—men romantically honourable and loyal, women faithful in love even unto death; both alike doing nobly with this life because they held it as a gauge of life eternal. And in her simple way of thinking she verily believed these shadowy forms to be portraits of the people whom she would one day meet with in the world. No one told her differently, for her own Mother had brought her up in that sweet pious simplicity which makes a woman charming in good fortune and patient and strong in the days of calamity.

As the marriage developed, young Mrs Elgar showed the strengths her husband lacked. It was common knowledge in the Leicester family:

Ann Elgar did not like W. H. E. walking alone to Worcester & came to church every Sunday, & eventually went to [the priest] for instruction & became a Catholic.[12]

---

[9] Quoted in Young, *Elgar, O.M.*, p. 32.
[10] Ibid. Young states (*Elgar, O.M.*, p. 20) that the relative at The Shades was not a sister but a brother. This may well be correct, and Young gives much ancestral information.
[11] Blake, MS notes.   [12] Blake, MS notes.

Roman Catholicism, in the Jesuit foundation of St. George's, offered an intellectual engagement to faith not easily available to a young woman of the lower middle class. But it was a brave decision for a tradesman's wife with no Catholic background in an English cathedral city in 1852. Her husband—whose organist's job had started it—cordially detested the whole business. At the time of his wife's conversion he wrote back to his own family in Dover of '. . . the absurd superstition and play-house mummery of the Papist; the cold and formal ceremonies of the Church of England; or the bigotry and rank hypocrisy of the Wesleyan.'[13]

Yet Ann Elgar was to raise all their children in her new faith. Her strong character showed in every arrangement made for the children who had already begun to arrive. Before the birth of the first in October 1848, the expectant parents had moved to a house in the centre of Worcester, No. 2 College Yard East (now called College Precincts), close by the east end of the Cathedral. And thus (as their eldest daughter Lucy wrote):

Three of us—Harry [born 15 October 1848], myself [29 May 1852], and Pollie [Susannah Mary, 28 December 1854]—had been born within the shadow of our dear, dear Cathedral.

But my Mother's wish always for a country life prompted Father to go to Broadheath [three miles north west of Worcester]. Broadheath itself is only a hamlet, with a handful of houses, a large clump of fir-trees by the entrance gate of 'Newbury House', and a wide stretch of heath known as the 'Common', dotted about with little pools and ponds for the cottagers' ducks and geese, and ponies and donkeys grazing.

The remote place was reached only by circuitous lanes, dusty in summer and bemired in winter. It was a real disadvantage to the father in his itinerant tuning work. But again the mother had her way—for a time. From 1856 they rented the cottage of Newbury House known as The Firs (after the tall trees at the corner of the cottage garden by the lane). The cottage measured only 20 by 21 feet overall, and comprised six tiny rooms on two floors. It would be a squash for a family with three children and some domestic help. Soon a fourth child was on the way. But in this mother's estimation, a country life for her children was next to religion itself. Her daughter Lucy recalled:

We were always taught to adore Him in the smallest flower that grew, as every flower loves its life. And we were told never to *dare* destroy what we could not give—that was, the life—ever again. There is a humanity in every flower and blade of grass . . .

Our summers had been lovely, but the winters, how we loved the winters. We did not fear the cold, but rejoiced in the clear sharp air: how beautiful it all was out of doors, then the brightness and warmth within! . . . I remember how eagerly we looked for snow—because upon the very first snow being seen on Malvern Hills we were allowed to have the 'warming-pan' in our little beds!

We were encouraged to go out in all weather during the whole of the year. Although we honestly loved the winter we welcomed the beautiful time of spring: the singing birds

---

[13] Worcester Papers, No. 6: Saturday 18 Sept. 1852 (MS at the Elgar Birthplace).

had come, hedge and heath, fields and forests were offering their gifts of flowers as a pledge that winter was over. The resurrection of sleeping nature with its yearly miracle awoke . . .

Our home life at Broadheath, as I grew older and understood, was *ideal*. Although Father in his profession was away many days in the week, the break was made up by the constant friendship between Mother and children . . . We indeed seemed to spell Life with one word, and that word was 'Mother'.

And to Mother we laid bare all the wounds of our hearts, she understood our little griefs. She saw in them—and made us see in them by explaining away, very tenderly, all sorrow. And yet with all her sweetness and charm she could, to we children, be very severe if we were in the wrong, and put on a forbidding tone and a look in her eye of microscopic power which was positively paralising and gave us horrid little shivers. But the scoldings never lasted long, and what we learnt we never forgot.

She taught us that as we all have our wearisome hours which cannot always be cast aside by an effort of the will, it is not wise to encourage the feeling of discontent—or as others may put it, *temper*. She would say, 'Be always busy, doing something that is useful and interesting.' She despised those who led a purposeless life . . . She wished us to be of the *crew* of this great ship we call the world, and not of its useless passengers.

She kept us from the contamination of the world as long as ever she could do so. It was too great a distance from the town for smuts to fly, and settle on our homely table-cloth! . . . If humanity has its sorrows, the green earth offers its compensations . . . She sought *natural* joy in her daily pin-pricks by taking long walks, and commun[ing] with Nature in its beauty—when the air was sweet with the breath of violets and all the tender flowers of the first months of the year, when the sense of awakening is everywhere . . . She loved an atmosphere peaceful yet glowing and vibrating with her own emotions.

Her eldest child, Harry, was already beginning to distil from his mother's world an essence for himself. Hubert Leicester remembered: 'Harry loved botany and the study of herbs and was always making concoctions of plants . . . It was always understood he would [be] a doctor and [make] his mark'[14]

The mother combined practicality with dreams. Hubert Leicester remembered her as

a very clever, ingenious, busy woman. Not good at ordinary domestic work. Did a vast amount of fancy work, painting &c.—always busy in her own way—extremely careful with every shred of silk &c. When her husband would not buy her a cradle for the baby, she put rockers on a washing basket.[15]

So a child of hers, like most children perhaps until recent times, would have the cradle-pulse of rocking from side to side.

Before birth, in the dark womb, the baby's first consciousness of any experience comes in the mother's heart beat, her breathing and walking, and the sound of her voice. So pulse and rhythm, movement and sound—all manifesting themselves through time—appear on the *tabula rasa* before there is anything to see. Medical opinion testifies that these experiences begin about

[14] [Blake] TS notes.
[15] Hubert Leicester in conversation with Fr. Driscoll, SJ, 28 Nov. 1909 (written down immediately afterwards by Leicester's eldest son Philip, who was present).

half-way through the pregnancy:[16] until the moment of delivery, they comprise almost the whole of experience.

A baby in the womb sometimes responds to patterned sounds outside with regular rhythmic movements of its own. The regularity proves that the response is positive—the happy recognition of principles already familiar.

Pulse and rhythm, movement and sound are also the elements of music. And there was music in the cottage at Broadheath whenever the father persuaded his friends to come for an evening or weekend. Then the expectant mother, when she could, attended St. George's Church in Worcester, where the big, organized sounds of the choral and instrumental Mass settings by the great composers were often to be heard.

The day of the new arrival at Broadheath, 2 June 1857, made a permanent memory for five-year-old Lucy:

How well I remember the day he was born! The air was sweet with the perfume of flowers, bees were humming, and all the earth was lovely. There seemed to be, to we little ones, a lot of unnecessary running about in the house, and Father came tearing up the drive with a strange man in the carriage. And before that, an old lady whom we had never seen there before arrived with a large bag, and we were told by the younger maid there was a baby in that bag! That was good enough for our weak comprehension, and so we were taken a scamper across the heath to be out of the way.

Very gradually in the newest consciousness, the larger world of sight and scent would begin to impress the pulses of its own times—day returning after night, a future summer after winter. These cycles of time shaped the countryside life of mother and children. But the baby who had come now held a special place in his family: he was the only child actually born outside the city. He was also to be the only child in his family to find genius. The potential bond of genius with the world of nature was the subject of a quotation which his mother later copied into her notebook from Emerson:

When Nature has work to be done, she creates a genius to do it. Follow the great man, and you shall see what the world has at heart in these ages. There is no omen like that.[17]

The childhood memory of Lucy emerged on another scene:

The first impression of my Mother . . . was as she stood talking to someone in the room. A beautiful little boy in a white dress, his hair all ashine from the light of the window, was perched upon a chair and looking up at his Mother, trying in his baby way to fathom and understand that beautifully minded Mother, for the charm of her face consisted in its great nobility. It was a pretty picture, and that little boy was Edward.

As with most babies, Edward's concentration began to centre on the sound of his mother's voice. The sense of this dawning insight was conveyed to a friend of later life, who wrote:

---

[16] I am indebted to J. A. Chalmers, FRCS, for a conversation on this point.

[17] From Emerson's essay 'The Method of Nature' (Ann Elgar's notebooks are at the Elgar Birthplace).

In his earliest years, perhaps before he could appreciate the full meaning of what he heard, he was entranced by the *sound* of his mother's voice as it rose and fell, the *sound* of the words as they flowed in well-turned phrases, and the *sound* of the rhymes and the metrical rhythm he quickly learned to appreciate in the poetry to which he listened.[18]

Lucy, being older than Edward, caught more:

I well recollect how beautifully she read and recited poetry, and would often gather us round her with the baby Edward on her knee, and tell us sweet stories—one of which we liked better than some was 'Speak gently'. And the poem of Mrs. Hemans', 'The Better Land', we never tired of listening to.

This poem, which Edward heard his mother reading from the time he could understand anything, was a dialogue between son and mother:

'I hear thee speak of a better land,
Thou call'st its children a happy band;
Mother! oh, where is that radiant shore?
Shall we not seek it, and weep no more?
Is it where the flower of the orange blows,
And the fire-flies glance through the myrtle boughs?'
  —'Not there, not there, my child!'

'Is it where the feathery palm-trees rise,
And the date grows ripe under sunny skies?
Or 'midst the green islands of glittering seas,
Where fragrant forests perfume the breeze,
And strange bright birds on their starry wings
Bear the rich hues of all glorious things?'
  —'Not there, not there, my child!'

'Is it far away, in some region old,
Where the rivers wander o'er sands of gold?—
Where the burning rays of the ruby shine,
And the diamond lights up the secret mine,
And the pearl gleams forth from the coral strand?—
Is it there, sweet mother, that better land?'
  —'Not there, not there, my child!

'Eye hath not seen it, my gentle boy!
Ear hath not heard its deep songs of joy;
Dreams cannot picture a world so fair—
Sorrow and death cannot enter there:
Time doth not breathe on its fadeless bloom,
For beyond the clouds, and beyond the tomb,
  —It is there, it is there, my child!'[19]

---

[18] W. H. Reed, *Elgar* (Dent, 1946), p. 3.

[19] There is at the Elgar Birthplace an MS entitled 'Six Songs in Six Languages', probably in the hand of Ann Elgar. No. 2 is a setting of 'The Better Land' by Miss Davis, an Irish composer of the day.

Death held no reality for the infant at Broadheath. 'The Better Land' could appeal to him by its bright images and rhythms touching the little countryside world of his experience, renewing itself in returning hours and days through seasons of change so slow as to be scarcely perceptible. So Broadheath itself might appear as 'The Better Land' in the poem the children 'never tired of listening to'.

The images of rhythm and recapitulation from Edward's mother were passive, contemplative. But the same images found an instant and active voice in the father's profession. For music offers an active, deliberate, human way of shaping things in pulse and cycle.

In the children's ears, the very approach of their father was signalled in sound. Lucy wrote:

I so well remember the intense pleasure and true joy we all felt when we at last heard the long-waited-for whistle Father always blew on his return home, at one angle of the lane, which could be heard distinctly across the meadows between. It was a very old-fashioned whistle in ivory which he always carried in his pocket—aye, and for years too after the Broadheath days . . .

One of us would run to open the gate, another the stable door, and Harry generally helped with 'Jack's' [the pony's] food, and each one thought ourselves very clever in helping to welcome Father. Then came the happy evening meal all together, and experiences to be related on each side, both Father's and Mother's. And of course we little ones had much to tell of what had happened during the time of absence—how many chickens had been hatched, how 'my garden' had flourished, how badly the 'ladies' avenue' wanted rolling (that was the long path in the sketch next the central one, which Father had made especially for we little girls to bowl our hoops and play upon, as it was thickly laid with sand purposely to dry quickly after rain). And then all the baby prattle of little Edward had to be listened to.

Verily the home of which he was head was a happy one because a simple one—and because he was blessed with a well-regulated mind, and undisturbed by that unbalancing element known as genius. He was a musician of great ability, and a distinguished man without having become so by the production of any work of distinction . . .

My Mother was not in a way called musical, though her calm, high-thinking mind could fully appreciate the nice little musical evenings we so often had in those early days, and her sweet voice blended delightfully with the others. For Father very frequently brought their friends from Worcester for an evening or for a week-end. And well I remember a dear man we all loved, and who was a constant visitor, named Allen, a solicitor—and another named Leicester, John Leicester, a tall man who sang in the choir at Church.

Well, they were thoroughly happy in our cottage home, and sang and sang for the very joy of singing. And Mr. Allen *always* (yes, every time he came) would sing to Father's accompaniment, 'Di Provenza il mar'[20]—oh! I can hear it now! We got to know every note of it by heart. Also we knew every note of the glee 'Mynheer Van Duncke'[21] with which they finished up and all sang together. It was lovely. And we three little ones sat on a sofa to listen, and of course thoroughly enjoyed ourselves—but dare scarcely move or blink our eyes, for fear of attracting attention and being sent to bed!

[20] From Verdi's *La traviata*, first produced in 1853.   [21] By Sir Henry Bishop.

Thus the music at Broadheath recapitulated not only by repetition and chorus; like the mother's reading of poetry, it also repeated pieces from one occasion to the next. So the children themselves could begin to have their own favourites to recognize and look forward to.

It is the sense of pulse—the source of life itself—that gives music its first and fundamental appeal. Even the most complex rhythm draws its power ultimately from some hint of simple pulse repetition—some pattern of events moving steadily, regularly forward in time. What is the repeating pulse of music if not a precise miniature of those larger pulses of nature, the cycles of hours and seasons returning. Music offers a means for venturing out from the babyhood of primal dark physical feeling toward the great world outside.

As these relations slowly found their places far below any actual consciousness in childhood's early experience, the consonance of the father's music with the mother's gifts offered to all the Elgar children a base of the firmest security on which to build their lives—as long as they should remain in the countryside. It was no accident that Lucy's recollections of Broadheath were shaped so strongly by thankfulness. But for Lucy and the older children Broadheath was a place to which they had moved: so it signified an early arrival. For Edward alone it was the point of departure.

* * *

The Elgar family's idyll in the country lasted for about two years after Edward's birth. But the presence of four growing children increased constantly the pressure of living in six tiny rooms. When a fifth child was expected, in the summer of 1859, the necessity of moving was clear.

Ann Elgar would have remained in the country. But her husband pressed for returning to Worcester. It was not only a question of his travelling convenience. Now there was another offer from his friend John Leicester. Leicester's nephew Hubert later recalled:

As [W. H. Elgar] got on as a tuner, called on to advise County gentlemen re: pianos & supply of them. Uncle John said:
'If you like, & want somewhere for people to try pianos [for sale], you can use my shop.'
Accepted, & 1st piano put in there (Leicester's shop). Sold 1 or 2 pianos. Saw necessity for developing.[22]

If there was to be any thought of expanding into a regular shop business, it would be vital to live on the scene.

So the parents decided to take their children back to the city. Lucy remembered all her life the sadness of parting:

Leaving the dear home—(that has even *now* kept as a nook the memory of those who had lived in it)—seemed at the parting inexpressibly beautiful and dear, and we had the

[22] Blake, MS notes.

longing desire to live on here for always. And what could come to anyone better than the beauty and peace of this place, in which all care and regret gave way to the mists of the mornings, the scented shadows of the nights, the breathless wonder of the days? The country—so fresh, so delicious: we wanted nothing in shape of distraction. Surely in no corner of beautiful England can anyone find more refreshment and tranquil pleasure than breathing the air of Broadheath Common.

And yet, as this woman recalling her childhood from the distance of fifty years had come to realize, Broadheath was really a state of mind. For the older Elgar children its memory was to remain forever associated with the time of their lives they had spent there. As those children grew up, the lengthening distance of time made it safe at last from any danger of corruption by later experience. From that vantage it appeared to Lucy Elgar finally as

a tiny world apart from the one in which people hurry and clamour and fret time away. There are many such villages in our sweet old England, and each one seems to be a little contented child that has stolen away from the noise and worry and all the tragedy wrought in the name of civilisation.

When Lucy tried to think of what had come after Broadheath, she found that 'the first few years of our town life I do not remember much about now . . .'

For Edward the life at Broadheath finished about the time he was two, so it would hardly have found a place in his permanent memory at all. Lasting memories do not survive from a baby's earliest months because there is as yet no frame of reference, no means of sorting what may be kept from what is let go. Even without conscious choice, the operation of any memory must signify some rough structure of values. The beginning of memory, then, is the sign that some first lines of individuality have been laid down. What has come before that—what can never be precisely recalled—has gone into shaping the matrix of that individuality.

For the two-year-old Edward that earliest experience had been at Broadheath. Though no conscious memory might ever reveal it to him, he would carry it with him always and everywhere: it was, together with what was in him of his parents, an aspect of himself. The knowledge of it could only be intensified by the fact that he stood apart from all his brothers and sisters in feeling Broadheath yet being hardly able to compare it with the clear and specific memories of his life. The boy questioning his mother in the childhood poem had anticipated Edward precisely:

> 'I hear thee speak of a better land,
> Thou call'st its children a happy band;
> Mother! oh, where is that radiant shore?'

From later childhood to old age he would go back to Broadheath again and again as if to consult it. When in his last years his achievements were commemorated in a baronetcy, he chose to become First Baronet of Broadheath. Before that he had told his daughter that he wished the cottage preserved there as his memorial before all the other places where his life was lived. But he came closest to defining his debt to Broadheath in a letter written to an intimate friend sixty years after he had left it:

So you have been to B.—I fear you did not find the cottage—it is nearer the clump of Scotch firs—I can smell them now—in the hot sun. Oh! how cruel that I was not there—there's *nothing between* that infancy & *now* and I *want* to see it.[23]

[23] Undated note to Alice Stuart of Wortley, perhaps written after her visit to Broadheath on 25 June 1920 (HWRO 705:445:7683).

2

Divisions

⬩⬩⬩

WORCESTER is divided from the countryside westward to Broadheath by the River Severn. A century ago the river drew the division of city from country more clearly than it does today. Up to its eastern bank crowded close buildings and narrow city streets. But from certain of those streets, and from Worcester Cathedral, you could look across the river to an almost uninterrupted landscape of field and countryside opposite—to the agricultural and pastoral world whose natural cycle still shaped so much of life then. The tall houses and narrow pavements of the city could not have made stronger contrast.

In 1859 the Elgars went to live at No. 1 Edgar Street.[1] Eighteenth-century brick houses of three and four storeys jostled one another in the shadow of the medieval Edgar Tower to form a south-eastern gateway to the Cathedral and its quiet, refined College Precincts (now called College Green), the site of the King's School and fine old ecclesiastical residences. The house in Edgar Street, just outside the Precincts, provided larger quarters for the Elgar family—now increased again by the arrival of 'Jo' (Frederick Joseph).

The Edgar Street house also provided a centre for W. H. Elgar's piano-tuning business. The business prospered so that he was quickly able to persuade his younger brother Henry to rejoin him.[2] They called themselves 'Elgar Brothers', though Henry was the employee and not a full partner. Their friend John Leicester wanted them to extend their trade to the selling of music. In 1860 he made them investigate a vacant shop at No. 10 High Street, north of the Cathedral. But by now there were two piano shops almost across the street—one of them run by the family of the eccentric former organist of St. George's Church—so the matter was not pursued.[3]

After a year in the Edgar Street house, the Elgar family with Henry moved round the north side of the Cathedral to No. 2 College Yard.[4] This was another

---

[1] The electoral Register of Householders published on 30 Nov. 1860 shows W. H. Elgar as living in Edgar Street. Seventy years later Edward Elgar identified the house for Edgar Day, assistant organist of Worcester Cathedral, as the building now known as No. 1 Severn Street.

[2] Letter of 11 Dec. 1859 (Elgar Birthplace).

[3] City Directories for 1855 and 1860 list George Baldwyn at 98½ High Street as a 'Dealer in musical instruments, pianoforte warehouse'. The census of 1861 shows 101 High Street occupied by Isaac Dimoline, 'Pianoforte maker and music dealer'. The same census shows the shop at No. 10 as vacant, the rooms above it occupied by a young family with two children.

[4] They were listed there in the 1861 census, which was taken in March.

old brick house, with no garden at all to the rear, but facing the churchyard lawn by the Cathedral. For the three-and-a-half-year-old Edward it opened a new world: '. . . I played about among the tombs & in the cloisters when I cd. scarcely walk.'[5] The Cathedral organist and lay clerks were friends of his father from the Crown Hotel Glee Club and other musical organizations round the city. Some of the Cathedral 'gentlemen' quickly became *habitués* of the Elgar household.

Yet the church of the mother and children was neither the Cathedral nor the old Parish Church of St. Michael at the eastern end of College Yard and close to their own front door. Instead they walked half a mile through the city to the small Roman Catholic church built thirty years earlier, at the time of the Catholic Emancipation Act.

There they came under the influence of Fr. William Waterworth, SJ, the priest who served as Superior at St. George's Church and led the Catholic community in Worcester for twenty years between 1858 and 1878—practically the entire period of all the young Elgars' childhood and adolescence. Fr. Waterworth was no ordinary parish priest. The Superiorship at Worcester was a late stage in a clerical career which would have taken him much farther if his health had been stronger. As it was, he had been more or less retired to Worcester in middle life by the Jesuits (as a colleague expressed it) 'that they might not lose so valuable a subject'. But Fr. Waterworth made his mark in Worcester:

On account of his learning and his kindness and zeal, he was much esteemed by Catholics and Protestants. The Catholics loved him and were proud of him . . . Dean Peel [of Worcester Cathedral] and many of the Protestant Rectors and Canons were his personal friends . . . His frequent lectures to the Worcester Catholic Institute, articles published in the *Dublin Review* and the *Rambler*, and sometimes in newspapers, caused him to be respected by Catholics and Protestants, and also to be feared by some of the latter, so that a Protestant would be afraid to publish anything, for example, respecting Worcester Cathedral without consulting him, lest he should prove him to be wrong in some point, by an article of three columns in a newspaper.[6]

Those abilities would have been of the greatest advantage in guiding the Catholic community of Worcester in the 1860s. In a cathedral city, prejudice was strong against Roman Catholics even more than against Protestant dissenters. As the Elgar children grew older, they would be made aware of living to some extent outside the regular community of their city. Even forty years later children going into St. George's Church were harassed by neighbourhood boys hanging on to the railings and calling:

> Catholic, Catholic, quack! quack! quack!
> Go to the Devil and never come back![7]

---

[5] Letter of 13 Nov. 1914 to Alice Stuart-Wortley (HWRO 705: 445: 7668).
[6] *Letters & Notices* (The Society of Jesus) XVI, 1883, pp. 150–3.
[7] Recalled by the late Brig. B. W. Leicester (third son of Hubert Leicester) in 1974.

A man of Fr. Waterworth's character and ability would not have been the priest to let slip any chance for turning adversity to advantage. If Catholicism in Worcester was surrounded with difficulties, that encirclement could be shown to the faithful as just the special strength of their special community. So the Faith itself might set its example of particular placement and uncommon election.

In Fr. Waterworth and his Church, Ann Elgar found the natural ally for her own aspirations. Her daughter Lucy remembered:

All necessary things were brought gently into our young minds—meaning, that we were never thrown upon our own resources for our amusements but were carefully watched and tended even until we were well through our teens, and our companions were carefully chosen, not leaving us alone to form what friendship we liked best.

Catholicism provided the manifest premise for her children's education. There was a Poor School at St. George's; but by the time the Elgar daughters were ready to begin their schooling, something more suitable to a family in trade had appeared.

Shortly before 1860 a 'Middle Class School for Girls' was started by Miss Caroline Walsh in her house at 11 Britannia Square. Britannia Square is an elegant, old-fashioned place three-quarters of a mile north of College Yard, beyond St. George's Church. The Walshes were a well-known family in the district, and Miss Walsh—like Ann Elgar herself—was a convert to Catholicism. She had made her confession in the 1840s, and then joined the 'Filles de Marie', a religious order whose members 'wore a Secular dress and mixed in Society as lay persons'. One of her Order wrote of Miss Walsh:

She used, we are told, to rise very early on Sunday morning for a round among the poor Catholics of the town. Thus she obliged the sluggards to leave their beds and prevented them from missing Mass. Both parents and children held Miss Walsh in great veneration. Her lessons were a source of enjoyment to the little scholars, whose hearts she filled with the love of God. The Jesuit priest who had charge of the Mission in Worcester said that the Catholic congregation in that town were largely kept together by the efforts of this Daughter of the Heart of Mary.[8]

Miss Walsh's Catholic 'Dame's School' was on a small scale. No. 11 Britannia Square is a narrow terraced house without any very large rooms. During the years of the school Miss Walsh seems to have been assisted entirely by one other lady of her Order. And in the Catholic congregation of

---

[8] Information from The Society of the Daughters of the Heart of Mary in England, and from a MS note about Miss Walsh in the archives of St. George's Church, Worcester. Miss Walsh lived originally with her family at Lower Wick House, beyond the western bank of the Severn. A Worcester directory of May 1855 shows her as Mistress of the Catholic Poor School at St. George's. The directories of Sept. 1860 and Sept. 1863 list her school at 11 Britannia Square. Two letters from Br. Henry Foley to Fr. Waterworth, both written 29 July 1878, show that the chapel at 11 Britannia Square was being furnished in Feb. 1862. (MSS in the Jesuit archives, Farm Street, London). A Worcester directory of Aug. 1868 shows Miss Walsh living at 11 Britannia Square, but there is no mention of the school. She died at the house in July 1878.

Worcester—a total of about five hundred—there would hardly have been more than twelve to fifteen girls eligible for such a school at any time. Among those girls were Lucy and Pollie Elgar. And along with them in September 1863 went little Edward, as he himself recalled for friends in later years : 'Told us . . . he was then in petticoats & with his sisters & other girls.'⁹

Thus Catholic education defined itself for Edward in a female world. One anecdote his memory preserved was noted down by a friend:

Miss W. was a dear old lady who told the girls to not only read books, but to ask her questions on anything they wished to know. After long cogitation a question was thought out & E's sister Lucy was chosen to ask it. She went up to Miss W. & said:

'Please, Miss Walsh, we want to ask a question, as you told us.'

'Yes, my dear,' said the old lady.

'Please, Miss Walsh, we want to know whether you wear ——' (an unmentionable lady's undergarment).

E. shook with laughter as did we all. Said he should never forget Miss W's face.¹⁰

The mother's Catholic conversion and all that followed had come from the father's employment as organist at St. George's. Yet the father himself would take no part in these arrangements except to criticize them. Hubert Leicester remembered: 'Old man a regular terror as regards the Catholicity of his family—used to threaten to shoot his daughters if caught going to confession.'¹¹ The pointed animus against confession before a priest could suggest that W. H. Elgar felt a threat to his own position as head of his family. Hubert Leicester added to his observation the significant words, 'Very tactful wife'.

Disagreements in the Elgar household spread more generally until they became common knowledge through the neighbourhood:

E.'s mother always worried as to where the money was coming from; his father taking life very easily, enjoying his rides into the country for piano-tuning, and the proceeds of each visit more than likely swallowed up by the unnecessary expense incurred in getting there.

Henry was persuaded to join W. H. E. somewhat against his will, for the two brothers never got on very well together. W. H. E. always found it impossible to settle down to work on hand but could cheerfully spend hours over some perfectly unnecessary and entirely unremunerative undertaking (a trait that was very noticeable in E[dward] especially in later life . . .) Henry was more practical and content to carry on with the business in hand . . .

The business should have been very good if [William] had only given his full attention to it, as his brother was always begging him to do. But anything that promised an hour's

⁹ Philip Leicester, MS notes on a visit from Elgar, 2 June 1914. The date of Edward Elgar's beginning at Miss Walsh's school is established by a flyleaf, detached from a book and preserved in an album at the Elgar Birthplace, which is inscribed:

Edward W. Elgar Decr 20 1865

Third Class Prize, Miss Walsh's School.

¹⁰ Ibid.

¹¹ Philip Leicester, MS note of Hubert Leciester's conversation with Fr. Driscoll, SJ, 28 Nov. 1909.

or a day's distraction always took him away. He once spent a whole day in the courts, as a customer who owed him a trifling sum was undergoing his examination. At the end of the day he came away having been paid 4d. Disgusted with the whole proceeding, he entered this sum in his accounts and added grimly to the entry: 'Spent the 4d. on beer.'

Arguments with H. then followed about the utter waste of time in thus spending a whole day. Then there would be a long silence while they sat smoking. Suddenly W. H. would break this silence with the words: 'Now Henry is a fool'—long pause—then: 'and a bloody fool.'[12]

Ann Elgar faced the difficulties of this household as well as she could. And when she could not face them, books and her imagination offered an escape which she never disguised from her children. Lucy recalled:

I used sometimes to wonder if she had come to the end of her illusions, and she said:

'No, and I hope they will last as long as I do; they give colour and variety to life and keep one's heart young.'

Those illusions found a special response in her Broadheath child, as young Hubert Leicester recognized: 'It was she who peopled the world with heroes and poets for E. and led his mind—which needed only the merest hint of things of beauty . . .'[13] Edward later gave his mother's encouragement the fullest recognition. In 1905, with his knighthood fresh upon him, he would stand up in the Worcester Guildhall to receive the Freedom of the City: '. . . He confessed that his position was owing to the influence of his mother . . . and many of the things she said to him he had tried to carry out in his music.'[14] A quarter-century after that, when two young visitors asked about his early life: '. . . he fishes out from a drawer some quaint family daguerrotypes, one of them showing himself as a little boy, resting his head against his mother's knee. "My mother", he says simply, "was one of the most beautiful women alive." '[15]

For the little boy in Worcester, that special attachment to his mother must have come partly through the disintegrating family ensemble. In the old days at Broadheath, where Edward's first experience was shaped, the father had been received as the family's most honoured member—whose music gave fulfilling complement to the mother's countryside world. Now in the city, the father's more constant presence broke the former harmony.

The places where they now walked made a contrast to the half-remembered country. A contemporary of the Elgar children recalled the city in the 1860s:

The Worcester of those days was a very different place from what it is now. It was then a quiet, somnolent, old Cathedral city, a stronghold of the clergy and their wives . . . The city was full of ancient churches, dedicated to such old-time Saints as St Wolstan, St Swithin, St Nicholas, and the like; and the artist or antiquary wandering off the main thoroughfares could still find quaint old streets, narrow and tortuous as those of any

   [12] [Blake], TS.   [13] Ibid.

   [14] 'Worcester and Sir Edward Elgar: Presentation of Honorary Freedom' (news cutting of 12 Sept. 1905, at the Elgar Birthplace).

   [15] Alan Webb, MS notes of a visit to Elgar, 13 Jan. 1931.

Flemish city, where the overhanging storeys of the old timbered houses almost met overhead.[16]

If ever there was a place to make a child conscious of the past and the influence of the past, Worcester in the 1860s was such a place. It could provide vivid illustration for the father's tales of old distinguished houses whose pianos he tuned, illustration also for the mother's readings and teachings out of a life 'peopled from noble books'.

In the city surroundings hours were measured by a heavy mechanism of bells in the huge Cathedral tower, resounding their strokes against walls and pavements and places that never saw the sun.[17] Renewing hours and reviving seasons here might only chime the imagination past pavements and buildings, across the river to a countryside where the cry of the child could echo again in the favourite poem:

> 'Mother! oh, where is that radiant shore?
> Shall we not seek it, and weep no more?'

A sense of vision first offered, then removed, and sought through all the world thereafter, can provoke creative impulse. For Edward, a primary ensemble had been made at Broadheath by both the parents together. In Worcester it might seem to be fading—equally through the parents' lives and actions. If the ensemble were ever to be found again, then, neither parent might show him the way. It would be up to Edward to teach himself.

The little boy found his way down to the bank of the river that divided Worcester from the countryside opposite—the countryside westward to Broadheath. And when he got to the river he saw something he would remember all his life. Addressing an audience fifty years later in the Worcester Guildhall: 'Said first letters he learnt were the names of the old tugs which he spelt out as he sat on the banks . . . They were "Enterprise", "Reliance", & "Resolution".'[18]

Those large and complex words would not have made a child's very first letters: separate sounds and syllables must have come first. But it was not the learning of small sounds that fixed itself in Edward's memory. What impressed him was the need to fit the sounds together to make something larger, longer, more complex. It might be another portent for the composer to come—the composer who would find his fulfilment in the writing of large works.

The big words put together at the bank of the river suggested the 'self-help' popularized by the Victorian writer Samuel Smiles. The three words all

---

[16] Mrs C. E. Jeffery, *One Road to Rome* (Burns Oates and Washbourne, 1927) p. 27.

[17] The Worcester Cathedral bells resound constantly through the pages of Mrs Henry Wood's novel *The Channings* (1862), written out of an intimate knowledge of the city in those days. *The Channings* was a great favourite with Edward Elgar and all the Worcester friends of his generation. (Information from Mrs Philip Leicester, 1978.)

[18] Philip Leicester, MS account of Elgar's speech at luncheon following the presentation of the Freedom of the City to B. W. Leader, RA, 3 June 1914.

described things for which one looked to oneself. The child's recognition of 'Enterprise', 'Reliance', and 'Resolution'—floating before his eyes on the dividing river—might hint again that any crossing of that river to return to the ideal country beyond it must come from within.

Where were his means for such a return? The clearest avenues lay through his father's music on the one hand, his mother's words and stories on the other. Psychologists now know that verbal and musical abilities are centred in opposite hemispheres of the brain.[19] That physical opposition underlies a basic distinction. Words and poems and stories are tied to particular things. But music need not be tied to any defined thing. So music must have a special appeal for a person who is emotionally isolated: the abstraction of music can offer a means of experiencing emotion without the threats and dangers always implicit in defined situations and conflicting characters.

When music does not describe anything in particular, the clearest way to relate the separate events within it is by some returning similarity. That notion underlies the classic musical pattern A—B—A: you make a statement, you go away from it, you come back to it again. That pattern shaped many of the musical compositions played in the household by W. H. Elgar and his friends. A similar cyclic pattern lay beneath Ann Elgar's understanding of nature and the countryside. For their little son, then, that simple notion of cycles might offer an ideal way to reconcile the diverging expressions of the two parents. It was the notion which would shape all the mature composer's most successful large works.

But which parent's mode of expression should the boy choose for himself? There was much evidence, then and later, that Edward shared his mother's keen interest in literature. But if he took words for his own medium, then his mother's strong, helpful character would direct him with a monitoring passion he might not be able to resist: so the words might not be truly his own. The case was different with music. Father would have little time to teach Edward, and Mother always pointed out that music was not her special skill. So in music both would leave him alone. And so there could emerge a remarkable paradox: a way back to the maternal countryside of the spirit across the river lay in cultivating his father's art of music.

All this remained below the surface of the child's consciousness. But the force of the huge dim equation came over Edward when his father at last took the vacant shop at 10 High Street and set up a music business in it. For what would come back in recalling his own earliest experience of the shop was the very image of the river. He said in 1904: '. . . I saw and learnt a great deal about music from the stream of music that passed through my father's establishment.'[20] It was no chance allusion. A dozen years later another conversation on the same subject yielded precisely the same image in expanded

[19] J. Z. Young, *Programs of the Brain* (OUP, 1978), p. 245.
[20] Interview with Rudolph de Cordova, 11 Feb. 1904, in *The Strand Magazine*, May 1904, p. 538.

form: 'A stream of music flowed through our house and the shop, and I was all the time bathing in it.'[21]

A reminder of the countryside life beyond the real river came when Edward began to be taken on his father's piano-tuning rounds. This was recalled vividly for a friend of later years, who wrote:

Piano-tuning was the outdoor department of the music-shop; and Elgar senior condescended to nothing less for his conveyance than a thoroughbred mare, until Edward was old enough to accompany him, when a pony and trap had to be substituted. When a piano had to be tuned at Croome Court or Madresfield the boy Edward was taken for what was practically a delightful day's outing. While the piano was being attended to he could roam about in the grounds until he was taken into the house and refreshed.

No detail or happening of those far-off times escaped him; he could tell me as we ambled about the lanes and passed these great houses, and many others too, the names of all the people who lived in them long ago, and relate to me the sayings of the members of the household, or the yarns spun for his benefit by the groom, or the old ostler who watered his father's horse.[22]

Once Edward recalled they were caught between houses in the rain, sheltered under a tree, and his father drew out a pocket notebook to write a musical idea.[23]

Back at the shop, the little boy tried to shape his own music. Hubert Leicester recalled:

At 4 or 5 he was seen trying to write down a tune, drawing the lines himself on a plain sheet of paper and very puzzled because he could not get it right. His father, who was thought to be a gruff kind of man generally, said to him:
'What are you doing there, you silly boy?'
'Writing music,' was the reply.
'Well, don't you know that music is written on five lines, not four lines?' was the father's comment on seeing that his little son (doubtless having seen some Gregorian music on four lines) had only ruled four lines on his piece of paper.[24]

Perhaps it was merely W. H. Elgar's anti-clerical temper showing itself again. But if this was a sample of the music-seller's tutelage, it could only confirm the son's impulse to teach himself.

Near the end of 1863 the family moved from College Yard into the rooms over the shop in High Street.[25] However much the mother may have felt a reduction in social status, for the children it made another adventure. Their

[21] Interview with Percy Scholes, 29 June and 5 July 1916, in *The Music Student*, Aug. 1916, p. 343.

[22] W. H. Reed, *Elgar as I Knew Him* (Gollancz, 1936), p. 45. .

[23] Young, *Elgar, O. M.*, p. 32.

[24] Unsigned newspaper articles of Nov. 1910 and 1934, identified as Hubert Leicester's from the original manuscripts in possession of Mrs Philip Leicester.

[25] W. H. Elgar is shown still at College Yard in the List of Householders published on 30 Nov. 1863 (probably compiled in Oct. of that year). But at the birth of their youngest daughter Helen Agnes on 1 Jan. 1864 the family are listed at 10 High Street.

young neighbour Hubert Leicester (the son of John Leicester's brother William, who ran the printing shop four doors away) remembered it all:

10 High Street was an old house in the narrow part of the High Street near the Cathedral. At the back there was a long courtyard containing the workshop where the piano repairs and polishing were done, and at the end of this long courtyard a cottage, the which was often let to . . . men employed by W. H. E. for special polishing who were very often foreigners, who were as a matter of course musical & played some instrument . . .

Ned Spiers, who had been stage carpenter with the touring companies, was their man who moved pianos and all the odd jobs. He met with an accident which left him with a limp but he was a tremendously strong man. He had seen all the great actors and quoted Shakespeare and speeches from all the repertory of the companies, and knew the whole of England, and was a wonderful companion—almost tutor to the children.

'Kit' was the nurse and faithful servant of the family and came from Herefordshire. The family adored her, and her extraordinary misuse of words was their delight all their lives: many of her sayings became household words . . . One of Kit's sayings was: 'Give me an item when the kettle is boiling.'

Then there was Tip, a mongrel black and tan terrier, the children's playfellow & devoted slave. The cat, also a family friend, learnt to ring the bell when he wanted to go into the house. The back door opened onto the courtyard, and over it there was a bell which was rung by a long cord, which reached almost to the ground. This cord the cat accidentally touched one day, causing the bell to ring. Naturally the door was opened, and the cat walked in. The cat never forgot this, and to the end of its life knocked the cord when it wanted to go in.[26]

Edward had passed his sixth birthday then. He had a special attachment to little Jo, two years younger and also quick at music though he was still halting of speech and so small he might be consumptive. Then there was the youngest brother Frank, born at College Yard in October 1861 and still almost a baby. For an elder brother to himself, Edward had to look all the way up to Harry, close to nine years his senior. In view of his father's temperament, Edward needed some slightly older boy as model and friend.

He found that boy in Hubert Leicester, more than two years older but soon as close to Edward as any brother. Edward could hold the elder boy's interest without effort. Hubert recalled him as permanently on the alert—notably to sounds and vagaries of speech from the fustian of Ned Spiers's old dramas to the malapropism of Kit:

E. was always alive to every impression. No one ever came to the house whose characteristics and tricks of speech were not noted and stored up for future reference and quotation—a habit which persisted all his life. He was . . . always on the watch for new experiences and fresh knowledge.[27]

When it came to an instrument for his music, one glance round 'Elgar Bros. Pianoforte & Music Warehouse' made the choice obvious, as he himself

[26] [Blake], TS.   [27] Ibid.

remembered: 'I was a very little boy indeed when I began to show some aptitude for music and used to extemporise on the piano.'[28] At this, paternal interest was persuaded to take a more agreeable turn. Music lessons were arranged—once again as a part of Catholic education. Seventy years later came this letter from the daughter of his first teacher:

My Mother was then Miss Sarah Ricketts, who gave you your first music lessons (piano). She was a singer in the Catholic Church. No doubt you have forgotten her—it is a long time ago. But she was a frequent visitor at your home in Worcester . . . Mother used to relate about your first attempts at composition. During a theory lesson Mr. Elgar, your father, came in, & suggested trying your hand at composition. Mother too, but she couldn't write a bar. But you soon had a few lines written, and you were only a young boy.[29]

A little later there were lessons from another young lady, Miss Pollie Tyler, at Miss Walsh's school.

Some of the ensemble music still used in Worcester, as elsewhere, was in eighteenth-century editions, which demanded skill in playing from figured bass. Edward recalled:

My father had the complete set of Vivaldi's Concertos & from the cembalo part of this I made my first infantile petticoated attempts to play from figured bass: *each part* (Violins, Violes, Bassi) has written [on] it 'This book belonged to Barrington Hall' signed Daines Barrington.[30]

The Hon. Daines Barrington, a well-known amateur of the previous century, had written an eyewitness account of the child Mozart's visit to London in 1764. Edward added: '. . . I seem to remember Mozart was asked to play from *Vivaldi's* concertos—invent an upper part to Vivaldi's figured bass . . . I always like to think we own the very book that the Wonder Child played from!'

It was a heady thought for a piano-tuner's son in Worcester. Realizing a figured bass at sight was like extemporizing one's own music, but under the harmonic guidance of a master-mind. He worked at it until he gained a fame in Worcester faintly reminiscent of the Wonder Child's fame in Europe. Years later he recalled it for a magazine editor, who wrote: 'If not quite a prodigy, young Elgar took to the pianoforte at an early age . . . Whilst still nothing more than a child, he could extemporise in such a way as to astonish those privileged to hear him . . .'[31]

Even in Worcester there was a rival. She was Maud Baldwyn, niece of W. H.

---

[28] *The Strand Magazine*, May 1904, p. 538.

[29] Letter from Florence M. Lawrence, 15 Apr. 1932 (MS at the Elgar Birthplace). Printed programmes of performances at St. George's Church and the Worcester Catholic Institute show Miss Ricketts as an occasional performer in the early 1860s—at least once in an ensemble of pianists including 'Master Elgar'.

[30] Letter to Edward Speyer, 13 Mar. 1916.

[31] [F. V. Atwater], 'Mr. Edward Elgar and his works', in *The Musical Courier*, Oct. 1896. Atwater's acknowledgement of information supplied by Elgar is preserved (HWRO 705:445:7540).

Elgar's eccentric predecessor at St. George's who now had an interest in a family music business in the High Street almost opposite the Elgars. Maud was widely advertised as 'The Infant Musical Genius, aged six years', and had appeared in London, Dover, Ramsgate, and Dublin.

That kind of exploitation would never have been countenanced by Edward's mother. But his father would take the boy to show him off before fashionable clients of his piano-tuning practice among the local gentry and clergy. Special interest was found in a house in the quiet old Precincts on the far side of Worcester Cathedral. Edward recalled:

One of the prebends [sic] was Dr. Davison, who lived with his two daughters on the other side of the College Green. As a boy I used to go across to play the piano there, and to extemporise.

Boys were not allowed to make the passage through the Cathedral, and it was a great delight for me, being 7 or 8, to dodge the Dean, and get through the North door and to the College Green.[32]

The Davison house was known as Castle House, because it was built along one side on a fragment of wall remaining from the medieval Worcester Castle. Most of the Castle had disappeared long since; but the Davisons' house, showing this surviving wall, rested on foundations as rich and storied as those of the Cathedral itself.

Inside the house this atmosphere of the past intensified. In the drawing-room, which contained the piano, Edward encountered a way of life as different from the life at 10 High Street as the contrasting positions of the two houses on opposite sides of the Cathedral could suggest. He recalled it in a preface drafted many years afterward for a collection of children's piano pieces.

Long ago, when I was a very little boy, smaller than you even, I used to be taken to play on an old piano, in an old house, in an old close, near an old cathedral, to two old ladies . . .

The drawing room was lined with books, books, books, all leather-bound & faded: the curtains also & the silk bands which held them, all were faded to a soft golden brown, as was the piano. When Broadwood sent it by long & weary road fifty years before, its mahogany had surely been a rich red: but time had shaded all the tints to one mellow tone—& I played in a golden-brown room.

Grown-up people came (I knew later it was to hear me play): also, amongst them old gentlemen—very courtly: people talk faster nowadays. And the two ladies received their friends with an old-world state that I loved to see.

The music was old-world too. Some of the pieces I played, and which I hope you may

---

[32] Address to the Friends of Worcester Cathedral, 18 June 1932, as reported in a Worcester newspaper of 20 June 1932. The newspaper prints the name as 'Davidson'. The prebendary was in fact John Davison (1777–1834), who had been appointed in 1826. After his death his widow Mary (who died at the age of seventy-seven in 1875), their daughter Grace (1825–85), and possibly another daughter lived in the house in College Green later known as Castle House. There they remained for the rest of their lives. See Michael Craze, *King's School, Worcester, 1541–1971* (Ebenezer Baylis: The Trinity Press, 1972), pp. 242, 252.

like, are now printed for you in this book. They were in many volumes—faded volumes, with red ribbon markers. Outside, the ribbon had no colour. But when the book was opened, the warm ruby-carmine tint flashed soberly out & the music seemed warm & alive, even coming from such wan paper & perished leather.

The old people talked gravely about the music, and my favourite old gentleman (who wore many seals & such a stiff collar) said: 'Kozeluch was more fiery than Corelli,' & that Schobert was 'artificial in his Allegros'. Dear Heart! how learned it seemed.

Once Miss Grace brought down a manuscript book—ever with a crimson marker—asking me to play a little piece by Quarles (it is printed for you here): 'It had once been a favourite,' she said, & the old gentleman smiled gravely. And as I played they looked very happy—I did not know why. I think now the music turned aside their faded lives, & roseate youth—(they must have been young once, only I never thought so)—then came up before them transiently. Music has done this magic to many.

And now *I* am not young. And sometimes, when wearied with great works by greater men than those whose pieces form this book, I play these formal things endeared to me by memories of the gentlest & courtliest of mortals. Long years since, at the opening of the old books, I saw the coloured ribbon still brightly where it had been most hidden. So, when I hear these pieces, I forget to be grey & feel the most of my heart's best feelings.

But I am older than you.

Ed. E.[33]

Coming into the Davison house in College Green from the family quarters above the shop, what had struck the boy immediately was the presence of 'an old-world state'. It was a response which could have drawn equal suggestion from his father's pride in the acquaintance of gentlemen and from his mother's life of the imagination 'peopled from noble books'. The son of these parents could feel encouraged to seek a natural aristocracy in himself. Yet the notion was strong, especially in the agricultural counties then, of a social order in which those at the top really did stand for the best. The ideal for a child of that day was not to beat them but to join them.

'An old-world state' was a way of life which had successfully stood out against time and change. The people who lived there would never have had to give up the Broadheaths of their earliest experience to be moved about the city from one house to another. So the old house in the old close near the old cathedral showed this boy something 'that I loved to see'.

And then it was clear that his own playing made a special appeal to the people in this old house. It appealed to them by reminding them of their own youth and past—of a private old world that must lie hidden inside each one of them, like the past inside Edward himself. Could the past, then, hold a special secret for Edward's music? Again the hint found its echo in the favourite poem, when the boy asked the question:

> 'Is it far away, in some region old,
> Where the rivers wander o'er sands of gold?'

It was a hint which might glimmer only dimly for a child of seven or eight. But if

---

[33] MS at the Elgar Birthplace.

it should be true, then it would follow that all the music Edward could make from out of his own experience might find a special response among people with their own experience of an old-world state.

The Elgars, whatever the cleverness of Edward or any child of the family, were tradespeople who lived over their shop. Social position divided this boy from his ideal audience as plainly as the Cathedral divided the High Street from College Green—as Catholicism divided mother and children from much of their city—as musical skill had begun to divide Edward from other children —as the Severn itself divided the boy's own present from the Better Land of his past.

*   *   *

In any family of the 1860s, the threat of death was never far away. Ann Elgar had begun as early as she could to prepare her children for the possibility. First there had been the lines concluding the favourite poem:

> 'Sorrow and death cannot enter there:
> Time doth not breathe on its fadeless blooms,
> For beyond the clouds, and beyond the tomb,
> —It is there, it is there, my child!'

Then, as Lucy remembered of their mother:

When Father once had a deep and serious illness she drew us together and said: 'Now children, this has to be faced. Be brave and pray always: if he goes, God will help us, and if he is spared to us, "Blessed be God."'

And he *was* spared to us—thanks to her constant care, love, and attention. And so we always learnt a lesson in some way or another from her sweet teaching.

One result of this teaching was to implant in the children almost an expectation of death. Lucy would find herself practically looking out for it in the life of their youngest sister:

In 1864, as the bells were ringing in the glad New Year, our darling mite Dott (Helen Agnes) was sent to us . . . I have often believed she was born a saint, and would not be allowed to suffer earthly ills for long: whose soul was already in Heaven, while the poor body is left to do good in this world till it shall wear itself out and melt away like a shadow.

In her baby life she had suffered a severe illness, such that her moments were counted. But Mother watched and prayed unceasingly, feeling that life *meant so much* . . .

Mother watched the whole of the night alone over the sweet baby, counting every breath, fearing each one may be the last, and the agonised soul of the dear patient watcher was on the verge of despair, when, as the dawn came, baby opened those large thoughtful eyes—one could almost imagine they possessed a super-human power, so beautiful were they, those wonderful eyes—and looked straight at Mother and smiled! dear Mother! she then knew her little one was given back to her—and so she recovered . . .

We felt that if Mother was taken from us, or had a serious illness, or was parted with us for any length of time, the world would be just four bare walls to us—and nothing else!

A few months after Dott came into the world, the threat of death appeared again—this time not to be denied:

In [May of] the same year came a great great sorrow, a very real sorrow—the death of my brother Harry, after an illness of four weeks only. This grief nearly cost my Father his reason, but by Mother's bravery and fortitude that awful calamity was averted.

At first it was an impossible sorrow. They did not, they could not believe in it. Why should such grief come to them? Life for them now was empty—nothing but withered hopes. For on that first boy they had naturally built all their bright anticipations for the future . . .

My remembrance of that dread time is like a confused dream. But I seemed to realise even then that when our joys die, they find no grave for us. We must live on, just as the rose-tree lives though all its flowers be broken off; and the Spring brings roses again.

And so, through many nights and days of pain and grief, they found that never was Christ's love dearer or His voice more sweet. Many times had grief brought good, and sorrow been full of strange joys and compensations: for, having bravely and heroically made a sacred offer of their first boy's life, was not the second boy given them as a sweet recompense?

The second boy, at the time of his brother's death from scarlet fever, was near his seventh birthday. Harry had been fifteen—the eldest child and the eldest son. Now Edward, from a middle position among the young Elgars, suddenly found himself cast for pre-eminence. His father began to refer to him as 'the governor'[34]—as if to show that Edward now was natural heir to the shop. On his shoulders rested the family's accumulated hopes for both its eldest sons.

Still Edward had his next younger brother to whom he could turn. And Jo was already a bright spark. Lucy wrote:

He was called the 'Beethoven' of the family, having very remarkable aptitude for music in every way from the time he could sit up in his chair. But alas! he was not strong, and a source of much anxiety to the dear dear parents.

Hubert Leicester remembered this too:

Jo, though very intelligent, was curiously undeveloped in many ways and never learnt to pronounce his words properly.[35] Jo went with family & Leicesters to Perrywood—Edward, Pollie. Jo began to cry as he was so tired—wanted to go home. Uncle H[enry] put him on his back & carried him home.[36]

Edward showed an elder-brotherly affection. It emerged in photographs of the two boys taken at about the time of Harry's death. In each photograph Edward stands close beside Jo with a protective hand on the smaller boy's shoulder, staring straight at the camera with a hurt defiance that would protect the smaller boy's weaker stance and open mouth. It is as if he knows already the darkening future. Nearly seventy years later Edward found himself reminiscing

---

[34] [Blake], TS.   [35] [Hubert Leicester,] newspaper article, 1934.   [36] Blake, MS.

with friends and colleagues in Worcester Cathedral: 'Pausing before a skull, underneath which are two fearsome wings, he told us how as a little boy the gruesome sight fascinated him, and he still remembered the thrill of fear with which he regarded it.'[37]

The Cathedral was the undoubted centre of music in Worcester. The lay clerks were constantly in and out of the Elgar shop and family rooms. W. H. Elgar supplied the Cathedral music, and Edward had been able to hear much of it rehearsed from so early a time that he could later say: 'My first music was learnt in listening in the Cathedral.'[38] The Cathedral organist William Done conducted both the Worcester Philharmonic Society concerts and the Worcester Three Choir Festivals in which W. H. Elgar took part as a second violin. So it was natural that the Cathedral doors should open a little wider for the music-seller's son. Edward recalled:

I drew my first ideas of music from the Cathedral, from books borrowed from the music library, when I was eight, nine or ten. They were barbarously printed in eight different clefs, all of which I learnt before I was 12 . . . I was allowed by Mr. Done to borrow them, and they were administered to me by friends who were lay clerks.[39]

But again his father stood aside, as Hubert Leicester recalled:

His mother was always ready to help and encourage him, but his father and uncle were merely amused and scoffed at these childish efforts—an attitude in which they persisted until E. really had made his way in the world. They failed to see, not only that they had an exceptionally gifted boy in the family, but even that he was moderately clever at music.[40]

Perhaps they did not want to see it.

Edward bided his time. One day the politics of his father's little world came into their house. The result would remain vivid in the mind of a family friend after more than sixty years:

As a boy Edward was very silent, rarely speaking unless spoken to. I have known him sit a whole evening in the family sitting-room in a corner behind the piano saying very little to anybody. Yet he was endowed with powers of repartee of a very precise kind. I remember one instance of this.

Amongst his father's friends was an amateur musician and composer named Mr. Spark. Mr. Spark always submitted his efforts to his friend Elgar pere, and on one occasion having received the flattering judgment of Mr. Elgar and several friends on the trial of a very successful anthem, played over at Mr. Elgar's house, he stooped down to Elgar junior at the corner of the piano, saying:

---

[37] Emmie Bowman Leaver, 'Some impressions of Sir Edward Elgar', in *Musical Opinion*, July 1934, p. 869 (written immediately after the meeting of the Union of Graduates in Music at Worcester, 30 July–1 Aug. 1931).

[38] Address to the Friends of Worcester Cathedral, 18 June 1932 (reported in local newspapers of 20 June).   [39] Ibid.   [40] [Blake], TS.

'And what does little Edward think of it?'

'If you puff a spark too much you will blow it out,' was the quiet but prompt reply.[41]

Another time Edward tried a pun in music. On 24 March 1866 he intersected four staves and gave each of the four a different clef, in such a way that a single note placed in their common centre was at the same time B, A, C, and (the German) H.[42]

Soon Edward began to look beyond the basis of his father's business, the piano. A sense of the piano's limitation came into his thinking, and it was to remain. Forty years later he would analyse it:

The rigid piano is capable of only two qualities of tone simultaneously in the hands of a moderate player—and most players are that. Better performers are able to produce further effects by allowing one part (one voice) to predominate in each hand, giving us four distinct weights of tone. Beyond this the piano cannot go.[43]

As a mature composer Edward described his inspiration in these terms:

'My idea is that there is music in the air, music all around us, the world is full of it and—' (here he raised his hands, and made a rapid gesture of capture) '—and—you—simply— simply—take as much as you require!'[44]

The sound of the piano, by contrast, started suddenly when a hammer struck a string: then it helplessly decayed. Its only sustaining was in the blurring pedal, its only reviving in repeating the process.

When the boy began to look for another instrument, it was still his father who held the key:

The organ-loft then attracted me, and from the time I was about seven or eight I used to go and sit by my father and watch him play. After a time I began to try to play myself. At first the only thing I succeeded in producing was noise, but gradually, out of the chaos, harmony began to evolve itself.[45]

Like the piano, the organ offered its player complete harmony. And the organ could sustain its tones to any length. But the other limitation was still there, as Edward expressed it later: 'The organ can sustain, but only with three or four varieties of the colour at once.'[46]

What he was looking for was an instrument which could vary infinitely while

[41] Anonymous article, possibly by Charles Pipe, in *The Worcestershire Journal*, 10 Mar. 1934. The amateur composer was Edward J. Spark (b. 1829), the second and least distinguished son in a musical family. In 1870 Edward Spark established his own music business in Worcester, where he was already organist of Holy Trinity Church.

[42] The scrap of paper, dated in the hand of Ann Elgar, was later pasted into an album now at the Elgar Birthplace.

[43] *'A Future for English Music' and other lectures*, ed. P. M. Young (Dennis Dobson, 1968), pp. 237, 239.

[44] R. J. Buckley, *Sir Edward Elgar* (John Lane The Bodley Head, 1905), p. 32, recounting a conversation of 1896.

[45] *The Strand Magazine*, May 1904, p. 538.

[46] *'A Future for English Music' and other lectures*, p. 251.

it sustained. He would find that only when he encountered the orchestra, as he afterwards realized:

If it be required to define the 'genius' of the orchestra in one sentence, I should say it is the 'simultaneous, varied sustaining power' . . . The modern orchestra is capable of an unending variety of shades of tone, not only in succession, but in combination. We see that a whole world of new harmonies is at the disposal of a composer for orchestra.[47]

Edward did not see it yet, but in the summer of 1866 (when he was nine) he had hopes. The only time a boy in Worcester could hear a really big orchestra was when every third year for a week in September the Three Choirs Festival came to the city and a sizeable band was recruited to accompany oratorio performances in the Cathedral. It was Worcester's turn in 1866: his father would be playing among the second violins and Uncle Henry among the violas.

The whole of that summer, however, was overshadowed by the decline of little Jo, 'the "Beethoven" of the family'. Lucy wrote:

After much suffering he passed away in the dawn of a September day, in the seventh year of his age. It was a great grief. But consolation seemed to come under the calm resignation which Mother taught us: after Jo came another little boy, Frank (Francis Thomas)—and his bright, happy little childhood made up to us in a way the loss we felt in giving up dear Jo.

But Frank was not yet five. Edward, the only other son, had now at the age of nine encountered death not only above him in the family but immediately below him as well. Of the three eldest Elgar sons, he alone survived. Their mother insisted it was the will of God working toward some benefit which only the future could reveal. So to the burden of Harry's inheritance was added the burden of Jo's musical promise.

Yet Jo's very gradual decline could not repeat the paralysing shock of Harry's death. Jo was buried beside Harry in Astwood Cemetery on 8 September, the Saturday before the Festival week. W. H. Elgar decided to go through with his engagement to play in the orchestra, and as a distraction was prevailed upon to see what he could do to get Edward into a rehearsal.

The Festival was to begin on Tuesday. Monday was the final day of rehearsal—when solo singers, choirs, and orchestra at last came together. One work in the programme was Beethoven's Mass in C. And the presence of nine-year-old Edward at the rehearsal on that Monday, 10 September 1866, made a life-long memory for Hubert Leicester:

H. says he can remember E. running down the street from 10 High Street to the Cathedral with a large score under his arm . . . His father had obtained for him admission to a rehearsal. He had never heard a big band until then, and when he came back it was such a revelation to him that he said:

'Oh, my. I had no idea what a band was like. Then I began to think how much more

[47] Ibid., pp. 251 and 239.

could be made out of it than they were making . . . If I had that orchestra under my own control & given a free hand I could make it play whatever I liked.'[48]

It was not the choral singing that attracted his attention. Despite the singers' preponderance of numbers their music was confined mostly to four parts, their tone uniformly the tone of the human voice. The ensemble which stirred him was the wealth of tonal and instrumental complexity that made the orchestra.

He responded to the orchestra just as he had responded earlier 'when I began to show some aptitude for music and used to extemporize on the piano': the orchestra was seen instantly as another means for playing 'whatever I liked'. But the complexity of this new ensemble also appealed to him. The size of the orchestra might actually have a bearing on the size of music which could be written for it. Looking at the development of the modern orchestra, Edward said later: 'Larger works have demanded larger orchestras, or larger orchestras have demanded larger works; I leave it to historians to decide, if it be worth discovering, which force was the moving power.'[49]

The Three Choirs Festival itself was such a gathering of size. Its triennial rotation among the cities of Worcester, Hereford, and Gloucester made a focal point for the society of all three counties. Local nobility and landed gentry, leading clergy and prominent laity, all came together there and then with important musicians. If Edward had the orchestra of that occasion under his own control and given a free hand, he would also have commanded the largest and most distinguished audience ever presented to his young experience. It would be like playing the piano for the old ladies and gentlemen at the house in College Green, only bigger in every way.

For the boy listening in Worcester Cathedral, the power to manipulate a large orchestra could show the importance of music itself in a way to which W. H. Elgar had never aspired. It could fulfil Ann Elgar's dreams for her son's achievement while preserving the son's originality. But Edward's encounter with the orchestra also came right after the death of little Jo. For all the sorrowing Elgars, this rehearsal of music by the great Beethoven could sound a second requiem for the little Beethoven who now would never appear. If there was still to be a Beethoven in the family, it could only be Edward.

It was close to this time[50] that Edward returned to the bank of the river. And there the child had another glimpse of the man who was to be. More than half a century later he recalled it:

[48] Blake, MS, and [Leicester,] Newspaper article, 1934. Leicester recalled this happening 'when E. & Hubert were 9 & 11—or a year or two older'. Edward and Hubert were 9 and 11 in 1866, and the date is confirmed by a note in Hubert Leciester's hand (MS now in the Jesuit archives at Farm Street Church, London): 'Sept. 1866. Elgar attended the first Festival—Beethoven in C.' The Beethoven Mass in C was not performed at any other Worcester Festival during those years.

[49] 'A Future for English Music' and other lectures, p. 251.

[50] The experience was placed 'at the age of nine or ten' by Elgar's biographer Basil Maine in Elgar: his Life and Works (Bell, 1933), Life, p. 17. Maine based his book partly on extensive conversations with Elgar, and the boy's experience at the river had not been mentioned by any previous biographer.

I am still at heart the dreamy child who used to be found in the reeds by Severn side with a sheet of paper trying to fix the sounds & longing for something very great – source, texture & all else unknown. I am still looking for This – in strange company sometimes – but as a child & as a young man & as a mature man no single person was ever kind to me.[51]

The loneliness might have astonished those who felt close to Edward. Perhaps his mother had an inkling of what her child suffered and strove for when she stumbled on another passage from Emerson and copied it into her scrapbook ten years later:

Nature is a mute, and man, her articulate speaking brother, lo! he also is a mute. Yet when Genius arrives, its speech is like a river.[52]

For Edward himself, the second experience at the river might have anticipated what he found long afterwards in a poem entitled 'The Music Makers':

> Great hail! we cry to the comers
>> From the dazzling unknown shore;
> Bring us hither your sun and summers,
>> And renew our world as of yore;
> You shall teach us your song's new numbers,
>> And things that we dreamt not before:
> Yea, in spite of a dreamer who slumbers,
>> And a singer who sings no more.[53]

One distant day this poem was to make the basis for a summation of his maturest art. In the reeds by Severn side, the dreamy child who faced the intimate, unknown country across the river might feel only the need to reunite his father's world of music with his mother's world of nature and the spirit. But the uniting this time should be permanent. The child had brought a sheet of paper to try to write it down.

[51] Letter of 13 Dec. 1921 to Sir Sidney Colvin (HWRO 705:445:3527).
[52] From 'The Method of Nature', entered in Ann Elgar's scrap-book 25 Mar. 1877 (MS at Elgar Birthplace).
[53] By Arthur O'Shaughnessy, first published in *Music and Moonlight*, 1874.

3

Ensembles

IN 1867, the year of his tenth birthday, Edward crossed the river and returned to Broadheath. There he spent a summer holiday with old friends of his family.[1] And this return to the Better Land of his own beginning brought about the real entry to what he had longed for in the reeds on the city side of the Severn. For it was linked ever afterwards in his memory with the first dated music he actually wrote. The music survives in a sketchbook of many years later with the note 'Humoreske a tune from Broadheath 1867':[2]

---

[1] On 13 June 1895 Elgar took his wife to Broadheath '& saw the house in wh. he was born & where he used to stay as a little boy' (MS diary, HWRO). According to local tradition among life-long residents of Broadheath, who had it from their parents and apparently from Elgar himself on late visits, he stayed with old Mrs Doughty at her farm beside Broadheath Common. The back of the Doughty farm ran nearly down to the cottage of his birth. (Conversations with Mrs Davis, who occupied the cottage for years before its purchase by the Worcester Corporation in 1935 as a Birthplace memorial, and with Mr Bowden, a grandson of the man who purchased Mrs Doughty's farm from her heirs around 1890. Mr Bowden's mother recalled Elgar coming to visit the farm about 1930.)

The youngest Elgar children also went to Broadheath. Years later Frank Elgar's widow wrote to Edward's daughter: 'When Grandma [Ann Elgar] was particularly busy in the shop in High St. she sent Frank & Dot out to Broad Heath to stay with some old people—Mr & Mrs Doughty; of course there was a well there & Mrs. D. would periodically come to the door & say "Master Alfred (Frank) be you in the well?" I believe the two children took tin whistles with them & played duets to the great joy of the village children—Also, being very lost without some sort of music, they drove a couple of nails into a wall & stretched elastic from one to the other, so that with their fingers they could produce *some* sort of sound.' (Letter of 24 Sept. 1940 from Mary Agnes Elgar to Carice Elgar Blake. MS at the Elgar Birthplace.)

[2] Sketchbook II, fo. 56ᵛ.

F ♯ minor Jape

The jig rhythm was familiar from eighteenth-century music heard and played. The pattern of notes going down a fifth and up a fourth to go down another fifth was a common bass figure in Corelli and Handel. Edward set his own 'tune' in the bass clef: yet as it was written in the later sketchbook, there was an extra stave left blank over each of the lines. The blank staves seemed to invite other voices to join in a rising ensemble—like the ensemble of voices that rose when the wind went through the reeds.

The jig rhythm made insistent repetition. Repetition is the simplest way to carry on—to avoid coming to a halt: so it can suggest the most basic and primitive need of self-teaching—the need to propel the expression. Training at the hands of others might direct a young composer's attention to more sophisticated things. Edward's love of repeating patterns was to go on and on, until it became an element of his mature style.

The same principle of repetition was carried into the 'tune' itself—in the sequential figure at its centre. Sequence-writing repeats a single shape in changing positions. Of all melodic devices, then, it most resembles and suggests the patterns of nature in the countryside round Broadheath: gentle undulations of field and hedgerow, copse and dell—fruit trees planted in rows to make an orchard—the linked chain of the Malvern Hills rising up suddenly out of the Severn Valley—and flowing through all that landscape, the curving and re-curving river. Sequential writing would become another of the most prominent elements in Edward's mature music.

The particular shape of sequence in this childhood tune was also suggestive. Each of its downward fifths made a small arrival: yet all the small arrivals together existed inside a larger pattern. Each repeating of the little sequence was like his own summer arrival at Broadheath: it was only temporary. The larger pattern of all the little arrivals projected the classic goal of musical excursion from tonic to dominant. So the little inner arrivals together opened something paradoxically new.

The whole movement of the sequence was not upward but down and down—as if searching endlessly for some harmonic firmness. Shaping his melody in downward steps was to be one of the most persistent ideas through all Edward's music of maturity even after his music had found the harmonic foundations he sought.

The 'tune' also gave an instrumental hint. Bass movement in fifths and fourths suggested lower instruments tuned to those intervals. He marked the tune to be played by 'Celli & C[ontra] B[assi]': the fifths and fourths would come easily across cello and double bass strings. And the quick repeating notes could be sounded more easily with a bow on a string than with a finger re-striking the same key of a piano. So it confirmed the earlier hint. Making his music without reference to the piano was to seem a basic tenet of his creative thinking as he recalled it for an interviewer years later:

There is one point Sir Edward makes in telling this. Music to him was never the keyboard. It was just 'music'. He never evolved his thoughts in a keyboard shape, as the early training of so many composers impels them to do. His thoughts came to him, and still do, as abstractions—just music, but clothed in one colour or another, determining their disposition on one line or another of the orchestral score.[3]

The second line of the 'tune from Broadheath' began a still larger repetition, matching the opening line note for note at first. But half-way through it, he started to introduce accidentals—semitones which subdivided the pattern of downward steps. It made a fresh variant of the sequential principle. At the end he forced the music back to G—and then suddenly concluded the line on F ♯. Physically the F ♯ was right next to G; harmonically it was far distant. So it could turn the tune from Broadheath into a 'humoreske'. He noted beside the second-line ending in the later sketchbook 'F ♯ minor—Jape'.

The surprise was very close to a pun. But sudden shifting of tonality was a natural ally of sequence-writing: where melodic shape repeated, tonal change must play a more important role. Although the 'F sharp minor Jape' led the child's tune no farther then, when the 'tune from Broadheath' was used years later as the basis of a movement in his first *Wand of Youth* Suite, the F sharp minor modulation there opened an entire new episode. In fact the so-called 'Neapolitan' relation of juxtaposing keys a semitone apart was to be another favourite device of his maturity.

* * *

When Edward outgrew Miss Walsh's Dame's School, the alternative lay two and a half miles outside the city. At Spetchley Park, along the road to Stratford, lived the Berkeleys, a distinguished old Catholic family related to the man who owned much of the land round Broadheath. The Berkeleys of Spetchley had built a small school for the children of their village and estate. The children were taught by the Sisters of St. Paul under their Superioress, Mother Mary. So Edward began trudging every day to Spetchley, which he would remember as 'the village where I spent so much of my early childhood—at the Catholic School house: my spirit haunts it much.'[4] There was a fragrant reminder of Broadheath in the large pine trees also at Spetchley.

---

[3] *The Music Student*, Aug. 1916, p. 343.
[4] Letter of 14 Aug. 1912 to Ernest Newman (MS copy by Lady Elgar, BL Add. MS 47908). The school building at Spetchley had been designed by the elder Pugin and built in 1840. The sisters of St. Paul taught there until 1870–1, when they went out to nurse the wounded in the Franco-Prussian War. Mother Mary was killed by a stray bullet on the battlefield. After that date the teaching at Spetchley was in the hands of seculars. The earliest pupils' records now preserved there begin in 1873 (after Edward Elgar had left the school): in August of that year there were 36 pupils enrolled, ranging in age between four and twelve.

The Spetchley school was named after the mother of the Virgin Mary, St. Ann. It could only make clearer still the power of feminine influence. For behind all these Catholic women stood the mother herself, and the mother alone. Ann Elgar found her own image of it one day in 'a beautiful thought' (as she described it) which she copied into her scrap-book from the Irish writer Thomas Moore:

'Among the voices I could distinguish some female tones, which, towering high and clear above all the rest, formed the spire, as it were, into which the harmony tapered as it rose.'[5]

While Edward pursued his education and his music, his mother was seeking to repair her own ensemble of family life after the loss of Harry and then of little Jo. Like many lower-middle-class parents in those years, she set store by the new documentation of the camera. Through the 1860s and 1870s the Elgar children were regularly trooped into the photographer's studio for a visual record of their life. But these images of the living children inevitably reminded their mother of those who were missing. She wrote of them later in a poem 'To my nephew in America, with the photograph of the group of "Mother's Five"':

> Aunty's picture of the jewels
> In her earthly diadem:
> Two have been removed for safety,
> So she cannot send you them.
>
> If perchance across the photo
> Shines a ray of lustre bright,
> Think it is a bit of radiance
> From the others out of sight.[6]

One of the earliest photographs of 'the group of "Mother's Five"' was taken in 1868—the year after Edward wrote his tune from Broadheath. The group pose shows a careful arrangement, and it distinguishes Edward from the rest. Lucy and Pollie—five and two and a half years older than Edward—stand at the back of the group. The youngest sister, Dot, is perched on a low seat at the left. To balance her on the right stands the younger brother, Frank. So they all together make a frame for eleven-year-old Edward, sitting alert at a table in the centre of everything. Before him is a sheet of music paper, and in his right hand is a pen poised at the inkwell.

Two motives could dictate this arrangement. The first was Edward's inherited position as the family's eldest son—'the governor'. So the ghost of Harry Elgar seems to inhabit the photograph at Edward's side. But the pose with music paper and pen suggests that Edward is in the centre also because of his talent. So the ghost of Jo, 'the "Beethoven" of the family', hovers there beside him too.

---

[5] From Moore's prose work *The Epicurean* (MS copy in Ann Elgar's scrap-book, now at the Elgar Birthplace).
[6] Copied into her scrap-book on 9 Dec. 1882 (Elgar Birthplace).

Close to the time of this photograph, Edward received a gift which would be kept for the rest of his life. It was a little religious engraving, inscribed by the Superior of St. George's Church: 'Eddie, from Father Waterworth Sept. 17 1868.'[7] It shows the Death of St. Joseph. There sits the earthly stepfather of God's son, full of years, calmly facing his own extinction. On the left kneels the Holy Mother. But her gaze goes past the dying husband to the divine Son standing on the right and lifting his arm toward a bright cloud. The legend is in simple French, which Ann Elgar could translate for her son:

YOU SHALL LIVE AGAIN FROM OUT OF THE TOMB
He who believes in me shall never die
and this passing death is nothing but a sleep
to make entry into Eternal Life.

On the other side was a text which said in part:

O death, precious before God and filled with ineffable consolations! . . . Jesus, wishing to console Joseph, fixes on him his eyes full of tears and says these words:
'Blessed are the pure in heart, for they shall see God!'
. . . The Angels come to gladden him with their presence, to console him with their songs . . .
JESUS, MARY, JOSEPH, PRAY FOR ME IN MY OWN AGONY.[8]

The death of a man grown old in faith was a favourite Catholic theme. Three years earlier the celebrated Father Newman of Birmingham had published his poem 'The Dream of Gerontius', incorporating the last words of this picture-text and showing the life after death guided by angels. 'The Death of St. Joseph' was a common subject in these little French engravings for children. But the giver of this picture, Fr. Waterworth, was also a Jesuit intellectual. He knew that Edward had experienced the death of his own Joseph in his favourite brother; and September was the anniversary of Jo Elgar's death. So the little gift could suggest again the importance of the past, even for a life which counted only eleven years.

The occasion of the priest's gift was Edward's beginning at another new school. This was at Littleton House, an old place surrounded by orchards and hop-yards, and again across the river to the west. There he would be taught for the first time by a man, a professional Catholic schoolmaster. Littleton House was advertised thus in the Catholic Directory:

Young gentlemen are prepared for Commercial pursuits by MR. REEVE, who has been engaged in tuition since 1848.

Francis Reeve had the respect of the whole Catholic community around Worcester. He had hoped to become a priest, but a fall in his youth while climbing a tree had caused the withering of his right arm. It was the right arm with which the Mass would be served, and only those perfect in body as well as mind could be eligible for priesthood. So he had married and set up a Catholic

---

[7] Elgar Birthplace.　　　　　　　　[8] Ibid.

Academy for boys. In 1867 he had moved his school to Littleton House. Hubert
Leicester was already attending there: he had conceived 'a tremendous respect'
for Mr Reeve and thought him 'a magnificent man'.[9]

Going to school at Littleton House would mean crossing the river for every
journey by the ferry from the Cathedral Steps. Ann Elgar asked young Hubert,
now thirteen, to go with Edward and show him the way. Hubert was the sort of
character who inspires respect on all sides. Edward remembered: 'My mother
said to me as a boy, "Wherever Hubert goes or whatever he does, you may join
in." That is a tremendous thing to say.'[10]

From the other end of their lives, Edward fondly recalled the journey with
Hubert to school, and two things especially. One was the elder-brotherly
presence of this more experienced, responsible friend—the boy who would
grow up to head his own firm of chartered accountants and become Mayor of
Worcester. The other was the magic suggestion of the journey itself: for
Littleton House, as it lay westward beyond the Severn, was on the Broadheath
side of things. Edward wrote:

. . . our walk was always to the brightly-lit west. Before starting, our finances were
rigidly inspected,—naturally not by me, being, as I am, in nothing rigid, but, quite
naturally, by my companion, who tackled the situation with prophetic skill and with the
gravity now bestowed on the affairs of great corporations whose accounts are harrowed
by him to this day. The report being favourable, two pence were 'allowed' for the ferry.
Descending the steps, past the door behind which the figure of the mythical salmon is
incised, we embarked; at our backs 'the unthrift sun shot vital gold', filling Payne's
Meadows with glory and illuminating for two small boys a world to conquer and to love.

Yet as they went, the younger boy watched for differences—things that
would show, in their companionship, some shape of himself:

Boys so inconveniently alive could not exist without speculating on the origin and
meaning of such delicious names as Newdix, Bedwardine and Diglis . . . I know that
while my own flibbertigibbet mind was concerned with the humourous possibilities of
such a word as Newdix, my school-fellow's more sedate intellect was busy with the
seriousness of derivation and signification . . . Nothing escaped our notice; everything
was discussed and adjudicated upon . . . I recall a report in the *Journal* of a committee in
which it was advised that an outlying street should be 'curbed and channelled'. This was
new to me and I read it with the joy of discovery; I was on a 'peak in Darien' certainly,
but was not likely to stand 'silent'. Sunday offered no easy opportunity to air the
acquisition, but on Monday on our way to school, in the hopyard then existing [near
Littleton House] I saw my chance. Casually kicking the turf, I said: 'Hubert, this ought
to be curbed and channelled.'

[9] Information from the *Catholic Directory* (editions of 1856 ff.) and from conversations with
Hubert Leicester's youngest son, Fr. Edward Dering Leicester. The school at Littleton House
(later called St. George's Academy) was listed in the *Catholic Directory* up to 1892. Mr Reeve was
born in 1825 and lived until 1912.

[10] Speech at the ceremony to confer the Freedom of the City upon Hubert Leicester, reported in
The Worcester Advertiser, 7 Apr. 1932.

Hubert (eagerly): '*You* saw that!'
E.: 'Yes,—and got it off first.'[11]

As Hubert ceded primacy where Edward deserved it, the older boy gained the lasting affection of the younger. Sixty years later Hubert's daughter-in-law was to see them together, still held by the old magnetism: Edward darting his eyes everywhere, talking about everything under the sun to entertain Hubert—Hubert basking in the incomparable brilliance of Edward's companionship.[12] The strength Edward drew from his friend was acknowledged again after they both had passed seventy, remembering the ancient sun that lit their way to school: 'In our old age, with our undimmed affection, the sun still seems to show us a golden "beyond".'[13]

For Hubert the fascination lay in the sheer maturity of Edward's company. Forty years after their school days he recalled: 'Nothing of a boy about him. One great characteristic—always doing *something*. When he stopped away from school (which he did about 1/3 of his time) it was not to play truant merely.'[14] The reason, as Hubert admitted, was often Edward's fragile health: 'He was a delicate boy, and the two miles to and from his school at Wick was, no doubt, a bit of a strain for him, so it frequently happened that young Elgar was absent once or twice a week.'[15]

The weakness offered another focus for Hubert's elder-brotherly solicitude. For the Elgars, after the loss of Jo, the weak health of this remaining elder son could only have been a constant anxiety.

Once Edward did play truant; and then he took Hubert with him. It was just the occasion to tempt Edward—a rare manifestation of old-fashioned pageantry. Describing the adventure in later years, Hubert's self-effacement was betrayed by his keenness of observation:

In those days the High Sheriff used to give a breakfast on the first morning of his shrievalty when he came in to meet the Judge. He used to have at least 12 javelin men, a carriage and four, two full-dressed footmen, a coachman in full uniform and two trumpeters, and all the gentlemen invited to the breakfast rode behind on horseback or in their carriages. Mr. Hornyold, of Blackmore Park, Hanley Swan, one of the wealthiest and most desirable of Worcestershire's landlords, was High Sheriff at the time.

Elgar was going to school with another lad who, rather suggestively, it would appear, remarked:

'Ted, the High Sheriff will be coming along this road to meet the Judge, and I should like to see the procession.'

[11] Foreword to *Forgotten Worcester*, by Hubert A. Leicester (Ebenezer Baylis, The Trinity Press, 1930). The quotations are from Henry Vaughan's poem 'The Fountain' (which Elgar set to music in 1914) and from Keats's 'On first looking into Chapman's Homer'.
[12] Conversations with Mrs Philip Leicester, 1975.
[13] Foreword to *Forgotten Worcester*.
[14] Conversation with Fr. Driscoll SJ, 28 Nov. 1909, noted down by Philip Leicester immediately afterward.
[15] Obituary notice written anonymously for a Worcester newspaper in 1934.

They had got well on the way to school, but had seen no sign of the procession. So Elgar said: 'What do you say to going to meet him?'

It was agreed, and they walked on until it was too late for school. Then he said: 'I don't suppose the governor (meaning the master) would mind.'

So they walked on as far as Powick Asylum, but there was no sign of the procession. There the two truants sat down under the hedge and ate their lunch, and then, having made up their minds that they had missed the show, they walked back to school.

The 'governor' naturally demanded an explanation, and young Elgar frankly replied that they went to meet the High Sheriff.

With an admonition the master forgave them.[16]

Hubert might have been the elder, but it was Edward who offered the cool justification of going to meet the High Sheriff—and got it accepted.

It recalled his first encounter with the schoolmaster at Littleton House. Asked his name on the opening day of term, the boy had answered:

'Edward Elgar.'
'Say, "Sir".'
'Sir Edward Elgar.'[17]

This famous anecdote has often been cited as a child's misapprehension. In the mouth of the self-possessed eleven-year-old Edward then was, it was almost surely deliberate. The barest hint of a pause made between the 'Sir' and the 'Edward' ought to have kept retribution at bay.

In the early summer of 1869, near the end of his first year at Littleton House, the Elgar family had the chance of a holiday together in a cottage on the Berkeley estate at Spetchley Park. It was the gardener's cottage,[18] standing close by the Spetchley schoolhouse in a grove of enormous pines. These family weeks together in a country cottage might almost recapture the old life at Broadheath. And the great evergreens surrounding the cottage at Spetchley would bring back for Edward the very scent of Broadheath as he recalled the earlier cottage 'nearer the clump of Scotch firs—I can smell them now—in the hot sun.'

So this recapitulation of earliest time could link itself with the boy's recent memory of his Catholic schooling at Spetchley. They came together powerfully enough to haunt his memory for thirty years, until at last they were given creative expression in music. A friend of his later life wrote: 'Elgar told me that as a boy he used to gaze from the school windows in rapt wonder at the great trees in the park swaying in the wind; and he pointed out to me a passage in

---

[16] Obituary article for a Worcester newspaper, 1934. John Vincent Hornyold became High Sheriff of the County in Feb. 1869. He attracted attention, as a Catholic occupying such an office was almost without precedent. In addition to this interest for the Leicesters and Ann Elgar, Hornyold's great house was just the sort of place to delight W. H. Elgar in his piano-tuning rounds.

[17] I am unable to find an early written source for this well-known anecdote. It is entirely in character and may well have emanated from the stories Elgar told in later years of his own youth.

[18] Information from Miss Juliet Berkeley at Spetchley Park, 1975. Miss Berkeley had it from her late father, Robert Berkeley, who was told by Elgar himself.

Gerontius in which he had recorded in music his subconscious memories of them.'[19] Visiting Spetchley Park nearly half a century after his days there, the distinguished composer would enter the library of the great house to find a copy of his own setting of Newman's 'The Dream of Gerontius', and open it to the page containing the music of the Soul's approach to God:

> 'The sound is like the rushing of the wind—
> The summer wind among the lofty pines.'

Next this he would write, 'In Spetchley Park 1869.'

It is rare to remember from such a distance of time the moment of starting so lengthy a process of private response. But then it is rare for a boy to be so absorbed in his own responses and their evolutions that he should be driven to find means of turning them to creative purpose.

*   *   *

Returning homeward from Spetchley in the summer of 1869, however, he had not yet found a way to develop his music. He set out to acquire further techniques for himself:

I am self-taught in the matter of harmony, counterpoint, form, and, in short, the whole of the 'mystery' of music . . . To-day there are all sorts of books to make the study of harmony and orchestration pleasant. In my young days they were repellant. But I read them and I still exist. The first was Catel, and that was followed by Cherubini.[20]

Using a theme from one of the exercises in Cherubini's *Counterpoint*, Edward began to write a fugue:

The fugue subject accelerated to double its opening pace. Acceleration is the simplest way of increasing rhythmic power and interest. It might be another portent of the future. Through much of Edward's later music, an accelerating pace would ally itself with sequential repetition to shape his melody.

A farther extending of his interest in acceleration emerged in the performances he would conduct. In the gramophone recordings he directed fifty and sixty years later, the mature composer often accelerated the pulse of his own figures and melodies toward a climax, afterwards dropping back the pulse to prepare for new acceleration.

[19] Ernest Newman, *The Sunday Times*, 30 Oct. 1955.

[20] *The Strand Magazine*, May 1904, p. 539. Elgar's copies of Catel's *Treatise on Harmony* and Cherubini's *Counterpoint* are preserved at the Elgar Birthplace.

[21] BL Add. MS 49973A fo. 75ᵛ, labelled in Elgar's hand 'about 1870'. Christopher Kent points out four bars of attempted continuation on fo. 74ᵛ.

The boy's fugue went through its four-voiced exposition. He made a fair copy, and on another sheet sketched a possible continuation.[22] But that was not as straightforward as following the rules to a ready-made fugal exposition. There at the end of Edward's fair copy was another bar ruled and ready, but nothing ever went into it. The fugue remained unfinished—stopped where the need to develop his ideas called for an assurance of knowledge Edward did not have.

The first real sort of friendly leading I had, however, was from 'Mozart's Thorough-bass School'. There was something in that to go upon—something human.[23]

A richer source of self-teaching lay in the 'stream of music that passed through my father's establishment'. Half a century afterwards the gift of some classic scores brought it all vividly back:

I renew my growth in reading some of the dear old things I played when a boy,—when the world of music was opening & one learnt fresh *great* works of every week—Haydn, Mozart and Beethoven . . . the feeling of *entering*—shy, but welcomed—into the world of the immortals & wandering in those vast woods—(so it seemed to me) with their clear pasture spaces & sunlight (always there, though sometimes hidden), is a holy feeling & a sensation never to come again, unless our passage into the next world shall be a greater & fuller experience of the same warm, loving & *growing* trust.[24]

Here were the people who might really be 'kind' to him.

One discovery made a special impression. It showed a tonal shift like the 'F sharp minor Jape' in his own 'tune from Broadheath'—but in the hands of Beethoven. Years later he recounted it for a critic who reported it thus:

The story of every mind susceptible to artistic feeling affords instances of sudden impressions of far-reaching influence. One of the earliest and most powerful of these came upon Edward Elgar on his first reading of Beethoven's First Symphony. Only the pianoforte score, but the effect was there. It came upon him like a lightning flash. The transition to the key of D flat, and back to C, in the minuetto, left him breathless—'sank into his very soul'—convinced him that counterpoint was not the last word of musical art: that Tallis and Byrd and Orlando Gibbons and the rest of the classic church-composers had not exhausted the possibilities: that, despite the dicta of the critics and university professors, the 'solid diatonic style' did not represent the Ultima Thule of composition; and, finally, that Mozart and Beethoven, having attained the highest plane of emotional expressiveness, were the best models for study.[25]

The same experience was recalled again for another interviewer in 1916 with an extended metaphor of water:

A new ocean was in view, and he longed to sail it. An ocean on which you might set forth from the safe harbour of Natural Quay, touch rapidly and momentarily at such adjacent

[22] BL Add. MS 49973A fo. 75ᵛ and 74ᵛ (continuation: identified by Christopher Kent). Elgar himself later dated the exposition fair copy 'about 1870'. It was reproduced in *The Music Student* (Aug. 1916), p. 343, where the composer's age was stated as 'fifteen'.

[23] *The Strand Magazine*, May 1904, p. 539. Elgar's copy of *Mozart's Thorough-bass School* (now thought to be the work of another hand) is preserved at the Elgar Birthplace.

[24] Letter of 15 Dec. 1909 to Edward Speyer.   [25] Buckley, *Sir Edward Elgar*, pp. 12–13.

ports as the Quays of One Sharp and Two Sharps, and then, with a sudden favouring breeze find yourself making for the Coast of Flats—passing quickly from port to port, resting nowhere and, almost before you realised the distance you have travelled from home, find yourself casting anchor for a time and swaying gently on the tide in the harbour of Five Flats.[26]

Beethoven had written this music for the orchestra. If Edward should write his own music for orchestra, he would need to know how to use all the instruments. Beethoven had left no book of instruction to help him. But as it happened, there was such a book written by a famous friend of Beethoven. It was Anton Reicha's *Orchestral Primer*, and it explained:

I. The Nature and compass of all the instruments used in the orchestra, and the distribution of them into two principal masses, stringed and wind instruments;
II. The Mode of arranging musical ideas for the orchestra;
III. The Treatment of the two orchestral masses, both separately and combined.

Edward acquired a copy of this book: it bears an inscription in his mother's hand: 'E. W. Elgar. March 7th, 1869'.[27] Reicha's opening description gave first place to the violin:

Of all the instruments used in the orchestra, the violin is the most useful and important . . . The violin may be played on in any key, and that with almost equal facility; this and the rapid execution which it admits of, has gained it the title of the *king of instruments*.

The year 1869 brought the Three Choirs Festival to Worcester again. No Beethoven was to be sung that year, but Edward extracted a promise from his father to get him into the rehearsal of Handel's *Messiah*—for which Elgar Bros were to supply orchestral parts. Anticipating this great moment gave another impulse to the boy's thinking:

I composed a little tune of which I was very proud. I thought the public should hear it, but my opportunities of publishing it were decidedly few.

I saw my opportunity when my father was engaged in preparing the Handel parts for the forthcoming festival. Very laboriously I introduced my little tune into the music. The thing was an astonishing success, and I heard that some people had never enjoyed Handel so much before! When my father learned of it, however, he was furious![28]

Yet once inside the big rehearsal, the interest of his own Jape was overborne by the sheer experience of Handel's music. He remembered it many years later for a group of musicians: a lady among them made notes of what he said.

One of his earliest recollections was of going into the Cathedral and hearing the orchestra rehearse 'O thou that tellest'. The opening greatly impressed him, and he thereupon resolved to learn the violin.'[29]

[26] Scholes, in *The Music Student*, Aug. 1916, p. 345, where Elgar's age at the time of this experience is stated to have been eleven.
[27] Elgar Birthplace.
[28] Newspaper cutting at the Elgar Birthplace.
[29] Leaver, 'Some Impressions of Sir Edward Elgar', in *Musical Opinion*, July 1934. The date of

This music made an ideal confirmation for Reicha's estimate of the violin as 'king of instruments'. The melody of 'O thou that tellest' contained a wealth of sequential imitations: the leaps between them would be difficult and unnatural to play on a keyboard, but were perfectly suited to moving across the strings of the violin. And the introduction and accompaniment of the aria was scored for all the first and second violins to join together in sounding their melody above the rest of the ensemble. The leaps over the strings would force all the violins to bow together—firsts on the conductor's left, seconds on his right, stretching across the entire front of the orchestra. The sight of them all moving together could match vision to the grandeur of sound:

He hurried home and begged his father to lend him a violin from the stock of the shop, when he at once retired to an attic and straightway endeavoured to pick out the beginning of 'O thou that tellest'. The first two notes, A and D, were easy on the open strings, but then came difficulties. However, the boy was not to be beaten, and at the end of a fortnight he had mastered the violin part of 'O thou that tellest'.[30]

There was no mistaking Edward's attraction for his new instrument. Hubert Leicester noticed that he became 'very thick with old tailor who did odd work and scraped the violin a bit at night'.[31] The interest was also noted by W. H. Elgar, as Hubert recalled:

When E. E. could scrape violin (at about 12) old man introduced him to Crown Hotel Glee Club.[32] They were short of a second violin, and his father said:
    'Well, I think the Governor would come.'
    'Do you think he can play?' Elgar's father was asked, and he replied:
    'He can play well enough.'
    . . . His father and Edward played the violins, his uncle the harmonium, and one or two other amateurs completed the orchestra.[33]

The instruments had been introduced to vary as well as support the singing. The Glee Club was an old institution even then. It seemed impervious to time, as described by Hubert Leicester's son fifty years later:

It is an autumn evening in a Midland city. The market carts have trotted out into the quiet country, the shops are shutting and queues are forming outside . . . the little theatre. We pass down a narrow side street, under the archway of an ancient hotel, such as Dickens loved, cross the cobbled yard and climb a dark and narrow staircase. At the top we find ourselves in a long, low room. Huge fires roar a cheery welcome. Round the walls and at numerous small tables are seated perhaps a hundred men, of all ages and

---

the experience is supplied by Maine, *Life*, p. 10, recounting the story from information supplied by Elgar himself. Elgar's copy of the 1869 Festival programme (now at the Elgar Birthplace) is inscribed 'My first festival', suggesting he may have heard more of the 1869 Festival than the one rehearsal he seems to have got into in 1866.

[30] Leaver, op. cit.
[31] Conversation of 28 Nov. 1909, noted by Philip Leicester.
[32] Blake, MS notes.
[33] Worcester newspaper article, 1934.

conditions, from prosperous city fathers to young clerks. Half a dozen buxom waitresses flit about with trays of tankards and glasses and the air is heavy with smoke. . . .

The Club has flourished since far back into the old coaching days. Matches are still prohibited and on each table is a bundle of spills and a lighted candle. The portraits of local worthies which line the walls have looked down upon the great-grandfathers of the present audience, upon pigtails and churchwarden pipes. Nelson's sailors and Wellington's officers have sat in these chairs . . .

At the top of the room, on a small platform, sits the chairman, a local alderman, and at his side the stout and genial secretary. In the corner beside them is the band—a piano, some violins, a couple of 'cellos, wood-wind and a solitary cornet. Near them sit two men in evening dress, artistes from the cathedral choir of a neighbouring city . . . The chairman looks at his watch, nods to the conductor and strikes a bell. At once the buzz of conversation stops, the maids disappear through a side door and the concert commences.[34]

The proceedings always commenced with 'Glorious Apollo', and seven other glees and two songs completed the programme. An additional accompaniment . . . was furnished by churchwarden pipes solemnly smoked by the senior Apollos.[35]

'And', as Edward remembered, 'they smoked all the time.'[36] Soon Edward himself began smoking. It was to give him pleasure all his life.

Corelli was largely drawn upon, Handel's Overture to 'Saul' was a favourite, and Haydn's symphonies were often heard. The rich store of our great glee writers furnished the vocal music, and they were very well done in those days. Not many songs were sung, and they were of a healthy, vigorous type.[37]

During the seasons of 1870–1 and 1871–2 the Glee Club instrumental programmes included the Overtures to *Zampa* (Hérold), *Norma* (Bellini), *Masaniello* (Auber), and *Maritana* (Wallace)—all in more or less *ad hoc* arrangements for the forces available—and occasionally a big selection from one of the operas—*Norma* or *Il trovatore*. There were Violin Sonatas by Beethoven and Schubert, and Haydn Symphony movements. The piece of Mozart's they often played was an arrangement of a trio (from the opera *Così fan tutte*) called 'My sweet Dorabella'. There was the Wedding March from Mendelssohn's *Midsummer Night's Dream*, and the Coronation March from Meyerbeer's *Le prophète*—with its daring triplet to open a measure in common time. That made a life-long appeal to Edward, as a friend of later years recalled: 'His eyes used to shine with excitement when we discussed this and hummed the strong rhythmic tune: the strength of that triplet on the first beat of the bar gripped him.'[38]

---

[34] TS article submitted to *The Musical Times* 30 Sept. 1921.

[35] [F. G. Edwards] 'Edward Elgar', in *The Musical Times*, Oct. 1900, p. 642, based on information supplied by Elgar himself.

[36] MS note on a galley proof of Edwards's *Musical Times* article: BL Egerton MS 3097A fo. 13–22.

[37] [W. Mann Dyson] in *The Musical Times*, 1 Oct. 1900.

[38] Reed, *Elgar as I Knew Him*, p. 86. A large collection of the Crown Hotel Glee Club's printed programmes for these years is preserved at the Elgar Birthplace.

Yet there was to be a moment when Edward met humiliation in the Glee Club: 'When a child, I once came in wrong with a second violin passage. I shall never forget my horror. I feel it even now. I did not analyse my sensations at the time, but I know that it was an artistic horror.'[39] The horror of breaking the ensemble might overtake him at any time, instantly and without warning, whenever performing music in the ensembles he joined with his father. The only musician free from its threat is the one who writes out his ensemble in advance. When Edward attempted to do that again, he used the ensemble of his mother—the family itself.

*   *   *

The Elgar children would present a play with music. The two parents would be spectators, with all the action and music provided by the brothers and sisters. Edward's later recollection insisted on the co-operative effort. Yet the drama they worked at reflected his own individuality and its private landscapes.

Some small grievances occasioned by the imaginary despotic rule of my father and mother (The Two Old People) led to the devising of [the musical play].

By means of a stage-allegory—which was never completed—it was proposed to shew that children were not properly understood.

The scene was a 'Woodland Glade', intersected by a brook; the hither side of this was our fairyland; beyond, small and distant, was the ordinary life which we forgot as often as possible. The characters on crossing the stream entered fairyland and were transfigured. The Old People were lured over the bridge by 'Moths and Butterflies' and 'The Little Bells'; but these devices did not please,—the Old People were restive and failed to develop that fairy feeling necessary for their well being. While fresh devices were making, 'The Fairy Pipers' entered in a boat and 'charmed them to sleep'; this sleep was accompanied by 'The Slumber Scene'; here we may note that the bass consists wholly of three notes (A. D. G.)—the open strings of the (old English) double bass; the player was wanted for stage management, but the simplicity of the bass made it possible for a child who knew nothing of music on any instrument to grind out the bass.[40]

Our orchestral means were meagre: a pianoforte, two or three strings, a flute and some improvised percussion were all we could depend upon; the double bass was of our own manufacture and three pounds of nails went into its making; the needless asperity with which one of the Old People enquired into the disappearance of these nails confirmed us in our resolution to produce our play. For the benefit of other young luthiers it is worth recording that the tone of the bass was poor, probably in consequence of a superfluity of metal. But we had the gorgeous imagination of youth and the ubiquitous piano became a whole battery of percussion, a whole choir of brass or an array of celestial harps as demanded by the occasion.[41]

[39] Buckley, *Sir Edward Elgar*, p. 22.

[40] Draft note for gramophone records, 1929 (HWRO 705:445:1304), quoted in Moore, *Elgar on Record* (OUP, 1974) p. 93.

[41] Draft note for *The Wand of Youth* Suite No. 2 first performance, 1908 (MS in possession of the writer).

To awaken the Old People, glittering lights were flashed in their eyes by means of hand-mirrors ('Sun-Dance'). Other episodes—'The Fountain Dance' in which the music follows the rise and fall of the jets; the water was induced to follow the music by means of the interior economy of a football. Other episodes, whose character can be deduced from the titles [—'Serenade', 'Fairies and Giants', 'The Tame Bear', 'Wild Bears'—] followed, and the whole concluded on the 'March'.[42]

Later recollection attached two dates to this effort. One was 1869—the year he was twelve, the year he began to play the violin. The other was 1871, when he was thirteen and fourteen. As a projection of Edward's experience, the play could have been written in either of those years. It hovered between childhood and maturity, looking uneasily forward, looking fondly back. So the nostalgia that was to shape so much of his mature style began to make itself felt. But the action reversed his childhood's experience in one vital respect: in the play it was the vision of childhood that guided the Old People, and not (as in real life) the other way about.

The play setting was a miniature of Edward's private landscape—a stream dividing the world of Old People from the Better Land of childhood. The foreground of the play evoked the countryside round Broadheath in two special ways. Close to the cottage of Edward's birth was indeed 'a woodland glade intersected by a brook'.[43] And in the garden of the cottage itself in the Elgars' time there was a real fountain: that could inspire the play's 'Fountain Dance'. Beyond the dividing stream lay the land of the Old People—the city world of maturity and dullness, 'the ordinary life which we forgot as often as possible'. In real life this was the foreground where the young Elgars spent most of their time; it was Broadheath that was really 'beyond, small and distant . . .'. So again the experience of reality was reversed.

The play's action showed why the countryside of childhood's past was a Better Land. The Old People who entered it were 'charmed to sleep' and then 'transfigured': when they had been 'awakened' by the glittering lights of 'our fairyland' flashing in their eyes, then they would find the world of inspiration. That reversal of old and new was re-enacted thirty years later in the setting of Newman's 'The Dream of Gerontius'. There his music would show again the death of an 'Old Man' from his mundane life to awaken the transfigured Soul in a blinding vision of the ideal life from which it had first come.

The children's play would need a lot of music. It needed more than an inexperienced composer could write all at once, especially leading a double life as a busy schoolboy. So Edward looked out the music he had already written. There was the tune from Broadheath. That became the basis for a scene in

---

[42] Draft note for gramophone records, 1929.

[43] There were two such places which the young Elgar could easily find during his boyhood visits back to Broadheath. One was near the eastern edge of the Doughty farm by the Common. The other, known as 'The Dingle', began just across the lane from the cottage of his birth: its stream was fed by a deep ditch for drainage running down the length of the cottage garden on its eastern side just beyond a hedge.

which Fairies and Giants were to enter. The Fairies entered to the old tune itself; then the three Giants made their appearance to a great distortion of it. At another point the action depended on Moths and Butterflies to lure the Old People over the bridge. Of the melody used for this he later wrote: 'I do not remember the time when it was not written in some form or other.'[44]

It anticipated the creative practice of Edward's maturity. Throughout his career he kept sketchbooks for noting musical ideas as they occurred to him—disconnected and without any particular use in view. Then when some project for a composition was formed, the sketchbooks could be consulted to see what material might be there, waiting to be 'used up' in a work unforeseen at the time the ideas were noted down. So an idea, once entered in a sketchbook, might ripen toward a maturity of its own. Then it could give its hint of the Better Land beyond the stream of time, where old ideas might be 'transfigured'—where the Older Person scrutinizing the notes of his former self might find those impulses transfigured through the Fairies and Giants, the Moths and Butterflies of his own recollecting.

At a critical moment in his creative life nearly forty years later, Edward sought out the music for his childhood play and re-scored it for full orchestra. He called it then *The Wand of Youth*. But the first making of the play and its music had already shown him one power of the talisman. When the boy organized his family into an ensemble for dramatizing the magic of his own inspiration, the Wand could show itself as a conducting baton.

[44] Draft note for *The Wand of Youth* Suite No. 2.

BOOK II

ENIGMA

4

The Man's World

ᶦᴎᴑᴊᴑᴥᶦ

THE children's play had sought to solve the problems of childhood at the expense of the adult world. But its hope fell victim to inexorable time. Like the tune from Broadheath and the attempted Fugue, the play and its music remained unfinished and unperformed as Edward grew older.

Outside the family, the Old Person who most helped him as he came in sight of adulthood was the schoolmaster at Littleton House, Francis Reeve. In his old age, Reeve recalled Edward for an acquaintance, who wrote:

As a boy at school he did not attract much attention. He was very shy and reserved, and was considered rather delicate; he had no taste for the rough games and rarely joined in them.

Mr. Reeve . . . was much struck one day, while watching Elgar play croquet, to hear the boy remark as he struck the mallet: 'Why, there is quite a note in it!'

He bore a very high character for obedience and attention to his work.[1]

At the end of Edward's first term in the school Mr Reeve inscribed a copy of Ormsby's *Autumn Rambles in Africa*: 'Littleton House. Christmas 1868. Prize awarded to Master Elgar for General Improvement. F. Reeve, Principal.'[2]

Edward recalled the schoolmaster's understanding when reminiscing long afterwards with Hubert Leicester:

Drifted on to Littleton House & old Reeve. Said he was a good teacher . . .

Said he [himself] never could learn anything in ordinary way by heart, so one day 'the dear old gaffer' called him up & questioned him. Found he could not repeat the work but knew it alright in his own way. Said Reeve:

'Why, you know it. You are not a stupid boy.'

'No Sir, certainly not,' said E. (He laughed loudly at this.)

Said he never did any 'prep'. Used to learn his lessons going to school, but as Reeve understood him, he always came out on top . . . Reminded [Hubert] how after he left, he (E.) was head of school for 9 months.[3]

Such a response to his own needs could not fail to win an intelligent boy's attention. And so it happened that a remark of Mr Reeve's bore fruit in the

---

[1] The Very Revd T. A. Burge, OSB, 'Edward Elgar', in *The Ampleforth Journal*, July 1902, pp. 26–7.

[2] Elgar Birthplace.

[3] Philip Leicester, Notes of conversation on 2 June 1914.

212564

future. Edward recalled it at the time of writing his oratorio *The Apostles* in 1903:

The idea of the work originated in this way. Mr. Reeve, addressing his pupils, once remarked:

'The Apostles were poor men, young men at the time of their calling; perhaps before the descent of the Holy Ghost not cleverer than some of you here.'

This set me thinking . . .[4]

The story of the Apostles could appeal to Edward also for its endlessness. It was a tale of inspiration progressing through a laying on of hands that repeated from one person to another—like repeating a melodic figure in a sequence.

Edward responded with an affection for the place of this encouragement which resembled his affection for the place of his life's beginning:

I remember when I was a boy of 13 or 14 that two of us . . . thought the communication between one side of the river and the other was rather feeble. We started to make a tunnel . . . You may have found a circle upon which I and another boy spent 10 days in digging two feet deep. We thought it might be an alternative route to school.[5]

If the Old People of his play had failed to cross the stream and re-enter the fairy glade of childhood, perhaps adolescence could make its own way to the Old Person who taught at Littleton House, beyond the western bank of the Severn on the Broadheath side of time.

Mr Reeve asked every boy for a letter each Saturday setting out the boy's activities and reflections for the week past. One of Hubert Leicester's letters to the schoolmaster survives:

Littleton House

May 28th, 1870.

My dear Sir,

Some of the would-be prophets stated that we were to have no rain for a considerable length of time, but I think it will not be long before we have some. The month of May has not been so pleasant as usual, I suppose it is owing to the weather. I have taken a walk after supper the last two nights hoping to hear the nightingales but I was disappointed. Elgar and I went to see the volunteers inspected on Monday evening. Mr. Butler's foreman was examined before the magistrates on Thursday, I think it was for embezzlement, but he was remanded till next week.

I remain

Your obedient pupil,
Hubert A. Leicester.

From this weekly letter-writing, the boys took to writing other letters, as Hubert recalled: 'When E.E. was 13 & H.L. 15 they used to write letters to papers on all sorts of subjects, frequently to the defunct 'Figaro' (an English imitation of the French one). Both prided themselves on their knowledge of

---

[4] Buckley, *Sir Edward Elgar*, p. 8.

[5] Speech on receiving the Freedom of the City of Worcester, reported in *The Worcestershire Echo*, 12 Sept. 1905.

English.'[6] Sometimes a letter published did not turn out as anticipated, but Edward could be resourceful in adapting to the unforeseen:

E. & Hubert (aged respectively 14 & 15 [sic]) once put an advertisement in the 'Evening Post' . . . requiring a *wife*, for a joke. They stipulated that she must have a cheerful temperament. The Post printed 'cheerful temperature'. E. thereupon cut out the avt. & posted it to the London 'Figaro', which paid him 5/– for it.[7]

But there were serious questions over which he and Hubert found they did not always agree. Edward remembered: 'We have differed and argued it out—instead of leading a sleepy existence, taking everything for granted and letting things slide—and respected each other afterwards. And we have seen eye to eye at last.'[8] Hubert wrote with his characteristic modesty of companionship with Edward: 'He and one or two other boys (now well-known citizens of Worcester) met almost daily, and were wont to talk of things which were far outside the humdrum philosophy of the old cathedral city's life.'[9] More than thirty years after his schooldays, Edward recalled: 'In 1871 I said literature must be provided for the newly educated people—a fortune could be made.'[10] It was the tradesman's son who spoke then, but it was also the child of a mother's imagination 'peopled from noble books'.

* * *

Just before his fifteenth birthday he ventured on the writing of a song. His tune from Broadheath had had only a single voice. His Fugue had four voices but no development. The music for his children's play had promise of development yet remained incomplete. But the music for a song could ride to completion on the back of words and stanzas already composed: the same music could even be repeated to fit as many stanzas as there might be.

Song words were usually by someone else. But Edward had no need for anyone else. For he also experimented with his mother's art: in addition to the children's play, later recollection was to bring back 'sundry lengthy poems, a novel and other literary exercises'.[11] Now if he wrote a poem in several stanzas of a single pattern and devised a tune to fit that pattern, there would be a finished work all by Edward Elgar.

Edward's poem took its subject from that favourite theme of his mother's, flowers and plants and how their cycle resembled the cycle of human life. The flower emblems traced the human cycle of conception, young innocence, mature fame and glory supported by a 'hidden sweetness', and the lasting memory.

[6] Conversation with Fr. Driscoll, SJ, 28 Nov. 1909, noted by Philip Leicester.
[7] Philip Leicester, MS notes of conversation with Elgar on 4 Nov. 1919.
[8] Speech on receiving the Freedom of the City of Worcester, 1905.
[9] Article written anonymously for a Worcester newspaper, 1910.
[10] *'A Future for English Music' and other lectures*, pp. 257 and 259.
[11] Note for *The Wand of Youth* Suite No. 2, 1908.

THE LANGUAGE OF FLOWERS

In Eastern lands they talk in flow'rs,
& they tell in a garland their loves & cares;
Each blossom that blooms in their garden bow'rs
On its leaves a mystic language bears.

The rose is a sign of joy & love—
Young blushing love in its earliest dawn—
& the mildness that suits the gentle dove
From the myrtle's snowy flow'r is drawn.

Innocence gleams in the lily's bell
Pure as the heart in its native heaven;
Fame's bright star & Glory's swell
By the glossy leaf of the bay are given.

The silent soft & humble heart
In the violet's hidden sweetness breathes,
& the tender soul that cannot part
A twine of evergreen fondly wreathes.

The cypress that daily shades the grave
Is sorrow that mourns her bitter lot,
& faith that a thousand ills can brave
Speaks in thy blue leaves, forget-me-not.

The[n] gather a wreath from the garden bowers,
& tell the wish of the heart in Flowers.

Most of these flowers and leaves, despite the poem's Eastern beginning, were part of the Worcestershire countryside. Most of their symbols were conventional, but one could touch his own experience. Finding a place for the evergreen and its 'tender soul that cannot part' made a small emblem of Spetchley and Broadheath.

As he set his music to stanza after stanza, Edward found he was not content simply to repeat the formula. He began to make variations. They started in the accompaniment, and soon reached up to change the vocal line. Variations offered an ideal means to extend musical expression: for variations harness an extemporizing impulse toward some sort of development.

With each stanza Edward's music became more complex. It suited the poem's theme of expanding life through the first three stanzas. It suited less well the violet sweetness and evergreen wreath of stanza 4. Faced with the choice between verbal and musical interests, the boy's loyalty went all to the music: through stanza 4 he increased the musical complexity again. When he reached the final stanza, however, he set its cypress shade and forget-me-not to a virtual repetition of the opening music. So the music as a whole still carried out the theme of life returning in cycles.

He dedicated *The Language of Flowers* to Lucy on her twentieth birthday, for she also responded to their mother's vision of nature. He wrote out the title in copperplate hand on the back of an old Beethoven Sonata cover:

THE LANGUAGE OF FLOWERS

The Poetry by PERCIVAL.

The Music composed and dedicated to his sister Lucy

by *Edward W. Elgar.* May 29, 1872.[12]

Later he signed his own name against the title's 'PERCIVAL'. But the music was what he claimed immediately. It is his earliest dated and finished composition to survive.

*   *   *

The boy facing his own fifteenth birthday in the spring of 1872 had another reason for wishing to show that he could finish a musical composition. The end of his fifteenth year would bring the end of his schooling at Littleton House, and he wanted to go to Germany, where all the best musicians went to study: 'My hope was that I should be able to get a musical education, and I worked hard at German on the chance that I should go to Leipsic, but my father discovered that he could not afford to send me away, and anything in that direction seemed to be at an end.'[13]

Whatever role W. H. Elgar played in this disappointment, Edward could now be reminded of the advertised terms of his education at Littleton House: 'Young gentlemen are prepared for commercial pursuits . . .'. In later years a journalist wrote after interviewing Edward: 'There was no encouragement held out to him to devote himself to music with a view to adopting it as his profession. As a matter of fact, he was intended for the law.'[14] The disadvantages of the music trade were everyday familiarities to the Elgar parents: the legal profession offered a clear upward step for their elder son.

The idea of the law as a career for Edward came from William Allen, the friend of Broadheath days who always sang 'Di Provenza il mar'. Mr Allen was himself a solicitor, with an office in Sansome Place, opposite St. George's Church: he was not only a member of the Catholic congregation but their solicitor as well. He was a familiar figure for Edward, who recalled for a biographer that the lawyer 'impressed himself upon the boy's mind chiefly by reason of his height, his gaunt appearance and his red dressing-gown'.[15] All in

[12] BL Add. MS 49973A fos. 1–4.
[13] *The Strand Magazine*, May 1904, p. 538.
[14] [F. V. Atwater], 'Edward Elgar and his works', in *The Musical Courier*, Oct. 1896.
[15] Maine, *Life*, p. 26.

all, it would have been hard to better Allen's suggestion (as Edward described it) 'that I should go to him for a year and see how I liked the law'.[16]

On 26 June 1872, having finished his last term at Littleton House, he entered William Allen's office at No. 7 Sansome Place. At the beginning it seemed to promise well. Soon the solicitor was telling Elgar senior that he found Edward a 'bright lad'.[17] In fact, as the new boy recalled one day for a friend: 'Allen intended to give him his articles and leave him all his property. But Arthur Beauchamp was already his pupil.'[18] Soon it began to appear that new abilities in that office met with small reward. This was also recalled for the biographer, who wrote: 'Edward applied himself to the Law with a serious mind, and made headway with a number of text-books. But so far from being encouraged in his studies he was made to wash the floors of the office.'[19]

It might have been only the standard office discipline for juniors. But in this world of men his reading was ill rewarded. So his reading was drawn round once more to the world of his mother's imagination. 'About this time a bookseller in Worcester, about to remove to another place of business, hired a loft over Mr. Elgar's stable and there deposited a large number of books for which he had no accommodation.'[20] Edward attacked them instantly:

There were books of all kinds, and all distinguished by the characteristic that they were for the most part incomplete. I busied myself for days and weeks arranging them. I picked out the theological books, of which there were a good many, and put them on one side. Then I made a place for the Elizabethan dramatists, the chronicles including Baker's and Holinshed's, besides a tolerable collection of old poets and translations of Voltaire, and all sorts of things up to the eighteenth century.

Then I began to read. I used to get up at four or five o'clock in the summer and read—every available opportunity found me reading. I read till dark. I finished by reading every one of those books—including the theology. The result of that reading has been that people tell me I know more of life up to the eighteenth century than I do of my own time, and it is probably true.[21]

He kept up his music as best he could outside office hours. Sundays were free, and through all the violin-playing he had continued to practise the organ. Again his father offered no help. He described it for the biographer, who wrote: 'Without supervision of any kind, he applied himself to the Organ Schools of Rinck and Best and so was able to relieve his father at St. George's by extemporising the voluntaries.'[22]

[16] *The Strand Magazine*, May 1904, p. 538.

[17] Reed, *Elgar as I Knew Him*, p. 46.

[18] Philip Leicester, Notes of conversation with Elgar, 16 Oct. 1932.

[19] Maine, *Life*, p. 26.

[20] T. A. Burge, OSB, 'Edward Elgar', in *The Ampleforth Journal*, July 1902, p. 27.

[21] *The Strand Magazine*, May 1904, p. 540. In describing the books for Basil Maine, Elgar again mentioned Baker's *Chronicles*, and added Drayton's *Polyalbion* and Sidney's *Arcadia* (*Life*, p. 10).

[22] Maine, *Life*, p. 18. Buckley (p. 16) also covers the point, as does Edwards's article in *The Musical Times*, Oct. 1900, p. 642. All were based on conversations with Elgar.

W. H. Elgar had now been organist of St. George's for twenty-five years, and he had his characteristic methods as Hubert Leicester observed when he joined the choir: 'Old E. always handed round the snuffbox before commencing the Mass, "damned" the blower, & began. Went out at Sermon for drink at Hop Market.'[23] The availability of an assistant to relieve him on occasions was therefore not a thing to be discounted. Three weeks after Edward entered Mr Allen's law office, Lucy noted: 'Ted played the organ at Church for Mass first time—14 July 1872.'[24]

Playing the organ brought immediate contact with the choir of St. George's. And there was another impression to be recalled for the biographer: 'As a boy, Elgar was struck by the effect if not by the significance of key-changes in some of Mozart's masses.'[25] Key-changing had made the Jape in the tune from Broadheath and the memorable effect in Beethoven's First Symphony.

Then came a suggestion in Mozart's sonatas for violin and piano. In the second movement *Allegro* from the Sonata in F (K.547), Mozart set two contrasted figures side by side. Beginning in diversity, sonata-writing can move contrasting themes toward some synthesis. Edward cast about for a way to adapt this sonata's *Allegro* to his own expression.

In those days it was common to draw on secular music of the masters to make church settings. It struck him that the Mozart *Allegro* might be fitted to the Catholic words of the *Gloria*. The piano part would make an organ accompaniment virtually as it stood. He devised a vocal ensemble in four parts to replace the violin solo, writing it out on narrow strips of paper pasted over the printed violin part.[26] It was a more sophisticated version of fitting his own tune to the music of Handel at the Worcester Festival. But W. H. Elgar could hardly object to a practical new piece for the choir at St. George's, saving him the trouble of writing something for himself.

September 1872 brought the Three Choirs Festival to Worcester again. Edward got leave from Allen's office to attend at least one rehearsal; and he recalled for a friend of later years

... how struck he had been by the way in which Handel treated the second violins when strengthening a lead for altos, bringing that part into prominence by throwing it into the upper octave, and he commented on the wonderfully resonant effect Handel obtained by the spacing of his chords in a chorus like 'Their sound is gone out'.[27]

Yet inevitably Edward's employment in the office stood in the way of his music. There was only an occasional chance for taking part in music-making of a secular sort in the evenings, after the office was shut. Thus the review of a concert at the Union Workhouse on 31 January 1873 showed 'Master Elgar' as one of the vocalists. Ann Elgar cut it out and pasted it into a little scrap-book of news cuttings she had begun to keep.

[23] Conversation of 28 Nov. 1909, noted by Philip Leicester.
[24] MS fragment cut from her diary (Elgar Birthplace).   [25] Maine, *Life*, p. 14.
[26] BL Add. MS 49973A.   [27] Wulstan Atkins, MS reminiscence.

A rare chance presented itself in the middle of April 1873. W. H. Elgar was engaged to play in an orchestral concert at Hereford. His son was all attention: 'At this time of my life opportunities to hear orchestras were very few; I seized every chance that offered. This was an afternoon concert; as it occurred in holiday time it was possible for me to get a day off from the office.'[28]

That day the energies pent up through months of office drudgery poured out in every kind of animation. And at the end of it all, the boy submitted his father a 'Bill of Costs' which recorded the day in ebullient parody of Mr Allen's legal language. As Edward wrote later of his employer:

His use of the words *'desirability'* and *'conferring'*, which appeared in his accounts more frequently than in those of other lawyers, is imitated in this 'Bill of Costs'.[29]

Bill of Costs. Mr. W. H. Elgar, to Edward Elgar Dr.

1873

April 17   Attending you, conferring as to the desirability of my accompanying you to Hereford on the 18th inst: being the date of the Philharmonic Concert, when, (after discussing the matter at length) you stated that if I could make it convenient to attend, you would be prepared to proceed thither with me.

6.8

„   18   Attending you at your residence conferring as to which train would be most desirable to go by, when (after some discussion on the point) you expressed an opinion that the train leaving Foregate Street Station at 11.29 would be best. & afterwards

6.8

Attending you accompanying you to the Station & also to the Refreshment Room—conferring with you and Mr. Henry Brookes[30] as to the rival qualities of draught and Bottled ale when we decided in favour of the latter

6.8

Journey to Hereford Conferring with you & fellow passengers as to the state of the weather & upon several other topics of general 'interest', & advising

6.8

Conferring with you as to the Tunnel near Ledbury when you stated that you considered it to be the most offensive one on the line in which view I concurred

6.8

Attending you at Hewitt's Eating House conferring as to the desirability of your having pork,[31] mutton, veal or other meat, of which there was a great variety, when, as I and Mr. Brookes had decided in favour of the first-named, you took our view of the case and ordered two porks on the spot

6.8

[28] MS note dated 18 May 1929 to accompany the 'Bill of Costs' (Elgar Birthplace).
[29] Ibid.
[30] 'Henry Brookes was a lay clerk (bass) in Worcester Cathedral and played the double-bass.' (Ibid.)
[31] 'The reference to "pork" was deliberate—my father detested it.' (Ibid.)

Attending you (after we had dined) accompanying you to the Cathedral Green conferring as to the antiquity of a portion of the Cathedral, which you were of opinion was of later construction than the bulk of the building, and after a lengthy discussion (as we were not experienced in architecture) we thought it best to let the matter rest

6.8

Attending you to the Shire Hall where the concert was to take place Conferring with the ticket taker as to the most desirable place for me to sit and giving the man in attendance instructions to furnish me with a Book of the words Engaged 2 hours

13.4

Attending you to the Hereford Railway Station accompanying you to your residence at Worcester and conferring on the business of the day and conferring as to the advisability of your washing yourself to renovate your spirits and to remove the depression which you stated had come over you, & advising

10.—
<hr>
£3–10–0.

Many years afterward Edward noted:

The last paragraph reverses the position,—a boyish attempt at humour as indeed is the whole 'Bill'. Excited over the music, the architecture of the Cathedral and the day's outing generally, I was talking volubly; my father, who had played second violin in the orchestra, was tired and wanted to rest; it was he who recommended *me* to wash and change for the evening 'depression'.[32]

If depression had overtaken the boy at the end of this day of release, it was because of a contrast with the office whose door would gape wider for its prodigal on the morrow.

There was also another purpose behind this elaborate exercise in bookkeeping. Its parody was a practical demonstration of skills acquired during the months in the law office. Years later he pointed out the motive to his biographer, who wrote:

The book-keeping of his father's business was far from regular. The business, it may be said, continued not because but in spite of the accounts. It occurred to Edward that, if he must work in an office, it was just as well that his father's business should receive the benefit of his industry.[33]

In fact the situation in Allen's office was coming to a head. The boy's quickness to gain his chief's good opinion had roused the jealousy of the upper clerk, Arthur Beauchamp. Edward recalled it many years afterward in conversation with Hubert Leicester's son Philip, who noted: 'A.B., seeing how the land lay and having his own views about things, made matters so unpleasant to E.E. that he could stand it no longer.'[34]

[32] Ibid.   [33] Maine, *Life*, p. 26.
[34] MS notes of a conversation on 16 Oct. 1932.

It was the mirror-opposite of life at Littleton House. There the older boy had offered himself as Edward's guiding friend, and the master showed a fatherly understanding of Edward's abilities. In the solicitor's office now, the boy who should have taken the elder brother's role used his position to thwart Edward's efforts, while their chief turned a blind eye. It was the man's world still, but with all the values reversed.

When Edward resolved to make his escape from this man's world, he turned to his mother: 'I told my mother once when I was young that I wouldn't be content until I received a letter from abroad addressed to "Edward Elgar, England."'[35] And to show how little he needed the law to gain that eminence: 'The only honour I ever coveted, & which I "arranged" with my mother (when I was fifteen) to obtain, is the Hon.D.C.L.'[36] The point of the Honorary Doctorate of Civil Law was precisely its honour: it was not awarded to an ordinary lawyer. In the end Edward obtained his parents' consent to leave the solicitor's office when he completed his year's agreement, and to work as his father's assistant in the music shop.

*   *   *

In June 1873—the month of his sixteenth birthday, Edward found himself able to walk out of the law office for the last time and into the world of music. This was his recollection thirty years later:

In studying scores the first which came into my hands were the Beethoven symphonies. Anyone can have them now, but they were difficult for a boy to get in Worcester thirty years ago. I, however, managed to get two or three, and I remember distinctly the day I was able to buy the Pastoral Symphony. I stuffed my pockets with bread and cheese and went out into the fields to study it. That was what I always did.[37]

The boy's impulse was to take it right away from the shop—just as he had taken away the sheet of paper to write the sound of the wind in the reeds. He took the Beethoven score to the churchyard at Claines, a quiet old place to the north of the city amongst fields and with no houses very close by.[38] Ann Elgar's family had lived at one time in Claines; and her parents, dead before the birth of her son, were buried in Claines churchyard. Many years afterwards Edward took his wife to see the spot, and she wrote in her diary: 'Saw his relatives' tomb & where he used to sit & read scores, years ago.'[39] There among the memorials to his ancestors the boy had sought his own relationship to Beethoven.

---

[35] Sir Compton Mackenzie recalling a conversation with Elgar in the 1920s (BBC broadcast, *The Fifteenth Variation*, 1957).

[36] Letter of 20 Jan. 1905 to Sir Hubert Parry, quoted in MS notes by Basil Maine (Royal College of Music).

[37] *The Strand Magazine*, May 1904, p. 540.

[38] So it survived as late as 1920, as recalled by Mrs Philip Leciester (conversation with the writer, 1975).

[39] Entry for 12 Aug. 1910 (MS diaries, HWRO).

Among the composers of the past, Beethoven must have made a special appeal to the boy just escaped from the solicitor's office. For it was Beethoven, of all the great composers, who had made the classic affirmation in music of the man's world.

The inspiration was irresistible. In July 1873, the month after his emancipation from the law, Edward chose themes from Beethoven Symphonies and combined them to make a setting of the *Credo*. Where the Mozart *Gloria* had followed the form of the master, this time the master's ideas would merely inhabit a form that was to be entirely of Edward's devising: he had only the shape of the text to help him.

The most striking aspect of Edward's setting was his modulation among the Beethoven themes. After opening with the beginning of the Seventh Symphony *Allegretto* in its original A minor, he sent Beethoven's A major second subject into F major—thus adding a flat to his opening key signature. The words 'descendit de coelis' brought more flats, taking the music suddenly to F minor and A flat major. So the opening section of Edward's *Credo* as a whole had moved from A minor to A flat—physically close but tonally remote. It recalled precisely the 'Jape' which had sent his G major 'tune from Broadheath' into F sharp.

For his 'Incarnatus' Edward transposed the Ninth Symphony slow movement opening to A flat major. His 'Crucifixus' then began on the note of A♭. Only now it was not A♭ at all, but really G♯—as it proved to be the major third in a new tonality of E major. So the 'Crucifixus' had interrupted 'Incarnatus' with a musical pun. The effect of that pun was that though Edward had avoided any direct movement to the dominant, he had none the less arrived there—ready to move toward flats yet again to regain his opening tonic.

Instead, he did something more interesting. He had the solo soprano sing 'Incarnatus' again, now in E minor. 'Crucifixus' returned a whole tone lower in D major. 'Incarnatus' replied in D minor. 'Crucifixus' reappeared another tone lower in C, to be overtaken by 'Incarnatus' in C minor. Again Edward's music echoed the 'tune from Broadheath', this time shaping a central sequence through descending steps.

So there emerged through his *Credo* the shadowy partnership of tonal descent in steps with movement into flats. That was not at all from Beethoven, who was before all else master of the dominant—of bright, aggressive, outgoing, climbing shapes for his developments of both melody and larger form. But Edward's impulses went all the opposite way—toward the dark, warm, inward-turning, descending development to flats. How deeply the attraction toward flats was to influence his music came to conscious recognition after half a creative lifetime, as he sent a fragment from his First Symphony to a friend in 1908:

This is a sort of plagal(?) [*sic*] relationship of which I appear to be fond (although I didn't know it)—most folks run through *dominant* modulations if that expression is allowable

& I think some of my twists are defensible on *sub*-dominant grounds. All this is beside the point because I *feel* & don't invent—I can't ever invent an *explanation—no excuse is offered* . . . after all I am only an amateur composer—if that means I compose for the love of it . . .[40]

Plagal movement was in fact to make the third term of Edward's style. Substituting subdominant for dominant development had been tried by almost all composers at some time or other; but few shaped their music with such consistency toward IV rather than V as this boy was to do. In the standard pattern of I V I, excursion is made to the fifth interval, while return is through a fourth. In I IV I, the fourth interval becomes the excursion; and the fifth (commonly associated with excursion) makes the return. So expectation is inverted, and the constantly recurring surprise creates a pattern of propulsion.

Therein Edward's subdominant development joined the other two basic terms of his style, rhythmic repetition and melodic sequences. All three make for propulsion. And propulsion can be of fundamental concern to an insecure personality—uncertain of its own place in the world, needing assurance and self-assurance to shore up private uncertainty.

Thus Edward's honesty with himself provided a basis for his music's strength. It would take him more than twenty years to forge his three terms—rhythmic repetition, melodic sequence, subdominant shaping of harmonies—to a consistent style. But when the forging was complete, the result was a style instantly recognizable. The style's recognizability—one indication of individual greatness in creative art—was based on the small number of its terms and the concord among those terms.

The subdominant impulse was also to give Edward's music its most celebrated quality. For where the excursion is redolent of return, there is nostalgia. It was the 'sunset quality' of which the great critic Ernest Newman was to write, and which Edward himself would come to prize perhaps beyond all else in his music. Posterity has agreed with him.

The *Credo* on themes by Beethoven written at the age of sixteen was still far from such knowledge. It was a structure of 306 bars, his earliest surviving achievement of large form in music. But the themes remained the themes of Beethoven. Edward did not set his own name to it, even as arranger. Instead he used his smattering of German (acquired the year before in hopes of going to study in Leipzig) to concoct a punning pseudonym:

Credo Beethoven arranged by Bernhard Pappenheim.[41]

'Pope's home'—that was the target at which this music was aimed. Did the German pun suggest a certain detachment? It was no more than the detachment already shown by the regular organist of St. George's Church

---

[40] 19 Sept. 1908 to A. J. Jaeger (Elgar Birthplace).
[41] MS written entirely in Elgar's youthful hand (Jesuit archives, Farm Street Church, London).

when he went out at the Sermon for his drink at the Hop Market. Edward's deputizing for his father at the organ had led very easily to this further deputizing, arranging the music for his father's choir to sing.

And sing it they did. The *Credo* manuscript survives with all sorts of conductor's notes showing the experience of practical rehearsal. So if the Pope's home had not immediately inspired Edward to melodic invention, it gave him the chance to shape his own form in music.

5

Hyperion

As Edward shaped his music, he followed his mother's taste in literature. One favourite author was Longfellow, whose romantic idealism enjoyed world-wide fame. Longfellow's prose romance *Hyperion* made a particular appeal to Mrs Elgar and her elder son. It was the story, as Edward later recalled, 'from which I, as a child, received my first idea of the great German nations'.[1]

Longfellow's modern Hyperion is a young man called Paul Flemming, feeling the cultural attraction of Germany just as Edward himself felt the wish to study music there. Flemming admires the German author Jean Paul Richter for writings that make an ideal marriage of 'tenderness and manliness'.[2] His favourite opera is Mozart's *Don Giovanni*. Yet Flemming himself is no Don Giovanni. His own attraction to women is subject to a stronger sanction, as Longfellow described it:

He resembled Harold, the Fair Hair of Norway . . . 'he who spent his mornings among the young maidens—he who loved to converse with the handsome widows.' This was an amiable weakness; and it sometimes led him into mischief.

Imagination was the ruling power of his mind. His thoughts were twin-born; the thought itself, and its figurative semblance in the outer world. Thus, through the quiet, still waters of his soul each image floated double, 'swan and shadow'.[3]

So every human attraction must embody a less-than-perfect fulfilment of what the imagination projected.

All this emerges in Paul Flemming's travels through Germany and Switzerland to forget an unhappy love affair. He spends a season with a friend, a young German baron, observing student life in Heidelberg. Then his travels bring him face to face with an English girl of such beauty and distinction that manly desires are stirred up again. Yet he cannot rest content with his reality, but soon insists to the young lady that 'Art is the revelation of man; and not merely that, but likewise the revelation of Nature, speaking through man.'[4]

Flemming introduces the girl to the glories of German literature. But his choice falls on Uhland's 'Ballad of the Black Knight', riding mysteriously to destroy all the flower of youth and the future. Flemming asks his lady to recite

---

[1] 25 Oct. 1899 to Hans Richter (MS, Richter family).
[2] Book I, ch. 5.
[3] Book I, ch. 3.   [4] Book III, ch. 5.

'The Black Knight' after him line by line in German. She naturally hesitates over such a subject; whereupon he insists on improvising a translation for her.

At last the hero recognizes the failure of his love-making. And he turns back again to the alternative feminine significance of Nature, shaping her cycle of seasons through magnificent Alpine landscapes. Then the young lady can appear to him only in a dream, saying 'I confess it now; you are the Magician!'[5] So the youthful hero, in the days of autumn, goes forth to pursue the search for himself.

Ann Elgar's interest in sharing this Hyperion with Edward might show no more than a mother's classic concern for her son's relations with other women. Yet it was undeniably the story of actual feminine attraction replaced by the feminine symbolism of Nature and season and countryside—of reality defeated by an ideal. If *Hyperion* was to become an exemplar for Edward, it would increase his difficulty in searching for a wife. In fact the idea was to hold him in thrall until another woman came to challenge his mother on her own ground.

The teenage Edward pursued his search for music. One medium offered itself in the organ of St. George's Church. The whole range of organ expression lay directly under the player's own fingers and feet, even though playing the organ still needed outside help: 'It was very difficult to get an organ-blower at the exorbitant sum of three halfpence an hour. But Hubert Leicester would come and blow the organ for nothing, but he would insist that I should play a certain piece of Mozart.'[6] After that the organist might choose for himself. And then the instrument showed its real usefulness: it offered a ready means for exploring orchestral music. Hubert Leicester remembered:

Was blowing organ one day for E. E. He played something quite new & strange. Ran round & said:

'Ted, what is that?'

'That, Hubert, is by a man who is not understood. His name is Wagner—a German—& you will hear more of him some day.' . . .

The piece was the Overture to *Tannhäuser*.[7]

The skill of Edward's organ-playing reached the ears of the Cathedral organist, William Done:

When the new organ, the gift of the Earl of Dudley, was being put in [to the Cathedral early in 1874,] one of the Cathedral clergy pointed to it and said:

'Mr. Done, there is room for display here.'

Done replied: 'Not for me. Leave it to these young men' (meaning [myself] among others . . .)[8]

---

[5] Book IV, ch. 8.

[6] Speech on the presentation of the Freedom of the City to Hubert Leicester, 6 Apr. 1932 (newspaper cutting at the Elgar Birthplace).

[7] Conversation with Mr Mason, Oct. 1909, noted by Philip Leicester.

[8] Speech to the Friends of Worcester Cathedral, June 1932.

I attended as many of the Cathedral services as I could to hear the anthems, and to get to know what they were, so as to become thoroughly acquainted with the English Church style. The putting of the fine new organ into the Cathedral at Worcester was a great event, and brought many organists to play there at various times. I went to hear them all.[9]

One was the great Samuel Sebastian Wesley of Gloucester, who improvised in a way that Edward remembered all his life. Wesley 'built up a wonderful climax of sound before crashing into the subject of the "Wedge" Fugue' of Bach.[10]

Another source of musical sound presented itself at St. Helen's Church, nearly opposite the door of the Elgar shop in High Street:

The curfew used always to be rung in those days at eight o'clock in the evening . . . I made friends with the sexton and used to ring the Curfew, and afterwards strike the day of the month. My enthusiasm was so great that I used to prolong the ringing from three minutes to ten minutes, until the people in the neighbourhood complained, when I had to reduce the time.[11]

Bells produced the clearest patterns of overtone and pulses of vibration. They also rang their measures of time. So Edward tried to extend control over them:

The day of the month used to be rung with the clapper of the seventh bell. Some little mechanical arrangement in the tower was prepared by myself with great skill—or rather effect, because I remember that night, after raising the bell, the curfew was so exuberant that we ran into the 42nd of September and then lost count.[12]

The best of bell-ringing at St. Helen's came with Sunday morning:

On Sunday the bells were supposed to go for half an hour before service, from half-past ten to eleven. The performance was divided into certain parts. With a friend [Hubert Leicester], I used to 'raise' and 'fall' the bell for ten minutes, chime a smaller bell for ten minutes or so, and at five minutes to eleven I would fly off to play the organ at the Catholic Church.[13]

With so many churches offering sounds for musical experience, the only real problem seemed that of running quickly enough from one to the next:

The services at the Cathedral were over later on Sunday than those at the Catholic Church, and as soon as the voluntary was finished at the Church I used to rush over to the Cathedral to hear the concluding voluntary.[14]

But it was the Catholic Church that offered music in which Edward could take an active part: 'The touring *English opera companies*, the *personnel* of which was mainly Catholic, gave their help, vocal & instrumental, at the

[9] *The Strand Magazine*, May 1904, p. 539.

[10] Leaver, 'Some Impressions of Sir Edward Elgar', in *Musical Opinion*, July 1934.

[11] Speech on receiving the Freedom of the City of Worcester, reported in *The Worcestershire Echo*, 12 Sept. 1905.   [12] Ibid.

[13] *The Strand Magazine*, May 1904, p. 539. The identification of the friend as Hubert Leicester was made by Mrs Philip Leicester, who often heard her father-in-law and Elgar discuss it.

[14] Ibid.

Sunday services. Many of the operatic artists were friends of my father.'[15] Edward's skill with his violin enabled him to help his father return the favour. For when the company mounted their operas at the old Worcester theatre, 'I was taken into the orchestra, which consisted of only eight or ten performers, and so heard old operas like *Norma*, *Traviata*, *Trovatore*, and, above all, *Don Giovanni*.'[16] *Don Giovanni* had been the favourite opera of Paul Flemming in *Hyperion*.

Vivid impressions of the theatre were always on the tongue of his father's handyman Ned Spiers. But then the theatre itself led to music again. He recalled this for a biographer, who wrote:

The visits of Mrs. Macready to Worcester enabled Edward to follow up the enthusiasm which had been kindled by Spiers. From time to time she gave readings of Shakespeare . . . At these readings, Mrs. Macready used to engage a German pianist, called Benjamin, to play groups of pianoforte solos. He also was an influence in the boy's developing life. They discussed music together, and from this pianist Edward Elgar heard a great deal of music that was either unfamiliar or entirely unknown to him. It was to Benjamin especially that he owed his real initiation into the works of Chopin.[17]

Yet the initiation into Chopin did not take Edward back to the piano, any more than serving as shop assistant in Elgar Bros' Pianoforte and Music Warehouse had done. He asked the Worcester Glee Club ensemble leader Fred Spray to give him formal violin lessons.[18] After six months Spray was 'honest & told him he could not teach him any more'.[19] It was strong confirmation of the boy's ability. Edward pursued his violin practice as if it held the key to every fairyland of imagination.

When a magic casement seemed to open one day, it opened directly out of his violin-playing. Practising in the showroom above the shop, he stood overlooking the busy pavements below. A factory girl on her way to the glove works nearby smiled up at him as she passed. He smiled back. Another day it happened again. The girl was interested. So when he saw her next, he called to ask if she would like to come up, and up she came. Yet her interest in his music turned out to be small. Perhaps she would respond to literature, as his mother responded with him. He took her to the loft where the old books were stored, chose a copy of Voltaire, and began to read out a favourite passage from *Candide*. But then she said something that sounded like 'You silly mutt!' and abruptly took her leave.[20] It was almost the re-enactment of Paul Flemming's

[15] 23 May 1928 to Hubert Leicester.

[16] *The Strand Magazine*, May 1904, p. 538.   [17] Maine, *Life*, p. 28.

[18] On a sheet of accounts drawn up later to show his 'Expenses, *in full*, for general and musical education', there is the item: 'Violin lessons, Mr Spray, Worcester, 2 q[uarte]rs 1874 £3.0.0.'

[19] Blake, MS notes of conversation with Hubert Leicester.

[20] Elgar told this story to Dr and Mrs Moore Ede in 1933 (recalled by Mrs Moore Ede in conversation with the writer, 1975). The late Sir Steuart Wilson also remembered hearing the story from Elgar with the word 'mutt' (conversation with the writer, 1954). Basil Maine, in his notes of a conversation with Elgar in 1932, recorded the girl's words as: 'I think you're a b——— fool.' (MS at the Royal College of Music.)

literary love-making in *Hyperion*. Ann Elgar saw her elder son in November 1874 as

> Nervous, sensitive & kind,
> Displays no vulgar frame of mind.[21]

*   *   *

September 1875 ought to have seen another orchestral Three Choirs Festival in Worcester. But the recent Cathedral restoration and the new Cathedral organ had been heavily financed by Lord Dudley on the understanding that the Cathedral would no longer be used for functions at which an entrance fee was charged. So there was no Festival orchestra that year—and hardly any Festival at all. Only a few extra anthems and choral services with the organ marked what might have been Edward's first chance for playing his violin in a big orchestra.

But the following months yielded several performances of a certain size. On 23 November the Worcester Musical Society gave Spohr's oratorio *The Last Judgment* under the direction of Alfred Caldicott, a well-known young Worcester musician who had been trained in Leipzig. Edward played in the second violins with his father—and found his name listed in the programme as 'W. Elgar, Jun.'[22] A month later he joined his father in the second violins again for the Worcester Philharmonic Society's Christmas performance of Handel's *Messiah* under the conductorship of the Cathedral organist Mr Done. Father and son were also in the second violins when the Musical Society gave a Mendelssohn evening early in February 1876, and they played side by side for two smaller concerts during April. But on 16 May, when Mr Done's Philharmonic Society performed Mendelssohn's *Elijah*, there was a change: W. H. Elgar remained in the second violins, but 'Elgar Junior' was placed among the firsts.

The tensions among family personalities told sometimes on the mother who tried to maintain the simpler peace of her own youth. Remembering her younger days as the youngest child by many years in a farmer's family, Ann Elgar could not avoid a rueful comparison with the vociferous individualities of her maturing offspring. She wrote in her scrap-book:

It is no joke to have men and women to rule, and keep peace between, and to keep *home* in something [of] order and comfort. With so many different dispositions, different ways and wants—each one wrestling for the mastery—or trying to have their own way in defiance of the rest—Yes! I own it at last, I have given them too much latitude, I have ruled by love instead of terror and the fetters are too weak for their stronger passions. I failed to see that they could be different to myself, forgetting I was *alone*, in my youth for so many years. I had no one to quarrel with—*or to play with*—but my parents, to whom I

---

[21] MS couplets on all her surviving children, in possession of Elgar's niece, the late Madeline Grafton.

[22] The programmes for this and other concerts cited are preserved at the Elgar Birthplace.

always gave implicit, blind obedience. I never remember stopping to question their authority, or their wisdom—hence, I suppose I expect too much—but I should be happier if they were a little more like I was.[23]

When she began the scrap-book in which that observation was written, the mother must have been thinking especially of Edward. For on its opening pages she copied out several passages from *Hyperion* bearing directly on the artist and his problems. The first was from a scene in which the young hero contemplates the ruins of a high old castle above the Rhine, and seems to hear it say:

> 'Beware of dreams! Beware of the illusions of fancy!
> Beware of the solemn deceivings of thy vast desires!'[24]

And then a conversation between Paul Flemming and his German host:

'After all,' said the Baron, 'we must pardon much to men of genius. A delicate organization renders them keenly susceptible to pain and pleasure. And then they idealize everything; and, in the moonlight of fancy, even the deformity of vice seems beautiful.'
'And this you think should be forgiven?'
'At all events it is forgiven. The world loves a spice of wickedness. Talk as you will about principle, impulse is more attractive, even when it goes too far. The passions of youth, like unhooded hawks, fly high, with musical bells upon their jesses; and we forget the cruelty of the sport, in the dauntless bearing of the gallant bird.'
'And thus do the world and society corrupt the scholar!'[25]

For the boy taking part in his first large-scale performances, however, recollection would preserve just what it preserved from his performances with the Glee Club—the sheer excitement of ensemble. This was noted in later years by his violinist friend W. H. Reed, who wrote:

He was immensely pleased when he heard those passages which he had noticed himself when he had first taken part as an orchestral player in such works as *Elijah* or *Messiah*. Year after year at the Three Choirs Festivals, his head would appear round a pillar in the cathedral to catch my eye when the altos and tenors enter, with their A^b and [C] respectively, in the concluding bars of 'All we like sheep' (adagio). He revelled in the sweeping passage for the violins in 'Thou shalt break them' . . .

The overture to *Elijah* he would never miss. He pointed out to me that Mendelssohn was the first composer to touch up the entries of the successive voices in the Fugue: the horns coming in on the last beat of the second bar of the exposition; then the clarinets and bassoons at the sixth bar, later the oboe: none of them taking any part in the statement of the fugue subject, but holding sustained notes against the moving parts of the strings, adding colour and great sonority to the whole texture.

Many other things in *Elijah* pleased him intensively, and he loved pointing out Mendelssohn's use of the trombone, and his cleverness in directing the first violins, in

---

[23] Written 10 Dec. 1874, copied into the scrap-book 31 Dec. 1876.
[24] Book I, ch. 3.
[25] Book I, ch. 7.

'Cast thy burden', to hold on to the top note of the chord long after the other notes had ceased, 'to keep the vocal quartet up to pitch'.[26]

All these passages juxtaposed sustaining, reinforcing tones with the motion of melody—yoking essentially opposed forces in wonderful ensemble.

And there was another paradox in Mendelssohn's use of melody. Several of the musical figures in *Elijah* were systematically related to the appearance of particular ideas or characters in the story. It was the system which Wagner had since used as *Leitmotiv*. Edward had never heard a Wagner opera. But Mendelssohn showed clearly enough how the systematic linking of melodic theme with dramatic event could make music develop with the story it espoused. As development was the most challenging part of writing music, here was a guide which a composer who wanted to handle large forms could not ignore.

An interest in the craft of writing music offered a healthier outlet for his energies than competing with his father. On the whole, the best avenue for his music lay in St. George's Church, for Catholic worship was not at the centre of W. H. Elgar's musical interest. Edward applied himself so assiduously to the study of Catholic music and its sources that one day a great expert on liturgical music, Sir Richard Terry, would write of Ann Elgar's son:

On the question of polyphony he used to embarrass me by his persistent attitude of a listener and a learner. I found out the depth of his knowledge (which I had long suspected), by the merest accident. Hearing that I had lost my volume of Rochlitz, he asked me to accept a copy which he had bought in early youth 'to try and get the hang of those old fellows' (as he put it). His notes in the book and across the music showed me that his had been no superficial study. He had noted all that was worth noting about the characteristics (contrapuntal and harmonic) of the old Polyphonists from Dufay and Josquin to Goudimel, Lasso, Palestrina and his school.[27]

The results of this study soon emerged. In the autumn of 1876 came Edward's earliest dated compositions for the choir of St. George's Church—a *Salve Regina* inscribed 16 September,[28] and a *Tantum Ergo* dated 27 November.[29] Though neither was on the scale of the Beethoven *Credo*, Edward's part-writing showed a big advance. Yet polyphonic interest remained firmly subservient to a continuing sense of harmonic poise and flow. The form of each piece revealed strength and purpose within the limits of the young composer's experience. In the *Tantum Ergo*, a central *Adagio* made a real chromatic development from the opening theme: this was short, but its thinner vocal writing was compensated by an elaborate organ part for Edward himself—of a dignity to match the rest.

[26] *Elgar as I Knew Him*, pp. 84–6.
[27] 'Elgar as I Knew Him', in *The Radio Times*, 9 Mar. 1934, p. 710. The Rochlitz Volume, *Sammlung vorzüglicher Gesangstücke*, 1834–40 (in 1981 in possession of Messrs Maggs, Berkeley Square, London), reveals only a few pencillings by Elgar, and his presentation inscription to Terry.
[28] BL Add. MS 49973A.
[29] MS parts now in the Jesuit archives, Farm Street Church, London.

That same autumn saw him also trying out his skill in orchestral writing. As his choral skills had first been tested in the Mozart *Gloria* and the Beethoven *Credo*, so now he arranged the Overture to Wagner's *Flying Dutchman* for the instruments in the Glee Club ensemble. Thus Wagner's music entered the Worcester Glee Club programme for the first time on 23 October 1876.

Elgar Junior was going from strength to strength in the Glee Club then. Soon his piano-playing skills were also in demand on Tuesday evenings. One of his contemporaries in the Club recalled: 'For about two years Edward was accompanist. In this his marked ability was at once manifest, though he would always insist that he was not a pianist. His accompaniments were a great delight to singers and audience.'[30] There was one member, however, who was not greatly delighted. It was W. H. Elgar, who had himself once played the Glee Club accompaniments and made the Glee Club instrumental arrangements. Hubert Leicester was to remember 'battles between father & son. Old man always great respect & fear of son's superior musical knowledge.'[31]

That superiority was maintained through the larger concerts in which the two Elgars took part. For the Philharmonic Society's performance of Haydn's *Seasons* on 22 November 1876, Edward took his place once more among the first violins while his father remained in the seconds. The roles were the same when the Musical Society gave Barnett's *Paradise and the Peri* on 12 January 1877. And the pattern continued in concerts beyond Worcester. On 7 February father and son played second and first violins in a small band to entertain the patients of the County Lunatic Asylum at Powick. Five days later the two Elgars occupied their positions again for a performance of Sterndale Bennett's *The May Queen* organized by C. H. Ogle at Pershore.

At last the younger Elgar worked himself up to leading the orchestra for Ogle's Pershore concerts. And then the driving concentration of Edward's violin playing made an incident which he himself would recount in later years for Philip Leicester with an easy comic retrospection:

They were all rehearsing one night at Pershore, in a room over a bake house. Ogle lived over the road & usually 'stood' supper. In the midst of a soul-stirring passage Bob Surman (who had slipped out) returned & whispered to E:

'Cut it short, Ted, there's a red hot leg of mutton gone over to Ogle's.'[32]

Whether W. H. Elgar was playing in the orchestra that night did not emerge from his son's recollection.

Then Edward found another reason for practising his violin. He began to give violin lessons on his own account. If his pupils were few and far between, his journeys to reach them took him out of the shop. One of the pupils recalled:

---

[30] W. Mann Dyson in *The Musical Times*, 1 Oct. 1900, p. 642.
[31] Conversation with Fr. Driscoll, SJ, 28 Nov. 1909. MS note by Philip Leicester.
[32] Notes of conversation on 3 June 1914.

At the age of 19 . . . Edward Elgar used to give violin lessons to the children of neighbouring villages.

He would come once a week by train from Worcester to Upton-on-Severn, walk a mile to the vicarage of Hanley Castle, where one of his small pupils lived, walk another two miles to my father's house to give me a lesson.

For this he received 5s and a glass of sherry. Then three miles back to Upton Station. I remember him as a very unassuming, shy youth, but painstaking and kind.[33]

Many schools in the neighbourhood gave opportunities for teaching. Thus Edward made acquaintance with the Revd William Wilberforce Gedge, Headmaster of the Wells House School, Malvern Wells. In addition to his pupils, Mr Gedge had a family of seven daughters, one of whom recalled: 'At the age of 18 or 19 Elgar used to come to our house at Malvern Wells every week to teach my sisters and others the violin.'[34]

Another teaching connection at the Worcester College for the Blind Sons of Gentlemen brought a pupil who one day would distinguish himself as an organist. William Wolstenholme studied piano and organ with the Cathedral organist Mr Done; but his violin lessons with Edward laid the foundations of a lifelong friendship.

Then Edward returned to give lessons at his old school, Littleton House. There Mr Reeve 'was often much amused to see him during meal time take a card out of his pocket and write down some musical idea that had struck him'.[35] But Mr Reeve's former pupil and head boy was in earnest. For on 16 March 1877 he signed and dated a composition of his own for violin and piano.[36]

The writing was as well conceived for the long lines of violin expression as the recent choral pieces were well conceived for choral singing. Again there was an overriding concern for harmonic address and harmonic shape. But the new piece went farther. It was entitled *Reminiscences*, and it brought together a number of Edward's musical attractions which had so far remained separate.

Its main theme recalled the favourite choral passages of Handel and Mendelssohn and the Beethoven Seventh Symphony *Allegretto* in setting long-held notes through shifting harmonies:

[33] K. M. Lechmere, in *The Times*, 7 Sept. 1934.

[34] Letter from Mrs A. M. Bowlby to Sir Landon Ronald, printed in *The Daily Telegraph*, 14 Sept. 1935.

[35] Burge, in *The Ampleforth Journal*, July 1902, p. 27.

[36] BL Add. MS 57912.

The end of the melody was a little figure repeated sequentially in a pattern of descending steps like the 'tune from Broadheath':

This was a pattern to appear and reappear in Edward's music—notably in the closing pages of his Second Symphony.

The piece's brief middle section repeated a figure of descending steps in a sequence that went on too long. Later Edward revised it to

This revision came closest of all to the child's tune from Broadheath—and so found rich significance in the music's title. The figure of descending steps was to haunt Edward's music to the end of his life. And so it hinted at the nostalgia which would become a central element in his creative maturity.

He dedicated this music to an older Worcester friend who was chained to a life of business when his spirit would have been elsewhere. Oswin Grainger was a grocer by trade, but he wrote to Edward years later: 'Though condemned to get my living by means not to my taste—nothing will quite crush the poetry out of me.'[37] A similar fate might well have hung over Edward's head then.

Two days after finishing and signing his new piece, a different goal for the future presented itself. A great violinist of world renown came to give a recital in Worcester. August Wilhelmj, the German virtuoso, was just then helping to organize a London Festival of the music of Wagner, to take place in May under the direction of the composer himself. The violinist included a 'Wagner Paraphrase' in his Worcester concert. But it was Wilhelmj's playing in his final piece of the concert that electrified Edward that night. Years later he described it for W. H. Reed, who wrote:

He imitated Wilhelmj, whom he had heard play with *Airs hongroises* by Ernst. From his account of this affair Wilhelmj must have had a colossal tone; and his attack on the

---

[37] In a letter of 2 May 1889 to Elgar addressed 'Dear Ted' (HWRO).

opening tenth on the G string must have been hair-raising. It excited Elgar to such an extent that he never forgot it; and when he showed how it was done I felt thankful he was content to perform upon an imaginary violin and not on mine; for the movement he made would have cut any ordinary violin in half.[38]

The new goal was revealed: 'Although I was teaching the violin I wanted to improve my playing, so I began to save up in order to go to London to get some lessons from Herr Pollitzer.'[39] Adolphe Pollitzer was one of the most celebrated violin teachers of the day, with large numbers of successful pupils. Amongst the friends of the eminent teacher's youth had been numbered both Mendelssohn and Ernst: in fact it was Ernst who had first suggested London as a place for Pollitzer to pursue his career.

By August 1877 Edward had raked together just enough money to take him to Pollitzer. It needed the strictest personal economy, as he himself was to remember: 'I lived on two bags of nuts a day.'[40] Many years later he recalled it for W. H. Reed, who wrote:

His ambition then was to become a famous violinist; and for this he worked unceasingly. At his first lesson, after he had played and been criticised, he was told to get a certain book of studies. At his next lesson, when he produced the book, Pollitzer said to him,

'Now, which one of these have you prepared for your lesson to-day?'

'All of them,' said the young student, having been at them night and day, in fact, until he had mastered them.[41]

He managed to stay in London long enough to have five lessons with Pollitzer. He described them for his biographer Robert Buckley, who wrote:

Pollitzer . . . ran him over his scales, showing the fingering of three octaves . . . The teacher also gave five pieces of music by way of solatium, and was surprised to find Elgar not only remembering the unmarked fingering, but also playing the five pieces from memory.[42]

And his eagerness did not stop there, as he himself recalled:

Students were led to invent passages for their special needs. Pollitzer . . . was much amused by the studies and exercises of my invention—I say invention, for the effort involved in 'making' these things can scarcely be called composition. Five of the studies

[38] *Elgar as I Knew Him*, pp. 73–4.

[39] *The Strand Magazine*, May 1904, p. 538.

[40] Maine, *Life*, p. 33.

[41] *Elgar as I Knew Him*, p. 46.

[42] *Sir Edward Elgar*, p. 23. The five pieces were a Mozart *Larghetto*, Raff's *Cavatina*, *Romances* by DeBeriot and Vieuxtemps, and a *Gigue* by Franz Ries. In 1927 Elgar wrote to the head of Schott's, the music publishers: 'It is fifty years since you, in Regent Street, purveyed five violin pieces to a "prescription" by Pollitzer for me. I remember the pride & pleasure I had in presenting the "order".' (Letter of 30 May 1927 to Charles Volkert, quoted in *Letters*, p. 297.) This has sometimes been taken to mean that the five pieces were actually arranged by the 20-year-old Elgar and published in those arrangements by Schott. There is no trace of such a publication, and the letter refers simply to a recollection of buying the music in Schott's shop.

(dedicated to Pollitzer), mainly for the 'poise' of the bow, although the left hand is not neglected, were published long after their inception . . .[43]

Pollitzer's interest was thoroughly roused, as Edward remembered: 'He suggested that I should stay in London and devote myself to violin playing, but I had become enamoured of a country life, and would not give up the prospect of a certain living by playing and teaching in Worcester on the chance of only a possible success which I might make as a soloist in London.'[44] The eminent teacher must have wondered at this pupil who drove himself in pursuit of opportunity—and then when opportunity offered would not take it up. But living in London would mean breaking the pattern of life with his parents. A violinist's career in the great city would register a triumph in one department of his father's art: but it left no quarter for the creative impulse of his mother's world. So economic prudence joined hands with other motives to lead him back to the country life of which he was 'enamoured'.

*　*　*

A few months earlier his friend Oswin Grainger had helped to found an Amateur Instrumental Society in Worcester. The new Society existed 'for the cultivation of Instrumental Music by means of weekly Rehearsals under the direction of an efficient leader'. It would also provide a regular, reliable orchestra to accompany local oratorio performances: rehearsals were directed by the Worcester Musical Society's conductor Alfred Caldicott.

The task of coaching all the orchestral players in their separate instruments, however, fell to the Society's leader. For this post they had tried to engage a prominent violinist in the city, but he did not turn up. Edward held his breath. He had his experience of the Glee Club ensemble, his reading of Reicha and other books about the orchestra, and boundless interest. Instructing for the Amateur Instrumental Society would give him a practical acquaintance with the full range of orchestral instruments. Grainger and one or two others used their influence, and in the summer of 1877 the Society announced that 'an efficient Leader and Instructor has been engaged'.

The playing membership of the new Society drew many of the city's young musicians. Edward's younger brother Frank, for instance, appeared with an oboe, Hubert Leicester and his brother William with flute and clarinet. But Edward quickly established himself as an effective and dominant force. Indeed his colleagues seemed to recognize more than that when after only a few months they presented him with a violin bow by the most famous English bow maker together with an address:

We, the undersigned, Members of the Worcester Amateur Instrumental Society, request your acceptance of a Tubb's Violin Bow, as a testimonial of the kindly feeling

[43] Letter to *The Daily Telegraph*, 24 Dec. 1920. The five *Études caractéristiques*, op. 24, were published by Chanot in 1892.　　[44] *The Strand Magazine*, May 1904, p. 540.

and regard we have towards you. We likewise take this opportunity of expressing the hope that you may by continual perseverance, eventually attain an eminence in the profession for which you have evinced so great a love and aptitude.[45]

The signatures included those of Oswin Grainger and Alfred Caldicott.

That Christmas a group of them went out to play 'Waits'. There were a dozen lads, most of them from the Catholic congregation. They were accompanied by Charles Pipe, a young merchant who was not a Catholic but who had none the less persuaded Mrs Elgar to accept his engagement to marry Lucy. Pipe recalled the 'Waits' musicians:

They were all friends of mine, amongst them Elgars, Leicesters, Griffiths, Box, Surman, Weaver, Coupe and Exton. One or two of us helped to carry the lanterns and desks.

One night the snow was so deep we could hardly wade through it, and we trudged with instruments and stands from the top of Battenhall to Henwick, playing about ten times in front of various friends' houses. The hot coffee which some provided was most acceptable.[46]

As a result of his instructing for the Amateur Instrumental Society, Edward carried a new instrument—a bassoon. Once he had a reception that memory treasured ever after:

I remember one moonlight night stopping in front of a house to put the bassoon together. I held it up to see if it was straight before tightening it.

'Hi!' said a voice, 'if you shoot that here it'll cost you five shillings!'[47]

From these players a smaller group had formed. They made a woodwind quintet to play what the Germans called 'Harmonie Musik'. But the Worcester quintet had an unconventional distribution—two flutes (Hubert Leicester and Frank Exton), oboe (Frank Elgar), clarinet (Willie Leicester), and bassoon (Edward):

There was no music at all to suit our peculiar requirements . . . so I used to write the music. We met on Sunday afternoons, and it was an understood thing that we should have a new piece every week. The sermons in our church used to take at least half an hour, and I spent the time [in the organ loft] composing the thing for the afternoon.[48]

(It was just the time when W. H. Elgar, on his own organ-playing Sundays, had gone across for his drink at the Hop Market Hotel.) 'The Brothers Wind', as they called themselves, rehearsed in one of the sheds behind the Elgar shop. With the ink barely dry on Edward's pages (Hubert recalled), 'he expected the

[45] Dated Jan. 1878 (Elgar Birthplace). James Tubbs (1835–1921) was one of the most celebrated English bow makers. The cost of a Tubbs bow in the 1870s was about 3 guineas. (See William C. Retford, Bows and Bow Makers [The Strad, 1964] pp. 72–7.

[46] MS Memoirs, a chronological selection of which was published in Berrow's Worcester Journal, 1972.

[47] The Strand Magazine, May 1904, p. 540, with the stranger's words as quoted by Young in Elgar, O.M., p. 41.

[48] Ibid.

band to play it on sight.[49] A favourite expression of displeasure to the Wind Quintette: "It's all spit & wind." '[50]

At first he had merely arranged the music of others—an *Allegro molto* from Beethoven's Violin Sonata in A minor, Op. 23, a march of Leybach, Glee Club songs ranging from Morley's *Now is the Month of Maying* to Barnby's *Sweet and Low*, selections from Lecocq's naughty operetta *La fille de Mme Angot* which was all the rage in London and Paris then. But already at Christmas 1877 Edward himself wrote a bright little *Peckham March* for the quintet.[51]

The new year 1878 saw further original works. On 4 April he completed an ambitious *Harmony Music* in sonata form. The six-minute movement was a bright and busy *Allegro molto* that wavered between Haydn and the sound of a German band, with sidelong glances toward French operetta. The most accomplished part of it was the scoring, which deftly varied the limited colours: here both Edward's organ-playing and his experience with the Instrumental Society stood him in good stead. And he displayed his own new-found virtuosity by entrusting the exposition's entire second subject to the bassoon, accompanied by the rest.

The exposition was repeated. There was a tiny development of the primary theme only. A recapitulation varied the scoring (with the second subject now entrusted to Hubert Leicester's flute). The whole finished in a coda, with a *stretto* (carefully marked in every copy) as long as the development itself. He dedicated the music to the second flute, 'Professor Exton'. Soon this *Harmony Music* was renamed *Shed* after the home of the quintet's rehearsals.

In May Edward wrote an *Andante* with variations, named after the nearby town of Evesham (where at the beginning of the month he played violin in a performance of the *Messiah* with his father and uncle). The *Andante* was a satisfied little tune; but the theme and every variation was punctuated with a two-bar postlude falling away through the triad in miniature farewell. There were variations for solo bassoon and oboe. One variation sent its accompaniment through daring chromatics in slow motion. The whole finished up with a *perpetuum mobile* in conventional semiquavers led by the flute of Hubert Leicester, to whom the piece was dedicated.

Hubert was the group's nominal director, and he took it with characteristic seriousness. He had them photographed at his own expense, and insisted that all the music written should come to him for safe keeping. Soon there was quite a pile.

Two other *Harmony Musics*, or *Sheds*, followed in quick succession. No.2 was dedicated to Willie Leicester, the clarinet. But in his own part Edward annotated it 'Nelly Shed' after Helen Weaver, the pretty and musical daughter of a shoe merchant nearly opposite the Elgar Shop in High Street. The new

---

[49] Obituary notice of Elgar in a Worcester newspaper, 1934.

[50] Conversation of 28 Nov. 1909, noted by Philip Leicester.

[51] BL Add. MS 60316. Subsequent references to pieces for the woodwind quintet are based on this source.

Shed was almost twice the length of its predecessor. Its themes had echoes of Weber, and its middle development was a big affair—in fact rather too big for its material.

Shed No. 3 promised to be still larger. Edward tried two versions. The first took a single theme—announced by the oboe, as it was dedicated to Frank Elgar—followed immediately by an extended development. The second version recomposed this in a more sophisticated way, with a really dextrous *fugato* in the midst. All was primary material: thus it foreshadowed the big orchestral expositions of Edward's maturity, where symphonic scale would be achieved through just such an assemblage of primary material before the appearance of any second subject. In *Shed* No.3 the exposition of primary material extended to a hundred bars before an oboe cadenza introduced a suave second theme. But after a few bars more the whole attempt was abandoned and left unfinished. Such a structure was beyond Edward's compass then. It stood only as a suggestion of the scale on which he would like to project his music.

On 21 May 1878 he wrote his name on the brown paper cover of a thin oblong sketchbook in which to work out preliminary ideas.[52] There, side by side with sketches for the wind quintet, were four-part hymn settings for St. George's Church (these were also arrangeable for the wind quintet) and an 'Introductory Overture' for an evening of Christy Minstrels at the Worcester Music Hall on 12 June: the Overture was scored for flute, cornet, and strings. On another page of the sketch book he wrote out all the old modal scales with their names in Greek characters.

June 1878 was the month of Edward's twenty-first birthday. The local life, whatever its chances found or made, was not yielding the scope he craved for his talent and his promise. He must try again to escape from Worcester. On his behalf an advertisement was placed that month in the Catholic magazine *The Tablet*:

To Musical Catholic Noblemen, Gentlemen, Priests, Heads of Colleges, &c., or Professors of Music.—A friend of a young man, possessed of great musical talent, is anxious to obtain partial employment for him as Organist or Teacher of Piano, Organ, or Violin, to young boys, sons of gentlemen, or as Musical Amanuensis to Composers or Professors of Music, being a quick and ready copyist. Could combine Organist and Teacher of Choir, with Musical Tutor to sons of noblemen, &c. Has had several years experience as Organist. The advertiser's object is to obtain musical employment for him, with proportionate time for study. Age 21, of quiet, studious habits, and gentlemanly bearing. Been used to good society. Would have unexceptionable references. Neighbourhood of London preferred; the Continent not objected to. Disengaged in September. Address Alpha Beta, TABLET Office.

---

[52] This book and its successors are at present (1983) in the keeping of Dr Percy Young.

Nobody responded.

He turned back to London for more violin lessons with Pollitzer.[53] And Pollitzer began to realize that this young man's promise might be fulfilled as something other than a violinist, for all his desperate application. The moment of the teacher's insight remained sharp in Edward's memory:

I was working the first violin part of [a] Haydn quartet. There was a rest, and I suddenly began to play the 'cello part.

Pollitzer looked up. 'You know the whole thing?' he said.

'Of course,' I replied.

He looked up, curiously. 'Do you compose, yourself?' he asked.

'I try,' I replied again.

'Show me something of yours,' he said.

I did so, with the result that he gave me an introduction to Mr. August Manns.[54]

This introduction to the conductor of the Crystal Palace Concerts secured a pass for Edward to attend orchestral rehearsals there. The excitement was recalled later for W. H. Reed, who wrote: 'While he was in London he took the opportunity of going to every possible concert; and he always spoke almost reverentially of those which August Manns conducted at the Crystal Palace. I think it was the attendance at those concerts . . . that fired his ambition and turned the scales on the side of serious composition.'[55] More than any other concert series in England then, the Crystal Palace Concerts were famous for producing large works of unfamiliar music—works whose scores were seldom available for study, but which any aspiring composer would want to know.

Back in Worcester Edward evolved a plan for making single-day visits to London which might yield him two hearings of each novelty:

I rose at six—walked a mile to the railway station;—the train left at seven;—arrived at Paddington about eleven;—underground to Victoria;—on to the Palace, arriving in time for the last three quarters of an hour of the rehearsal; if fortune smiled, this piece of rehearsal included the work desired to be heard; but fortune rarely smiled and more often than not the principal item was over. Lunch,—Concert at three;—at five a rush for the train to Victoria;—then to Paddington;—on to Worcester arriving at ten-thirty. A strenuous day indeed; but the new work had been heard and another treasure added to a life's experience.[56]

At home he began a series of small *Promenades* for the wind quintet. They were less ambitious than the *Sheds*, mostly in simple A–B–A form but full of deft touches. One, entitled 'Mme Taussaud's' [sic] or 'Waxworks', reflected Edward's London experience. A slower one, for which he invented the word 'Somniferous', might show the sleepiness at home. And to the *Promenades* he

---

[53] The account of Elgar's 'Expenses, *in full*, for general and musical education' includes as its final item: 'Violin lessons, Mr. Pollitzer 1877–8 £15.15.0'

[54] *The Strand Magazine*, May 1904, p. 538.

[55] *Elgar as I Knew Him*, pp. 46–7.

[56] MS draught for speech given at 'His Master's Voice' reception in London, 16 Nov. 1927 (Elgar Birthplace).

added an *Adagio cantabile* named after a patent medicine, 'Mrs. Winslow's Soothing Syrup'. Between the mellifluous opening and close, a smooth development slipped down chromatic phrases as easily as the egregious product itself might do. Thinking back to his childhood that summer, the composer just turned twenty-one looked out the music for the old children's play and copied 'Moths and Butterflies' into his new sketchbook—as if to test the potency of that early inspiration amid all the newer effort.

One of the hymns he wrote for St. George's Church had the same idea:

> Hear Thy children, gentle Jesus,
>   While we breathe our evening prayer,
> Save us from all harm and danger,
>   Take us 'neath Thy sheltering care.

Edward's setting of the hymn was dated 21 July 1878. Three days earlier had come the death of old Miss Walsh, his first schoolmistress in Britannia Square. And at St. George's it was announced that Father Waterworth, now aged and ill, was to leave Worcester after a Superiorship which had lasted virtually the length of Edward's childhood.

Yet the summer was also full of anticipation, for 1878 was another Three Choirs Festival year at Worcester. Despite the preventive efforts of Lord Dudley, the Worcester Festival was to be revived in September with the full orchestra. Edward accepted an engagement to play in the Festival orchestra, even though it was only in the second violins with his father.

One of the works the orchestra rehearsed for the 'Wednesday Evening Secular Concert' at the College Hall, in College Green, was Mozart's G minor Symphony (K.550). The experience of rehearsing this music now focused an opinion which had been forming in Edward's mind ever since the Mozart Violin Sonata *Allegro* had guided the making of his boyhood *Gloria*: 'Mozart is the musician from whom everyone should learn form.'[57]

Of all musical forms, the symphony was held in highest regard at that time. The Mozart G minor Symphony showed with great clarity how themes could be contrasted and developed to present a big structure. Years later Edward described the work as 'amongst the noblest achievements of Art. We have to marvel that with such a selection of instruments, variety and contrast can be found sufficient to hold the attention for thirty minutes.'[58]

So once again he submitted himself to the discipline of Mozart.

I . . . ruled a score for the same instruments and with the same number of bars as Mozart's G minor Symphony, and in that framework I wrote a symphony, following as far as possible the same outline in the themes and the same modulation. I did this on my own initiative, as I was groping in the dark after light, but looking back after thirty years I don't know any discipline from which I learned so much.[59]

[57] *The Strand Magazine*, May 1904, p. 539.
[58] *'A Future for English Music' and other lectures*, p. 273.
[59] *The Strand Magazine*, May 1904, p. 539.

In a second sketchbook started during the Festival rehearsals in August, Edward set out his own G minor Symphony in full score: 'Flute, Oboes, Bassoons, Horns (1 & 2 G, 3 & 4 B^b), Tympani (F B^b), Clar[ine]t B^b, Violin I & II, Viola, 'Cello, Contra Bass'.

He scored fully 17 bars of an opening *Allegro* and carried it on through several instruments for another 25 bars. But it was a difficult business seeing to all the problems of composing and orchestrating a big sonata movement—even when the framework was all Mozart. The movement went no farther.

A simpler design was offered in the minuet. He began again with a 3/4 *Allegro molto*. The sections of this Minuet were shorter than Mozart's: Edward wrote 8 bars with a repeat, and 20 bars with a repeat. Then he started a Trio; but before that was finished the music stopped again.

At the Festival in September he played in the G minor Symphony, Handel's *Messiah*, Mendelssohn's *Elijah* and *Hymn of Praise*, Part I of Haydn's *Creation*, the Mozart *Requiem*, Spohr's *Last Judgment*, and several modern English works. An oratorio entitled *Hezekiah* had been written by Dr Philip Armes of Durham, who came to conduct his piece; and there were choral settings by Sir Frederick Gore Ouseley and Sir John Stainer. Edward remembered Stainer 'walking up and down, in a nervous sort of way, until the composition was over'.[60] But much of the new music did not impress the young second violinist:

We had been accustomed to perform compositions by Sir Frederick Ouseley, Dr. Philip Armes and others of the organists and professors of music who furnished meritorious works for festivals, but they lacked that feeling for orchestral effect and elasticity in instrumentation so obvious in the works of French, Italian, and German composers.[61]

Again it was the use of orchestra, not chorus, on which his judgement concentrated.

The experience of playing in the Mozart Symphony found its first result in the week after the Festival. On 17 September Edward completed a new large *Shed* for the wind quintet—No. 4, entitled 'The Farm Yard'. The atmosphere of its themes and contrasts was Mozartean in a way none of the previous *Sheds* had been. But the new *Shed* also solved the problem abandoned in *Shed* No.3—an immediate big development of the first subject, contrasting with a second subject to fill a large exposition. In *Shed* No. 4 the large exposition was completed, and there followed a finely accomplished formal development section. After full recapitulation, a coda neatly dovetailed fragments of first and second subjects in a satisfying conclusion. It was by far the best big structure Edward had yet achieved.

He went straight back to his own 'Mozart' Symphony. In a third sketchbook dated 1 October 1878 he began the Minuet again. This time he gave it a

---

[60] Address to the Friends of Worcester Cathedral, June 1932.
[61] 'The College Hall', dated 3 June 1931, in *The Three Pears* annual magazine (Worcester, 1931).

suggestion of Beethoven with the title 'Minuetto (Scherzo)', and set out a more elaborate orchestration with the instrument names all in proper Italian: 'Flauti, Oboi, Clarinetti, Fagotti, Cor I° "C", Cor II° "G", Trombi, Tromboni, Tympani, Violino I, Violino II, Viola, Cello, C.Basso'. This time he finished the music. The achievement spurred further study: on 7 October he wrote his name in a copy of Ebenezer Prout's treatise on *Instrumentation*.[62]

His own symphonic Minuet, however, was no mere copy-book exercise. He added to it a part for the harmonium, or reed-organ. When his Uncle Henry played the harmonium for the Glee Club ensembles and elsewhere, he filled in parts whose instruments might happen to be missing. (Indeed there were times when the entire 'orchestra' for a local choral performance consisted of Uncle Henry at the harmonium and Edward at the piano.) Adding a harmonium part to the Minuet score bettered its chance for performance.

Then came a further chance for Edward's orchestral skill. It came from an odd quarter. The Elgars had continued to play in occasional concerts given for the inmates of the County Lunatic Asylum at Powick. But the Asylum doctors attached such importance to musical entertainment for their patients that they had recruited an ensemble of players from their own staff. They had even hired a director to give instrumental instruction to the players, conduct the band, and compose and score lancers, polkas, and quadrilles for the patients' Friday dances. This post was to fall vacant at the new year. So on 21 December Edward scored another Minuet in G minor for the instruments available at Powick: flute, clarinet, 2 cornets, euphonium, bombardon, violins 1 and 2, bass, and piano.

The demonstration succeeded. Edward received his appointment as musical director, to commence from January 1879. He was to spend a day a week at Powick. His salary would amount to about £30 per annum; in addition he would receive 5s. for each quadrille or polka he composed, 1s. 6d. for every Christy minstrel accompaniment he might arrange. It was his first chance to compose and conduct on a regular basis.

He proposed to give up none of his other musical activities. His day a week at Powick was fitted into a schedule which included violin solo and ensemble engagements (as far afield as Gloucester in an orchestra conducted by the Cathedral organist Charles Harford Lloyd), constantly practising the violin himself and teaching pupils scattered over many miles, leading and instructing for the Amateur Instrumental Society, leading and playing the piano for the Glee Club meetings, playing the organ at St. George's Church, rendering a modicum of service in his father's shop, and going to London when a concert there contained something he especially wanted to hear. His composition sketchbook showed the considerable beginnings of a string quartet.

And there was the wind quintet. At Christmas 1878 he arranged Corelli's *'Christmas' Concerto* for them. In February 1879 he arranged the March from

62 Now at Elgar Birthplace.

his old children's play music. Then came another big effort. On 2 March, as he sat in the organ loft at St. George's Church through the long sermon, he began setting out a new *Shed Allegro*.[63] It was the opening of a work for the wind quintet that was to put every previous effort in the shade—in effect, a four-movement symphony of twenty minutes' length. It was entitled 'The Mission'.

Its first-movement exposition followed the lines of the big *Allegro* primary development and secondary contrast realized six months earlier in *Shed* No. 4. The polyphonic accomplishment in the new writing was greater still. But the melodic ideas, though fresh inventions, continued Mozartean: the personal note was now more distant than in the violin *Reminiscences* of two years earlier, more distant even than in the music for the children's play.

Some sense of this facility without depth may have affected the coda to the new *Shed Allegro*. The coda extended and extended—as if seeking for some personality of its own. But everything remained obdurately Mozartean to the end—reached after 324 bars (not including the exposition repeat) on 24 March 1879.

For a second movement he drew on a previously written Minuet. There was reaction in his choice, for the theme of this Minuet was more personal than anything in his previous music for the wind quintet. Its melody brought together again many traits of *Reminiscences*. Here was the down-turning shape, the initial subdominant goal, and a sequence shifting dextrously to the dominant from which another return would be made:

Then on 28 March (the day after finishing the first movement) he sketched an *Andante* in A minor promisingly reminiscent of a pensive Schubertian night-piece. Yet as soon as it was submitted to development, the Mozartean atmosphere came back.

Edward turned away to work at smaller pieces for the quintet—on 3 April a *Gavotte* named after Hubert Leicester's sister Alphonsa, on 5 April a *Sarabande* and *Gigue*. There was not much thematic originality in any of these. On 7 April he finished the *Shed Andante* and gave it the private title of 'Noah's Ark' in his own bassoon part.

He completed 'The Mission' *Shed* with a Finale again in sonata form, with some chromatic progressions in the second subject but little else to distinguish it. The whole work was signed and dated 4 May 1879—a four-movement piece

---

[63] MS in the writer's possession.

approaching the scale of Mozart's G minor Symphony. But its lack of individuality suggested that Mozartean form had somehow caged Edward's expression.

His restlessness during the completion of *Shed* No. 5 emerged also in the writing of five tiny *Intermezzi* for the wind quintet. Here in the small forms there was some reaching out. The first two *Intermezzi* were full of syncopations, odd and whimsical intervals, suprising key-changes, and more daring harmonies than the ambitious *Sheds*. Edward noted in his bassoon part: 'I like the Shed on the whole but the "Intermezzi" are "mine own".' And he gave them feminine names: *Intermezzo* No. 1 was entitled 'Nancy'[64] and No. 2 'Mrs. & Miss Howell'. By the time he finished the series with No. 5 on 5 May, the style was somewhere between Schubert and the homely strains of a composer Edward had yet to encounter—Dvořák.

Two days later he sketched an *Andante arioso* which looked in a new direction. It brought the best of his melody-making into a sort of slow march with dark dignity:

Here was an utterance which could combine nostalgia for the past with real aspiration for the future. With the help of careful subsidiary themes he sustained the new piece to some length and finished it. It became the slow movement in a new *Shed* No. 6.

He added an opening *Allegro molto* in G—another sophisticated sonata structure of some scale. On 12 June he started a Minuet in the wind quintet part-books, but it went no farther. *Shed* No. 6 remained a two-movement work—a sonata *Allegro*, and an *Andante* which could not be followed by a Minuet. So the new *Shed* raised squarely the problem facing him: it was all very well to learn classic form from Mozart, but the style of Mozart's music did not hold the expression Edward needed for himself.

Near the first idea for the *Andante arioso* in the sketchbook, he noted another idea still more spendid:

But for this there was then no development.

*   *   *

[64] Stated in some sources to be No. 3. Elgar's manuscript, recently discovered, is however quite clear.

He went to concerts in London whenever his finances could be made to meet the interest of the programme offered. He had begun to make acquaintance with the music of a composer whose example was to offer him private sympathies of a kind he had never yet encountered—Schumann. He had heard the great piano *Fantasy* played by Hans von Bülow in one of the Monday Popular Concerts at St. James's Hall in November 1878. Other London concerts during the winter yielded several performances by the Joachim String Quartet, culminating on 10 March 1879 in the Schumann Quartet in A minor.

Amongst provincial musicians such as Mr Done, Schumann was still looked on as a radical. For Edward, however, his music evoked a strange sense of recognition. Schumann's life had ended in Germany a year before Edward's own had begun at Broadheath. Yet Schumann had been drawn to obsessive rhythms and to sequence-writing before Edward. Moreover, that impulse to rhythm and sequence seemed even to touch Schumann's way of writing melodies. Again and again he would shape a melody as if from the inside out—a first phrase cast harmonically as if it were returning, followed by an answering phrase that was like an excursion: it was the sequence of surprising expectation which Edward himself had sometimes touched in his own melody-writing. Coming hard on the thraldom to Mozartean form, the heady freshness of Schumann offered rich promise.

Further illumination came from the big *Dictionary of Music and Musicians* which George Grove began to publish in 1879. It contained articles by many authorities: 'but the articles which have since helped me the most', Edward said later, 'are those of Hubert Parry . . .'[65] Parry's articles in the new *Dictionary* addressed the formal and technical aspects of music with a rare combination of practical experience and generous lucidity. His article on 'Symphony' cast its net wider still when it came to Schumann. For Parry showed that Schumann, in addition to all the musical affinities which Edward could feel for himself, had also been a kind of auto-didact:

Schumann seemed to have developed his technique by the force of his feelings, and was always more dependent upon them in the making of his works than upon general principles and external stock rules . . . Schumann developed altogether his own method of education. He began with songs and more or less small pianoforte pieces. By working hard in these departments he developed his own emotional language, and in course of time, but relatively late in life as compared with most other composers, he seemed to arrive at the point when experiment on the scale of the symphony was possible.

Beginning with songs and more or less small piano pieces was practically what Edward had done. He was now nearly twenty-two—relatively late in the life of a Mozart. If he worked hard, as Schumann had worked, would he too succeed in developing his own emotional language?

The career of the Elgar family's elder son seemed committed to music in a way

---

[65] *The Strand Magazine*, May 1904, p. 539.

from which there would be no turning back. The younger son Frank was likely to follow him. For the ageing mother pursuing her family duties at 10 High Street, the creative urge that formerly made a special bond with Edward had taken the son where his mother could not follow.

Early in 1879 Ann Elgar finished a suggestive poem:

> Into the belfry tower one night
> My wandering fancy winged its flight
> For the dear voices of the night
>   Had lured me with their spell—
>
> I roamed the gloomy chamber round
> And all so stark and drear I found
> When lo! methought I heard a sound
>   A voice as from a bell—
>
> Saying, 'Why come you here to see
> The secrets of our minstrelsy?
> Mortal! what are our ways to thee?
>   Be silent, or away!'—
>
> I fancied then the tower shook,
> But I crept close within a nook
> And waited anxiously to look
>   Upon this strange array—
>
> On the huge frame of each bell to be rung
> With an airy lightness a figure sprung,
> A goblin, elflooking fairylike thing,
> And grasped from each clapper a tiny string,
> Then lightly, gracefully poised on a toe
> Each lifted the string and struck one blow,
> Orderly, quietly, each in its place
> The little forms moved with exquisite grace.
> Their task accomplish'd, the chimes and the hour
> Had been struck with precision and wonderful power,
> They were moving away, when a fierce desire
> Came o'er me to speak before they retire.
> I cough'd, and I Hemm'd but no word could I say
> And saw the last little one turning away
> When, there came a loud knocking! Dot opened the door!
> Saying 'Ma, you sleep soundly, I've called you before!'[66]

If the energies that shaped Edward's music remained enigmatic, the call back to mother's duties still sounded clearly enough through the voice of her youngest child—for Dot was only fifteen.

---

[66] The version in Lucy's 'Reflections'. An earlier draft, written in Ann Elgar's scrap-book, was dated Nov. 1878.

The elder Elgar daughters were already women in their own right. Lucy had been engaged to Charles Pipe, the young tradesman in Broad Street, but as he was not a Catholic, the outcome was in doubt. Pollie, the second daughter, was married in the Catholic Church on St. George's Day, 1879. Her husband could not have been more welcome: he was William Grafton, the eldest son of a well-known Catholic family who were close friends of the Leicesters.

Edward cultivated his band at Powick. There was no formal challenge in any of the music he could write for the Lunatic patients' concerts and dances, which must be always of the simplest. The scoring depended on the talents and instruments available among the staff: so the balance was peculiar and also subject to change. It challenged him once more to find every possibility in every instrument at his command, as had the wind quintet and the children's play music before that.

The mainstay of the Powick band—as of the old improvised band for the children's play—was the piano. The Asylum pianist was a young lady, Miss Holloway. Soon her presence was leading Edward's thoughts in several directions. On 7 March 1879 he inscribed a *Polonaise* 'for J.H. with esteem'.[67] Like a Polonaise by Wieniawski which he was playing then, this was for piano and violin: an ensemble for Miss Holloway and Edward. But the sketch was not quite finished. What Edward did finish was a set of quadrilles for one of the Asylum dances in May. He dedicated them to Miss Holloway—'*with* permission', as he wrote in the piano copy.[68] And he gave them the title *Die Junge Kokotte*. Perhaps his relations with the dedicatee of these continental trifles were finding no more fulfilment than the friendship with the factory girl over Voltaire.

The marriage of his sister Pollie—the first in the family—produced a change in his own life. Before the year was out Edward left his parents' home over the shop in High Street to join Pollie and Will Grafton in the little house they had taken at the far end of Chestnut Walk, north of the Catholic Church. It might be seen as the act of a dutiful elder son considering the burdens of the remaining family upon parents nearing sixty. But even when Lucy and then Frank were ready to leave the old home, Edward never went back to live with his mother and father.

Two other marks of maturity made their appearance. About this time he began the moustache which was a prominent feature ever afterwards. And he acquired another personal adjunct of consequence:

When a young man, Elgar had a 'turnip' silver watch which he took to a Swiss watchmaker well known as a character in Worcester—Fritz Friedrich of Broad Street.

F. said to E.: 'If you really want to get on in the world you ought to have a gold watch & a top hat.'

---

[67] BL add. MS 49973A fo. 61.

[68] MS, Elgar Birthplace. 'Die kleine Coquette' was an alternative title jotted down in his sketchbook.

E. took this advice to heart & saved up & bought a gold watch, of which he was always extremely proud.[69]

He might laugh at his father's snobbery in riding a thoroughbred to grand houses. Yet he himself was showing more determination to get on in the world than ever his father had.

On 21 June 1879 the fiftieth anniversary of St. George's Church was celebrated with a special service which included two of Edward's anthems, the *Tantum ergo* and a *Domine salvum fac* completed that spring. Yet there had also been an unfinished work of special private suggestion. At a time when most of the surviving Elgar children were beginning to find their ways in the world, it must have been impossible not to think sometimes of Harry, the eldest son and brother—or of Jo, whose musical promise would now have been approaching maturity. In a new sketchbook started in April, Edward began another anthem with the words 'Brother, for thee he died'.

The past came up again during that summer. Arriving for a concert he was to lead at the Powick School on 20 July, he found an ensemble lacking any bass instrument. 'Played Double Bass', Edward noted on his programme. It recalled the family play music, when he had devised the home-made double bass to support the ensemble. At the end of the summer Edward looked again at his old music for the children's play, and once again drew on it to serve present need. He recalled:

Two eight-bar sections of ['Wild Bears'] were taken to complete a dance [the final number in a set of Quadrilles for Powick entitled *L'assomoir*, dated 11 September 1879]; a portion of the 'March' also 'assisted' in a . . . trio for 2 violins & piano . . .[70]

In the midst of young manhood, however, the past could not dominate for long. Edward's immediate outlook showed itself in a set of *Valentine Lancers* dated 14 February 1880, and a Polka he called *Maud* for one of the Asylum concerts at the end of May. He asked Charlie Pipe to share a summer holiday to France. Young Pipe was about to solve the question of his engagement to Lucy Elgar by joining the Catholic Church. The happy couple could look forward to getting married in the following spring.

On 15 August 1880 Pipe was duly received into the Catholic Church. Immediately afterwards, as he recalled:

My future brother-in-law, Edward Elgar, and myself went to Paris for a few days. We went to an American house, the Hotel Buckingham, and were very comfortable . . .

Of course with E.E.'s mind already musical, he went to hear M. Saint-Saëns at the Madeleine and I expect one or two other organists at different churches.

E.E. had the laugh on me and my French one evening. We went to an American bar

---

[69] MS note by Carice Elgar Blake, *c.*1935: 'Told me by Mr. C. Bryant of Cooper's Jewellers, The Cross, Worcester. E. aged about 20–22.' (MS, Elgar Birthplace).

[70] MS notes on *The Wand of Youth* Suite no. 2, 1908. The Powick music is now at the Elgar Birthplace. Émile Zola's *L'assommoir*, a novel of lower middle class tradespeople whose lives move inexorably downward, had made an instant *succès de scandale* on its appearance in 1876–7.

for one of their drinks, but did not know what to ask for. We fixed eventually on 'Grog americaine', and were asked by the waiter, 'Chaud ou froid?' I forgot the right word and said 'Chaud.' The summer lightning was playing round us, and so were the fools of waiters, who thought it fun to watch us drinking hot rum on so hot a night. We said it was an English custom to cool us!

As I had been in Paris three times before I was able to cicerone E.E. about, and we did not miss many of the principal sights. We were very amused at the Haulon-Lees at the Chatelet Theatre, especially at one scene which I cannot well write about.[71]

Edward's recollection of this first international adventure preserved a different experience—a memory to come wafting back to him one day after more than fifty years: 'In passing through the pine-scented forest of Fontainebleau [in 1933] I had come to a turn of the road leading to Barbizon. The scent recalled a romance of 1880, and I nearly—very nearly—turned to Barbizon.'[72] The scent that kept the romance through all those years was also the evergreen scent of Spetchley and Broadheath.

Returning home, the French holiday was drawn on to provide titles of new music for the Powick Asylum. A set of quadrilles written in October was given the collective title *Paris*, with the separate numbers reflecting particular impressions: 'Chatelet', 'L'hippodrome', 'Alcazar d'été (Champs Elysées)', 'Là! Suzanne!', and 'Café des ambassadeurs' (introducing a French song, 'La femme d'emballeur'). The impressions were brought home in the music's dedication once again to Miss Holloway. But there was nothing to galvanize further composition on the scale of the last *Sheds*.

He was to make one final attempt in the following spring. But *Shed* No. 7 was again only a single *Allegro* of the dimension of *Sheds* written three years before it. Its primary subject

bore a coincidental homely relation to the primary subject of another work being completed at that moment in Germany, whose sounds would one day reach Edward's ears—the *Tragic Overture* of Brahms.

He was being recognized in the ways a small city recognizes its clever young men. One well-to-do citizen whose interest he attracted was E. W. Whinfield, the head of a local organ-building firm and vice-president of the Worcester Musical Society. Mr Whinfield often gave musical evenings at his large house, Severn Grange, near Claines. There Edward and his violin were more and more regularly invited, and there he met a wide range of musicians.

The pianist for the Severn Grange evenings was Mrs Harriet Fitton of

[71] 'The memoirs of Charles Pipe', *Berrow's Worcester Journal*, 1973.
[72] 'My Visit to Delius', *The Daily Telegraph*, 1 July 1933.

Malvern. She had studied in Germany, and in London with one of Chopin's pupils, Henry Brinley Richards. Her musicianship was in great demand and might have supported a professional career, were it not for her large family. Mrs Fitton's musicianship was matched by outstanding personal qualities:

She brimmed over with vitality. She was seldom tired, and hardly ever showed a sign of it; she was never bored. With a child's eagerness she flung herself into whatever new enterprise was being started. And with this vitality was linked a rare shrewdness. If you put some project before her, in a flash she saw both its strong and weak points. With the same intuition she read character . . . No one in genuine distress ever went to her without finding the most sympathetic and practical of friends. But I doubt if the cleverest imposter ever managed to get the better of Mrs. Fitton's shrewdness.[73]

Soon Edward found himself invited to the Fittons' home at Malvern, Fair Lea. There he and Mrs Fitton played through the repertory of violin and piano sonatas. For trios, the violoncello part was taken by either of two young clerics who were also *habitués* of the house—Edward Capel-Cure, recently arrived in Worcester as curate at one of the Anglican churches, or Fr. Henry Bellasis. Bellasis was the youngest son of Cardinal Newman's great friend Serjeant Edward Bellasis, who had enrolled his sons as the very first pupils at Newman's Oratory School. Two of the sons, Henry and his elder brother Richard, had become priests at the Birmingham Oratory. They shared the old Cardinal's musical interest, and he bore them a fatherly affection.

Then the Worcester Musical Society's conductor Alfred Caldicott approached Edward:

The Royal Academy of Music instituted their local examinations in music . . . At the earliest of these test occasions—in 1881—the local representative of the Academy at Worcester (Mr. A. J. Caldicott) came to [Edward] and in anxious tones said to him at the last minute:

'I am one candidate short in order to make a centre—that is, to ensure the visit of an examiner from London—for the Academy examination. Will you go in for it?'

'What are the pieces?' asked [Edward].

Raff's *Cavatina* and Kreutzer's Concerto in D minor,' replied Mr. Caldicott.

'I think I can play those, and will oblige you,' said [Edward], and his name was duly entered as a candidate for a local examination in music.

When the eminent musician from the Academy [Brinley Richards] came to Worcester in his examining capacity, he asked the stop-gap candidate this question:

'How many quavers in a double-dotted minim?'

'If it's a joke,' replied the examinee, 'I don't quite see it.'

'There is no such thing as a double-dotted minim,' gravely remarked the examiner.[74]

If that was a sample of academy teaching, it could confirm again the wish to teach himself. Yet at the end of the day, the name of Edward Elgar appeared at

---

[73] A. C. D., *In Memoriam—Harriet Margaret Fitton* (privately printed leaflet, 1924).

[74] Obituary article in a Worcester newspaper, 23 Feb. 1934, probably written by Hubert Leicester.

the top of the Royal Academy honours list with a certificate for 'violin and general musical knowledge'. Also on the list was William Wolstenholme from the Blind College.

There was further experience gained outside the city. During the previous autumn, Edward had played in the Herefordshire Philharmonic Society concerts (where Mrs Fitton also acted as pianist). The programmes included two Mendelssohn Symphonies. A London weekend in February 1881 had yielded the *Hebrides* Overture at the Crystal Palace and Mendelssohn's E minor String Quartet played by the Joachim Quartet at St. James's Hall. And each of these concerts had offered something else of note. At the Crystal Palace Edward heard Berlioz's Overture *Les francs juges* with its virtuosity of orchestration. At St. James's Hall he heard Clara Schumann play her husband's *Études symphoniques*: here was a monument to the musician who had truly found a wife of sympathetic temperament.

During the early months of 1881 Edward finished three small compositions. On 1 March the Glee Club performed two of them in the same concert—a part-song entitled *Why So Pale and Wan?* and a March *Pas redouble*. Each was repeated at a subsequent Glee Club concert. On 17 May the Amateur Instrumental Society under Alfred Caldicott brought out a new *Air de Ballet*. Edward appeared that evening in three capacities—as composer, leader, and solo violinist.

His violin-playing was sometimes held up against him as a composer:

Several young musicians . . . myself among the number, were continually repressed and even reprimanded for an alleged ignorance of harmony . . . It was believed by most professors and amateurs of that day that no one could understand harmony unless he played the organ or the pianoforte: 'Elgar's boy (meaning me) cannot possibly know any harmony because the instrument he plays (the violin) sounds only one note.'[75]

Still he secured his place in the Worcester Festival orchestra for September 1881: he was now at the last desk of the first violins. Local music was represented in several programmes. There was a *Magnificat* and a *Nunc dimittis* composed especially for the Closing Service by Edward Vine Hall, the Precentor of the Cathedral. Alfred Caldicott had responded to a Festival commission with a cantata called *The Widow of Nain*. Edward offered to supply a part for the harp when one became available, and noted in his programme 'harp—E.E. orchestrn'.

But the Festival event which electrified him that year was the production of a cantata entitled *The Bride* by a composer only a decade older than himself. His name was Alexander Mackenzie, and he had mastered what the British composers for the previous Worcester Festival had not—the secret of writing for the orchestra. Fifty years later Edward recalled the impact of Mackenzie's new music:

The coming of Mackenzie then was a real event. Here was a man fully equipped in every

---

[75] 'The College Hall', *The Three Pears Magazine*, 1931.

department of musical knowledge, who had been a violinist in orchestras in Germany. It gave orchestral players a real lift and widened the outlook of the old-fashioned professor considerably. *The Bride* was a fine example of choral and orchestral writing, [and] had a rousing reception . . .

I had the honour to meet the composer the following morning and actually shook hands with him at Sansome Lodge.[76]

As soon as the Festival was over, Edward returned to his own music. He had just begun another sketchbook—his fifth—and on 14 September started *Pas redouble* No. 2. But the duties at Powick beckoned, and three days later he was working on another quadrille. In October there was a Polka which Edward attributed to his brother Frank—it was said to encourage Frank's creative hand. Hubert Leicester recalled: 'Frank was clever and would have made his name as a conductor or instrumentalist, had he been able to overcome a kind of lethargy which dogged him all his life.'[77]

The Polka was entitled *Nelly*, after the neighbouring shoe merchant's musical daughter Helen Weaver; but the ascription to Frank suggested ambivalence. All Edward's efforts to write music had so far produced no finished work more individual than the violin *Reminiscences* completed nearly five years ago. Some vital ingredient was still missing from his creative life.

In October he went to London to hear two larger works by Berlioz. At the Crystal Palace, August Manns directed the first English performance of *Lélio* (written to follow on from the earlier *Symphonie fantastique*). The programme note stated quite categorically that Berlioz's greatest genius was in his use of the orchestra. Soon Edward acquired a copy of Berlioz's treatise on Instrumentation.[78]

The other concert of October 1881 contained a work in which Berlioz had directly addressed the matter of melody. It was the song cycle *Nuits d'été* for solo voices and orchestra. The words of the songs were taken from several poets: each had appealed to the composer in a different way. So the collection as a whole might show some new and larger image of the composer and his music—suggesting one way of shaping separate inspirations toward a larger result.

The programme which included *Nuits d'été* began a new season of orchestral concerts directed by Hans Richter at the St. James's Hall. Richter had first come to England in 1877 in the company of Wagner, for the Wagner concerts which Wilhelmj helped to arrange. The year before that Richter had conducted the world première of the complete *Ring des Nibelungen* at the new Wagner Festival Theatre in Bayreuth. Now it was announced that he would preside over a London production of the *Ring* cycle in May and June 1882, as part of a

[76] Ibid.
[77] [Blake] TS. In 1903 Elgar allowed the *Air de Ballet* to be published as a '*Pastourelle* par Gustav Francke'—still in the hope of encouraging Frank Elgar as a composer.
[78] Elgar's signature in the copy is dated 27 May 1882 (Elgar Birthplace).

season of German opera to include also the English premières of *Tristan und Isolde* and *Die Meistersinger*.

The *Meistersinger* Overture in fact opened the Richter concert in which Edward heard *Nuits d'été*. The range and power of Wagner's orchestration equalled that of Berlioz. But his melody and harmony fired Edward's keenest interest. Wagner drove Mendelssohn's use of characteristic motives much farther, applying them in a consistent way to designate specific characters and ideas throughout the drama. When ideas in the drama were related, Wagner's themes showed melodic relations, and often harmonic relations as well.

The Overture to *Die Meistersinger* was none the less a powerful instance of pure form. In it Wagner used three independent themes in succession, and developed them; but then his recapitulation combined all three in a sudden polyphony which revealed them as aspects of a new and larger unity. If Edward could blend his own disparate ideas in such a way, his music would gain real unity.

Back in Worcester, his *Air de Ballet* was repeated 'by desire'—and encored—at the Amateur Instrumental Society's concert on 20 February 1882. The programme also included the 'New March *Pas redouble*'. Both were repeated when the Society gave a concert at the Powick Asylum in March. Through the spring he worked at two pieces for the Catholic choir and a violin study.[79] There was nothing approaching the scale of the *Shed Allegros*.

Yet the *Air de Ballet* attracted the spontaneous interest of a complete stranger when it was included in the Instrumental Society's concert given with members of the Catholic choir on 11 August to entertain a convention meeting of the British Medical Association in Worcester. One of the members attending was a thirty-year-old doctor from Yorkshire who was also a keen amateur cellist, Charles William Buck. When the solo violinist and leader at the Worcester concert stood up and conducted the orchestra in his own *Air de Ballet*, Dr Buck was sufficiently impressed to seek him out and introduce himself.

The new acquaintance was helped by the fact that the two young men had a friend in common—John Beare, a London dealer in musical instruments who often came to visit Elgar Bros' shop in Worcester. The two men soon became friends. Buck was introduced at High Street, where Ann Elgar told him she kept Edward's early manuscripts safe in a box.[80] Then Buck asked Edward to spend a holiday with him in Yorkshire. The Doctor lived in the village of Giggleswick near Settle with his mother, who played the piano well enough to accompany her son. If their guest would bring his violin to add to Dr Buck's cello, the three of them could attack the repertory of trios.

[79] A fragmentary *O Salutaris Hostia* for bass solo is dated 17 Apr. 1882 (Elgar Birthplace). The fifth sketchbook contains a violin study dated 30 Apr. and a *Benedictus* with nearly complete string parts dated 7 May.

[80] Letter of 22 Aug. 1932 from Buck to Elgar (HWRO 705:445:4513).

When he went to Yorkshire at the end of August, Edward took his current sketchbook. One day he found himself sketching new music for the trio. It was a simple melody and accompaniment, set down partly in ink and partly in pencil:

It was one of the earliest pieces to show something of Edward's mature style to come. The violin sounded its melody on the first and third beats of the bar, where the natural accents of rhythm are strongest. His friend's cello filled in the weaker second beat, giving propulsion to the violin's longer note. It made a repeating pattern of complement and contrast. The melody began with a small step upward and moved gradually into upward intervals of larger assurance. But it returned always to the A from which to renew its subdominant impulse. The mother's piano only provided background support. In a second and answering phrase of eight bars, the cello followed the violin's figure an octave lower, while the piano took over the former cello figure of two crotchets and rest. So the old children's play theme of youth inspiring its parents emerged again in the textures of this latest music. Altogether the Yorkshire trio showed a mirror-reflection of Edward himself to a degree that no finished music of his had yet achieved. Many years later it was to be looked out and arranged and published with the title *Rosemary* — *'That's for remembrance'*.

Yet the young man of 1882 was not quite satisfied. He recast the music with the title *Douce pensée*—'Sweet thought'. The French language translated its thought to the feminine gender. And so it could whisper again that the ideal of his music might still remain in the air all around him, as it had seemed to waft through the pine-scented forest of Fontainebleau.

*   *   *
*   *
*

Returning from Yorkshire at the summer's end in 1882, Edward tried to settle to solo violin-playing at home. A miscellaneous concert at Woodford on 3 October found him playing Miska Hauser's *Hungarian Rhapsody*. And he secured a testimonial letter from Alfred Caldicott which distinguished him from other musical Elgars:

I can strongly recommend Mr. *Edward* Elgar as 1st Violinist or Solo player. The terms would be *Two* Guineas if he leads or £1.11.6. otherwise. His brother, *Frank* Elgar, oboe, will come for a guinea . . .[81]

But on 31 October Edward confided to Dr Buck:

I am *'so busy'* really get no time to myself. This is, of course, all very nice from a commercial point of view, but oh! my fiddling; I never touch it now, save to give lessons or scrape at a Concert.[82]

In fact he was looking for a change. Alfred Caldicott was leaving Worcester, and the Amateur Instrumental Society needed a new conductor. Edward grasped the chance and was successful, as he wrote to Dr Buck: 'I was elected conductor to our Society: I had 33 votes, my oponent 3.' The opponent was Herbert Wareing, a young Birmingham musician of exactly Edward's age. Wareing had studied at Leipzig, and he had just taken his Mus.B. at Cambridge. Edward had only the advantage of local residence. Conducting at Powick had stood him in good stead.

When a way opened to orchestral experience that was more professional, however, its vehicle was still the violin. He found a place in William C. Stockley's Concert Orchestra in Birmingham. The fifty-four-year-old Stockley was an interesting figure: he too had studied his music at home, he too had earned his money by means of a music shop—which even now he carried on as his business. Yet Stockley had risen to become conductor of the Birmingham Festival Choral Society and chorus master to the Birmingham Festival, one of the most important musical events in the country. Then he had founded his own

---

[81] Quoted in Ralph Hill, 'When Elgar played for £2–12–6', in *The Radio Times*, 19 Nov. 1937.
[82] *Letters*, p. 4. The MSS of the Elgar letters to Dr Buck are in possession of the recipient's daughter, Mrs Monica Greenwood.

Concert Orchestra to give Birmingham a regular series of winter concerts. For the second concert in the current series, on 30 November 1882, Edward played in a miscellaneous programme finishing with a Suite from Delibes's ballet *Sylvia*.

Among Stockley's players he met new friends. One was Alfred Probyn, representing the latest generation in a family of horn players whose fame went back into the previous century. And seated with Edward in the first violins was Charles Hayward of Wolverhampton, the son of Henry Hayward who was known as 'the English Paganini'. The younger Hayward, a charming fellow of many talents, quickly became Edward's boon companion. Their adventures at local concerts were fondly recalled for the violinist friend of Edward's later years, W. H. Reed, who wrote:

I never met Charles Hayward; but his name was always cropping up, when Sir Edward would stand up to his full height and play an imaginary violin with the wildest abandon, singing some phrase out of one of the old pieces at the top of his voice, moving his imaginary bow about at incredible speed with a white-hot enthusiasm, and telling me that this was how he and Charles Hayward electrified the natives on their old rounds. I always thought that he was acting the old pictures of Paganini on these occasions.[83]

Electrifying the natives was not a game that could hold Edward's imagination for long. That autumn he laid plans for a trip to Germany—to realize his schoolboy dream of going to Leipzig. It could not be for a proper term of study now, but only a visit of a few weeks in the Christmas holidays to hear all the music he possibly could. There was also another reason: Helen Weaver, the High Street girl who had already given her name and initials to several of Edward's compositions, was studying music at Leipzig. On 15 October Edward inscribed another Polka: 'H J W   vom Leipsig gewidmet.'[84] It was entitled *La blonde*. For his Leipzig visit, Edward arranged to be at the *pension* where Helen was staying.

His journey was made on the last day of 1882. In Leipzig he duly joined forces with Helen Weaver and her friend, a seventeen-year-old piano student named Edith Groveham. Edith's quick admiration gave Edward a taste of female hero-worship—as the worshipper herself recalled:

The day after he arrived in Leipzig, Elgar was anxious to collect his mail at the Post Office but assured us he might find it difficult, not speaking German. We at once offered our services, & having both been in Leipzig for six months, thought we spoke fluently.

My friend marched up to a clerk with E.E.'s visiting-card saying, 'Haben Sie Briefe für dass.'

Elgar's eyes had a very merry twinkle as he came up behind & said, 'I am not a dass, I never was a dass, & I never will be a dass,' & in his usual courteous manner asked for his letters in perfect German.

For the rest of his visit he made great fun of our *fluent* German.

There were one or two important musical events which he was anxious to hear,

---

[83] *Elgar as I Knew Him*, p. 73.   [84] MS, Elgar Birthplace.

amongst them Anton Rubinstein's opera [*Die Makkabäer*], which the composer was conducting . . . Being a young female with time upon my hands, he asked me to get the tickets for the first performance. I took two or three places in a box—& when he asked what time one ought to eat, I said,

'Oh, before six, as we must be at the theatre at 6.30.'

I remember how very astonished he looked—but we duly had supper at 5.45 & were on the doorstep at 6.15.

I was greatly excited about it all (only 17) so suggested a *Droschke*—that we should not be late. Off we drove, & were at the Theatre door just before 6.30.

No doors open—not a soul about—& I shall never forget the twinkle in [his] eyes as he said, 'Edith, what have you done?'

The doors slowly opened & I *tore* up the steps to the 2nd range, where our box was. Suddenly I heard peals of laughter—& looking round saw [Edward] supporting himself against a pillar—& with the others literally in 'fits' of laughter. When at last he could speak he said,

'Infant, you will be the death of me.'

We had an *hour* to wait in that box, but I received an education from him. As each member of the orchestra came in, he said, 'Now Infant—that is so & so—listen to his tuning up.'[85]

Edward recalled:

We used to attend the rehearsals at the Gewandhaus; 9 a.m.! Most of our pros: are not up by that time. There the violinists play 3 over one desk, except the principals! The first thing I heard was Haydn Sym. in G—the Surprise [—] fancy!! I was so astonished. I thought it strange to go so far to hear so little—After that I got pretty well dosed with Schumann (my ideal!) [*Overture, Scherzo, and Finale* on 4 January, Symphony No. 1 and songs on the 9th, Piano Concerto on the 11th], Brahms [Songs on the 9th, String Quintet op.88 on the 13th], Rubinstein [*Die Makkabäer*, and '*Ocean*' Symphony on the 11th], & Wagner [*Parsifal* Prelude on the 9th—only a few months after the opera's première, *Lohengrin* on the 12th, *Tannhäuser* on the 14th], so had no cause to complain.[86]

The days in Leipzig overflowed with experiences. One of them could seem for a moment to cast an odd light over several facets of the German visit:

I saw two dancers . . . in Leipzig who came down the stage in antique dress dancing a gavotte: when they reached the footlights they suddenly turned round & appeared to be two very young & modern people & danced a gay & lively measure: they had come down the stage *backwards* & danced away with their (modern) faces towards us—when they reached the back of the stage they suddenly turned round & the old, decrepit couple danced gingerly to the old tune.[87]

It was reminiscent of the Two Old People in Edward's childhood play.

Yet the teeming days at Leipzig hardly gave time for much reflection. Every minute was full of experience, and there was music even at the *pension*. Edith Groveham remembered:

[85] Letters of 1937–8 to Carice Elgar Blake (Elgar Birthplace).
[86] 1 July 1883 to Dr Buck.
[87] 4 Feb. 1899 to A. J. Jaeger (HWRO 705:445:8330).

One evening before supper I heard, in the small drawing room, the sound of a violin—someone playing Hauser's *Rhapsody*. I felt sure it was Edward Elgar—so did not go into the room.

When the sound ceased I imagined he had left the room—so myself sat down to the Concert Flügel in the large room & played Schumann's *Nachtstück* from op.23.

Elgar came into the room, asking what I was playing: he did not know it, but thought it delightful. He then discussed Schumann with me—& asked for some other of my favourite bits—I played the *Kinderscenen*—& after thanking me, said he would send Jessie Fothergill's *First Violin*. He did not forget, & wrote in a little note: 'I know the Infant will revel in this.'

I remember how much he loved Longfellow's prose—& as we did not even know Longfellow wrote prose, he sent Helen the book containing *Hyperion*—& being all about the Heidelberg student life we loved it.[88]

So the book Edward shared with his mother now went to Helen. It was the appeal to literary romance which guided Paul Flemming in the book itself, and which Edward had tried already to reproduce when he read Voltaire to the factory girl. But Helen was more understanding.

The book he sent to Edith Groveham also made its suggestion. *The First Violin* was a recent novel of love and music in Germany. Its English heroine was seventeen—just Edith's age. And the hero could have stood for an ideal translation of Edward himself in German surroundings. Tall, with dark eyes and a moustache, his brilliant violin-playing hides a secret nobility. For he is really a Count in adversity. When at last his fortunes are restored, he marries the English heroine and carries her off to live happily ever after in his castle on the Rhine.

Edward wrote to Edith several times. One of his letters said how much he regretted not being in Leipzig to see the score of *Parsifal* being exhibited there round the time of Wagner's death in February 1883—and Edith duly went to see it: 'I gazed in awe, & left the shop actually in tears—deciding that I knew nothing about music & never should.'[89] With Helen Weaver events would shape a different course.

Edward's German adventure was brought to an end by the need to return to play in Stockley's next concert on 18 January 1883. That programme included two vocal solos by the distinguished tenor Edward Lloyd: 'Love in her eyes sits playing' from Handel's *Acis and Galatea*, and the 'Prize Song' from *Die Meistersinger*.

There was more of *Die Meistersinger* in a Wagner Memorial Concert at the Crystal Palace on 3 March, together with selections from *Tannhäuser*, *Lohengrin*, and *Parsifal*, *Siegfried Idyll*, Siegfried's Death from *Götterdäm-merung*, and the Prelude and 'Liebestod' from *Tristan und Isolde*. Against the

[88] 19 Apr. 1938 to Carice Elgar Blake (Elgar Birthplace).
[89] Ibid.

words of the 'Liebestod' Edward wrote in his programme: 'This is the finest thing of W.'s that I have heard up to the present. I shall never forget this.'[90]

The contrast of his own life was pointed by affairs at home in the early months of 1883. His sister Pollie was moving her young family closer to her husband's employment at the Stoke Works near Bromsgrove. So the little house in Chestnut Walk which had been Edward's home for more than three years was given up. He did not take the chance to return to his parents at 10 High Street, even though he used the showroom over the shop for giving violin lessons. Instead he went to the house of his eldest sister Lucy, who had been married for a year to Charley Pipe and was living at 4 Field Terrace, off the Bath Road. Pipe recalled: 'E.E. had a room at [the] top of the house & a pianette, & so very often had friends there.'[91] One was young Wareing, whom he had defeated for the Instrumental Society conductorship.

That spring Edward completed a piece for full orchestra. The new music evoked distant scenes, but not those of the German visit. Instead it had the kind of 'Spanish' subject which was highly popular just then—and which could appeal to Edward through its chances for insistent rhythm and syncopation. He called the piece *Intermezzo moresque*.[92] Almost immediately came an offer of performance. Vine Hall, the Cathedral Precentor, also conducted the Worcestershire Musical Union. He produced it on 4 April, as Edward wrote to Dr Buck: 'I had an exciting time over my composition, but it was successful so that is so far satisfactory.'[93]

Among the audience for the première was Herbert Wareing. Since taking over the Instrumental Society Edward had included one of Wareing's pieces in a concert in January. Now Wareing returned the favour with interest by recommending Edward's music to W. C. Stockley in Birmingham. Stockley recalled:

Mr. Elgar played in my orchestra for some little time as a first violin. But my first real knowledge of him came from . . . Herbert Wareing, who told me that Elgar was a clever writer, and suggested that I should play one of his compositions at one of my concerts. At my request Wareing brought me [the *Intermezzo moresque*], and I at once recognised its merit and offered to play it.[94]

Edward wrote to Dr Buck: 'Thanks for your congratulations re the Intermezzo. It is accepted for the Birmingham orchestral Concerts (band of 80) & may be heard in London: but I do not intend to push it.'[95]

The contrast of his own small work with larger music was inescapable. On 19 April Stockley's Orchestra was conducted by Frederic Cowen—a man only five years Edward's senior, but with German training and wide success in London and the provinces. In Birmingham Cowen conducted two of his own compositions. One was a symphony—an achievement beyond Edward's reach

---

90 Elgar Birthplace.
92 Apparently lost.
94 *The Musical Times*, 1 Oct. 1900, p. 645.

91 'A few anecdotes of E. Elgar'.
93 13 May 1883 (*Letters*, p. 7).
95 1 July 1883.

then. The other was a new suite for strings, *In the Olden Time*, made up of old-fashioned dances, recalling his own experience of the old-and-new gavotte in Leipzig.

Edward took the hint. He made his own Gavotte, wherein 'old' and 'new' styles would be set side by side as variations on the same theme. The opening 'old' Gavotte he based (he said) on a theme by Corelli, a favourite composer of W. H. Elgar and his friends. Edward's 'old' Gavotte had a solid, square-toed rhythm, began its minor tune on the tonic, and proceeded to a canon:

The 'masculine' assertion of this music was then balanced with a softer 'feminine' figure moving in parallel thirds:

A little development brought back the opening canon figure.

The 'new' Gavotte shot away at high speed in G major. Yet it was still the offspring of the two 'old' figures—with a special closeness to the 'feminine' strain:

The 'old' figures returned to recapitulate, but the offspring seized on the coda. So 'old' and 'new' remained as contrasts.

Helen was returning from Germany to spend the summer at home. When another invitation came from Dr Buck for a northern holiday to include chamber-music playing, Edward made clear that it must take second place to the arrival of his 'Braut'—his bride. He wrote on 1 July 1883:

If you are still in the same mind about my coming to you for a day or two I need not say how delighted I shall be. The vacation at Leipzig begins shortly; my 'braut' arrives here on Thursday next; remaining 'till the first week in Septr; of course I shall remain in Worcester 'till her departure. After that 'twould be a charity if you could find a broken hearted fiddler much trio-playing for a day or two.[96]

The broken heart would come only after Helen's departure. For that summer they were engaged. There was difficulty in the way of the marriage because the Weavers were Protestant chapel-goers. But Edward and Helen were happy for the summer, and did not borrow trouble from the future. Only Helen had to be a great deal at home, as her mother was not well.

There was also music to attract Edward. At the end of May he had gone to the Crystal Palace to hear the first performance in England of Berlioz's gigantic *Requiem*. Another Crystal Palace concert in August brought three movements from the *Symphonie fantastique*. The movements played were 'Un bal', 'Marche au supplice', and 'Songe d'une nuit de Sabbat'. Its brilliance may have inspired him, for in August Edward informed Dr Buck: 'I am engaged on a "gigantic" orchestral work.'

At the end of the summer, Helen went back to Leipzig and Edward took himself north to his holiday. With Dr Buck he visited the Lake District. The romantic scenes seemed to spread before Edward the manifest subject for his big orchestral work, an overture. Dr Buck recalled it long afterwards for an interviewer, who wrote:

When [Edward Elgar] went with him on his first visit to Lake Windermere, the effect was extraordinary upon the composer. Not a word could be got out of him, and then suddenly he began to write furiously. When he had finished he said that he had never known quite the same sensation before, and that he was simply obliged to write.[97]

After this burst, the Overture did not progress so easily. Back in Worcester after his holiday, Edward toyed with the idea of a Suite. What he achieved was another Polka, which he signed 'Edward Elgar, composer in ordinary to The W[orcester] C[ity] & C[ounty] L[unatic] A[sylum] Octr. 1, 1883'. How much he missed his 'Braut' emerged in the Polka's title—*Helcia*.

Then Helen herself arrived from Leipzig—in unhappy circumstances, as Edward wrote to Dr Buck on 11 November.

Well, Helen has come back!! Mrs. Weaver is so ill, dying in fact, so the child thought it best to return & nurse her; so we are together a little now & then & consequently happy . . .

I was playing at Hereford on Wednesday & at Birmingham (Stockleys) on Thursday. Splendid Concert. My piece is down for Decr. 13 . . . I get no time for composition now, altho' I have about finished my Suite in my head if I could but get time to write it down.

96 *Letters*, p. 8.
97 *Yorkshire Weekly Post*, July 1912 (quoted in *Letters*, p. 14).

The programme for Stockley's concert on 13 December described the *Intermezzo moresque* as 'one of a suite of three pieces by Mr. E. W. Elgar of Worcester, who is a member of Mr. Stockley's Orchestra'. In his own copy of the programme Edward crossed out the 'three' and wrote '4'. But if those other movements were 'about finished' in his head then, they were nowhere else. And when it came time for his *Intermezzo* performance, Edward's uncertainty showed itself on the orchestral platform in Birmingham. Mr Stockley remembered: 'On my asking him if he would like to conduct, he declined, and, further, insisted upon playing in his place in the orchestra. The consequence was that he had to appear, fiddle in hand, to acknowledge the genuine and hearty applause of the audience.'[98] In that audience were Ann Elgar and Helen Weaver, who had travelled up together from Worcester for the occasion.

The performance of Edward's music in Birmingham also brought the exposure of criticism written by strangers for large newspapers. One of the critics went out of his way to congratulate and encourage Mr Elgar to 'go on in a path for which he possesses singular qualifications'.[99] Others were less enthusiastic. And in the ebb of emotions that followed, there came news of a London friend going to Leipzig to study music, and a rumour of Dr Buck's engagement to John Beare's sister Emma. Every dream but his own seemed to find fulfilment then as Edward tried to frame his congratulations to send to Yorkshire.

4, Field Terrace   Worcester.

Jany. 14. 1884

My dear doctor,

If not too late—a happy new year & many of them. I have been loth to write before, for I have had nothing worth telling you. I had a good success at Birm. despite what the papers say; the man who wrote the slighting article is a Mus:Bac: who had sent in two pieces & they were advertised—and withdrawn because the orchestration wanted so much revision as to be unplayable!

Enough of this—I had a characteristic letter from Pollitzer—he asked for the parts & is trying to introduce the sketch in London—I don't anticipate a performance tho' I will let you know if it comes off.

I was sorely disappointed at not going to town—but 'tis no use going there to sit in the house all day & well—I have *no money*—not a cent. And I am sorry to say have no prospects of getting any.

We have had a very quiet time; my father was ill just before Xmas which made it dismal; the younger generation at the Catholic Ch: have taken an objection to him & have got him turned out of the Organist's place; this he had held for 37 years!! He thinks a great deal of this & I fear 'twill break him up.[100]

    [98] *The Musical Times*, 1 Oct. 1900, p. 645.
    [99] Newspaper cutting at the Elgar Birthplace.
    [100] According to the manuscript diary of Will Grafton's sister Agnes, on Sunday 19 Nov. 1883 old Elgar had 'got cross'—a situation apparently not without precedent. The following evening the new Superior at St. George's Church, Fr. Foxwell, called a meeting which had the result of dismissing the organist (MS in possession of Paul Grafton).

Frank gets on: was playing 2nd to Horton at Birmingham on Decr.26 & it seems to me that the only person who is an utter failure in this miserable world is myself.

I have heard from Arthur & am glad he is going to Leipzig. I fancy he will get good lessons; there are few, comparatively, for the 'cello.

What have you gone & done? He says 'the Doctor is in great glee over his last business!' Surely 'tis not matrimony: I am more than anxious to know all about it. I suppose I begin regular work next week, but I don't look forward to it: . . . I am disappointed, disheartened & sick of this world altogether.

Well—'tis time I was retiring—I am afraid I have sent you a dismal epistle—sorry for it—better next time . . .

Always sincerely yours,
Edward Elgar.

P.S. Miss Weaver is remaining in Worcester & the little Music &c. that we get together is the only enjoyment I get and more than I deserve no doubt.

Here were hints aplenty that things were not well with his own engagement. Mrs Weaver had died, and since then Helen seemed less certain than ever that she wished to marry a Worcester musician of Edward's prospects.

He tried to improve those prospects with the composition of a new 'Spanish' piece entitled *Sevillana*, for which he wrote a detailed programme note:

This sketch is an attempt to portray, in the compass of a few bars, the humours of a Spanish *fête*. It consists of three principal themes, which may be briefly characterized thus—1st, an imitation of a Spanish folk-song, played by the Violins on the fourth string; 2nd, a softer strain in the major, which may (or may not) be taken to represent 'un passage d'amour', for which, as in England, such gatherings are supposed to lend opportunity; and 3rd, a brisk Valse measure in D major. Something very like an *émeute* takes place during the progress of this, missiles are freely thrown, and at least one stiletto is drawn—but these are only modern Spaniards, and no tragic result follows; 'Çela était autrefois ainsi, mais nous avons changé tout çela.' Quiet is restored—the itinerant resumes his song—the Valse continues, and somehow or other all ends happily.

It is not assumed that this little piece embodies an accurate representation of all the above; suffice it to say that amidst some such scene, and as a souvenir thereof, was it written.[101]

The music owed less to Spain that to Delibes. The opening theme was in G minor, and the second 'love' theme made a variation of sorts in G major—after the fashion of the 'old' and 'new' Gavottes. The central 'Valse' led to the little development of conventional drama. But the earlier melodies were quite fresh enough to welcome recapitulation and a brief coda. It was all briskly orchestrated.

He dedicated *Sevillana* to W. C. Stockley in Birmingham. But the response came this time from London: for when Pollitzer showed the new score to August Manns, Manns promptly set it down in a Crystal Palace programme for 12 May. Worcester was not to be outdone. Mr Done placed *Sevillana* in his

---

[101] Quoted in the Worcester Amateur Instrumental Society programme for 9 Apr. 1885.

Worcester Philharmonic Society programme for 1 May—almost a fortnight before London—with the legend 'Composed expressly for this Concert'.

For Edward's own hopes it came too late. The engagement with Helen was broken. At first he could not bear to tell Dr Buck (whose own engagement was recently announced) but wrote to his friend on 21 April 1884: 'My prospects are about as hopeless as ever. I am not wanting in energy I think; so, sometimes, I conclude that 'tis want of ability & get in a mouldy desponding state which is really terrible.' To Buck's enquiries over the 'Braut', Edward responded: 'Miss Weaver is very well. I do not think she will remain in Worcester much longer now.' Two months more passed before he could bring himself to confess the truth—when he received an invitation to his friend's wedding:

You ask me to let you know 'soon' whether I can visit you & also be at the marriage feast: With many thanks & many regrets I must say 'nay' to both.

I will not worry you with particulars but must tell you that things have not prospered with me this year at all, my prospects are worse than ever & to crown my miseries my engagement is broken off & I am lonely.

Perhaps at some future time I may come out of my shell again but at present I remain here; I have not the heart to speak to anyone.

Please give my kind regards to Mrs. Buck & all friends; & once more accept my good wishes for your happiness, these I can give you the more sincerely since I know what it is to have lost my own forever.[102]

*   *   *

The emptiness of his loss must have resounded through his sensibilities as he listened alone to the music of others. In London at a Richter concert on 5 May he heard an orchestral work by Alexander Mackenzie on Keats's *La belle dame sans merci*. The programme also contained two German masterpieces. One was Schumann's Symphony in E flat, which grew from the composer's impressions of his own countryside round the Rhine. The other was a first performance in England. It was a choral ode by Johannes Brahms entitled *Gesang der Parzen* ('Song of the Fates'), set to a text of Goethe showing sorrow stricken dumb before destiny. Nothing was lacking to remind him of his loss.

What could not escape his notice was the announcement in the programme that in the next concert a week later Richter was to direct the first English performance of Brahms's new Third Symphony. So this première on the evening of the 12 May would follow by a few hours the first hearing of his own music in the environs of London, when Manns directed *Sevillana* at the Crystal Palace.

Brahms was the most highly regarded of living symphonists. Parry wrote of him in Grove's *Dictionary*:

[102] 20 July 1884.

The greatest representative of the highest art in the department of symphony is Johannes Brahms. His first two examples have that mark of intensity, loftiness of purpose, and artistic mastery which sets them above all other contemporary work of the kind. Like Beethoven and Schumann he did not produce a symphony till a late period of his career, when his judgment was matured by much practice in other kindred forms of instrumental composition . . .

If that hint of slow development touched Edward, what could touch him equally was Parry's description of Brahms as the defender of an old-world state in formal tradition:

He seems to have set himself to prove that the old principles of form are still capable of serving as the basis of works which should be thoroughly original both in general character and in detail and development, without . . . falling back on the device of programme . . . In all these respects he is a thorough descendant of Beethoven, and illustrates the highest and best way in which the tendencies of the age in instrumental music may yet be expressed.

When the hour came for the newest Brahms Symphony, Edward resolved to be there.

On 10 May Manns rehearsed *Sevillana*. Edward sent a post card to 10 High Street:

Heard my 'Sevillana' played at the Crystal Palace by the Orchestra in the Centre Transept! E.W.E.[103]

The card was addressed to W. H. Elgar: now surely the father must join in his son's recognition. Two days later, at the public performance, the success of Worcester was distinctly repeated. But neither parent was present.

That evening came the Richter Concert which included the new Brahms Symphony. It made a deep impression: twenty years later Edward devoted to it an entire university lecture, in a series which gave such attention to only one other work—the Mozart G minor. But an equal impression was made by the conductor of the Brahms Symphony that night in 1884.

On the European continent, controversy divided the followers of Brahms from the followers of Wagner. Brahms, it was said, stood for the tradition of abstract form in music, Wagner for a revolutionary marriage of musical theme and literary idea. Yet here was Hans Richter, the apostle of Wagner, showing himself equally the apostle of Brahms. Brahms had entrusted to Richter the world première of this new Symphony at Vienna in the previous December.

So Richter, representing the two greatest modern orchestral composers, could make a profound appeal to Edward. Younger than both Wagner and Brahms, Richter was still nearly a generation older than Edward himself. More than a quarter century later Edward acknowledged what began that night when he signed himself in a letter to Richter 'Your affectionate son'.[104]

Back in his Worcester life, on 22 May 1884 he inscribed another dance for the

103 HWRO.   104 29 July 1912.

Lunatic Asylum 'Blumine Polka von Eduard Wilhelm.' And when *Sevillana* had been liked well enough to earn a second performance at the Crystal Palace, he wrote to Dr. Buck:

Did I tell you Manns has been playing one of my pieces through the summer? Joke, isn't it.[105]

During one visit to London, Pollitzer sent him to hear Wilhelmj play the Beethoven Violin Concerto. Edward found himself marvelling again over the great violinist's tone, as he had marvelled in Worcester in 1877. But hearing Wilhelmj now only seemed to emphasize the inadequacy of his own playing. Years later Edward analysed it for his biographer, who wrote: 'he was dissatisfied with his own tone-quality, which he considered to be too thin to permit any outstanding achievement.'[106]

After hearing Wilhelmj this time, Edward asked Pollitzer the question direct:

'Shall I ever be first-class?'
'You will be very good.'
'Shall I ever be first-class?'

Pollitzer had been forced to answer no. And that hint of weakness in projecting his violin-playing might hint at a weakness of personal force in other phases of Edward's life. Back in Worcester, Hubert Leicester was forming an attachment to Agnes Grafton, the youngest sister of Pollie's husband Will. Soon there would be another engagement to set beside Dr Buck's marriage.

In the summer holidays that followed, Edward journeyed alone to Scotland. There he projected his own emptiness on the grand and wild landscapes. He explained to Dr Buck:

About my Scots excursion—I got into a very desponding state (you ken what happened) and it behoved me to do something out of the common to raise my spirits—the continent was in a measure closed to me—too much quarantine about—so I thought of Scotland and—went from here to Glasgow—thence to Rothesay . . .[107]

He began to record the days in short diary entries:

Monday Augt 11 1884. By Columba to Oban via Ardrishaig—encamping on grass, boat late (Linnet) to Crinan.

Tuesday Augt 12. Mull & Oban—round Mull to *Iona, Staffa going south.*[108]

Against the photograph of Fingal's Cave in his guide book,[109] Edward pencilled two themes from Mendelssohn's *Hebrides* Overture—as if joining the picture to the music its landscape had once inspired might suggest a way to something for himself. Next day he repeated the trip in the opposite direction:

---

[105] 28 September 1884.   [106] Maine, *Life*, p. 34.   [107] Sunday 8 Mar. 1885.
[108] HWRO.   [109] Ibid.

Wedy Augt 13. Round Mull to Staffa & Iona *going north* . . .

Friday Augt 15. *6.15* a.m.! Oban to Balachulish (& Corpach) thro' Glencoe & back to Balachulish—Boat to Oban Met E.E. on L[och] E[tive] went with her to Corpach.

The coincidence of the girl's shared initials made a beginning, and before they parted they arranged to meet again the following evening at Inverness.

Saturday Augt 16. Oban (6.15 a.m.) to Inverness Canal.

As the ship passed Fort Augustus early in the afternoon, Edward heard the bells of the Monastery. Here were actual sounds from the landscape to invite his response: he noted their tune in the guide book below the photograph of the Monastery: 'Chimes ½ past 2. Augt 16 1884.'

Arrived Inverness at Omnibus Station Hotel. Meet E.E.

Sunday Augt 17. Out early. Cath[olic] Ch. general larks, islands . . .

At the Free Kirk, Inverary, Edward scrawled two lines of a hymn tune across the back of an envelope which he would carefully keep.

Monday Augt 18. Saw off E.E. & party to Edinbgh not well. Walked about—Cemetery &c.

They had arranged to meet once more in Edinburgh:

Tuesday Augt 19. Inverness to Stirling . . .

Wednesday Augt 20. Stirling to Edinburgh Old Waverley Rd met E.E.

Thursday was shared at Holyrood House and the National Gallery of Scotland. Then it was over:

Friday Augt 22. Adieux! Flowers. Edinburgh to Glasgow.

And so home.

    This holiday that began and ended in loneliness produced another piece for violin and piano: but the new melody opened a kind of self-expression never yet realized in Edward's music.

The melody, in Edward's old key of G, moved along the borders of those 'sub-dominant grounds' whose magnetism he had felt since at least the time of the Beethoven *Credo*. And the opening of the new melody was instantly echoed in harmonic movement away from sharps as G major slipped to A minor in bar 4. These plagal darkenings gave the music a quality of reflection—almost of some noble renouncing—that was clearer than ever before in Edward's music. It marked another step forward in self-discovery.

The contrasting middle section had a smaller, quicker, more nervous figure climbing through the music's centre:

It was like the nervous quickness in Edward himself. Then the grander tune emerged again, singing its way to a coda which sought to join the two impulses in an ideal marriage.

Thus it was that the Hyperion journey to a place of fresh romantic magnificence had brought him face to face with a different E.E.—a female image. Yet meeting the girl who only happened to share his initials had been so casual. Instead of giving his fine melody a measured deliberation of self-assurance then, Edward quickened its tempo marking to *Allegretto*. Instead of facing any implications of meeting the other E.E. he called it merely *Idylle*. And to that title was appended the further description 'Esquisse façile'. When Dr Buck saw the new music in print with a dedication to 'E. E., Inverness', he wrote to ask about it. Edward replied:

Miss E. E. at Inverness is nobody—that is to say that I shall ever see again. I wrote down the little air when I was there & dedicated it to her 'with estimation the most profound' as a Frenchman would say, that's all.[110]

*   *   *

He returned home from Scotland to encounter a new influence on his music. At the Worcester Festival in September 1884 Antonin Dvořák came to conduct his *Stabat Mater* and his recent Symphony in D major.[111] Thus during the days of the Festival a composer of world renown, who enjoyed the friendship of Brahms, walked the pavements and streets of Edward's home career. Edward, sharing with Charles Hayward the third desk in the Festival orchestra's first violins, rehearsed and played under the great composer.

When he heard Dvořák's works, he heard a strength and deftness which were quite new to him. He wrote to Dr Buck: 'I wish you could hear Dvořák's music. It is simply ravishing, so tuneful & clever & the orchestration is wonderful; no matter how few instruments he uses it never sounds thin. I cannot describe it; it must be heard.'[112]

[110] 8 Mar. 1885.   [111] Now numbered 6.   [112] 28 Sept. 1884.

On the Wednesday evening, just before the performance of Dvořák's Symphony, Charley Pipe gave a small 'Festival' dinner at Field Terrace. Edward invited the most eminent composers of his acquaintance: Alfred Gaul, whose cantatas were produced in Birmingham; Charles Swinnerton Heap, a brilliant pianist and conductor of the Choral Society at Wolverhampton; and Herbert Wareing, who 'very often came to see E.E.'[113] The great composer was elsewhere in Worcester that evening.

The visit of Dvořák, coming hard upon the late summer reflection in Scotland, was a challenge. Where was Edward's composing and conducting in 1884? Its oldest appointed platform was the Lunatic Asylum band. In October—a month after the Festival but without any new expectation—he resigned the Bandmaster's post at Powick.

In London he negotiated the publication of a *Romance* in E minor for violin and piano, dedicated to Oswin Grainger. The publishers were the German firm of Schott, whose London office in Regent Street he had known ever since Pollitzer had sent him there to purchase music for his violin lessons in 1877. Thus it was that this *Romance* became his Opus 1. The contract he signed conveyed its copyright to Schott for 'the price of One shilling & 20 Gratis copies'.[114] In the terms of the day, it was a fair exchange for a young composer whose name was all but unknown in London.

Another chance seemed to open in London—possibly through the interest of Frederic Cowen, to whom Edward sent some of his compositions in hopes of performance.

The directors of the old Promenade Concerts at Covent Garden Theatre were good enough to write that they thought sufficiently of my things to devote a morning to rehearsing them. I went on the appointed day to London to conduct the rehearsal. When I arrived it was explained to me that a few songs had to be taken before I could begin.

Before the songs were finished Sir Arthur Sullivan unexpectedly arrived, bringing with him a selection from one of his operas. It was the only chance he had of going through it with the orchestra, so they determined to take advantage of the opportunity. He consumed all my time in rehearsing this, and when he had finished the director came out and said to me:

'There will be no chance of your going through your music to-day.'

I went back to Worcester to my teaching . . .[115]

In February 1885 *Sevillana* was played in Birmingham at a Stockley concert. Edward's thoughts moved again to larger things: he would write a 'Scotish' Overture to embody the experience of the past summer. His conducting energies were poured into the Amateur Instrumental Society—which now enjoyed the participation of aristocracy and gentry. In March he reported to Dr Buck:

[113] Pipe, 'A Few Anecdotes of E. Elgar'.
[114] Copy of the original assignment, dated 20 Jan. 1885 (Schott archives).
[115] *The Strand Magazine*, May 1904, p. 541.

Our orchestra (in which you played once) has developed alarmingly under my disabled guidance & are blossoming into a grand Concert April 9. that's a long time ahead, you will say, but we have to take time by the forelock.

Programme. Fest Marsch—Raff
    Entr'actes [*Manfred*] Reinecke
    Gavotte ['Stephanie'—Czibulka]
    Symphony B min.—Schubert
    Sevillana—Elgar
    Overture Haydée —Auber
     ,, Alfonso—Schubert
    &c. &c.

Vocalists—Honble. Mrs. Lyttleton (Miss Santley *as was*)
    Lord W. Compton &c.

Conductor, E.W.E.!!! There ye gods! Now, my most gay Settleite, digest that & bow down before me—

 . . . The lakes overture is done with—I am on the Scotish (with one t) lay just now & have a big work in tow. Of course all these things are of no account —but they serve to divert me somewhat & hide a broken heart.[116]

The 'Scotish' overture made no rapid progress. When he completed some music, it was on a smaller scale, still springing from a memory or even a name. In his once-a-week teaching at Mr Gedge's school in Malvern Wells, two of the Gedge daughters had violin lessons; another, Ethel, he instructed in the skills of piano accompaniment.[117] He used the notes G E D G E to concoct a Schumannesque melody, and made it the basis of another *Allegretto*. But this personal borrowing brought no very personal expression.

In August 1885 he sold to Schott an elaborate *Gavotte* for violin and piano. It was dedicated to Dr Buck 'in memory of the old days' they had shared before Buck's marriage. Yet again Edward's music had little to offer beyond conventional virtuosity.

The dedication to Dr Buck looked forward to another summer holiday in Yorkshire. Now there would be the addition of Buck's wife. At home in Worcester, Hubert Leicester was engaged to his Agnes. There was no such ensemble for Edward. On the day he wrote to dedicate the *Gavotte* to Dr Buck, 10 August, he wove music round a poem by the American writer John Hay:

> Through the long days and years
> What will my lov'd one be,
> Parted from me?
>
> Always as then she was
> Loveliest, brightest, best,
> Blessing and blest.

[116] 8 Mar. 1885. Edith Santley was the daughter of the famous baritone Charles Santley. Her husband was R. H. Lyttelton. Lord William Compton was brother of the Dean of Worcester Cathedral.

[117] Ethel H. Davis, 'Some memories of the first fourteen years of my life at the Wells House, Malvern Wells', in *The Wells House Magazine*, No. 123 (December 1960).

> Never on earth again
> Shall I before her stand,
>   Touch lip or hand.
>
> But, while my darling lives,
> Peaceful I journey on,
>   Not quite alone.

The music for this poem showed little more individuality than the recent violin pieces. And soon Helen Weaver planned a final departure, as Edward wrote to Dr Buck just after the 1885 visit to Settle: 'Miss W. is going to New Zealand this month—her lungs are affected I hear & there has been a miserable time for me since I came home.'[118]

At the end of the Yorkshire visit that summer, the Bucks had pressed him to accept their pet collie dog. It was an opportune moment. Back in Worcester, the old family dog Tip had died in July at High Street. Lucy's husband Charley Pipe was also missing a little dog he had lost—a mongrel they had named 'Beckmesser' after the character in Wagner's *Meistersinger*. So the Pipes agreed to receiving the newcomer at Field Terrace. He was named, 'Aesculapius' in honour of the Doctor's profession. And when he arrived, 'Scap' turned out a great success at High Street as well. On 7 October Edward wrote to Dr Buck:

I shut him up at the shop this morn: while I went to give a lesson, he whined so his grandmother [Ann Elgar] let him out, he darted down the stairs, caught his leg in *twenty concertinas* that are piled on the staircase & rolled over with all the lot into the middle of the shop! There were some ladies there & my old father enjoyed it awfully.

We live in an atmosphere of 'Scap', the creature's advent has curiously changed all our relationship—my father who had been a respectable citizen for 40 years is now no more than Scap's grandfather. Charles is an Uncle of his—if we refer to you & Mrs. Buck 'tis only as Scap's father & mother & I hide my important personality under the style & title of Scap's Master—I am no more & bear my effacement with equanimity . . .

At the end there was news of the long-delayed big effort: 'Scotch over: progress favourably & the score begins to look important.'

The new concert season brought encounters with larger works by more distinguished composers. On 8 October 1885 he played with the Stockley orchestra in two choral works written for the Birmingham Festival—which was now under the direction of Hans Richter. One was a short cantata by Frederick Bridge, who was born in Worcestershire a dozen years before Edward, educated at Oxford, and was now organist of Westminster Abbey. The other was an oratorio, *The Three Holy Innocents*, by the Irish composer Charles Villiers Stanford—less than five years Edward's senior, and a graduate of Trinity College, Cambridge (of which he was now organist) and Hon. D.Mus. of Oxford. In November he encountered the music of Herbert Wareing again, as he wrote to Dr Buck: 'I went over to Birmingham on the 14th. to lead (in

[118] 7 Oct. 1885.

private) a new String IVtett by Wareing The mould is on it, but it is not bad.'[119]

This example of a friend writing music in larger form none the less registered. The very next remark in the same letter to Dr Buck reported the fate of his own large work: 'I shall not bother further with my Scottish overture. Old Stockley is afraid of it.' Later he would add:

I showed it to old Stockley & he candidly said he could not read the Score & it sounded to him disconnected. So I have retired into my shell & live in hopes of writing a polka someday—failing that a single chant is probably my fate.[120]

He had agreed to take over the organist's duties at St. George's Church from his father's successor. W. H. Elgar was not happy about it, as Edward wrote to Dr Buck: 'The old man does not take quite kindly to the Organ biz: but I hope 'twill be all right before I commence my "labours".'[121] It was not all right—either at the Church or anywhere else, as the new organist confided to his friend early in January 1886:

I am very lazy now—have nothing to do—am *not doing* it. Begin again on the 25th. Fear our orchestral Society is going—to pot. People leaving; &c.—I am a full fledged organist now &—*hate* it. I expect another three months will end it; the choir is awful & no good to be done with them.

The choirmaster's duties at St. George's were in the hands of Hubert Leicester. Together they drafted a set of regulations which were sent to the choir with a stiff printed note:

In order to put the Choir on a satisfactory footing, and to ensure regular and punctual attendance at the Services and Rehearsals, we find it necessary to request you to sign the enclosed Conditions of Membership and return the same to the Choirmaster, before the end of the present month, if you wish to remain a Member.

                     EDWARD ELGAR, *Organist.*
                     HUBERT LEICESTER, *Choirmaster.*[122]

Amongst the regulations for punctuality and diligence was one less expected:

The Members to conform in all Musical matters to the instruction of the Organist, whose authority on all questions of Music shall be supreme . . .

He helped the assistant priest, Fr. Knight, in arranging non-liturgical services to fill hours when the Mass itself was not said. One such service over which Edward took special pains was the 'Bona mors: devotions for a good death'. It was from this text that Cardinal Newman had taken many lines for his 'Dream of Gerontius'. After a 'Kyrie eleison', the sung words began in English:

Holy Mary! Pray for us.
All ye holy Angels and Archangels, Pray for us.

A long succession of prayers culminated in:

[119] 29 Nov. 1885.   [120] 8 Jan. 1886.
[121] 29 Oct. 1885.   [122] Broadside in possession of the Leicester family.

Be merciful unto us, Spare us, O Lord.

· ·

From the peril of death, O Lord, deliver us.

Near the end came a chant of many verses to recount the Crucifixion. For certain of these verses, Edward re-harmonized the chant to make it echo the 'De profundis'—with which 'Bona mors' was often placed.

He wrote and arranged short Litanies in a book which had once held sketches for the wind quintet.[123] The choir sang them week by week, and in the end there were as many Litanies as there had been hymn arrangements in the summer of 1878. One Litany copy he finished with the words 'Ora pro Edwardus Elgariensis'.

When the limitations of liturgical music-making seemed too close, he applied commentary from the concert-room. Charley Pipe remembered him playing Gounod's *Funeral March of a Marionette* as a voluntary. Edward himself recalled another instance for a friend, who wrote:

. . . he officiated at a baptismal service, and as at that time he was greatly attached to Berlioz's *Symphonie fantastique*, he extemporised the 'March to the Gallows' as the worshippers were leaving the church.

All left but one man, who lingered, and approaching young Elgar said, 'Excuse me, I am the father of the boy that was baptized, and I should like to know the name of the beautiful tune you were playing.'

Elgar thought rapidly and cautiously replied, 'Oh! that was a march by a French composer.'[124]

On 10 April 1886 he went to the Crystal Palace to attend a performance in honour of the seventy-five-year-old Franz Liszt. Liszt was visiting England for the first time in many years, and Manns's orchestra surpassed itself. The old Abbé shook hands with Manns and bowed many times to a cheering audience. Edward recorded his own judgement on this music with a cipher pencilled into his programme. It was an early instance of his interest in this arcane art, more secret than a pun yet with its own show of cleverness. Many years later the cipher on Liszt's music was decoded to read: 'GETS YOU TO JOY, AND HYSTERIOUS.'[125] 'Hysterious' was a portmanteau invention: it married 'hysteria' with 'mysterious' through a 'tear'.

Writing to Dr Buck in July, Edward struck the pose of a world-weary cynic content to enjoy his position as a very local paladin:

I have been coming out of my shell lately; went to a large picnic with sister Lucy last week; high jinks; a sequestered spot by the river 9 miles out. I helped to boil the kettle &c. &c. flirting (out of practice), dancing (stiff in the joints), &c. &c.

But, perpend, I am *great* at Tennis now. I am a big pot in our set & have to teach 'em!!

---

[123] In possession of the writer. Elgar's Litanies were still in use at St. George's in 1916, when Hubert Leicester made a fresh copy of the entire book.

[124] F. W. Gaisberg, *Music on Record* (Robert Hale, 1946), pp. 246–7.

[125] Communicated by Anthony Thorley, 1977.

I can't play a bit as I haven't touched a racquet since I was at Settle 3 years ago, but I am coming on & can easily boss all the lot we have.

Anticipating another summer holiday with the Bucks, he had been working at a Trio for violin, cello, and piano. A slow introduction led to an *Allegro* exposition of Schumannesque themes. But as with the other big projects commenced since the loss of his 'Braut', the Trio did not progress. He lacked the inner drive to realize large structures—structures such as he had achieved in the wind quintets of nearly ten years earlier.

He took the Trio fragment to his holiday with the Bucks in August 1886, but left the manuscript in Yorkshire when he returned to begin his Worcester teaching in the autumn. He had fewer pupils now: a German violinist from Stockley's Orchestra had established himself in Malvern and was taking a number of Edward's students away. The appearance of a new pupil in his own roster then seemed an event worthy of making a special note: 'Miss Roberts. 1st lesson. Oct. 6th.'[126]

*   *   *

*   *

*

Miss Caroline Alice Roberts was a short woman in her late thirties with a quiet voice and a definite, assured way of speaking. Her features tended more to character than to conventional beauty: a high forehead, china-blue eyes under arched brows, a prominent nose, and firm lips were set off with abundant light-brown hair swept up and plaited in the fashion of the day.

She desired tuition in pianoforte accompaniment, so as to take a larger part in the musical activities of her friends. Miss Roberts was a lady. She lived with her elderly mother at Hazeldine House at Redmarley, below Malvern on the southern borders of Worcestershire.[127] It had been her home for most of her life—ever since the retirement of her late father, Major-General Sir Henry Roberts, from a distinguished career in India.

Miss Roberts already knew something of music, having studied the piano as a girl with a well-known teacher in Brussels named Kufferath. More recently she had studied harmony with the Gloucester Cathedral organist Charles Harford Lloyd. But all that was as nothing in the face of the interest she now evinced. The son of the Roberts's coachman at Hazeldine recalled long afterward:

When Miss Alice met Sir Edward first, [my father] used to drive her to Malvern in an open Phaeton to take *music lessons* and on one occasion (when returning home at night) I heard him say to my mother he thought there was more in it than music lessons . . .[128]

It was almost a year since Helen Weaver had disappeared over the horizon, two years since Edward had met and parted from the other E. E. in Scotland.

---

[126] MS chronology in Elgar's hand (HWRO).
[127] Following revision of county boundaries, Redmarley is now in Gloucestershire.
[128] T. A. Rouse, letter of 30 Oct. 1938 to Carice Elgar Blake (Elgar Birthplace).

One of his friends at about that time saw him as 'a spare, dark, shy young man, standing by a piano and looking at me with a gaze that was at once difficult and aloof—a look that I was to see many a time in after years —as if he was half here and half in some other place beyond our ken.'[129] There were other young ladies. One was Laura Cox, a soprano from Stratford whom Edward met at concerts in the area. He wrote a *Laura Valse* in her album. But her family discouraged the attentions of an obscure musician, and the girl obeyed her family.

With Miss Roberts of Hazeldine the case was different. The fact that she was almost nine years older than himself almost precluded the sort of instant attraction he had felt for Helen Weaver. On the other hand, Miss Roberts's background brought her a good deal closer to the society which had filled his own childhood dreams than any other lady who took an interest in him. Moreover, for a son who had grown up under a kind and powerful mother, the interest of an older woman must evoke an almost familiar sense of comfort. And there was something else at least equally important for a man once disappointed in love: an older woman's interest in himself would be far less likely to wane. So she offered a security that Edward's earlier relations with women had never known.

At the end of the month in which Miss Roberts's lessons began, Edward finished the sketch of a song to words translated from the poetry of French chivalric romance.

> Is she not passing fair,
>   She whom I love so well?
> On earth, in sea, or air,
>   Where may her equal dwell?
> Oh! tell me, ye who dare
>   To brave her beauty's spell,
> Is she not passing fair,
>   She whom I love so well?

It was a poem to engage Edward's private 'old world'. It drew his music once again to the 'sub-dominant grounds' which had cast the melody of *Idylle* two years earlier. The new melody added a propulsive rhythm of 3/4 *Allegro molto*.

Yet he did not publish his song then, and the manuscript remained in a semi-finished state for more than twenty years.[130] At the end of *Is she not passing fair?* he drew over the final chord the arcane design of a cryptogram. Its verbal hints remained possibilities, committing the cryptographer to nothing. Some possible interpretations of the *Is she not passing fair?* cryptogram are ANON (signifying 'nameless' or 'soon' with equal likelihood); NONE, ONE, or LONE; NEW or NOVEL; and LOVE.[131]

---

[129] A. B. L-W., quoted in Diana McVeagh, *Edward Elgar, his life and music* (Dent, 1955) p. 14.
[130] BL Add. MS 58053.
[131] Communicated by Anthony Thorley, 1977.

On 7 October 1886 Mr Stockley opened his orchestral season in Birmingham with a concert which included Schumann's Piano Concerto played by Fanny Davies. Miss Davies was a young woman of twenty-five who had studied with the best teachers in Birmingham and Leipzig before spending two years with Clara Schumann. Dvořák appeared in the same concert, again conducting his D major Symphony. The next Stockley concert on 18 November included an Overture by Herbert Wareing: it was part of a Cantata which had just won him the Cambridge Mus.D.

What touched Edward's deepest interest was the announcement of a concert of chamber music by Brahms, to take place in Malvern in late January 1887. Anticipating the event, he wrote an article on Brahms for the local paper which concluded:

My presence, by great good fortune, at the first production in England of several of his larger works has given me an enthusiasm for the advancement of Brahms' music which I would fain communicate to others. It is to be hoped that Malvern, which has shown itself so ready to appreciate compositions bearing well-known names, will do itself honour by duly appreciating the classical composer *par excellence* of the present day; one who, free from any provincialism or expression of national dialect (the charming characteristic of lesser men: Gade, Dvořák, Grieg) writes for the whole world and for all time—a giant, lofty and unapproachable—Johannes Brahms.[132]

The prospect of Brahms chamber music encouraged Edward's own. He asked Dr Buck to return the little melody written during the first visit to Yorkshire, together with a Trio slow movement written since. Edward acknowledged it on 7 January 1887:

The Trio came safely to hand & we have played the slow movement several times in private & the verdict is—*gigantic*! It's too monstrous however for three instruments, Fiddle, 'cello & piano all seem to distend themselves frog-wise 'till they burst in a vain endeavour to represent an orchestra. Since I received it (the M.S.) I have added a 'Scherzo & Trio', wh: is *jam*. The Scherzo is new & very difficult, but goes with a swing wh: would make you dance to Pen-y-Ghent & carry a cage of parrots as well (do your remember our exploit?)[133] the Trio, (mea culpa) I have transferred bodily from the little thing I wrote for you on my first visit to Giggleswyke; and I humbly ask pardon, but think it too good & sugary to be left to your tender mercies only. I hope to play this with you some day.

In the same letter he sent news of his younger brother Frank's engagement. Hubert and Agnes Leicester were about to celebrate the first anniversary of their marriage. But for Edward himself as he entered the year of his thirtieth birthday, there was no commemoration. 'These holidays have been dull; as

---

[132] *The Malvern Advertiser*, 21 Dec. 1886. Elgar recalled that the editor printed his article in the form of a letter to avoid paying for it.

[133] Dr Buck's daughter recalled the parrot-adventure: 'I have often heard my father laugh over this same incident. He and Sir Edward were carrying the parrot in a cage which was slung on a stick between them. As they crossed the Settle Bridge the bottom dropped out and they walked on, leaving the parrot behind them.' (*Letters*, p. 28.)

they are nearly past, I look back & find a miserable record of wasted time. I have read three of Scott's Novels—written 3 Ballet airs & partly scored 'em—& coloured a clay-pipe—beast! This saddens me.'

One of the Scott novels, *Old Mortality*, presented a hero who could stand for Edward himself. The novel was set in the seventeenth century, and the young man was dependent on an uncle. But here again was individuality self-taught yet divided against itself. Scott described his youthful hero thus:

The base parsimony of his uncle had thrown many obstacles in the way of his education; but he had so far improved the opportunities which offered themselves, that his instructors as well as his friends were surprised at his progress under such disadvantages. Still, however, the current of his soul was frozen by a sense of dependence, of poverty, above all, of an imperfect and limited education. These feelings impressed him with a diffidence and reserve which effectually concealed from all but very intimate friends, the extent of talent and the firmness of character . . .[134]

The hero feels himself caught between two powers. The strong and mysterious leader of low-church rebels claims domination through ancient comradeship with the young man's father. But the dashing royalist commander of high-church ideals eulogizes the old chivalric virtues epitomized in the Chronicles of Froissart. For Edward, they were the ideals associated with his mother. His music had sought them recently in *Is she not passing fair?* His musical imagination responded to those ideals again in January 1887 when he wrote a second letter to the newspaper following the Brahms chamber-music performance. In it he characterized the Finale of the String Quartet Op. 51 No. 1 as 'noble and chivalrous as a chapter of Froissart'.[135]

Time's passage sounded another tolling in the death of his boyhood employer, the solicitor William Allen. Edward sketched a choral setting of the *Pie Jesu* from the Requiem Mass, and inscribed it:

Offertorium
In Memoriam—W.A.
obit Jany.27:1887.[136]

A child-like tune opened in descending steps to recall the 'tune from Broadheath' and *Idylle*. Years later he would reset it to the words of *Ave Verum*; then it achieved some fame. In 1887 the Offertory was only one more small work.

Within a month he had the chance to measure his small effort against one of the choral masterpieces of the century. On 24 February Stockley produced for the first time in Birmingham the *Manzoni Requiem* by Verdi. Edward played in the performance, and more than forty years afterward he would describe the Verdi *Requiem* as a 'work I have always worshipped'.[137]

[134] *Old Mortality*, ch. 13.   [135] *The Malvern Advertiser*, 29 Jan. 1887.
[136] BL Add. MS 49973A fo. 90. Part of the melody was based on a *Kyrie* he had sketched on 19 Aug. 1886.
[137] 18 Jan. 1931 to Fred Gaisberg (*Elgar on Record*, pp. 127–8).

In March the little song of 1885, *Through the long days*, found a publisher in Stanley Lucas (a London firm who hired music to the Herefordshire Philharmonic Society). When the first printed copies arrived, Edward inscribed one of them: 'Miss Roberts from Edward Elgar, Mar.21 1887.'[138] Lucas also took *Sevillana*, but only in a piano version.

Miss Roberts invited him to a tea-party at Hazeldine House, Redmarley, to meet her mother Lady Roberts. Hazeldine House was filled with relics of the late Major-General's Indian career. Many of the Roberts's friends were ladies of similar comfortable position who devoted themselves to good works, and to intellectual and artistic pursuits. Miss Roberts herself was a case in point. Music was only one of her interests. In years past she had written a long poem and a two-volume novel, both of which had been published in London.

In London on 25 May Edward attended another Richter concert. It opened with Brahms's *Academic Festival Overture* and concluded with another first performance in England—the Seventh Symphony by Anton Bruckner: 'Fine intro.', Edward noted in the margin of his programme, but nothing else. Between Brahms and Bruckner, Richter placed two extracts from Wagner's *Die Walküre*. They drew Edward's deepest interest. At the end of the 'Ride of the Valkyries' he wrote: 'Brass telling through all the strings—most weird & witchlike hurry scurrying through the air & the battle &c & rushing wind "embodied".'[139] Here, as with Berlioz, a brilliant orchestrator used instrumentation itself to tell his story. And it was Wagner's story that caught Edward's special attention that night. In his programme he set a mark against the translated lines of the other *Walküre* extract, the Love Duet:

> Siegmund the Volsung
> Stands revealed!
> For bride-gift
> He brings thee this sword;
> And fearless woos
> A wife sweet and fair:
> From foeman's house
> He flies with his bride.
> Far from hence
> Follow his steps,
> Forth in the smiling
> Softness of Spring:
> There shields thee Needful, my sword,
> And Siegmund but lives in thy love.

When a few days later Lady Roberts died quite suddenly at Redmarley, Edward lent Miss Roberts his own copy of Cardinal Newman's poem about Catholic death and immortality, 'The Dream of Gerontius'. Miss Roberts was not a Catholic. But Newman's poem had recently aquired a wider familiarity through connection with the late national hero, General Gordon of Khartoum.

---

[138] Elgar Birthplace.   [139] Ibid.

Edward's copy of the poem had Gordon's markings in it. How they came there made a curious story.

On the eve of Gordon's last departure from England in 1884, a copy of 'The Dream of Gerontius' had been placed in his hands by a young admirer.[140] Gordon was a high Anglican, but his response to Newman's poem was clear from the many markings and underscorings he had made in the copy. One of the most significant of these was on the dedication page: seeing 'Fratri desideratissimo Joanni Joseph Gordon' (a young convert and priest who had died supreme in the poet's friendship), the later Gordon had drawn a circle round his own surname in the days immediately before his own death.

When General Gordon met his hero's death at Khartoum, there was a national eruption of grief extending from the old Queen to the Gordon Boys' Clubs that sprang up all over the country. Gordon books, poems, and pamphlets poured from the presses. Gordon's copy of 'Gerontius' was saved and sent to the old Cardinal at Birmingham, where news of its arrival created a stir in the Catholic community. Copies of the Gordon markings circulated from hand to hand all over the Midlands, and soon Edward had his own copy. On 12 June 1887 Miss Roberts re-copied all the Gordon markings from Edward's book into her own.[141]

Edward himself sketched another song on 1 July.[142] It was on a poem of ageing feminine despair, the Lute Song from Tennyson's *Queen Mary*:

> Hapless doom of woman happy in betrothing,
> Beauty passes like a breath and love is lost in loathing:
> Low! my lute: speak low, my lute, but say the world is nothing.
>   Low! lute, low!
>
> Love will hover round the flowers when they first awaken;
> Love will fly the fallen leaf, and not be overtaken;
> Low, my lute! O low, my lute! we fade and are forsaken.
>   Low, dear lute, low!

An elegant curve of melody wove itself over a slowly pulsating 6/8 through minor and major and minor. The voice, when it entered, sounded almost subsidiary to the instrumental expression. So it might sing of what could not be surmounted. But again this song found no publisher.

September 1887 brought the Three Choirs Festival to Worcester once more. Edward took his place in the first violins. Frederic Cowen conducted his 'Scandinavian' Symphony and the première of his oratorio *Ruth*. There was music by Sir Arthur Sullivan and the Cambridge composer Dr Charles

---

[140] E. T. Maund, who wrote to Elgar about it in 1930 (HWRO 705:445:6807).

[141] Elgar later recalled being given a copy of 'Gerontius' with the Gordon markings at the time of his wedding in 1889. However that may be, Alice Roberts's copy with the markings noted as copied from E. E.'s copy is clearly dated (Elgar Birthplace).

[142] BL Add. MS 58053.

Stanford. Stanford also conducted at Stockley's first concert of the new Birmingham season in October: there he produced a Symphony.

That autumn brought another encounter with the world of academic music. The blind William Wolstenholme was sitting for the Mus.B. degree at Oxford. He needed the help of a sighted musician to prepare for the examination. Edward offered himself, as he recalled many years afterward for a gathering of musicians. One of his audience wrote down this summary:

He was very friendly with the blind organist, and many of his Saturday afternoons were spent in helping the latter with his examination work . . . At the examination, Wolstenholme was to have an amanuensis provided by the University, in this case one of the Cathedral lay clerks, and the blind candidate went down to Oxford some days before to practise working with this man.

A little later Elgar received a frantic message from Wolstenholme beseeching him to come to Oxford at once and act as his amanuensis. Wolstenholme had obtained permission for this from the examiners, on pointing out to them how impossible it was for him to work with the individual provided, who required to have everything explained to him, such as the length of notes, tempi and various points which Elgar had been in the habit of immediately writing down as Wolstenholme dictated.

Elgar consented to help his friend, and on going into the examination room was greeted by [Dr] Frederick Bridge (one of the examiners) with some such words as these:

'We do not usually have anyone as knowing as you acting in this capacity, but of course we rely on your honour to put down only what the candidate dictates.'

The examination papers were given out, [Dr Bridge] keeping in the room with the blind candidate.

On reading over the first counterpoint paper, Elgar immediately detected, in a subject given for an exercise, either a misprint or a note which the examiner writing the canto had not noticed to be unworkable. [Dr Bridge], on this being shown to him, first said in his breezy way, 'Nonsense, nonsense;' but on Elgar persisting, he said:

'Really, Elgar, you must realise that an amanuensis cannot speak to an examiner like this!'

However, the latter would not yield the point, feeling himself (as he told us) to be absolutely in the right and also to be a greater authority on counterpoint than any of the examiners.

[Dr Bridge] at length, owing to Elgar's persistence, looked at the canto and left the room to consult the other examiners. On returning, he announced that the candidate would not be required to work that particular exercise.[143]

Young Wolstenholme gained his Oxford degree—the first blind musician to achieve it since John Stanley in the eighteenth century. But for Edward the revelation of clay feet in academia repeated the experience of the RAM examination he had taken side by side with Wolstenholme six years earlier. The mistrust of academic things would remain through his life: after being showered with degrees and honours of every kind, he still recalled in old age

---

[143] Leaver, 'Some Impressions of Sir Edward Elgar', in *Musical Opinion*, July 1934. The MS diary of Henry Baldwyn shows that Wolstenholme left Worcester for his Oxford examination on 26 Oct. 1887.

'the crudities which one of the many—why are there so many?—unbrilliant university men has used in reference to myself.'[144]

Edward himself, as his thirtieth birthday year drew toward its end, seemed in thrall to inaction of every kind. It was reflected in another text he chose to set as a song—another chivalric romance, versified in archaic spelling by the nineteenth-century writer 'Thomas Ingoldsby' (R. H. Barham). Some of the stanzas recalled the masculine joy of *Is she not passing fair?* They alternated with others which framed again the feminine despair of *Queen Mary's Song.*

> As I laye a-thynkynge, a-thynkynge, a-thynkynge,
> Merrie sange the Birde as she sat upon the spraye!
>   There came a noble Knyghte,
>   With his hauberke shynynge brighte,
>   And his gallant heart was lyghte,
>   Free and gaye;
> As I lay a-thynkynge, he rode upon his waye.
>
> As I laye a-thynkynge, a-thynkynge, a-thynkynge,
> Sadly sang the Birde as she sat upon the tree!
>   There seem'd a crimson plain,
>   Where a gallant Knyghte lay slayne,
>   And a steed with broken rein
>   Ran free,
> As I laye a-thynkynge, most pitiful to see! . . .

The music Edward wrote for this poem was mellifluous, but it showed hardly a trace of personal synthesis.

It was none the less clearer every day that he had attracted a feminine attachment of faith and persistence in the person of Miss Roberts. Finding herself after her mother's death alone in the world as she was nearing forty, she let Hazeldine House and took a succession of rooms and smaller houses in Malvern. Later she had a year's lease of a new semi-detached residence called Saetermo in The Lees, close to the town centre.[145] It would facilitate her lessons in piano accompaniment with Mr Elgar. One of her friends remembered:

Dear Alice! how hard she worked at it. She nearly wore her fingers to the bone practising and I couldn't think what for. She would never have made a fine player.[146]

If that was clear to Alice Roberts's friend, how much more clear it must have been to her teacher. None the less he inscribed for her a copy of *As I laye a-thynkynge* when John Beare published it at the new year 1888. Beare was not an established publisher, but he printed a number of Edward's pieces, including the *Allegretto* on G E D G E, as a friendly gesture when the well-known firms would not take them.

144 Foreword to *Forgotten Worcester*, p. 9.
145 For information about Alice Roberts's residence in Malvern I am indebted to Nigel Edwards.
146 Mary Frances Baker, quoted by her step-daughter Dora Powell in *Edward Elgar: Memories of a Variation* (Methuen & Co., 1949) p. 1.

Edward's energies had seemed to take a new lease of life. He wrote to Dr Buck: 'Jape! I have started a Ladies' orchestral Class & have sixteen fair fiddlers all in two rows & I direct their graceful movements; it is doing well & I think will pay very well & flourish another season.'[147] On 1 November 1887 he opened a new sketchbook with a Sonata for violin and piano.[148] The beginning was set out in bold ink. The following music was written more tentatively in pencil, so it could be changed if need be. Then uncertainty overcame it, and it faded out altogether. Turning the book upside down, he began from its other end the Sonata slow movement—a rich melody rising from its tonic note in rhythmic sequence. Again it was unfinished: but that rising melody was a few years later to make the basis of a big instrumental *Sursum Corda*.

On another page of the new sketchbook he projected a work for a different ensemble: there appeared two separate schemes for a Suite in four movements, one for orchestra, the other for strings. This was what he had boasted four years earlier he had 'about finished . . . in my head'. For the composer who was still failing to complete any large movement, the Suite offered a way of approaching big composition by degrees.

He consulted Mr Stockley about the chance of playing an entire suite at a Birmingham concert. There was some hesitation, but at length a four-movement Suite by Edward W. Elgar was put down for performance. The first movement 'Mazurka' derived pretty closely from Delibes, except for a little secondary descending figure:

Then came a new 'Moresque'. It opened with another descent:

After an 'exotic' primary subject, a second subject descended in a wistful sequence:

A clever middle section combined primary and secondary ideas in major-key polyphony. The third movement was the 'Fantasia Gavotte' inspired by the reversible dancers in Leipzig: here were more descending figures. The Finale was the old March *Pas redouble* which had been played at the Glee Club concerts.

---

[147] 12 Dec. 1887.   [148] BL Add. MS 49974D.

The Suite fared well when Stockley's Orchestra gave its performance on 23 February 1888. This time Edward conducted. But the account he sent to Dr Buck showed defensiveness:

I had a good success at Birmingham with my Suite, but the critics, save two, are nettled. I am the only local man who has been asked to conduct his own work—& what's a greater offence, I *did it*—and *well* too; for this I must needs suffer.

Glad the bairn is flourishing & I pray earnestly he may not be musical—it is *no* blessing but is a big C[URS]E I believe to have a good ear.[149]

But then he finished the other Suite in his sketchbook—a Suite for strings. In the end there were three *Sketches*—'Spring Song' (*Allegro*), 'Elegy' (*Adagio*), and 'Finale' (*Presto*). They were played on 7 May 1888 at a Worcestershire Musical Union concert under the conductorship of the Revd Vine Hall. The successful composer wrote again to Dr Buck:

Thanks for congratulations re Birmingham Suite; that's ancient history now; I have since written, for the Society here, three movements for String Orchestra, classical Style; they were played on May 7 & took well. *I like 'em.* (The first thing I ever did). Also there's a terrific (!!!) song in this month's Maga[zine] of Music.[150]

Edward's new song appeared in the *Magazine of Music* because it had won a prize in the publisher's competition. It was his setting of a poem by Alice Roberts entitled 'The Wind at Dawn'—'on a country saying that such & such weather will ensue if the wind goes out to meet the sun':

> And the wind went out to meet with the sun
> At the dawn when the night was done,
> And he racked the clouds in lofty disdain
> As they flocked in his airy train.
>
> And the earth was grey, and grey was the sky,
> In the hour when the stars must die;
> And the moon had fled with her sad, wan light
> For her kingdom was gone with the night.
>
> Then the sun upleapt in might and in power,
> And the worlds woke to hail the hour,
> And the sea streamed red from the kiss of his brow,
> There was glory and light enow.
>
> To his tawny mane and tangle of flush
> Leapt the wind with a blast and a rush;
> In his strength unseen, in triumph upborne,
> Rode he out to meet with the morn!

Miss Roberts's diction might not match that of the greatest poets, but her conviction was clear enough. Two forces of nature, each inhabiting a different

---

[149] 30 Mar. 1888 (*Letters*, p. 33).

[150] 8 July 1888. This Suite appears no longer to be extant. It may have formed some basis of the *Serenade* for strings (1892).

sphere, could move in concert to create brilliance. The wind added nothing to the inherent power of the sun, but when it rose at his approach it could open a way for him to shine over the world in fullest glory.

Her thoughts had drawn a compelling response from Edward. He set the stanzas in a pattern of variations which moved gradually into the melody to alter its basic expression. The early stanzas, set in F minor, pursued the hesitant conventions of *As I laye a-thynkynge*. But half-way through its course the music shifted from minor to absolute major—never to return to the dark opening. The completing stanzas set in the major realized their music in great march-like strides through big intervals to sound an assurance absolutely new in Edward's music.

The whole expression, moreover, was shared in perfect equality between voice and piano. The vocal line was subtle and heroic by turns, and it was matched with a piano part of daring virtuosity. So *The Wind at Dawn* was a true ensemble. It was that ensemble which won Edward's music its prize.

Alice Roberts had written the poem in 1880—years before she had heard of Edward Elgar. Those were the years when she was writing and publishing her long poem and her novel. Both of those works disclosed an interest in classes of people less fortunately placed than her own, and in what might be done to help them. So the act of offering *The Wind at Dawn* as a vehicle for Mr Elgar's music could also help a romantically dark young man who had grown up the son of a tradesman.

The long poem Alice Roberts had published in 1879 showed a daughter left alone by the death of an aged parent, trying to fulfil herself in schemes for the betterment of the poor. Yet the heroine says to the young man who is already in love with her:

> 'Think not because some women must go forth
> And tread the world alone, slow pave their way
> To slender means or e'en an useful life,
> Think not they deem such path their highest good
> Or glory in their solitude; not so;
> Full oft they long for touch of shelt'ring hand,
> Full oft, beholding as they pass alone
> Some sister woman soul whose ev'ry step
> Is shielded and o'erwatched by guardian eye,
> There comes across their thought a sickening sense
> Of something they have lost or never known.'[151]

In the novel which appeared three years later in 1882, Miss Roberts was quite explicit in defining her ideal relation of wife and husband. The novel's title was *Marchcroft Manor*, and it drew the picture of two separate couples. Julian de Tressanay, who unexpectedly inherits Marchcroft Manor, vows that his love and approaching marriage will not deflect him from his avowed aim of

---

[151] *Isabel Trevithoe*, by C. A. R. (The Charing Cross Publishing Co., 1879) p. 40.

ameliorating the lot of the poor. And indeed, so far from overwhelming Julian's purpose, the marriage with Olive provides his greatest strength.

Olive sympathized in all his ideas; her strong, direct nature untroubled by subtle intricacies of thought, her perception of things in a clear direct view unaffected by changeful side-lights, and above all her unwavering reliance on the wisdom and purity of his intentions, ever re-established in times of trial or depression, that confidence in himself and in the goodness of his cause which was the main spring of his working powers.[152]

In the case of the novel's second couple, the wife's contribution to her husband's strength was clearer still. The second hero, Roger Osborne, has always considered his lawyer's income as insufficient to support a marriage—until he met his Ella.

Notwithstanding the unexpected sufficiency of his resources, Roger took it into his head now and again that he would do great things and become a distinguished member of his profession. He thinks that so remarkable a member of society as Ella, ought to have a celebrated man for her husband; but in reality it is Ella who is ambitious for his sake, and who stirs up the sparks of activity which flicker fitfully through his *fainéant* tendencies.

'It is a great deal of trouble, Ella,' he would say, after a few days of more arduous application than usual.

But she would stroke his hand gently and reply—'It is very good for such a clever idle man to exert himself.'

Roger's friends shook their heads and smiled at the idea of his making such unwonted exertions, and they were quite incredulous as to his carrying on the effort very long. 'You do not know Mrs. Osborne,' they would say, 'what a hard life it is to be in the front ranks of the profession.'

But Ella would look up proudly with her clear, sweet eyes, and answer—'Yes, I daresay it is for most people, but my husband can do anything he pleases so easily.'

Then they would smile again and say that it was no use to argue with so infatuated a lady.

But we do not know if Roger Osborne ever became Lord Chancellor.[153]

Here was the wife defining heights for her husband undreamt by the man himself.

And this authoress nearing forty could make another appeal to Edward Elgar in the spring of his thirty-first year. The symbolism of her writing drew deeply on the world of nature. At an early point in *Marchcroft Manor*, before the patterns of engaged love and marriage are fully revealed, the feminine presence of Nature is recognized as the initiator of insight.

It was a long time since there had been such a beautiful autumn as there was this year when Julian De Tressanay first came to Marchcroft. One fine day rolled slowly on after another; the deep golden sunlight hung over the corn fields, and shone on the ripples of the river as it took its placid course through the valley; the bright fruits and ripening berries made the whole land appear like one smiling garden. These were days when to

[152] *Marchcroft Manor*. A Novel in two volumes By C. A. Roberts (Remington & Co., 1882), Vol. II, p. 280.   [153] Ibid., pp. 283–5.

those who were well and strong it seemed impossible to stay in the house; it was the time for lingering on the hill side amidst gorse and bracken, for seeking the shade of the woods, for a closer union with Nature, who, in such a season, veils her stormy and chilling aspects, and appearing to us like a gentle and beautiful being, woos us to her side; and so intimately are we connected with this strange being, Nature, that we must not be surprised if we find that the lights and shades which we see varying and changing in the sunlight, enter into and work strange changes in the lives of some of us as well as play over the surface of the waters and hills.[154]

Thus Alice Roberts reflected the landscape of Edward's own Worcester-shire. It was the vision which his mother had sought to define in herself. Yet Miss Roberts answered equally the aspirations of Edward's father. Her status as the daughter of a major-general and a knight identified her as one of the gentry whose society W. H. Elgar courted: Edward's success in attracting this lady's interest was a different measure of the son's ability to go beyond his father.

In July 1888 Edward finished a new piece of music. Because of Alice Roberts's fluency in German, he called it *Liebesgrüss* ('Love's Greeting'). When it was published later, it appeared with the French title by which its fame would go round the world—*Salut d'amour*. Like many of his smaller pieces, *Salut d'amour* appeared in several versions—for piano, for cello and piano, for orchestra. But there remained a special suggestion in the music's tonality of E major. The E string of the violin embodied the highest and brightest expression of 'the king of instruments'. So the music's performance as a violin solo with piano would evoke Edward's meeting with Alice—the violinist's ensemble with his pupil in piano accompaniment.

The primary, 'masculine' melody of *Salut d'amour* opened with the major sixth intervals of *The Wind at Dawn*—wedded instantly to the inward-turned descending steps which had haunted Edward's music ever since the 'tune from Broadheath':

The secondary, 'feminine' subject (whose G major was covertly related to the opening E major through the one-sharp implication of E minor) was made from those descending steps as Eve from Adam's rib:

But again the exultation climbed:

154 Ibid., Vol. I, pp. 94–5.

The climactic F♯ was harmonized:

THE PUN OF B♭–A♯ brought suddenly together the distant triads of G minor and F sharp major. It enacted the meeting of two individualities from distant backgrounds in a quick recognition of secret sharing.

Edward thought it all over during his September holiday with the Bucks in Yorkshire. On his return he set on his *Salut d'amour* the dedication 'à Carice'—a punning double on the Christian names Caroline Alice—and laid it before her. And on the day after the autumn's equinox he could write:

September 22nd 1888. Engaged to dearest A.[155]

[155] MS chronology in Elgar's hand (HWRO).

6

The Black Knight

ⁱₙₑₒₘₐₗ

ANY hope for an easy acceptance of Edward's engagement to Alice vanished as soon as the families heard of it. In later years their daughter wrote:

The news brought cries of horror from the Roberts [aunts] & Raikes [cousins] (with the notable exception of Mr. & Mrs. W. A. Raikes and her mother who lived in Norwood and were always most helpful). They said he was an unknown musician; his family was in trade; and anyway he looked too delicate to live any length of time. One of the aunts was going to leave Alice a substantial sum, but on hearing the news, she forthwith cast her out of her will altogether.[1]

The current attitudes of polite society both to shopkeeping and to the arts were described by the heroine of *The First Violin*—the novel Edward had given Edith Groveham at Leipzig.

In our village at home, where the population consisted of clergymen's widows, daughters of deceased naval officers, and old women in general, and those old women ladies of the genteelest description—the Army and the Church (for which I had been brought up to have the deepest veneration and esteem, as the two head powers in our land . . .) looked down a little upon Medicine and the Law, as being perhaps more necessary, but less select factors in that great sum—the Nation. Medicine and the Law looked down very decidedly upon commercial wealth, and Commerce in her turn turned up her nose at retail establishments, while one and all—Church and Army, Law and Medicine, Commerce in the gross and Commerce in the little—united in pointing the finger at artists, musicians, literati, *et id omne genus*, considering them, with some few well-known and orthodox exceptions, as Bohemians, and calling them 'persons'—a name whose mighty influence is unknown to those who never were and never will be 'persons'.

They were a class with whom we had and could have nothing in common; so utterly outside our life, that we scarcely ever gave a thought to their existence. We read of pictures, and wished to see them; heard of musical wonders, and desired to hear them—*as* pictures, *as* compositions. I do not think it ever entered our heads to remember that a man with a quick life throbbing in his veins, with feelings, hopes, and fears and thoughts, painted the picture, and that in seeing it we also saw him—that a consciousness, if possible, yet more keen and vivid produced the combinations of sound which brought tears to our eyes when we heard 'the band'—beautiful abstraction!—play

---

[1] Quoted in Percy M. Young, *Alice Elgar: Enigma of a Victorian Lady* (Dennis Dobson, 1978) pp. 95–6.

them. Certainly we never considered the performers as anything more than people who could play—one who blew his breath into a brass tube, another into a wooden pipe; one who scraped a small fiddle with fine strings, another who scraped a big one with coarse strings.[2]

Robertses did not consort with shopkeepers. The Roberts family were devout supporters of the Church of England: their orthodoxy would hardly countenance an alliance with Roman Catholicism. Even supposing this violinist could actually have fallen in love with Alice, who was by so many years his senior, everyone else would be quite certain to see in his shabby background the outlines of a bounder and a fortune-hunter. And if the man had grandiose notions of trying to compose music, as Alice seemed to think, that could only contribute to uncertainty both of income and of temper. The prospect of such a marriage could not have been more unsuitable.

Edward knew what they thought and said. For the remembered childhood haunted by intimations of 'old-world state', the most hurting part of all was the 'tradesman' label. Years later, when fame and real success had begun to come, he could still write to an editor preparing an article on the history of his career:

Now—as to the whole '*shop*' episode—I don't care a d—n! I know it has ruined me & made life impossible until I what you call made a name—I only know I was kept out of everything decent, 'cos 'his father keeps a shop'—I believe I'm always introduced so now, that is to say—the remark is invariably made in an undertone . . .[3]

If Alice's relations disapproved, even Edward's family might wonder whether such a gulf would be bridged by the simple ceremony of marriage. Ann Elgar could have found little comfort in the prospect of her son's marrying a woman who was so many years older than himself, whose background and religion were entirely outside the Catholic faith.

All in all it seemed best to have the wedding as quietly and as far from Worcestershire as possible. Alice agreed to a Catholic ceremony, and they chose the Brompton Oratory in London—where they knew almost no one. The day was fixed for 8 May 1889, and to ensure that marriage would make a clean break with the past, they resolved to pursue their life together in London.

The intervening months were got through as well as might be. During the autumn of 1888 Edward nursed an eye injury sustained while visiting Pollie's family at Stoke. Yet he kept on his full complement of teaching and music-making. For a visit of the Archbishop of Birmingham to St. George's Church on 9 October, he wrote an anthem, *Ecce Sacerdos Magnus*, based on the 'Benedictus' melody in Haydn's *Harmoniemesse*. It was accepted by a small publisher of Catholic music, Alphonse Cary of Newbury, who also issued four of Edward's *Litanies*.

He went to play in the first violins at a 'North Staffordshire Festival' which Swinnerton Heap had organized at Hanley. But Edward's Wolverhampton

---

[2] Jessie Fothergill, *The First Violin* (Richard Bentley and Son, 1882) pp. 82–3.
[3] 19 Sept. 1900 to F. G. Edwards (BL Egerton MS 3090 fo. 39).

friend Charles Hayward was not in this orchestra. From the eminence of his own experience Edward informed Dr Buck: 'Charles Hayward, in answer to my last letter, has turned up teaching & taken to the "road",—with a little concert (costume) party! I always thought he would.'[4]

As for himself, he had little trouble in selling *Salut d'amour* to the publisher Schott. The sum involved was two guineas[5]—a fair price for a small work from an unknown composer. During another visit to his sister Pollie's family in January 1889, he wrote a little two-movement Sonatina to encourage the eight-year-old piano-playing efforts of her eldest daughter May.

Alice took him to meet her old friend Mary Frances Baker (affectionately known as 'Minnie') at Miss Baker's large house Covertside in the hamlet of Hasfield, near Redmarley. A friend remembered: 'Miss Baker . . . really went out of her way to help and befriend [Edward]. A woman of great personal charm and character, she had known Alice for many years and, so far from frowning on the marriage, gave it every encouragement.'[6] But on this first visit, as Miss Baker herself recalled, 'he was terribly shy and quiet'.[7]

Edward had begun to make his preparations. In February 1889 he interviewed Heinrich Sück, a German violinist in Stockley's Orchestra, with a view to passing on his own violin-teaching connection around Worcester. Through late winter and early spring there were sortings through accumulations of past years. Some bits of old composition were given to his pupil Frank Webb with a deprecating note. Young Webb's letter of thanks revealed an affection that many friends now came forward to express:

We shall ever look back with great pleasure to our quartett lessons and 'musical evenings' as some of the happiest hours of our lives, and we shall all miss you terribly, I know. But there, it's no use my trying to express my feelings on paper . . . I only know that there will be a sort of blank when you're gone, which will never be filled up again, as far as *we* are concerned.

Forgive my selfishness—I know I ought to be congratulating you, and I do, indeed, most heartily, and may you be enabled to do as much for the cause of music, and to promote as much happiness, as you have been able to do here.

Believe me to remain ever

Your friend & affecte pupil
Frank W. Webb[8]

A record of the days that followed was set down in a diary where both Edward and Alice had begun making entries—in hands often hardly distinguishable.

April 17, 1889. To London at 7.0. Called at Oratory & Registrar's to arrange. Home 4.45. Ch[oir] rehearsal.

---

[4] 13 Oct. 1888.   [5] Contract dated 10 Dec. 1888 (Schott archives).
[6] Rosa Burley (and Frank Carruthers), *Edward Elgar: the record of a friendship* (Barrie & Jenkins, 1972) p. 94. The text of the book, based on memories, diaries, and notes, was written many years later with the help of a professional journalist.
[7] Powell, *Edward Elgar: memories of a variation*, p. 1.   [8] 2 May 1889 (Alan Webb).

April 18. Maundy Thurs Mass at 9 Malvern: A. & I walked down Wells Road & back thro' fields . . .

April 22. Packing! . . .

April 26. Mr. Whinfield's for last time—3 to 10 . . .

May 1. Ladies class *last*. Ch: rehearsal *very last*.

There was a final session of chamber music with friends in Malvern:

May 2. To Mrs. Fitton's for Trios with Mr. Capel-Cure. Presentd. with dressing bag [by] Messrs. Wall & Weaver[9]

That was Frank Weaver, the brother of Helen.

On the final morning in Worcester, Edward gave a parting gift to his sister at Field Terrace, where he had lived. It was a copy of Longfellow's *Hyperion*, which he inscribed:

Lucy, May 4: 1889.
From her affectionate Brother Edward
(In Memory of our six years companionship).[10]

So the tale of the young man whose ruling imagination had kept him from the love of women could be given up at last to the sister who had watched over the final difficult years of his bachelorhood.

Despite Edward's long residence in their house, however, neither Lucy nor her husband was invited to the wedding. 'No invitations issued,' wrote Charley Pipe in his diary,[11] and it was almost true. Even Edward's mother and father were not to attend. But Uncle Henry would come: 'he was passionately fond of E.E.', Pipe recalled. And Pollie and Will Grafton would be there to sign the marriage register as witnesses for Edward.

May 4 1889. To London at 2.15. Dad saw me off . . .

May 7 . . . Saw House   3 Marloes Rd. [which they took as a London residence until the end of July]. Met H[enr]y. Will & Pollie at S.Pancras. To *Yeomen of the Guard* with Uncle.

May 8. Wedding at Oratory 11 o'c—Fr. Garnet.

Alice's cousin Cyril Raikes signed the marriage register with her. At the last moment Dr Buck arrived from Yorkshire. Hardly anyone else was present to see Edward and Alice married in one of the side-chapels of the Oratory. The ribbon from Alice's bouquet was sent home afterward to Ann Elgar at 10 High Street.[12]

After a wedding breakfast, the newly-weds departed for a honeymoon in the Isle of Wight.

---

[9] The Elgar diaries are in HWRO.
[11] 'Memoirs', Instalment 14.
[10] Now at the Elgar Birthplace.
[12] Preserved in an album at the Elgar Birthplace.

May 14. Shanklin. Alice & I by coach to Freshwater & back: Cave & arched rocks—had to wade: kissed her wet foot.

From Ventnor on the 25th Edward wrote to Dr Buck: 'This is a time of deep peace & happiness to me after the vain imaginings of so many years & the pessimistic views so often unfolded to you on the Settle highways have vanished! God wot!'[13]

At the end of the honeymoon their money was nearly gone—until Alice's small income should provide for them again. On 28 May they travelled back to London—'3rd class' he noted in the diary. But the house in Marloes Road stood waiting to receive them. They moved into it with the help of Sarah Allen, Alice's maid from Redmarley days. Sarah was to remain a pillar of the Elgars' domestic arrangements through much of their married life.

Kensington spread the feast of London music before them. It was the season for the Richter Concerts at St. James's Hall. The first one they attended contained the Brahms Third Symphony. A fortnight later Richter conducted Dvořák's *Symphonic Variations*. The programme note stressed the significance of variation-writing for modern music:

On speaking of a work of this class, one cannot but revert to the fact that in no other direction, unless we except that of orchestration, has music made greater strides than in the art of variation-making . . . From at first being a purely mechanical exercise, variation-making, or more strictly speaking, 'thematic development', has become one of the most important factors in the symphonies and music-dramas of later times, and has been employed as an adequate means of representing circumstances, personages and their emotional feelings . . .

It was what Edward himself had attempted in the *Fantasia-gavotte*, *The Wind at Dawn*, and more subtly in *Salut d'amour*. But Dvořák had given his variations a 'symphonic' character by presenting a long series capped with 'an extended finale'. It showed the possibilities.

The next Richter Concert, on 24 June 1889, brought an all-Wagner programme. A week after that Richter conducted the closing scene of *Götterdämmerung*, and the first performance of a new Symphony by Hubert Parry—No. 4 in E minor—written at Richter's special request. Thus the great German conductor had virtually brought this English symphony into existence. It made another example for the future.

At Covent Garden there was *William Tell*, Gounod's *Romeo et Juliette*, *Carmen*, Verdi's *Otello* in its first English season, and Edward's boyhood favourite *Don Giovanni*. But now a new favourite supplanted it. In July Edward attended three performances of *Die Meistersinger* within a fortnight. The story of the young knight, brought by the love of his lady to submit himself to learning the rules and mysteries of music, could suggest an array of possibilities now. Alice carefully kept all their concert and opera programmes,

[13] *Letters*, p. 43.

to supplement Ann Elgar's old newspaper cutting-books as a continuing record of Edward's life in their marriage.

Edward took the score and parts of his string *Sketches* to the Crystal Palace for August Manns to look at. He visited John Beare and one or two possible publishers. There were no striking results. Alice cultivated all the London friends she could find: high on the list were the Raikes cousins from Norwood. Edward knew few people in London, but one of their first dinner guests was his old violin teacher Pollitzer. One day Lucy and Charley Pipe came up from Worcester. On Sundays Alice often accompanied Edward to one of the Catholic churches within range of Kensington.

The short lease of the house in Marloes Road terminated at the end of July 1889. Further house-hunting was forestalled when William Raikes and his wife offered to lend their house in Upper Norwood for the coming winter. Oaklands was on a grander scale than anything Edward and Alice could have afforded, and it was close to the Crystal Palace. To fill the time until they could move in, they decided to spend the summer back at Malvern, where Alice's lease on Saetermo had still a few weeks to run. Returning to home scenes in this way was only a temporary arrangement: London beckoned again in October.

They went into Saetermo on 29 July. His mother sent flowers, and a few days later she came for an afternoon's visit. On the first Sunday Edward found his way to the Catholic Church at Little Malvern, overlooking a wonderful prospect of fields and villages through the Severn Valley. He took Alice to see Blackmore Park, the residence of Mr Hornyold—whom he and Hubert Leicester had *not* met in his High Sheriff's procession when they played truant from Mr Reeve's school. Another day they climbed the Herefordshire Beacon above Little Malvern—to the 'British Camp' where the ancient hero Caractacus had tried to defend all this countryside against invading Romans. On 8 August he wrote a little Schumannesque piano piece for the twenty-first birthday of Mrs Fitton's second daughter Isabel.[14]

If the summer in Malvern raised memories, it also raised difficulties. One local acquaintance wrote:

The Elgars were not very happily placed for forming friendships. The rigid ideas of caste which Alice had naturally taken over from her Anglo-Indian father made many of Edward's friends unacceptable to her—she hardly ever visited the old Elgars at their music shop—and a good many of her own friends had cut her for marrying a man who, in their pathetically limited view, was 'unsuitable'.[15]

Yet the summer life with Alice amid old scenes brought to focus Edward's first strong impulse for big composition since the 'Lakes' and 'Scotish' overtures faded out in the mid 1880s. The new impulse went towards a setting of words. It was a practical notion for a composer hesitant in handling big forms: a framework of actions and words gave a ready-made structure to guide

[14] BL Add. MS 60315.
[15] Burley, *Edward Elgar: the record of a friendship*, p. 93.

the shaping of music. A choral work, moreover, would address all the amateur choral societies abounding in the Midlands. Choral standards were higher than those of the relatively few orchestras then, and a big work for choir stood a better chance of performance.

The text of Edward's choice for his choral setting was the ballad of 'The Black Knight'—the poem which Longfellow's Hyperion had chosen to attract his lady.

'Twas Pentecost, the Feast of Gladness,
When woods and fields put off all sadness,
    Thus began the King and spake:
'So from the halls
Of ancient Hofburg's walls,
    A luxuriant Spring shall break.'

Drums and trumpets echo loudly,
Wave the crimson banners proudly,
    From the balcony the King looked on;
In the play of spears,
Fell all the cavaliers
    Before the monarch's stalwart son.

To the barrier of the fight
Rode at last a sable Knight.
    'Sir Knight! your name and scutcheon, say!'
'Should I speak it here,
Ye would stand aghast with fear;
    I am a Prince of mighty sway!'

When he rode into the lists,
The arch of heaven grew black with mists,
    And the castle 'gan to rock.
At the first blow,
Fell the youth from saddle-bow,
    Hardly rises from the shock.

Pipe and viol call the dances,
Torch-light through the high hall glances;
    Waves a mighty shadow in;
With manner bland
Doth ask the maiden's hand,
    Doth with her the dance begin;

Danced in sable iron sark,
Danced in measure weird and dark,
    Coldly clasped her limbs around.
From breast and hair
Down fall from her the fair
    Flowrets, faded, to the ground.

To the sumptuous banquet came
Every Knight and every Dame.
   'Twixt son and daughter all distraught
With mournful mind
The ancient King reclined,
   Gazed at them in silent thought.

Pale the children both did look,
But the guest a beaker took:
   'Golden wine will make you whole!'
The children drank,
Gave many a courteous thank:
   'Oh, that draught was very cool!'

Each the father's breast embraces,
Son and daughter; and their faces
   Colourless grow utterly.
Whichever way
Looks the fear-struck father grey,
   He beholds his children die.

'Woe! the blessed children both
Takest thou in the joy of youth;
   Take me, too, the joyless father!'
Spake the grim Guest
From his hollow, cavernous breast:
   'Roses in the spring I gather!'

It was another picture of the medieval and chivalric world which had already provided the subject for several of Edward's songs. Yet here was ambiguity itself. The Black Knight rode to destroy youth—but it was the youth of the old kingdom. The Knight's season was Pentecost, the season of confirmation. For Edward, finding music to set this ballad of masculine force might offer a victory over the hegemony of his wife's social world. And Alice persistently encouraged him to gather his ideas into a large-scale work.

He wrote out these ideas in an eight-page sketch,[16] where the music gathered once again round the major and minor of G. To begin it, he used an idea which seemed new to him, though it was reminiscent of *Tannhäuser*:

[16] BL Add. MS 47900A, fo. 58–61.

He set it out in purely instrumental terms—showing again the springs of his invention in abstract musical thought. The whole melody was then repeated to make the vocal entry: so the opening words, ''Twas Pentecost . . .', would be sung to music already returning.

Then came a second figure:

Its upward swing reversed the primary subject's descent, and the shorter compass invited a sequence. No words were written against it in the sketch. But it led back to a full reprise of the opening—to set the beginning of stanza two, 'Drums and trumpets echo loudly . . .' The rest of the sketch page showed only a question mark in pencil.

On the next page Edward wrote the line, 'In the play of spears'. Below it he sketched a small triadic variant for repetition '&c—end in G'. The third stanza's opening line, 'To the barrier of the fight', was set to another short figure, defined by minor sixths in the bass. Further lines suggested small variants.

The fourth stanza did not readily suggest music. But then there was the beginning of a 3/4 counterpoint based on the dominant D major triad:

This music was already worked up elsewhere: it was in fact the piano opening of an easy piece for violin and piano set out in the 1887 sketchbook.[17]

For the end of stanza five the ghostly dance was evoked in an elaborate rhythm alternating G minor and A flat major—another pairing of tonalities physically close yet harmonically distant:

Above it elements from the primary and secondary opening themes combined in a close-stepping minor variation—which also invited sequential repetition:

17 BL Add. MS 49974D fo. 35. The identity was drawn to my attention by Christopher Kent.

And then the minor sixth of the dance bass figure suddenly reached into the melody to join Edward's old idea of descending steps in a bold shape that was ultimately to evoke the old king's grief:

An aspiratory leap upward to sink gradually down: it was a combination which mirrored Edward himself, and it was to characterize his melodies through all his music to come. This very figure in *The Black Knight* sketches haunted his music for more than twenty years, until it was incarnated at last in a motto-theme of dying 'Delight' in his Second Symphony.

Near the end of *The Black Knight* sketch, the opening G major theme was noted for sinister return at the deaths of the children in stanza nine. After it he wrote:

&c, (deeper & deeper)
immer schwacher und schwacher

—dreaming already perhaps of performances in Germany. Then he signed the entire sketch:

C. Alice & Edward Elgar
Augt.29:1889.

The writing was entirely Edward's. Alice's portion was the contribution of intangibles.

Yet it was all so far away from any finished and performable score. Except at the opening, most of the ideas were mere thematic tags, which might or might not be developed. The sketch called for '&c. *cresc.*', '*accell°.* &c&c', '&c till . . .', 'later', at all the points which would need development. But it was precisely the developing of ideas that would measure the personal assurance behind them. Two days after signing his sketch, Edward wrote in the diary the words 'blank hopeless'.

Alice was indomitable. She organized an expedition down the Malverns towards Redmarley, and afterwards wrote in the diary for Edward to see:

September 5, 1889. E. & A. to Midsummer Hill all day. Loving & lovely!

Two days after that they went over to Hasfield to lunch and tea with Minnie Baker.

Then the first printed copies of *Salut d'amour* arrived from Schott in London. On 13 September 'E worked hard all day'—and at the end of it sent off a final version of the two-year-old *Queen Mary's Song* to a small firm of London publishers called Orsborn and Tuckwood, asking whether they would consider printing it. They replied that it might be possible.

There were sessions of chamber music with his old pupil Frank Webb and Webb's sisters. There were trios with Basil Nevinson and his pianist friend Hew Steuart-Powell. Nevinson was a London gentleman whose family connections often brought him to Malvern: as a keen amateur cellist he had heard of Mr Elgar and sought him out. On 28 September Mrs Fitton came with her family to Saetermo and they joined Edward for a programme which included Trios by Raff and Schumann, the G major Violin Sonata of Brahms, and Gade's Novelettes. A friend recalled the Fittons as 'perhaps the most cultured and . . . interesting members of the local society of that time. Musically alone they were an immense asset for they provided amongst themselves a thoroughly competent piano quartet, but their capacity for wise and understanding friendship was an even more valuable gift.'[18] Ann Elgar came to stay over one night, and mother and son went for a long afternoon walk together. So the time at Saetermo drew to a close.

Until Oaklands at Upper Norwood should be ready for them in October, they had planned to visit Minnie Baker's sister and brother-in-law, the Townshends, in Oxford. But Edward's eye that he hurt a year earlier suddenly became inflamed, and they had to go into rooms in a house close to Saetermo until he could recover. Perhaps it was only coincidence that this particular form of ill health should show itself just when he could not see how to go on with his big composition, or how to return to London without it.

Alice's understanding saw them through once more. It might have seemed some final confirmation of the marriage itself, as Edward wrote to Dr Buck on 6 October:

And now (after all our talks about the mystery of living), I must tell you how happy I am in my new life & what a dear, loving companion I have & how sweet everything seems & how *understandable* existence seems to have grown: but you may forget the long discussions we used to have in your carriage when driving about but I think all the difficult problems are now solved and—well I don't worry myself about 'em now![19]

Four days later, with Edward's eye improved, they went to London and moved into Oaklands.

Settled in south London for the winter, Edward could believe that his dreams of a composer's life would be realized. Freedom from provincial tyrannies, mastery of house and household in the metropolis, the companionship of a sympathetic and sophisticated wife—all held their promise. The proximity of the Crystal Palace, with its constant rehearsals and concerts, might almost do the rest.

---

[18] Burley, *Edward Elgar: the record of a friendship*, p. 94.
[19] *Letters*, p. 46.

Nearly every day found Edward at the Crystal Palace, often with Alice by his side. A rich repertoire spread before them: selections from *Die Meistersinger* (including the Overture) and from other Wagner operas, two performances of Brahms's Second Symphony, and another work that might have found a special response in those first days and weeks of the London autumn—Mendelssohn's Overture *Calm Sea and Prosperous Voyage*, which Manns conducted on 26 October. One theme was particularly noted in the programme:

A spell comes over the scene, and we are made aware of the presence of deeper and more human feelings than the mere excitement of the breeze can awake. The melody in which Mendelssohn has embodied these sentiments, whatever it may be intended to depict, is one of the most lovely he has ever written . . . In fact so characteristic is it, that at the time of the composition of the work, the phrase was used by Felix and his most intimate friends as a signal or call . . . Schubring asked Mendelssohn if it was not intended to embody the accents of love, fulfilling itself as the goal of the prosperous voyage approached.[20]

So this music might sing to Edward now.

At Oaklands, before their tenancy was two weeks old, a grand piano arrived from Worcester. A few days after that a small organ was installed in the music room of the house. At the end of the month W. H. Elgar came up to tune both instruments for his son. Edward met his father in central London, and together they went to Orsborn and Tuckwood's offices to sign the agreement for publishing *Queen Mary's Song*. The price for the copyright was £4.

A reminder of practical necessities came early in November when Edward had to go down to play his violin in orchestral concerts at Birmingham and Hereford. But when he returned to London on the 11th, Alice met his train at Euston and together they went to hear August Manns conduct *Salut d'amour* in an afternoon concert at the Crystal Palace. So again this music could offer its consummation. Though Alice was still officially a member of the Church of England, Edward that day inscribed a copy of *The Catholic's 'Vade Mecum'*:

C. Alice Elgar
Nov: 11: 1889
from her loving Husband, E.W.E.[21]

Of the hours that followed, he wrote in the diary: 'a Happy Evening'.

He began two new projects, each taking up a different side of the challenge of big ensemble-writing offered by *The Black Knight*. One was a Suite of *Vesper Voluntaries* for just such a small organ as he had now at Oaklands.

---

[20] C. A. B[arry], Programme notes for the Crystal Palace concert of 26 Oct. 1889. The Elgars' copy is preserved at the Elgar Birthplace.
[21] Elgar Birthplace.

When the sketches were assembled, they provided material for eight short pieces (arranged on two staves for playing on a harmonium, but with indications for an independent pedal part). They made simple designs of mostly A–B–A form, full of fresh melodic invention and deftly sketched harmonies. In several cases the middle notes of a primary theme were shaped to make a second subject. All the Voluntaries ranged round D major and minor. The Introduction, Intermezzo (between the fourth and fifth Voluntaries), and concluding coda were cast from a single pair of themes. So they hinted at the big cyclic structures for which *The Black Knight* had raised his hopes.

The other composition project in the early days at Oaklands addressed the vocal aspect of *The Black Knight*—in a first mature attempt at writing part-songs. There was a setting of verses written by Alice in February 1888, before their engagement. The lines began:

> O happy eyes, for you will see
> My love, my lady pass today;
> What I may not, that may you say
> And ask for answer daringly.

The text pictured secret hesitancy; Edward's music was assured.[22]

The other part-song came from a 'Romance' by Andrew Lang.[23] Again the atmosphere was medieval. Now the viewpoint was feminine, but again there was contemplation in the place of action—together with fate's denial of youthful promise.

> My love dwelt in a Northern land,
> A dim tower in a forest green
> Was his, and far away the sand
> And grey wash of the waves was seen
> The woven forest-boughs between:
>
> And thro' the Northern summer night
> The sunset slowly died away,
> And herds of strange deer, silver-white,
> Came gleaming thro' the forest grey,
> And fled like ghosts before the day.
>
> And oft, that month, we watched the moon
> Wax great and white o'er wood and lawn,
> And wane, with waning of the June,
> Till, like a brand for battle drawn,
> She fell, and flamed in a wild dawn.

[22] BL Add. MS 47900A, fo. 16. There is also a notation of *O happy eyes* on the fairly complete sketch of *My love dwelt in a northern land* (49973B, fo. 2), indicating the composition of the two side by side.

[23] *The Century Magazine*, May 1882. Lang at first refused permission to use his words, and later assented (as Elgar recalled) 'with a very bad grace'.

I know not if the forest green
Still girdles round that castle grey,
I know not if the boughs between
The white deer vanish ere the day:
The grass above my Love is green;
His heart is colder than the clay.

In this Romance, love and life were at night. The dawning of day destroyed its fragile world, and with it the young master who commanded for a season his own house and home. So it could show the obverse of what Alice and Edward found in *The Wind at Dawn*. Yet the clean beauty of the words, and the delicate harmonic fantasy of Edward's setting sprung between A minor and D major over a 3/4 rhythm, could seem denial enough that the man who made this music might ever find himself half in love with easeful death.

*   *   *

The denial came again in two further conceptions—whose progress would keep close parallel through the months to come. One was identified on 10 December: 'Dr. Eccles called to see darling A (Baby).' The baby's projected arrival in the following August placed its conception close to that 'Happy Evening' in early November when Edward had returned from his time away to give Alice her *Catholic's 'Vade Mecum'* and to share the first performance of their *Salut d'amour* in the environs of London.

The other expectancy began when it was agreed that Edward should write an orchestral work for the Worcester meeting of the Three Choirs Festival in September 1890.[24] So his first sizeable work would be played by the Festival orchestra which had so moved him in 1866, when the child came running home to exclaim, 'If I had that orchestra under my own control and given a free hand, I could make it play whatever I liked!'

The acceptance of the Festival commission carried the clear commitment to finish a significant composition against a specified date. So this work must not share the fate of the 'Lakes' and 'Scotish' Overtures and *The Black Knight*. Edward took the responsibility seriously as he addressed a letter to the powerful critic of *The Daily Telegraph*, Joseph Bennett.

I have been requested to write a short orchestral work for the Worcester Festival. I am afraid my name is unknown to you, but I thought you would forgive my writing (not wishing to prejudice any criticism of the committee's choice you may make) to tell you that I am a native of Worcester and have written several orchestral works (including a suite of four movements for full orchestra and a set of pieces for string orchestra) . . .

I hope you will not think I am writing from egotistical motives, but it is a crucial time in the career of a young musician and I was afraid you might question the committee's action in asking me to contribute to their scheme.

[24] See the letter of 1 Jan. 1890 from William Done (HWRO 705:445:2903).

The fact of my being a professor of music in my native town till last May and my having produced many things somewhat successfully locally has been the cause of their choice.[25]

The Daily Telegraph duly announced Edward's Festival commission on 12 December 1889.

In the first days of January 1890 the *Vesper Voluntaries* were finished and sold to Orsborn & Tuckwood for £5. They appeared with an inscription to Mrs Raikes—a music-room present to commemorate their winter at Oaklands. Then Edward left his part-songs at the offices of Novello, the leading English publisher of choral music. He enclosed a careful letter to a member of the firm acquainted with his father through supplying music to the shop in Worcester:

Oaklands, Fountain Road, Upper Norwood, S.E.

Jan:13:1890

Dear Mr. Neale,

With this I send two part-songs & should be exceedingly grateful if you could bring them quickly under the notice of your firm. You may remember I have been requested to write for Worcester Festival & am very anxious to get some things introduced & published before that event.

I heard this morning from Worcester & my people beg to be kindly remembered to you.

With many thanks & kind regards

Believe me

Vy truly yours

Edward Elgar[26]

Eventually one of the songs was accepted. Novellos would print *My Love Dwelt in a Northern Land* only if Edward could accept 100 printed copies in full exchange for the copyright. Edward accepted the copies, and thus began what was to become the longest and most fruitful publishing association of his career. But the only public performances of his music that winter took place when August Manns conducted the orchestral Suite at the Crystal Palace on 20 February and again four days later.

The term at Oaklands was approaching its end. After some searching Edward and Alice found a house at 51 Avonmore Road, West Kensington. The Crystal Palace would no longer be close, but the whole of the West End was practically on the new doorstep. The drawback was that the Avonmore Road house had to be taken for a considerable tenancy. Alice's resources were smaller since her expectations had been diminished by the marriage.

March 5 1890. E. to Spink & Co. Sold dear A.'s precious pearls.

It was the first of several such visits.

Before the end of the month they had moved into Avonmore Road. If the terraced house was less commodious than Oaklands, at least it was their own

[25] Quoted in J. A. Westrup, *Sharps and flats* (OUP, 1940) pp. 91–2.
[26] Novello archives. All correspondence with Novellos is from this source unless otherwise noted.

home. Yet strangely, after less than a week, Edward had another bout of eye trouble—just as he had suffered before they went into Oaklands in the autumn. After a few days he settled down, and on 6 April a note was entered in the diary: 'Began Overture for Worcester.'

On 9 April he had to go down to play his violin in a Herefordshire Philharmonic concert. He would be staying in Worcester for several days, while Alice remained in London. Her friend Minnie Baker came up from Hasfield to keep company in Avonmore Road during Edward's absence.

In Worcester Edward stayed this time with his parents at 10 High Street. He visited Fr. Knight at St. George's Church, looked in on a choir rehearsal directed by Hubert Leicester, and went back to supper afterwards with Hubert and Agnes. He joined his brother Frank for a visit to Pollie and Will Grafton at Stoke. He saw Vine Hall. And when his time in Worcester was over, his mother promised to pay a visit to Avonmore Road.

So the old world of the past again filled Edward's experience—just at the moment of beginning his work for the Worcester Festival. He thought of his reading in Scott's novels three years earlier. The youthful hero of *Old Mortality*, caught between the causes of tradition and revolution, had sought to define himself as Edward sought to now. Then Scott's young man had been addressed by the brilliant and sophisticated Royalist commander:

'Did you ever read Froissart? . . . His chapters inspire me with more enthusiasm than even poetry itself. And the noble canon, with what true chivalrous feeling he confines his beautiful expressions of sorrow to the death of the gallant and high-bred knight, of whom it was a pity to see the fall, such was his loyalty to his king, pure faith to his religion, hardihood towards his enemy, and fidelity to his lady-love!—Ah, benedicite! how he will mourn over the fall of such a pearl of knighthood, be it on the side he happens to favour, or on the other. But, truly, for sweeping from the face of the earth some few hundreds of villain churls, who are born but to plough it, the high-born and inquisitive historian has marvellous little sympathy . . .'[27]

Froissart had written his *Chronicles* to honour a rich heritage. His subject was a time when the faith of Europe was Catholic. When the faith was pure, then every other facet of life should find perfect accord. Loyalty to the temporal sovereign would be an ideal expression of that faith. Hardihood towards every enemy must draw strength from it. Death itself could come more nobly to the gallant and high-bred knight in the knowledge that his very knighthood stood witness to the just proportion of his world. And fidelity to his lady love might ultimately lead to a perpetuation of that world.

As Alice progressed in her conception through the spring of 1890, so did Edward in his. Without the marriage to Alice and without her encouragement, he might himself have remained one of the 'hundreds of villain churls' whose ploughings of the earth never adorn the pages of any history. A fortnight after

---

[27] *Old Mortality*, Chapter 35. This passage was identified by Elgar as the inspiration for *Froissart*. See Percy Pitt and Alfred Kalisch, *Analytical and Descriptive Notes [for the] Covent Garden Elgar Festival Vocal and Orchestral Concert, Wednesday, March 16th 1904*, p. 3.

the first anniversary of their marriage, the knightly theme connected itself explicitly with the Overture for Worcester. On 25 May its ideal title was confided to the diary in a recognized new beginning: 'Commenced *"Frois-sart"*. He headed the music with an epigraph from Keats:

> . . . When chivalry
> Lifted up her lance on high.[28]

Twice more in the weeks that followed Edward took himself to performances of Wagner's *Meistersinger*. The young hero's search for mastery in music and love in its mediaeval setting now coupled with the musical model of the *Meistersinger* Overture. Edward followed Wagner in making three principal themes for his own Overture. And when the *Froissart* themes were finished, they bore shadowy resemblances to the *Meistersinger* Overture themes:

Yet the *Froissart* themes were fragmented as Wagner's were not. The *Froissart* primary theme was subdivided by a change of octave in bar 2, and subdivided again in bar 3 by the overtaking C quaver starting a new voice. Again (as in *Salut d'amour*) an inner phrase of the primary theme generated later themes: Ib+c made the basis of 2A and 2B—which in themselves projected 'masculine' and 'feminine' variants.

Years later the process of Edward's generative thinking was described by one of his closest friends:

Like Beethoven, he allowed an idea, which may have occurred to him as a short phrase, to germinate and transform and throw out branches.[29]

The Beethoven comparison would have pleased Edward, as the metaphor of springtime nature would have pleased him. But when he himself described the process for another friend, his own metaphor emphasized the effort and heat of the craft:

. . . There was a great mass of fluctuating material which *might* fit into the work as it developed in his mind to finality—for it had been created in the same 'oven' which had cast them all.[30]

[28] *To* **** (beginning 'Hadst thou liv'd in days of old') from *Poems*, 1817.

[29] Reed, *Elgar as I Knew Him*, p. 129.

[30] Charles Sanford Terry, MS notes written on visits to Elgar during the composition of the Second Symphony, October 1910 and January 1911 (The Athenaeum library).

In the casting of *Froissart* the process which had generated the main themes 2A and 2B also produced further variants—short sub-themes which could be of use in any process of combining and developing.[31] The first took the

down-up-down shape of 1b  and compressed it

to half-step chromatics:

The second was contrapuntal, contrasting close chromatics with a questing reference to the primary theme, and extending the whole formula in sequence:

The third covered eleven of the twelve notes in the chromatic octave.

Chromatic courage took him no farther: the repetition of this figure was not sequential but literal.

Even with this variety of material, Edward still felt uncertain of sustaining his music to the necessary length. He elaborated four more figures from the main *Froissart* themes. They were all sequential, but less chromatic. The two most vigorous of them joined to make a striking opening for the Overture. The first was in fact the most vigorous and propulsive of all the *Froissart* ideas. Its running arpeggios were set off with bold rhythmic contrasts, as the music raced toward the sub-dominant:

---

[31] No evidence appears to survive about the order in which the *Froissart* themes were invented. Their ordering here is designed to show inter-relationships.

C minor in turn moved toward its own sub-dominant F: only it was not F minor but F major—thus making a dominant base from which to launch another 'return'. So the ground was prepared for a first-subject entry. Instead Edward played his second introductory figure:

Of the remaining extra figures, one appeared as a rhythmic development of Intro 2:

The other followed in tandem with it:

The chromatic beginning promised adventure, but soon it came back to an older, diatonic world—all in a single bar, sequentially repeated. So it hinted at a yearning after what might be envisioned but not realised. Its ardour suited perfectly a vision of Froissart's world from the distance of 1890, and the ardour of Edward Elgar to write great music.

Another phase of composition began with the shaping of his ideas into a continuous argument. Tracing this phase of the composition-process in eye-witness accounts needs quotation from later years of the composer's life. But the process, once evolved in *Froissart*, remained fundamentally the same. Sketches surviving from his maturity show that the settled method was to begin with a thematic idea harmonized through two or three staves, often in ink. The idea was typically developed through twenty or thirty bars (with later stages often in pencil, which could be erased and changed) to cover much of a twelve-stave sheet on one side. Alternatives were written on separate sheets. W. H. Reed recalled in later years:

He wrote innumerable repetitions of the same section in the music; I have seen a matter of twenty or thirty bars written in short score eight or nine times without very much change except for a little twist here or an interpolation there. He liked to see how it shaped—how it presented itself to the eye as well as to the ear. Fugitive phrases he would redraft and play with by inversion, augmentation, and other devices as if they haunted him.[32]

They haunted him because they were (as he was to say later of his *Gerontius* music) out of 'my insidest inside'.[33] Until he could see and feel and hear all their subtle multifarious implications, he could not be satisfied that they really reflected himself.

A single surviving sheet of sketches for *Froissart*[34] shows how hard he worked through the late spring and early summer of 1890. The sheet contains two main sketches, in this case one on either side. Both sketches refer to moments near the end of the exposition when the vigour of 2A would give way to the gentleness of 2B. The recto sketch begins with repeating and sequential figures of descent, leading gradually toward an opening for 2B in the dominant. The music is in short score, with tempo and dynamic markings and some instrumental indications. So his thoughts were already coming to him (as he later described it) 'clothed in one colour or another, determining their disposition on one line or another of the orchestral score'.

In the *Froissart* sketch, all these things are set out firmly in ink: there seems a degree of certainty. Yet the sequential figure has not yet acquired the rhythm it would achieve in the final score. One of the repetitions is crossed out in ink. Then a single note is crossed out in pencil. Many accidentals are added: each yields some new point of interest.

The sketch arrives at its dominant goal by the middle of the page. The staves lower down are used for smaller sketches of special points. The main line of musical argument is continued again from the top of the verso page. Here 2B is set out with some extending. Again the foundations are in ink, with all the signs of later considering and reconsidering in pencil.

So Edward slowly devised all the horizontal and vertical patterns through the whole of *Froissart*, mentally testing every phrase in every context—its rhythm, its melodic shapes and harmonic implications for the past, the present, the future. Such solicitude made the only difference between workaday elaboration and that special glow of love and interest which must shine through this music to attract the attention he was not certain to command. It was the sedulous interconnecting of every strand and fragment of vertical and horizontal expression, to make them all sing together, that shaped a real style.

In the mature compositional process, these elaborations of half a page or a page were not necessarily written in order. W. H. Reed described the heterogenous work on the Third Symphony right at the end of Edward's life:

He began to get all these fragments—in some instances as many as twenty or thirty consecutive bars—on paper, though they were rarely harmonically complete. A clear vision of the whole symphony was forming in his mind. He would write a portion of the

---

[32] *Elgar as I knew Him*, pp. 130–1.
[33] 20 June 1900 to A. J. Jaeger (HWRO 705:445:8426).
[34] MS in possession of Oliver Neighbour.

Finale, or of the middle section of the second movement, and then work at the development of the first movement. It did not seem at all odd to him to begin things in the middle, or to switch off suddenly from one movement to another.[35]

His mind was here, there, and everywhere over the entire surface of the work, evolving developments and cross-references as they struck him, gradually shaping the wealth of interconnections for concatenating in a final pattern.

Through all this phase his piano-playing skills could be drawn into the process. Piano-playing did not usually serve actual thematic invention: he had not (as he was at pains to make clear to the interviewer in 1916) 'evolved his thoughts in a keyboard shape, as the early training of so many composers impels them to do. His thoughts came to him, and still do, as abstractions, just music . . .' But having assembled a number of ideas, he could use the piano to weave them speculatively into webs of possible structure.

In playing over his themes in this way, the presence of other people could stimulate him. It is an experience that comes to many of us—most often perhaps in conversation. Edward became adept at responding to silent attitudes and measuring silent responses. The very magnetism of his quickness and his insecurity would become the encourager of both.

When the music had been elaborated to a certain point on separate sheets, then came the task of arranging their final order. Separate sheets of short score sketch made possible every sort of arrangement and rearrangement. Reed was to see it vividly when Edward sought his assistance at this point in the writing of the Violin Concerto in 1910:

I found E. striding about with a lot of loose sheets of music paper, arranging them in different parts of the room. Some were already pinned on the backs of chairs, or struck up on the mantelpiece ready for me to play . . . He had got the main ideas written out, and, as he put it, 'japed them up' to make a coherent piece.[36]

Having set the sketches round the room, he could survey the whole landscape of any proposed formal arrangement in a single sweep.

Froissart gradually took its form. The Overture opened with Intro 1+2 *Allegro moderato*, followed by a new tempo *Andante* for the primary theme. That was given a small extension, but not enough to prepare the music for the second subject. So Edward introduced the extra material X 1+2, and extended it in a clever counterpoint with the little sub-theme A. Then came the second subject (2A) *Allegro moderato*. It was extended with the second sub-theme B. The primary theme returned *fortissimo* in the tonic, repeated a third higher, and again higher in the dominant. The music subsided through a sequence of la (this was the passage being sketched on the recto page of the surviving manuscript sketch). Both tonality and mood suggested the advent of development.

[35] Reed, *Elgar as I knew Him*, pp. 170-1.
[36] Ibid., pp. 23-4.

Instead Edward used the dominant tonality to introduce 2B. It was fully repeated, and was extended with the third sub-theme C, another novelty. The whole exposition now extended to 142 bars—made from eleven separate ideas, most of them used more or less complete and set side by side as pieces in a mosaic. Taking this final phase of exposition to the dominant had obscured a distinctive feature of sonata form—the tonal change associated with moving to development. Blending late exposition with development, however, reduced the responsibility to develop—always the most daunting part of sonata-writing.

Formal development in *Froissart* consisted mostly of pairing off separate ideas. 1a linked with 2B; their unit repeated sequentially; then a shorter sequence appeared from 2B alone; and the latter half of 2B made a still shorter sequence. So the idea drained away. Intro 2 linked with sub-theme A: again the full sequence ran its course, again the elements reduced. Then X 1+2 reared up *fortissimo*—a much-needed injection of energy—but it combined with nothing else. So the music returned to sequential diversion. The development as a whole added little to the work beyond the one vital requirement of form—to hold off recapitulation long enough for it to be welcome.

The entire primary subject returned (augmented) in the dominant, raising the question of whether it was ending development or beginning recapitulation. Intro 1+2 followed in portentous crescendo. It led not to the primary theme but to 2A in the tonic (followed by sub-theme B, just as in exposition). Then came 2B, also starting in the tonic (again with its follower, sub-theme C). The 'extra' X 1+2 found no place at all: if its presence had made the exposition diffuse and development exciting, its disappearance from the recapitulation contributed to a vital, headlong finish. The coda led back to the first full statement of Intro 1 since the opening of the Overture. At last the primary subject entered in the tonic—showing Edward's cleverness in holding this most important recapitulation in reserve for the final climax. A neat dovetailing of 1c/2B then combined with Intro 1+2 to bring *Froissart* to an exciting end.

The central feature of *Froissart* was its almost continuous pattern of melody and variation. From this pattern followed the principles of Edward's orchestration. One principle was safety. It was vital to protect the melody: so he almost always gave it to more than one instrument at a time. In 1890, and for many years after, the standard of orchestral playing in the English provinces was so low that it was unsafe to count on always having a player capable of sustaining a long solo on any given instrument. It was the problem which Schumann had faced in Germany. But Edward's experience of instruments saved him from the charge of thickness sometimes levelled at Schumann's orchestration. For Edward introduced into his doublings a constant variety.

Using just a few instruments at a time (as Dvořák did) was the best way to achieve variety—because there were always others to be brought in. Certainly there was no variety in using the entire force continuously—a tendency Edward criticised in other composers: 'It is possible to see what Strauss is driving at when he superimposes one mass on another: but his imitators frequently blaze the whole orchestra in an immature way . . .'[37] In *Froissart* no solo or combination of instruments was allowed to persist for more than a few bars.

In fact Edward's orchestral experience led him to use his constantly varied doublings to solve a practical problem in a unique way.[38] Orchestras never had enough rehearsal time, especially to prepare a new work. So he applied the principle of continuously varied doublings to shape the actual phrasing he wanted—using different instruments to touch in points of emphasis at the beginning and end of a phrase, or to add fresh colour in the middle. In this way the scoring itself phrased the music, and so made its interpretation easier.

It is also possible to see a private reflection in Edward's varied doublings. Orchestral doublings especially in the melody meant that the sound of his orchestra might not be so highly coloured as some. But as the sequences in Edward's melody could give back the repeating shapes of his own countryside, so the subtly changing orchestral variety gave soft atmosphere and muted colours. (It was the atmosphere of English countryside as seen to perfection in the watercolours of one of Edward's favourite artists, Paul Sandby.[39]) Thereby his own orchestration added one more element to the 'Englishness' of his music.

If the sounds and shapes of *Froissart* reflected its composer's private wishes, none the less the achievement of his first big work was a monument to his new-found security in marriage. The final page of the orchestral score acknowledged this ensemble with its inscription:

Edward Elgar

Fine Kensington

C.A.E.[40]

*   *   *

In the last days of July 1890 the *Froissart* Overture stood complete—a structure of 340 bars scored on 17 separate instrumental staves—altogether nearly 6,000 bars of Edward's expression. But the years of self-doubt and lost time had left their mark. Writing to his old pupil Frank Webb on 29 July, his fears turned upon his friends:

My Overture is finished & I do not think will be liked but that must take its chance: I find in my limited experience that one's own friends are the people to be most in dread of: I could fill a not unentertaining book with the criticisms passed on my former efforts: when I have written anything slow they say it ought to have been quick—when loud, it shd. have been quiet—when fanciful—solemn; in a word I have always been wrong

[37] '*A Future for English Music*' *and other lectures*, p. 253.
[38] Several points in this and the next paragraph were suggested by Bernard Keeffe.
[39] See Elgar's letter of 17 August 1917 to Sidney Colvin (HWRO 705:445:3466).
[40] Worcester Cathedral Music Library B.1.20.

hitherto—at home. I am truly sorry you will not be there to stand aside with me & watch the simple Vigornians execute a dance of triumph on the grave of my Overture!

But before taking *Froissart* to the professional copyist for making orchestral parts, Edward addressed the publisher Novello.

51, Avonmore Road, West Kensington, W.

July 27: 1890

Dear Sirs,

I have written for the Worcester Festival an overture for Full Orchestra & should be glad to know if I might submit the Full-score for your inspection with a view to your publishing it: the overture is not excessively difficult, this might be somewhat in its favour should you think of taking it up.

You may remember that two months ago you were good enough to publish a Part-song of mine.

I am, dear Sirs,

Faithfully yours
Edward Elgar.

The reply was not very encouraging, but they asked to see the manuscript. Edward took it to them, and after due consideration their offer arrived. Novellos would publish *Froissart*—using the manuscript as conductor's score, printing string parts, hiring their copyist to make manuscript wind parts. In exchange Mr Elgar must assign his copyright to them without payment.

When he opened the post containing this offer, Edward sent off a jubilant postcard to his mother in Worcester:

Friday [8 August 1890] 9.30

Letter from Novello to say *all right*.

I am to say you may think of *A's* pleasure. Bless her: Bless me. Bless us all. Only orchestral parts as yet to be done.

Hallelujah!

E.E.[41]

In a letter written later that day, Edward informed his mother: 'One of our wealthy friends wishes to pay anything to advance my music but *of course* I will not hear of it: I intend to *make* my way as *I have done.*'[42]
The contrast with his bitter letter to Frank Webb written ten days earlier measured again his neurotic swings of self-estimation.

Alice's baby was overdue. On 5 August Edward wrote in the diary: '*Poult-day—no poult*'. Already he had written to Frank Webb: 'It is of course a very trying anxious time but, God willing, we hope it may pass safely by & be a blessing to us.'[43] And despite Alice's age of nearly forty-one, all was to be well.

41 HWRO 705:445:4594.   42 Elgar Birthplace.   43 29 July 1890.

Before dawn on 14 August, after an unsettled night, Edward rose to fetch the doctor, and at 10.15 that morning a girl was born. They named her after her mother—or rather after her father's dedication of *Salut d'amour* to her mother—Carice. She would remain the only child of their marriage—except for the music, whose first notable achievement had already preceded her.

On 1 September Edward's parents came to London to see their granddaughter. Next morning Ann Elgar went with Edward as he took Carice for her Catholic baptism at Brook Green, and afterwards W. H. Elgar joined the little party to luncheon. On the day after that he was taken by Edward to the London rehearsals for the Worcester Festival, where Edward had a first run through of *Froissart*. For the rest of the rehearsals, Edward played in the orchestra's first violins. His father had now retired from Festival playing.

When he went to Worcester for the Festival, he came alone as Alice was not yet considered well enough to make the journey. He stayed with Lucy and Charley Pipe. On Monday 8 September there was a second rehearsal of *Froissart* in the Public Hall, in preparation for the première two days later.

The Wednesday Evening Secular Concert offered a long programme including Parry's *St. Cecilia Ode* conducted by the composer, as well as vocal solos by Edward Lloyd and the young Irish baritone Harry Plunket Greene. The moment came for *Froissart*. It was vividly remembered by a young man of twenty-one who was to become a close friend of Edward's later years. His name was Ivor Atkins. He was assistant to the Hereford Cathedral Organist G. R. Sinclair, and this was his first Three Choirs Festival:

Never before had I heard such a wonderful combination of a first-rate Chorus and Orchestra. I was naturally specially interested in Elgar, knowing that he was to produce a new Overture whose very title attracted me, for I had just been reading Froissart's *Chronicles*. Sinclair pointed Elgar out to me. There he was, fiddling among the first violins, with his fine intellectual face, his heavy moustache, his nervous eyes and his beautiful hands.

The Wednesday Evening came. I had no dress clothes with me, having come over from Hereford for the day, so crept up the steps leading to the back of the Orchestra and peeped from behind those on the platform. The new Overture was placed at the end of the first half of the programme.

The great moment came, and I watched Elgar's shy entry on to the platform. From that moment my eyes did not leave him, and I listened to the Overture, hearing it in the exciting way one hears music when among the players. I heard the surge of the strings, the chatter of the wood wind, the sudden bursts from the horns, the battle call of the trumpets, the awesome beat of the drums and the thrill of cymbal clashes. I was conscious of all these and of the hundred and one other sounds from an orchestra that stir one's blood and send one's heart into one's mouth.

But there was something else I was conscious of—I knew that Elgar was the man for me, I knew that I completely understood his music, and that my heart and soul went with it.[44]

[44] Quoted in Wulstan Atkins, 'Music in the Provinces', in *Proceedings of the Royal Musical Association*, 1957–8, pp. 28–9.

The music which could gain such a response as that was music which had every chance of reaching far.

Froissart made a good impression generally. The *Daily Telegraph* critic Joseph Bennett went out of his way to suggest more than that:

The work . . . is one of considerable interest, arising rather from promise than actual achievement . . . But Mr. Elgar has ideas and feeling as well as aspiration, and should be encouraged to persevere. Mr. Punch's memorable advice to persons about to marry [i.e., 'Don't!'] is that which true charity dictates in nine cases out of ten when young men propose to write overtures and symphonies. I regard Mr. Elgar as an exception. Let him go on. He will one day 'arrive'.

Alice collected all the reviews for the cutting-books she had now taken over from Ann Elgar. Edward reported to Frank Webb:

I have had very good notices in nearly all the papers especially in those most to be feared & the overture was much liked by the *musicians* present. Novello is publishing it & in fact printed the string parts in time for the performance.

    After the Festival I came up to town & fetched Mrs. Elgar away to pay some visits to Herefordsh: & at Severn Grange (Mr. Whinfield's).[45]

They stayed for a week with Minnie Baker's sister and brother-in-law, the Townshends, at their house near Redmarley. The host was described by a young relation:

R. B. T. was, by any reckoning, a very unusual person. He was a classical scholar of Cambridge, a first-class rifle shot, had been cattle rancher and gold prospector in Texas and Colorado, and had written books and stories about it . . . His brother-in-law, W. M. B[aker of Hasfield Court], called him a wild Irishman and objected strongly to the strange clothes he generally wore.

    He had a curious didactic manner of speaking . . . and he had a trick of finishing up a rather tall-sounding story with an impressive 'I'm telling *you*' to convince you of the truth of it. Retailing more lurid stories of Mexican life to men friends, I have been told that R. B. T.'s blue eyes blazed with exasperation and excitement when the audience did not seem sufficiently impressed: 'Damn and blast it, man! Can't you understand what I'm telling you?'[46]

Edward relished these eccentricities so far removed from his own sphere. He drew keen pleasure from Townshend's company.

Behind the Festival success and country-house visiting, the practical difficulties of life were exacerbated by insecurity. His own summary, made a year or two later, remained vivid in the memory of a friend to whom he talked:

It appeared that his wife had an income of her own which she was only too willing to dedicate to the cause of freeing him for his true vocation. [But] the income which had

45 28 Sept. 1890 (*Letters*, p. 50).
46 Powell, *Edward Elgar: memories of a variation*, p. 104.

been sufficient for one proved inadequate for two, and his wife's inexperience as a housekeeper had been a further handicap. He told me of his wearisomely unsuccessful journeys in search of engagements, of the further complication of their difficulties by the birth of a child, above all of the crushing discovery that, even had he been able to sell his compositions, the anxieties of the home had completely deprived him of the power to produce them.

Finally, hearing from his friend D'Edgville that the teaching connection in Worcester and Malvern had been abandoned by [Heinrich Sück] and was therefore to be had for the taking, he had thrown in his hand and had returned once more to the hated routine . . .[47]

In November 1890 he began going down to Worcester once a week to give violin lessons again. He 'practised Violin very much'[48], and even sketched a violin concerto. Yet the net result of it all was to be sent back to the provincial trade from which he had seemed at last to be rising clear. When he and Alice paid another visit to the Bakers at Hasfield in December, Edward several times had to absent himself from the comfortable house party to give his lessons.

The return to London just before Christmas was grim:

December 22 . . . Train late. Fog at Paddington. Roads very bad. No coal (Coal strike at W.End wharves).

On Christmas Eve Edward found some coal at last, but their holiday was a lonely one. The weather grew worse.

December 30. A. to dressmaker & Registry Office. E. practiced. A. thought this the coldest day she ever felt (& cried with the cold).

December 31. E. wrote for permission to use names as reference on circular abt Vio. lessons.

In the first days of the new year 1891 he took a notice of his violin teaching for insertion in the advertising columns of *The Musical Times* (published by Novellos), and sent out circulars for pupils in London. There were no replies.

January 24. Thinking of leaving London. Spoke to landlord (Potter) about letting 51 Avonmore Rd.

A performance of *Froissart* by Stockley's Orchestra in Birmingham on 5 February only emphasized Edward's unbreakable tie to the Midlands. Back in London afterward, he wrote to Frank Webb: 'The winter has been truly awful: the fogs here are terrifying & make me very ill: yesterday all day & to-day until two o'clock we have been in a sort of yellow darkness: I groped my way to church this morning & returned in an hour's time a weird & blackened thing with a great & giddy headache.'[49]

The weekly journeys to Worcester for giving lessons also included orchestral engagements in the district. He played in the orchestra at Worcester on 14

[47] Burley, *Edward Elgar: the record of a friendship*, pp. 39–40.
[48] Diary entry at 1 Jan. 1891, referring to late 1890 and early 1891 generally.
[49] 8 Feb. 1891. MS in possession of Alan Webb.

April when Hubert Parry conducted his *Judith*. Before the performance was due to begin, Edward stole up to the conductor's desk and laid on it his own baton. In later years he gave the whole affair a stout protection of jocularity by pointing out all the knocks the baton had received: 'Oh! Parry did all those . . . I played first fiddle then and put my stick on his desk. I wanted to make it immortal. He did not break it!'[50]

In the spring of 1891 Edward had no music of his own to notice. He might have found time for composition during his days in London, or even possibly odd hours in Worcestershire. But the need to resume his trade as an itinerant violin teacher in the provinces had dealt a heavy blow to self-esteem and so to self-expression. The offer of three old but still unpublished songs to the publisher Tuckwood elicited no definite response.

Before the end of April 1891 the notice boards 'To Let' were up at Avonmore Road, and in May Edward and Alice were house-hunting around Malvern. Malvern was the geographical centre of the violin teaching. It also offered the protection of eight miles' distance from the familiarities of his youth in Worcester. They found a semi-detached stone house off a quiet road in Malvern Link, between Malvern and Worcester. The house was named Forli.[51] There had been an early Italian painter of angel-musicians named Forli: it seemed a sign, and they took the house on a year's lease.

On 15 June they left Avonmore Road and journeyed down to Malvern. Through the next days Alice arranged furniture at Forli while Edward gave his lessons in Malvern and Worcester. Carice arrived with her nurse from London, W. H. Elgar tuned the piano, Mrs Fitton came with her daughters for chamber music, and soon the scene was as if they had never left it.

*   *   *

One of Edward's teaching connections was at a girls' school in Malvern called The Mount. Here he spent half a day a week giving lessons one after another to a succession of girls, always accompanied by an elderly female 'dragon' to assure the proprieties. The ownership of the school had recently changed hands, and a new headmistress arrived at The Mount, though this fact had not been brought to Edward's attention. But the headmistress—Miss Rosa Burley, a young woman of twenty-five—made her own enquiries about the violin master.

From various sources I had learned that he was not always good-tempered and that in consequence the girls were afraid of him. Thus it was the custom for each pupil at the end of her lesson to telegraph the state of the emotional atmosphere to her successor; and there was one child who enraged him to such an extent that the others had begged that she might be placed last on the list in order to prevent her from making things impossible for them . . . The violin lessons were unpopular and the girls who took them

---

[50] *The Musical Times*, 1 Oct. 1900, p. 647.
[51] The diary shows that the house already bore this name when the Elgars first saw it.

a dreary little company who sawed away to the general discomfort in distant rooms . . .

One afternoon, however, when the violin master was in attendance, I heard sounds coming from the music-room which made me pause and listen. Apparently waiting for a pupil, who was lingering upstairs in order to shorten the time of tribulation, he was happily playing the piano and—as I should guess in the light of later knowledge—hoping that she would never turn up . . . What I heard that afternoon seemed to detach itself from the mill of sound in the school with a suddenness and strangeness that I have never forgotten. Hardly able to wait for the first full close I pushed open the door and went in.

A tall slight young man with a pale face and dark eyes rose hastily and rather awkwardly from the piano and stood scowling at me. Clearly the interruption was not welcome. Feeling rather chilled, I told him my name and said that I hoped his pupils were working satisfactorily. This elicited some sort of mumbled reply which struck me as, if anything, more hostile than his silence. He then stood blinking at me for what seemed a very long time until my struggle to think of some further contribution to this not very promising conversation was mercifully interrupted by the arrival of the pupil and her dragon . . .[52]

Miss Burley soon identified the violin master as 'one of the most difficult problems that faced me in the management of the school.'[53] The difficulty was that her interest was deeply engaged:

I was struck by the conflicting elements in his character. One never quite knew what he would say next or how he would say it . . . I have never known anyone, indeed, who changed so abruptly and completely. In some degree these changes may have been due to nervousness for he had to exert great control, I believe, in order not to stammer when he was excited, with the result that he would often speak in a stilted, rather measured style . . .

He had at this time, and indeed never wholly lost, a marked Worcestershire accent and was not then a young man of any particular distinction, yet he had a habit of speaking of Malvern in the condescending manner of a country gentleman condemned to live in a suburb . . .

One day he asked me with the greatest formality if I would grant him a favour . . . He would be particularly grateful if I would give him a bunch of the syringa which grew in the drive . . . While I was cutting it and adding a few other flowers he explained that he wanted to take something to his wife which would remind her of the gardens and parks to which she was accustomed. It appeared that she felt unconscionably cramped in the small houses to which they had been condemned since their marriage.[54]

That autumn, in fact, Alice decided to sell Hazeldine to provide some needed capital.

Miss Burley asked Mr Elgar for violin lessons on her own account. She would marshal many reasons:

It seemed possible that I might at once help myself and encourage a warmer interest in the instrument, might possibly serve as a buffer between the master and his somewhat intimidated pupils, and might certainly be in a better position to observe . . .

With the start of the violin lessons came my first opportunity of studying Edward

[52] *Edward Elgar: the record of a friendship*, pp. 18–21.
[53] Ibid., p. 31.   [54] Ibid., pp. 28, 37, 25–6, 29.

Elgar at close quarters. The impression made on me at the outset was that he was extremely shy but that his shyness masked the kind of intense pride with which an unhappy man attempts to console himself for feelings of frustration and disappointment. I remember that as he entered the long room in which the lessons were given he assumed a simply tremendous dignity and spoke about all sorts of trifles, but that when he came to the actual business of teaching he stammered, picked up ornaments, quickly replaced them, and appeared almost unequal to the task in hand. He seemed to be a man whose emotional reactions were out of all proportion to the stimulating causes.[55]

The renewal of life among its old scenes produced fitful creative impulse. 'E. writing music', the diary notes on 11 August 1891—almost as if to chronicle a special event. A month later he wrote out the parts of his old unfinished Trio, to play it with Basil Nevinson (who was visiting Malvern again) and Steuart-Powell.

On a two-day visit to London, Edward looked in at the violin dealer Hill's shop in Bond Street. There he tried a fine old instrument made by Niccolo Gagliano in eighteenth-century Naples. With Alice's encouragement he decided that its purchase would make an investment—though the sum involved ran to three figures.[56] The same visit also yielded a performance of Mascagni's new opera *Cavalleria Rusticana*. This he would condemn, in talking it over with Miss Burley, 'not on musical grounds . . . but because the characters of the opera were persons in low life'.[57]

'I am not writing much now but hope to sometime,' wrote Edward to Dr Buck on 20 December. After the acquisition of the Gagliano he finished a new violin piece. He called it *La capricieuse*, from the character of its primary theme. The contrasting second subject recalled the rich reflection of the 1884 *Idylle*. But now Edward's strain of nobility was hidden inside capriciousness. Together the juxtaposed themes made a piece which was to win a popularity faintly reminiscent of *Salut d'amour*—and bring in almost as little cash.

Over the New Year 1892 they went to stay with Minnie Baker's brother and his wife at Hasfield Court. William Meath Baker had inherited the Court through an uncle who had bought it with a fortune made in the Potteries. The younger Baker had qualified as a barrister, but was more than content to occupy his position as a country gentleman. He had a few close friends, but was not deeply concerned with people: when riding with the Hunt his abrupt manner often bordered on rudeness.[58] Yet his own family were devoted to him. A younger friend would recall him as

a small, wiry man, very quick and energetic. He had an incisive way of speaking—and of laying down the law sometimes. He was an excellent host, and the Elgars called him 'The Squire'. He was always well turned out . . . I used to like seeing him in the Hunt

[55] Ibid., pp. 21–3.
[56] Elgar's diary entry follows the £-sign with three squiggled marks.
[57] *Edward Elgar: the record of a friendship*, p. 25.
[58] Conversation with Mrs Brooksbank (W. M. Baker's daughter-in-law) 3 June 1977.

coat and knee-breeches which he usually wore at dinner when there was a house-party at Hasfield. He used to say that it was a good way of wearing the coats out! . . .

He took me to the Opera at Covent Garden (Wagner only: no other was worth listening to—according to him!) *Die Meistersinger* was his favourite . . .

On one of my visits to Hasfield, a day that I particularly remember was when the whole party staying there was taken over to a Point-to-Point meeting. The ladies of the party were in the library after breakfast and we heard a quick step coming along the hall. 'There's Bill,' someone said, 'Now we shall get our orders for the day.'

He came in, shut the door sharply behind him, and stood against it. 'Oh, here you all are! That's all right. Now, about this business to-day,' consulting a card in his hand, 'the Brougham will take three, five can go in the brake—and someone will have to drive over with me in the dog-cart.' His eye fell on me, sitting on the floor near the large wood fire. 'Dora, will *you* come?'

'Oh *please!*' I cried joyfully. Now that really was delightful and I knew that I was in for a jolly day.

'Well, look here,' he went on, 'we must get over to this place by twelve sharp and I've arranged the start for eleven-fifteen.' And he disappeared as suddenly as he had come, pulling the door to behind him with a loud clap.[59]

The Squire fascinated Edward, as the Squire's brother-in-law Townshend had fascinated him. Through the visit they were inseparable. The evenings, according to Alice's diary, were given to 'music & coruscations'.

From Hasfield Alice negotiated the final dispersal of family furniture at Hazeldine—with the result that a quantity of the old Major-General's Indian effects followed them home to sound their exotic notes amid the humbler harmonies of Forli. In the days and weeks after their return, Edward suffered from throat illness and influenza which kept him out of everything.

Early in February, as soon as he was mending, Mrs Fitton brought over the young cellist Edward Capel-Cure to play through the Trio again. Then Alice took Edward back to stay a week with Minnie Baker at Covertside, close to Hasfield Court. They were seeing so much of Miss Baker that Edward dubbed her 'The Mascotte' after the popular comic opera by Audran. A few weeks later The Mascotte turned the tables by inviting Alice and Edward to accompany her on a journey that summer to Bayreuth for the Wagner operas.

The stimulation was immediate. On 31 March, two days after receiving news of the Bayreuth plan, Edward's own writing started again. This time it was a *Serenade* for string orchestra. The three movements may have been based on the *Sketches* for string orchestra which the Revd Vine Hall had conducted in 1888. Now the movements linked hands so intimately that not a note was out of place.

The opening *Allegro piacevole* found the mood for the whole work. At its centre a warm idea leapt up to descend gradually

pp espress.

[59] Powell, *Edward Elgar: memories of a variation*, pp. 104–6.

It was another essay in the shape already sketched for the old king's grief in *The Black Knight*

—yet its mood now was entirely the *Piacevole* of this *Serenade*. The transfiguration suggested a skill of development which could bring range and power to Edward's expression.

The slow movement *Larghetto* was at the centre of the *Serenade* in every sense. Reflection, yearning, and aspiration complemented one another with a completeness utterly satisfying. Superficial critics might hear influences of Mendelssohn or Schumann or Dvořák in this or that aspect of the music, but as a realized ensemble it found an individuality to make a lasting impression. It was the first such slow movement Edward had achieved, and it showed the most promising path his music had yet discovered.

The tiny *Allegretto* Finale brought the first movement E minor to G major—before returning to close the cycle with the opening *Piacevole* music, including the aspiratory leap to steps descending. In the coda a final idea

descended in fourths reminiscent of *Parsifal* (which Edward had begun to study in anticipation of hearing it at Bayreuth). Yet this closing *Serenade* idea showed also a ghost of the child's 'tune from Broadheath', transmuted now by his marriage. Finishing an arrangement of the *Serenade* for piano duet, Edward acknowledged the role of his wife in his music-making when he noted on the manuscript: 'Braut helped a great deal to make these little tunes.'[60] Now Alice enjoyed the name once given to Helen Weaver.

When the *Serenade* was sent to Novellos, they refused it:

We have given your 'Serenade' our attention, & think it is very good. We find however that this class of music is practically unsaleable, & we therefore regret to say that we do not see our way to make you an offer for it.[61]

So Edward's only way to hear his *Serenade* was to rehearse it with the Ladies' Orchestral Class which he had restarted in the previous autumn. The present class included Miss Burley, who attended regularly with a friend.

One afternoon at the orchestral class Jessie and I found ourselves playing in a work which was unfamiliar at any rate to me. I think I must have arrived late and commenced hurriedly, for I do not remember looking at the title. But I do remember the profound impression its rather Mendelssohnian slow movement made on me.

'What is this?' I asked.

⁶⁰ BL Add. MS 57989.   ⁶¹ 30 July 1892 (HWRO 705:445:8277).

'Oh, it's a thing he wrote himself,' she said. *'Serenade for Strings.'* She spoke casually and quite without enthusiasm.

'Wrote it himself?' I could scarcely believe it.

'Oh yes. He's always writing these things and trying them out on us.'[62]

In their meetings at The Mount, Miss Burley made a more conscious effort to impress Mr Elgar. She would later describe the relation as

an encounter between two extremely conceited young people, each of whom was desperately anxious to impress the other. I was very young and wished to make it clear that in addition to being the headmistress I knew something about music and that I had what I fondly supposed to be a wide listening experience. He on the other hand was equally concerned to make me realize that he was no mere provincial but the musical equal of anyone I had heard or met. It is a tribute to his quality that I was impressed by this despite the fact that I had heard practically every artist of note who was before the public at that time . . .

[But] he regarded executants, those gods of the great public, as creatures of an altogether inferior breed. In view of the fact that it was, after all, as a trainer of executants that he had appeared on our scene this came as something of a shock. He was quite firm on the point however. 'The artist, the true artist is not the executant but the composer.'

He then went on to tell me, a little haltingly at first, that this was the branch of music in which he felt that his own future lay. He told me of various small successes he had already made, of performances here and there, of a struggle against almost insuperable obstacles raised by his determination to write what he wished rather than what was demanded . . .

It was a story, none the less vivid for being disjointed and incoherent, of frustrated ambition. The one thing he wanted to do in life, the be-all and end-all of his existence, was to write great music. For this he had lived, worked and suffered, had abandoned the shelter of safe employment, had risked the displeasure of his parents and had plunged headlong into a difficult and uncertain way of life. The creation of music had become a veritable passion with him. He felt himself swayed by an urge which, if obeyed, must carry him to a goal undreamt of by the journeymen-composers of his time. And yet, in spite of this urge, in spite of the glorious possibilities which he knew to be almost within his reach, he had come to a standstill and could do nothing.[63]

The entire conversation showed graphically how he depended on outside encouragement.

One of Edward's encouragers was Hugh Blair, the twenty-seven-year-old assistant to Mr Done at Worcester Cathedral. Mr Done was now in his middle seventies, and young Blair was already Cathedral organist in all but title. He conducted the Festival Choral Society, which provided the Worcester contingent for Three Choirs Festivals and gave intermediate concerts on its own account. Edward led the Festival Choral Society's orchestra: 'Blair (of the Cathedral) & I are pulling together & making things lively here,' he wrote to Dr Buck on 20 December 1891. And the diary recorded: 'Blair visited us much

[62] *Edward Elgar: the record of a friendship*, p. 31.   [63] Ibid., pp. 27–8, 38.

just now.'[64] One day they played over the old sketches of *The Black Knight*. Blair's response was immediate: 'If you will finish it I will produce it at Worcester.'[65] It was what Edward needed. He started to work again.

On 18 April he was looking through old sketchbooks. What he found gave a surprise. For there in the sketchbook of 1879–80 was the very G major theme he thought he had invented ten years later to open *The Black Knight*. 'How strange,' he noted as he added now below the old notes the opening words from Longfellow's translation, ' 'Twas Pentecost . . .' It showed how the same impulse could persist through varied, developing experience.

Before beginning in good earnest, he made a preliminary trial by setting for chorus and orchestra a *Spanish Serenade* from Longfellow—'muted strings, tambourine & all sorts of games!' as he described it for Dr Buck.[66] When he submitted this to Novellos, they met his own request of 5 guineas for the copyright. It gave a lift all round. The old violin studies written for Pollitzer were dusted off and sent to the violin makers and publishers Chanot, who printed them as *Études caractéristiques*. Chanot also wanted a set of *Very Easy Melodious Exercises in the First Position* which Edward had written for his niece May Grafton's violin study. He wrote again to Tuckwood, offering to present the copyright of three songs submitted earlier if they could be brought out soon: and so it was arranged.[67]

Then he began his new sketch for *The Black Knight*—in chastened mood, as he inscribed the opening page:

The Black Knight
(proposed!) Ballad for Chorus and Orchestra
words by Longfellow   Music by Edward Elgar
if he can.

Side by side with the old thematic ideas were fresh developments. Together they made possible a more continuous musical argument. The new sketch, covering the whole structure, extended to 28 well-filled pages of piano or 'short' scoring.[68] During intervals in his teaching, he worked at the full piece with Alice's constant encouragement. On 5 May she wrote in the diary: 'Met Isabel [Fitton] who came in & had tea & heard some of the Black Knight—A. recited with E's music (piano) (beginning).'

The first of the four Scenes into which Edward divided the poem depicted the castle setting and the tournament. His music juxtaposed two themes—the old one descending

Allegro maestoso

[64] 10 Nov. 1891. Many references to Blair's visits follow through subsequent months.

[65] *The Strand Magazine*, May 1904, p. 542.   [66] 13 Nov. 1892.

[67] Correspondence in the archives of Chappell & Co. The songs were *Like to the Damask Rose*, *The Poet's Life*, and *Song of Autumn*.   [68] BL Add. MS 47900A fo. 38f.

and the second generating a sequence

Each opened with an instrumental statement, to be simplified drastically for setting to words. It confirmed again the abstract, instrumental springs of his musical inventing. He later described *The Black Knight* as 'a sort of symphony in four divisions . . . it's not a proper cantata as the orch: is too important'.[69]

In place of any close development for Scene I came a series of variants, each extended sequentially in turn. At the end came a brief recapitulation of the opening theme and words—but now with chromatic harmonies sounding through the lower orchestra, looking toward the sinister events to come.

Scene II covered the appearance of the Knight at the tournament and his triumph over the king's son. Both of its principal themes were derived from the second subject of Scene I. The first was squarely diatonic

The second, evoking the malign Knight, was more chromatic:

The contrast of diatonic and chromatic was to be used throughout Edward's creative life as a paradigm of good and evil, hope and doubt. Their dialogue shaped the musical form in Scene II. The Knight's approach spelt doom for the old kingdom. Edward's choice of this poem as a subject for a major work in 1889, the year of his marriage, might have reflected his own desire to upset old dispensations. Then he had lacked the assurance to develop it, and the recapitulating form of *Froissart* had followed instead. Now he pursued his metamorphosis with a vengeance.

Scene III was divided into two distinct parts. The opening, to show the hall prepared for the dance, was all diatonic innocence: one of the many fresh ideas was a dotted figure with tambourine sounds and other devices from 'Spanish' music. But then came the dance itself, where the Black Knight embraced the king's daughter. Its music, oscillating between G minor and A flat, was set

[69] 1 Mar. 1898 to A. J. Jaeger (HWRO 705:445:8305).

almost entirely in the orchestra, punctuated by the chorus with short chanted phrases. Edward's orchestral invention was formidable, and the syncopated rhythm never let up to the end of the scene.

By 23 July there was enough vocal score to enable him to leave it at Novello's office as he and Alice passed through London on their way to Bayreuth. They joined their hostess, Minnie Baker, at Margate and crossed to Ostend; thence to Cologne, Bonn (where they paid a visit to Beethoven's birthplace), up the Rhine to Mainz, on to Nuremberg, and so at last to Bayreuth.

Edward had made careful preparation for this visit. He had acquired printed analyses of the three operas they were to see and hear—*Meistersinger*, *Tristan*, and *Parsifal*.[70] In those days *Parsifal* could be seen complete only at Bayreuth, and Edward carefully collated every thematic motive in the analysis with its page-reference in the vocal score. The story of the guileless outsider winning his way to spiritual knighthood contained many of the same elements that were in *The Black Knight*, with a fundamental difference of purpose.

He saw *Parsifal* for the first time on 28 July. The next evening brought *Tristan*, and 31 July *Meistersinger*. Then on 1 August there was *Parsifal* again. The Bayreuth performances began at 4 in the afternoon, with generous intervals. After the First Act, Edward and Alice wandered away from the Festival Theatre and found themselves in a pine forest. It was the scent of Broadheath, of Spetchley, and the romance of Fontainebleau once more.

After Bayreuth their holiday with Minnie Baker continued. She took them to Nuremberg, on to Munich (where they stayed at the famous Hotel Vier Jahreszeiten), and for an enchanted week to the Alps on the Bavarian border with Austria. They joined in the life of the villages, went for long walks, climbed mountains, and in the hot afternoon of 9 August high up at Einödsbach 'E. & Min walked on to the Lavine & touched snow.'

A reminiscence lay waiting at Heidelberg, for here was one of the main settings of Longfellow's *Hyperion*. Edward wrote to his mother on 12 August:

Dearest Mother:

I must send a line from *here* about which we have read & thought so much. I have marked with a cross our hotel which is *above* the Castle: it is exquisitely lovely here & we are just going exploring.

Last night we accomplished a good slice of home journey—Lindau to Heidelberg—you will see on your map—then when driving up here we suddenly had to stop & make way for a great procession of Students—torchlight—the three duelling guilds with a brass band & marching—all their faces wounded (silly fools) & many with bandages on—gay uniforms & no end of torches: it did remind me of Hyperion . . .

Immer und immer
Edward[71]

On 15 August they crossed to Dover and stayed the night in London. Edward called at the offices of the German publishers Breitkopf & Härtel, who agreed

---

[70] In possession of the writer.   [71] HWRO 705:684:BA5664.

to publish the *Serenade* for strings. Then he went to Novellos to discover their reaction to *The Black Knight*. The reply was a cautious encouragement: if he would finish it, they would give it their best attention. Through the remaining days of summer back at Forli, Edward pressed his music to a conclusion.

The Fourth and final Scene revealed another striking instrumental idea:

This opened the banquet. But the swift-moving action now demanded a more vital role for the chorus. Adapting such an instrumental figure to the words led to some awkward shapes and rhythms for singing. Edward redoubled his studies of counterpoint, as Miss Burley noted: 'That this discipline had not been relaxed even when he was thirty-five I can testify, since it was I who lent him some of the books (for example Cherubini's *Counterpoint and Fugue* . . .).[72]

In place of any direct development for the Scene IV music came a device which was to take a central place in many of Edward's Finales to come. The music reached back to recall themes of a previous movement. As Edward grew older, these recursions would grow more and more literal to underline a growing nostalgia in his music. In *The Black Knight* Finale they were new developments of the older themes. First an elaboration of the chromatic motive for the Knight himself from Scene I accompanied the sickness of the King's children. Then a vigorous development of the Scene III dance set the Knight's offer of the fatal cup.

There followed two more notable anticipations of Edward's music in later years. At the words 'he beholds his children die' came a choral model for the opening of 'Nimrod' in the *'Enigma' Variations* half a dozen years later in almost the same E flat harmonies. Shortly after that both chorus and orchestra traced the leap upward to descending steps for the king's grief: it closely foreshadowed the figure of 'dying Delight' which was to haunt the Second Symphony of 1910–11.

The biggest return in *The Black Knight* was reserved for the end. At the mention of youth there came rolling back the opening primary theme: now it surrounded 'Woe! the blessed children both Tak'st thou in the joy of youth'. The joyous character of the old melody made it wrong for this unhappy end, and no amount of musical qualification (of which he applied a good deal) could

---

[72] *Edward Elgar: the record of a friendship*, p. 46. Elgar had known the Cherubini volume, however, from his teens. See above, p. 41.

blacken it sufficiently. Faced with a choice between words and music, it was perfectly clear where this composer's impulse would ultimately lead him.

The completed vocal and piano score was posted to Novellos on 30 September 1892. The response was an invitation to call on the firm's music editor, Berthold Tours. Edward made a special trip to London on 12 October—only to hear that several passages in the accompaniment would be difficult for the average choral-society pianist. He brought the manuscript home that evening.

Forli, Malvern.

Oct:16:1892

Dear Sirs:

'The Black Knight'

Since my interview with Mr. Tours on Wednesday last I have most carefully gone through the P.F. accompaniment of the above-named Cantata & have removed all the difficulties which he was so kind as to point out. I now return the M.S. & shall be extremely obliged if you will consider the question of accepting it for publication.

I may repeat that the leading Society of this district ('Worcester Festival Choral') are announcing the first performance of the work early in the year at their second concert so that an early reply will be most welcome.

Believe me,

Faithfully yours,
Edward Elgar

P.S. The arrangement now fairly represents the orchestral effects but should any of the passages be found intrinsically too difficult I would be willing to alter them rather than anything should stand in the way of the acceptance of the work.

E.E.[73]

A further project for choral setting suggested itself in Newman's 'The Dream of Gerontius'[74]—which might answer the challenge of *The Black Knight* with a vision of specifically Catholic after-life. But the season of violin teaching had begun again. Monday was Edward's day for private lessons in Worcester. On Tuesdays he was at The Mount. On Thursdays he gave private lessons at the Haynes Music Shop in Malvern. Friday or Saturday was devoted to the Ladies' Orchestral Class. And in the previous spring he had taken on another commitment at the Worcester Girls' High School: every Wednesday afternoon of the school term he would teach pupils there.

The headmistress of the Girls' High School was Miss Alice Ottley, a lady of considerable experience in education, and a friend of Lewis Carroll. Miss Ottley might not have the personal interest in music displayed by Rosa Burley at The Mount, but music was given an important place in her curriculum. The

---

[73] Novello archives.
[74] Near the production of *The Dream of Gerontius* in Oct. 1900 Elgar said: 'The poem has been soaking in my mind for at least eight years.' (*The Musical Times*, 1 Oct. 1900, p. 648.)

piano mistress, Mary James, was a pupil of the famous German pianist Nathalie Janotha—who made occasional visits to the school. When Edward joined Miss Ottley's staff, he joined the best musical establishment offered by any school that employed him.

His enthusiasm was captured, and he proposed to give a series of lecture-recitals at the Girls' High School in addition to his regular teaching. The School Logbook recorded the plan.

Oct. 3rd 1892. It has been arranged to have some Pianoforte & Violin Recitals in the Big Hall at the High School on Oct. 19th, Nov. 2, Nov. 16, 30 & Dec. 14th when Miss James & Mr Elgar will play straight through Beethoven's P[iano] & V[iolin] Sonatas, taking two each time.[75]

Edward found himself so nervous for the first recital on 19 October that he would not allow Alice to come. But it went off well enough, and Alice proudly attended the rest of the series:

Nov. 2nd. 2nd Piano & Violin Recital. Mr. Elgar gave a short lecture after the 1st Beethoven Sonata (No. 3) on the growth of the Sonata, to illustrate which several short Sonatas by Corelli, Tartini & Leclair were played, and followed by Beethoven's Sonata No. 4.

Nov. 16th. Third Piano & Violin Recital. Mr. Elgar gave a short lecture on the form of the first movement of a Sonata, illustrated by little pieces of a Sonata by Emmanuel Bach. [Then followed the Beethoven Sonatas 5 & 6.].

Nov. 30th. 4th Piano & Violin Recital. Mr. Elgar gave a capital explanation of the form of the first movement of Beethoven's Sonata Pathétique [together with performances of Sonatas 7 and 8].

The fourth recital had been given in the aftermath of a visit from Lewis Carroll to the School two days earlier. After the final recital in December, Miss Ottley entertained the Elgars to tea to express her appreciation. The recitals were in fact almost Edward's valediction to solo violin-playing. Miss Ottley's gift of appreciation was, presciently enough, a conducting baton.

These successes did not go unnoticed at The Mount. Miss Burley had already placed herself under Edward's tuition. Now she resolved that the girls of her school should follow suit.

Taking lessons myself had been the first small step, but something more was needed and it was at this point that I hit on the idea of a school orchestra. I was in some doubt as to how he would react to the suggestion for he was, as I have said, extremely difficult and uncertain. To my surprise he welcomed it . . .

We soon found that Mr Elgar was far happier with the ensemble class than he had ever been with individual pupils. For one thing the personal relation was easier with a class which could be cursed roundly as a body without its bursting into tears, and there was another even more important . . . In our early days when we could muster only a few weak and dubious strings it was necessary to have a sort of continuo to hold us together,

75 MS in the archives of the Alice Ottley School, Worcester.

and this Mr Elgar provided at the piano in a particularly attractive way . . . supporting our stumbling efforts and decorating the simple works we played with impromptu counterpoints and figurations. It may be noted, however regretfully, that he was not above adding these to a Haydn quartet . . . If there were only the nuclear four [strings], we would grind through a Haydn or a Mozart quartet or Mr Elgar himself might play the piano part with us in the Schumann Quintet. Later, as our numbers increased, we began to study such works as Schubert's Fifth Symphony . . . or even Mozart's Symphony in G minor. [Then] he filled in the missing parts and did so with immense skill . . .

Despite his lifelong contempt for the piano and for those composers whose orchestral writing showed a keyboard influence, it was nearly always the piano rather than the violin that he chose to play at our meetings.[76]

His use of the piano might seem to contradict his oft-expressed aversion from the instrument. But the keyboard extemporizing which was an essential part of his creative process also offered a private escape. Miss Burley noted of the latest extemporizing:

He loved doing this, and often became so absorbed in the intricacies of his invention that he seemed to forget our presence.

Sometimes when one of us produced a particularly painful squawk, he would look round with a petrifying scowl and say, 'What do you think you are doing?' . . . On one occasion he was so annoyed by the quality of the playing that he jumped up and walked out of the house. I got the girls together after this and rehearsed them thoroughly to avoid the repetition of such an incident. The next time he came the music went with a flourish and he was amiability itself . . .

Very occasionally, when a subject really interested him, he would become animated and really helpful. At these times light broke through the clouds indeed and shone upon the poor little class. 'Now then, second violins and violas,' he would say, 'this crescendo passage depends on you. Don't forget that the higher strings are already doing their utmost and that it is you who must swell the body of sound.' . . .

We were always willing to provide him with whatever score he might wish at the moment to study. Again and again works which were wildly beyond the scope of our players would be hired from Goodwin and Tabb's for purposes which were not at first clear to us . . . More complex scores would be eagerly scanned on arrival and exhaustively analysed . . . By 1892 . . . we had already nicknamed him The Genius.[77]

Alice, at home, was writing verses. On 6 December she wrote two *Mill-wheel Songs*—'Winter' and 'May (a rhapsody)'. The first began:

> On and on the water flows,
> Stops the mill-wheel never,
> Crashing on and on it goes,
> Hearts may ache for ever.

It echoed almost too closely the fateful love in Schubert's *Schöne Müllerin*. That afternoon Edward started a setting.[78] The same manuscript also contained

[76] *Edward Elgar: the record of a friendship*, pp. 32–4.
[77] Ibid., pp. 34–7.
[78] BL Add. MS 57995 fos. 67–72.

notes for another water setting, 'The High Tide on the Coast of Lincolnshire (1571)' by Jean Ingelow—the story of a mother and a son, and the son's wife lost in rising waters. But then Alice wrote on the manuscript of her second *Mill-wheel song* 'In case my Beloved might like it'.

> And I've touched both her hand and brow
> Oh! the mill wheel may thunder now,
> Let it splash, let it dash at will
> Our hearts' song is louder still.

December 9 1892 . . . E. wrote the 2nd Millwheel Song.[79]

Novello's response to *The Black Knight* was delayed. An enquiring letter from Edward on 1 November 1892 had brought no reply. Perhaps they were waiting to see whether the work would be included in the Worcester Festival programme for the coming September. The best that Hugh Blair could promise was the performance by the Worcester contingent of the Festival Chorus at their own concert in April. Not until Edward sent a 'reply paid' telegram on 10 November was an answer forthcoming. But the answer was affirmative—if Mr Elgar could accept a royalty of 2d. after the sale of 500 copies of the vocal score. Only the vocal score was to be printed: orchestral material would be duplicated in manuscript copies.

Forli, Malvern.

Nov. 11:1892

Dear Sirs:

'The Black Knight'

In answer to your letter of yesterday's date I write to say I shall be glad to accept the terms therein contained.

I should esteem it a favour if you would let me know how soon the vocal score can be ready as it seems to me (I am inexperienced in such matters) that time is very short: *can* it be issued at Christmas? The Chorus will require copies for practice early in the new year.

I propose to complete the orchestration during the holidays in January & shd. require a copy to work from as I have only my rough sketch . . .

I enclose a copy of the announcement of the first performance & awaiting your reply am

Vy faithfully yours
Edward Elgar

P.S. I need not say that I am extremely obliged & gratified by your acceptance of the work.[80]

The printed proofs were delayed. The first of them arrived only in time for Edward to take them away to Hasfield Court, where he and Alice spent a few days at Christmas with the William Bakers. They returned home to Forli on 27

---

[79] MS not traced.   [80] Novello archives.

December to find more *Black Knight* proofs and orchestral proofs of the *Serenade* for strings from Breitkopf & Härtel. On the last day of the year Edward and Alice each wrote a diary summary for the other to see:

End of our lovely year. Thank God for my sweet life with my sweet wife Braut. E. E.

Thank God for one beautiful year. May He grant us more happy ones. It has been more beautiful than ever with my beloved. [C. A. E.]

That day he had begun the orchestral score of *The Black Knight*. He had undertaken to finish it within a month. He could be so quick over what Arnold Bax was to call 'the Egyptian labour that is orchestration'[81] because his instrumental thinking was an aspect of the actual composing. As the interviewer of later years described it: 'His thoughts came to him, and still do, as abstractions—just music, but clothed in one colour or another, determining their disposition on one line or another of the orchestral score.'[82] The quickness of Edward's instrumental thinking would become proverbial among his professional friends. One of them wrote:

Elgar was extremely modest about his music; but he was rather proud, and rightly so, of his prodigious skill in laying it out for the orchestra. He knew unerringly what he wanted in the way of orchestral or choral tone, balance, and colour . . . Meyerbeer's plan of writing alternative scorings in differently coloured inks to find out how they sounded was a favourite subject of derision with Elgar.[83]

Another friend remembered:

I once watched him orchestrating something, the 24-stave music paper held at the bottom by his left hand, the first finger at a bar on the lowest line, the right hand and pen running up to the top to do a passage for the flutes, coming down to put in something for the brass, lower for the harp, and below, a whole cascade of notes for the violins.[84]

The Black Knight orchestration was posted complete to Novellos on 26 January 1893. It was close to a thousand bars of music, scored for four-part chorus and full orchestra.

The beginning of the new year's teaching was relieved by another diversion—golf, first shown to Edward during the Christmas visit to Hasfield by the Bakers' brother-in-law R. B. Townshend. In March 1893 Edward joined the Worcestershire Golf Club. It was the beginning of long-lived interest. A decade later he responded to a query from a sporting paper thus:

Golf—call it a game, a sport or what you will, no one can define golf—is the best form of exercise for writing-men, as it involves no risk of accident, is always ready to hand without waiting for a 'side' . . . and it has the inestimable advantage of being seldom worth seeing and rarely worth reading about.[85]

In the spring of 1893 he could already begin to value his privacy. When he

[81] *Farewell, My Youth* (Longmans, Green, 1943), p. 31.
[82] *The Music Student* (August 1916) p. 343.
[83] Reed, *Elgar as I Knew Him*, p. 149.
[84] Powell, *Edward Elgar: memories of a variation*, p. 127.
[85] TS, winter 1902–3 (Elgar Birth place).

invited Miss Burley to attend the first performance of the *Spanish Serenade* on 7 April in Hereford, her attention was caught by a conversation in the audience:

The two ladies behind me seemed less interested in the music than in its composer. What appeared to strike them as remarkable, however, was not that he could write music, but that he had married a member of their own social circle . . . They did not wish to cut Alice Roberts—though apparently a good many of them had done so—but they naturally felt that they could not be expected to meet a man whose father kept a wretched little shop in Worcester and even tuned their pianos. It was all *most* awkward.[86]

Yet the music itself had gone well, as Alice wrote in an intimacy of childish diction now beginning to appear in the diary: 'My darling Star's Spanish Serenade given—*most* lovely—A. vesy pwoud & everyone admiring.'

Final rehearsals of *The Black Knight* promised a much bigger success. Coming away from one of them Alice noted: 'Tune of Black Knight heard in the street'. With the actual performance on the 18th, she could see herself as Edward's 'Braut' in every sense: 'The Black Knight heard for the first time. Quite glorious & splendid reception. Star most beautiful & Braut's—had the proudest happiest evening in all ser lives.'

Only the local press noticed the event (except for a brief mention in *The Musical Times*, published by Novellos). But there was no doubt that a corner had been turned in the young composer's career. *Berrow's Worcester Journal* wrote of *The Black Knight* on 22 April 1893:

It was the greatest effort of composition Mr. Elgar has made. He has done himself eminent credit with instrumental suites and other trifles, but he has not previously ventured upon so large a work as a cantata.

On the same date *The Herald and Chronicle* noticed the special prominence of orchestral writing in the new choral work:

It emphasises the characteristics of Mr. Elgar's orchestral writings—sumptuous examples of orchestration, relieved by charming melodies and sparkling with bright and picturesque passages. Mr. Elgar had a very cordial reception as he took the baton to conduct the performance of his work, and there was no mistaking the enthusiastic cheering which followed its conclusion.

For the rest of the concert, Edward returned to his place at the leader's desk in the orchestra. But afterwards Alice collected all the reviews and proudly pasted them side by side in the cutting-book which was beginning to fill more quickly now. Nearly four years had come and gone since the visible beginnings of *The Black Knight* in the first days of their marriage.

*   *   *

*   *

*

---

[86] *Edward Elgar: the record of a friendship*, pp. 51-2.

Teaching continued unabated. Outside the school room, with a less overweening student than Miss Burley, Edward's instruction could be successful. One of his private pupils was Mary Beatrice Alder, who was to recall her lessons after eighty years.

When I was about twelve, my parents began looking about for the best teacher of the violin they could find in the neighbourhood of Malvern. Elgar was clearly the best, and he was invited up to our house for an interview . . .

He came in and had about five minutes' conversation with my parents, and then they went out, leaving me to get to know him myself. I was very much struck with his looks . . . Instead of being dressed in black striped things (like my piano man, who didn't interest me at all) I never saw him dressed in anything but plus-fours and gaiters: that was a new thing to me . . . I very much admired his long thin hands—sensitive hands—and very expressive eyes. I should think he hated giving lessons . . .

He put me at my ease by taking me up to a picture of the Sistine Madonna above our mantelpiece, and saying: 'The German students, you know, make paper hats for the two cherubs at the bottom of that wonderful picture.'

That cheered me up.

And then he took me to a mirror and said: 'Now I want you to look at yourself when you are playing, so that you stand quite properly.'

. . . He said: 'As far as you possibly can, have an accompanist the whole time.' . . . I told him I had five brothers and sisters and a mother who all played the piano, so I thought I could do that.

. . . I remember his quizzical, amused look when I calmly handed my violin to him whenever I wanted it re-tuned. I can see him now, very amused and slightly critical.

. . . We soon advanced from simple music to Beethoven's Sonatas. And he was rather shocked I think to find that, having the whole book of Beethoven, I hadn't run through the lot . . . When he thought I was mature enough, we went to those earlier people like Nardini, and then on to a man I had never heard of called Rode—very advanced in technique . . . The lessons appear to me now to have been really more like recitals, in which he occasionally made remarks. He was never impatient, and always lured me on to do things . . .

As I grew into the teens, he invited me to join a school class orchestra or band of about six pupils and two mistresses, which was most enjoyable. He played the piano, and we played some of his music unpublished . . . When I went home after my lessons, my ribald elder brothers used to say in mocking voices, 'How was the great man today?'

. . . His remarks were most unusual, and he said to me one day: 'Great musicians are things to be ashamed of.'[87]

During a week in London between teaching terms, he left *The Black Knight* for August Manns to look at in hope of a Crystal Palace performance. The hope was not realized. But the excitement then was the prospect of a second holiday in Germany—this time on his own resources. Though there was no Bayreuth Festival in 1893, a Wagner Festival was offered in Munich.

Again there were careful preparations. On the day before Edward's birthday, his own copy of the piano-vocal score of *Tristan* arrived. He wrote in it:

[87] BBC interview with Michael Pope, recorded in 1973.

This Book contains the Height,—the Depth,—the Breadth,—the Sweetness,—the Sorrow,—the Best and the whole of the Best of This world and the Next.[88]

Throughout June he was teaching himself German. On the 28th he and Alice snatched a single night in London to attend a performance of *Tristan* at Covent Garden.

Edward's happy anticipation showed itself during another visit to Hasfield. Minnie Baker remembered him singing at the top of his voice the vigorous figure from *The Black Knight* where the magic drink is offered: he called it his *Perrier Jouet* theme. 'He sat in the strawberry bed and wished that some one would bring him champagne in a bedroom jug.'[89] His hostess brought him a suitable substitute, a curtsey, and a 'Sir'. The appellation stuck.

One day at The Mount Miss Burley mentioned that she was planning a Munich holiday with one of her older pupils.

He plunged at once into an exciting lecture on the theories behind the new music-drama [of Wagner], its divergence from the older Italian opera, its use of leading themes—which he illustrated on the piano with the 'gaze' motive from *Tristan*—and the welding which was attempted of musical, plastic and dramatic elements into one art-form. His enthusiasm for the subject . . . was at once a stimulus and a challenge to one's power of understanding.

I asked him where he proposed to stay in Munich. He answered that over this he was in some difficulty. The only possible place, he felt, was the Vier Jahreszeiten, but it would be altogether too expensive—and in any case almost certainly full. So I told him that my own plan was to go into rooms, of which I had a number of addresses, supplied by my German governess, and that, if he chose to do so, he could make use of my list. He thanked me and suggested that his wife, who had a good knowledge of German, should call on me. Thus it was that I first met her.

. . . No one had told me what, at first sight, was the most striking thing about Alice Elgar—namely that she was considerably older than her husband. Indeed, as at that time he was an unusually youthful thirty-six and she a rather mature forty-[four], she seemed almost to belong to a different generation.[90]

Miss Burley's observations were inevitably shaped by her own dreams and wishes.

The meeting ended in an invitation for Miss Burley to dine at Forli and a few days later, on 20 July, she came.

When the evening of my visit arrived it brought a number of fresh surprises, of which the first was that my host and hostess were in full evening dress—something of a reproof to the demi-toilette which I had supposed suitable to the occasion. However they gave me a very kind welcome to their home and showed me over it.

It was the inside of the house that gave me my second surprise that evening, for the furniture and decoration suggested a taste and a culture that could not have been guessed from its somewhat suburban exterior. The dining-room was on the right of the

---

[88] Preserved at the Elgar Birthplace.
[89] Powell, *Edward Elgar: memories of a variation*, p. 2.
[90] *Edward Elgar: the record of a friendship*, pp. 56, 58.

door as one entered, the drawing room (bigger because of its bay window) on the left. The little study where so much music was to be scored was on the first floor at the back, to the left of the staircase. Each contained a number of fine pieces of Indian carved furniture . . .

I had never seen the Elgars together before, and the outstanding impression of that first evening was of the strange disparity between them . . . She had all the vagueness of manner which was then considered a mark of feminine refinement, and all her utterances tended to float off into space. Even her most impressive pronouncements had to be finished by a wave of the hand rather than by anything so definite as the completion of a sentence.

Compared with his fluttering and perhaps rather affected wife, the Genius seemed to belong to a world of cold realities, and indeed I had a feeling that a certain impatience with her affectation was making him rather more downright than he might otherwise have been. Thus when she complained that since her marriage she no longer had had access to the Army & Navy Stores, he said abruptly, 'No; because I don't make it my business to kill my fellow men.' . . .

Despite the softness of her voice and the gentleness of her manner, there was a certain firmness—almost hardness—in the look of Mrs. Elgar's china-blue eyes which convinced me that in the ordering of their home, at any rate, it was she who drove the chariot even though her husband might occasionally flourish the whip. [Yet] one fact was evident and could not be questioned. This was that Mrs. Elgar worshipped her husband with a devotion so absolute as to make her blind to his most obvious faults . . .

Conversation over dinner (to which the Genius formally 'took me in', having formally offered me his arm) turned on a variety of subjects which showed the range of the Elgars' interests. At that time I had not met many people in Malvern who had much knowledge of, or interest in, foreign languages or literature; and it was pleasant to find that Mrs. Elgar shared my own taste for French and German. She was reading Sudermann at the time and offered to lend me *Frau Sorge*, which I did not know and was glad to borrow . . .

The main subject of our evening's discussion was of course the projected visit to Germany. For various reasons it was not possible to travel out together. And in the end we did not stay in the same house. But it was arranged that we should meet in Munich and join forces for most of our outings.[91]

Through the final days of July preparations were made. Carice was got ready for packing off with her nurse to stay out the time of her parents' absence in Miss Baker's house at Hasfield: both mother and father would be absent for her third birthday, as they had been absent for her second.

Poor Carice: she came always last in the list of considerations—except when her own self-expression threatened her father's composing. Miss Burley's first impression of the child would remain with her ever after:

She was in fact a very beautiful little girl with flaxen hair and a roseleaf complexion. But the expression on her face troubled me, for it was one of profound sadness. She never smiled or laughed; and when I learned that from the first she had been taught never to make the least noise for fear of disturbing her father, I understood her unnatural look of resignation.[92]

[91] Ibid., pp. 59–62.   [92] Ibid., p. 30.

In her maturity Carice would speak of her mother's sacrifices and not of her own. Only once in conversation with a close friend she said quietly that she thought she was the only woman of whom her mother might ever have been jealous.

At the time of the child's conception in the first autumn of the marriage, even Alice's faith in Edward's future might have needed some tangible outcome. Any woman who had given up as much as Alice had given up for marriage at the age of forty might long for some present talisman even though the marriage had been made in expectation of the music it was to produce. Now with *Froissart*, the *Serenade*, and *The Black Knight*, it was clear that the marriage had begun to bring forth its ideal progeny. From that point onward the physical child was of smaller importance. If Alice's strength could seem heartless, Edward acquiesced.

* * *

On 2 August 1893 they were off. The holiday began with a fortnight in Garmisch at the guest-house of an English family they had met the previous year, the Bethells. The diary gathered a harvest of impressions.

Prinz Regent at Partenkirchen . . . Splendid uniforms to be seen . . .

Swallows feeding young *in houses*, flying in & out over doorways left open for their pretty sight . . .

Rain—thunder—retired to cottage—peasants played & sang *in parts*, fresh from the hay-cutting &c. Shook hands all round . . . Walls of room covered with horns . . .

Garmisch: Schuhplatt'l Tanz . . .

Haymakers remain on hills all night—fires lighted up in evening—cows coming home—bells, light wagons—old man in village with immense heap of fir-tree trimmings . . .

On 17 August they bade farewell and journeyed to Munich, where Miss Burley and her pupil Alice Davey had already arrived:

The Elgars were full of their experiences at Garmisch, we of our journey up the Rhine . . . On the night of the Elgars' arrival . . . we all went to *Die Meistersinger* . . . The Genius had an immense admiration for the part-writing of the quintet . . .

The next day they came round from their rooms in the Gluckstrasse to have tea with us . . . He was delighted with the Biedermeyer furniture of our rooms, the view from the window, the Münchener Kind'l holding up his beer mug, with the ancient tea-caddy and its green glass sugar box which Frau Würmer, our landlady, had brought out in great excitement over the arrival of these English visitors—and particularly with a pointless but endearing little family joke which began that afternoon and was to last for many years. A reference was made to the *Herrschaft*, and the Genius—not fully understanding the meaning of the word—supposed it to refer to himself. 'This Herrschaft', he said, 'is enjoying himself.'

In a quieter way Mrs Elgar also seemed to be happy. The German culture was one with which she was clearly in deep sympathy . . . Graciously—for she was always gracious—she spoke in her excellent German to Frau Würmer, complimenting her on her tea and on the arrangement of her rooms. But it was noticed that when she referred to the Genius as 'mein Mann', the old Frau seemed slightly taken aback . . .

In many ways her behaviour towards the Genius was, as Frau Würmer had hinted, that of a doting mother of a gifted son rather than of a wife. She had all the doting mother's fussiness and anxiety, especially over his health, with the result that she sometimes irritated him. Moreover, as sons frequently do, he often refused to play the part for which she had cast him . . .

But despite these little brushes and the irritation they sometimes caused, one impression remained from that Munich holiday—and this was of Mrs Elgar's devotion to her husband. She had married him in defiance of the opinion of her friends, and it was clear that she would stand by him whatever misfortune might befall. Hers was truly a great love.[93]

Through the holiday, Alice changed over to German for addressing her 'Eduard'.

On four nights they attended all the *Ring* operas in succession. Then came Wagner's early opera *Die Feen*, *Tannhäuser*, and on 29 August *Tristan*. Miss Burley recalled that

The Ring impressed me chiefly by its interminable length, and I was quite unable to understand the Genius's enthusiasm. But *Tristan* was a shattering experience—Mrs Elgar, always deeply affected by romantic music, was the most touched . . . but on all of us the heavily erotic melodies worked such a spell as to make sleep impossible for the whole night.

The performances were arranged on the leisurely Bayreuth plan . . . and began in the late afternoon. Then after an act or so came the Grosse Pause, in which one retired to the theatre restaurant to repair the emotional ravages made by the music with beer, Schnitzel, Semmel Brot and Sauerkraut, enlivened by an exchange of views on the performance.

The real discussion, however, came after we left the theatre and adjourned to the Hofbrauhaus. There the whole opera would be reviewed, the playing criticized in detail, and the technical means by which each effect had been obtained carefully analysed. At that time the Genius had a great regard for Wagner, but . . . he did not jettison his critical faculty. I remember that, despite his great admiration for the orchestration and the direction [of Herman Levi], he was very displeased with the coarse quality of the brass. The fact is that he had begun to understand very fully how the new music was put together, and was realizing that he could convert this knowledge to his own use.

Throughout the holiday he took copious notes of what he had heard, and spent many hours over them at his rooms. In this he was encouraged by Mrs Elgar, who sometimes when we called to take him out would put her finger to her lips and tell us he was too busy. 'My word,' said Alice Davey, 'Doesn't she keep him at it?'[94]

Before the end of the visit Edward bought a large engraved portrait of Wagner. When it came home to be hung in the study at Forli, it would demand such a place of honour that Alice was to write: 'Moved all the pictures all the evening!'

---

[93] Ibid., pp. 64–7.   [94] Ibid., pp. 68–9.

To the end of the German holiday, however, there were things to defy the understanding of Miss Burley:

It was one of Edward Elgar's peculiarities never to speak, even in the early days, of the teaching and playing by which he really earned his living . . . Although the Genius was returning from Munich to play . . . in the orchestra of the Three Choirs Festival at Worcester, he did not once mention it during the holiday.[95]

The return home from this holiday meant a re-fastening of shackles. On his copy of the Worcester Festival programme for September 1893—listing no performances of music by Edward Elgar—he wrote: 'I played 1st Violin for the sake of the fee as I cd. obtain no recognition as a composer. E.E.'[96]

If his music was winning a local celebrity, the social advantages were double-edged. The Elgars might find themselves less frequently 'cut' by people reacting against Edward's tradesman origins. But there was embarrassment over golfing with more comfortably established friends and acquaintances— R. B. Townshend at Hasfield, Hugh Blair, Basil Nevinson, R. P. Arnold (a son of the poet Matthew Arnold), the Malvern bank manager H. Dyke Acland—when their foursomes had to be regulated to Edward's appointed hours for teaching the violin.

In October 1893 he toyed again with ideas for *The High Tide*. In November he revised the old part-song to Alice's words of hidden affection, *O Happy Eyes*. In December he looked out the Violin Sonata fragments of 1887 and began to recast the slow movement melody as an *Andante religioso*: but he did not finish it. Teaching ground forward term by term. It was, he once said, 'like turning a grindstone with a dislocated shoulder'.[97]

In the early days of January 1894 he worked at three solo songs. One set a 'Rondel' by Froissart in Longfellow's translation. So it brought together two favourite authors—but in a personal statement less certain than *O Happy Eyes*:

> Love, what wilt thou with this heart of mine?
>   Nought see I sure or fixed in thee!
>   I do not know thee, nor what deeds are thine:
> Love, what wilt thou with this heart of mine?
>   Nought see I fixed or sure in thee!
>   Shall I be mute, or vows with prayers combine?
>   Ye who are blessed in loving, tell it me:
>   Nought see I permanent or sure in thee:
> Love, what wilt thou with this heart of mine?

Neither the Rondel nor its companions[98] found any publisher in 1894.

---

[95] Ibid., p. 73.   [96] Elgar Birthplace.   [97] Buckley, *Sir Edward Elgar*, p. 43.
[98] The other two songs were listed by Alice as 'The Wave' and 'Muleteer's Song'. The latter may have been *The Shepherd's Song*, for which permission to use the poem by 'Barry Pain' was obtained on 3 March 1894 (Chappell archives). Of *The Wave* nothing further emerged.

Just as the new term's teaching was to begin, an alarming throat illness stopped all Edward's activity for a month. It was his first prolonged illness since the move to Forli. When he recovered Alice persuaded him to re-submit the revised *O Happy Eyes* to Novellos, and on 23 March the diary could show: 'E. heard from Novello that they wd. publish "Happy Eyes" & *pay* for it.' The purchase price of 3 guineas made a contrast to the exchange of its companion, *My Love Dwelt in a Northern Land*, merely for 100 printed copies four years earlier.

Then Hugh Blair wanted the *Andante religioso* if it could be finished and scored in time for a special Cathedral service early in April honouring a visit to Worcester by the Duke of York. (This was the future King George V, who was later to confer upon Edward the Order of Merit, Mastership of the King's Music, and a Baronetcy.) Edward only just managed to paste together a score and parts for the piece now to be entitled *Sursum Corda*. The scoring was for organ, strings, timpani, and brass. The brass writing especially, inspired by the recent experience of Wagnerian orchestral colours, made a sustained dark colour quite new in Edward's music.

An introductory figure leapt up to descend in steps—tracing again a basic shape at the centre of the *Serenade* for strings first movement. The main theme of the *Sursum Corda* (from the Violin Sonata) lent itself well to the new purpose, lifting slowly to descend again in steps.

From the descent the organ elaborated a variant,

which repeated antiphonally with the strings as its sequences rose through the triad to a big contrapuntal climax. Then the primary theme returned, followed by the introductory figure, to close the music in slow farewell. So the *Sursum Corda* sounded for the first time an ensemble of traits that would haunt the centre of Edward's maturest music—aspiration ennobled within darkening nostalgia.

The spring of 1894 remained a wretchedly uncertain time. There was illness again as the hour of the *Sursum Corda* première approached on the evening of 8 April. *The Worcester Herald* reported:

Mr. Elgar rehearsed his piece with the band in the afternoon, though the doctor's veto prevented him assisting in the performance of it in the evening. However, under the able conductorship of Mr. Hugh Blair, Mus. Bac., it suffered nothing by his absence.

The work bears the impress of a master of composition, and one inspired with much religious and poetic feeling. Without attempting an analysis of the Adagio—which takes about ten minutes to perform—it may be remarked that evidence is present of a grasp of the capabilities of brass as well as stringed instruments. Strings and brass are admirably blended, and in a secular building the performance would certainly have had the inevitable accompaniment of hearty applause.[99]

During the inter-term holiday after Easter Edward and Alice left Carice with her nurse, and spent a week in London as guests of Basil Nevinson at his large house in Tedworth Square, Chelsea. Edward called at Novellos to leave the *Sursum Corda*, a 'solemn march' arranged for the organ from *The Black Knight* Scene II, and some piano accompaniments he had devised to go with Kreutzer's Studies for unaccompanied violin. Ten days of golf at Littlehampton in Sussex brought him back to the beginning of summer-term teaching.

Earlier in the year Edward had half promised to arrange the Good Friday Music from *Parsifal* in such a way as to make possible its performance in a concert at the Worcester Girls' High School. When Novellos declined *Sursum Corda* and temporized over the Kreutzer accompaniments, the only thing left was to busy himself with *Parsifal*. Over another weekend at Hasfield in the beginning of June he produced a version of Wagner's music for the High School forces with the dexterity of his old scorings for the Powick Band. The arrangement was duly rehearsed, together with the other items for the school concert. They included a violin solo played by Edward himself and Mendelssohn's D minor Piano Concerto in which the soloist was Ethel Gedge. On 13 June the School Log-book recorded:

The long looked-for Matinée Musicale took place . . . the music to which all were looking forward was the little piece from Wagner's Parsifal at the end of the programme, which had been so splendidly arranged by Mr. Elgar with express permission from abroad for performance at the School. It was played by Miss Perks at the Organ, Miss James and A. Wigram at the two grand Pianos, Mr. Elgar 1st Violin, A. Woodward 2nd Violin, O. Webb 3rd Violin and Mr. Fitzherbert 'Cello . . . Our warmest thanks are due to Mr. Elgar for his kind help in getting up the concert and conducting rehearsals.

Wagner was in prospect again with another excursion to Germany later in the summer. So for a moment Wagner's greatest achievement in setting to music the Nibelung Sagas seemed to offer an example: 'July 15 1894 . . . E. wrote Sagas all day—booful.'

Before the saga-writing had begun that Sunday morning, Alice had accompanied Edward to Mass at the Catholic Church. She had been taking

[99] 14 Apr. 1894. Review probably by Revd E. Vine Hall.

lessons in Latin at The Mount, and all the spring she had gone to Fr. Knight in Worcester for instruction in the Catholic faith. When she was received into Roman Catholicism at St. George's Church on 21 July with Edward by her side, she wrote in the diary:

A. & E. Worcester 10.48. Fr Knight at St George's—very quiet & peaceful—just zu & me.

So again in her conversion Alice fulfilled the example of Edward's mother.

Then the end of summer teaching was at hand, and the familiar rush of preparations. Carice was left with Miss Burley, who was staying at home that year. The precious Gagliano violin, known familiarly as 'Messer Niccolo', was entrusted to the safe keeping of the violin dealer Hill in London, and a modern instrument by Chanot (the firm who published Edward's Violin Studies) was taken as a travelling substitute. On the last day of July they boarded the boat-train at Liverpool Street and caught the night boat from Harwich.

Again they made for Garmisch, where they shared a few days with Mrs Fitton and her daughters who were touring Germany that summer. But for most of their seven-week holiday Alice and Edward roamed about familiar scenes and explored new ones by themselves, equipped with rucksacks. They went for a few days to Innsbruck, and on the way back one Sunday morning saw the Procession of the Blessed Sacrament through an Alpine village. Edward wrote down the notes of the chimes sounding then—as he had written the notes of the monastery bells at Fort Augustus in Scotland ten years earlier almost to the day.

The return journey in mid-September 1894 took them once more through Munich, where they attended performances of *Götterdämmerung* and *Die Meistersinger*. Then on to Frankfurt (with a visit to Goethe's house), Bruges, Ostende, London—carefully collecting Messer Niccolo from Hill's—and so home to Forli, where Carice was already re-installed by Miss Burley. By the end of the month another year's teaching had begun.

One of Edward's new pupils was a nine-year-old girl named Norah Masefield, whose eldest brother John was soon to acquire a reputation as a poet. Norah had begun her lessons on a half-size violin with a teacher close to her home in Ledbury, on the Herefordshire side of the Malvern Hills. As she progressed her family made enquiries for a teacher of larger reputation: Mr Elgar had been recommended to them. Norah would need a full-size violin now, so Mr Elgar sold her family the Chanot which he had taken abroad during the summer and which he had liked so much.[100]

---

[100] The Chanot violin was given by Norah Masefield Parker to the Elgar Birthplace, where it may be seen together with this bill in Elgar's hand:

        Forli, Malvern.
        IIIrd term 1894

Mrs. Masefield to [Edward Elgar]	
Violin lessons	3. 3. 0
Violin	10.10. 0

The lessons were an unmitigated trial, as Norah Masefield recalled after more than eighty years. She used to come over with her governess in the train to Malvern, climb the hill up to the Haynes Music Shop, go through the shop and down some steps at the back leading to a little room where a grand piano was crowded in with an upright piano. The previous lesson finished, Norah's three-quarters of an hour began.

The first book he gave me was Mendelssohn. He would tune my violin himself—he wouldn't have allowed me to tune it, not thinking it adequate. He would say what he wanted me to play, and would begin at the piano.

He sat at the grand piano and played it all the time, hardly ever stopping. His hands moved all over the place, all up and down the piano keyboard, with great facility—in that tiny room with the second piano and I and my governess tucked into a corner. He seemed at times hardly to listen to my playing, but was absorbed in his own.

Then he'd stop and say, 'You didn't do that right,' or 'Go and do that again.' Sometimes he would take the violin and show me.

It wasn't awfully good. I didn't look forward to the lessons. Small children were of no use to him: he never laughed or joked. Imagine a child of nine years old and this very handsome man playing away. Oh, I was frightened. After a time I went alone, but it was always the same.

Once he gave me *Salut d'amour*. I didn't realise it was his. My impression was that he was put out.[101]

The lessons were to go on for three years.

On 1 November 1894 he burnt the fire-motive from *Die Walküre* in poker-work over the fireplace in his study. What he achieved in the month that followed was only the orchestration of Hugh Blair's *Advent Cantata* due for performance in December. Then Alice offered him another pair of lyrics. They were from her long poem *Isabel Trevithoe*, published in 1878. Again they were of winter and spring, and again they inspired Edward's music.

The winter poem was entitled 'The Snow'. It suggested how a quiet season served to conceal the life persisting underneath—how the appearance of death might be in reality a snowy innocence covering an understanding of time's passage in cycles:

―――――

Music		
Mendelssohn net	1.	6
Händel net	3.	6
Exercises	1.	0
	£13.19.	0

Received Jany 10 1895
Edward Elgar with many thanks.

―――――

[101] Conversation with the writer, 1975. Norah Masefield later studied with Arthur Quarterman, the teacher who succeeded Elgar in Worcester, and ultimately attained a level of playing that enabled her to take the solo part in local performances of Mendelssohn's Violin Concerto.

O snow, which sinks so light,
Brown earth is hid from sight,
O soul, be thou as white
   As snow.

O snow, which falls so slow,
Dear earth quite warm below;
O heart, so keep thy glow
   Beneath the snow . . .

Then as the snow all pure
O heart be, but endure;
Through all the years full sure,
   Not as the snow.

The poem's innocence was emphasized by Edward's scoring of it and its companion for female voices and two violins, with piano. An opening theme in E minor moved *Andantino* in small intervals over spare vocal harmonies to evoke the cold—all set off with slowly whirling semiquavers for the violins. For the second-stanza warmth within, the melody turned to E major, repointed and with upward intervals widening toward an octave leap. At its centre the music entered G major with a new idea rising up to a big climax. Then the original E minor melody returned against a redoubled persistence of falling semiquavers; and a short coda set the minor melody once more in the major. The simple form made a perfect frame for melodic and harmonic invention as fresh and sure as the best he had ever achieved.

Both *The Snow* and its companion were dedicated to Mrs Fitton. The writing for women's voices again suggested feminine inspiration—from Mrs Fitton and her daughters, from Miss Burley's counterpoint books, from Alice's constant encouragement. Alice's spring poem moved with quicker passion to send its cry after a music for which both husband and wife waited:

Fly, singing bird, fly,
From the woods where lies shelter'd thy nest,
From the tree whence thou pourest thy song,
Fly away, far away to the west,
Tell my love that I wait, Ah! too long
   And lonely, I sigh.

His music for this poem moved in a 6/8 *Allegro* from its G major opening through a rising sequence. At 'Fly away, far away to the west', a descending sequence lightly touched a memory of the tune from Broadheath. A smaller figure rose in hesitating sequence through the long and lonely wait. Two further stanzas extended the tension of headlong instrumental impulse against the hesitant human voices.

Novellos asked for some simplification of the instrumental parts. But they accepted Edward's suggestion of 12 guineas for the copyrights. It was the largest fee his music had ever earned. Vine Hall expressed interest in performing the new songs with the Worcestershire Musical Union. Then the Midlands musician Charles Swinnerton Heap sent an invitation for the composer to conduct *The Black Knight* with the big Choral Society in Wolverhampton. Heap himself had conducted it with a smaller society in Walsall. Now he was so enthusiastic that he wished to encourage Edward to write a really big work.

Through December 1894 Edward considered two projects. One was a setting from St. Augustine's *Civitas Dei*. St. Augustine interpreted the past, present, and future of the City of God through a vast assemblage of quotations from the Old and New Testaments. It was akin to the way in which the libretto of Handel's *Messiah* was constructed from a selection of biblical sources. Minnie Baker (now engaged to the vicar of Wolverhampton) agreed to make a selection from the *Civitas Dei* texts, and on 16 December the selection arrived.

The last three books of *Civitas Dei* concluded St. Augustine's picture of the future. There was 'The Last Judgment' (based on the Book of *Revelation*) and the coming of Antichrist, 'The Punishment of the Wicked', and 'The Eternal Felicity of the City of God'. But such an ultimate division of good from evil needed a personal security that was beyond Edward and the art that reflected him then.

The other possibility was the Saga project. On 14 December 1894 he had brought home several books of sagas from The Mount, for Miss Burley was eager to help. A month later there was further definition: on 15 January 1895 Edward wrote his name in a volume of Longfellow which contained 'The Saga of King Olaf'.[102] Longfellow's version of this saga was part of his *Tales of a Wayside Inn*, gathering travellers from many walks of life to tell stories in the fashion of Chaucer's *Canterbury Tales*. The central figure in Longfellow's *Tales* is not the Innkeeper, however: he is the Musician.

Like Edward himself, Longfellow's Musician is a violinist. He is also a Norseman: so if 'Elgar' was really 'Ælf-gar' or 'fairy spear', here was another likeness. Longfellow described his Musician thus:

> He lived in that ideal world
> Whose language is not speech, but song;
> Around him evermore the throng
> Of elves and sprites their dances whirled
>
> . . . .
> Voices of eld, like trumpets blowing,
> Old ballads, and wild melodies
> Through mist and darkness pouring forth,
> Like Elivagar's river flowing
> Out of the glaciers of the North.

So this Musician could enchant the whole company gathered at the snow-bound inn, creating with his music an atmosphere for all their tales.

[102] Elgar Birthplace.

When the Musician himself settles to story-telling, the tale he chooses is the Saga of King Olaf—or as much of it as survives. In fact the broken character of the separate saga narratives gives the Musician his own chance:

> And in each pause the story made
> Upon his violin he played,
> As an appropriate interlude,
> Fragments of old Norwegian tunes
> That bound in one the separate runes,
> And held the mind in perfect mood,
> Entwining and encircling all
>
> . . . .
>
> As over some half-ruined wall
> Disjointed and about to fall,
> Fresh woodbines climb and interlace,
> And keep the loosened stones in place.

If ever there was a group of stories to invite music, here it was.

The setting for this saga was something like the old chivalric world of Froissart and The Black Knight. And the saga hero offered his identity to Edward in other ways. First, he was an exile from his rightful kingdom, and his young life gave heady example: from obscurity he journeyed to triumph, vanquishing all who stood in his way. Then Olaf—alone among Norse saga-heroes—was a Christian: it had been the faith of his own choosing, and it guided his actions. Yet here was also a mirror of Edward's uncertainty: for as Olaf grew older, he found his triumph in the world only fleeting despite his faith. And there was a final suggestion. Olaf was attracted to women; but each of the women he cared for betrayed his vision, and the last of them accomplished his death. At the end there was only the hero's mother to witness the faith by becoming a nun.

The proposal to set 'King Olaf' to music was met by Miss Burley with headmistress superiority:

I confess that my heart sank when I realized that Longfellow was once more to serve as librettist, and my hopes were not encouraged by a study of the poem. *Olaf* is a wretchedly muddled story, in which there is neither consistency of character nor unity of plot.[103]

Edward meted out a punishment subtle enough that Miss Burley seemed unaware of any reprimand. For as she continued:

. . . He was fascinated by what may be called the Longfellow movement. Indeed he asked me to read it up for him, and I spent a fortnight of my holidays in the British Museum Library studying the literary group who had surrounded Longfellow at Cambridge, Massachusetts, and made a sheaf of notes on the subject.[103]

Swinnerton Heap's invitation to write the big work for a definite Festival performance was long in coming. Alice again stepped into the breach in a most

[103] *Edward Elgar: the record of a friendship*, pp. 87–8.

resourceful way. She reasoned that Novellos had refused two of Edward's instrumental works, but had accepted part-songs—including several set to her words. Some further verses to recall their shared experience might offer him a guiding hand. One ideal experience they had shared in the Bavarian holidays, and a musical focus presented itself in the *Schuhplatt'l* dancing and part-singing remembered so happily.

Alice wrote six poems imitating the spirit of those dances. Each was subtitled with the name of a place they had visited, each captured a different aspect of the picturesque life: 'The Dance' (Sonnenbichl), 'False Love' (Wamberg), 'Lullaby' (In Hammersbach), 'Aspiration' (Bei Sanct Anton), 'On the Alm'—to which Alice gave the added title 'True Love' (Hoch Alp), 'The Marksmen' (Bei Murnau). As all the verses were made for Edward's music, they could be cut, extended, or adapted to suit whatever his musical ideas might suggest.

Like the marksmen of her final poem, Alice scored a direct hit. By 15 February 1895 Edward found himself deep in his biggest composition since finishing *The Black Knight*—yet with each of its sections quite separable for individual planning and constructing to make his task easier. The triumphant Wolverhampton performance of *The Black Knight* on the 26th hardly interrupted the new work.

Despite the demands of running a household for a composer in full spate, Alice kept up a thread of social life. She went to tea with a new friend, Mrs Harry Acworth, whose husband had literary interests. She entertained Hugh Blair and Ann Elgar: the old lady made them a visit over three days of summery weather in mid-March. On the morning of the 16th, before she arrived, Edward finished the sketch of his latest Bavarian part-song.[104]

As their outlines emerged, the new songs began to show a unity. Alice's poems might be separable, but Edward's keenness for larger expression led him to interrelate and develop themes from one song to the next—just as he had done from scene to scene of *The Black Knight*. The innocence of Alice's poems suggested the simplest musical beginnings. He opened 'The Dance' with a simple down-and-up tune moving to sequential repetition:

It was an idea to interlace with each of the contrasting figures to come. The second-stanza figure repeated:

[104] BL Add. MS 47900A fos. 115–18.

And that opening inverted to open stanza 3:

Down the path the lights are gleam-ing,

The music then followed a rondo pattern to round out its picture of peasant invitation.

The second song, an *Allegretto* in F major, was made from two developments. The first was based on I.1:

Now we hear   the Spring's sweet voice

while the second combined elements from I.2A and 2B:

E - ver   true was I to thee

The 'Lullaby' opened with an instrumental tune, through which the voices gently reminisced over I.2A:

Sleep,   my   son,   Oh   slum - ber   soft   -   ly,

A second theme reminisced over I.2B:

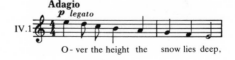

Far - a-way Zith - ers play,

The fourth song, 'Aspiration', extended I.2A slowly down the entire scale:

O - ver the height the   snow lies deep,

and then contrasted a theme which combined suggestions from all three of the opening 'Dance' ideas set in slow motion against a fresh 'Snow' figure:

'On the Alm' ('True Love') opened with another instrumental variant:

presently set off with yet another vocal variant:

The Finale, 'The Marksmen', returned to G major to pour out a cornucopia of fresh variants, beginning with

and culminating in a figure whose double extension of the rising and falling pattern could seem to gather everything that went before in an embrace of new importance:

Often in these *Scenes from the Bavarian Highlands* the vocal figures betrayed their instrumental origins. Yet the writing of so many spirited variations and developments made a new measure of Edward's power to expand his music from the simplest beginnings. Vocal counterpoint and orchestral polyphony engaged one another without overreaching the innocent happiness which had brought them together. So Edward's music filled all the sketches Alice had drawn.

Less than six weeks after the project's first mention in the diary, the entire musical design was approaching completion. On 28 March 1895 he took the first five *Bavarian Highlands* songs in vocal score to London and laid them on the desk of Novello's music editor Berthold Tours. The Finale's complex structure needed a little more time. Back at Forli he soon finished it and sent it off with a letter on 10 April:

Dear Sirs:

By this post (registered) I send you No. 6 of the set of six partsongs for Chorus and orchestra completing the set . . .

Nos. 1, 3 & 6 wd. make a *very* useful Suite for orchestra alone, or for *piano*—duet or solo.

I submit the work to you with confidence as I believe from the character of the music it will find easy acceptance in many quarters.

I shall be much obliged by an early reply.

Believe me

Faithfully yours

Edward Elgar

P.S. The words are partially arranged (imitated) from Volkslieder—the music is my own.[105]

His manuscripts of *Froissart*, the *Serenade*, and the first sketch for *The Black Knight* had all made special acknowledgement of his debt to Alice. Now Edward's achievement alone was to be flourished. Alice's determined understanding never faltered. When the first London performance of *The Black Knight* was reviewed slightingly at the end of March 1895, she silently withheld the cutting; but she kept it, and left a space for it in the cutting-book. Six years later it was duly entered in its place with a pencilled note: 'Not put in book till 10 Ap 1901—when such foolish remarks can have no sting. (C.A.E.)'[106]

* * *

On the very day of posting the final Bavarian song, Edward began work on another big project. Hugh Blair had asked for a new organ voluntary to play in a Cathedral service for a party of American musicians who were to visit Worcester in July. Edward resolved that the piece must take the form of a four-movement Sonata on symphonic scale. Writing such a work for the organ at this moment could offer a way of carrying on the musical development sketched through the four scenes of *The Black Knight*. The organ, large and varied as it was for a single instrument, did not have the bewildering variety of the orchestra: the mere three staves on which organ music was written allowed him to focus on the vital matter of horizontal development.

[105] Novello archives.   [106] Elgar Birthplace.

The movement sketched on 10 April 1895 was not the beginning of the Sonata, but an Intermezzo.[107] It was not naturally shaped for easy keyboard playing. In fact the sequential opening recalled the slowly whirling figure for two violins of *The Snow*:

The real definition of this figure came with the upward leap of a fourth repeated through every group in the opening bar. That suggested kinship with an older theme, which had been set down as an *incipit* for a Suite movement in the sketchbook of 1887,[108] and which was now to make a slow-movement subject for the Organ Sonata:

The concentration on fourths emerged again in the subject which opened the Sonata's first movement. Once again the opening was in the key of G, and in fact this primary subject started as a close relation of the theme that opened *The Black Knight*:

The likeness showed how Edward's thinking could extend from one work to another. The contrast showed his progress in self-expression: where the *Black Knight* theme projected a balanced and complete melody, the Sonata theme raced headlong into emotional volatility.

The pressures of composition were exacerbated by events during the spring of 1895. A throat infection kept Edward low during much of May. The beginning of summer-term violin teaching brought a really promising little girl called Marie Hall—who would one day become a celebrated virtuoso; but she

[107] BL Add. MS 57993 fo. 16.
[108] BL Add. MS 49974D fo. 33, where the title noted for the movement is 'Traumerie' [sic].

soon returned to her regular professor. Then Novellos rejected the *Bavarian Highlands* songs with a polite letter casting doubt on their saleability in any form. Edward sent three solo songs to the publisher Tuckwood, offering to present the copyright just to get them into print: they were *The Shepherd's Song*, *Through the Long Days*, and the Froissart-Longfellow *Rondel*.

On 2 June he passed his thirty-eighth birthday: time to realize that half of life was gone. As if to draw out the irony, the editor of the *Dictionary of National Biography* chose this moment to write to Alice requesting an article on the distinguished career of her late father, Major-General Sir Henry Roberts, KCB. On 21 June Edward left off whatever work he was doing on the Organ Sonata to write a new song to a poem by Philip Bourke Marston called 'After'—which might make its comment on all the expectations that Alice had sought to enlarge with her share in the Bavarian songs:

> A little while for scheming
> Love's unperfected schemes;
> A little time for golden dreams,
> Then no more any dreaming.
> . . . .
> But long, long years to weep in,
> And comprehend the whole
> Great grief that desolates the soul,
> And eternity to sleep in.

Morbidity had emerged more than once in the lonely letters written to Dr Buck before Edward's marriage. It had emerged again in the first months of the marriage with *My Love Dwelt in a Northern Land*. Now it recurred just when the ensemble with Alice had raised expectation to a new height.

As the date of the Worcester Cathedral service on 8 July drew closer, Hugh Blair appeared more and more frequently at Forli. On the day that followed the writing of *After*, he listened to Edward playing the Sonata sketches. Then he repeated and reinforced the encouragement given three years earlier over *The Black Knight* and *Sursum Corda*. Edward's response to it all was a pattern described by Miss Burley as

. . . a sequence of moods that was to be repeated with nearly every work I watched him compose. To begin with there would be a period of great exaltation over the conception of the work and the commission . . . This was always followed by a period of black despair over the intractability of the material and the utter impossibility of ever getting it into a satisfactory shape. An immense amount of encouragement, accompanied by assurances that he was the only person able to do it, and reminders that it must at all costs be done, had now to be expended in order to shift him into the next phase—which was one of increasing hope and enthusiasm.

Tunes, contrapuntal patterns, and sequences had begun to suggest themselves for various sections of the work, and would start to build up almost without his conscious control into something like the completed whole.[109]

---

[109] *Edward Elgar: the record of a friendship*, p. 89.

All available time through the final week of June 1895 went into developing the sketches.

June 22 . . . Mr. Blair to dine & stay. Heard Organ Sonata.

June 23. Mr. Blair left at 9.30. E. to S. Joseph's [Catholic Church, Malvern Link] at 8.30—A. to S. Joseph's at 10.30 very misy. E. wrote his Sonata all day.

June 24. E. wrote his booful Sonata—to Worcester [for lessons] at 1.46. Intensely hot. A. posted her notes to Dictionary [of National Biography]. Mr. Blair to dinner & stay.

June 25. E. into Malvern [for lessons]—wrote Sonata in the afternoon—Intensely hot.

June 26. Mr. Blair over early—putted [golf] most of the morning. E. writing his Sonata—to Worcester at 1.46 . . .

June 27 . . . Mr. Blair to lunch.

The next evening after dinner, the shaping of the Sonata's final form began.

The first movement contained almost as many ideas as *Froissart*. But the Sonata ideas were presented in a clear sequence of emotions. In the wake of the explosive opening theme, two figures followed one after the other. The first slowed the pace with heavy insistence

The second descended in steps

Its mood foresaw the lyric second subject:

The second-subject 'answer'

anticipated again the sequence of dying 'Delight'. Then another sub-theme

brought the exposition to a lyrical close.

The development was another measure of Edward's progress in the five years since *Froissart*. In place of the lame repeated sequences of the *Froissart* development, there was now a polyphonic combining of elements from the primary theme to shape something new in Edward's mature style—pastoral trio writing:

sonore

It recalled Bach; and Edward told an interviewer a year later: 'I play three or four preludes and fugues from the "Well-tempered Klavier" every day. No. 33, in E major, is one of my favourites. No. 31 is another, and No. 29, a wonderful masterpiece, is constantly before me.'[110] There had been hints of such pastoral writing in the youthful wind quintets nearly twenty years earlier: but then he had no framework of personal style into which it could be fitted. The trio writing rediscovered in the Organ Sonata was to find its place in the orchestral works of his creative maturity—notably in the *'Enigma' Variations* (1898–9), *In the South*, (1904), the First Symphony (1907–8) and *Falstaff* (1913).

The Sonata first-movement development exactly equalled the length of the exposition. Recapitulation followed, and a coda which climbed in brilliance toward a vital, masculine end.

The second and third movements were in simpler A–B–A forms. The opening of the second movement *Allegretto* (the Intermezzo sketched on 10 April) extended the polyphonic trio style. A contrasting middle section used the triads of C and F to revive memories of the first movement:

[110] Buckley, *Sir Edward Elgar*, p. 31.

The slow movement was in B flat major:

All polyphonic busy-ness vanished, and the old melody of 1887 was set with rich harmonies and chromatic movement to recall the orchestra of Wagner. Yet the old-world dignity of the melody itself carried forward the mood of Edward's own *Sursum Corda*. A middle section moved enharmonically to F sharp major for a lyric theme derived from the first movement:

Recapitulation led to a coda which joined primary and secondary themes in polyphony:

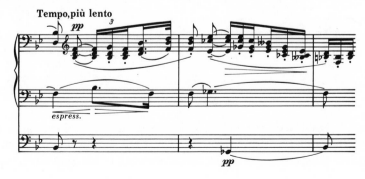

The Finale was in G minor—so the Sonata traced again the darkening progress of *The Black Knight* from G major to minor.

After forty bars, two descending sequences led to the second subject:

Its opening fourth recalled the opening subjects of both the first movement and the slow movement. The Finale development (just as in *The Black Knight*) brought another reminiscence. This was the slow movement first subject, but now in the Finale rhythm: indeed it soon combined in polyphony with the Finale primary theme. Recapitulation was deftly compressed.

Then the coda did what *The Black Knight* and the *Bavarian Highlands* songs had only hinted at. It drew all the preceding ideas together in a final figure suggesting that here at last was the real theme to which all the rest had been leading—the answer to the downward fourth with which the Sonata had begun, the apotheosis of the slow movement's reflection:

The triumph was stated once only, just before the end; so it did not quite carry the weight to cap a thirty-minute work. But it did show a pattern of moving from smaller things to larger within the framework of recapitulating themes.

And so it looked forward to a masterpiece now less than four years in the future—the *'Enigma' Variations*.

The Sonata was completed just as Hugh Blair was due for luncheon on 3 July 1895. Turning back to the opening page of his manuscript, Edward wrote:

> Friday June 28, 9 pm.
> Finished the Sonata July 3, 1 pm
> (one week's work).[111]

The 'week' covered only the formal putting together; but it was a pardonable exaggeration, because it was precisely the putting together that held the difficulty and the triumph.

Even now the Sonata existed only as a rough score, full of inked additions and pencilled corrections. Blair had only four days to learn a work of virtuoso difficulty on the largest scale. On Sunday morning, 8 July, the Elgars went over to Worcester accompanied by Miss Burley. At the conclusion of the Cathedral Matins the Sonata was given its première. Miss Burley rendered a sharp verdict on the player, whose personal habits she thought 'erratic':

> He had on one occasion been engaged to spend a day examining the girls at The Mount, and had neither turned up nor sent any explanation. His performance of the Sonata showed that he had either not learned it or else had celebrated the event unwisely, for he made a terrible mess of poor Elgar's work. I was present at this débacle and commiserated with the Genius. But with a splendid flash of loyalty he refused to blame the murderer who, he said, had not had time thoroughly to study the victim.[112]

It was not the first time or the last that Miss Burley would oppose her own judgement to an explanation of Edward's. But the significance of the Organ Sonata première was to emerge only gradually. This was the first instance of a fate which overtook later premières of several of his large works written against a deadline. Edward depended on increasing pressure of an approaching performance to force the final constructive effort. It left no time for rehearsal.

Still, the Organ Sonata stood now as a solid achievement, finished just as Alice finished her article on her father's career for the *Dictionary of National Biography*.

> July 13 1895 . . . E. & A. . . . walked to Broadheath & saw the house in wh. he was born & where he used to stay as a lissy boy.

It marked a distance in his own career.

*   *   *

During the spring of 1895 Edward had played golf with the Revd T. Littleton Wheeler, the Worcester Secretary of the Three Choirs Festival. And the chance emerged of a choral commission for the Worcester Festival of 1896.

---

[111] BL Add. MS 57993.
[112] *Edward Elgar: the record of a friendship*, p. 86.

Edward asked Mrs Fitton's Anglican cleric friend Capel-Cure to suggest possible subjects. One centred on the Magi, another on St. Barnabas. But Edward chose the first of all—the story of the blind man whose eyes were opened by Christ, related in the ninth chapter of St. John's Gospel. Capel-Cure proposed the title *Lux Christi*. [113]

When the summer term's violin teaching was done and Carice deposited once more with Miss Burley, Edward and Alice set out for another holiday in Bavaria. This time they made a wide circle of approach, going down the Danube to Linz and back through Salzburg and the Austrian Alps. They climbed mountains and explored countryside so generously that there was only a fortnight to spend in Garmisch.

There Capel-Cure's draft libretto for *Lux Christi* reached Edward. The libretto drew heavily on direct quotation from St. John, but these passages were interspersed with original stanzas to provide a broader basis for musical set pieces. And so, instead of seeking inspiration in the *Schuhplatt'l* dancing, this year Edward used his days in the picturesque summer heights to sketch music for the blind man's story. One of the first figures for the new project[114] echoed the high notes of a repeated passage in the Organ Sonata Finale:

But again, as in *Froissart*, there was a source in Wagner—the dying Tristan's motive of bitter yearning, 'Still no ship in sight':

It was the 'Weary wait' of Alice's 'Fly, singing bird'.

As the holiday drew to its close, Edward had another throat infection. At Munich they managed a single performance of *Der Fliegende Holländer*. Then Alice nursed him all the way home through Strassburg, Paris, Dieppe, and across the Channel. In London they consulted a well-known specialist, Greville MacDonald, who gave assurance that the trouble was not organic. And so in the evening of 12 September they arrived back at Forli.

[113] Letter at Elgar Birthplace. Miss Burley (p. 87) also claimed credit for this title.
[114] Buckley, *Sir Edward Elgar*, p. 34.

In Birmingham that summer Swinnerton Heap had taken over as conductor of the Birmingham Festival Choral Society. Heap immediately set about plans for a Birmingham performance of *The Black Knight* in December. It would be the first large-scale Elgar performance in a major city. Heap also said that if Edward offered a new large choral work he could now arrange its performance at the North Staffordshire Festival in October 1896—only weeks after the Worcester Festival for which *Lux Christi* was promised. Edward accepted the challenge of writing two big works within little more than a year.

For the North Staffordshire subject, he returned to the King Olaf Saga. Through the autumn of 1895 many thematic motives were shaped toward a continuous design. The locus of the new music was once again G, now in the minor. The opening figure was the old step-wise descent, transfigured in a slow unfolding dignity to represent 'Sagas'[115]:

The orchestra would sound the descending scale, and the chorus replied in soft octaves just opening to bare harmonies over chanted Gs. Out of it emerged a first development—another descent with an upward leap—to stand for 'Legends':

It was introduced by a solo bass and extended slowly, majestically through the chorus. Then a triplet-and-triad variant

rounded out this opening evocation of misty northern strength. Its suggestion of power held in reserve had no parallel in Edward's previous music.

The thunder god Thor was summoned by a new variant:

He was evoked in a canonic development with further triplets:

---

[115] The nomenclature of the *King Olaf* motives follows that of the *Analytical Notes* written by Joseph Bennett with Elgar's silent collaboration (Novello, 1899).

And suddenly out of this sprang up a chorus of prodigious invention and resource. Muttering at first over a B flat minor *ostinato*

it erupted in a major triadic variant through the men's voices:

etc.

A new variant sounded Thor's Hammer:

A reprise of the *ostinato* chant redoubled the chorus's opening energy with a Fire figure crackling above. Then a reference to the ruler of the gods brought chromatic descent reinforced in iron rhythm:

Briefly recognizing the opposition of Christ over an *idée fixe* of the Hammer-motive, the chorus roared final defiance in a huge, brief B flat minor return. It dwarfed everything in Edward's previous music or in any English music of the time.

    The descending motives of pagan forces were answered by rising figures to evoke the Christian masculinity of King Olaf. These were welded into an heroic tenor solo to answer 'The Challenge of Thor' at even length. First came a vigorous, airy 'Sailing' motive:

This generated several ideas. There was a chordal motive to embody Olaf's heroism:

Out of 'Heroism' would arise the later blowing of Olaf's bugle

(which had a precursor in *The Black Knight* when 'Drums and trumpets' sounded:

)

When it came to a motive for Olaf's personal beauty, the dominant fifth of 'Heroism' initiated a new ghost of the 'tune from Broadheath':

Where the child's tune could only repeat its descent, however, the *King Olaf* motive would be answered with a climbing figure to denote the hero's Christianity:

Near the end of this heroic solo, the romantic dream found quieter and stronger expression in a motive representing the hero's mother:

Thus Edward's new hero moved from energy to reflection, just as the Organ Sonata first-movement exposition had done. It made an arresting answer to the challenge of heroism.

Edward showed his sketches to Swinnerton Heap, who thereupon proposed the production of *King Olaf* to his North Staffordshire Festival committee. The Committee pointed to several other works deserving attention, said that they had not heard of Edward Elgar, and recalled that Grieg had already written a series of choral and orchestral scenes on the Olaf Saga. 'Yes,' said Swinnerton Heap, 'but the composer who in years to come will stand head and shoulders above Grieg is Edward Elgar.'[116] And he won the day. Edward's *King Olaf* would be produced at Hanley in the following October, though the committee could not afford any composer's fee.

The new music drew enthusiastic response from closer friends. On 12 October Edward brought Richard Arnold back from the golf links and played the *Olaf* sketches for him. Arnold was staggered, as his wife wrote to Alice that evening:

My dear Mrs. Elgar,
Dick has just come in & told me the deeply interesting news that Mr. Elgar has been asked to write for the Staffordshire Festival, & I cannot help writing off at once to tell you how *truly* delighted I am. It will give the world a chance of hearing that there is great musical genius *yet* to be found—Dick has come home perfectly *possessed* by Mr. Elgar & his wonderful cleverness, (that is hardly the word)—He says Schumann was a babe compared to him, & can think & talk of *nothing* else . . .
                                                                    Affect[ionat]ely yrs.
                                                                    Ella C. Arnold.[117]

Soon the *Olaf* music was overflowing into Edward's violin lessons. Some of his pupils at the Worcester Girls' High School lived with neighbouring families, and sometimes their lessons came after school hours. Two pupils had their rooms with Dr Crowe's family in Foregate Street. Dr Crowe's daughter Norah, also a pupil at the High School, was curious about her friends' teacher:

I'd heard a lot about him, you know: he was just beginning to get known in Worcester . . .
He rang the bell absolutely at the moment that he was expected, and a parlourmaid let him in and escorted him up our very beautiful old staircase to the upstairs drawing room where the upright piano and his two pupils awaited him . . .
I crept in: there were deep window-seats in the drawing room looking over the main street, and I know I seated myself by the curtain, and I was present at one lesson at least . . .
He put his violin in its case on the top of the piano, and struck a note, and he said, 'Now we'll have scales.' He was very quiet, very very shy and diffident the way he went, and as he got more at home he accompanied them a little bit on the piano. And he ended up with what I was told afterwards was *King Olaf*.
[The pupils] stopped, and they asked him what he was playing, and he told them that

116 Burley, *Edward Elgar: the record of a friendship*, p. 90.
117 Elgar Birthplace.

he was composing a choral work which he hoped to dedicate to the King of Norway. He was just—you know—fingering the notes while they were doing their scales.[118]

Edward regularly extemporized on themes in his study at home—where the presence of sympathetic listeners often helped to stimulate new variants and developments. Now he had found a way of turning even the hated time of teaching to the same end—by extemporizing on the themes in his mind within the framework of his pupils' scale-playing. It showed the quickness of his thinking. And if the pupils stopped their scales to listen to the teacher's new music, that was better still. For then there was no more lesson, but only young enthusiasm to reassure him of the value of what he was creating.

At The Mount, Miss Burley was fascinated by the process of his mind moving from smaller toward larger patterns:

He did not begin at page one and write the work in the order in which it would be played. He was more likely to start with a finale and build up each section in reverse order until he reached the opening. But more generally he wrote as fancy and the inspiration of this or that passage of the libretto dictated.

He would now begin to be comparatively happy. His pupils—always the principal sufferers during the period of gestation—began to retire into the grateful shade of his disregard, and he would excitedly play me thematic scraps and chord progressions with which he proposed to illustrate points in the text of the cantata. 'What do you think of this for Gudrun?' he would say, or 'This is the Challenge of Thor.'[119]

On 4 December Edward and Alice went up to Birmingham for the rehearsal of *The Black Knight* with Swinnerton Heap's Festival Choral Society. Before the performance next day, they made a visit to the Oratory at Edgbaston founded by Cardinal Newman: there Newman had written 'The Dream of Gerontius', and there he had died at a great age in 1890. So the day of Edward's first large choral performance in Birmingham brought him to the place intimately associated with a poem that was already 'soaking in my mind' for setting to music. But the whole matter of death had still to find its place in his thinking. *King Olaf* would be his first large work to show the hero's death.

Following the performance of *The Black Knight* that evening, *The Birmingham Daily Gazette* hailed the work with an enthusiasm worthy of Swinnerton Heap himself:

The 'Black Knight' is no merely ingenious vamping-up of stale and worn-out platitudes. From first to last the work bears the impress of strong and original thought. There is little or none of the quality known as elegance, but in its place is a rugged power combined with a richness of imagination and a fertility of invention which remind us of Richard Wagner or Thomas Carlyle. Without being affectedly eccentric, the themes are novel

---

[118] Conversation with the writer, 2 Feb. 1977, recorded for the BBC. The two pupils were Mabel Ward and Mary Douglas: both married early and did not pursue their music. Miss Crowe celebrated her hundredth birthday before she died in 1980.

[119] *Edward Elgar: the record of a friendship*, p. 89.

and striking, their development masterly, their harmonic treatment and orchestral colouring of a great and noble type, as well as modern in the extreme sense of the term.

The writer was Robert Buckley, an enthusiastic young critic who had known Edward from the days of Stockley's Orchestra.

Before returning home, Edward and Alice went on to nearby Wolverhampton to visit Minnie Baker. In the summer 'The Mascotte' had married the Revd Alfred Penny, Rector of the Parish Church. The old house at Hasfield had been given up, and many Covertside furnishings found new places in the Wolverhampton rectory. Mr Penny was a widower, and his twenty-one-year-old daughter Dora was at the station with her stepmother to meet them. When she was nervous Dora had a decided stammer.

The train came in and, of course, not having seen one another for an age, the two friends fell upon each other and Mr. Elgar was left for me to look after. I quickly found out that music was the last thing he wanted to talk about. I think we talked about football . . .

He came into the drawing-room before luncheon: 'Hullo, there's the black piano! Let's see how its inside has stood the move.'

Although I had not left school very long I had heard a number of good pianists, but I had never heard anything quite like this. He didn't play like a pianist, he almost seemed to play like a whole orchestra. It sounded full without being loud and he contrived to make you hear other instruments joining in. It fascinated me . . .

After luncheon that first day we all went to the drawing-room for coffee and he took hold of a high-backed wooden chair to bring it forward—and its back came off.

'Here's another old friend and its back still comes off. Why don't you mend it?'

I said it was a job which got put off to another day.

'Well, this is the day. Got any tools?'

So, after coffee, bearing with us the chair, we departed to my sitting-room and started on it.

'Now clearly understand,' he said, 'if this is a success *I* mended it; if it's a failure *you* did it.'

That, I think, sealed our friendship.[120]

Back at Forli, he returned to *King Olaf*. Even Longfellow's version (which formed the basis of the libretto) abounded in adventures and characters. Some abridgement was needed, and perhaps some reshaping. One day in November Alice had gone to tea with their neighbours the Acworths, and Edward looked in afterwards to fetch her home. The difficulty with the libretto was mentioned, and Harry Acworth (who had some experience of collecting and editing ballads during a career in India) offered his help.

Soon they were hard at it. First they cut out a number of stories altogether: many of the excisions showed Olaf's brutality, which did not suit Edward's

---

[120] Powell, *Edward Elgar: memories of a variation*, pp. 2–3. The Elgars had met a 'Miss Penney' in Bavaria in August 1893, but I can find no evidence that she was connected with the Wolverhampton Pennys.

ideal hero. Other stories were revised and rewritten to emphasize the more visionary, lyric, and feminine elements of the hero.

Acworth's verses, labouring under Edward's evolving and sometimes contradictory instructions, had none of Longfellow's fitful inspiration. They occasioned Edward no qualms: despite his love of literature, he clearly regarded the collaboration with Acworth as a mere vehicle for his music. So little did the needs of drama move him that he cast each of his solo singers in two overlapping roles, explained in a score note:

In the following Scenes it is intended that the performers should be looked upon as a gathering of skalds (bards); all, in turn, take part in the narration of the Saga and occasionally, at the more dramatic points, personify for the moment some important character.

It was a comparable half-drama to that which shaped *The Black Knight* in Edward's description: 'where the "picture" is fixable for a little time the words are repeated—in dramatic parts the words "go on".'

Dramatic conflict was reduced again in the stories which Acworth rewrote for Edward's libretto. The first of these was the hero's encounter with Iron-Beard, the leader of the old pagans. Longfellow had made much of Iron-Beard's earthy vulgarity; Acworth suppressed all that, raising Iron-Beard to a nobility which nearly matched Olaf's own. It focused their conflict on the spiritual issue. As for Olaf himself, where Longfellow had made clear his savagery in murdering Iron-Beard and forcing the pagans to choose between baptism and the sword, Acworth muted the killing and followed it with a vision of the Cross for all to see. The entire scene was renamed 'The Conversion'. So it anticipated another project with conversion at its centre which lay in Edward's future—the Apostles of Christ. 'The Conversion' in *King Olaf* was to make the longest scene.

Musically 'The Conversion' was divided in three parts. An opening section contrasted opposing motives for Olaf and Iron-Beard, based on rising and falling figures. A second section of quicker movement began with the blowing of Olaf's bugle and went on to the battle. Forced at last to show a conflict, Edward's music dealt with it as indirectly as possible: here, as in *The Black Knight*, the fight was set as brief choral description observed at a distance. The third and final section returned to the opening slow tempo to show the Conversion itself fulfilling the scene's opening motives in a big majestic coda. Once again Edward's inspiration was the flowering of lonely vision in an adverse world.

He worked at 'The Conversion' through the final days of 1895. On New Year's Eve Alice noted in the diary: 'E. writing. E. & A. to call at the Acworths after tea. Many beautiful things—Deo Gratias.' Among the year's blessings, side by side with *Olaf*, there was now an official invitation to write *Lux Christi* as a short oratorio for the Worcester Festival. On 9 January 1896 Edward sent a letter about both works to Novellos and asked for an interview. He went to

London a few days later—to be told that the firm must see substantial portions
of both scores before making commitment.

He returned to *Olaf* through the January holiday from teaching. Next came
the tale of Olaf's political marriage with Iron-Beard's daughter Gudrun—who
then tried to kill her husband on their wedding night. The burden of drama
through the short scene was given to solo singers. They again revealed their
composer's lack of dramatic impulse: his music for this scene, despite some
acute motivic references, was on a distinctly lower level of inspiration.

Next to 'Gudrun' was placed another story which could be treated in a purely
choral way, and Edward's music responded instantly. 'The Wraith of Odin'
made a choral ballad. A jolly G minor tune developed from the Olaf music
showed the men feasting and drinking. Across it came the chromatic Odin
motive for the entrance of the Wraith. Both were laced with a tuneless chant of
refrain: 'Dead rides Sir Morten of Fogelsang'. The ensuing conversation and
story-telling were all given to the choir. Opposing motives mixed and mingled
as Olaf's company fell asleep when the Wraith departed. But where Olaf
decided that the Wraith's appearance must mean the death of the old gods,
Edward set his hero's Christian insistence amid sinister Thor and Odin motives
without losing for an instant the ongoing energy of the ballad.

Then came adventures with two further women. Sigrid, the old Queen of
Svithiod, would marry Olaf but refused to give up her paganism. Longfellow
had shown Olaf offering Sigrid a ring of false gold, dealing her a cruel physical
blow, and crying forth her age and ugliness. Acworth cut out the false ring,
leaving only the religious contest. So Olaf emerged like Paul Flemming in
Hyperion—ready to sacrifice a woman's love for his spiritual ideal.

Again Edward's music softened the drama. Sigrid's motive descended once
more from the old gods, but its rapid syncopation obscured the emotional
expression:

(The instrumental origin of this idea is emphasized in its close relation to the
Organ Sonata, p. 191, ex. 2.) The Sigrid music was given largely to the women's
chorus, and Sigrid's quarrel with Olaf was another largely choral description
from a distance. Sigrid achieved real character only at the end, where she
vowed vengeance in a figure which would come to stand for Olaf's death:

The third woman of the libretto was Thyri. She had been married to another
king against her will, and escaped to find solace with Olaf. Longfellow had
written a gossiping ballad to report her chequered career. Edward set it as

another purely choral section—a gay antiphonal up and down, but ending always in a refrain of the old descending steps:

Hoist up your sails of silk, And flee, flee a - way from each o - ther,

Then Longfellow had shown Olaf bringing the fruits of summer fields, but Thyri angrily insisting instead that he reconquer her dowry-lands from her former husband—and the conquest duly followed. Acworth softened all this, uneasily equating Olaf's summer fruits with his new wife's lands. The pastoral theme drew fresh invention from Edward, but again the writing for solo voices was unsatisfactory. At the conclusion of their duet the two voices sang parallel octaves for nearly 30 bars. It was a familiar operatic device: but in Edward's instrumental vein the vocal octaves were too wooden to express a dynamic of mutual love.

Almost the last scene showed Olaf's death in a sea battle fought against the forces of the vengeful Sigrid. Acworth rewrote this for two purposes. One was to convert another conflict into choral description. The other was to alter Longfellow's lines so that Edward could fit them to his old unpublished setting of Alice's first *Mill-wheel Song*. In fact Edward took his rough *Mill-wheel Song* manuscript right into the *King Olaf* assemblage[121]—where the rising song theme perfectly extended Sigrid's vengeance. So again an old inspiration served him, and behind that inspiration again stood Alice.

He used the rest of the Death scene as a huge coda to revive all the leading motives at appropriate references in the text—Olaf's 'Sailing' and 'Heroism' and his 'Bugle' gradually overwhelmed by the music of Sigrid and the old gods until the hero's passing beneath the waves was signalled in the tolling of a single bell.

There remained Longfellow's final picture of the hero's mother surviving to pray for acceptance of her son's vision. She was answered by a mystic choral voice, which likened lonely human vision to the inspiration of the countryside that Edward shared with his own mother:

> As torrents in summer,
> Half dried in their channels,
> Suddenly rise, though the
> Sky is still cloudless,
> For rain has been falling
> Far off at their fountains;
>
> So hearts that are fainting
> Grow full to o'erflowing,
> And they that behold it
> Marvel and know not

[121] BL Add. MS 57995 fos. 67–72.

That God at their fountains
Far off has been raining!

The sketch for Edward's setting of this Epilogue was dated 9 February 1896.[122]
For it he had saved a final variant, unique in its unaccompanied choral
simplicity:

It was a musical fulfilment as the last variants in the *Bavarian Highlands* and the
Organ Sonata had been fulfilments.

At last, as in the Introduction, the G minor 'Saga' and 'Legends' motives
were interwoven to complete the musical circle with Longfellow's final words
about his Musician's achievement:

> A strain of music closed the tale,
> A low, monotonous funeral wail,
> That with its cadence, wild and sweet,
> Made the long Saga more complete.

The revival of the opening music suggested perhaps that Edward's inspiration
had been greater at the outset than at the end. But the structure of the whole
was strong and accomplished:

	metre	key
Introduction	4/4	Gm
Chorus I: Challenge of Thor	3/4	Bᵇm
Tenor: King Olaf heard the cry	3/4 frame	Bᵇm
Chorus—drama—chorus: Conversion	3/4 frame	C
Women I: Gudrun	9/8 basis	Bᵇ
Choral ballad II: Wraith of Odin	2/4	Gm
Women II: Sigrid	9/8, 4/4	Bm
Choral ballad III: A little bird	3/4	G
Women III: Thyri	6/8 frame	C
Chorus: Death of Olaf	12/8, 4/4	C
Epilogue	4/4 frame	G; Gm

(episode I: Chorus I through Women I, 'relative' to Gm — men; episode II: Choral ballad II through Epilogue, 'absolute' to Gm — women)

On 21 February 1896 this first really big work, to occupy more than 90 minutes
in performance, stood complete in vocal score.

* * *

[122] Ibid., fo. 92.

He turned instantly to *Lux Christi*. Early in February he had spent several days with Capel-Cure settling the final form of the libretto. Edward's latest hero was another young man, this time affected with blindness. So he could appeal to Edward's old eye trouble. Since the marriage with Alice, eye trouble had beset him just often enough to suggest that he might seek reliance on the insight of others where his own should not suffice. But when the Blind Man of this libretto went to wash his eyes in the pool of Siloam, there was a ghost of the child who found his own way to the river which promised him vision.

The Blind Man has no wife, yet once again the hero's mother would play a prominent role. This mother recalls her son's conception in sin, yet she cannot believe that such a circumstance should be punished by blindness. She puts her case to Christ, who agrees with her. So the mother initiates her son's healing, his inspiration, and his ultimate apostlehood. The likeness to Edward's mother, and to the mother of King Olaf, was manifest.

Yet the Blind Man's recovery of sight was only half the libretto's story. After the miracle is performed and the apostle self-discovered, he finds that his inspiration has isolated him from his fellow men—who were quite content to have him blind but find it less convenient to cope with a man of sight and insight. The price of vision, then, is the sacrifice of any chance for practical influence in the world. The saga of the outsider, begun in the Black Knight's victory and continued in King Olaf's ambiguous battles, here found another chapter.

As in *King Olaf*, here again the mother prays for her son's failed vision; but where in *Olaf* the mother's prayer virtually finished the work, now it opened the action. Here also was a conflict of old and new gods. But the issues were more complex: and in place of the simple *Olaf* opposition of upward and downward motives was a development of contrasts less clear. The *Lux Christi* themes breathed the diatonic innocence of Edward's earlier music. But they had a slow-paced mastery that seemed to grow out of a nostalgic insight.

The main themes of *Lux Christi* were exposed in an opening orchestral 'Meditation' which encapsulated the drama itself. The music began with a motive from which all else would spring. It represented the Levites' Psalm 'O Give Thanks unto the Lord'. Once again Edward's opening theme incorporated a descending triplet:

Out of it came a motive for the Blind Man's prayer:

Then the descending steps of the melodic 'answer' shaped a lyric theme of
'Blindness'

and a more forceful 'Darkness'

Here again was a ghost of the dying 'Delight'. Finally two answering motives
rose. One stood for 'Christ the Healer and Consoler':

The other was a superb invention to evoke Light itself:

Yet this could be heard as only a development of *The Black Knight*:

Without a break the music went into the opening chorus of Levites within the
Temple Courts (4/4 in G major) contrasted with the solitary voice of the Blind
Man outside (3/4 in G minor). A contralto narrator's recitative then introduced
the high-reaching figure Edward had written in the Bavarian Alps the previous
summer: now it described Christ passing by, and throughout *Lux Christi* would
be used to signify narrative progress. The chorus of Christ's Apostles asked
'Who did sin?' over and over in a figure of three descending steps. Then the
Mother made her appeal in an aria which developed the opening Chant-motive
toward the culminating Light:

Be not ex – treme, O Lord ___

And Christ answered with a rising variant of his own:

Nei-ther hath this man sinned, nor his pa - rents,

The full chorus entered (female choral voices now sounding for the first time in the work) singing the Light-motive in a big *Allegro* of irresistible impulse, to be superbly scored. Yet the joy was not sustained, nor the certainty: in a fugato over the words 'Within the shadow of Thy Cross Now burns a light', the musical figure of the narrative recitative shrank to

Thou hast borne___ The sin - ners sen - tence

At the end of this chorus came the first pause in the music. It marked a division one-third the way through—the ending of an exposition.

The central section began the healing process. It was approached in a canonic duet for women, 'Doubt not Thy Father's care', set over the descending steps of Blindness. There ensued a fascinating choral contrast of the sexes: where the women had faith, the men questioned. It could have stood for Edward's own marriage—as he recalled long afterwards, for a friend, the atmosphere with which Alice surrounded his composing of *King Olaf* and *Lux Christi*:

It seems strange that the strong (it is *that*) characteristic stuff shd. have been conceived & written (by a poor wretch teaching all day) with a splitting headache after dinner & at odd, sustained moments . . . You, who like some of my work, must thank *her* for all of it, not me. *I* shd. have destroyed it all & joined Job's wife in the congenial task of cursing God.[123]

At the end of this chorus men and women came together in a vocal counterpoint spreading slowly through 'And the eyes of the blind shall see' into eight parts. Not all the mastery of *Lux Christi* was in its orchestra.

At the centre of the central section the Man who was Blind described the coming of sight. Yet curiously, the theme of his aria was again the descending motive of Blindness:

---

[123] 30 Dec. 1922 to Ivor Atkins.

The remainder of the central section mirrored its opening. The men and women fought another choral battle about the significance of what had happened: the men shouted over vicious orchestral descents, the women developed the rising hopes introduced earlier by the Mother and Christ. At last the contralto narrator found a new variant of rueful rest for the words 'Make a silence in my soul, Where only Thy true voice shall sound.' It brought the end of the second section and another pause in the music.

The final section retraced the pattern of appearances which had opened the work, but each element registered a change. Now the Mother's faith was beset by the men shouting something very like the Heroic motive of *King Olaf* but caged inside one tonality. So they cast out the son to whom sight had been given. Then the Mother (in place of the pleading solo she had sung at this point in the opening part) led the women in an angry 'Woe to the shepherds of the flock'. Again her character showed the greatest musical definition of any personage in the work. And Christ replying, where formerly he had raised the Mother's hopes, now comforted the son in still another rising figure for the vision which had cost him his place in the world: in the midst of Christ's comfort, the orchestral accompaniment returned to the figure which had set 'Thou hast borne the sinner's sentence' in the opening section of the work.

So *Lux Christi* went toward its ambiguous end. Unlike *King Olaf*, this story showed only a fragment of its hero's life. The concluding chorus, instead of a grand review of earlier motives as in *Olaf*, pursued fresh ideas. The doom-laden chromatics of the *Olaf* score were largely absent, as they had been absent everywhere in *Lux Christi*: where heroism had reaped its own destruction in the world, the rewards of grace were indeterminate. Yet again there was symmetry in the entire design:

		key
1.	'Meditation.	Gm, E^b, G
2.	Men (Levites) vs Blind Man	Gm vs G
3.	(Alto recit); Men (Disciples)	Fm (A^b)
4.	Mother	Fm→A^b, Fm→F
5.	Christ	G
6.	Full chorus	G + A^b
7.	Alto recit & Christ	
8.	Women	
9.	(Alto recit); Men vs women	Dm
		Cm→C, coda
10.	Man who was Blind	*divisi* in G
		E^b (Dm)
11.	(Alto recit); Men vs women	Dm vs B^b
12.	Alto	F
13.	Alto recit & Mother; Men (Levites) vs Man who was Blind	Bb, Cm, Dm, D
14.	Mother & women	Dm
15.	(Alto recit); Christ	B, E^b
16.	Full Chorus	G

Despite all the musical recapitulations and symmetries of *Lux Christi*, the Blind Man was the hero of change in a way King Olaf had never been. When the waters closed over Olaf's head, the world was still what it had been before he came into it. But the Blind Man's vision had changed the world permanently—at least for himself.

The question of whether or not experience changes the individual is as old as the notion of human individuality itself. The answer for most of us, whether pro or con, is determined in early life. A man who shows the will to frame both sides of the question in alternative major essays on the nature of heroism reveals a remarkable ability and a remarkable need. Moreover, the opposition of *King Olaf* and *Lux Christi* repeated the opposition of the recapitulating *Froissart* and the evolving *Black Knight*.

More than for most composers, perhaps, Edward's music was his spiritual autobiography. His personal insecurity, his constant need to show his musical thinking to others and secure their approval, his extreme sensitivity to his surroundings—all suggested an artist in uncertain touch with his own philosophy. All his heroes inhabited a world of shadowy, knightly aspiration. But the sequence of heroes from *Froissart* to *Lux Christi* recognized more and more clearly the remoteness of this aspiration from the ways of the world. Whether the stories of these heroes came back to the points of their beginnings or went forward to occupy new ground, the heroes of Edward's major works found themselves more and more clearly outsiders.

*   *   *

The two works were written with astonishing dexterity. The vocal score of *King Olaf* was virtually completed through a full schedule of teaching within four months. On the day after finishing it, 22 February 1896, Edward had turned to *Lux Christi*. Then a week had to be given to work for smaller publishers: the German firm of Breitkopf & Härtel had agreed to publish the Organ Sonata and sent proofs for correction, and Edward quickly orchestrated the *Bavarian Highlands* songs to be published by the small house of Joseph Williams in time for a première in the spring. By early March he was back with the composition of *Lux Christi*.

Later in the month he revised *King Olaf*—before taking both scores to Novello's music editor Berthold Tours on 31 March. Tours asked him to remove passages of music linking the *Olaf* scenes, so that individual numbers might afterwards be printed separately. And the editor also wanted some excisions to reduce the length. He could not be so specific about *Lux Christi*, for it was not yet finished. If Edward met these difficulties, the firm would still demur about publishing without a subvention. At last it was agreed: £40 must come out of the small royalty to be paid on the sale of *Lux Christi* vocal scores and £60 out of *King Olaf* royalties: if the royalties did not meet these sums within three years, the composer must pay the difference. Alice guaranteed the money, and the bargain was struck.

Back at Forli, Edward finished the vocal score of *Lux Christi* on 6 April—six weeks after beginning it—and Alice posted it to the publishers the same day. For performance in an English Cathedral, Novellos suggested the title *The Light of Life*, and so it was agreed. Two days later a final version of the *King Olaf* vocal score followed, cut down by some thirty pages.

But Edward's clear annoyance over the changes soon caused Novellos to transfer his affairs to a younger editor. He was August Johannes Jaeger, three years younger than Edward, born in Germany but settled in London for many years. Jaeger's enthusiasm for the music of his adopted land and his warm-hearted encouragement of Edward soon opened a new era in relations with the firm. It was Jaeger who saw both the new works through the press.

In Worcester Edward conducted the Festival Choral Society in *Scenes from the Bavarian Highlands* for a successful but very local première on 21 April 1896. Next day he received vocal score proofs for *The Light of Life*. He corrected them at speed, and by 5 May commenced the orchestration. To save time, Alice set out the framework of the score. A friend recalled:

She made herself a good copyist, since an amanuensis would have been too expensive a luxury. She 'laid out' his scores, copied in the voice parts, planned the barring—all this for several [hundreds] of pages of 40-line score . . . She would ask overnight what size his orchestra was to be, and . . . would say, 'With a bass clarinet, I suppose?' or 'Aren't you going to have a *cor anglais*?'; and he would come down next morning to find as much of the form ready as he could fill in during the day with orchestral parts.[124]

His orchestrating went forward rapidly as always, for the instrumental sounds had been clear in his mind from the moment of the music's invention.

The seventh anniversary of his wedding with Alice found him busy as he had never been busy before. With the arrival of the fine summer weather, he decided to carry on the work out of doors. He ordered a tent, set it up on the lawn, and carried on the orchestration inside. A flag raised above the tent denoted that he was not to be interrupted. The summer term's teaching hardly impeded progress with *The Light of Life* score, which was posted off to Novellos section by section as it was finished. On 20 June the 250-page orchestral score was complete. Two days later he began to score *King Olaf*.

When the Birmingham journalist Robert Buckley came to interview Edward on 31 July, he was given a warm welcome.

It was the riotous summer. The hedge of the lawn before the house was in flower, and the wicket opened amid poetic blooms. Close at hand was a larger lawn, a pleasaunce of sloping banks and smooth-shaven turf, whereon was a sunny tent, the opening of which commanded a glorious valley, extending to the purple horizon.

'Forty miles and never a brick!' ejaculated mine host, as we took our seats in this ideal retreat, where were easy chairs, a table and a couch which reminded me of Rossini dashing off operas in bed. There, too, was a proof copy of *Lux Christi*, afterwards called *The Light of Life*, concerning which we held sweet converse together. The Worcester Festival was due in a few months, and the composer felt that much depended on the success of this, his first choral work to be heard at an important meeting.

Overflowing with enthusiasm, he spoke rapidly and continuously of the state of musical art in England, deploring the fate of works commissioned for festivals, which, after painstaking and elaborate production, were heard no more. His bearing was that of one in deadly earnest, not wholly inaccessible to the jocular, but too intent on his aim to waste time on anything not directly leading to the goal. He laughed but rarely, and his mirth was soon checked. In the heat of the early struggle, and with the winning-post in sight, his mind seemed occupied with a fixed resolve to make the world aware of the power he believed to be his own.

King Olaf was in hand, and the tent was littered with sheets of music-paper bearing myriad pencil marks, undecipherable to the stranger as the hieroglyphics on a blackbird's egg, and, like the proverbial lost pocket-book, of no use to any one but the owner . . .

Of a fugue in *The Light of Life* he said: 'I thought a fugue would be expected of me. The British public would hardly tolerate oratorio without a fugue. So I tried to give them one. Not a 'barn-door' fugue, but one with an independent accompaniment. There's a

---

[124] Obituary article [by A. H. Fox-Strangways] in *The Times*, 8 Apr. 1920.

bit of a canon, too, and in short, I hope there's enough counterpoint to give the real British religious respectability!'. . .

He would have enjoyed working on opera, but wanted both subject and libretto . . .

[Going into the house] the composer revealed himself as a book enthusiast, a haunter of the remoter shelves of the second-hand shops, with a leaning to the rich and rare. In the sitting-room was a grand piano, in the study a smaller instrument, surrounded by books, and books, and more books.

. . . Other features of the study spoke his many-sidedness. A large portrait of Wagner was conspicuous, and a board over the fireplace displayed in poker-work an ascending flash of chromatic semi-quavers. 'The Fire-motive', he said, 'from *The Ring of the Nibelungen*; one of my own attempts at decoration.' A cosy room, with quaint bric-a-brac from foreign lands; bits of carving from the Bavarian Highlands, then his annual summer resort.

He showed the silver buttons of his waistcoat as specimens of Bavarian handicraft, described the character of the people, and pointing to the score of *The Light of Life* said he wrote the beginning of number three recitative and chorus, 'As Jesus passed by', six thousand feet above the sea-level. 'It has at least that claim to be called high art,' he remarked airily.

Tacked lightly to the wall was an uproarious illustrated joke cut from a German newspaper, and in a dim corner a photograph of a thirteenth century panel sculpture of the Crucifixion from Worcester Cathedral. 'It shows a wonderful feeling,' he remarked, as he looked upon it lovingly.

Presently he spoke of recreations, and declared a liking for golf, remarking that if not of the first force he was certainly animated by the best intentions. He was for some time a follower of the American craze for kite-flying, with its aerial photography and its scientific aims, desiring to invent a compensating kite that should adapt itself to whatsoever currents it might meet in its celestial course . . . Nothing came of it except the fall of his neighbour's spouting, and the occasional employment of a powerful navvy to pull down the rebellious thing from the central blue.[125]

During the first fortnight of August he corrected orchestral parts for *The Light of Life* side by side with ongoing orchestration of *King Olaf*: by the 16th he was at the love-duet with Thyri. Next day the Mascotte came over from Wolverhampton for the afternoon with her stepdaughter Dora, who remembered:

There was a fine view of the hills from the front of the house, and the North Hill stood up like a huge hump and seemed a good deal closer than it really was . . It was a hot day and on the lawn in front of the house was a small bell-tent. E.E., in his shirt-sleeves, was writing at a little table.

'You can't come in here—it's private.'

Hot and stuffy too, I thought, but he seemed to like it.

After luncheon he suggested a walk and we spent the afternoon on the North Hill. How lovely it was up there! The wonderful air and the view—I had never been to Malvern before. He pointed out various places and landmarks and I said admiringly:

'You're as good as a map!'

[125] *Sir Edward Elgar*, pp. 29–36.

'Better,' he said. 'We'll do the Worcestershire Beacon next time you come, only you must stay, not flit like this.'[126]

A week later, with the *Olaf* orchestration finished, Edward journeyed to Leeds to rehearse *The Light of Life* with the Leeds Choral Union, one of the choirs providing a contingent for the Worcester Festival. He met the Choral Union's patron and secretary Henry Embleton—one of the great enthusiasts for choral music in the north, and with a big industrial fortune to back his enthusiasm. And there were two days at Settle, renewing memories with Dr Buck, before returning for the Worcester Festival orchestral rehearsals at the Queen's Hall in London.

Arriving with Alice on 3 September for the *Light of Life* rehearsal, they found a large group of interested musicians including his father's old acquaintance Sir Walter Parratt (once organist at Witley Court in Worcestershire, now Master of the Queen's Music at Windsor), Charles Harford Lloyd (Alice's old teacher, formerly organist of Gloucester Cathedral and now Precentor at Eton), and Alberto Randegger (conductor of the Queen's Hall concerts and director of the Norwich Festival). The solo singers included Edward Lloyd to sing the Blind Man and Anna Williams for the role of the Mother. All was enthusiasm, as the press reported of the new score before its première:

Mr. Edward Lloyd says it is one of the finest English works composed for some time, and that the instrumentation is particularly fine. The tenor part is the finest he has had presented to him for many years . . . Signor Randegger, after hearing the first rehearsal in London, said he thought it was the best English work that had been produced within his knowledge for certainly twenty years.[127]

On Monday 7 September came the final rehearsal at Worcester, and on Tuesday evening the performance. The London press were present. *The Sunday Times* summarized:

Seldom does the dip into the 'local art' lottery yield a prize so conspicuously promising as Mr. Edward Elgar. Here is a musician of whom Worcester has perfect reason to be proud, and the place accorded his short oratorio, 'The Light of Life', in Tuesday evening's programme, was eminently justified by the critical verdict of the following day.

The young Malvern teacher has uncommon talent. He knows his Wagner well—sometimes, perhaps, a trifle too well and he has turned his experience as an orchestral player to good account; hence the marked superiority of his scoring as compared with his vocal writing—his choruses as compared with his solos. But his sense of proportion and tone colour, and his knowledge of effect are quite exceptional . . .

He might have done more with his representative themes, but otherwise his handling of the various situations betrays little of the novice or paucity of bold ideas. The best number in the work is the chorus 'Light out of darkness' [No. 6], and this is of such

126 Powell, *Edward Elgar: memories of a variation*, p. 6.
127 *Worcester Daily Times*, 7 Sept. 1896.

excellence that I cannot help looking to Mr. Elgar for a really fine work when he comes across a 'book' which appeals in every sense to his strong artistic temperament. Meanwhile his present achievement will suffice to gain him a ready and grateful hearing in quarters where his name had hitherto been wholly unknown.[128]

A week later the distinguished composer Charles Stanford and his wife, staying in the neighbourhood, called on the Elgars. He and Edward spent several afternoons together: one day they played through Stanford's new *Requiem* (which the composer was endeavouring to have produced at a Three Choirs Festival), and another they spent walking up the north end of the Malvern Hills to a high farming estate called Birchwood. Then the new teaching term began.

Through these days Edward had been correcting orchestral parts for *King Olaf*. On 8 and 17 October he went to Stoke-on-Trent to rehearse the North Staffordshire Festival chorus. On one of those evenings a young man happened to pass the end of the street leading to the Town Hall. He was a twenty-year-old organist named William Havergal Brian, and he wanted to be a composer:

A choral rehearsal was on and I heard music that held me spellbound—it was unlike any known to me.

At that time I was organist of Odd Rode Parish Church in Cheshire. On the following Sunday I told my choirmaster of my experience with this unknown strange music. He said:

'Oh, I was there—it was the chorus of the North Staffordshire Triennial Festival and we were rehearsing a work by a new chap named Elgar—called *King Olaf*. It is strange music and we don't understand it . . . We are not supposed to lend out copies, but I will bring mine for evening service and you can look at it during the sermon.'

I carried that copy home. By the time of the performance I had borrowed it incessantly and knew it backwards. I regarded Elgar as a phenomenon and spent my time gaining converts.[129]

Back in Malvern, Edward met his pupils through the following week. On Monday 27 October, after a morning's teaching at The Mount, he and Alice went up for the première. At Hanley they toured the Minton Porcelain Works, where a special cup was afterwards made to commemorate the production of *King Olaf*. Edward rehearsed the chorus and orchestra over the next two days without the title character, as Edward Lloyd was occupied elsewhere until the day of the final rehearsal.

But even for the last rehearsal Lloyd did not turn up, owing to an error of trains. Edward's agitation turned upon Swinnerton Heap. After all he had done to get *King Olaf* produced, Heap could not keep his wounds to himself, and it became the gossip of The Potteries.[130] Finally Lloyd did arrive amid

---

[128] 15 Sept. 1896.   [129] Quoted in Reginald Nettel, *Ordeal by music* (OUP, 1945), p. 11.
[130] Ibid., p. 22. Nettel suggests that the rift was irreparable, but letters written afterward by Elgar to Heap in 1897 and 1898 (partly quoted in Christie's Catalogue of *Autograph Letters and Manuscripts* sold on 6 Apr. 1977) as well as subsequent Elgar performances conducted by Heap convey a different impression.

apologies and assurances—without rehearsal but in time for the performance. In the audience for *King Olaf* was young Havergal Brian:

Its production was on a cold October morning—I was there with two others whom I had fired with my own enthusiasm . . . The soprano was Medora Henson—Edward Lloyd was at the top of his form and Ffrangcon Davies was a fine looking fellow with a mop of curly hair. Willy Hess was the leader of the orchestra.

The unknown Elgar walked on to conduct his work in a light woolly suit and was obviously fidgety and nervous . . . When Lloyd sang 'And King Olaf heard the cry', something went wrong—subsequently Hess saw that Elgar was losing his grip. Hess jumped to his feet and straightened the thing out by his presence and his bow. He saved *King Olaf*. Elgar admitted it to me years afterwards: 'But for Willy Hess, what a fiasco that performance would have been!'[131]

Many in the audience discerned no difficulty. Dora Penny recalled:

Quite a large party of us—relations and friends—went up for that Hanley Festival and heard *King Olaf*, and I think it was a really fine performance . . . Part of the entry [in my diary] for Friday, 30 October 1896 reads:
Mr. Elgar came and saw us in the interval, Went to Choral Symphony in the evening.
Mr. E. sat with me most of the time.
Anyone who has sat next E.E. during a performance which he was enjoying knows what it is to be thrilled, and also knows what it is to have an arm black and blue with bruises next day. I have done it many times. With practice one learnt to shift imperceptibly now and then so that the position of the grip varied.[132]

Again it was left for Alice to reaffirm the priorities:

October 30 1896 . . . Glorious King Olaf a magnificent triumph. D[eo] G[ratias].

The judgement was confirmed in every review. The *Daily Telegraph* critic Joseph Bennett wrote:

Behind all his work lies the power of living talent, the charm of individuality in art, and the pathos of one who, in utter simplicity, pours forth that which he feels constrained to say . . . Almost every number was applauded fervently . . .[133]

One enthusiast wrote that Edward Elgar was 'the greatest English genius since Purcell . . . English music is lifted to the highest place of contemporary art.'
What such a position might mean in private terms, however, was recognized by Harry Acworth the day after the performance:

Sherborne, Malvern Wells.

31 Oct. 1896.

My dear Elgar
Having now seen the two Birmingham papers & most of the London papers I must write a line to offer you my warmest congratulations on a grand and triumphant success. We heard K.Olaf through yesterday but to catch our train had to miss the final plaudits

[131] Ibid., pp. 11–12.
[132] Powell, *Edward Elgar: memories of a variation*, pp. 4–5.
[133] 31 Oct. 1896.

& the double call for you. My wife was quite carried away by the beauty of the music & I should say the same only my musical judgment is worthless . . .

In one respect I am sorry for you. You have now the obligation of living up to a great reputation, a reputation which if I read the critics aright places you at the head of living composers, & high wrought expectations are difficult to satisfy—& such all your work will henceforth evoke.

I really felt for Mrs. Elgar, who sat just in front of me, under the overpowering weight of feelings which such a splendid triumph caused her.

With renewed congratulations

Yrs Sincy
H A Acworth[134]

It was what Edward felt for himself when he reported the triumph to Miss Burley: 'Truly success is harder to bear than adversity.'[135]

His old mother sent a poem from 10 High Street 'to my boy':

> I will not praise as others praise—
> Thou needst it not from me,
> The Genius has won its much,
> And Fame is crowning thee.[136]

But Edward, when he saw her alone, put his head in her lap and said he could not face the exposure.[137]

He addressed a letter of painstaking thanks to Swinnerton Heap, from 'one who would have remained in outer darkness' but for the North Staffordshire commission. Yet now he proposed to re-enter that darkness voluntarily:

My work is done and I feel I have proved myself a man! but I cannot afford to write any more: but I cannot thank you sufficiently for allowing me to appear under your auspices & (if all the rest of my life will only be 'it might have been') I shall always reverence you for the way in which you gave me a helping hand.[138]

It was the first of many such 'farewells' to composition. After a time, with the growth of his fame, no one took them very seriously. But they showed the extent of his dependence on constant encouragement from every quarter—and on Alice's rock-like fortitude.

*   *   *
*   *
*

After the première of *King Olaf*, Edward and Alice had some days in London. One object was to interest August Manns in performing *King Olaf* at the Crystal Palace, and Edward spent considerable time going through the score at

---

[134] HWRO 705:445:3916.   [135] *Edward Elgar: the record of a friendship*, p. 91

[136] 1 Nov. 1896 (HWRO 705:445:1117).

[137] Recalled by Miss Winifred Whitwell in 1975. Miss Whitwell's mother was a close friend of Ann Elgar, who told her the story.

[138] 4 Nov. 1896. (Quoted in Christie's Sale Catalogue of 6 Apr. 1977, p. 41.)

the piano for the old man. At last an offer was forthcoming: *King Olaf* could have a performance at the Crystal Palace, but only with financial guarantees. Novellos would stand some of the guarantee, but not all. On 23 November Edward returned to London, interviewed the firm, and arranged that any extra sum would come out of his own pocket without Alice knowing. It was to be the first performance of a major Elgar work in the environs of London.

With the Crystal Palace performance in prospect and chances of several provincial productions as well, Novellos felt that the value of Elgar's music might be on the rise. They suggested that he consider writing two smaller works to celebrate the forthcoming Diamond Jubilee of Queen Victoria's accession. One was to be an 'Imperial March': it might be played at all sorts of occasions through the Jubilee year. The other was a short cantata of moderate difficulty which they wished to have set to words about St. George and the Dragon then being written for the firm by a Mr Shapcott Wensley of Bristol.

Edward had no experience of writing to this sort of commission. But the moment was right. Writing to someone else's order must reduce the burden of creative decision which had been constantly his through the last year and more. Then there was the prospect of immediate substantial fees. And on top of all else, the occasion of the Jubilee found a specific response in the promptings of his own music.

The celebration of sixty Victorian years would almost certainly be the final great anniversary in a reign which had begun before the majority of its subjects had been born. Many people, both at home and in the Empire, remarked on a general feeling that the best was with them then, or had even begun to pass: the mood was soon to be expressed in Rudyard Kipling's poem 'Recessional'. Edward's upward leaps to downturning figures perfectly met this feeling in his project for an *Imperial March*:

And a second subject turned to gentler reflection:

As for 'St. George', though he had as yet only a fragmentary libretto, the story offered some similarity to *The Black Knight*. Here the old kingdom was invaded by two outsiders—one for evil, the other for good. Evil was vanquished by Good; but then Good renounced the lady he had won, showing the same difficulty in embracing love which all Edward's heroes encountered. A final section of the libretto paid tribute to the heritage of St. George as symbolized in the modern British flag—'It comes from the misty ages'.

Forli, Malvern.

Nov:30:1896

Dear Sirs:

St. George

By this post I send three sketches for the above work & shall be glad to hear if the music is suitable for the purpose.

I have treated the opening as a separate number & in my own style, it requires revising but I think it good—for me.

Following this would be the narrative which I should propose to treat in freër Volkslieder style but I have not the libretto except the descriptive lines 'approach of St. George'

The Epilogue I propose to treat as a separate number (the last portion of the narrative wd. run into it but it wd. be detachable) in a broad march movement: I enclose my first sketch which is effective, but I have not proceeded until I hear from you: the length of this wd of course depend upon the time occupied by the narrative.

I shall be glad to hear your views as early as possible.

If I have the libretto I could finish the whole work early in January.

Believe me
Vy faithfully yours
Edward Elgar:

P.S. . . . If the sketches are not sufficiently clear I would gladly amplify them.[139]

The plan was again symmetrical: a short opening scene of lamentation, to balance the triumphant Epilogue, with the central action and battle framed between them. Once again it was to be a purely choral narrative, with optional soprano solo for the Princess but the battle description and St. George's words sung by the male voices of the chorus. As in *King Olaf* and *The Light of Life*, the malign force (in this case the Dragon) was represented by a descent in steps. St. George's motive was a marching rhythm rising through a triplet. And once more the best melody was saved for last—another descending figure whose clearest contrast with the Dragon-motive was its rousing melody and rhythm:

Novellos liked the *St. George* sketches, sent the rest of the libretto, and offered £50 for the copyright of the completed score. They also liked the *Imperial March* sketch. But their editor registered an early query about Edward's habit of elaborating his ideas in short sequences:

. . . It might be adversely criticized on account of the fact that it contains so many short phrases of two bars & even one bar: & we are of opinion that it would be enormously

---

[139] Novello archives.

improved if you could remodel the march with a view to including in it phrases say sometimes of eight bars.'[140]

Edward did what he could, and sent a revised version on 9 January 1897 with a covering letter:

The March is now concise & effective—it will be particularly so for orchestra: if you should wish any alteration as to the difficulty in the p.f. arrangement I will gladly see if I can meet your wishes.

I shall be glad to hear your views as to the above points & as to terms: I should not have mentioned the last item, but that the Term will shortly recommence & it will depend entirely upon this matter whether I return to teaching or continue to compose—or try to.

Novellos responded with an offer of 20 guineas for the *Imperial March*, together with a letter of encouragement from the firm's chairman, Alfred Henry Littleton. It was not sufficient to consider giving up teaching.

Through January 1897 and the first half of February he worked at *The Banner of St. George*. The libretto offered no real incidents beyond the battle with the Dragon. Edward's unadventurous setting showed how little his inspiration could be drawn by a subject not of his own choosing. He added no distinguished melodic ideas to those already submitted. The Princess and her women were given a theme that echoed the Gudrun-motive in *King Olaf*; the old King seemed to overhear Olaf's Bugle-motive; and they were set in a comfortable diatonic style that showed less of Edward's individuality than any other work of his on a large scale.

But it was clearly designed for broad appeal, and the publishers were happy. Edward devised the masterly orchestration with an eye to economy in performance, as he noted in the score:

The instrumentation of this work has been so arranged by the composer that a small orchestra (String Quintet, 1 Flute, 1 Oboe, 1 Clarinet, 1 Bassoon, 2 Horns, 2 Cornets, 1 Trombone, and Drums) will be effective. These instruments may be supplemented by any or all of the other instruments indicated in the Full Score.

Amid teaching and some depressive bad health, the scoring was completed on 15 March 1897.[141]

The Crystal Palace performance of *King Olaf* took place on 3 April. Edward received a tremendous ovation from the performers. But the hall was far from

[140] 8 Dec. 1896. It is not clear whether this is the opinion of Berthold Tours or the man who succeeded him as music editor early in 1897, John Ebenezer West.

[141] W. H. Reed (*Elgar*, p. 42) states: 'There is a note in one of the diaries to say that *The Banner of St. George* was given by Miss Holland's choir at St. Martin's Hall, Trafalgar Square in London, on 14th March [1895]. (This must have been a preliminary sketch of the work, which was eventually completed and performed in 1897.)' The diaries I have seen (all I know to exist) contain no such reference; but a programme at the Elgar Birthplace shows that there was a performance of *The Black Knight* by these forces on 28 Mar. 1895. In view of the history of *The Banner of St. George* as it emerges from the Novello correspondence, Reed would seem to have been mistaken.

full, and there were heavy expenses to offset. The net loss was £57 19s. of which Novellos had agreed to cover only £21. Edward wrote to the chairman:

Ap 11:97

Private
Dear Mr. Littleton:
   As you spoke to me in the first instance as to contributing to the expense of the C. P. Concert I send the enclosed account—recd. last night—to you.
   I am sorry it has turned out so very badly: if you will kindly let me have the amt. of your contribution I will send it on with my share to the Manager.
   I write to you as I may say that I do not wish to worry my wife with seeing these details of expense—*just now*— & I shd. be much obliged if you wd. send anything in connection with this matter to me 'The Club' Malvern. This I could scarcely put in a business letter.

Believe me
Vy sincerely yours
Edward Elgar.[142]

   But on 19 April the *Imperial March* was given its première at the Crystal Palace in a concert of massed bands. Six days later it was heard in the Queen's Hall. Then it was down for performance at a Royal Garden Party during the week of the Jubilee celebrations in June, and for a State Concert in July. Novellos were also arranging performances of *The Banner of St. George* with smaller choral societies in London and elsewhere.
   Lux Christi was revived at Hereford on 23 April. On the 24th *The Snow* and *Fly, Singing Bird* were given in Worcester. On the 27th, far in the north at Bishop Auckland, County Durham, an amateur conductor named Nicholas Kilburn produced *King Olaf* with the fervour of an Old Testament prophet recognizing the Messiah. Three days after that Swinnerton Heap revived it at Hanley to 'terrific applause'. On 4 May Edward conducted its first performance in Worcester—a 'tremendous success'. Frederic Cowen wrote to say that he was planning performances at Liverpool and Bradford.
   It seemed the fulfilment of much that Alice envisioned, even if some of the expectations remained dark. On Sunday 30 May (responding to the request of a small publisher), Edward copied out the music he had written to a little poem of Alice's:

Closely cling for winds drive fast,
Blossoms perish in the blast,
   Love alone will last.

Closely let me hold thy hand,
Storms are sweeping sea and land,
   Love alone will stand.

Kiss my lips and softly say,
Joy may go and sunlit day,
   Love alone will stay.

[142] Novello archives. The Crystal Palace account is preserved at Worcester (HWRO 705:445:2837).

That afternoon Carice came home from The Mount for an afternoon's visit. (She had been sent there as a boarder: it was only a few minutes' walk from Forli, but safely out of earshot of her father's composing.) She was brought over by Miss Burley—who found the new song, picked it up and sang it. It was three days before Edward's fortieth birthday.

The Hereford organist George Sinclair had decided to ask him to write a new setting of the Anglican *Te Deum and Benedictus* for the Three Choirs Festival Opening Service in September. Edward cast it on a large scale (with orchestra), combining current notions of English Church music with the style of his own early works for the Catholic choir at St. George's: but now the diatonic themes were developed with persistent chromaticism.

On 5 June he took his sketch over to Hereford and stayed the weekend with Sinclair. Sinclair's pupil Percy Hull was in the house:

I was privileged to hear Elgar play over his *Festival Te Deum and Benedictus* in Sinclair's house to see whether the work would be acceptable for the programme of the Festival at Hereford . . . He was as nervous as a kitten and heaved a huge sigh of relief when Sinclair said: 'It is *very very* modern, but I think it will do; you shall play it again after supper when Hull and I will give you our final verdict.' All this in Sinclair's stammering and somewhat patronising fashion.

During the same visit, the three of us took turns in sight-reading Tchaikovsky's *Pathetic Symphony* arranged as a piano duet . . . Musicians were all agog about the 5/4 movement. Elgar tackled the bass part but he soon got into a mess and had to stop. After another attempt he eventually reached the end and remarked: 'The violin is my instrument, not the piano. I can read any old or new rhythmic patterns on the fiddle.'[143]

With Sinclair's approval, he sent the completed vocal score of the *Te Deum and Benedictus* to Novellos, who offered 15 guineas for the copyright.

But Tchaikovsky's 5/4 melody stayed with him, contrasting against the world of cathedral music. When the new editor of Novello's *Musical Times*, Frederick George Edwards, printed several skits on the National Anthem in the Diamond Jubilee year,[144] Edward sent him a scrap setting the National Anthem against the Tchaikovsky 5/4—with a covering note of mock musicology: 'This setting has been declared suitable for cathedrals as well as profaner areas and, (perhaps a doubtful advantage) the conductor might join in the vocal part as it has been found to require no conducting.'[145] The new editor was a notorious punster, and a fast friendship began.

In mid-July Edward and Alice went up to Wolverhampton for a long weekend with the Pennys. Young Dora was ready for him.

We had music at all hours: *Lux Christi, Scenes from the Bavarian Highlands* . . . I remember how much I liked the 'Bavarians' and, after he had played the Lullaby ('In Hammersbach'), I could not help interrupting:

[143] 'Elgar at Hereford', in *The RAM Magazine*, 1960, p. 6.
[144] F[rederick] C[order], 'A Lost Opportunity', in *The Musical Times*, 1 Oct. 1897, pp. 666–8.
[145] BL Egerton MS 3090 fo. 2.

'That's lovely—I should like to dance to that.'

'I wish you would: I'll play it again.'

An interruption fortunately gave me time to escape and slip into another frock and also to think out something. (I used to amuse myself by inventing dances rather in the Maud Allan style of later years.) When I got back into the drawing-room . . . we tried it out. He seemed much pleased and we did it again, trying bits where steps had not quite fitted in and so on. So much did he like it that I was called upon to 'come and dance Hammersbach' on several occasions at Malvern.[146]

Returning to Forli to complete the *Te Deum and Benedictus* orchestration, Edward had another bout of eye trouble. But when he sent the finished score to Novellos, it drew a letter from Jaeger full of his own gratitude for Edward's music. Edward responded instantly:

Forli, Malvern.

Aug. 4 1897

Dear Mr. Jaeger:

I send you my very sincere thanks for your kind and appreciative letter: I feel after reading it that I shd. like to 'go on': but cui bono?

You praise my new work too much—but you understand it;—when it is performed will anyone say *any*thing different from what they wd. say over a commercial brutality like the 'Flag of England' for instance: naturally no one will & the thing dies & so do I—

All the same hearty thanks for your sympathy: I told you that I wd. never put pen to paper when I had finished this work: but shall I?

Vy sincerely yours
Edward Elgar[147]

When Jaeger commiserated over his firm's payment for the *Te Deum and Benedictus*, Edward replied:

Please do not think I am a disappointed person, either commercially or artistically— what I feel is the utter want of *sympathy*—they i.e. principally critics, lump me with people I abhor—mechanics. Now my music, such as it is, is alive, you say it has heart—I always say to my wife (over any piece or passage of my work that pleases me) 'if you cut that it would bleed!' *You* seem to see that but who else does?[148]

*   *   *

Many people were beginning to see it, in fact there might be a chance for a new major work at the Leeds Festival, which attracted nation-wide attention. The

---

[146] Powell, *Edward Elgar: memories of a variation*, pp. 8–9. She said that she also heard 'lots of sketches from *Caractacus*'. This is doubtful: in trying to correlate memories and diary dates in later years, Mrs Powell often supposed she had heard sketches of works which could not yet have existed in any form. The suggestion of 'lots' of *Caractacus* sketches in July 1897 remains unsupported by any evidence known to the writer, and it is contradicted by much other evidence (e.g. Elgar's letters to Novellos).

[147] MS at the Elgar Birthplace. Novellos had published *The Flag of England* with words by Kipling and music by Frederick Bridge side by side with *The Banner of St. George* early in the Jubilee year.   [148] 6 August 1897. MS at the Elgar Birthplace.

next Leeds Festival was due in October 1898. During the winter and spring of
1897 Edward had been thinking about subjects. One idea, perhaps inspired by
Berlioz's *L'Enfance du Christ* (which he had heard at the Crystal Palace), was
'The Flight into Egypt'. He sketched out a three-part plan: the Journey, the
Holy Family at rest, and the Return to Jerusalem seven years later when Jesus
was preparing for the 'Work of My Father'.[149] The mother Mary would
explicate the Child's words, her husband Joseph would not understand them.

Then there was another suggestion—for a setting closer to home. It came
from Edward's golfing companion, the bank manager and amateur cellist H.
Dyke Acland, who proposed the subject of St. Augustine converting
Britain—or perhaps something earlier still:

> What do you think of 'St Augustine' as a subject?
> Ancient Britons
>   wood, battle axes
>   Druids—Mistletoe (Oh fie)
> Ancient Rome
>   Martial tum tum
>   Christian strain, Monks chant, Angeli non Angli . . .[150]

There had been Druids in the Malvern Hills at the time the Romans defeated
the British leader Caractacus in the first century. Such a theme offered the
chance to draw inspiration directly from the home landscapes of his music. Six
months later this theme received sudden suggestion from a powerful quarter.

In the late afternoon of 4 August 1897 Edward and Alice went over the
Malvern Hills to the village of Colwall. There Ann Elgar was spending a
holiday: it was close to the parents' fiftieth wedding anniversary, but the old
lady still liked to get away by herself. Later she wrote:

When I was staying at Colwall E and Alice came to see me—on going out we stood at the
door looking along the back of the Hills—the Beacon in full view—
    I said Oh! Ed. Look at the lovely old Hill. Can't we write some *tale* about it. I quite
long to have something worked up about it; so full of interest and so much historical
interest.
    I said to write some *tale*, and you *can* 'do it yourself Mother' He held my hand with a
firm grip 'do' he said–
    No I can't my day is gone by if ever I could and so we parted . . .[151]

The 'British Camp' on top of the Herefordshire Beacon near the southern
end of the Malverns is still marked with earthworks—made (according to
legend) by Caractacus against the Roman invaders. The patriotic story could
continue the theme of Edward's recent music for the Diamond Jubilee. The
figure of the old Briton himself held a more personal interest: his was yet
another story of an outsider and his apostles. But more clearly than any

---

[149] MS draft dated 7 Feb. 1897 (Elgar Birthplace).
[150] 9 February 1897 (HWRO 705:445:2302).
[151] 11 Dec. 1898 (quoted in Young, *Elgar, O.M.*, pp. 80–1).

Elgarian hero who had yet appeared, the story of Caractacus was the hero's defeat.

Four days after the visit to Colwall, Edward decided suddenly that he must get away from home scenes altogether and go again to the mountains of Bavaria. There was barely a month before the Hereford Festival. Alice hurriedly packed, and on the morning of 10 August they were off. In Munich they heard *Tristan* directed by the thirty-three-year-old Richard Strauss, whose works already enjoyed European fame. They had a fortnight in Garmisch, visiting friends and returning to Hammersbach and St. Anton: it was to be the last sight of Bavaria for many years. On the return journey they attended a Strauss performance of *Don Giovanni* at Munich, and arrived in London in time for Edward to correct orchestral parts for the *Te Deum and Benedictus* at Novello's offices.

At Forli on 10 September, Alice merely repacked for a week's stay in Hereford through the Festival. The *Te Deum and Benedictus* was rehearsed on Saturday 11 September, and performed at the Festival opening service next day with the *Imperial March* as orchestral voluntary. Through the service Edward sat in the organ loft with the tenor Edward Lloyd. Among the large congregation below was Jaeger of Novellos, and he wrote to Edward a few days later:

I hunted for you high & low during the Service (awfully long!) & afterwards, but you were not to be seen. Never mind, I spent a most delightful 22 hours in the delightful cathedral town & I *have* heard your finest, most spontaneous & most deeply felt & most effective work & I was *very* happy . . . You see I am conceited enough to think that I too can appreciate a good thing & see genius in musicians that are *not* yet dead, or even not yet well known . . .[152]

Other new faces were seen at Hereford that year. Sinclair's young pupil Ivor Atkins had become organist of Worcester Cathedral after Hugh Blair departed unhappily (due, it was said, to his taste for drink). Charles Lee Williams's retirement from Gloucester Cathedral had brought up a local man, Herbert Brewer, in his place. And Nicholas Kilburn had come down with his wife from Bishop Auckland to attend the Festival.

Afterwards came the return to hated teaching. From that autumn Edward limited his violin teaching to Malvern. He was restless, and there was talk of taking a new house. Alice spent several days going over some premises with Basil Nevinson's brother Edward—who had now set up an architectural practice in Malvern—and his young assistant Arthur Troyte Griffith. But then Edward had another bout of eye trouble. He met Jaeger's further encouragements with heat:

Yes! I have some ideas: but am about taking a new house—*very noisy* close to the station where I *can't write* at all—but will be more convenient for pupils (!) to come in—I have no intention of bothering myself with music.

[152] 15 Aug. 1897 (Elgar Birthplace).

Look here! in two years I have written
Lux Xti
King Olaf
Impl. March
S. George
Organ Sonata (big)
Te Deum
Recd £86.15. Debtor £100 [to Novello's for publishing *Lux Christi* and *King Olaf*]
After paying my own expenses at two festivals I feel a d—d fool! (English expression)
for thinking of music at all. No amount of 'kind encouragement' can blot out these
simple figures.[153]

During a visit to London in late October to conduct three of the *Bavarian
Highlands* Scenes at a Crystal Palace Promenade Concert, he left with Novellos
a new violin piece—another slow G major melody with a contrasted episode of
darkly descending steps. Provisionally he called it 'Evensong'; the publishers
persuaded him to *Chanson de nuit*, and they offered 10 guineas for the
copyright. Edward replied:

I *wish* you could arrange terms for it which would leave me some interest in it: the last
Violin piece I wrote [*Salut d'amour*], which unfortunately I sold some years ago for a
nominal sum, now sells well—I understand 3000 copies were sold in the month of
January alone. An orchestral arrangement of this piece no doubt materially helped the
sale & the piece you now have would arrange satisfactorily for a small orchestra.
    In any case I accept the terms you offer.[154]

When published, the new piece approached the popularity of *Salut d'amour*.

Alice let it be known that their search for houses was extending again toward
London. Thereupon local friends responded by organizing a choral and
orchestral society for Edward to conduct. He wrote of it to the *Daily Telegraph*
critic Joseph Bennett: 'The people here are much afraid I was leaving so they
have started a brand new soc[iet]y "The *Worcestershire Philharmonic*" which I
am to conduct—a sort of toy I suppose for a petulant child: I think we may do
well in time.'[155] So again—as with Hubert Leicester's wind quintet and Miss
Burley's 'Little Orchy' at The Mount—Edward's friends banded themselves
together in a musical group to encourage him.
    The new enterprise was on a large scale. The Secretaryship was shared by
two admiring ladies—Miss Martina Hyde (the daughter of a prominent
Worcester solicitor) and Miss Winifred Norbury (whose family lived at
Sherridge, an old country house between Worcester and Malvern). They
gathered an impressive committee headed by the Earl of Dudley and Lord
Hampton, and including Lord Beauchamp's sister Lady Mary Lygon. The

153 19 Oct. 1897 (Elgar Birthplace).
154 28 Oct. 1897 (Novello archives).
155 15 Nov. 1897 (MS in possession of Denis Mellor).

three Cathedral organists Atkins, Sinclair, and Brewer accepted honorary membership, and many local magnates lent their support. Miss Burley took a great interest in the new Society:

Recruitment to its ranks, vigorously undertaken by a few enthusiasts, was fairly easy since those who could play an instrument readily joined the orchestra, those who sang could be pressed into the choir and those who did neither swelled the audience and were glad of an excuse for driving into Worcester, lunching with friends, and driving home again . . . With a little beating up, we managed to collect a very large number of subscribers.

Not that everyone was as hopeful as we of success. Old Mrs Fitton pointed out from the start that Edward had not the precise qualities to hold such an organization together. 'He'll throw it up as soon as he gets tired of it,' she said. But most of us were hopeful and enthusiastic.[156]

The orchestra of fifty players contained numerous friends, pupils, and colleagues: Miss Burley herself, Hilda Fitton, Dyke Acland, Frank Ehrke, Bob Surman (now an established dentist in Worcester), Canon Claughton's son Alban, Heinrich Sück, and Alfred Probyn from Birmingham. The choir numbered over a hundred, including everybody from Lord and Lady Hampton to Miss Holloway from Powick and Hubert Leicester. The Society's musical motto was the 'Wach auf!' chorus from Die Meistersinger: it was to be sung to open each meeting and it was quoted on the Society's letterhead, worked into an engraving of the sun rising along the Malvern Hills.

Edward suggested he conduct for a year without remuneration. He began voice trials for prospective members in November 1897. Fortnightly rehearsals were commenced for a first concert to take place the following May. Miss Burley recalled them:

We started with immense enthusiasm. Edward had been given a completely free hand in the choice of the programme and, although he had selected Humperdinck's Die Wahlfahrt nach Kevlaar for the first concert, and insisted on doing it in German—a language of which ninety-nine percent of the choir knew nothing—the practices were well attended and the work attacked with good humoured determination. That the result was more satisfactory musically than linguistically is not surprising when one remembers that the conductor knew little more German than his choir. Alice of course spoke it fluently as did Miss Norbury (one of the secretaries), but for the most part strange chewing noises were produced that sounded like no known European language. Nevertheless a success was made with the local public.[157]

One of the orchestra evaluated Edward's musical leadership thus:

[M]r. Elgar is hopeless as a teacher, but is a fine conductor. Those who need to be taught orchestral playing must go elsewhere. If the band is experienced, and knows how to allow itself to be played on, he will play on it to some purpose. But he is at the mercy of his moods: and rarely does a thing twice alike. At preliminary practices with an

[156] Edward Elgar: the record of a friendship, pp. 103–4.
[157] Edward Elgar: the record of a friendship, p. 107.

incomplete band, he plays missing wind parts [on the piano] with the left hand, and beats time with the right. He can bear much provocation with patience, and little provocation with no patience at all. If a violin player drops her mute there is a 'rumpus'; on the other hand, he is unsparing in his care for detail, and will repeat a passage many times until he is satisfied.[158]

On 11 November Edward again met the wealthy secretary of the Leeds Choral Union, Henry Embleton. The immediate matter in hand was an engagement to conduct the *Bavarian Highlands* with the Choral Union. But Embleton was also a member of the Music Sub-committee for the Leeds Festival. The question of a new Elgar work was raised, and Embleton undertook to carry the matter with his colleagues: 'I feel sure', he wrote afterwards, 'that they will fall in with our wishes.'[159]

Still Edward felt uncertain over the subject of Caractacus. There was Dyke Acland's first suggestion of St. Augustine and the Conversion of Britain: but that might prove too Roman Catholic for a Festival Committee. Then Edward's thinking went again toward something purely orchestral. His own interest, in everything from melodic invention to versatile scoring, was always in the orchestra. And orchestral music eliminated all the specific doctrine and interpretation of a libretto. He thought now of shaping the stories of Caractacus, St. Augustine, and others like King Canute into 'perhaps a series of illustrative movements for orchestra with "Mottoes" from English history'.[160] Such an anthology might offer more personal musical expression than following any one story exclusively. He suggested the orchestral suite for Leeds.

But Embleton said that Leeds wanted a cantata. So Edward returned to 'Caractacus', as he wrote to the chairman of Novellos on 26 December 1897:

I hear that nothing save the merest accident will prevent my being asked to contribute a Cantata to the Leeds fest:—I have hinted at other things but it seems they wish a Cantata. Now, I do not know if they will offer me any fee or whether it is usual for them to do so, but I thought it wd. simplify matters if you could give me an idea,—nothing official, but an expression of opinion,—if your firm wd. be likely to purchase a work: I cannot say it will be easier than K. Olaf—it seems to be less involved so far as written: the subject is 'Caractacus'.

A conference at Novello's offices early in the new year 1898 produced a contract: they would pay £100 for the copyright. It contrasted happily with the agreement for publishing *King Olaf* and *The Light of Life*. But once again the composer had signed away his copyright.

The legend of Caractacus rested on just two historical facts—an unavailing

---

[158] Florence Fidler in *The Musical Standard*, quoted in *The Worcester Echo* c.Jan. 1903.
[159] 16 Dec. 1897 (HWRO 705:445:3159).
[160] 29 Sept. 1897 to Nicholas Kilburn. Elgar later offered the Canute idea to Edward German: see W. H. Scott, *Edward German* (Chappell, 1932) pp. 178–9.

defence of his homeland, a pardon (with his family) in Rome. Could these two facts be connected in the hero's love for his native land—a love so powerful as to move even his captor? This was a theme to touch Ann Elgar's son. Yet if it was so, then more than ever the hero's nobility was defined in his defeat.

It would all need most careful handling by a librettist sympathetic to Edward's thought as it developed, and close at hand for constant consulting. The obvious choice was Harry Acworth. After the success of *Olaf*, Acworth readily agreed.

In consultation with Edward, he manufactured a family for Caractacus. Once again there was to be no wife. But there would be a daughter—whom they named Eigen after a young neighbour in Malvern, Eigen Stone. Some love-interest could shape a secondary plot: they made Eigen's lover a young Bard of the Druids. And so they could bring in the entire holly-and-mistletoe side of Dyke Acland's idea.[161] The tree-worshipping Druids personified the love of native countryside. Caractacus would consult their powers of divination about his battles. Only there again was the unanswerable fact of defeat.

Edward had begun already to shape the music. He made a motive for 'Britain'[162] from his own descending steps, with a great rise in the midst, and all repeated in sequence:

The contrasting presence of 'Rome' was defined in four-square assertion enclosing a triplet—exactly as 'the Roman hosts' of Acworth's libretto 'Have girdled round our British coasts':

Both of these basic motives were purely instrumental shapes. Whether or not they had been first designed for the orchestral scenes of British history, their instrumental character promised a strong orchestral presence in *Caractacus*.

Acworth's libretto divided the Caractacus story once again into scenes. But they were altogether longer than the *Olaf* scenes, and Edward resolved to fill each with continuous music in such a way as would make subsequent

---

[161] One source for the Caractacus libretto was probably James Mckay's *The British Camp on the Herefordshire Beacon*: Fifteen short essays on scenes and incidents in the lives of the ancient Britons (Malvern: Published at the *Advertiser* Office, 1875). Many of the essays deal with the Druids and their life in the Malvern Hills, suggesting connections with Caractacus. The volume was dedicated to the Revd W. S. Symonds of Pendock, near Redmarley, who had taught local geology to Alice Roberts and Minnie Baker.

[162] The nomenclature follows that of the *Analytical Notes* written by Herbert Thompson for the first production and subsequently published by Novellos. No direct evidence of Elgar's close supervision of Thompson's writing has come to my notice, but he would certainly have been consulted—especially as to the representative themes.

fragmentation impossible: *Caractacus* would involve the biggest structures his music had yet essayed.

Scene I was set in the British Camp on the Malvern Hills at night. He was uncertain whether to write an overture. But the opening cry of the Watchmen was set into an instrumental figure which made the Roman motive climb up and down as if the enemy were already swarming through the lower Hills:

And this entire night-time opening was framed by a figure that plunged, twisting the sequence of Edward's old 'tune from Broadheath' in strange ways:

The 'Roman' theme reduced to a 'Desolation' motive evoking the violated countryside:

Of the 'Britain' theme there were alternating developments for women:

and for men:

At the end of this opening section of the scene, words describing the landscapes of home suddenly raised again the old shape of dying 'Delight':

Straight to it came the hero himself: his portrait occupied the central third of the opening scene. He was given a plunging motive, purely instrumental:

This orchestral definition of Caractacus would make him less a dramatic character than a natural force. Like King Olaf, the new hero was also given separate motives to evoke the lyric and heroic sides of his character. But the heroic motive for Caractacus lacked the melodic distinction of the *Olaf* themes: when he addressed his soldiers, Caractacus struck rather the note of elegy. His lyric theme told of no personal beauty, for this was a significantly older man; instead it evoked again the countryside peace of home:

The final third of the scene introduced the hero's daughter Eigen with another instrumental motive:

This variant of the hero's lyric motive of the countryside invested his daughter with almost maternal mystic qualities. Her Druid lover Orbin followed, but Edward's music did not characterize him with any clarity. A culminating trio for the principals set their united aspiration to unadorned octaves. But when the chorus, as Spirits of the Hill, chanted a new variant of the 'Britain' theme to greet the dawn, Edward's inspiration came instantly back. The distant 'Watchmen, alert!' sounded softly again to end this scene of more than twenty minutes' length. Despite the weak solo writing, it was the most imposing opening Edward had yet achieved.

He steeped himself in the *Caractacus* landscape, as Miss Burley recalled: 'Elgar walked all over the ground. He tramped over the hills and went along the Druid path from end to end, along the top of the hills.'[163] At the north end of the Malverns, on the Birchwood estate, he found a tiny lodge cottage unoccupied. It was perched nearly on top of the Hills in woods, but with spectacular views over miles and miles of the Severn Valley. In the midst of major composition, the notion of moving house had dropped. But here was an ideal summer retreat. And Birchwood Lodge itself made a private appeal: built on a familiar Midlands cottage plan, it almost reproduced the old cottage at Broadheath. He consulted Edward Nevinson's architectural assistant Troyte Griffith, who knew the Birchwood squire, and a short lease was soon

---

[163] Quoted in *Letters to Nimrod*, ed. Percy M. Young (Dennis Dobson, 1965) p. 21.

negotiated. Through the next weeks, as Alice began to equip the cottage, Edward got together his ideas for the second Scene in *Caractacus*, to show the Druid world.

Acworth set this Scene in 'The Sacred Oak Grove by the Tomb of Kings'. He supplied an Arch-Druid to direct the proceedings with pagan dance and sanguinary incantation, and made Eigen's lover Orbin a Druid bard who could read the future. For dramatic complication, when Orbin foresaw the hero's disaster in battle, the Arch-Druid was to conceal the adverse prophecy from Caractacus in the hope of encouraging a braver fight for Britain. At the end of the Scene Orbin would be banished for trying to tell the truth: so Edward's music could show another outsider.

The music for Scene II was designed as a huge slow movement to follow the big opening *Allegro* of Scene I. Scene II introduced a motive to represent oracular power in the Druids

Motives for the Mistletoe and Sacred Dance were less successful. Each had a diatonic innocence and rhythmic filigree reminiscent of the *Bavarian Highlands* songs: in the dark Druid setting they sounded coy. So *Caractacus* further exposed its composer's dramatic limitations. It was not surprising that he had failed to find any subject for an opera.

When the Arch-Druid invoked the pagan god, the music instantly regained power. It sounded an awesome variant of the childhood invocation to nature in the 'tune from Broadheath':

roared the chorus in octaves to be reinforced by brass instruments and full organ, while the rest of the orchestra swung out a vicious off-the-beat counterpoint. It exposed a raw compulsion beyond anything in *King Olaf*. 'Caractacus frightens me in places,' Edward wrote to Jaeger on 1 March 1898.[164] Jaeger replied:

That Caractacus 'frightens' you is a *good* sign; for directly you become familiar with the 'visions' that now 'frighten' you, they will inspire you & become your friends! . . . *Don't* hurry over your magnum opus, whatever you do. I hope your Health is good & your spirits as high as I suppose those of your Hero when he *didn't* get a 'licking'.[165]

Jaeger had a way of putting his finger on things.

[164] HWRO 705:445:8305.   [165] 8 Mar. 1898 (HWRO 705:445:8300).

The remaining two-thirds of Scene II was less inspired. Orbin read his divination of doom to an unmemorable phrase, the chorus responded wrongly with their coy dance, and the Arch-Druid produced his false encouragement to a noble extension of the 'Britain' theme when nobility was belied by the situation. The appearance of Caractacus was marked by a welcome 'Sword Song' *vivace e con molto fuoco* with brilliant regular flashes through upper strings and woodwind, and solo resolution amplified by the chorus of men: but the animation was brief. Within two minutes the music had returned to Orbin's unmemorable accents, the Arch-Druid's sham nobility, and a chorus response of stock opera phrases. On 29 March Edward wrote to Nicholas Kilburn 'I have just arrived at hating what I have done & feeling a fool for having done it—but my wife says I always do that at certain stages: anyway there are some gorgeous noises in it—but I can't say how much music—but it "flows on somehow" like the other best of me.'

Yet the best of him was elsewhere, as he had already written to Joseph Bennett on 17 March about the work *Caractacus* should have been and was not:

I hope some day to do a great work—a sort of national thing that my fellow Englishmen might take to themselves and love—not a too modest ambition! I was going to write to *you* to ask if ' S. Augustine' might form the basis of such a work . . .?'[166]

Bennett, in addition to his position as the *Daily Telegraph* critic, was known as a skilled libretto-compiler. But he did not seem very eager to fall in with Edward's idea.

The Leeds Festival Secretary, Frederick Spark, began to ask when the Leeds Chorus could expect to have their copies of the completed *Caractacus* for rehearsal. Still Edward had not decided whether to write an overture or merely an introduction to Scene I: so he could not send its music to the publishers. He obtained Novello's permission to send in Scene II for engraving first, and posted it on 4 April with a query: 'Do you ever send out "sections" of works for fest. choirs for rehearsal?' With the première only six months away, this was agreed.

Then he turned to Scene III. Acworth had set it in 'The Forest near the Severn: Morning. In the distance youths and maidens sing while they weave sacred garlands.' Here Edward's impulses were back on home ground. The soft chorus now echoed melodic shapes and vocal textures of the *Bavarian Highlands* music with the greatest happiness; while in their midst Eigen sang a song whose theme was strongly reminiscent of King Olaf's 'Beauty' motive. If this had been a symphonic third movement, it would recall a Brahms *Allegretto* rather than a Beethoven *Scherzo*.

As the rest of Scene III contained only solo and duet music, and the Leeds Chorus was waiting to rehearse, Edward turned instead to Scene IV. It would show the climactic battle of Britons and Romans. Yet Edward's music made no climax. As with battles in *The Black Knight* and *King Olaf*, the crucial

[166] Elgar Birthplace.

encounter was taken off the stage of action. 'Wild rumours shake our calm retreat,' sang the women to diatonic figures that recalled the gossiping 'Thyri' chorus in *King Olaf*. Acworth's libretto then had the women describe the battle from afar. But Edward cut out those words, leaving the entire account to the remnant of British soldiers returning in defeat. The soldiers' account did not even mention Caractacus: it was as though he had stood watching over Edward's shoulder.

The actuality came when Caractacus sang a Lament in 7/4 metre (perhaps in memory of the Tchaikovsky 5/4), its melody superbly combining the 'Rome' and 'Britain' themes:

It gave finer focus to Edward's music than anything in the foregoing action, for here at last the imperial theme found the heart of his own nostalgia. Action had given way to reflection; the past was better than the present. Yet the hero had still to seek his reassurance—not at the hands of women, nor indeed of any men who remained in the world at all:

> Oh, my warriors, tell me truly,
>   O'er the red graves where ye lie,
> That your monarch led you duly,
>   First to charge and last to fly;
> Speak, ah! speak, beloved voices,
>   From the chambers where ye feast . . .
> And the god shall give you heeding,
>   And across the heav'nly plain,
> He shall smile, and see me leading
>   My dead warriors once again!

Where self-pity could ennoble, there indeed this hero of home countryside might show Edward a vision of himself.

On 19 April 1898 he posted the completed vocal score of Scene IV to Novellos, together with Scene I. For now he had determined to write no overture. Instead there were 26 bars of terse *Allegro* to plunge directly into action. It differed from every previous Elgar opening: thematic fragments flashed briefly across a darkened scene before rising arpeggios rang up the curtain on the British Camp at night with 'Watchmen, alert!'

He showed further uncertainty. Sir Frederick Bridge, who conducted the Royal Choral Society, had asked Novellos for early proofs of *Caractacus* with a view to a London performance. To Jaeger's query Edward replied:

I don't want Bridge to see *early* proofs of Caractacus. I can't prevent his seeing the thing when ready for the chorus: you are the best judge, but if he didn't like it, his remarks, *altho' not unkind*, might prejudice me . . .

Proofs? You might send on that second scene.[167]

Within a fortnight proofs began to arrive, and the remaining composition of *Caractacus* went forward side by side with proof-correcting of earlier portions.

On the day he had written to Jaeger, Edward and Alice went over to Hereford for a performance of the *Bavarian Highlands*. They stayed the night with George Sinclair, whose bachelor house in the Cathedral close was shared with his bulldog Dan. Edward had taken to writing in the Visitors' Book there little musical inventions to describe 'The Moods of Dan'.[168] A figure which seemed to echo the *Caractacus* Lament now went into the book to show the big dog as 'He muses (on the muzzling order)':

Back at Forli, his thoughts were still elsewhere. Joseph Bennett had not replied over 'St. Augustine'. But on 3 May Edward wrote to him again about an enquiry from the great Birmingham Festival, whose next meeting would take place in the autumn of 1900: 'I do not expect a reply to this or my former enquiry as to St. Augustine: But the Birmingham people have more than hinted & I will call & see you one day.'[169]

When Herbert Brewer requested a short orchestral work for the Gloucester Festival in September, Edward asked him to apply to Samuel Coleridge-Taylor, a young composer over whom Jaeger had enthused. And it seemed a very local stir that was created by the successful début of the Worcestershire Philharmonic Society on 7 May. Edward conducted the Humperdinck cantata twice in the programme, which was filled out with extracts from Massenet's *Le Cid* and from *King Olaf*.

All the time he was being pressed hard to finish *Caractacus*. Frederick Spark, the Secretary at Leeds, demanded as much of the music as possible for the Festival Chorus to rehearse before their month's holiday beginning on 15 July. Scenes I, II, and IV were now with Novellos. Edward wrote an orchestral opening to Scene III from an old G major idea in the 1887 sketchbook to breathe dewy innocence in a 'Woodland Interlude':

[167] HWRO 705:445:8308.   [168] Now at St. Michael's College, Tenbury (MS 1510).
[169] MS in possession of Denis Mellor.

So the oldest idea in the new score made one of its freshest moments. Now he was in a position to send to Novellos all the choral music of Scene III, as he wrote on 10 May:

I should be so much obliged if you could let the Chorus have Scenes I. II. III & IV to go on with: I have been—a month ago—laid up with cold—& am rather anxious that the latter portion of my composition—which I am now completing—shd not be hurried without due consideration.

Scs. I & II are in type

Of Sc III I send M.S. down to the point where the solo voices enter.

Sc IV you already have (M.S.) complete.

The Committe wd. like to have this much by June 1.

I trust the extra expense in sending these sections first wd. not be great & I must bear the cost if any: the great consideration your firm has always shewn me leads me to hope that the extra trouble will not stand in the way of this plan which will make things easy all round—giving the Chorus something to do & giving me time to complete my work without hurry.[170]

So that the Leeds Chorus could have the rest of their music, he turned to the final scene. This showed Caractacus and his people as prisoners in Rome. 'The Triumphal Procession' began—as every other good thing in *Caractacus* began—with the orchestra. It started with the big brilliant 'Rome' theme. Soon a counter-melody brooded over the situation with Edward's finest sensibility:

The chorus entered casually voice by voice, its placement and the necessary simplification of the vocal writing under instrumental figures emphasizing its inferior position. Caractacus, Eigen, and Orbin appeared to another orchestral motive denoting 'Captivity', heavily dragging its rhythmic sequence through minor keys:

It was quietly juxtaposed with an 'Eigen' memory of British woodlands, rising in the major:

170 Novello archives.

Then the chorus gathered toward ever weightier celebration of the Roman achievement.

At the entry of the solo characters the temperature dropped perceptibly—until Caractacus could answer the Roman Emperor Claudius with the self-defence of loyalty to his home countryside. Edward wrote to Jaeger: 'I made old Caractacus stop as if broken down on p. 168 & choke & say "woodlands" again because I'm so madly devoted to my woods.'[171]

They were the woods round the Birchwood cottage, where Edward and Alice began to spend more and more time between teaching and engagements. In the tiny upstairs bedroom which he had turned into a study, he worked on his music on an old square piano they had secured for the cottage.[172] When Dora Penny came to visit Birchwood she laughingly challenged Alice over it:

I played a few notes on it and it made a tinny little noise rather like a spinet.
   'Surely he's not going to use this?'
   'It does sound rather funny, dear Dora, but I assure you dear Edward makes it sound beautiful!'
   'That's uncommonly clever of him!'[173]

It showed how little his musical thinking relied on the piano sound.

After the hero's self-defence, the chorus of Romans roared 'Slay the Briton', as Bach had made his *Passion* choruses roar for the Crucifixion of Christ. And Edward cast his hero's reply in the mould of his own Christ's 'I am the good shepherd' in *The Light of Life*. The ensuing solo exchanges tried again to reduce complex instrumental figures to clear vocal lines—again without success.

At last the chorus returned to draw the opening themes from Scene I into a big coda. The goal was to be the 'Britain' theme scored for all forces. Musically it was the ideal ending, but it demanded an awkward volte-face in the libretto: where a few moments earlier the chorus had impersonated Roman victors, now they must suddenly return to being British—yet without the stigma of defeat. Acworth did his best by calling the future to witness:

> The clang of arms is over,
>   Abide in peace and brood
> On glorious ages coming,
>   And Kings of British blood.
> The light descends from heaven,
>   The centuries roll away.
>
> . . . .
> For all the world shall learn it—
>   Though long the task shall be—
> The text of Britain's teaching,
>   The message of the free . . .

[171] 21 Aug. 1898 (HWRO 705:445:8316).
[172] The square piano is now in the Broadwood collection.
[173] *Edward Elgar: memories of a variation*, p. 10. Dora Penny said further that the next time she

To it Edward brought an immensely skilful montage of earlier motives building toward a vastly imposing 'Britain'.

The completed final scene went off to Novellos on 4 June. When libretto verses applying the words 'menial' and 'jealous' to non-British nations met the eye of the German-born Jaeger, he protested. Edward replied:

Good! by all means will I ask Acworth to eliminate the truculent 'note' in the lines; any nation but ours is allowed to war whoop as much as they like but I feel we are too strong to need it—I *did* suggest we should dabble in patriotism in the Finale, when lo! the *worder* (that's good!) instead of merely paddling his feet goes & gets naked & wallows in it . . .[174]

But then he defended the truculence after all: 'I knew you wd laugh at my librettist's patriotism (& mine)—never mind: England for the English is all I say—hands off! there's nothing apologetic about me.'[175]

There was none the less an intolerable contrast between the hero's Lament at the end of Scene IV and the brazen Roman opening of the Finale. So he and Acworth added a brief Scene V to show the captive Britons embarking from the banks of the Severn for transportation to Rome. The new lines (using the old name of the Severn) could touch the memory of the child in the riverside reeds longing for something very great:

> They shall ne'er return again!
> Lap their bark with sob and sigh,
> Sombre Habren, swirling by;
> For they never more shall see
> British heav'n, or land, or thee.

Edward set the lines over a rolling arpeggio to a slow descent breaking in sequences

It was the same figure which had filled the centre of *Sursum Corda* four years earlier. Now in *Caractacus* it ushered in an orchestral transition which gradually mingled themes of farewell with the victor's triumph, until the entire voyage was synthesized in a brief orchestral arc of deeply-felt music. It showed yet again where Edward's power really lay. For now more clearly than ever his heroism had abandoned dramatic contest with the world, to pursue instead an inner renunciation through nostalgia.

He turned at last to the solo and duet music which would finish Scene III. Here Orbin would emerge a fugitive from the Arch-Druid's wrath. So his came to Birchwood the square piano had been changed for an upright. But the square piano contains a note in Elgar's hand to the effect that *Caractacus*, *Sea Pictures* (1899), and *The Dream of Gerontius* were all written with its assistance.

[174] 21 June 1898 (HWRO 705:445:8311).
[175] 12 July 1898 (HWRO 705:445:8313).

love-duet with Eigen could sound again the music of farewell, with the hope of meeting only in another world:

> Where all is peace under summer suns,
> And clear of battle the river runs,
> And in placid waters the lilies float,
> And the sweet birds sing an untroubled note;
> Where never are heard the sounds of strife,
> But all is radiant, joyous life,
> > When this sad life is o'er.

It was the Better Land evoked more powerfully than ever from the landscape of home. Now it lent inspiration to Edward's solo and duet writing, for the Woodland motives which had opened Scene III here gained a recapitulation more memorable than anything else he had written for the lovers.

Perhaps the inspiration of the British countryside in this music suggested its dedication to the old Queen whose reign had preserved the peace at home through more than sixty years of social and industrial change. Edward wrote to his father's former acquaintance Sir Walter Parratt, once the organist at Witley Court in Worcestershire and now Master of the Queen's Music at Windsor. Parratt replied on 20 July 1898: 'I will certainly bring the matter before the Queen's notice. It will not be my fault if your request is not granted. I hope you are aware that I use your music constantly [for private State Concerts], and the Queen likes it.'[176] The royal assent followed for Edward to dedicate *Caractacus* to his sovereign.

After finishing the last of the vocal score on 12 June, the orchestration of *Caractacus* went forward at speed, with Alice again preparing the paper for his scoring. Weekends brought visits from the architect Troyte Griffith, who was rapidly becoming an intimate friend. His tall, stooping figure inspired Edward's epithet, 'the Ninepin'. Together they rambled through long summer afternoons in the Hills.

By 21 August the score was finished. Edward wrote to Nicholas Kilburn:

I have to-day written the last note of the score of 'Caractacus', commenced on June 21, & feel free. I am *bursting* to talk to you but when I come to hold the pen it seems to want to write 'Please be lenient': I feel frightened at my score (which is big) & if I ask for justice, surely I shall hang? Anyway, I *tried*, & shd. like to please you . . . I have thought of you often & my thoughts of you & Mrs Kilburn & your kindness to me are woven into many a bar of my score—*you* will understand how that can be . . .

If Edward's music reflected its composer, by that token it reflected his most valued friends.

Scoring done, he turned to correcting the shoals of vocal-score proofs and orchestral parts now arriving almost daily from Novellos. In the correcting he had devoted help from friends close at hand—particularly from Mrs Fitton's

[176] HWRO 705:445:1970.

daughter Isabel and from the Norbury sisters at Sherridge, just a mile below Birchwood. A young Norbury niece recalled a late evening walk to the cottage:

I remember going up with my two aunts after dinner (a great treat to stay up so late) . . . It was so dark thro' the wood we had to hold on one behind the other, and the tiny path was lit up on both sides by glow worms.

W[inifred] N. was very sedate and calm, rather like a kind governess with him, but had a sense of humour and I believe he purposely kept on being tiresome till he had got the laugh—rather like a deep bell—[that was characteristic of her]. I remember her saying 'We didn't come up here at this time of night to spend the time giggling' with a very straight look at him, and he got up and bowed and they went to the piano room while the rest of us stayed behind . . .

W. N. and Lady Mary Lygon (who was a lively intelligent creature) could keep him in order and *make* him work as well as amuse him. My little aunt [Florence Norbury] and Dora [Penny] could run with him, bicycle, climb hills, fly kites . . . Their part was to get him ready for work.

Mrs. Elgar was a wonderful woman. His work and well-being was everything to her, and I believe she *made* these friendships with other women—all young and attractive—who could do the parts she couldn't always manage. She was almost *too* sweet with him . . . She did everything she could, and effaced herself completely and did the chores when she had got the others going. My aunts always said he was lazy and would never have done anything with his music but for his wife.[177]

Alice's encouragement of younger women to stimulate Edward, in the summer of her own fiftieth year, was the proudest testament of her will to spur on Edward's music. Yet no woman who loved her husband as Alice loved Edward could contemplate welcoming possible rivals without some assurance of ultimate control. Edward's regard for Alice was totally involved in the large-scale music-making which their marriage had initiated and which she constantly mothered. Whatever parts the other women might play or seek to play, when it came to the music only Alice was of final importance.

Twice he went up to Leeds for Chorus practices. On 9 July he rehearsed the early scenes of *Caractacus*; on 27 August, after the Chorus holidays, he rehearsed the later sections. At Leeds he met the Festival conductor, Sir Arthur Sullivan, and told him of the contretemps at the Covent Garden Promenade rehearsal all those years earlier, when Sullivan had unknowingly taken up the time designated for the unknown Elgar. Edward described the meeting at Leeds:

When we were introduced, he said, 'I don't think we have met before.'

'Not exactly,' I replied, 'but very near it,' and I told him the circumstance.

'But, my dear boy, I hadn't the slightest idea of it,' he exclaimed, in his enthusiastic manner. 'Why on earth didn't you come and tell me? I'd have rehearsed it myself for you.'

They were not idle words. He would have done it, just as he said.[178]

[177] Letter of 9 Dec. 1948 from Mrs Gertrude Sutcliffe to Roger Fiske, published in The Gramophone, July 1957, p. 54.   [178] The Strand Magazine, May 1904, pp. 541–2.

The Three Choirs Festival at Gloucester in September 1898 contained very little Elgar. Only the 'Meditation' from *The Light of Life* was to appear as an orchestral voluntary at the Opening Service. None the less Edward attended London rehearsals for Gloucester, held at the Queen's Hall on 2 and 3 September. There Jaeger introduced Samuel Coleridge-Taylor, whose Ballade in A minor had resulted from Edward's recommendation.

But the presence of Jaeger himself seemed to call forth Edward's need for reassurance all over again. He remembered it himself: 'I was very down in the dumps; everything seemed to be going wrong. I was feeling pretty wretched . . . and told him I was going to give it all up and write no more music.'[179] Jaeger had leapt to the defence: 'He said that Beethoven had a lot of worries, and did *he* give it all up? No. He wrote more, and still more beautiful music— "And-that-is-what-*you*-must-do." '[180] Then Jaeger sent a letter to chase Edward back to Forli before the Gloucester Festival began: 'He wrote me such a screed, reams and reams of it, all about my ingratitude for my great gifts, as he called them, and he abused me for my wickedness—and I don't know what else!'[181] Edward's reply acknowledged the growth of intimacy in a pun:

Forli, Malvern.

[8th or 9th September 1898]

My dear Jaeger, (The Mister-y is soluted)
Very many thanks for your letter which soothed me immensely: a 3-choir festival always upsets me—the twaddle of it & mutual admiration. It was a real refreshment to me to see C. T. & know him . . .
I am afraid I gave you a bad impression of my temper the other day—I am not really bitter & my heart warms to anything like naturalness & geniality (C.T. e.g.) but I detest humbug & sham & can't talk it (well). Q's Hall seemed reeking with it those two *daze*.

Yrs.

Ed Elgar[182]

For the week of the Gloucester Festival he and Alice went over to stay at Hasfield Court with the Bakers. The Squire was his peppery self, assigning members of the party for various expeditions and slamming doors for emphasis. The Mascotte and Mr Penny were also in the party. Young Dora was elsewhere for the week, but she received a vignette of Edward and her stepmother:

He took her in to dinner and one of the first things he said was:
 'How is my sweet Dorabella?'
 'Oh! So it has got to that, has it?' said my stepmother.
 'That's a quotation from Mozart's *Così fan tutte*, don't you know it?'[183]

In those days of restricted Mozart performance, Edward himself might hardly

---

[179] Quoted in Powell, *Edward Elgar: memories of a variation*, pp. 110–11. See also Elgar's account of their 'long summer evening talk' in *My Friends Pictured Within*: IX. Nimrod.
[180] Ibid.   [181] Ibid.
[182] HWRO 705:445:8318.   [183] Powell, *Edward Elgar: memories of a variation*, p. 11.

have known it but for the memory of old Worcester Glee Club programmes that included the 'Comic trio, "My sweet Dorabella"'.

Back at Forli after the Festival, Edward played through every one of the *Caractacus* orchestral parts with the help of John Austin, the local violinist who led the Worcestershire Philharmonic. Then on 25 September he and Alice took the corrected parts to Novellos for the London rehearsal of orchestra and soloists before the Leeds Festival. The principal solo singers were the same as for the Crystal Palace *King Olaf*—Medora Henson, Edward Lloyd, and Andrew Black.

The *Caractacus* rehearsal excited those present, and there were frequent interruptions of applause. But Edward's rehearsal technique came in for criticism in the press:

It is difficult to say what would have happened if all the novelty-composers [for the Leeds Festival] had been fidgety and exacting in the same degree as Mr. Edward Elgar. Somehow Mr. Cowen with his 'Ode to the Passions', Dr. Stanford with his 'Te Deum', Dr. Alan Gray with his 'Song of Redemption', Herr Humperdinck with his 'Moorish Rhapsody', and M. Gabriel Fauré with his Ode 'La Naissance de Venus', contrived to make satisfactory progress and finish off their various works within an approximately just space of time. Not so Mr. Elgar. The Malvern musician is one of those composers who understand what they want a great deal better than the art of getting it. His idea, apparently, is to worry his forces into a comprehension of his intentions. He stops at every third bar and calls for repeats until the band fairly loses its temper (without perhaps showing it), and there ensues a general feeling of impatience, and dissatisfaction.[184]

Edward's special difficulties in rehearsing a first performance could be seen as another manifestation of insecurity—a reluctance finally to give over the creative phase and resign his work to its performance.

After London they went directly to Leeds. Saturday evening 1 October brought the final rehearsal, with orchestra and chorus together at last. Alice noted: 'Very enthusiastic. Tremendous applause. Mr. Spark interfered.' The Festival Secretary evidently objected to such demonstrations as a waste of valuable rehearsal time. But for Alice the applause was 'very great & wonderful to hear'. Parry was there again. And a number of friends had come especially for *Caractacus*: Basil Nevinson and Steuart-Powell, the Kilburns from Bishop Auckland, Sinclair from Hereford, Lady Mary Lygon, Miss Hyde, and Miss Burley—who contrasted the Leeds atmosphere with that of a Three Choirs Festival:

The Leeds Festival was a secular meeting . . . The audience were a good deal less clerical, less County and a good deal more fashionable and opulent . . .[185] The Lord Mayor (Mr. C. F. Tetley) entertained a very distinguished party at luncheon and Elgar and his wife were placed close to Sullivan . . . A. J. Balfour was at this luncheon and it would be a very interesting group that was assembled there. The solo artists were invited

184 *The Sunday Times*, 2 Nov. 1898.
185 *Edward Elgar: the record of a friendship*, p. 113.

to luncheon. The Elgars were very shy and rather frightened by the hearty Yorkshire people.[186]

The opening day of the Festival was reported in *The Leeds Mercury*.

The policemen's gloves never looked whiter, their bearing never more lofty or responsible, their helmets never so metropolitan. All day long, in gentle, persuasive accents, they implored, entreated, and directed. Now it was, 'Please, ladies, not so close in'—to the poor folk who had no tickets and less money—or 'Not this way!' to the cabby who would take the shortest route—in flagrant violation of all the printed regulations that ever were concocted. As the carriages, and the cabs, or whatnot, began to roll up for the opening concert in the morning, the crowd and the interest grew in volume. Women in discoloured attire, with shawls over their heads, divided their attention 'twixt the opening and closing of carriage doors and the disinterested cries, groans, gurgles, and shrieks of countless unhappy-visaged babies . . . Young girls, hatless, and arm in arm, stood patiently [in] that crowd outside the Town Hall all day long.

But it was nothing to the evening . . . It was here that the crowds—for there was a crowd outside all three main entrances—gaped most, and stared and jostled and pushed in their energetic attempts to see the Festival from without. And in sooth it was a brave sight to see the gaily dressed ladies tripping from the steps of the broughams and up over the red-carpeted steps, with the gentlemen behind shouting directions to the Jehus and giving a final pat for the satisfactory adjustment of their white tie . . . Hither and thither hurried people—greybeards, youths, elderly ladies, maidens—with 'scores' rolled under their arms . . .

With every seat again occupied, as it appeared, the Victoria Hall presented an even more brilliant spectacle [within]. There was, too, a certain pleasant air of expectation and excitement observable, for the first of the specially commissioned works stood awaiting judgment. This was Mr. Edward Elgar's 'Caractacus', which promised to be the most important of the novelties brought forward during the week . . .

It was a triumph, and everybody admitted it. Exclamations were to be heard all over the crowded hall after the conclusion of every scene. 'Tremendous, isn't it?' said one delighted lady to a bosom friend. 'Yes, that it is, and so dramatic,' was the gushing response. But, in spite of everything, in spite even of the noise of the brass and the crash of the cymbals, perhaps in consequence, hands were clapped as seldom they are in evening-dress circles, and the chorus rose en masse and cheered Mr. Elgar for all he was worth.

The composer was most modest. His bow was hurried, almost nervous, and he seemed only too glad to be able to get away from it all.[187]

In those days of gargantuan programmes, the ninety-minute *Caractacus* made only half an evening's music. The second half began with the Theme and Variations from Tchaikovsky's Suite No. 3.

The papers were loud in praise of the new work. *The Court Journal* struck a very official note: 'Mr. Elgar has not inaptly been dubbed 'the Rudyard Kipling of the musicians.'[188] Only one critic, E. A. Baughan, writing in *The Westminster Gazette*, wondered about the quality of the music itself. He made an

---

[186] Quoted in *Letters to Nimrod*, p. 21.   [187] 6 Oct. 1898.   [188] 8 Oct. 1898.

observation which could not have been applied to any of Edward's previous major works:

He has strength of execution, and his ideas, though tending towards the sentimental, and even now and then to the commonplace, are dramatic enough. It is only when we expect really original and strong invention that Mr. Elgar fails. His power of making music covers up his poverty of direct and appropriate thematic invention. All through the work we are brought to a standstill by a theme which has not risen to the dramatic and poetic situation, and the music as a whole gives one the impression that a second-rate mind has, by some freak of nature, been endowed with a capacity for expressing itself which we do not find in anyone who is not a genius.[189]

It confirmed a private reaction from Nicholas Kilburn—shared with Alice at the time, but not with Edward until several weeks had passed:

Your dear Lady & I exchanged a word or two at Leeds, when I said that C[aractacus] did not appeal to me to the same extent as Olaf. But in certain respects it seems to go far beyond that work, & this very bigness may perhaps stand in its way. That its thematic material does not give me as much satisfaction as 'Olaf' may be my own fault, but so it is.[190]

The quality of thematic material—of what Baughan called 'direct and appropriate thematic invention'—made the first measure of a composer's response to his subject; and Caractacus had been commissioned at a time when Edward himself wanted to write for the orchestra, where his inventions would not have to be reduced from their origins in his instrumental thinking to simpler shapes for voices. It was true that the subject of Caractacus had come out of the very 'air all around' Edward's home countryside. Yet as it emerged in his music, it was a story of heroic action defeated by dreams—by everything except the importance of squarely facing the task at hand. After the première, Miss Burley remembered Edward leaving Leeds 'with the air of one who has fought—and is inclined to think he has lost—a heavy engagement.'[191]

*　*　*

After Leeds Edward and Alice spent ten days in London. Three orchestral Dances arranged from the Bavarian Highlands were played at the Queen's Hall on 11 October. But that was the date of a Worcestershire Philharmonic rehearsal. Edward interrupted the London visit to go down to Worcester: he stayed the night with the Richard Arnolds in Britannia Square before returning to London. Then came two Crystal Palace performances of 'The Triumphal Procession' from Caractacus—again in an orchestral version.[192] Alice noted: 'Great success. E. had a real ovation. Gott sei Dank.'

[189] 6 Oct. 1898.   [190] 20 Nov. 1898 (HWRO 705:445:6544).
[191] Edward Elgar: the record of a friendship. p. 115.
[192] The programme of 14 Oct. lists Elgar's 'Grand Festival March in C', while that of the 15th specifies the Caractacus 'Roman Triumph' in a 'first performance at these concerts'. The

He called at Novellos to consult about his commission for the Birmingham Festival. The publishers said that if he fulfilled the commission with a simple work such as *The Banner of St. George* of wide appeal to amateur societies, they would be disposed to consider it favourably. It was what he least wanted to hear. And back at Forli (to which he and Alice returned on 19 October) there was a letter from Alberto Randegger requesting 'a short choral work, sacred or secular at your choice, for the [Norwich] Festival wh. will take place in the 2nd week of October 1899'.[193]

Nobody asked for the big orchestral work he had wanted to write instead of *Caractacus*. So he himself suggested to the Worcester Cathedral organist Ivor Atkins that he might write a symphony for the Three Choirs Festival in September 1899. He had a prescient idea for its theme: a symphony could be written round the subject of General Gordon—the hero who had gone to his death having just made acquaintance with Newman's 'The Dream of Gerontius'. Edward sketched a new figure of descending steps intersected by a great upward leap and extended in ways more interesting than literal sequence-making:

The entire figure could repeat and continue in a concentration of melody that had no parallel in *Caractacus*. Edward's sketch of it filled nearly an entire page.[194]

But that day he answered a letter from Jaeger with something like complete unhappiness:

Malvern

Oct. 20. 1898

My dear Jaeger:

Ja! I'm here, tho' I only arrd. last night & found your note—letter rather. No—I'm not happy at all in fact never was more miserable in my life: I don't see that I've done any good at all: if I write a tune you all say it's commonplace[195]—if I don't, you all say it's rot.—well I've written Caractacus, earning thro' it 15s/-d *a week* while doing it & that's all—*now* if I will write any *easy*, small choral-society work for Birmingham, using the fest. as an advt.—your firm will be 'disposed to consider it'—but my own natural bent I must choke off. No thank you—no more music for me—at present . . .

explanation was a shortage of rehearsal time, observed a few years later by young Adrian Boult: 'In this connection I may recall what Sir August Manns did in the old Crystal Palace days. Every day for a week he played a work which he variously called Rondo in F, Symphonic Sketch, etc., etc., by Richard Strauss. Finally on the Saturday afternoon he announced the first performance in London of Strauss's *Till Eulenspiegel*. That is one way of solving the rehearsal problem!' ('The Orchestral Problem of the Future', in *Proceedings of the Musical Association*, Session 49, 1923, pp. 50–1.)
    Letters to Elgar from Manns's programme-note writer Charles Barry (14 and 21 Sept. 1898: HWRO 705:445:3032–3) request information only about the *Caractacus* 'Triumphal Procession'.
    [193] 20 Oct. 1898 (HWRO 705:445:2927).
    [194] Sketch dated 20 Oct. 1898, later mounted in Jaeger's copy of *The Dream of Gerontius* vocal score (A. Rosenthal).
    [195] Referring to newspaper criticism of the recently published *Chanson de nuit*.

'Gordon' sym. I like this idee but my dear man *why* should I try?? I can't see—I have to earn money somehow & it's *no good* trying this sort of thing even for a 'living wage' & your firm wouldn't give 5£ for it—I tell you I am sick of it all: why can't I be encouraged to do decent stuff & not hounded into triviality . . .

Yours ever,
E.E.[196]

Next morning he replied to a request from F. G. Edwards for news of further Elgarian projects to go into *The Musical Times*: 'No: please not: it would be nearer the mark to say that "E.E. having achieved the summit (or somewhat) of his amb[itio]n, retires into private life & bids adieu (or a diable) to a munificent public." '[197] Then he went out to his day of teaching at The Mount.

Was there anything in that day, 21 October 1898, to mark a new beginning? It was hardly likely. The day of weary teaching, the return home with the little whistle of a third [musical notation] he always gave to tell Alice he had come in—all was dispiritingly usual, as he himself recalled it: 'After a long day's fiddle teaching in Malvern, I came home very tired. Dinner being over, my dear wife said to me, "Edward, you look like a good cigar," and having lighted it, I sat down at the piano . . .'[198]

He did not look to the piano for real invention. The piano's customary task in his creative process was the secondary, more 'ordinary' business of developing ideas already invented: so Edward's piano skill served its master as his father's piano skill had served the gentry among his clients. If he extemporized on the piano at the end of a weary day, it was usually nothing but aimless rumination.

In a little while, soothed and feeling rested, I began to play, and suddenly my wife interrupted by saying:
'Edward, that's a good tune.'
I awoke from the dream: 'Eh! tune, what tune!'
And she said, Play it again, I like that tune.'
I played and strummed, and played, and then she exclaimed:
'That's the tune.'[199]

[196] Elgar Birthplace.
[197] BL Egerton MS 3090, fo. 67.
[198] J. A. Forsyth, 'Edward Elgar: True Artist and True Friend', in *The Music Student*, Dec. 1932, p. 243. Though quoted from Elgar's memory many years after the event (like Basil Maine's account below), this agrees with the story published in *The Sheffield Independent* and other papers on 12 June 1899, a week before the première of the *'Enigma'* Variations.
[199] Combined from Forsyth and Basil Maine (*Works*, p. 101). Maine told me that his quotation was probably found in a drawer of source material in Elgar's study made available for his research in 1931–2. Later attempts to identify the source have been fruitless, and it may be that the words were quoted from what Elgar himself said. Much of Maine's material was gathered in this way from their conversations: but he was a careful writer, devoted to his subject, and would have made every attempt at accuracy from any source.

Once more Edward's music was juxtaposing G minor and G major. What he was playing now echoed the last juxtaposition of minor and major in the final scene of *Caractacus*, where 'Captivity'

was set side by side with the memory of Woodland freedom

Now the minor idea dragged slowly upward in successive thirds, with the first beat of every bar in the treble line a blank:

Then the idea turned to the major, smoothly filling those opening rests and gathering the thirds together in a half-memory of *Caractacus* 'Woodlands' with an augmented inversion through the last two bars tracing Edward's old descending steps:

---

[200] Quoted here from the first known draft of the music, (BL Add. MS 58003 fo. 2ᵛ).

It was as though the two ideas embodied the most simple, fundamental distinctions—major *v*. major, leaps *v*. steps, the hint of triple things in the top line of one *v*. the quadruple actuality in the other. So this latest juxtaposition could suggest a return, at a moment of deep uncertainty, to scrutinize first principles.

He said afterwards that through this music 'another and larger theme "goes", but is not played'.[201] It was widely assumed that the hidden 'theme' must be a deliberately concealed melody. Yet Edward never said the hidden theme was a tune. (In fact, Dora Penny recalled, 'we always spoke of the hidden matter as "it", never as tune or theme'.[202]) And at the moment of first playing the music, his mind had been so far from deliberation that it needed Alice's interrupting to show him an entity at all. This was not consistent with any deliberate concealment by counterpoint, polyphony, or cipher. Whatever it might mean, then, its significance was yet to emerge.

> The voice of C. A. E. asked with a sound of approval, 'What is that?'
> I answered, 'Nothing—but something might be made of it . . .'

And again Edward's uncertainty showed itself. The 'Nothing' theme had been noticed by Alice, not himself. So for its development he looked again not to himself but to others. Who?

Playing the piano suggested other pianists. Long afterwards it was thought that the G minor theme was an unwitting reminiscence of the 'Benedictus' in the *Requiem* which Stanford had played to Edward on this same piano more than two years earlier.[203] Yet there was as much similarity in the 'Credo' from Beethoven's *Missa Solemnis*—or in Tchaikovsky's orchestral Variations in G which had followed *Caractacus* at Leeds.

It was the memory of Leeds that now provided Edward's mind with the intimate pianistic relation it sought. One of the friends gathered there for *Caractacus* a fortnight earlier was Hew Steuart-Powell, the pianist of their old trio evenings with Basil Nevinson. Before each of those sessions Steuart-Powell had warmed his fingers with 'a characteristic diatonic run over the keys before beginning to play'.[204] Edward's quick brain instantly rearranged the descending steps of the 'Nothing' theme's B-inversion in a travesty 'chromatic beyond H.D.S.-P.'s liking',[205] and he said to Alice:

'Powell would have done this':

---

[201] Quoted in Charles Barry's programme note for the first performance of the *'Enigma'* Variations, 19 June 1899.

[202] Powell, *Edward Elgar: memories of a variation*, p. 119.

[203] See *The Times*, 20 Aug. 1977 and following correspondence.

[204] *My Friends Pictured Within* (Novello, 1949), reprinting Elgar's notes written to accompany player-piano rolls of the *Variations*. Later quotations about *Variations* subjects are from this source unless otherwise noted.   [205] Ibid.

Against it in the bass he brought a ghost of the A-theme. Thus he briefly developed the 'Nothing' music through another's medium.

Who else might be used in this way? There was the cellist Nevinson, who had also been at Leeds. Edward turned to Alice again: 'or Nevinson would have looked at it like this.'[206] And his fingers traced a slowly singing variation in the tenor register setting A

side by side with B ... etc.

The reflective Nevinson music took Edward's mind beyond the matter of simple imitation to another idea, as he himself described it: 'The variation is a tribute to a very dear friend whose scientific and artistic attainments, and the whole-hearted way they were put at the disposal of his friends, particularly endeared him to [me].' So the whole matter of friendship emerged as a central concern in this piecemeal self-development through other personalities. But that as yet glimmered dimly beside the amusement of evoking them in his 'Nothing' theme.

Could the suggestive personality be evoked so clearly that he would be recognized directly from the music? It would need a very definite character. Edward thought of the Squire at Hasfield Court, played the A and B themes through *Allegro di molto* with a bang at the end, and turned to Alice:

'Who is that like?'

The answer was, 'I cannot quite say, but it is exactly the way W. M. B. goes out of the room. You are doing something which I think has never been done before.'[207]

After identifying the theme for him, Alice was now encouraging him to realize

---

[206] Quoted in Maine, *Works*, p. 101.

[207] Quoted in Maine, *Works*, p. 101. In fact the idea was not new. Mozart wrote to his father of shaping a movement after the character of a lady. Schumann repeatedly fashioned music round the personalities and initials of friends. And the following reminiscence shows a similar interest in the music of Chopin: 'Chopin liked and knew how to express individual characteristics on the piano. Just as there formerly was a rather widely-known fashion of describing dispositions and characters in so-called "portraits", which give ready wits a scope for parading their knowledge of people and their sharpness of observation; so he often amused himself by playing such musical portraits. Without saying whom he had in his thoughts, he illustrated the characters of a few or of several people present in the room, and illustrated them so clearly and so delicately that the listeners could always guess correctly who was intended, and admired the resemblance of the portrait.' (Stanislas Count Tarnowski, quoted in James Huneker, *Chopin: the Man and his Music*, Dover Publications, 1966, pp. 57–8).

himself in every way he could. If the inspiration of these other personalities would help, it was no more than the friendships of young and attractive women 'who could do the parts she couldn't always manage'.

Several friends of the Hasfield connection offered personal traits for new variations. Dora Penny had the grace of youth and a stammer in her speech: it suggested a little hopping figure round the A theme's middle phrases

Underneath it presently came a form of the B theme in smooth counterpoint.

Then there was R. B. Townshend, the Baker brother-in-law who had acted an old man in amateur theatricals at Hasfield—'the low voice flying occasionally into "soprano" timbre'. His variation turned both A and B themes to quirky rhythms and wide skips.

Back in Worcester there was Richard Arnold, whose characteristic laugh, '*HA*-ha-ha, ha-ha-*HA*-ha-ha!'[208] readily shaped itself to the B theme's counterpoint of descending steps:

It was set within a dark minor opening. Juxtaposed together, the two ideas gave Edward's impression of Arnold at Worcestershire Philharmonic Society tea parties: 'His serious conversation was continually broken up by whimsical and witty remarks.'

Thoughts of the Worcestershire Philharmonic suggested the Misses Norbury and Hyde, and he began a variation whose sketch was first labelled '2 sec[retar]ys'. Then, on Sunday 23 October, when he walked up to Birchwood with Troyte Griffith, they were caught in a thunderstorm and sheltered in the Norburys' comfortable old house Sherridge in the foothills below.[209] So the '2 secretarys' idea came to focus at Sherridge as a sort of minuet evoking the old house. It was finally initialled 'W.N.'

The thunderstorm itself recalled Troyte. Edward had dubbed him 'the aged Ninepin', and the storm's noise reminded him of bowling over ninepins. But these violent sounds also brought back another shared experience, which Edward described after the 'Troyte' variation was orchestrated:

The uncouth rhythm of the drums and lower strings was really suggested by some maladroit essays to play the pianoforte; later the strong rhythm suggests the attempts of

---

[208] Powell, *Edward Elgar: memories of a variation*, p. 107.
[209] MS notes by W. M. Baker's eldest son.

the instructor (E.E.) to make something like order out of chaos, and the final despairing 'slam' records that the effort proved to be vain.

Yet again Edward's music had caught the subject's broader character. An acquaintance recalled:

Troyte Griffith could be direct to the point of abruptness and just as 'staccato' as Elgar portrayed him—in action, speech, and temperament. He could be difficult if anyone suggested a variation of his opinions or plans.[210]

What about other friends who had gathered at Leeds for *Caractacus*? Several were musicians. Sinclair's virtuosity on the organ pedals might suggest a variation. Or Ivor Atkins. Or Lady Mary Lygon, who had founded a Competition Festival for local musicians living round her brother's stately home at Madresfield. Could Parry or Sullivan be brought in? When it came to framing other composers, however, Edward found his own music merely parodied their styles and gave him nothing back. So he gave them up.[211] But there was Kilburn—to whom he had written two months earlier about *Caractacus*: 'My thoughts of you & Mrs. Kilburn & your kindness to me are woven into many a bar of my score—*you* will understand how that can be . . .'[212]

Thoughts of intimate encouragement inevitably evoked Alice. So he inserted the little figure of his own whistled greeting into the theme she had identified:

and wrote of the result many years later: 'The variation is really a prolongation of the theme with what I wished to be romantic and delicate additions; those who knew C. A. E. will understand this reference to one whose life was a romantic and delicate inspiration.'

When it came to purely musical encouragement, there was one more inevitable friend. It was Jaeger, whose Beethoven comparison had touched Edward deeply. He turned the G minor theme to a quiet E flat major

[210] Bertha Flexman, letter to the writer, August 1983.
[211] Maine, *Works*, p. 101.
[212] Traces of a 'Kilburn' variation can be found in the sketches.

where it found affinity with the slow movement of the 'Pathetique' Sonata. As 'Jaeger' in German meant 'hunter', Edward concealed the music's almost too strong emotion under the classical allusion of 'Nimrod'.

On Monday 24 October he wrote again to his friend. It was just five days since the return from London—four days since the depressed letter about the 'Gordon' symphony—three days since the evening in which this new music had begun.

Malvern

Oct. 24 98

My dear Jaeger,

Here is the 'grecian ghost which unburied remains inglorious on the plain' or on the hills . . . Our woods look lovely but decidedly damp & rheumaticky—unromantic just now.

Since I've been back I have sketched a set of Variations (orkestry) on an original theme: the Variations have amused me because I've labelled 'em with the nicknames of my particular friends—*you* are Nimrod. That is to say I've written the variations each one to represent the mood of the 'party'—I've liked to imagine the 'party' writing the var: him (or her) self & have written what I think *they* wd. have written—if they were asses enough to compose—it's a quaint idee & the result is amusing to those behind the scenes & won't affect the hearer who 'nose nuffin'. what think you?

Much love & sunshine to you.

Ed. Elgar[213]

The one element common to all the friendships—and thus to all the variations—was Edward himself. So the collection of them all together could hold the hope of some self-discovery—self-discovery of the sort that the defeated Caractacus had denied him. His own friendships, after all, enjoined no single plot: the result of them all would be entirely what he himself made of it. So these variations offered a perfect vehicle for the impulse seen already in the shaping of several earlier works—a progress from small diversities toward some unifying generality. Edward projected a Finale for these variations to suggest himself—'merely to show what [the composer] intended to do'.[214] This Finale could be a goal for the theme Alice had discovered in his extemporizing. But then the Finale initials should not be 'E.E.', the man who was: rather they should be 'E.D.U.'—the 'Edu' of Alice's intimate vision.

Other friends were consulted. Friday 28 October brought the next weekly teaching session at The Mount. Miss Burley recalled it:

. . . He came to The Mount in the rather excited state which usually indicated some new inspiration and played me a sixteen-bar tune on the piano. I thought it wistful but hardly of outstanding interest and asked what it was. So far as I remember he did not answer the question but continued to play, apparently extemporizing, a set of variations each of which, he said, represented a friend . . . He was far more concerned with the variations than with the underlying theme and constantly challenged me to guess whom they represented . . .

---

[213] Elgar Birthplace.　　[214] Elgar, *My Friends Pictured Within.*

In many cases, however, the portraiture was astonishingly accurate and the translation of physical or mental characteristics into musical terms wonderfully ingenious. As the work progressed, indeed, and variation was added to variation, I realized that the complete set, when illuminated by Edward's brilliant orchestration, might show an enormous advance on anything he had previously written.

I believe it is true to say that Edward enjoyed the writing of the Variations more than that of any other work. At any rate he seemed happier to me. The fact is that for once he was not writing on commission but for the pleasure of doing so. I doubt indeed if he foresaw at the beginning that he had begun an important work.[215]

But Alice recognized it. And she recognized again the need for other encouragers. She wrote to Dora Penny, describing 'some wonderful and most *exciting* music, dear Dora! You simply *must* come soon and hear. I have promised H[is] E[xcellency] not to say a *word*.'[216] ('His Excellency' had been Dora's reply to the 'Dorabella' sobriquet.) An opportunity for Dora's next visit would shortly present itself in the second concert of the Worcestershire Philharmonic—to include the *Bavarian Highlands*—on 1 November.

Meanwhile Edward took himself off to Hereford, where George Sinclair was to give his fiftieth recital as organist of the Cathedral. He stayed the weekend, and one afternoon walked the banks of the Wye with Sinclair and his bulldog Dan. As they walked, Dan wandered too far from the path and rolled down the steep bank into the river. But he was a strong swimmer, paddling downstream until he could get a foothold and scramble up the bank with a bark. Sinclair, remembering the 'Moods of Dan' which Edward wrote from time to time in his Visitors' Book, challenged him: 'Set that to music.'[217]

It took the idea away from making a variation out of the organist's pedalling. The dog could roll down the bank, fall in, paddle a steady quaver version of A, scramble out to B *fortissimo*, and bark. None the less, one of Sinclair's later friends found that the music also hit off precisely G.R.S.'s 'prancing' walk—'a quick pace but with a stiffness in it'.[218]

Returning to Malvern on Tuesday 1 November, he was joined by Alice and then Dora Penny on the way to Worcester for the Worcestershire Philharmonic rehearsal and concert—separated by luncheon. Dora recalled:

The composer was in high spirits but I could get nothing out of either of them about the new music. All he said was:

'You wait till we get home. *Japes!*'—taking up a spoon and conducting something with it.

And I had to be content with that.[219]

After the concert the Elgars took Dora Penny back with them to Forli:

[215] *Edward Elgar: the record of a friendship*, pp. 116, 117, 128.
[216] Powell, *Edward Elgar: memories of a variation*, p. 12.
[217] Elgar, *My Friends Pictured Within*.
[218] Olive Gosden, letter to the writer, 1982.
[219] *Edward Elgar: memories of a variation*, p. 12.

No sooner inside the door than E. E. fled upstairs to the study, two steps at a time—I after him, the Lady following at a more sedate pace.

'Come and listen to this,' and he played me a very odd tune—it was the theme of the *Variations*—and then went on to play sketches, and in some cases completed numbers, of the *Variations* themselves. I turned over and saw the next page headed 'C. A. E.', the Lady's initials, something dedicated to her, evidently. Very serene and lovely—and in some curious way *like* her.

Then he turned over two pages and I saw No. III, R. B. T., the initials of a connexion of mine. This *was* amusing! Before he had played many bars I began to laugh, which rather annoyed me. You don't generally laugh when you hear a piece of music for the first time dedicated to someone you know, but I just couldn't help it, and when it was over we both roared with laughter!

'But you've made it so *like* him! How on earth have you done it?'

'Go on, turn over.' And the next piece was called No. IV, W. M. B., another connexion and a great friend; very energetic and downright. Why did it remind me of him so?

I think he then played Troyte, and a shout of laughter followed. 'What do you think of that for the giddy Ninepin?'

After that, 'Nimrod'.

'That must be a wonderful person, when am I going to meet him?'

A voice from near the fire-place: 'Oh, you *will* like him, he is the *dearest* person.' . . .

Then I turned over and had a shock. No. X, 'Dorabella'. Being overcome by many emotions I sat silent when it was over.

'Well, how do you like *that*—hey?'

I murmured something about its being charming and rather like a butterfly, but I could think of nothing sensible to say . . . (I had no idea what it really meant. It was not until many years afterward that it dawned on me that I had been as much the victim of E. E.'s impish humour as had R. B. T. I stammered rather badly at times . . . Elgar exploited his humour at my expense with such marvellous delicacy that no one could help laughing with him—if they understood it.) . . . My mind was in such a whirl of pleasure, pride, and almost shame that he should have written anything so lovely about *me*.[220]

Dora Penny's reaction—that the music was about her—was to be echoed again and again by the Variations' subjects. Dora particularly came to feel that she must hold a special clue in the search this music was only beginning to pursue. Yet one misunderstanding that evening might have warned her:

When 'E. D. U.' was first played to me on the piano, hilarity knew no bounds. E. E. shouted with laughter. But I had not then grasped who E. D. U. was and I remember thinking what a determined and forceful person this must be. Then something, not far from the end, caught my attention; where *had* I heard that? But I could not remember and I didn't ask.[221]

After Dora's visit, Edward sketched one more variation on a young lady—Isabel Fitton. Her tallness and her viola lessons with Edward both suggested splaying the A theme across viola strings

---

[220] Ibid., pp. 12–13, 112.   [221] Ibid., pp. 115–16.

—the reminder of an exercise he had once devised for her. A friend recalled:

> She was very tall & graceful, and seemed to 'float' rather than walk—was nearly always late for orchestra, and then sat amidst various scarves and belongings which she had discarded and strewn about her . . .
> She was full of fun. She stopped her viola lessons with him at one stage, & when he asked her to reconsider her decision we were told she said:
> 'No, dear Edward, I value our friendship much too much!'[222]

So 'Ysobel' (as Edward's eccentric spelling rendered her) brought the variation inventing full circle: the memory of her remark could remind him again of how it all began—with friendship.

On 7 November 1898 Edward met Charles Beale from the Birmingham Festival committee, '& we settled, verbally, that I am to have the principal place in the Birmingham Festival' of 1900.[223] The world of the national choral festivals could offer no higher distinction. He and Alice went up to Birchwood for a late autumn stay, and from there he wrote on the subject of women and music to Jaeger, who was engaged to marry a brilliant young violin student at the Royal College of Music.

Birchwood Lodge (Deo gratias!)

Nov 11

My very dear Jaeger,

I was sorrowing for a line from you & now comes a nice, *human* letter to me here, where you know I find the only really happy times now: thank you a thousand times.

We came up here two days ago & are in fog & the leaves are falling too rapidly but it's jolly nevertheless away from all feuds, intrigues & cranks . . .

I am so very very happy thinking of your new life, because I've seen Miss D[onkersley] & can, thank God, congratulate you & believe you *will* be happy—one can very seldom say this to men who shew you the 'modern young women' they are going into partnership with: Lord pity 'em! . . .

The Variations go on slowly but I shall finish 'em some day.

Private. I have agreed to do the *PRINCIPAL* novelty for Birmingham.

Randegger asks me about Norwich wanting, of course, something new—I suggested Black Knight amongst other old things but have not heard anything since the first committee meeting.

Then, *un*officially, poor old Worcester wants a symphony! . . .

Now as to Gordon: the thing possesses me, but I can't write it down yet: I *may* make it

222 Olive Gosden, letter to the writer, 1977.
223 17 Nov. 1898 to Alfred Littleton (Novello archives).

the Worcester work if that engagement holds. So don't, please, pass on the idea to anybody else just yet. I would really ask [F.G.] Edwards to announce it [in *The Musical Times*] but I must first get to know if the Dean & Chapter wd. object to the subject in an English Cathedral! . . .

Yrs ever
Ed. Elgar[224]

He wrote to the Novello chairman Alfred Littleton about the possibility of publishing the new works. No answer came back. A month later, near the end of the year, depression tightened its grip. He traced it back six weeks: that was virtually to the day he had agreed to take the principal place at the next Birmingham Festival.

Forli, Malvern.

Dec. 17

private

My dear Jaeger:

I put this on another sheet & I shd. not have bothered you with it—only you ask: for the last six weeks (about) I have been very sick at heart over music—the whole future seems so hopeless. I wrote to Mr. L.—because I had talked to him previously—about the Bir: fest: work & he does not reply: also I have asked how my egregious debt to the firm (K[ing] O[laf]&c) stands & they tell me nothing.

Now I have worked steadily & honestly till I am offered all the festivals & then the firm seem to have had enough of me. I can quite understand that my big works don't pay—i.e. shew any good return but I shd. have hoped that on artistic grounds the very small remuneration I ask shd. be forthcoming for things which at least interest the better portion of the musical public. No! the only suggestion made is that the Henry VIII dances [by Edward German] are the thing—now I can't write that sort of thing & my own heartfelt ideas are not wanted: why K. Olaf shd. be worthless when it's done often is a mystery to me when things by, say Mackenzie, which are never touched, shd. be good properties—You see I want so little: £300 a year I must make & that's all—last year I subsisted on £200. It seems strange that a man who might do good work shd. be absolutely stopped—that's what it means.

Now you see how things are. do *not* tell anyone all this or any of it. I did not intend to write as it may seem disloyal to the firm but apparently this is the end of all things so it doesn't matter.

Ever yrs.
E.E.[225]

It was not that he and Alice failed to make ends meet: teaching and conducting together with her income could always make a subsistence. The money from compositions was needed to reassure Edward that his music was valued in terms that could be measured more clearly than in laudatory reviews and patronizing encouragements: it should be his means to the life of gentry and aristocracy, with bespoke suits of finest cloth and first-class railway travel and hotel accommodation.

Worcester accepted the 'Gordon' Symphony for the Three Choirs Festival in

---

[224] HWRO 705:445:8320.   [225] HWRO 705:445:8324.

September 1899—without a commissioning fee, as was the custom—and it was announced in the press. And Randegger proposed a change of plan which made the Norwich project more attractive to a composer whose thoughts were more and more orchestral: 'We want a short work not *entirely* Choral—as we must employ on one evening Miss Clara Butt, Mr. Ben Davies, & Mr. D. Bispham . . . If you have not found a suitable "*libretto*"—could you write a "*Scena*" for either Contralto, Tenor, or Bass?'[226]

Clara Butt was a young contralto who had made a success before going to study in Paris. She had recently returned to become the rage of all the concert platforms. Edward sketched some songs for her, and included the little setting of Alice's 'Love alone will stay'. On 9 January he brought the song sketches to London to show to Randegger—who expressed approval provided Clara Butt would sing them.

So Edward took himself to call on Clara Butt at her flat in Hyde Park Mansions. Years afterward she would recall the scene for her biographer, who wrote:

Her companion, Madame Snella, went to the door one day in answer to a knock, and came in to report that some one had called with some songs, and wanted to show them to her mistress. Clara was splashing contentedly in her bath at the time, and said she wasn't coming out to see anyone!

Madame Snella came back, evidently impressed, after delivering this message to say, 'Oh, "Baby" dear, *do* see him! He's such a nice man, and he says he's got a whole *cycle* of songs to show you!'

'I don't care if it's a *bi*cycle! If he wants to see me he must call back another day!'[227]

Edward complained to Randegger. Randegger remonstrated with her manager, Narciso Vertigliano—who had long since abridged his name for English purposes to 'N. Vert'.[228] When Edward called again at Hyde Park Mansions on 14 January, he was cordially received. Ideas were canvassed, interest shown, promises given. So there was another Festival commission—again without any commissioning fee.

Yet contact with Vert opened a different prospect. He acted as concert manager for Hans Richter, and through his agency it was now decided to approach Richter with the idea of giving the first performance of the *Variations* at one of his London concerts.[229] Thus introduced, Edward's most private music might place him in the royal line of Richter's great composers directly after Wagner and Brahms.

[226] 3 Jan. 1899 HWRO 705:445:2925.

[227] Winifred Ponder, *Clara Butt, her life-story* (Harrap, 1928).

[228] Vert began his career in London in the 1860s as assistant to Charles Dickens's manager. By the 1890s he acted as agent for many international musicians. The earliest known record of Elgar's association with his firm was in Sept. 1897, when Vert undertook to act as a forwarding agent (HWRO 705:445:3211).

[229] Elgar wrote to Vert on 13 Oct. 1901: 'I have to thank you for the introduction to Richter which led to the first performance of the work—you may have forgotten this—but I have not & shd. like you to understand that I am very grateful to you for your kindness *when I much needed it*.' (Ibbs & Tillett archives).

On the evening of the day he saw Clara Butt, Edward and Alice went to call on Jaeger and his bride in their West Kensington house. Edward showed the *Variations* sketches, and then told Jaeger about the prospect of interesting Richter. Jaeger resolved to help matters on. He took the news to Parry, who had influence with Richter. And Parry, with characteristic generosity, put in his own word for a work as yet unfinished by a man he scarcely knew.[230]

Returning to Malvern on 17 January 1899, Edward threw himself into finishing the *Variations* with such energy that in less than three weeks the main outlines emerged. One advantage of variation-writing is to allow separation of creative thinking into distinct phases. The composer's whole thought can be first given to inventing separate variations in any order; then later he can turn fully to the question of ordering them in a larger structure. As the final sequence of Edward's theme and variations emerged, it traced an extraordinary pilgrimage.

The 'tune' which Alice had identified to start these *Variations*—with its alternatives of minor and major, leaps and steps, triple hint and quadruple reality—posed a fundamental problem: how to synthesize the opposite impulses? Where those impulses came from was an inscrutable, unknowable thing. Edward called this theme 'Enigma', placing the word only over the theme in the score. It was a vital point, persistently ignored or confused by later listeners: the 'Enigma' is nothing more or less than just the seventeen bars of music out of which the Variations grew—the 'nothing-but-something-might-be-made-of-it'.

Here was not a riddle to be solved, as Edward himself realized:

The Enigma I will not explain—its 'dark saying' must be left unguessed.[231]

It was the inchoate blackness out of which creation comes. Years later he would write of the 'Enigma' theme:

. . . it expressed when written (in 1898) my sense of the loneliness of the artist . . . and to me, it still embodies that sense . . .'[232]

---

[230] Harry Plunket Greene (*Charles Villiers Stanford* (Edward Arnold, 1935, pp. 157–8) stated that Jaeger showed the score of the *Variations* to Parry, who took it enthusiastically to Richter on a rainy night. Events could not have happened thus, for the score was sent on completion directly from Malvern to Vert and thence on to Richter. Jaeger's first sight of the *Variations* score was only when it reached Novello's offices early in Apr. 1899, by which time Richter had accepted the work.

Yet Plunket Green was Parry's son-in-law, and his story will have had some basis in fact. As Elgar later noted that Vert (as Richter's orchestral manager) 'did the work as a favour' (letter of 1 Sept. 1899 to Jaeger) it seems likely that Parry put pressure on Richter. It would be typical of Parry to fight for an unknown work, whose success might well prejudice his own *Symphonic Variations* produced eighteen months earlier. When the *'Enigma' Variations* were produced Parry wrote 'a nice rapturous letter' to Elgar—who replied with a presentation copy of the score when it was engraved and published in Jan. 1900.

[231] Quoted by Charles Barry in the première programme note for the Richter Concert on 19 June 1899.

[232] Notes on *The Music Makers* enclosed in a letter of 14 Aug. 1912 to Ernest Newman (Elgar Birthplace).

The notion of developing this Enigma through the personalities of friends had lightened the mood—but only at first, he recalled:

This work, commenced in a spirit of humour & continued in deep seriousness, contains sketches of the composer's friends. It may be understood that these personages comment or reflect on the original theme & each one attempts a solution of the Enigma, for so the theme is called.[233]

Each of the variations juxtaposed elements of both the A and B figures from the original theme. The problem addressed in each variation, then, was that of finding out how the fundamental differences of musical discourse represented by those two figures could be reconciled. This was the problem which 'E.D.U.' must solve—finally and convincingly.

The variations between the theme and the 'E.D.U.' Finale were all experiments—essays which contained hints of what or what not to do. Recognizing in others the traits that appeal to oneself is the beginning of individuality. It is the process by which a child first learns to make his own way—by imitating only those things in others which appeal to him. So these variations might enact the very experience of character building that leads toward identity. And not only for the composer. For the listener too, the real subject of the *Variations* would be the creation of a self in music.[234]

After the 'Enigma' theme, the music continued without a break into the first variation, 'C.A.E.' The link delicately acknowledged Alice's recognition of the theme. Her music's incorporation of Edu's whistling figure suggested the wish to synthesize. And the polyphonic combination of the climax suggested synthesis again:

Thus 'C.A.E.' emerged not so much as a portrait as a vision of ideal relationship. It made at once an epigraph and a goal.

'H.D.S.-P.' mixed the B theme with chromaticism. In the midst of it a ghost of A rose obscurely in the bass; then the chromaticism reasserted itself.

Next came 'R. B. T.', cleverly counterpointing A in the major with a summary of B:

---

[233] MS statement in Elgar's hand, dated 'Torino, Ottobre 1911', drafted for the programme of a concert he was to conduct there (Elgar Birthplace).

[234] In his *Portrait of Elgar* (OUP, 1968, p. 59) Michael Kennedy drew attention to the fact that Elgar signed a letter to Dora Penny with the notes of the *Variations* 'Enigma' theme, and added 'the name Edward Elgar "goes" [with the theme's opening notes] in almost natural speech-rhythm.'

But it was all so whimsical, with grumbles from the bassoon, that this music could be no more serious than the amateur theatricals it recalled at Hasfield.

It was a sly stroke to follow the elaborate fussiness of 'R. B. T.' with the forceful Hasfield brother-in-law 'W. M. B.' His rough handling of both the A and B figures closed a first paragraph in the *Variations*, where the opposition of A and B had been gradually reduced to jocularity.

'R. P. A.' opened new complexity in a new tonality, C minor. His music reset the B-inversion in 12/8, against A in 4/4:

But then came Richard Arnold's laugh round the B-inversion: '*HA*-ha-ha, ha-ha-*HA*-ha-ha!' For the first time two entirely separate moods were contrasted within a single variation. And just as in Arnold's conversation, the two remained separate through repeated recapitulations.

A last *pianissimo* C minor ushered in the 'pensive and, for a moment, romantic'[235] viola of 'Ysobel' in C major. The solo viola spread its little exercise of the A figure across the strings. The B figure moved gracefully away to a graceful sequence. Again the viola tried its gentle rhetoric of A, now with 'romantic and delicate additions' of its own. Again B stepped lightly aside. That left the solo viola alone at the end to play its A exercise *pianissimo* all by itself. Yet the touch was so light throughout this music that any sense of cross-purposes almost disappeared.

Cross-purpose emerged again in 'Troyte' with thunderous contrast. Its rhythm was 'uncouth' because it set A triple-time in the bass against B in the treble:

[235] Elgar, *My Friends Pictured Within*.

In reply, B moved into the bass to insist on strict duple metre. The instruction was in vain: 'Troyte' came banging back with his uncouth rhythm under cover of a treble run *brillante*. The encounter ended in brass *fortissimo* with the 'despairing "slam"'[236] of A.

After the 'Troyte' thunderstorm came the shelter of Winifred Norbury's old house. 'W. N.' returned the music for a moment to its home G, but full of witty subtlety. Here the A figure was shaped

and reshaped

Then came B trilling in a clear 6/8 as 'a little suggestion of a characteristic laugh'—upon which the B inversion was joined with a further reshaping of A:

So again the elements of the problem were mixed with humour.

The resolving G metamorphosed suddenly to a major third in the totally new landscape of E flat major for 'Nimrod'. So this second paragraph of variations would end as it had begun in three flats. Unlike the other variations of the second paragraph, however, 'Nimrod' entertained no subtle contrasts of mood. A great melody of A in the major projected nothing but manly nobility. B made a brief hushed appearance, spreading from a single note through the whole ensemble in 8 bars. Then A returned, moving majestically toward a coda which concentrated the full power of melody and harmony *fortissimo*—only to fall back to *pianissimo* in a single bar. It was to be acclaimed as the most memorable of all the individual variations. Yet as 'Nimrod' stood for single-minded devotion, his near-total absorption with the A side of the theme left the last answer still to seek.

[236] Ibid.

The girlish stammer of 'Dorabella' brought an 'Intermezzo'. It began in the midst of the A figure, giving each note a little pirouetting shake. Underneath it presently came B singing on a solo viola:

A sly reminiscence of the 'R. B. T.' bassoon grumbles sketched a small middle section. The pirouettes returned with a flute echo of the lovely viola melody. But with each return the echoes diminished until the pirouettes ended as innocently as they had begun.

The final paragraph of variations opened with the 'G. R. S.' bulldog rolling down the bank *Allegro di molto* to paddle in the water of G minor. It recalled the G minor *Allegro di molto* with which 'W. M. B.' had closed the first paragraph, and it had the same rapid concentration: there was the bark, another little canon, another bang at the end.

Then the violoncello of 'B. G. N.' sounded—as near to the end of the *Variations* as his friend 'H. D. S.-P.' had been near the beginning. The solo cello sang a slow, long-breathed song: it was the same inner phrase of A that 'Dorabella' had pirouetted through, but now reminiscent of 'Nimrod'. The cellos of the orchestra explored both the A and B figures in slow triplets within the quadruple metre, as if to hint at some metrical reconciliation. Yet at the end of the solo cello was left again to play its lonely inner phrase of A.

It led without a break to the penultimate Variation XIII, which also brought back Edward's old descending fourths:

Moderato

This music had first been suggested by the charming Lady Mary Lygon of Madresfield.[237] Then late in January 1899, just as the whole work was taking shape, it was announced that Lady Mary would accompany her unmarried brother Lord Beauchamp to a colonial governorship in Australia. Out of Edward's memory came the phrase from Mendelssohn's *Calm Sea and Prosperous Voyage* noted in the Crystal Palace programme all those years ago:

[237] Miss Burley (*Edward Elgar: the record of a friendship*, pp. 126–7) insisted that Variation XIII really represented another lady, whom she declined to identify. But on an early sketch sheet of Variation *incipits* this music is clearly labelled in Elgar's hand 'L.M.L.' (BL Add. MS 58003 fo. 6).

A spell comes over the scene, and we are made aware of the presence of deeper and more human feelings than the mere excitement of the breeze can awake. The melody in which Mendelssohn has embodied these sentiments [is] so characteristic . . . that at the time of the composition of the work, the phrase was used by Felix and his most intimate friends as a signal or call.

The descending steps of Mendelssohn's phrase linked hands with the B-inversion of the Enigma theme. Suddenly there opened an immense horizon of friendship and love and longing. The Mendelssohn phrase went into Edward's music in A flat major—a mere semitone from the music's G major yet aurally a world away. A recursion to the G major opening brought the *Calm Sea* motive in E flat, the tonality of 'Nimrod', only a moment before 'E. D. U.' himself was to appear.

So this penultimate music took its composer too far from his innocent friendship and simple regret at Lady Mary Lygon's departure. He deleted her initials from Variation XIII (and later even contemplated cutting out the Mendelssohn phrase altogether[238]). But he planned a recompense. Remembering Novello's suggestion of writing something like German's *Henry VIII Dances,* he looked out the old orchestral Suite of 1888 and interrupted work on the *Variations* to spend 23 and 24 January 1899 revising three of the four old Suite movements: these he sent in to the publishers as *Three Characteristic Pieces* Op. 10, with a special request to get them printed for Lady Mary before her voyage in April. (Novellos were delighted and agreed to pay Edward no less than 50 guineas for the three.)

On 25 January he was back with the *Variations.* The initials of Variation XIII were replaced with three asterisks, and the music was now headed 'Romanza'. So it might sing of aspiring, vaulting imagination both light and dark before the arrival of 'E. D. U.'

The 'Romanza' G major fourths opened 'E. D. U.':

After its opening in G, the same figure repeated at B, C, E, A, and D—together projecting a near-augmentation of the 'Romanza's'

---

[238] 2 May 1899 to Jaeger (HWRO 705:445:8346).

Enigma A appeared forcefully in the 'Nimrod' E flat major; the descending steps of the *Calm Sea* motive were echoed

a new version of the B figure moved *piu tranquillo* closer to A

Then (at cue 68) came the 'E. D. U.' solution of the Enigma. It united the B-inversion with an augmented A in the major:

Here the descending steps which had haunted Edward's music from childhood were wed to the opening 'Enigma' theme which Alice had identified from his extemporizing. It was just the combination of A with B that 'C. A. E.' had discovered, in the minor, in the first variation of all. Had her music then foreseen the whole of E. D. U.'s pilgrimage of self-discovery through all the other friendships?

One insoluble problem remained. In the *Grandioso* combination, the top line of the A figure, without its rests, suggested triple metre; the B figure remained unambiguously quadruple. So their combination in polyphony would never finish together. The overlapping sequences in 'E. D. U.' began to spiral inward, contracting and contracting until they came down to a single unresolving point.

The 'E. D. U.' opening recapitulated. And then Edu's little whistling figure called 'C. A. E.' to witness what he had done, to help him finish it. So she reappeared with her own version of A incorporating the whistle—retrospective, gentle, but gradually gathering the entire orchestra to play her music, to close his cycle. A flourish, and the portrait was done: a many-sided man who had wit and grace and courage to discover himself in the friends that surrounded him, and most of all in the woman he had married.

The *Variations* were scored between 5 and 19 February 1899. The music was dedicated 'To my friends pictured within'. But there was no mention of the Enigma: the title-page specified merely 'Variations on an original theme'.[239] Only at the end of the score he set a quotation from Tasso (which might have been found by Alice or by Miss Burley, who was a great linguist, for Edward knew no Italian then):

'Bramo assai, poco spero, nulla chieggio.'

He appended the translation:

'I essay much, I hope little, I ask nothing.'

On 21 February the score was posted to the concert manager Vert, who next day sent it on with his own recommendation to Richter in Vienna. A fortnight later Vert relayed the first intimation of good news: 'I have heard from Dr Richter acknowledging the score of your Variations. He is at present travelling in St Petersburg, but on his return he will look through the work, and he says "that he shall be only too pleased to promote the work of an English artiste".'[240]

Edward busied himself with smaller things. He attended an inaugural meeting for the Folk-Song Society in London, which included addresses by Parry and Mackenzie. He completed an old violin and piano sketch which suggested itself as a companion-piece to *Chanson de nuit*: he called it *Chanson de matin* and foresaw, correctly again, that it would attain immense popularity. At the request of Sir Walter Parratt at Windsor, he contributed a madrigal to be included in a volume of *Choral Songs* in honour of Queen Victoria's eightieth birthday. The leading poets and composers of the land were represented. The words assigned to Edward were by one of the lesser poets, Frederic Myers, and when he completed his work he wrote in banter to F. G. Edwards at *The Musical Times*: 'I've just finished a *Partrigal* (S.A.T.B) *to order* & feel weak.'[241]

At last the time had come to move from Forli. While he was finishing the *Variations*, Alice had been prospecting. She had found a new house south of Great Malvern along the Wells Road, set well up in the Hills. It was a detached house, altogether more comfortable than Forli, and offering a big first-floor front room for Edward's study with views all over the Severn Valley: 'I can see across Worcestershire, to Edgehill, the Cathedral of Worcester, the Abbeys of Pershore and Tewkesbury, and even the smoke from round Birmingham.'[242] He made the letters of E., A. and C. ELGAR into an anagram to name the new house Craeg Lea.

[239] The word 'Enigma' appears pencilled on the opening pages of music in Elgar's full score manuscript; but the pencilling is in another hand, possibly Jaeger's. It would certainly have been done with Elgar's approval. But the manuscript did not reach Novellos until April 1899.

[240] 9 Mar. 1899 (Novello archives).

[241] 1 Mar. 1899 (BL Egerton MS 3090 fo. 14ᵛ).

[242] *The Strand Magazine*, May 1904, p. 544.

But the actual task of moving house was not at all to his taste. Ten days before the move in March 1899 he wrote to Jaeger:

I am awfully worried with this moving & do anything to escape—I *fled* out yesterday straight across country to think out my thoughts & to avoid everyone—will you believe it? I had walked 9 miles & was on the road and a man rode silently (on a bicycle) behind me & said 'Oh! Mr. Elgar! can you tell me if *Novello's have any performing right in &c. &c.*' I was speechless.[243]

The move itself would coincide with Edward's first rehearsal for the London première of *Caractacus* with Frederick Bridge's Royal Choral Society. In the end Alice sent him up to London four days early, while she supervised the move alone.

In London he agreed with the Novello chairman Alfred Littleton that the *Variations* would be published with a royalty for their composer, who would thus retain a permanent interest in the success of his work. After his rehearsal he returned to Malvern to enter the new home. Alice had managed arrangments at Craeg Lea so well that within three days they were able to entertain Lady Mary Lygon to tea.

It was then, on 24 March, that he asked permission to dedicate to her the *Three Characteristic Pieces*, and received a gracious assent. That evening he sent back the proofs of the *Pieces* with a note to Jaeger: 'I want to know if you could get the *title* [with the dedication] done *very soon* as Lady Mary is going away & I should like her to see it first . . . She is a most angelic person & I should like to please her—there are few who deserve pleasing . . .'[244] Early in April, after another Madresfield Musical Competition, he headed the Worcester delegation at Foregate Street Station to see her off with her brother Lord Beauchamp to Australia.

More rehearsals with the Royal Choral Society preceded the London première of *Caractacus* on 20 April—noted by Alice as a 'great success'. A fortnight later Edward conducted the spring concert of the Worcestershire Philharmonic Society. The programme included Mendelssohn, Wagner, Delibes, his own *Spanish Serenade*, and Alexander Mackenzie's cantata *The Dream of Jubal*. Mackenzie's libretto was by Joseph Bennett, who had failed to respond to Edward's proposal for something about St. Augustine for Birmingham.

Both Bennett and Mackenzie, however, accepted invitations to become honorary members of the Worcestershire Philharmonic. Parry was invited and felt unable to accept. Sir Arthur Sullivan acquiesced, as Edward announced in advertising the *Dream of Jubal* concert.

Dora Penny came from Wolverhampton for the concert, and afterwards went back with Edward and Alice to Craeg Lea. In the months since her last visit in November, the allurement of the *Variations* had grown—especially the

---

[243] 10 Mar. 1899 (HWRO 705:445:8336).   [244] HWRO 705:445:8339.

mystery of 'E. D. U.': 'When I went to Craeg Lea for the first time in May 1899, I felt I *must* get to the bottom of this puzzle.'[245]

That evening E.E. played the whole of the *Variations* and played the 'Intermezzo' again afterwards for me to dance to; but I would rather have sat still and heard him play it, I should not have cared how often; the thrill of it was still upon me.

'You wait till you hear it properly played by a decent orchestra; that'll make you sit up!'

Looking through the music and counting up how many of the 'Variations' I knew and how many I had only heard of, I remembered how puzzled I had been by seeing 'E. D. U.' over the Finale.

'Who on earth is E. D. U.?' I asked.

'Well, I should have thought you'd know *that*.'

But I was quite stupid and said, 'I don't know any friend of yours whose name begins with "U".'

'It doesn't begin with "U".'

As he put the slightest possible emphasis on 'begin' my wits at last woke up. 'Oh! of course,' I said rather shyly, knowing that 'Edu' was what the Lady called him—and no one else, so far as I knew. 'It's you.'

'That's a secret. Will you remember?'[246]

On 9 May he was in London again for a performance of the 'Meditation' from *The Light of Life* conducted by young Henry Wood in a concert of the 'London Musical Festival' at Queen's Hall. The evening's soloist was Paderewski. The Queen's Hall manager Robert Newman entertained them all at a reception on the Saturday evening: Newman had vainly tried to secure the *Variations* première for his London Festival, and clearly had his eye on Edward.

Another reception was given by a well-known patron of the arts, Frank Schuster, in his elegant house overlooking St. James's Park. Schuster was later described by his young friend Siegfried Sassoon:

He had 'the soul of an artist', as far as appreciation of music goes, but his limitations were expressed by his clumsy hands, which were almost grotesque, like his slightly malformed feet. Unable to create anything himself, he loved and longed to assist in the creation of music . . . He was something more than a *patron* of music, because he loved music as much as it is humanly possible to do. In the presence of great musicians he was humble, bowing before them in his semitic way, and flattering them over-effusively because he knew no other way of demonstrating his admiration.[247]

Such admiration was a new thing to Edward.

Ten days later came Queen Victoria's eightieth birthday. Sir Walter Parratt had arranged an 'Aubade' to take place on the birthday morning, 24 May, in the Great Quadrangle of Windsor Castle below the room in which the Queen

---

[245] *Edward Elgar: memories of a variation*, p. 116.

[246] Ibid., pp. 15–16.

[247] *Siegfried Sassoon Diaries 1920–1922*, ed. Rupert Hart-Davis (Faber and Faber, 1981), pp. 293–4.

breakfasted. He assembled 250 singers from the Choral Societies and Chapel Choirs of Windsor and Eton to sing a short programme which included the first performance of Edward's madrigal. Edward found himself summoned by telegram to Windsor. The press reported:

Quite a burst of sunshine flooded the scene as Sir Walter Parratt gave the sign, and the words 'God Save our Gracious Queen' floated up to Her Majesty's window in the Victoria Tower whilst all heads were bared . . . The concluding item of the programme to which Her Majesty listened were a couple of madrigals. The first of these, entitled 'To the Queen', was written by Mr. Frederic W. H. Myers, the music being by Mr. Edward Elgar . . .

Then the quadrangle became alive with enthusiasm. The young Etonians waved their shiny silk hats, the volunteers lifted their helmets on high, and all cheered and cheered again, as only British boys can. In response to this loyal outburst the Duke of Connaught appeared at the windows of his Royal mother's apartment, and intimated that the Queen would express her own acknowledgments of the attentions so liberally showered upon her. In a few moments her Gracious Majesty was seen at the window, to which she had been wheeled in her chair. She raised herself and bowed repeatedly, whilst all below stood uncovered. Then the Sovereign spoke a few words, which were distinctly heard by the choir and those nearest her window. She said, 'I am very pleased with all I have seen and heard, and I thank you all very much.'

So it was that Edward saw and heard the dedicatee of *Caractacus*.

Novellos were preparing the *Variations* for the first performance at the Richter Concert on 19 June 1899. The score had reached the publishers from Vienna only in April, and various small changes had been suggested and carried out. Then Jaeger said he thought the 'E. D. U.' Finale should be made more of. Edward replied:

If you *really* think it wd. be better pray do add Finale to the title—I of course should prefer simply

Variations.
Op. 36
Edward Elgar.

That's modest & becoming but if necy I will sacrifice my own bruisèd feelings on the altar of Mammon (High Priest—Nimrod).[248]

So the idea dropped.

A proof of the printed piano arrangement had been sent to the writer of the programme notes for the Richter Concerts, Charles Barry. He had got wind of the 'Enigma', though it was nowhere specified in the proof, and he pressed for an explanation. Edward delayed and delayed before sending the reply which Barry would print in the programme:

It *is* true that I have sketched for their amusement and mine, the idiosyncracies of

[248] 28 Apr. 1899 (HWRO 705:445:8344).

fourteen of my friends, not necessarily musicians; but this is a personal matter and need not have been mentioned publicly.

The Enigma I will not explain—its 'dark saying' must be left unguessed, and I warn you that the apparent connexion between the Variations and the Theme is often of the slightest texture; further, through and over the whole set another and larger theme 'goes', but is not played . . . . So the principal Theme never appears, even as in some late dramas—e.g., Maeterlinck's 'L'Intruse' and 'Les sept Princesses'—the chief character is never on the stage.[249]

The chief character who was never on the stage in both the Maeterlinck plays was Death.

In the same month he was forced to acknowledge that the 'Gordon' Symphony could not be written for the Worcester Festival in September. He broke it to Ivor Atkins:

I am awfully sorry to tell you that I have had to write to Canon Claughton as Chairman of the Music Committee—& withdraw my new work from the scheme—the reason is merely the pecuniary one & this is insurmountable. Its withdrawal need not be announced—just let it drop out of the scheme & no one will notice it.[250]

But it was noticed. One newspaper printed a story that the withdrawal of Elgar's 'Gordon' Symphony had been caused by the refusal of the Worcester Festival Committee to pay £100 for it. That elicited a letter from Canon Claughton, stating that a fee of £100 had neither been offered nor requested.[251] Edward had complained of poverty to cover his inability to fulfil the commission.

Richter's first rehearsal of the *Variations* was down for St. James's Hall on Saturday 3 June. It was well in advance of the concert, but the manager Vert said that the great conductor wanted an extra preliminary rehearsal to make certain that all would be well for the new work. Edward took himself up to London for the weekend, met Richter, listened to the rehearsal, and was 'much pleased'.

On the Monday he went back to St. James's Hall to show the tenor Edward Lloyd the sketch of a new song called *The Pipes of Pan*. The words by the popular verse-writer 'Adrian Ross'[252] began:

> When the woods are gay in the time of June
>   With chestnut flow'r and fan,
> And the birds are still in the hush of noon,—
>   Hark to the pipes of Pan!

---

[249] Quoted in the programme note for the Richter Concert of 19 June 1899, p. 206. The four dots preceding the final sentence appear thus in the programme. The original letter to Barry has not been traced. Barry's letters repeatedly requesting information were written between 10 Apr. and 26 May 1899 (HWRO 705:445:3034–8).

[250] 9 May 1899.

[251] *The Worcester Echo*, 7 Oct. 1899.

[252] The pen-name of Arthur Ropes.

> He plays on the reed that once was a maid
> Who broke from his arms and ran,
> And her soul goes out to the list'ning glade—
> Hark to the pipes of Pan!

So Pan himself realized his music through another and feminine personality. Lloyd was enthusiastic, and the song promised a popular success.

On 16 June Edward returned to London with Alice. The *Variations* were given their second rehearsal on Saturday 17 June, with Jaeger, Barry, and F. G. Edwards in attendance together with Mr and Mrs Charles Beale of the Birmingham Festival. A third rehearsal on the morning of 19 June made final preparation.

In the première performance that evening, a few passages fell short of the ideal. But the quality of applause at the end made it clear that the *Variations* had been a real and resounding success. Richter brought Edward on to the platform to acknowledge his triumph, and afterwards carried him off to supper with Ivor Atkins and Sir Alexander Mackenzie, who recalled: 'Seated opposite to him at supper after the concert at which Elgar's *Enigma Variations* were so successfully introduced to the public, I heard the enthusiastic terms of admiration, shared by us all, which the conductor addressed to the composer.'[253] For Alice there was a letter from old Ann Elgar at 10 High Street: 'What can I say to him, the dear one—I feel that he is some great historic person—I cannot claim a little bit of him now he belongs to the big world . . .'[254]

Some of the reviews complained about secrets which seemed to hedge the work about, but it was generally acknowledged that a new voice had been heard in London, and an important one. One critic felt such an acuity in the feminine variations that he called Edward 'in the best sense a "feminist" in music'.[255] Jaeger saw to it that *The Musical Times* did full justice, and then he voiced his own hope:

Effortless originality—the only true originality—combined with thorough *savoir faire*, and, most important of all, beauty of theme, warmth, and feelings are his credentials, and they should open to him the hearts of all who have faith in the future of our English art and appreciate beautiful music wherever it is met.[256]

But then Jaeger asked whether Edward could rewrite the Finale to make more of it. He adduced the views of several people who seemed dissatisfied with 'E. D. U.' A second performance was in prospect for an all-Elgar concert planned by Granville Bantock, the young composer who conducted concerts at New Brighton near Liverpool: there should be time to alter the score before then.

[253] *A Musician's Narrative* (Cassell, 1927) p. 240.

[254] HWRO 705:445:1132, dated merely 'June 23rd'. It was attributed in Young's *Elgar, O.M.* (p. 105) to 1901, but references to an illness of Carice's and other matters show it to belong to 1899.

[255] *The Referee*, 25 June 1899.   [256] 1 July 1899, pp. 464–5.

Edward took several days to reply to Jaeger's renewed charge that 'E. D. U.' somehow failed to make his mark.

Malvern

June 27

My dear Jaeger,

I waited until I had thought it out & now decide that the end is good enough for me . . . You won't frighten me into writing a logically developed movement where I don't want one by quoting other people!

If 'E. D. U.' was not logically developed, that lack of logic might be the closest portraiture. Yet at the end of his letter Edward did just leave the door open:

As to engraving the score of the Vars: hadn't it better wait until after New B[righton]—that is on the 1[6]th.

I think that's all—I've a frightful headache to-day—sunheat.

Yrs ever
Ed. E.

How wd. this do—use the Skoughre & parts just as they are (Variations) at Bantock Concert—I would touch up the thing *after* & let you have it say on the 20th of July? Commonsense ain't it?[257]

But Jaeger pressed for a considerable lengthening of the Finale. He said that Richter himself had been dissatisfied. For Edward it could imply that 'E. D. U.' had not really solved his Enigma. In the composer's mind it might call in question the whole issue of his own music.

Malvern

June 30

My dear Jaeger:

As to that finale—it's most good of you to be interested & I like to have your opinion—I have my doubts as to some of the rest 'cos it's generally *suggested* to them.

Now look here, the movement was designed to be concise—here's the difficulty of lengthening it—I *could* go on with those themes for $\frac{1}{2}$ a day—but the *key* G is exhausted—the principal motive (Enigma) comes in grandioso on *p. 35* in the tonic & it *won't do* to bring it in again: had I intended to make an extended movemt. this wd. have been in some related key reserving the tonic for the final smash. In deference to you I made a sketch yesterday—but the thing sounds Schubertian in its sticking to one key. I should really like to know *how* you heard that Richter was disappointed—he criticised some of it but not the end—the actual final flourish was spoilt in performance as you know by the insts. going wild. You see there's far too much of this sort of thing said: somebody wants to find fault & in course of conversation says 'the end did not please so & so—I find it very poor—*don't you?*'   the other chap hadn't thought of it at all but says 'Yes it's very abrupt'—& so it goes on.

This sort of thing is of no value to me—what *you* say is your own opinion & wd. be given on *any*body's work. All the other fellows wd. never have made a remark if the work had been written by any great man.

[257] HWRO 705:445:8353.

If I find, after New B[righton,] that the end does not satisfy me, I may recast the whole of the last movement but it's not possible to *lengthen* it with any satisfaction I fear.

If I *can* find time to make a readable copy of my 'end' I'll send it to you & then you'll see how good

E. Elgar

is at heart.[258]

When Jaeger read this, he took the measure of Edward's difficulty. In his reply he asked his friend not to worry himself over what the Germans call a *Nebensache*—a very secondary thing.

Yet something in Edward must have told him that Jaeger was right—that the Enigma was not solved by recapitulating 'C. A. E.'. However he might pay homage to the feminine beginning, the *Variations* traced his own journey of discovery. After all the other friends had been consulted and the self-portrait drawn, he must not simply return to the point of its departure. If Jaeger saw a possibility beyond Alice, he was a man. So it began to appear that the Enigma held yet one more revelation. On 7 July Edward wrote to Jaeger again: '*Your* opinion is not a "nebensache"—some people's are: I am hoping to send you a sketch of a proposed extended finale: there's one phrase wh: I can use again.'[259]

The sketch left all the 'E. D. U.' music as it stood up to the last few bars: there were the 'Romanza' developments, the attempted solution, the whistle, and the returning 'C. A. E.' And then, in a hundred new bars, a real solution appeared. The opposed elements in his Enigma had failed of their first reconciliation because their opposing metres forebade combining one above the other in any polyphony. The alternative was to couple the two elements not vertically but horizontally—side by side, in tandem—to create a single melody. It was what the sustained melody of 'Nimrod' had suggested. Now Edward joined the opening of A with the germ of B in a new and greater extension of melody:

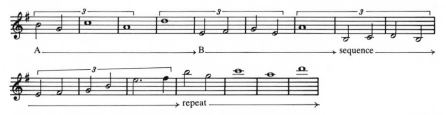

Here at last was the right solution. For it was melody and its allies that provided the readiest vehicle for Edward to carry his self-expression in music to great lengths. And it was melody that was to be the truest expression of Edward's individuality in music. The *Variations*, with Jaeger's help, had brought him at last to the consummation of that individuality. He was forty-two.

---

[258] HWRO 705:445:8354. Page-reference is to the piano score, which had already been printed.
[259] HWRO 705:445:8357.

Through his most recent decade, the quest had been pursued largely through a succession of outsider-heroes. Those heroes had dominated works in which he had sought to side-step a full development of his own melody by use of the shorter Leitmotiv. It was just the sort of motive which had made the A and B figures of his Enigma—which had, after the defeat of his last hero, Caractacus, sent him back to extemporize over the opposing elements in his own music. The rest of the story was chronicled in the *Variations* he had now completed.

Alice had identified his Enigma for him, Jaeger had helped to solve it. Yet both the Enigma and its solution were the expressions of himself. So 'the principal Theme' that never appeared in the original *Variations* emerged at last—or partly emerged. It was a famous victory, to resound through the annals of English music. Whether the victor was to find himself ultimately as Maeterlinck's Intruder, The Black Knight, or some wholly new manifestation, the years ahead would discover.

BOOK III

SYMPHONY

7

'Child Rowland to the dark tower came.'

THE words spoken in the storm of *King Lear* became a title for Browning's poem of wandering in desolation. They were set down again in one of Edward's sketchbooks near the turn of the century. The Enigma solved with a triumphant 'E.D.U.' did not after all resolve inner doubts. It made only another commitment to the future foreseen by Acworth at the time of *Caractacus*: 'In one respect I am sorry for you. You have now the obligation of living up to a great reputation . . . High-wrought expectations are difficult to satisfy—& such all your work will henceforth evoke.'

There was also the fact that greater success would lead him into places where manners and behaviour must seem more and more important to the tradesman's son who had taught himself. Yet an artist could only develop by exploring and revealing what was within himself. When high-wrought expectations were then taken to heart, the deepest doubts went home with them.

Edward's first work after the '*Enigma' Variations* was the cycle of songs with orchestra for the Norwich Festival in October 1899. Early in the year he had looked at the little 'Lute Song' written in 1897 to Alice's words. He consulted Novellos, who sent back the 'Lute Song' with the suggestion that it might be expanded with a different setting for each stanza.[1] That would not produce the large work to which he had committed himself.

Yet Alice had identified the Enigma that led to the triumph of 'E.D.U.' Her words to the 'Lute Song' might give another hint. If 'Love alone will stand', those words could suggest thoughts and feelings wandering elsewhere—when 'Storms are sweeping sea and land'. So Edward's thoughts could drift towards the spring voyage of Lady Mary Lygon—'the pretty lady' attracted by his music—whose social rank was far above Alice's, whose age was a dozen years less than his own. If Alice's 'Lute Song' made a 'Sea Song', then other poems drawn from other poets (in the manner of Berlioz's *Nuits d'été*) might comment and reflect on Alice's 'Lute Song' as the other 'friends pictured within' had expanded her perception of the Enigma.

Still he was uncertain, as he wrote to Nicholas Kilburn on 22 February: 'I wish you could give me another idea for a song cycle (Clara Butt) for Norwich

---

[1] 11 Feb. 1899 (Elgar Birthplace).

Festival. I have sketched out a plan & music but you may have an idea . . .'[2] Later in the spring of 1899, as if to exorcize sea-imagery, he proposed taking out the 'Calm Sea and Prosperous Voyage' figure from the Variation which had begun as Lady Mary's.[3] But Jaeger easily persuaded him to let it stand. And as the Norwich Festival drew closer, the sea metaphor came in with it.

Alice altered her 'Lute Song' words to make a 'Sea Song' with the title 'In Haven'. She suggested it as a reminiscence of a long-distant visit to Capri before her engagement to Edward; but the new lyric was made too poignant for the innocent music of Edward's setting:

> Closely let me hold thy hand,
> Storms are sweeping sea and land;
> Love alone will stand.

> Closely cling, for waves beat fast,
> Foam-flakes cloud the hurrying blast;
> Love alone will last.

> Kiss my lips, and softly say:
> 'Joy, sea-swept, may fade to-day;
> Love alone will stay.'

Love had here the permanence of death itself.

Death might be the theme of Roden Noel's 'Sea Slumber-Song', where the sea sang a mothering lullaby. Its dream recalled the 'caves & arched rocks'[4] of the honeymoon in the Isle of Wight ten years earlier:

> Isles in elfin light
> Dream, the rocks and caves,
> Lulled by whispering waves.

Those waves diffused the music of Edward's own instrument:

> Sea-sound, like violins,
> To slumber woos and wins.

The sea-mother offered again the oblivion that overwhelmed King Olaf, that might have drowned all the anguish of Caractacus embarking on his beloved Severn to exile:

> 'I murmur my soft slumber-song,
> Leave woes, and wails, and sins,
> Ocean's shadowy might
> Breathes good-night,
> Good-night!'

This 'Sea Slumber-Song' took its place at the head of the new cycle, like the 'Gute Nacht' that opened Schubert's *Winterreise*.

---

[2] 22 Feb. 1899.   [3] 2 May 1899 to Jaeger (HWRO 705:445:8346).
[4] See above, p. 133.

Alice's 'In Haven' would come next. Yet human passion would be left far distant in Edward's choice of another poem for the cycle, Richard Garnett's 'Where Corals Lie':

> The deeps have music soft and low
> . . . .
>   When night is deep, and moon is high,
> That music seeks and finds me still,
>   And tells me where the corals lie.
>
> Yes, press my eyelids close, 'tis well
> . . . .
> Yet leave me, leave me, let me go
>   And see the land where corals lie.

In Elizabeth Barrett Browning's 'Sabbath Morning at Sea', a new day's dawning showed the lonely voyager his own separation from those left behind in their haven:

> Love me, sweet friends, this sabbath day.
>   The sea sings round me while ye roll
>   Afar the hymn, unaltered,
>   And kneel, where once I knelt to pray.

The voyager himself finds in the sunrise at sea a brilliance too high for contemplation even by saints:

> And on that sea commix'd with fire
>   Oft drop their eyelids raised too long
>   To the full Godhead's burning.

The goal of the lonely voyage, then, lay elsewhere.

    The cycle would find its close in a poem by Adam Lindsay Gordon. He was a distant cousin of General Gordon of Khartoum: but where the General's death was ambiguous, this Gordon had committed the clearest, most wilful suicide. His poem, 'The Swimmer', matched single human strength against a challenge of nature hurled through overwhelming desolation:

> With short, sharp, violent lights made vivid,
>   To southward far as the sight can roam,
> Only the swirl of the surges livid,
>   The seas that climb and the surfs that comb.
>
> A grim, grey coast and a seaboard ghastly,
>   And shores trod seldom by feet of men—
> Where the batter'd hull and broken mast lie,
>   They have lain embedded these long years ten.

Ten were the years in Edward's own life since the honeymoon with Alice, and the poem drew its own contrast of distant companionship on these same shores:

Love! when we wandered here together,
Hand in hand through the sparkling weather,
From the heights and hollows of fern and heather,
    God surely loved us a little then.

Now the grim coast showed only a spectre of the sabbath sunrise:

One gleam like a bloodshot sword-blade swims on
The sky line, staining the green gulf crimson,
A death-stroke fiercely dealt by a dim sun
    That strikes through his stormy winding sheet.

I would ride as never a man has ridden
In your sleepy, swirling surges hidden;
To gulfs foreshadow'd through strifes forbidden,
    Where no light wearies and no love wanes.

A friend of later years recalled: 'E.E. used to say that it is better to set the best second-rate poetry to music, for the most immortal verse *is* music already.'[5] Thus the poems gathered for this cycle could invite again the music for which the lonely child had listened in the reeds by Severn side. But the ocean which he now faced could show no comforting opposite bank with its promise of Broadheath just beyond. These *Sea Pictures* opened the prospect of a voyage into the self: it was the dark face of that self-discovery which had been the triumph of 'E.D.U.'.

The Lute figure from his setting of Alice's poem became a source from which to draw the rest. Its arpeggio up and down

was slowed to make an E minor opening for the 'Sea Slumber-Song':

⁵ Vera Hockman, 'Elgar and Poetry' (TS, 1940).

The descending triplet steps appeared in the tenor almost as a hallmark.

The descending half of the Lute figure slowed further still to make a slow profundity of sea-motion, while over it the sea began its song to an echo of the Enigma:

The Lute figure was reduced to shape a central section in C major:

And that opened another development:

Then the opening E minor and major music returned to complete a quiet symmetry.

The gentle innocence of Alice's 'In Haven' continued the C major voyage. And C major opened again 'Sabbath Morning at Sea', to a new version of the Lute's upward arpeggio:

A recitative, accompanied by the 'sea moving', moved to a new variant of the 'Foam' figure (from 'Sea Slumber-Song') in the harmonic distance of B major:

Another enharmonic change to D flat opened the second half of a binary form whose goal this time was a triumphant C major. But just before the coda the descending steps went back into E minor to counterpoint the opening music of the 'Sea Slumber-Song':

'Where Corals Lie' moved to B minor with a new form of descending arpeggio:

A transparent vocal line above almost static accompaniment distilled the 'sea moving' figure. Only the third line of the stanza recalled the rising shape which had opened the first three songs

before that hint of passion cooled in renewed descending steps

reaching down and down to touch 'all the land where corals lie'.

    The Swimmer, instead of recapitulating the cycle's opening E minor, moved to the new tonic key of D major. Therein it took the cycle's images of sleep and waking dreams to a new goal. A furious descent of steps opened on a big swinging sequential figure

The solo voice interrupted, shouting across a troubled expanse. Downward pressure from E flat

met upward pressure from a sequence beginning in D

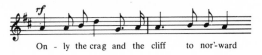

It drove the prayer of 'Sabbath Morning' through 'Corals':

At the centre of this heroic Finale came reminiscence: the central C major 'Isles in elfin light' music from the opening song rose up again at 'The skies were fairer and shores were firmer'. But this glimpse of the past only showed the Swimmer where he was now. After more turbulence of E flat and D major, the big sequential figure sounded its final triumph in the Swimmer's vision of his lonely fate. Thus the music itself moved away from the close development which shaped the earlier songs in the cycle. And with it Edward's music left the haven of its older tonalities and moved into a locus of D. This was the key that would frame the new work for Birmingham in the months ahead.

*   *   *

When Richter accepted the *Variations* in the spring of 1899, Edward agreed with Novellos for a 'Royalty according to published price'. But the rate of royalty had not been fixed. Just before the première, the publisher's offer had come through: nothing on orchestral performances, 4*d*. a copy on sales of the piano arrangement of the whole work, proportionately smaller royalties in case of publishing separate movements.

Craeg Lea, Wells Road, Malvern.

June 9: 1899

Dear Sirs:

Variations op 36

I am greatly disappointed by your offer of 4*d*. on the piano arrgt. of the above work & trust you will see your way to increase the royalty to at least 6d. I think also that, in the event of any excerpts being published, the smaller reprints shd. not be a *proportionate* royalty: it surely shd. be somewhat higher

Believe me
vy faithfully yrs
Edward Elgar

The publishers drafted this reply:

14.6.99

We are favoured by your letter of the 9th inst. & are sorry to note that the propositions contained in ours of the 8th inst. are not acceptable to you.

We very much regret that the question of terms was not discussed before we put the work in hand, as we should have preferred, rather than disagree as to terms, to give you the opportunity of making other arrangements, had you wished to do so.

Then on reflection that hint for Mr Elgar to approach other firms was struck out.

We thought however that we might assume that the terms which have proved acceptable to Sir Alex: Mackenzie   Mr. Edward German & many others would have proved acceptable to you, & we are sorry to find that we are mistaken.

We regret, however, that we are unable to make any better offer & can only suggest that you employ us as your publishers retaining the copyright in your own hands: by this means the whole profit derived from the publication after paying the expenses will be yours.

Craeg Lea, Wells Road, Malvern.

June 15: 1899

Dear Sirs:

Variations op 36

Many thanks for your letter of Yesterday's date.

You now tell me that the offer contained in your former letter is usual: therefore I accept it . . .

May I point out that I read your letter as a question—asking my opinion: as smaller firms are glad to give me 4d. on a piece nominally selling at 2/– I not unnaturally imagined a higher priced work wd. bear a higher fee.

It is only just to myself to say that I have never cavilled at the terms you have offered me on any occasion so far as I remember: I am rather pained therefore at the wording of the second paragraph of your letter, which seems to me to contain an innuendo which I feel is not called for—but I hope I am mistaken

Yrs v faithfully
Edward Elgar

Again the publishers drafted their reply:

Dear Sir

We are favoured by your letter of yesterday's date. Ours of the 14th inst. was certainly not intended to convey any innuendo to which you could object. We merely wished to remark that anything like a disagreement as to terms might have been avoided had we observed the most usual course of discussion of terms before arranging to publish; &, to a certain extent, that is a reflection upon ourselves!

The Royalties we named are usual with every well-known composer who publishes with us; & the length of the composition is a drawback rather than an advantage seeing that it adds to the expense of production (especially when there are band parts) & at the same time restricts the sale.

The man who drafted Novello's letters was the company secretary, Henry Reginald Clayton, a son-in-law of the chairman Alfred Littleton. Clayton was a graduate of Trinity College, Cambridge, where he had captained the University rugby fifteen and headed the law tripos. He was a qualified barrister, rapidly gaining recognition as an authority on international copyright law. He was described by a colleague thus: 'A certain brusqueness of manner, due to an isolated temperament, was a misleading index to his nature. He was kindly and generous to a degree unsuspected by acquaintances and known only to the circle of his friends.'[6] It seemed a well-kept secret: one of his juniors was to recall Clayton as 'a Justice Avory sort of man—a fierce and savage judge'.[7] In the negotiating just concluded, Edward had been overmatched.

He had not regained equanimity before encountering a further example of what could seem close methods. Richter's manager Vert received Novello's bill for another 30s. on account of hire of parts for the extra rehearsal in advance of the first performance. As the purpose of that rehearsal was only to ensure a successful launching for the work in which the publishers themselves had an interest, the added bill enraged Edward, as he wrote to Jaeger: 'Vert has lost heaps over the old gang's orchl. attempts & did the work as a favour—an *extra rehearsal* cost him £40 . . . & the publishers "try it on" to get 30/– more out of him—this certainly roused my disgust.'[8]

By now the need to finish the *Sea Pictures* for Norwich was pressing. But Edward's mind, as he finished the new coda for the *Variations* at Jaeger's urging, was elsewhere. In Hereford on 8 July for a weekend with Sinclair, he wrote in the Visitors' Book a new 'Mood of Dan'—'triumphant (after a fight)'—in a key unrelated to anything before him at that moment:

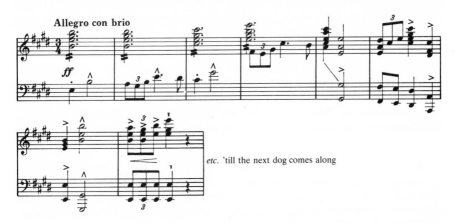

etc. 'till the next dog comes along

Then he and Alice went up to New Brighton to prepare the concert with Bantock's orchestra. It was one of a series devoted to modern English composers and their work. The thirty-year-old Bantock's miracle of turning a municipal band into an orchestra for serious modern music at regular Sunday concerts had gained the admiration of a wealthy Liverpool merchant and

[6] Anonymous obituary in *The Musical Times*, Jan. 1933, p. 80.
[7] Harry Fowle, who joined the firm in 1920, in conversation with the writer, 1976.
[8] 1 Sept. 1899.

patron of music named Alfred Rodewald. Rodewald was a disciple and friend of Hans Richter: in fact he was Richter's only conducting pupil.[9] He was founder and conductor of the amateur but highly competent Liverpool Orchestral Society. Rodewald and Bantock were the best of friends, and they joined in welcoming Edward into their midst.

Edward and Alice stayed with Bantock and his young wife Helena. When Edward felt depressed or unwell, Alice's protective ministrations created a formidable impression in the household, as the Bantocks' daughter wrote:

Elgar was somewhat delicate and many arrangements were necessary for his comfort, including an apparatus for his nightly tea-making. Elgar's wife was absolutely devoted to him and surrounded her husband with a ring-fence of attention and care that was almost pathetic . . .

The still unemancipated wives of that period were, I think, more devoted than women today. There was a tendency among some to treat ill-health in their husbands or children as a kind of cult, every little ache or sneeze being regarded with grave suspicion and an excuse for additional coddling. My newly married mother was, I am sure, awed by Mrs. Elgar, with her array of rugs, shawls and cushions, extra body-belts and knitted bedsocks for Edward's comfort. One evening Helena noted with astonishment no less than seven hot water bottles being filled for his bed, on the occasion of Elgar complaining of a slight chill![10]

At the concert on 16 July, Edward none the less conducted the *Imperial March*, an orchestral version of a *Minuet* he had written for Nicholas Kilburn's musical son Paul,[11] the *Variations* (with their original ending as the new coda was not yet engraved), the first performance of the *Three Characteristic Pieces* dedicated to Lady Mary Lygon, the *Serenade* for strings, and extracts from *The Light of Life*, *King Olaf*, and *Caractacus*.

Next day he journeyed across to Sheffield to take a preliminary choral rehearsal of *King Olaf*, down for performance at the second triennial Sheffield Festival in October. The conductor of the Sheffield Choral Union, round which the Festival was planned, was a self-educated director of great ability, Henry Coward. He wrote:

For the 1899 Festival . . . I asked the choir to think of the things which a very small select or professional choir could do, but which by a choir of 350 had hitherto not been attempted, because it was deemed '*impossible*'. These 'impossibles' included coalescence or unity of tone quality in each vocal part, mobility, clearness of *every* word, well-graded crescendos and diminuendos, extremes of force—thrilling fortissimos and bewitching pianissimos; and, to secure brilliant attack with oneness of movement, every piece must be memorized.[12]

[9] Recalled by Hans Richter's daughter, Mrs. Mathilde Loeb, in conversation with the writer, 1976.

[10] Myrrha Bantock, *Granville Bantock: a Personal Portrait* (Dent, 1972) pp. 47–8.

[11] W. H. Reed (*Elgar*, p. 49) said that Elgar conducted the *Minuet* at New Brighton on 1 Oct. 1898. On that day the Elgars were in Leeds for a *Caractacus* rehearsal, and the diary records no visit to New Brighton at that time.

[12] *Reminiscences* (Curwen, 1919) pp. 137–8.

The local press attended Edward's rehearsal, and afterwards he found himself being interviewed for *The Sheffield Independent*. He enthused over Coward's choir:

'The entire chorus tells the tales that are given in "King Olaf" as though it believed them. The basses and tenors shout out their determination with realism that is startling in its vehemence and fierceness. A chorus that can sing as the Sheffield chorus has done to-night when the Festival is three months ahead is capable of great things.'

All this uttered with the candour of a man who would fain speak with reserve, but who felt that the occasion demanded acknowledgment.

The Norwich song cycle for Clara Butt was brought up:

In fact, the composer is at present devoting his attention to the work, and had with him in his portmanteau the framework of the songs with which the great-voiced contralto will in a few short months thrill thousands of hearers.[13]

He stayed the night with Coward, who recalled:

Just before retiring he hinted that he would like a candle. This was supplied at once, although we thought it strange, seeing he had only to switch on the light. In the morning only a small portion of the candle was left, and as I knew he was busy with his masterpiece, I have always thought that some sudden inspiration was written down at Western Bank, Sheffield.[14]

Back in Malvern on 19 July, Edward and Alice soon went up to stay at Birchwood. A stream of friends and visitors followed. The last day of the month brought Dora Penny.

I had bicycled from Wolverhampton, forty miles, and arrived, rather warm and dusty, at the cart-track leading up through the woods to the house. When I was nearly there I thought I would rest, out of sight, and get cool. I heard the piano in the distance and, not wishing to lose more of it than I need, I soon went on.

In a moment I came in sight of the Lady sitting on a fallen tree just below the windows. She had a red parasol. I think she sat there partly to warn people off—particularly people with bicycles who had been known to commit the awful crime of ringing a bell to announce their arrival.

Leaning the bicycle against a tree, I went and sat down by her without speaking . . . Soon, however, the music ceased and a voice behind us remarked: 'Are you two going to have your photographs taken, or what?'

That afternoon the Lady said she was busy, and we went into the woods and sat down on the ground. After a bit E.E. said: 'If we are perfectly quiet perhaps someone will come and talk to us.'

In a few minutes a robin came, and then a little love of a fieldmouse. It ran towards us in jerks and came quite close, within touching distance, without a sign of fear.

[13] 18 July 1899.
[14] *Reminiscences*, p. 282. Writing of the incident for his memoirs twenty years later, Coward speculated that Elgar had been at work that night on *The Dream of Gerontius*. But his work was almost certainly on the *Sea Pictures*. There is no reliable evidence for any connected work on the *Gerontius* music before January 1900.

Later on E.E. lay down and went to sleep, and I felt very like dozing, what with the effect of my forty-mile ride, the hum of the bees, and the sheer beauty of it all.

'Tank-y-tank-tank,' said the sheep-bell distantly. Lo! it was tea-time.

We went back to the house and had tea, and afterwards he settled down to the piano and I to my usual work of turning over. The music was in manuscript [and] he went straight through it . . .

At last I came down to earth and realized with dismay that my time was up and that I had a train to catch at Worcester—and a seven mile bicycle ride. How the time had flown! I collected my things, said good-bye to the Lady and tore myself away. H[is] E[xcellency] walked with me to the lane which led down to the main road. It had clouded over and had become very sultry; it was pretty obvious that a storm was coming up.

'I wish you hadn't to go off like this,' he said, looking rather anxiously at the sky; 'come again soon.'

I looked back when I got to the bottom of the hill. He was still standing there; I threw up an arm and got an answer.[15]

On 11 August he took the *Sea Pictures* to London to go through them with Clara Butt. But the trouble with Novellos still simmered. Next morning he had a 'stormy interview' with Clayton at the publisher's office in Berners Street. It rankled further during the next few days, as he orchestrated *Sea Pictures* at Birchwood. On 21 August he went to London again to tell the firm's chairman Alfred Littleton of his 'disgust'—as he described it privately to Jaeger:

I would have withdrawn the expression &, if need be, apologised as far as A.H.L. was concerned, *but,—this annoyed me more than anything—he* said V[ert] ought to have got the extra rehearsal out of his men *for nothing*!! I was shocked at the sheer brutality of the idea but he repeated it: I confess the prospect of a rich man seriously considering the fleecing of those poor underpaid, overworked devils in the orchestra *quite* prevented me from feeling Xtian. If that is 'business'—well *damn* your business—I loathe it.[16]

And he stormed out of the office.

Novellos might be the best publisher for choral music, but their leading competitor specialized in songs and ballads. This was Boosey & Company, almost as old as Novellos and with a better address at the top of Regent Street. Edward had reconnoitred before breaking with Novellos: on 22 August, the day after his quarrel with Littleton, he settled with Booseys to publish the *Sea Pictures*. Booseys bought the copyright for £50, but any publication of the separate songs would bring a royalty of 3*d.* a copy.

---

[15] *Edward Elgar: memories of a variation*, pp. 17–19. At the time of writing this recollection some thirty-five years after the events, Mrs Powell identified the music heard in the woods as the Prelude to Part II from *The Dream of Gerontius*, and the music played to her later as the entire chorus 'Praise to the Holiest'. But there is no reliable evidence of any work on the *Gerontius* music as such before Jan. 1900; and Elgar could not have played 'straight through' 'Praise to the Holiest' (as she said he did) before the end of May 1900, when he finished constructing the great chorus.

The most probable explanation is another mis-connecting of a memory to a date. In July 1900 Dora Penny paid her second visit to Birchwood in similarly hot weather, just as Elgar was orchestrating Part II of *Gerontius*: he would then have had every reason to be playing through both the Prelude to Part II and 'Praise to the Holiest'.

[16] 1 Sept. 1899.

If Booseys were not equipped to deal with a succession of large and serious works, their publication of the *Sea Pictures* made Edward's point at Novellos. When he came up to London again for the Worcester Festival orchestra rehearsals in early September, there was a reconciliation at Berners Street. For their part, Novellos agreed to print the full score of the *Variations* and all the orchestral parts, rather than rely on manuscript copies. It would facilitate performances everywhere. Afterwards Edward wrote to the chairman:

Craeg Lea, Wells Road, Malvern.

Sunday [10 September 1899]

Dear Mr. Littleton:

Here, in the quiet of my own home, I feel I must send what perhaps you will think a wholly unnecessary note to say how very glad (& grateful to you) I am that our short misunderstanding is at an end: my long, quiet Sunday here on the hills is more bearable & happier on this account than last week & I feel I should like you to know this & that your forbearing & courteous way is really appreciated by me

V sincerely yours

Edward Elgar:

At the Worcester Festival on 13 September he conducted *The Light of Life* (having revised some of the solo writing). Edward Lloyd repeated the tenor role of the Blind Man which he had created three years earlier, and the part of the Mother was taken by the famous soprano Emma Albani. That evening came the first performance of the *Variations* with the new coda. Bakers, Townshends, and Pennys were all there. For most of them it was a first hearing of the music, and Dora Penny was especially stirred.

I shall never forget the gauntlet that I had to run afterwards, at the Star Hotel, Worcester, where we were all staying for the Concert. [W. M. B.] wanted to know what each variation meant including his own, and I was put in a very awkward position, with my father and stepmother (W. M. B.'s sister) there as I could not possibly have told what I knew about them, they would not have understood.

Then of course they demanded to know what the Intermezzo meant and I suggested—rather lamely—wasn't it a charming dance tune? Of course my father rose to that bait at once: 'What did he want to give you a dance tune for?' But everyone said how charming it was and further criticism was stayed . . . No. X was so lovely that I felt—that first time—that I wanted to hide somewhere.[17]

It recalled her first reponse to the 'Dorabella' music when Edward played it to her on the piano: 'My mind was in such a whirl of pleasure, pride, and almost shame that he should have written anything so lovely about *me*.' Yet 'Dorabella' figured as only an Intermezzo in the larger work.

Dora's separating of the 'friends pictured within' from Edward's Enigma was to go on. It caught the notice of Miss Burley:

---

[17] *Edward Elgar: memories of a variation*, pp. 20–1. Alice Elgar's diary for 13 Sept. shows that the Pennys were in fact staying at the Great Western Hotel, close to the Worcester main line station at Shrub Hill.

The fame which was soon won by the *Variations* cast a certain reflected glory on the subjects, some of whom exasperated him by the airs they began to assume in consequence. I remember that at a party at the Hydes' shortly after the *Variations* had been produced and the initialled headings were still being discussed, one of the 'variants' came to me and said:

'Well, Miss Burley, I'm a variation. Are you?'

'No,' I answered gravely. 'I'm not a variation; I'm the theme.'

Edward was much amused when I told him of this.[18]

But he would never dedicate any piece of music to Miss Burley.

* * *

After the Worcester performance of the *Variations* on 13 September 1899, the Elgars went home to Malvern for the night. They were to go back next day to hear a new oratorio by the American composer Horatio Parker entitled *Hora Novissima*. In the morning, Alice was driven to the station while 'E. walked with Father Bellasis', who was staying with the Fittons for the Festival. It was pure coincidence that on his way to *Hora Novissima* Edward should fall in with this old acquaintance whose entire life had been shaped by the author of 'The Dream of Gerontius'—the poem whose central expression of Catholic dying and rebirth sounded through the cry 'Novissima hora est'.

As a possible subject for his music, *The Dream of Gerontius* had been 'soaking' in Edward's mind (as he would describe it) since at least the time of finishing *The Black Knight* in 1892.[19] In fact the theme of the dying old man reached back to childhood, when Fr. Waterworth had given him the little picture showing 'The Death of St. Joseph'. And within the past year Newman's poem might have recurred in Edward's consciousness through a litany of Gordons: the poet of 'The Swimmer', the hero of Khartoum, and the young priest John Joseph Gordon whose death had precipitated the writing of 'The Dream of Gerontius' with the dedication 'Fratri desideratissimo'. Therein the young brother who had died might touch the memory of Edward's own childhood Jo.

The subject of 'Gerontius' was right for Edward now. Its 'old man' sought death and a new life as ardently as the Swimmer had sought his own. But for Gerontius it was the 'Bona mors' of Catholic benediction. So it could re-enact on the bigger stage of mature belief the children's play in which the 'Old People' were 'charmḗd asleep' by the Fairy Pipers of a boy's music to emerge 'transfigured'. In his forty-third year it might reconcile the vision in *The Light of Life* with the dark horizon of the Swimmer—to open the 'Nightflower of Belief' described by Jean Paul Richter in a favourite passage that Edward copied out:

[18] *Edward Elgar: the record of a friendship*, p. 131.
[19] *The Musical Times*, 1 Oct. 1900, p. 648.

When in your last hour (think of this) all faculty in the broken spirit shall fade away and die into inanity—imagination, thought, effort, enjoyment—then at last will the Nightflower of Belief alone continue blooming, and refresh with its perfumes in the last darkness.[20]

Yet such self-exposure might well give him pause. The possibility of setting 'Gerontius' to fulfil his commission for the Birmingham Festival was discussed with Miss Burley:

He was afraid, however, that the strong Catholic flavour of the poem and its insistence on the doctrine of purgatory would be prejudicial to success in a Protestant community. He told me in fact that Dvořák, who had planned a setting of the work for the 1888 [Birmingham] Festival, had been discouraged from making it for this very reason.[21]

No Festival objections were voiced against the great Masses of Bach and Beethoven or the Requiems of Mozart and Verdi, and Dvořák himself had produced a Requiem at the Birmingham Festival in 1891. A very different explanation of Dvořák's difficulty over 'Gerontius' was recollected by one critic, who wrote: 'The matter, indeed, was about fifteen years ago discussed between [Cardinal Newman] and Dr. Dvořák, who afterwards found the subject too placid and lyrical for his special style.'[22] Edward never at any time said that the subject of *Gerontius* failed to meet the needs of his own expression. His hesitation could suggest, on the contrary, that the fitting might be too close. The end of Gerontius was a soul in Purgatory.

After the Worcester Festival, Edward corrected orchestral parts of the *Sea Pictures*. Then on 3 October 1899 he and Alice went to Norwich to stay with their hosts for this Festival, Mr and Mrs James Mottram. Nervousness exacerbated a heavy cold. A doctor was summoned, and the composer was confined to the Mottrams' house on the eve of his première. It all made an impression that old Mr Mottram recalled for the family biographer, who wrote:

During the night of the 4th he gave J. M. the fright of his life. At 2.0 a.m. Mrs. Elgar came knocking on J. M.'s door, begging him to fetch a doctor, her husband was so ill. J. M. bundled into his camlet cape and trousers and was half-way downstairs when she came after him, charmingly apologetic. Mr. Elgar felt better. J. M. bowed politely holding tight to his trousers, and went back to bed. Fanny thought it must have been an attack of 'nerves', But J. M., who had never had any, naturally didn't know how that could be.

However, the morning performance [of the 'Meditation' from *The Light of Life*] went off smoothly enough . . . Then in the evening, after Thomas' 'Mignon' overture, Elgar's distinguished, hollow-chested figure, rather that of a hawk dreaming poetry in captivity,

[20] From Richter's *Levana*. Elgar inscribed the quotation in Frank Schuster's copy of the *Gerontius* vocal score in the German edition (in possession of Sir Adrian Boult).

[21] *Edward Elgar: the record of a friendship*, p. 134.

[22] Newspaper cutting at the Elgar Birthplace.

took the baton, and there stood up beside him the majestic figure of Clara Butt, in a wonderful dress, the material of which, it was whispered, indicated appropriately the scales of a mermaid's sinuous form. She created at the same time a record, and inaugurated an era. Until then all the stars, Albani, Patti, Clara Novello, Nevada, had appeared corseted 'from the knees to the nose' as irreverent rebels put it. But when Clara Butt rose to sing, a dowager aristocrat in the patron's stalls saw what was impending, and remarked audibly to the next bejewelled dowager: 'Look, my dear . . . guiltless of all confinement!'[23]

Yet as Edward wrote to Jaeger: 'The cycle went marvellously well & "we" were recalled four times—I think—after that I got disgusted & lost count. She sang *really well*.'[24]

Two days later Clara Butt sang four of the *Sea Pictures* with the composer at the piano to a packed St. James's Hall in London. Jaeger tried unsuccessfully to get in; and even Basil Nevinson, the Elgars' host in London, was turned away when he drove Alice to the Hall. She wrote in the diary: 'Wonderful scene & thrilling when C. Butt repeated "In Haven".'

Edward had not really recovered his health when they left London for the Sheffield Festival. There the confinements of Norwich were more or less repeated; but again he was able to conduct, and *King Olaf* scored its success with Coward's brilliant choir despite a bad orchestra. Two days after that came a concert organized by Sir Walter Parratt at the Royal Albert Institute, Windsor, in the presence of Princess Christian: Edward conducted and played accompaniments for no fewer than eleven of his own works. In London he sat to the sculptor Percival Hedley for a life-sized bust in bronze to adorn the music room of the wealthy patron and admirer Frank Schuster.

On 23 October at St. James's Hall, Richter repeated the *Variations* in a programme shared with a Piano Concerto by the youthful Hungarian composer Ernö Dohnanyi. In the audience that night was an aspiring English musician fifteen years Edward's junior, Ralph Vaughan Williams:

I had been advised by a friend to go to a Richter Concert and hear a work by Dohnanyi, of all people. So I went. The Dohnanyi was all right. But the *Variations*: here was something new yet old—strange yet familiar—universal yet typically original, and at the same time typically English.[25]

That judgement was echoed in many places now, and it was in the air again when Dora Penny came to Craeg Lea a week later. Now it was Alice who had fallen ill, and Edward telegraphed to ask if Dora could help them for a few days.

Fortunately I was able to go and set off next day, arriving at Craeg Lea in the forenoon. I found the poor little Lady looking like nothing on earth and obviously feeling

[23] R. H. Mottram, *Portrait of an Unknown Victorian* (Robert Hale, 1936) p. 253.
[24] 6 Oct. 1899 (HWRO 705:445:8365).
[25] BBC broadcast, *The Fifteenth Variation*, on the centenary of Elgar's birth in 1957.

wretchedly ill, but still up and dressed. I did my utmost to get her to bed and succeeded early in the evening, and she was really thankful to be there.

As for E. E., he was in high spirits. The [*Variations* orchestral] proofs were arriving in batches every few days with voluminous letters from Mr. Jaeger, some of which I had to read aloud while E. E. checked the corrections, with a constant flow of interjections and comments. 'What does he say? The crazy old Moss-head! I'm not going to alter that for him or anyone else.'

. . . Next day I had little difficulty in keeping the Lady in bed. She said: 'I *do* like hearing you both laugh. Then I know that H[is] E[xcellency] is happy and that makes me feel better!'

. . . Then came Saturday and another batch of proofs . . . That evening the proofs, which were nearly ready to be returned, were put on one side and E. E. settled down to the piano. I think we had the *Sea Pictures* all through . . . Then he went back to the *Variations*, and I asked about the 'Enigma' and what *was* the tune that 'goes and is not played'?

'Oh, I shan't tell you that, you must find it out for yourself.'[26]

(He did not—in her later recollection of the scene—trouble to correct her assumption that the Enigma hid a tune.)

'But I've thought and racked my brains over and over again.'

'Well, I'm surprised. I thought that you, of all people, would guess it.'

'Why "me of all people"?'

'That's asking questions!'

. . . Then we moved over to the fire and began talking about some of the *Variations* and how splendid they sounded on the orchestra.

'Why, I haven't seen you since Worcester! Didn't they play well? I saw you sitting next W. M. B.; how did he like it? Do tell me what he said.'

And so we went on and on, E. E. standing with his back to the fire and I sitting in a big arm-chair. Suddenly he took me by my two hands and half lifted me up:

'And how did you like *yourself*, my Dorabella?'

Then I tried to tell him how wonderful I thought it, and how it was far too delicate and lovely for the likes of me.

'Well of course it is! We all know that.'

But I wouldn't be put off and I said how marvellous it was to feel oneself part of the music which had been acclaimed by half the world as being his greatest work.

'You dear child,' he said, and kissed me on the forehead.

Thus he responded perhaps to her hint of understanding that her variation was part of something larger.

Through continuing violin lessons at The Mount and preparing the

[26] *Edward Elgar: memories of a variation*, pp. 21–4. This passage of Mrs. Powell's reminiscence is heavily confused with later events. She thought that the proofs arriving were those of *Gerontius*, which is disproved by Elgar's correspondence with Novellos. She cites a memory of Elgar playing the trombone, which he began to study only in November 1900. Her memory of hearing 'sketches and bits of *Cockaigne*' during this visit must also be suspect, for there is no independent evidence that either idea or music was in Elgar's mind before the autumn of 1900.

Worcestershire Philharmonic for a performance of Berlioz's *Childhood of Christ* in December, the question of a subject for Birmingham was still unresolved. Of alternatives to 'The Dream of Gerontius', the conversion of Britain by St. Augustine had failed to tempt the librettist skills of Joseph Bennett. But now there came back to him the original Christian conversion—the theme of the Apostles which had first attracted him in the schoolmaster's comparison of the boys at Littleton House thirty years earlier.

The Apostle who had been strong during Christ's days on earth was Judas: and there Edward found identity. Judas had done the dreadful thing. But why? He was an intelligent man: he was the only non-Galilean among the Apostles—a cut above the others in every way, perhaps an aristocrat. They had made him their treasurer. Afterwards they had all written disparagingly of him. In fact all the descriptions of Judas came from those others—who in the aftermath would have every reason to find a scapegoat. Who more ready to hand than the man of superior abilities, the outsider? Edward wrote:

I [see in] Judas Iscariot a much more terrible 'lesson' than the ordinary acceptation of his character by the unthinking, the unreading, and the invincibly ignorant allows.[27]

The 'lesson' lay in Judas's intellectuality:

I was always particularly impressed with Archbishop Whately's conception of Judas, who, as he wrote, 'had no design to betray his Master to death, but to have been as confident of the will of Jesus to deliver Himself from His enemies by a miracle as He must have been certain of His power to do so, and accordingly to have designed to force Him to make such a display of His superhuman powers as would have induced all the Jews—and, indeed, the Romans too—to acknowledge Him King.'[28]

Judas, in this view, wagered that he could force Christ's hand. When the wager failed, Judas sought his end in suicide. So, as the Swimmer had replied to the triumph of E.D.U., the nightmare loneliness of Judas could haunt the unachieved dream of Gerontius.

Edward had begun to make musical sketches, some of which he had played to Jaeger during the London visit in October. But he described them so vaguely that Jaeger assumed them to be part of the long-deferred Symphony. On 7 November 1899 Jaeger asked:

How *is* that Gordon Symphony getting on! *You Sphinx*!!
Why Dontcher answer???
    Are you lazying over your work?'[29]

Edward responded ambiguously:

All composition is a Dead secret but I say I *have* written a *theme*, alas! orchestral & it's no good on the piano.[30]

---

[27] 'Notes for an Article (sketch) on the Character of Judas Iscariot' (TS at the Elgar Birthplace).
[28] *The Strand Magazine*, May 1904, p. 542.
[29] HWRO 705:445:8386.   [30] 8 Nov. 1899 (HWRO 705:445:8371).

Of course Jaeger wanted to see it, and it came—headed with a biblical quotation: 'Then Jesus said unto him—That thou doest do quickly—he (Judas) went immediately out: *and it was night!*'

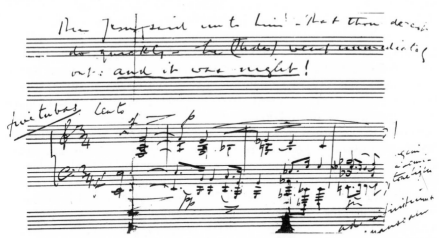

—'again a semi-tone higher—ad infinitum & nauseam'.[31] The bass was once again a sequence of falling fourths. Over its opening E came a clashing F major chord to start step-wise descent. Thus one primal shape of Edward's expression challenged another in 'Judas'.

Jaeger responded with enthusiasm. But Edward hesitated over showing more of such a conflict: 'I'm glad you like my idea of Judas. I'll send you another wildly expressive bit but it's very hard to try & write *one's self* out & find that one's soul is not *simple* enough for the British choral society.'[32]

As he considered the whole subject of 'The Apostles', it grew and grew until it suggested a trilogy of works perhaps on the scale of Wagner's *Ring*. It would need a theologian's mastery of the Bible to help select the stories and quotations which could best shape the great work: and for Birmingham that theologian should be Protestant. He consulted his librettist for *The Light of Life*, Edward Capel-Cure. On 5 December 1899 Capel-Cure came up from his vicarage in Dorset to stay two nights at Craeg Lea. He met Edward's idea of 'The Apostles' with dismay.

To work out the libretto with a penetrating character analysis and write all the music would take manifestly longer than the ten months now left before the Birmingham Festival. And if somehow the whole gigantic trilogy plan could be worked out and its first instalment finished in time, there was still the prospect of new financial negotiation over the whole project with Novellos—the only real publisher for a major choral work. Edward wrote to the chairman of the Birmingham Festival Orchestral Sub-Committee, George Hope Johnstone, to

---

[31] Pasted into Jaeger's copy of the *Gerontius* vocal score.
[32] 18 Nov. 1899 (HWRO 705:445:8373).

resign the commission. Excusing himself to Jaeger, he retreated to the financial plea he had used over the 'Gordon' Symphony: 'I gave up Birmingham as I could not really afford to go on writing . . .'[33]

*   *   *

On New Year's Day 1900, G. H. Johnstone and his wife came to Craeg Lea. He was a man of formidable business ability, and he was an experienced patron who understood musicians' problems, having dealt at close quarters with Gounod, Grieg, and Dvořák and entertained them in his home. Now he made two proposals. First, in addition to a payment from Birmingham for the first performance, he would make himself responsible for Edward's publishing negotiations over the new work. Second, the choice of subject was at this point a matter of sheer practicality. 'The Dream of Gerontius' offered a complete poem, on which Edward had already done some of the work needed to shape it as a libretto: it was the only one of his schemes which stood a chance of completion in time. Johnstone made no difficulty over his committee's acceptance of the Catholic subject, and there was to be no difficulty. The bare comments in Alice's diary expressed her relief:

January 1, 1900. Mr. & Mrs. Johnstone came to lunch[34] & arranged for E's work Birmh Fest. Deo Gratias

January 2. E. sent telegram accepting terms. Began again at former libretto.

Edward himself was very unsure. On 4 January he wrote to Jaeger: 'All goes slowly here: I'm working like a —— fool & get kicked for my pains &—or pleasures—I don't know which they are.'[35] He gave no news of the subject. When Jaeger asked the question direct, he answered: 'Judas is dropped! I *may* have some news for you concerning my works very shortly—but I am sick of the whole thing & wd. never hear, see, write or think of music in shape, form, substance or wraith again.'[36]

On 12 January he went to the Birmingham Oratory to consult with Cardinal Newman's friend and executor Fr. Neville about permission to abridge 'The Dream of Gerontius' for the libretto. The poem ran to 900 lines, divided into seven sections: a prologue on earth showed the dying dreamer surrounded by a priest and 'assistants', the other six parts traced the Soul's immortal progress through Judgment to Purgatory. Edward's libretto shortened the earthly prologue by only a little to form his Part I. But the 730 lines of the poem's other six sections were cut down to 300 lines to form his Part II. It sharpened the dramatic contrast between the two states: life and death, death and life.

---

[33] 5 Feb. 1900 (HWRO 705:445:8407).
[34] This word in Alice Elgar's diary has been misread as 'church'. 1 Jan. 1900 was a Monday, and the diary says nothing about church. Moreover Johnstone was not a Roman Catholic: he attended the Swedenborgian 'new Church'. See Adelina de Lara, *Finale* (Burke Publishing Co., 1955) p. 65.
[35] HWRO 705:445:8402.   [36] 10 Jan. 1900 (HWRO 705:445:8403).

The opening scene on earth showed the solitude of the dying man—as the solitude of Judas would have been shown in 'The Apostles'. Its central expression was Gerontius's long and lonely statement beginning in faith (with the words from the Good Friday liturgy, 'Sanctus fortis, Sanctus Deus') but moving toward 'horror and dismay'. At the end of Part I his life was dismissed by a priest in the final words of the Burial Service, 'Proficiscere, anima Christiana, de hoc mundo'.

In the other-worldly setting of Part II, the hero's loneliness found answer. Through Part II the central expression was to be choral, moving from demons to choirs of 'angelicals'. The massed singing contrasted with solo statements from the Soul of Gerontius, the Angel now guiding his progress, and the Angel of the Agony. This last awesome figure, preparing the Soul to go before God, answered the seeming finality of the earthly priest with prayers that were continuing.

Thus *The Dream of Gerontius* might embody Edward's growing sense of isolation, as success demanded yet more creative success, in the Catholic faith of his mother and of Alice. *Gerontius* was not on its surface a story of feminine recapitulating in cycles, but of masculine forward striving. Yet it offered a synthesis of rejection and acceptance in the Catholic prospect of Purgatory. The Angel's Farewell at the end of Part II echoed the farewell of the earthly priest in Part I, of Caractacus embarking on the Severn, of the hero's mother at the end of *King Olaf*—but now with a vital difference:

> Farewell, but not for ever! brother dear,
>   Be brave and patient on thy bed of sorrow;
> Swiftly shall pass thy night of trial here,
>   And I will come and wake thee on the morrow.

Was it such a morrow that the lonely child had been seeking all those years ago in the reeds by Severn side as he longed 'for something very great'? The companion Angel of Edward's *Dream* would be sung by a woman.

The abridgment of Newman's poem suited his music in another way. The 435 lines remaining made the shortest libretto he had yet used for a full-length work. These words would neither confine nor guide the development of his music as the *King Olaf* and *Caractacus* libretti had done. It would be for the music itself to discover and define its salvation and his own.

On 21 January 1900 Edward sent a letter to lay the project before Alfred Littleton at Novellos, and next day the chairman had a visit from G. H. Johnstone of Birmingham. Littleton wrote to Edward:

1, Berners Street, London, W.

Jan 23 1900

Dear Mr. Elgar

Your letter reached me yesterday and today I had a visit from Mr. Johnstone. He showed so much interest in the whole matter and put everything in such a nice way that I could do nothing but agree to his proposition. It is therefore agreed that we shall publish your work 'The Dream of Gerontius', which piece of information will I hope be satisfactory to you . . .

> With kindest regards,
> Yours sincerely
> Alfred H. Littleton[37]

The chairman had made a note of the terms for his files:

23 JAN 1900 Agreed this day between Mr. Johnson [sic] of Birmingham & Mr. Alfred H. Littleton to publish Elgar's new Birmingham work *'The Dream of Gerontius'* & to pay Two hundred pounds for all rights.[38]

When Johnstone wrote to Edward, however, he named an extra provision not in Littleton's note:

They take the whole responsibility of publishing & pay you Two Hundred pounds for the work & will, if there is any profit on the Sale after the Two Hundred pounds has been cleared pay you a fair proportion.

I hope this will be satisfactory to you & that you feel the benefit of freedom from financial responsibility. I trust the work will soon be in Novellos hands to print so that we can begin rehearsing it early.[39]

But there was no question of that. The main lines of musical assemblage had barely begun to emerge. In fact it was 5 February before Edward felt able to identify even the subject for Jaeger:

I am setting Newman's 'Dream of Gerontius'—awfully solemn & mystic.

　. . . Now I must go on to my Devils' chorus—good! I say that Judas theme will *have* to be used up for death & despair in this work—so don't peach.[40]

The 'Judas' theme would be given to the Angel of the Agony—a suggestive exchange of characters.

The Demons' Chorus held its mirror in a different direction. Edward used his skill at ciphering to make a Demon figure out of a name he disliked. Charles Stanford combined an Irish bluffness with cocksure and patronizing familiarity that grated on Edward's susceptibilities. To the end of his life he would recall Stanford's habit, whenever he caught sight of two people talking together, of walking up and butting in on the assumption that they were talking about him.[41] It was said that Stanford's politicking at Leeds had got the ailing Sir Arthur Sullivan turned out of the Festival conductorship there after an association of nearly twenty years.[42] And Edward disliked Stanford's music, as he had written to Jaeger: 'Anything "genuine" & natural pleases me—the stuff I hate and which I know is ruining any chance for good music in England is stuff like

[37] Elgar Birthplace.　　　　　　　　　　[38] Novello archives.
[39] 30 Jan. 1900 (Novello archives).　　　[40] HWRO 705:445:8407.
[41] MS notes of Elgar's conversations with Basil Maine (Royal College of Music).
[42] Sullivan's diary, 18 Sept. 1899. 'Received letter from Spark saying that my ill health had caused committee so much anxiety that did not intend running "risk" again. Rubbish. Of course I know what it was.' (Yale University.)

Stanford's which is neither fish, flesh, fowl nor good red-herring!'[43] In the Demons' Chorus, 'STANFORD' was ciphered to 'SATANFORD'.[44]

On 18 January Edward had conducted four *Sea Pictures* for Clara Butt at a Hallé Concert in Manchester. The invitation had come from Stanford, who conducted the remainder of the concert, and who seemed to take the occasion to right a bit of misunderstanding.

Early in February Edward and Alice were in Manchester again for the second of two Richter performances of the *Variations*. Alice wrote to Nicholas Kilburn: 'Dr. Richter is simply devoted to the work. It was a great pleasure to be in his company the rest of the evening after the Concert & hear him talk. All his words & thoughts seem on so noble & high a plane, & his appreciation of my Husband so truly touching.'[45] It was a good augury for Birmingham, where Richter was the Festival conductor. At supper after the Manchester concert, he introduced the leader of the Hallé Orchestra, the distinguished Russian-German violinist Adolf Brodsky, and his wife: since the death of Charles Hallé, Brodsky had served also as Principal of the Manchester College of Music. The Brodskys were becoming keen admirers of Edward's music. So was the cellist of the Brodsky Quartet and Professor of Violoncello at the Mancester College, Carl Fuchs. Fuchs extracted from Edward a promise to write something for the cello one day.[46]

Another ten days of hard writing at Craeg Lea completed a rough outline for the whole of the *Gerontius* music. In London to conduct the *Sea Pictures* at the Crystal Palace on 24 February, he went to tea next day, Sunday, with the Jaegers and played for them the Angel's Song which was to conclude *Gerontius*. The Jaegers were so entranced that he promised to write it out for them to enjoy through the weeks before the work should be finished and this music in print.

While in London, Edward also called on Parry's son-in-law, the baritone Harry Plunket Greene, with two songs—*After* (sketched in 1895) and a setting of Christina Rossetti's *A Song of Flight*. Plunket Greene was to give their first performance at a concert with the pianist Leonard Borwick in early March and the singer's interest was advertised on their title pages when Booseys brought them out. The meeting had a further importance: Plunket Greene was engaged for the Birmingham Festival, and he was in view for the 'Priest' and 'Angel of the Agony' solos in *Gerontius*. Returning to Malvern on 26 February Edward threw himself into writing out the short score.

The *Gerontius* music would begin with a lengthy Prelude. So his orchestral impulse—immeasurably strengthened in the *Variations*—here found expression at the head of a major choral work. The Prelude opened in a simplicity,

[43] 11 Dec. 1898 (HWRO 705:445:8290).   [44] Communicated by Anthony Thorley.
[45] 12 Feb. 1900 (HWRO 705:445:BA8089).
[46] Recalled by Fuchs in a letter written to Elgar on 24 Nov. 1900 (HWRO 705:445:1411).

almost a nakedness, unlike the opening of any previous work of Edward's
maturity. An unaccompanied A dropping to G ♯ suggested one harmony, a
second drop from A to G ♮ suggested another:

This darkening shape sent a chill across the music's beginning. Only in the
second half of the phrase was the tonic of D minor implied:

There followed more darkness and cold, as a bass-figure struck in with the
slowest, softest descent of pedal steps in all Edward's music:

At length the music returned to its unaccompanied beginning. In the thematic
analysis[47] which Jaeger was to prepare under Edward's supervision, all this
music was denoted as 'Judgment' and described as the most important motive
in the work.

It was followed immediately by two contrasting motives. The first had been
written earlier on a sheet headed 'Sketches 1896', containing material for *The
Light of Life*.[48] It fitted into the Gerontius texture now as a sequence of rising
'Fear', with the descending steps sounding chromatically in the bass:

[47] *Analytical and Descriptive Notes* (Novello, 1900).
[48] BL Add. MS 47904A fo. 120. This point and several others in the following discussion I owe to
Christopher Kent.

The two opening harmonies of D and A flat measured the interval of a tritone—the ultimate opposition of which the medieval theorists had written, 'Mi contra fa est diabolus in musica.' The figure then moved deliberately through F♯ and E to emphasize the whole tones between Ab and D.

The contrasting motive stood for 'Prayer'. Its first four notes were practically those of the old chant in the *Bona mors* service, lingering in Edward's memory from his organist's days at St George's:

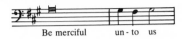

It was one of many musical quotations and near quotations from Catholic sources in the *Gerontius* music to parallel the quotations from the Bible and Missal in Newman's poem. Most recently, this melodic shape had evoked the deep ocean movement in *Sea Pictures*. Before that it had been set down in Sinclair's Visitors' Book to describe the bulldog Dan as 'he muses (on the muzzling order)'. Now it entered perfectly the *Gerontius* texture as 'Prayer':

Then the alto at the end of that motive generated another. It inverted the opening 'Judgment', to float uneasily over a bass *ostinato* of 'Fear':

When Jaeger called this 'Sleep' in his analysis, Edward wrote: 'I meant "to be lying down weary & distressed" with your poor head buzzing & weak &—have you ever been really ill? Sleep will do but it's the ghastly troubled sleep of a sick man'.[49] The 'Sleep' motive gave way before a new development of 'Judgment' and 'Prayer' to chromatic descending steps supported still by the 'Fear' *ostinato*. Jaeger called it 'Miserere':

Another variant—a chromatic sequence aspiring upward but slipping inevitably down—portended 'Despair':

'Prayer' replied by thundering through a C major augmentation passionately repeated. Its long sequence now suggested that the drama of Gerontius's loneliness had already begun in this Prelude.

Suddenly the music was wrenched to A flat major to show the old pattern of four descending steps. The crisis led through 'Despair' toward a synthesis of all these transformations—the grand melody Edward had set down fifteen months earlier on the day of writing to Jaeger about the 'Gordon' Symphony. That hero's dead hand seemed to grip the motive now as it found its place in *Gerontius* as the figure of 'Committal'. Its essence was again descending steps, set amid big upward leaps:

The entire 'Committal' theme was repeated in sequence through fifty bars of climax: in Edward's previous music, only the 'Nimrod' melody approached it. Then 'Sleep' and 'Fear', 'Miserere' and 'Judgment' re-sounded in a near symmetry that closed the Prelude in the soft chill of its unaccompanied opening.

[49] 28 Aug. 1900 (HWRO 705:445:8401).

There was no pause as the music went to the first hushed utterance of the mortally ill Gerontius. Between its set pieces, the solo writing would extend through the work in recitative. Quantities of surviving sketches show how hard he worked to fit this recitative to the poem's verbal accents. It meant abandoning any notion of finished tunes which might be fitted to the words. Some sketches began with just the words written out in ink, the music emerging round them in tentative pencil. When an instrumental idea was set down first, the words were introduced with a skill that was tireless in seeking accent and balance. Through draft after draft, the music gradually became a sort of dramatic plainsong.

The arioso recitative had been tried in parts of *King Olaf* and *Caractacus*. But the words of *Gerontius* were of a different order of dignity. And many of them were thrice familiar through the Catholic liturgy they echoed. The old chants could suggest their pulses and rhythms at deeper levels of Edward's consciousness. And thus a new sort of variation-writing began to emerge— where a single figure in music might be endlessly extended through changing accents of speech. It raised the subtlest possibilities for musical-dramatic integration.

The opening words of Gerontius were the opening words of the poem: 'Jesu, Maria—I am near to death . . .' Immediately there appeared a variant of the great 'Committal' theme which Jaeger would call 'Christ's peace':

After the D minor of the Prelude, this opening in B flat moved farther into flats for the first choral entry of the Priest's Assistants, 'Kyrie eleison', in E flat major. Here began the setting of the 'Bona mors' material in Newman's poem. Like most of the other early choral utterances in *Gerontius*, Edward set this 'Kyrie' to a slow fugato. This one started from the single upward step which had inverted 'Judgment' to 'Sleep'.

Gerontius returned to his E flat (now in the minor), and when the chorus came again they had moved farther into flats to sound a fugal 'Be merciful' in F minor. Here the old 'Bona mors' chant gave a hint for the more dramatic fall

There followed nearly fifty bars of fugal development to prepare the great central solo of Part I, 'Sanctus fortis'.

Here Edward paused. On 2 March 1900 he sent this much of the finished

vocal score to the publishers for printing in a first instalment: once again the
instalment plan offered the best hope for meeting the demands of choral
rehearsal in the time remaining before the Festival première. He enclosed a
covering note to Jaeger:

Let me know by wire, if you can afford it, that this has reached you—I'll be in a fit 'till I
know . . . Deal tenderly with it   it's the *only* copy there is . . .'[50]

Next morning Jaeger telegraphed:

44 PAGES GERONTIUS RECEIVED HURRAH AND GLUCK AUF FOR THE
REMAINDER[51]

> Sunday [4 March 1900]

My dear Jaeger:
   Many thanks for your nice telegram—get on with it fast as I want to see some in print
to 'kinder' encourage me—I'm wofully nervous about my powers—or want of them—as
you know . . .

> Yours ever
> Ed: Elgar[52]

But with the nervousness went a feeling of intense personal commitment, as he
wrote to Nicholas Kilburn on 11 March:

I like what I have done—I am bold & have shirked nothing—I've made my own
'atmosphere' & stuck to it—Much modern music annoys me thus:—when we come to
the 'high' situation the composers too often let the voice recite on one note
&c.&c.—This seems to me that the musician confesses his art is unworthy—Perhaps it
shews the composer is 'unworthy'—I don't know—but I do know that I won't write
so—if I fail it's not the fault of music per se—but of me!

   The writing of *Gerontius* went forward. The real demonstration of Edward's
new recitative writing came with the big 'Sanctus fortis' solo. He set the
opening words in a repeated figure of F-Db—to suggest the still further
flat-relation of B flat minor, though the main solo was once more in the major.
Through a 3/4 *Allegro* of 160 bars, utterance and melody responded to each
other in total naturalness. The overall musical form followed the form of
Newman's verses: a 'Sanctus fortis'/'Miserere' quatrain at beginning, middle,
and end, separated in each case with 12 lines of aspiration to belief. Later
appearances of the 'Sanctus fortis' music varied the opening to

This echoed the 'Judgment' opening of the whole work. In a multitude of
elements, one melodic pattern stood out: it culminated in upward leaps
opening to greater and greater breadths of assurance:

50 HWRO 705:445:8409.   51 HWRO 705:445:8457.   52 HWRO 705:445:8399.

And I hold in ve - ne - ra - tion, For the love of him a - lone,____

Afterwards Gerontius shouted his fear—his voice as lonely as the Swimmer's, engulfed in savage orchestral chromatics. His final words emerged to a wrenching distortion of 'Christ's peace':

In Thine own ____ a - go·ny__

In the following chorus, 'Rescue him', the E flat major 'Kyrie' music moved to minor—as far into flats as the dying Gerontius himself had penetrated—and then still further, to an unadorned chant set in the remote subdominant distance of A flat minor, telling over the old prophets rescued 'by Thy gracious power'. For this Edward adapted the Gregorian 'De profundis'—(closely linked with a harmonization he had once made for the chant 'Have mercy upon us' in the 'Bona mors' service at St. George's). Then came Gerontius's dying 'Novissima hora est', set to the shape of 'Christ's peace' and the 'Agony' music but now rising by a fourth. Its four-sharp signature equalled eight flats, bringing the tonal progression with perfect logic out on the other side.

Thus Part I came to its final section, the Priest's 'Proficiscere'. The solo began with repeated tones through descending steps, to project an atmosphere of chant:

Moderato solenne e con elevazione

Proficiscere, anima Christi - a - na,   de hoc mun-do!____

With the words 'Go, in the name of God', the music emerged in D major to answer the D minor of the Prelude. And on the word 'God' there began the steady tread of a slow march—a final development from the 'Sleep' motive to support a great triadic melody. The chorus thundered responses to the Priest, descending in majestic steps to the 'Committal' music and moving gradually through a huge softly-dispersed ensemble of Priest, Assistants, chorus, and orchestra sounding the slow march of D major to the end.

On 20 March 1900 Edward sent this concluding half of Part I to Jaeger, addressing him with a pun that took 'Nimrod' to the world of Shakespeare's Falstaff:

My dear Corporal Nym:

By this post come pp. 44–99 incl. up to the end of Pt I—the final chos. is godly effective &, I think, not quite cheap: if the score required

viz 1 [stave] solo, 4 semich[orus]; 8 full ch, & 2 acc[om]p[animen]t 15 staves in all— is unmanageable the *semi*chorus might go in short score on *two* staves—but don't say a word unless the printer takes the Button off his foil & shews real fight.

It was another pun: H. Elliot Button was Jaeger's colleague on the printing side at Novellos. And then Jaeger's latest suggestion—that Edward should write a virtuoso piece for the violinist Eugène Ysaÿe—provoked the most elaborate pun of all:

As soon as this 'Dream' (nightmare) is done—I'll write such a lot of things for Ysaiah—& all the profits . . .

Wire me that the new 'batch' arrives safely there's a dear, 'cos I get angkschzsuszcs over it sometimes.

Hirschen Sie auf! (Buck up!) translation, with the remainder, 'cos if the Bir[mingham] people want something to go on with they can have Pt I. as soon as you've done with it.[53]

But printing was nowhere near any stage of providing the individual chorus parts used in those days for rehearsal. First proofs had only begun to appear in mid-March. Jaeger wrote when he saw the new music: 'It looks 'shivery' & awe-inspiring to me . . . *When* is that Angel's Song coming??????'[54]

Then Edward wrote out and sent for his friends' private enjoyment what he had played at their house in February. Its security of D major would return a final answer to all chromatic questions. And its instrumental shape was easily simplified for singing:

When Jaeger saw it, he wondered whether Clara Butt—if indeed she was to sing it at Birmingham—could do justice to this 'uncanny' music:

I say 'uncanny', because it has a character unlike anything else in music as far as I know. Simple as it is, its very simplicity is its wonder, for it is a kind of simplicity I have never met with before, so aloof from things mundane, so haunting & strangely fascinating. It has been running through our heads ever since we played & sang it several times to make ourselves a little familiar with it . . . You are a genuine wizard to thus play with our emotions.

. . . Since 'Parsifal' nothing of this mystic, religious kind of music has appeared to my

---

[53] HWRO 705:445:8396.   [54] HWRO 705:445:8456.

knowledge that displays the same power & beauty as yours. Like Wagner you seem to grow with your greater, more difficult subject & I am now most curious *and* anxious to know how you will deal with that part of the poem where the Soul goes within the presence of the almighty. *There* is a subject for you![55]

But this time Edward showed his mind made up to avoid everything big and dramatic when the Soul goes before God. He replied to Jaeger:

Please remember that none of the 'action' takes place in the *presence* of God: I would not have tried *that* neither did Newman. The Soul says 'I go before my God'—but *we* don't—we stand outside . . .[56]

It was the significance and secrecy of the confessional.

Every phase of daily life was drawn into the writing. Miss Burley remembered:

Throughout the early months of 1900 we simply lived *Gerontius*. We talked of little else on our walks and Edward seemed to think of nothing else. Again and again manuscript fragments would be brought to The Mount on the lesson days, tried over and discussed. On these occasions Edward usually played the piano since only he could understand the complexities of his much-corrected score, my own share being to add on the violin counterpoints he had written and of which he wanted to hear the effect. Even the Worcestershire Philharmonic Society was pressed into the service and were asked to play the . . . introduction to Part II . . .[57]

This music, revealing the time after death, was scored for strings *pianissimo*. Again the beginning was in unaccompanied tones, but now in the major:

The rising fourths took up the hint of Gerontius's 'Novissima hora est'. Edward wrote: 'I intended all this peaceful music to be the "memory (remembrance) of the soul"—an utter childish (childlike) peace . . .'[58]

The Soul of Gerontius sang a quiet arioso, and an Angel appeared with an 'Alleluia' set to a Gregorian 'Ite missa est'.[59] Questions from the Soul and answers from the Angel led to a short duet. Here ecstacy was realized in a polyphony which far outstripped the love-duets of *King Olaf* and *Caractacus*. Yet this immortal love-making gave Edward pause. He wrote to Jaeger:

I want you to look very carefully at Pt. II of Gerontius, as it comes from the printer, &

[55] 13 Apr. 1900 (HWRO 705:445:8458).   [56] 17 Apr. 1900 (HWRO 705:445:8395).
[57] *Edward Elgar: the record of a friendship*, p. 135. The Worcestershire Philharmonic Society rehearsals during this period took place on 13 and 27 Feb. and 13 and 27 Mar. 1900.
[58] 4 July 1901 to Jaeger (HWRO 705:445:8500).
[59] Quoted in correspondence with Elgar by the music critic Vernon Blackburn (HWRO 705:445:2819 and 3088).

tell me if you think those *conversations* between the Angel & the Soul are wearisome
. . . Between ourselves this is the only part of the work that I fear or even think twice
about—if the words are sufficiently interesting the music will do . . .[60]

Progress with Part II was delayed by a severe cold. On 17 April he wrote to
Jaeger:

I've been really ill with awful chill—throat—everything in fact & *all* has perforce been at
a standstill for a fortnight, alas!
   I could send you another batch but strictly entre *nous* am waiting to hear definitely
who is to sing who.[61]

It might be better to entrust the music of the Angel-companion to the other
contralto engaged for the Birmingham Festival, Marie Brema.
   Directly after the duet came the Demons' Chorus. A chromatic introduction
descended through enharmonic cleverness which practically destroyed tonal
identity. Then the Demons entered to a sort of deformation of the three-note
fanfare for 'E.D.U.':

They avidly imitated from one voice to another, but made no real
development. Their expression lay in gross contrasts—leap against half-step,
fortissimo again sudden *piano*, complex harmonies against open octaves and
lame single notes. A climax came with what Jaeger described in his analysis as
'four giant strides' to a 'towering eminence':

It was the Demons' vulgarization of the simple four-note figure which had
opened the Prelude to Part II.
   That fantasy was followed by a furious fugal opening:

---

[60] 7 May 1900 (HWRO 705:445:8417).   [61] HWRO 705:445:8395.

But instead of any proper fugal development came *fortissimo* shouts of empty fourths and fifths, and another perversion of the Part II introduction:

Then the Demons compressed the figure of Gerontius's aspiration from 'Sanctus fortis' ('And I hold in veneration, For the love of Him alone') to

At last the whole box of pseudo-Elgarian tricks emptied into an ultimate reduction of the 'Judgment' motive

It was all to be awesomely orchestrated, and when it met with a chorus trained to master its formidable difficulties it made a terrific noise in turn-of-the-century cantata-writing. But it was the one section of *Gerontius* that was not to wear well. The passage of time has blunted its daring and left its vulgarity exposed. Yet after all perhaps that was what the subject needed. Before the Demons began, the Soul had said that their noise 'would make me fear, Could I be frighted.' Now the listener heard as the Soul heard. Immediately after the Demons had disappeared the chromatics were drained out of the 'Mind bold and independent' counterpoint to set the Soul's question about God: 'Shall I see my dearest Master . . .?'

'E. writing vehemently,' noted Alice on 26 April. Three days later he was setting out a fair copy of the music to introduce the Angelicals' chorus 'Praise to the Holiest'. With the performance at Birmingham just five months away, time was short. He wrote to Jaeger on the 29th:

I want you to push on the printing of what I sent & shall send today or tomorrow: then, if the Bir. people like, they can have copies up to the end of the Demons' chorus. The *end* of the work requires some consideration & I don't want to send the M.S. until I've been thro' it—(as usual)—again—again—again—again & after that once more![62]

But Jaeger had already warned him that Novellos were putting *Gerontius* behind all the new music they were printing for the Hereford Festival in September, because Birmingham was not until early October.

The first Saturday in May 1900 brought another Worcestershire Philharmonic concert. Edward conducted a mixed programme including Wagner's *Rienzi* Overture, the Beethoven G major Piano Concerto with the solo played by a young lady from Malvern, choral music by Brahms, and an orchestral scene from Granville Bantock's *The Curse of Kehama*. Jaeger came down for the concert, and Dora Penny arrived from Wolverhampton.

After the concert we had a most hilarious tea at the Hydes, and when tea was over we said our farewells and went to Foregate Street station on our way back to Malvern. Mr. Jaeger and I became fast friends at once. He *was* a most delightful person. His English was fluent, not to say voluble, but with a strong German accent. He asked me if I knew Gilbert and Sullivan well and we sang 'Oh, Captain Shaw!' on Foregate Street platform while waiting for the train. E. E. in the background remarked, 'Now they're off.'

That journey was one of the noisiest I have ever made. Our party filled a compartment, and all the way Mr. Jaeger was telling me with great volubility and much gesture how wonderful *Gerontius* was. E. E. was trying to stop him and was calling him a whole string of comic names, and the rest of the party were in fits of laughter.

On Sunday E. E. and Mr. Jaeger shut themselves into the study all the morning. When I came back from church, having called for Carice at her school on the way, I found them still busy.

The Ninepin [Troyte Griffith] came to luncheon and it was all most amusing. When they came into the dining-room E. E. saw Carice, standing behind her chair, waiting.

'Hullo, Fishface! Quite well?'

'Yes thank you, father.'

'"Yes thank you, father,"' imitated E. E. in a high sort of squeak, after which the unfortunate child was expected to say Grace![63]

The little girl's features did not arrange themselves for classical beauty; and with parents so mutually absorbed and over-powering when they suddenly turned in her direction, self-assurance was hardly possible during those few Sunday hours each week when she was allowed home from The Mount. Yet the situation was not always so brutal. As Carice began to read and write easily, her parents' frequent absences from home seemed to make their hearts grow fonder. Her mother used the baby language to write of their distant doings. Her father turned his quick brain to delighting her in his letters. From an interest shared in animals, Carice became his 'Duck's-eye'; and his letters were adorned with little sketches, especially of cats and mice, as she herself recalled: 'The "mice" originated in his writing of figures on a cheque; he always put a tail

---

[62] HWRO 705:445:8415.
[63] *Edward Elgar: memories of a variation*, p. 25.

to the final nought, and it was easy to add the ears—*not* on the cheque—but on bits of paper for my amusement as a child.'[64]

Cats and mice also decorated some of his letters to Dora Penny—who did not share Carice's embarrassment on that Sunday in May 1900:

When E. E. was at the top of his form meals used to be exciting. He kept up a running fire of absurd remarks, comments, chaff, and repartee. I often laughed so much that I could hardly eat and was positively afraid to drink. Also it did not help matters to have the Lady, at the bottom of the table—not always completely approving, particularly if Carice was present—putting in remarks to try to check the flow:

'Oh, Edward dear, how *can* you?' or 'Oh, Edward, *really!*'

'Cheer up, Chicky!' was all she got for her pains.

On this occasion I think the fun was even more than usually fast and furious, because Mr. Jaeger was there and E. E. was in the best of good spirits. The climax came when E. E. started conducting with a carving knife!'[65]

Jaeger's weekend at Craeg Lea ended in a wild rush for the train, owing to Alice's mistaken reading of a timetable. She apologized by letter: 'Forgive such mistakes, it must be owing to the "Martha" like duties I have to accomplish while wishing to dwell in the IDEAL.'[66] Jaeger replied:

I can quite appreciate how two 'idea angeleyte' people like yourself & E. E. must suffer occasionally when coming into contact with the commonplace in things or men—(e.g. Philistines like 'Nimrod'!). Though I am made of commoner clay, I can sympathise with you twain most thoroughly, & wish you could be spared all unpleasantness. But it has ever been a law, I daresay, that genius must suffer (—we might alter the saying to—'il faut souffrir pour etre génie!') But so long as your *Health* remains fairly good, the lovely Home you have built yourselves on the beautiful Malvern Hills, and the love & admiration of your many friends should help you to make life happy & enjoyable.[67]

However carefully he might cast himself as their inferior, Jaeger saw things very clearly.

In London on 10 May Edward conducted for the first time at a Royal Philharmonic Society concert: once again it was the *Sea Pictures* with Clara Butt. Next day he consulted Plunket Greene about the Angel of the Agony solo to come after the Angelicals' Chorus 'Praise to the Holiest'. Back at Craeg Lea, he passed temporarily over the big chorus to write the following solo music.

The Angel of the Agony solo was approached through the 'Fear' motive. As the Soul heard the voices of friends left on earth, the treading figure of the Priest's 'Go in the name of God' sounded again: it linked the earthly Priest with the Angel of the Agony. This magisterial figure was introduced with an 'appalling chord' that Edward recalled having set down on a 'tiny scrap, my first

[64] *The Musical Times*, 1 Oct. 1942, p. 298.
[65] *Edward Elgar: memories of a variation*, pp. 25–6.
[66] 7 May 1900 (HWRO 705:445:8816).
[67] 8 May 1900 (HWRO 705:445:8807).

note (made probably on the golf links) on a sorry shred of paper and pasted into
my note-book

Only later he saw that it was a close development of the 'Agony' figure from
Part I. That seemed a sign of real and present inspiration, as he wrote months
afterward to Jaeger of the motivic connections in *Gerontius* as a whole:

I really do it without thought—intuitively, I mean. For instance, I did not perceive till
long after it was in print that—
    (p. 34) 'In Thine *own agony*'
    & the appalling chords
        I last bar p 150   ⎫ introducing & dismissing
        II 3rd line, bar 2, p. 154 ⎭ the *Angel of the Agony*
were akin but they are, aren't they?[69]

The upward seventh leap to introduce the 'appalling chord' was a familiar
gesture through Edward's music of recent years. And the descending steps
brought the Angel of the Agony clothed in the mantle originally designed for
Judas:

In the middle of this chromatic intensity the Angel of the Agony found a
moment of diatonic comfort. But a huge slow crescendo through his rising
'Hasten, Lord, their hour' brought back the chromatic agony and a final
repetition of the 'appalling chord'.

    All this, together with the solo utterances to precede the Angelicals' chorus,
was now ready for the printer. On 21 May Edward wrote to Jaeger:

I send m.s. down to big chorus—this said chorus I want to '*dwell on*' for a space—but I
think you cd. go on with the soli parts [following the chorus] (also enclosed) putting in
dummy paging . . .[70]

When Jaeger saw this music printed, he wrote:

I have just spent an hour over your latest batch of proofs (pp 103 to 111 & pp A to F,)
and, Oh! I am half undone, & I tremble after the *tremendous* exaltation I have gone

---

[68] Quoted in Marjorie Ffrangçon-Davies, *David Ffrangçon-Davies: His Life and Book* (John
Lane The Bodley Head, 1938) p. 34.
    [69] 28 Aug. 1900 (HWRO 705:445:8401).   [70] HWRO 705:445:8393.

through. I don't pretend to know *everything* that has been written since Wagner breathed his last in Venice 17 years ago, but I have not seen or heard *anything* since 'Parsifal' that has stirred me, & spoken to me with the trumpet tongue of genius as has this part of your latest, & *by far* greatest work . . .

But that solo of the 'Angel of the Agony' is overpowering & I feel as if I wanted to kiss the hand that penned those marvellous pages. Those poignant melodies, those heart-piercing, *beautiful* harmonies! I recognise the chief theme as having belonged to 'Judas'. *Nobody* could dream that it was not originally *inspired* by *these very words* of *Newman's*.

You must not, *cannot* expect this work of yours to be appreciated by the ordinary amateur (*or* critic!) after one hearing. You will have to rest content, as other great men had to before you, if a few friends & enthusiasts hail it as a work of genius, & become devoted to its creator.[71]

Edward now was deep in the Angelical chorus 'Praise to the Holiest'. First came the semi-chorus (which had represented the 'Assistants' in Part I) over the treading bass-figure from the 'Proficiscere'. It culminated in a sharp crescendo-diminuendo for the full chorus on a single chord of 'Praise'. Jaeger wrote of this in his Analysis: 'It is as if one of the gates of heaven were opened, and we heard for one moment the full, harmonious hymning of the Angelicals.'

The Angelicals began their first full statement in A flat major—the four flats of 'Be merciful' in a melodic shape which seemed to combine that figure with 'Christ's Peace' from the beginning of the drama:

An elaboration

projected essentially

and so looked back to the falling 'Judgment' and 'Sanctus fortis' openings.

As the Angel announced the Soul's entry in the House of Judgment, the treading bass began again below 'Judgment'. The semi-chorus tenors sounded a crescendo-diminuendo of 'Praise' on E flat, the full chorus basses echoed the 'Praise' a 3rd below. This time the doors opened on Edward's own childhood. As the Soul sang

> The sound is like the rushing of the wind—
> The summer wind among the lofty pines . . .

[71] 29 May 1900 (Elgar Birthplace).

the scent of the long-ago summer at Spetchley came wafting back, with the oldest memory of firs at Broadheath.

As the Angelicals sounded their 'Praise' in E flat major, the music began a slow movement away from flats (to answer at last the movement into flats of the dying Gerontius in Part I). But any orderly progression toward sharps was suddenly overtaken by 'a grand mysterious harmony'; and the poem raised the other landscape of Edward's beginning—the river that divided him from Broadheath:

> It floods me, like the deep and solemn sound
> Of many waters.

Now, as in Wagner's *Götterdämmerüng*, the river overflowed its ancient banks. The 'Fear' music, running dark below, rolled resistlessly upward through A flat major—E flat major—B flat—through four huge rising chords to engulf the 'towering eminence' of the Demons: then 'Praise to the Holiest' thundered in a grand C major.

It opened a magnificent culmination extending to nearly forty pages of vocal score. One figure

answered finally the Demons' 3-note beginning of 'Low-born clods'. The vast movement of choral sound went finally toward a coda *molto accelerando* to finish on a gigantic chord of C major held by all forces through nine bars.

Malvern

Thursday [31st May 1900]

My dear J.

By this post comes the great Blaze: as soon as I return the proofs already here everything will be straight.

There's still some more M.S. to come but not much.

I can't tell you how much good your letter has done me: I *do* dearly like to be *understood*.

No time for more,

Yrs ever,

E. E.

. . . Please wire receipt of M.S. I'm angshuss![72]

With the music now complete to the end of the Angel of the Agony solo, there was only the moment of the Soul's Judgment to finish before the Angel's song ended the work. 'Praise to the Holiest' was the climax, and the rest would shape a long and gradual *diminuendo*—to balance the long crescendo through Part I up to its central 'Sanctus fortis', and the crescendo again from the

---

[72] HWRO 705:445:8429.

beginning of Part II to the Demons. So, as in the original ending of the *Variations*, the forward-moving masculine impulse would be softened by feminine rounding in a cycle.

The *dimenuendo* after the Angel of the Agony led to a hushed chorus of the friends left on earth: their 'Be merciful' prayer, revived from Part I, was now built on an almost literal quotation of the old chant in 'Bona mors'. The Angel described the Soul 'Consumed, yet quickened by the glance of God.' Then a chorus of Souls in Purgatory sang 'Lord, Thou hast been our refuge . . .' to an echo of the Assistants' 'Holy Mary, pray for him' in Part I. Finally the Soul whose sight of God had left only Purgatory for him ended his singing as quietly as the dying Gerontius had opened Part I:

> Take me away, and in the lowest deep
> There let me be
> . . . .
> There will I sing my sad perpetual strain,
> Until the morn.

Edward recalled entertaining the possibility of making a big moment of the Soul's sight of God: 'But I thought it too much for the fat mind of the filistine.'[73] Moreover, the artistic problems involved in directly representing such a supreme thing might lead to vulgarity. The present understatement avoided that by returning its music to the safety of cycle and symmetry.

Then came the Angel's Farewell. After the main melody for the Angel, the tenors and basses of the chorus entered with a reprise of 'Lord, Thou hast been our refuge'. The Angel recalled the music of first meeting the Soul of Gerontius near the beginning of Part II. Sopranos and altos re-echoed 'Praise to the Holiest'. And the full chorus of Souls in Purgatory sang:

> Come back, O Lord! how long: and be entreated for Thy servants.

It was the most masterly ensemble he had ever achieved—without violent dynamics or paradoxical accents, answering 'Praise to the Holiest' with perfect quietness. On 6 June it was finished. Writing to Nicholas Kilburn, Edward identified the entire achievement with the landscape of his home:

My work is good to me & I think you will find Gerontius far beyond anything I've yet done—I *like* it—I am not suggesting that I have risen to the heights of the poem for one moment—but on our hillside night after night looking across our 'illimitable' horizon (pleonasm!) I've seen in thought the Soul go up & have written my own heart's blood into the score.[74]

In London for a visit Edward saw *Götterdämmerung* at Covent Garden, lunched with Frederic Cowen (who was taking up the conductorship of the Royal Philharmonic Society), attended a Richter Concert on 11 June and

---

[73] 11 July 1900 to Jaeger (Elgar Birthplace).   [74] 27 June 1900.

supped afterward with Jaeger and Percy Pitt, a young composer who played the organ for the Queen's Hall Concerts and wrote programme notes.

That day had seen a fateful event in Birmingham. Swinnerton Heap, who was to have guided the Festival Chorus through the new score, died suddenly of pneumonia at the age of fifty-three. His enthusiasm for Edward's music would have been invaluable now when time was short. With the Festival less than four months away, ten major works to prepare, and the conductor-in-chief Hans Richter not available until shortly before the performances were to begin, the chorus-master was the crucial figure.

In the circumstances the Festival committee turned to old W. C. Stockley, who had been chorus-master of the Birmingham Festival for forty years before Swinnerton Heap took over in 1897. But Stockley, for all his encouragement of Edward's early orchestral music, was said to have anti-Catholic views. And he was by now an old stager. It remained to be seen whether at the age of seventy he still had the energy and insight to lead a big amateur chorus through the chromatic and contrapuntal complexities of the Demons' Chorus and the subtle balances needed everywhere else.

The printing of chorus parts for *Gerontius*—despite Edward's constant pleas for speed—was still in its early stages. On 15 June Jaeger told Dora Penny that the chorus parts would not be in the hands of the singers 'for another 3 weeks or more'.[75] In the event his estimate was optimistic.

That evening none the less, as Jaeger sat in his London home going over the final section of *Gerontius* in proof sheets fresh from the printer, almost everything seemed to be well. His one criticism was just the moment he had felt anxious over—when the Soul goes before God.

16, Margravine Gardens, West Kensington, W.

June 15. 1900   11 p.m

My dear Elgar

I have this Evening spent 1½ Hours over the Finale to Gerontius, & as before *must* write a line to let off steam, lest I explode during the night! Do I like your Stuff? There is one page (159) I can make nothing of, i.e. nothing adequate to Newman's words or the situation, though I have played & sung it over & over again & imagined all sorts of orchestral 'dodges' to make it good to this critical ass!

But all the rest is *most beautiful, exquisite, Ethereal* . . . To you I send my warmest congratulations on the completion of the 'Dream'. It almost seems a dream *to me*, for with all my appreciation of English music I did not expect such a work to come out of this tight little Island, & I *felt quite* sure it would not—much as I admired him—come out of E. E.! All the greater seems the marvel & all the intenser my *delight*!! much love to you & warmest greetings to yours,

Ever thine
A. J. J.[76]

---

[75] *Edward Elgar: memories of a variation*, p. 29.

[76] HWRO 705:445:8463. Jaeger's copy of these proofs are preserved in his copy of the vocal score.

Edward—once again after finishing a large work—was depressed. But there was the orchestral score to be done. Alice set out instrument names and bar-lines on the large sheets of scoring paper. Above the full-score title Edward wrote 'A[d] M[ajoram] D[ei] G[loriam]: Birchwood, In Summer, 1900', and then a quotation from Virgil

> 'Quae lucis miseris tam dira cupido?'[77]

with Montaigne's adaptation in the English translation of John Florio:

> 'Whence doth so dyre desire of Light on wretches grow?'

Jaeger wrote again on 16 June—a letter full of praise, small suggestions, and the one serious criticism renewed:

Page 159   I have tried & tried & tried, but it seems to me the *weakest* page in the work! *Do* re-write it! Surely, you want something more dramatic *here*!! It seems mere weak whining to me & not at all impressive . . . Have you tried p. 159 on any other *outspoken* critical friends? Remember, that not everybody loves, I mean *cares* to criticise a friend's work adversely . . .[78]

Malvern

June 20.

My dear Jaeger:

Now I've had time I've been all through your suggestions for which a heap of thanks: I'm truly glad you like the thing 'cos I've written it out my insidest inside. Some of your remarks are just and I've endeavoured to meet 'em: some I *can't* see the force of & will be pigheaded & sit tight! . . .

P. 159   You must read the poem: I cannot rewrite this: the Soul is shrivelled up & voiceless—& I only want on this page a musing murmur & I've got it—it 'wakes up' later—but I can't do this better if I try for fifty years: I've played it to several people & they like it: in Music if you—as you wd. do—point out a passage & say 'Don't you think this weak?' people always say *yes*. If you don't point it out they don't notice it: then after pointing out a passage to ten people, the pointer says—'nine out of ten people don't like it.' then you get an average against a thing which if let alone wd. pass &, if not admired per se, wd. be thought fit & good enough. I would bet you all I possess that if you don't draw attention to p. 159 nobody else wd. say anything against it. Try.

It was the same argument by which he had attempted to persuade Jaeger a year ago that there should be no definitive revelation for 'E.D.U.' Then he went on to the Soul's utterance, where Jaeger was also unconvinced by the transition at cue 126 to the Angel's Farewell.

Now 126   The last of your grumbles: I don't think you 'appreciate' the situation—the soul has for an *instant* seen its God—it is from that momentary glance shrivelled, parched & effete, powerless & finished & it is condemned to Purgatory for punishment or 'purging'—he *sighs* and if you prefer it 'whines' I've given him some of his 'aspiratory' tunes in the middle but the situation is as sad as can be—deepest dejection & sorrow—the present unworthiness for heaven is awful: therefore my final cadence,

---

[77] *Aeneid* VI 721.   [78] HWRO 705:445:8864.

which is good—then comes the Angel 126   peaceful & soothing & in this 'tone' the work ends: I can't see how you *can* ask for the Soul to have a 'dramatic' song here: he is in the most dejected condition & sighs 'Take me away' *anywhere*—out of sight.

No! I can't alter that.

Now I'm *awfully obliged* to you for all you've said . . .

> Much love
> Yrs ever
> Ed E.

Write soon.[79]

Proofs of this final section in the vocal score had been at Craeg Lea for nearly a week, but Novellos had neither returned Edward's vocal-score manuscript so that he could correct the proofs nor sent any corrections of their own. Until this had been done, there could be no chorus parts. On 24 June, when there was still nothing to hand, Edward wrote again:

Aren't we wasting time sadly? . . .

I hope to send about 60 pages full score tomorrow or Tuesday & the remainder as usual a bit at a time.

Don't forget about the *full sc*[ore] fair copy—that is to say, if it is to be made at all why not now?[80]

New works were not usually engraved and printed in the full score. The publishers' practice was to use the composer's full-score manuscript as the conducting score for hiring out. But all of Edward's major works so far had proved so popular that copy-scores had soon to be made to meet the demand. His own suggestion now was for an immediate copy of the orchestral score. A second full score of *Gerontius* would also make it easier to correct orchestral parts near the time of performance—when composer, publisher, copyists, and conductor would all be needing access to the score. But the Novello chairman Alfred Littleton declined the expense, as Jaeger reported on 27 June:

As to *Duplicate* Full score, I spoke to Mr. L. about it some days ago. He says, quite rightly I guess, that you must supply *one* Score to the publishers to work with—whether that be the original or the copy he careth not. No doubt if the work is in demand (—and I am sure all the *good* Societies will wish to do it—) we shall require a duplicate & you can then send us the copy you have had made. It is the usual 'being-frightened' at the scope of the really difficult work . . .

Now for your last long letter, which I'm sorry I have not been *able* to answer before. Don't think you have convinced me, & don't imagine I have been delayed because I wanted to ask other peoples' opinions. Nothing of the kind. I have opinions of my own (& prefer them (often) to other peoples').

To begin with: A musician has no business, in setting a poem, to become:

1 Dull

2 clumsy

3 ugly

[79] HWRO 705:445:8426.   [80] HWRO 705:445:8427.

4 tied by the leg
5 unimaginative
&c &c &c.
even if the poem seems to *'invite'* such delightful attributes.

Now I *had* read the poem well, & appreciated the Situation at the end (Soul after seeing God) *well*. *But*, *surely*, the first sensations the soul would experience would be an *awful*, *overwhelming agitation*!; a whirlwind of sensations of the acutest kind coursing through it; a bewilderment of fear, exitation[sic], crushing, overmastering hopelessness &c &c, 'Take me away!!' *Your* treatment shirks all that &, if you will allow one to become for a moment downright vulgar & blasphemous, *your* view, as expressed in the music, suggests to me nothing so much as an: 'O Lor! is that all? What a poor show, Take me away, it gives me the *miserables*.' Yes, a whine, I called it.

You may take it for Gospel that Wagner would have made this the *climax* of *expression* in the work, especially in the *orchestra* which *here* should surely shine as a medium for *portraying emotions*! Wagner *always revelled in* seemingly *'Impossible'* situations & this one would have brought forth his most splendid powers. He always surmounted the most appalling difficulties like the giant he was, 'rejoicing'. You must not forget, that if Wagner had not been the *poet* musician he was he would never (to mention but one 'Impossible' situation) have given Isolde that *glorious* Death Song (Liebestod) to sing, for no woman that ever was, or ever will be born, would sing or make speeches over the body of her lover at *such* a moment.

No, the poet amongst musicians does *more* than the 'necessary', the 'likely', the 'natural'. He does the *un*-likely, the *un*-necessary, *Im*possible, but—the *great*, the *beautiful*!

I *don't* want your 'Soul' to sing a 'dramatic "*Song*"'. Heavens! But what is your gorgeous orchestra for? & why should you be *dull* & sentimental at such a *supremest* moment? Don't tell me you can't do it in 50 years! Here is your *greatest* chance of proving yourself *poet*, seer, doer of *'impossible'* things—and you shirk it. Bah! I see *your* point of view *quite*. It's alright for the despised Brixton Baker & Bayswater Butcher,—*not* for an inspired poet-musician like E. E., though.

There! now come & flay me. Quite frankly, I *looked forward* to this situation, as treated by you and, as I expected, you have shirked it, & sorely disappointed this wretched fellow, 'who is an ass' &c &c &c. Other People will like it, no doubt!! The Brixton Baker, for instance.

You, no doubt, believe in dealing with such a supreme situation, *almost* introducing the Almighty himself in quite a subdued, quiet way. But Wagner isn't *dull* in Parsifal, where he deals with the Supernatural, the divine. Don't tell me yours is 'only a Cantata'. No! it's a *great drama* with the supremest moment shirked.

Did I not hold your splendid powers in such reverence & admiration, & were the rest of the work less superb, I should hold my peace & be *content* with a 'mere English Cantata'!

But you 'can't alter that passage', so I will merely sing to *myself* 'my sad perpetual Strain'

Much love & admiration!
Ever yours
A. J. Jaeger[81]

[81] HWRO 705:445:8466–7.

Edward replied by return of post:

I've kept p. 159 back for consideration—but all the time know I'm right & that you're wrong. However I'll see—one thing does annoy me—you say I've 'shirked it'—now I've shirked nothing—I've only set the thing as I feel & see it — wh: is not shirking at all, at all.

I can't stay to refer to the rest of your sermon which is very Nimrodisch as pictured &c

Two days later he sent Jaeger a doggerel poem:

Craeg Lea, Wells Road, Malvern

June 29: 1900

Poem.

Old Jaeger preached
   (as is his wont)
In Nimrodishest way;
And Elgar heard,
   And blushed & squirmed,
And—took another day.

So Jaeger hoped
   And thought it o'er
And almost prayed the while;
   —Alas! the proofs
   Came back untouched
(Malvèrn is barren sile).

On the verso he wrote out 'Nimrod—revised up to date', with the variation opening set against the treading figure of *Gerontius*—E flat against E:

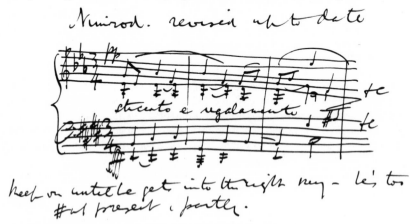

Keep on until he gets into the right key—he's too ♯ at present, partly.[83]

[82] 27 June 1900 (HWRO 705:445:8428).
[83] MS preserved in Jaeger's copy of the *Gerontius* vocal score.

1, Berners Street, W.

June 30 1900

My dear Poet (now Laureate)

Good! So that Poem of N[ewman]'s has beaten you after all in its superb climax! And you are *no* Wagner! . . . [Your] poem is worthy of any of the minor poets now before an undeserving public . . .

Do you know, I wanted you to suggest, in a *few* gloriously great & effulgent chords, given out by the whole force of the Orchestra in its most glorious key, the *momentary vision* of the Almighty. A few Chords! (remember those wonderful chords at Brunnhilde's Awakening in Siegfried, Act III) & then for a few bars the Soul's overwhelming agitation with a quasi-choked, suppressed 'Take me away' molto *agitato*, & *then* as miserable a whine as you like. The dejection to come *after* the first agitation.

No, it need not have been done 'theatrically' at all, at all! And to suggest the *glory* of the momentary vision need not have been blasphemous either. But I grant you, it wanted a Wagner or R. Strauss to do that, nobody else could dare attempt it. No!, as I know now, not even E. E. Only don't say I preach! I thought it would do you good to come across a really enthusiastic fellow *occasionally*, who believes in your powers more than you do yourselves & who only applied the most elevated standards to your work.

Verb Sap!

Yours ever
A. J. Jaeger.[84]

It was 'E.D.U.' all over again: Edward, uncertain of himself, underplaying his own powers in the interest of returning his music in a safe cycle—Jaeger insisting on the expansion of those powers to force the creative spirit forward to a new goal. And once again Edward yielded at last to Jaeger's masculine advice. When he made the decision, Edward carried out his friend's prescription almost to the letter.

After the Angel's description of the Soul 'consumed yet quickened', a slow sequence of 'Judgment' began sounding in the orchestra. It grew louder through bar after bar. There was an instant's pause: and then '"for one moment" must every instrument exert its fullest force' on a single chord the like of which had never detonated through his music. The Soul shouted 'Take me away!', and a development *con gran espressione* returned the music gradually to the soft singing of the original. The 'Souls in Purgatory' sang 'Lord, Thou has been our refuge' as a perfect transition to the Angel's Farewell.

It was a further Enigma which Jaeger had helped him to solve—the need to face an ultimate test of his musical expression and everything it stood for. But this time Edward could not bring himself to rejoice in the achievement. It was almost as if he had rather have been left alone.

Craeg Lea, Wells Road, Malvern

Sunday [1 July 1900]

My dear Jaeger:

Very well: here's what I thought of at *first*—I've copied it out & send it—of course it's biggety-big . . .

84 HWRO 705:445:8468.

Now: *important.* immediately you get this tell me if the printers can understand it. if they cannot, send me a wire at once here & I'll come up at once—*Don't delay* 'cos we're going to Birchwood on Tuesday & it's very difficult to get away to trains. It'll relieve my mind—(what *you have* left of it)—to get a wire anyhow.

I hope you'll like the emendation.

But wire-wire-wire or I shall be gone into the woods.

Yrs ever
Ed E.

The rehearsal figs: will want correction later. Hurry this up, all of it. I'm quite tired of seeing it.[85]

Persuading Edward to alter the *Gerontius* climax had taken further precious days. Now there were only three months to the première. Chorus rehearsals would be delayed still further while the new music near the end was set up, printed in proof, sent for correction, returned, and the individual chorus-parts drawn out for separate printing. And the month of summer holiday for the Birmingham Chorus was drawing near. On 5 July Edward wrote from Birchwood:

I am very anxious to get all the Chorus copies out now, & a voc[al] sc[ore] as *soon* as possible 'cos I must consult the Birmingham people *early* about *semi-ch.* &c.

Can you tell me if *Dodd* is solely employed by your firm, or may I ask him to copy the score for me—if he has time &c. &c.—let me know this as time is short enough.[86]

William Dodd was the professional copyist employed by Novellos. It had been arranged for Edward to send finished portions of his full score direct to Dodd: after Dodd had drawn out the separate orchestral parts from each portion, he would then send that portion of score to the publishers. Dodd was now so busy making parts for *Gerontius* as well as new works by Coleridge-Taylor and Horatio Parker for Hereford and Parry's *De Profundis* for Birmingham that he could not undertake a second *Gerontius* score for Edward.

Yet already the want of other copies was being felt. On 11 July Edward wrote to Jaeger:

Do let me have a complete copy [of the vocal score]—*two*—as soon as you can—You see, *your!* confounded alterations have left me with nothing to work with—for instance the new bar on p 64 (old paging)   I've no copy at all & haven't an idea what I put in, do send me this quick—I'm supplying Dodd fast & am well into Part II. It's a giddy skourrghe, I can tell you.

Now dear man, do keep the things going & let's have those chorus parts ready. I am to go over to Birmingham this week to arrange . . .[87]

When the day came, there were neither chorus parts nor proof copies. Edward

[85] HWRO 705:445:8430.   [86] HWRO 705:445:8431.   [87] Elgar Birthplace.

could only settle with G. H. Johnstone that Jaeger should do a full-scale analysis of *Gerontius* for the première programme.

Returning to Birchwood that evening, he found Dora Penny arrived for a visit, and showed her parts of *Gerontius* on the piano.

He went straight through ['Praise to the Holiest'] and after the nine bars tied chord passage at the end, he stopped, sat back in his chair and got a pipe out of a pocket.

I had been so absorbed, so amazed and overwhelmed by the music that I could say nothing, and there was silence. At last I murmured: 'How perfectly wonderful!'

More silence. Then he said: 'How does that strike you?'

What a question! What *could* I say? An idea had come into my mind while he was playing; should I tell him? I summoned up my courage. 'It gives me the impression of great doors opening and shutting.'

He turned round in his chair and looked at me. 'Does it? That's exactly what I mean.'[88]

Throughout those days he was immersed in finishing the *Gerontius* score. He put the last note to it on 3 August 1900. The feeling of achievement was commemorated in a quotation from Ruskin which Edward set above his own signature on the final page of *Gerontius* manuscript:

This is the best of me; for the rest, I ate, and drank, and slept, loved, and hated, like another; my life was as the vapour, and is not; but *this* I saw and knew: this, if anything of mine, is worth your memory.[89]

That day Alice wrote to Mrs Kilburn:

It seems to me that E. has given a real message of consolation to the world & that dread & terrors are soothed by the infinite sweetness & mercy which penetrate one's soul with intense emotion. I think human hearts will owe him a deep debt of gratitude. I cannot help writing this because I feel it so intensely & as if a new path opened out . . .

E. loves orchestrating here in the deep quiet hearing the 'sound of summer winds amidst the lofty pines'. I wish you & Mr. Kilburn cd. see the *lovely* scenery here. It is a nice little cottage on the edge of woods.

We have both been learning to bicycle, E. can now go beautifully & I am just beginning . . .[90]

Soon the long rides were expanding the range of his intimacy with home countryside. The distance available was so much greater than for walking, yet the pulse of pedalling was not unlike the walking pulse. He recognized it later when he answered an interviewer's question about his 'inventing': 'That comes anywhere and everywhere. It may be when I am walking, golfing, or cycling . . .'[91] Thus the bicycle itself might confirm the opening of creative horizons that came through *Gerontius*. When he bought a bicycle of his own, sheer anticipation named it 'Mr. Phoebus'.

Then the *Musical Times* editor F. G. Edwards arrived in Malvern to write a

---

[88] *Edward Elgar: memories of a variation*, pp. 18–19. See above, p. 288, footnote 15.

[89] *Sesame and Lilies*, p. 20. Elgar's copy was the 4th edition (Smith, Elder, & Co., 1867) (now in possession of the writer).

[90] HWRO 705:445:BA8089.   [91] *The Strand Magazine*, May 1904, p. 544.

big article on Edward's career, for publication at the time of the *Gerontius* première. He was immensely struck with the creative surroundings, and at Edward's suggestion headed his piece with a quotation from the Middle English poem *Piers Plowman*:

> In a somer seson · whan soft was the sonne,
> I shope me in shroudes · as I a shepe[rd] were,
> In habite as an heremite · vnholy of workes,
> Went wyde in þis world · wondres to here.
> Ac on a May mornynge · on Maluerne hulles,
> Me byfel a ferly · of fairy, me thouȝte.

It gave the keynote to the editor's opening impression of his Elgarian visit:

The Malvern uplands are to be seen, not described. No appreciative mind can fail to be impressed with the bold outline, the imposing abruptness, and the verdant loveliness of these everlasting hills. Nature has left the impress of her smile on this favoured region . . . The enjoyment of a quiet stroll along these grassy heights is greatly enhanced by the companionship of one who habitually thinks his thoughts and draws his inspirations from these elevated surroundings.[92]

Full of his impressions, the editor went away to write his piece.

Orchestral parts for *Gerontius* had begun to arrive, but Edward could not correct them as he did not have the full score. He wrote to Jaeger on 15 August: 'I *must* have the score to read the parts by or with & Richter wants it—he said—first week in September. I am to go through it with him then.'[93] At last the score of Part I came back, and before the end of August he was able to send some corrected parts. But the full score was not available for Richter's use.

The printed chorus parts had now arrived in Birmingham, where they were given out to the Festival Chorus on their return from their summer holidays. The press attended the first rehearsal under Stockley on 20 August—just six weeks before the première:

The chorus were in their places betimes, and Mr. Stockley, before starting the rehearsal, gave an account of the early days of Mr. Elgar as a composer. One of his very first works to be heard in public was produced at Mr. Stockley's orchestral concerts a good many years ago, the composer occupying a place among the first violins. Of course these remarks were the prelude to the work of the evening, when the chorus essayed the principal Festival novelty, Mr. Elgar's setting of Cardinal Newman's highly-imaginative mystic poem, 'The Dream of Gerontius'.

It would be premature at this stage to attempt any description of the music, or to deal with the method of treatment. The Choral parts are so interwoven with the work of the soloists that even with the graphic aid of Mr. Stockley—who takes up the different voice parts with the readiness of old times—it is difficult to gain any definite idea of the work . . . The chorus had to work in faith that all would in time become clear . . .[94]

[92] *The Musical Times*, 1 Oct. 1900, p. 641.
[93] HWRO 705:445:8438.
[94] *The Birmingham Post*, 21 Aug. 1900.

After another rehearsal a week later, some clarity seemed to emerge:

The Festival Chorus is making great progress with the new work 'The Dream of Gerontius', which is winning golden opinions from all who hear it . . . Even without the orchestra, the greatness of the music is discernible. The foot of Hercules is there.[95]

Inside the Chorus itself, with the Festival now only five weeks away, there was uncertainty. Dora Penny's aunt, in the Wolverhampton contingent, said: 'How lovely this music is! but shall we *ever* learn it in time?'[96] Another of the Wolverhampton ladies, Mrs Evans, told Miss Burley that old Stockley seemed hardly competent:

Stockley, Mrs Evans describes as a pathetic figure unable to bear the strain even of standing for the long periods required. Again and again he had to rest and eat sweets to keep himself going. The rehearsals were so ruthlessly shortened that sometimes the Wolverhampton party felt resentful at having taken the twelve miles' journey for so little result.[97]

A third Chorus member thought he saw another reason for the short rehearsals: 'I am sorry to say it, but the dear old Chorus master who took Dr. Heap's place was not able to grasp this new and different work, he was too old-fashioned . . . There is no doubt he was not so interested as circumstances demanded . . .'[98]

Stockley invited Edward himself to take a rehearsal. It fell on 12 September, during the week of the Hereford Festival, and Edward travelled with Alice to Birmingham. At his request, Henry Coward had come down from Sheffield to give an opinion. The press also attended:

Mr. Edward Elgar was the recipient of a most enthusiastic welcome, and, after the applause had subsided, he thanked the choristers for the cordial reception, and said:

'I am not coming to a strange place, as I shall always cherish my associations with Birmingham, where I have learned all the music I know when a member of Mr. Stockley's orchestra'.

The composer took much pains, and many important suggestions were made during the rehearsal . . .[99]

There were, of course, many repetitions of important passages, notably at the beginnning of the 'Kyrie', which opens pianissimo for tenors. He entreated them not to sing as if they were in a church. He wanted 'more tears in the voices' in the Kyrie, 'as if they were assisting at the death of a friend.' They must sing as if they felt each strain, not conventionally.

. . . At the close of the rehearsal he thanked the choristers and said that he was extremely pleased with what they had done.[100]

[95] *The Birmingham Gazette*, 28 Aug. 1900.
[96] 'The First Performance of "Gerontius"', in *The Musical Times*, Feb. 1959, p. 78.
[97] *Edward Elgar: the record of a friendship*, p. 139.
[98] W. T. Edgley, letter to Dora Powell, *c.* 1950.
[99] *The Birmingham Mail*, 13 Sept. 1900.
[100] *The Birmingham Post*, 13 Sept. 1900.

But Henry Coward was nowhere to be found:

[I] slipped away rather than meet him and give a depressing verdict. The singing reminded me of an automaton—shape and movement, but lifeless.[101]

Only three more Chorus rehearsals of *Gerontius* were scheduled before the première.

Yet Alice had been immensely moved, as she wrote to Jaeger:

We had a thrilling experience hearing the Choral rehearsal for the first time at Birmingham; like me, I am sure you wd. have been transported into a heavenly region of peace—& touched to choking too—& anything like the *sustaining* power of 'Proficiscere', I never heard. *Please* be there [at the première] to hear.[102]

Jaeger replied:

I sent the other day a Score of E's Variations to Professor J. Buths of Düsseldorf, one of the *Ultra moderns* of Germany, great propagandist for Richard Strauss, & I enclose his remarks, which *I* won't attempt to put into English. The Firm at my suggestion, have invited him to B'ham to hear some English music. He is conductor of the Lower *Rhenish Musical Festivals*! Do you guess something? But mum's the word!

. . . I wish I had been there to hear the B'ham Rehearsal; but I *shall* be (D.V.) at B'ham with Buths.[103]

The galley proofs of the *Musical Times* article on Edward's career arrived at Craeg Lea in mid-September. Following the *Piers Plowman* quotation and the eulogy of surrounding countryside, there was a sketch of early experiences, a list of compositions, and then the article concluded with anecdotes and choice extracts from several punning letters to the editor F. G. Edwards.

Alice went carefully over all the galleys and sent a long letter to Edwards explaining her emendations:

Every word concerning him seems momentous to me—& I *trust* you will agree with my suggestions which are to me *very* important. You will see that they in no way touch the body of your work.[104]

First, the editor had described Craeg Lea as a villa. Alice wrote:

Please not *villa*, if it is 'not villainous'. We cannot reconcile ourselves to the name! & as you allude to E.'s library that is a thing not found in a villa, in its mordern [*sic*] English adaptation.

Then as E. has *nothing* to do with the business in Worcester [at 10 High Street] would you please leave out details which do not affect him & with which he has nothing to do—His interests being quite unconnected with business.

Edward sent a note of his own:

Now—as to the whole 'shop' episode—I don't care a d—n! I know it has ruined me & made life impossible until I what you call made a name—I only know I was kept out of

---

[101] *Reminiscences*, p. 152.   [102] 17 Sept. 1900 (HWRO 705:445:8822).
[103] 18 Sept. 1900 (HWRO 705:445:8806).   [104] 18 Sept. 1900 (BL Egerton 3090 fos. 35–8).

everything decent, 'cos 'his father keeps a shop'—. . . but to please my wife do what she wishes.[105]

Alice was also worried by one line describing the end of their first year in London: 'Ill health, doubtless aggravated by disappointment, compelled him to leave . . .'.[106]

I cherish the London episode, but strongly think it wd. be *far* better to leave out details, this I feel strongly & will explain many reasons when we meet.

Then my father was a K.C.B. . . . & please! Robert Raikes was my *great* grandfather, please not relegate my birth into nearly the last century!

. . . Please be very kind & forgive my suggestions & *please* think as I do about them.

With many thanks & kindest regards

Sincerely yours
C. Alice Elgar

And the editor did.

The printing and copying of *Gerontius* orchestral parts was being delayed all along the line by the fact that there was only the one full score. The copyist needed that score to draw out manuscript parts for single instruments in the orchestra. The printers needed it to draw out string parts for engraving. The publishers needed it to collate their work. And the composer needed it to check and correct all the results. Through the middle days of September orchestral parts had been arriving in parcels from Novellos, with Jaeger's urgings for speed over their return. But they were still incomplete. Edward wrote:

I am doing all I can with the parts—but where are the rest & where are the M.S. things. I'm anxious to correct 'em.[107]

On 14 September Jaeger wrote again:

I send you a Batch of *Second* proofs unread, as West [Novellos's music editor in succession to Berthold Tours] has not time to go through them & time is getting precious. More D.V. tomorrow. Send 'em back *quickly* please, that we may proceed to print 'em. I have told Dodd to send all Wind parts to you direct, so that the Full Score may be released for Richter.[108]

Two days later Edward replied:

I'm tired out.

I return *all* I can correct with score.

I *must* have wind on Tuesday early somehow or I cannot correct the stuff—my man (helper) [John Austin] comes on that day.—Wire Dodd to send Horns & transposing things at least.[109]

[105] 19 Sept. 1900 (BL Egerton 3090 fo. 39).
[106] Galley proof with the Elgars' emendations (BL Egerton 3097A fos. 13–22).
[107] Undated letter of probably 14 Sept. 1900 (HWRO 705:445:8441)
[108] HWRO 705:445:8479.   [109] 16 Sept. 1900 (HWRO 705:445:8389).

Tuesday morning brought John Austin from Worcester to help Edward—but no wind parts. After wild telegraphing back and forth, some brass parts arrived in the early afternoon. But in London, with less than a week to go before the orchestral rehearsal, Jaeger had now taken real alarm:

1, Berners Street, London, W.

18/9 1900

My dear E

I'm nearly off my head with your & Parry's & Taylor's things. *All behind*! For God's sake return all *proofs* at once or we shall be landed in a fine quandary. Poor Brause [the engraver] is crazy!! & so is (nearly)

Yours ever
AJJgr[110]

Two days later Edward brought the outstanding proofs—still unfinished—with the score to London when he and Alice came for the Birmingham Festival orchestral rehearsals at the Queen's Hall. They stayed just across Portland Place at the Langham Hotel. The Langham was expensive, but its convenience made it the Elgars' headquarters in London through the next years.

Richter arrived from Birmingham, where he had met the Chorus at last during the week just ended, and on Sunday evening 23 September Edward took the full score to him. It was the conductor's first sight of the complete work he was to direct in ten days' time. The first and only rehearsal for the orchestra alone would take place the following day.

Monday morning found the press gathered at Queen's Hall in force. Arthur Johnstone wrote in *The Manchester Guardian*:

So far as Dr. Richter was concerned the rehearsal was a veritable *tour de force*, for the gifted conductor had had no opportunity of studying the full score until the previous evening . . .[111]

But E. A. Baughan of *The Morning Leader* saw the fault in the whole system:

A work such as Elgar's 'Dream of Gerontius' requires more orchestral rehearsal than it has been given, and more than one full rehearsal of soloists, chorus, and orchestra, which is all it will get at Birmingham. The orchestral rehearsal on Monday at the Queen's Hall which I attended was one of the most primitive order. Dr. Richter had a fine grasp of the score, and was indefatigable in his endeavours to obtain light and shade from his band, but four hours in the morning, with a couple in the afternoon for the soloists with orchestra, are not sufficient. Practically the orchestra was only reading its music, and the soloists were in not much better plight.

From this primary stage the soloists and orchestra will join forces with the chorus in the final rehearsal at Birmingham. The new work is scheduled with Parry's 'De Profundis' for rehearsal in the afternoon. That means, I suppose, that it will be given a couple of hours', or at most three hours', rehearsal. The chorus, of course, has been rehearsing its music for some time, and Mr. Elgar himself has directed its final [sic] rehearsal, but these rehearsals are not the same thing as full rehearsals.

[110] HWRO 705:445:8480.   [111] 25 Sept. 1900.

Of course, composers always publicly express themselves as satisfied with the performances of their new works, but if I were Mr. Elgar I should want my composition played and sung with more delicate appreciation of its light and shade than the band and soloist rehearsals promise.[112]

The only combined rehearsal of soloists, chorus, and orchestra was at Birmingham on the afternoon of Saturday 29 September: it was virtually a public preview with the hall full of interested guests and spectators. Edward, Alice, and Carice took their places mid-way down the room. Part I went without disaster, though the choral singing was rough. It became known that the Chorus had had only seven rehearsals with this music; and it was recalled that Richter, for all his orchestral prowess, was no trainer of choruses. In later years Henry Coward remembered: 'The worst performances of *Messiah*, *Faust*, and *Gerontius* were under his baton, and though he was excused on the grounds of his lack of sympathy with or knowledge of the idiom of the works, this did not make for the musical success of a Festival.'[113]

In Part II things began to go wrong. First Richter stopped the orchestra, to repeat several times the passage at the Soul's first meeting with the Angel. But it was the Demons' Chorus that showed the dimensions of a disaster. Edward took matters into his own hands. *The Sheffield Independent* reported:

Having tolerated much, he suddenly left his seat in the hall, hurried to the orchestra, and stopped the chorus, declaring the rhythm was wrong, the time wrong, and all unsatisfactory. 'There is', he exclaimed, 'no accent. It is nothing. The whole colour is gone. It is like a ballad in a drawing room!'

For a moment it seemed that the privileged audience were standing on the brink of a volcano. There was consternation among the choralists and orchestra. But subsequently the choralists, who for the moment were representing demons, shook off their drawing room manners, and infused a little more diablerie into their work, for which the composer thanked them.[114]

Yet Henry Coward was shocked:

I was present at the historic final rehearsal, when the composer, in the presence of thousands, gave that flaming outburst against the performers because of their lack of insight . . .[115]

The Birmingham Post was strongly critical:

We venture to think that this was a mistake, and showed want of tact, more especially as the baton was in Dr. Richter's hands.[116]

That elicited an official reply from G. H. Johnstone:

Dr. Richter desires me to say that it was at his request that Mr. Elgar addressed the choir, so as to give them his view of how the work should be rendered.[117]

[112] 29 Sept. 1900.   [113] *Reminiscences*, pp. 237–8.
[114] 8 Oct. 1900.   [115] *Reminiscences*, p. 153.
[116] 1 Oct. 1900.   [117] 2 Oct. 1900.

To which the *Post* critic rejoined:

. . . There was no evidence or intimation of the fact at the time.[118]

What was quite clear was that some of the Chorus took it as an affront to their powers, as Baughan reported afterwards in *The Morning Leader*:

I have received a letter from a correspondent who was a member of the Birmingham chorus . . . The composer . . . 'grossly insulted' the singers by telling them it was 'all wrong', and by saying that the chorus of Demons was sung like a drawing-room ballad; 'And that', adds my correspondent, 'after we had fairly shouted ourselves hoarse.'[119]

The effect of Edward's presence beside the rostrum that day was to remain in the memory of another Chorus member for a half a century:

He was alongside Richter most of the rehearsal, prompting him and trying to explain what he required. I must record, however, that things got very chaotic and everyone worked up to a high pitch and unfortunately E. E. more than anyone, naturally. He seemed desperate, with whom I cannot remember, but it was not all 'the chorus' . . .[120]

In the end, Richter, seeing an impossible situation, cut short the rehearsal. *The Birmingham Gazette* gave a carefully laundered account:

Two hours and a half, instead of the regulation five hours of the final three days—then Herr Richter laid down the enchanter's wand, and everybody made for home. There was a glimpse of Mr Elgar in a long brown waterproof; a sight of Sir John Stainer surrounded by a crowd of admirers; a vision of charming ladies from Worcestershire, come up to hear the first of Gerontius—strictly with a hard G, if you please![121]

Over the weekend Richter closeted himself with the score, trying to perfect his understanding now he had possession of it at last. 'I can see him now,' wrote Henry Wood, 'pacing up and down his bedroom with the score on the mantelpiece.'[122] On the Monday—traditionally a free day before the opening of the Festival on Tuesday—Richter called an extraordinary special chorus rehearsal 'to make all safe'. 'As a matter of fact,' said *The Spectator*, 'they had six hours' rehearsal on the Monday, and were thoroughly stale when they began their labours.'[123] The Festival opened on Tuesday morning, and that evening Edward conducted four of the *Sea Pictures* for Clara Butt. 'Glorious, great reception,' wrote Alice in the diary. But even after the concert, 'E. to see Richter at 9.30.'

Next morning at 11.30 came *Gerontius*. Just before they went on, one of the Chorus sopranos recalled: 'Dr Richter came and stood on the steps leading to our dressing-rooms and with unforgettable voice and gesture besought us to do our very best "for the work of this English genius".'[124] It was no good. Miss Burley's chief impression was of Richter taking the music so slowly—trying no doubt for safety—that it 'seemed to continue for an eternity':

[118] 3 Oct. 1900.   [119] 11 Oct. 1900.   [120] W. T. Edgley, letter to Dora Powell, *c.*1950.
[121] 1 Oct. 1900.   [122] *My Life of Music* (Gollancz, 1946), p. 249.
[123] 7 Oct. 1900.   [124] Quoted in McVeagh, *Edward Elgar*, p. 31.

Before the end of the *Kyrie* it was evident that the chorus did not know the parts they were trying to sing, and as the music became more chromatic they slipped hideously out of tune. [Part II was] hopelessly wrecked by the choir, whose pitiful stumblings indeed remained the outstanding impression. There were times when they seemed to be a whole semitone out and when the orchestra, disregarding the directions on the score, would play fortissimo in order to drag them back to the true pitch. The whole thing was a nightmare.[125]

The *Sunday Times* critic Herman Klein was amazed:

A more perfunctory rendering of a new work it has never been my lot to listen to at a big festival. The tenors began flat in the very first semi-chorus, and set an example of doubtful intonation that prevailed through most of the many passages in the cantata where awkward intervals and trying dissonances lay a trap for these unwary choristers. Nay more; the attack was rarely unanimous, and their rendering of passages requiring the utmost delicacy often offended the ear by a grating harshness of tone and slovenliness of phrasing . . . The probabilities are that an additional month of preparation would not have strengthened the weak spots in this particular Birmingham choir.[126]

Baughan in *The Musical Standard* thought it was deliberate sabotage:

Doubtless Mr. Elgar's protest [at the combined rehearsal] took the spirit out of the chorus, and it seems to have so seriously injured the pride of many that it is quite a question if they did their loyal best with the composition on Wednesday morning.[127]

Plunket Greene recalled his own quandary when he came to the Priest's solo 'Go Forth':

The choir, audibly dragged down by a single tenor in the first chorus, had flattened by degrees until in the last number of the first part there was half a tone between them and the orchestra. I can guarantee the truth of this as I, being the bass soloist and having to sing a part above the chorus, had to choose between them and 'was not happy with either.'[128]

Young Ralph Vaughan Williams, who had come to Birmingham expecting a success to equal the *Variations*, was 'bitterly disappointed'. If Plunket Greene had 'lost his voice', Vaughan Williams felt that Marie Brema as the Angel 'had none to lose'. And Edward Lloyd sang the title part 'like a Stainer anthem, in the correct tenor attitude with one foot slightly withdrawn'.[129]

After such a rendition, the vigorous and prolonged applause of the audience came as a shock. Applause was strictly against the Festival rules governing morning performances. *The Worcester Herald* reported:

Had it not been for a very determined effort on the part of the audience to see Mr. Elgar at the conclusion of the cantata he would not have appeared, although the applause was most enthusiastic. It certainly showed that the appreciation was no mere formal

[125] *Edward Elgar: the record of a friendship*, p. 142.
[126] 7 Oct. 1900.   [127] 13 Oct. 1900.   [128] *Charles Villiers Stanford*, p. 74.
[129] Quoted in Michael Kennedy, *The Works of Ralph Vaughan Williams* (OUP, 1964) p. 389.

recognition of a great work when an immense concourse of people, who have only the short space of thirty minutes in which to procure refreshment, deliberately curtail this period in order to manifest the delight which they have felt . . .[130]

Edward wrote afterwards to F. G. Edwards at *The Musical Times*:

People said I was long in coming forward to acknowledge the call: previous to the performance they told me, (nay—'twas printed in the books also) that 'no applause was allowed at the morning performances'—so I should not be 'wanted'—when the folk *did* applaud I was hopelessly bescronged & could not move—when Richter signalled to me I made a rush & was 'held up' on the stairs by a man who had his back to me applauding furiously: I modestly said 'Will you kindly allow me to pass?' He said soothingly 'Now *just wait* a minute or two—we want Elgar up, & mean to have him' I was really too shy to say 'I'm 'im' But at last I said 'I think it would really expedite the matter if you'd let me go on to the orchestra' Which he then did. I fled as fast as I cd. but not sufficiently fast to save myself from the reproach that I was intentionally rude to the audience.[131]

But the bowing composer was intentionally rude in another direction, as was noticed from the stalls: 'He quite properly put on record his opinion of the choral shortcomings by resolutely refusing to acknowledge the applause of the choir by the conventional bow.'[132]

Despite the disasters of the performance, the critics almost unanimously seconded the Birmingham audience's recognition of a great work. The critic of *The Star* wrote:

'The Dream of Gerontius' appeals to me as by far the strongest and most beautiful work Mr. Elgar has yet given us. In two respects particularly it shows a wonderful advance even on 'King Olaf' and 'Caractacus'. First of all, there is no longer that tendency to over-elaboration of detail which was sometimes allowed to obscure the broad outlines of Mr. Elgar's music. It is a temptation to which such a master of fine workmanship easily succumbs, and therefore the overcoming of it is the more admirable. Secondly, the writing for the solo voice is immeasureably more grateful and expressive without any attempt at mere effect making.[133]

Several critics recognized the special influence of the subject. Baughan wrote in *The Musical Standard*:

It is music which could not have been written by any man who had not felt the beauty of Newman's Angel and had not longed for a tender charity within which the bruised human soul could at last fold its wings and sleep.[134]

The suggestion of morbidity disturbed Joseph Bennett of *The Daily Telegraph*:

Those who know the composer of 'Gerontius' are aware of a peculiar artistic temperament in which, of course, extreme susceptibility, uplifting enthusiasms, and a tendency towards more or less mysticism have a place. Such a man when armed with

---

[130] Undated cutting at the Elgar Birthplace, possibly written by Hubert Leicester.
[131] 19 Oct. 1900 (BL Egerton 3090 fos. 45–6).
[132] E. A. Baughan in *The Daily News*, 4 Oct. 1900.
[133] 4 Oct. 1900.   [134] 6 Oct. 1900.

technical skill, and able to put a healthy restraint upon himself, may go far and do much.[135]

Bennett was an old stager, and he had his doubts about the motivic plan and the advanced harmonies of the new work. But he was resolutely just:

It must now suffice, while expressing from my point of view serious disagreement with its method, to declare that 'The Dream of Gerontius' advances its composer's claim to rank amongst the musicians of whom the country should be proudest.[136]

The Times critic Fuller-Maitland, a notorious reactionary and apologist for Stanford, contrasted the dramatic writing of 'Sanctus fortis' unfavourably with solo writing by Parry; yet he admitted that Birmingham's choice was just.

Most of the other critics went much farther. *The London Musical Courier*:

For the first time in history, it seems to me, a provincial festival commission has been successful in bringing to birth a choral work by an Englishman worthy in any way of arrogating to itself the title 'a great art work' . . . I, for one, felt all my preconceived antagonism to the poem melt away before Elgar's setting, and left the hall throbbing with emotions that no English work has raised in me heretofore.[137]

Baughan in *The Morning Leader*:

I can honestly and frankly say that no composition by an Englishman equals it in sheer technique, to say nothing of real poetic feeling. It is music of the heart, and appeals to the heart, and yet the head that worked it is no scamper of workmanship. In the orchestral score you will find an ingenuity beside which the attempts of other Englishmen seem but clumsy imitations of this or that dead master—Brahms or Wagner or even Dvorak. Elgar has studied these giants—but his music as a whole has a tone of its own—and most of its ingenuities are of his own conception.[138]

The Morning Advertiser on the same day:

Those who have a belief, a sanguine faith in the development of English musical art will take a solid pleasure in Mr. Edward Elgar's new work, 'The Dream of Gerontius', and the more so because, coming at the end of a time-expired century, they can point to it as the cornerstone of musical thought and progress of the Victorian era.

Herman Klein in *The Sunday Times* agreed:

If this cantata does not belong to the type of works that live and flourish in the full light of day, then I am greatly mistaken concerning the present trend of musical feeling and opinion in this country.[139]

And the judgment was confirmed abroad as well by Jaeger's German friend Otto Lessmann, visiting Birmingham with Julius Buths. Lessmann's article in *Allgemeine Musik Zeitung* was translated for *The Musical Times*:

If I mistake not, the coming man has already arisen in the English musical world, an artist who has instinctively freed himself from the scholasticism which, till now, has held

[135] 4 Oct. 1900. [136] 4 Oct. 1900.
[137] 12 Oct. 1900. [138] 4 Oct. 1900.   [139] 7 Oct. 1900.

English art firmly bound in its fetters, an artist who has thrown open mind and heart to the great achievements which the mighty tone-masters of the century now departed have left us as a heritage for the one to come—Edward Elgar, the composer of 'The Dream of Gerontius'.[140]

Vernon Blackburn, the Catholic critic of *The Pall Mall Gazette*, went farther still:

I am about to speak words which may seem exuberant and enthusiastic; but I have thought over them carefully before setting them down for the public eye, and I will venture to say that, since the death of Wagner, no finer composition (I am quite remembering Tschaikowsky and his great Symphonies) has been given to the world.[141]

At the end of the day, the position was fairly summed up by Arthur Johnstone in *The Manchester Guardian*:

We shall doubtless hear of plagiarism from 'Parsifal', and there is indeed much in the work that could not have been there but for 'Parsifal'. But it is not allowable for a modern composer of religious music to be ignorant of 'Parsifal'. One might as well write for orchestra in ignorance of the Berlioz orchestration as write any serious music in ignorance of Wagner's symbolism . . . The symbols, though employed in the Wagnerian manner, are nevertheless thoroughly original, taking us into an atmosphere and a world absolutely remote from all that is Wagnerian . . .

I am more than usually troubled by the sense of utter inadequacy in these notes, and can only hope that I may have some opportunity of doing better justice to a deeply impressive work.[142]

The critics unanimously castigated the disastrous performance, and there could be no doubt that it had its effect. August Manns, who was considering a full London production at the Crystal Palace, was so appalled by what he took to be the work's insurmountable choral difficulties that he decided to play only the Prelude in a miscellaneous orchestral programme. Henry Coward was approached at Birmingham by W. G. McNaught (the well-known editor, choral conductor, and adjudicator who knew Edward from Lady Mary Lygon's Madresfield Competition Festivals): 'McNaught said to me, "Coward, you must give this work at Sheffield, or it will be lost." I reassured him by saying, "I have already decided to perform it at the first opportunity." '[143] But that opportunity would not come until the autumn of 1902.

For one man only the failure seemed utterly complete. Edward himself saw in the accumulated mischances of the première a consummation of all his deepest doubts. Six days after the performance he wrote to Jaeger:

I have worked hard for forty years & at the last, Providence denies me a decent hearing of my work: so I submit—I always said God was against art & I still believe it. anything obscene or trivial is blessed in this world & has a reward . . . I have allowed my heart to open once—it is now shut against every religious feeling & every soft, gentle impulse *for ever*.[144]

[140] 1 Jan. 1901; quoted also in *The Daily Graphic*, 5 Jan. 1901.   [141] 4 Oct. 1900.
[142] 4 Oct. 1900.   [143] *Reminiscences*, p. 153.   [144] 9 Oct. 1900 (Elgar Birthplace parcel 343).

It showed how he had secretly used the inspiration of *Gerontius* as a wager against his own insecure faith. A fortnight later this thinking showed its logical end:

I really wish I were dead over & over again but I dare not, for the sake of my relatives, do the job myself.[145]

*   *   *

There was further trouble over *Gerontius* with the publishers. Novellos sent their payment together with an assignment for Edward's signature conveying the copyright to them absolutely. Those were the terms of the chairman's note from his conversation with G. H. Johnstone in January 1900. But Johnstone had told Edward that the payment was an advance on royalties. Edward now disputed the assignment, and Johnstone asked Alfred Littleton to amend the document to meet his recollection. Littleton refused:

Immediately after our interview I had a memm. of terms written down & this memm. contains no reference to the payment of a Royalty. You may have raised the question but I am certain that I could not have agreed to it. I should never have agreed to pay so much as £200 for the work unless it had been introduced by you [at the Birmingham Festival] in the way in which it was: & I have been much blamed by my co directors for agreeing to pay so large a sum.

Unfortunately for us the work is a commercial failure & we have no anticipation of ever realizing our initial expenses.[146]

38, Northampton Street, Birmingham.

Nov 7/00

Dear Mr Littleton

I am sorry you have forgotten the arrangement made with me as to 'Gerontius'. When I asked you as to a royalty you said that when your expenses were covered you should be pleased to give Mr. Elgar something extra & I presumed it would be a Royalty on the copies sold. I thought 2d a copy might be the extent & I hope you will see your way to carry this out. As the work, as you say is a 'financial' failure it at any rate can do no harm . . .

I am
Sincerely yours
G H Jonstone[147]

Littleton refused to move:

In purchasing any copyright we always object to paying a sum down and a Royalty in addition: we do one or the other, not both . . . We have paid a very large sum for the work; which we should not have done but for your intervention & with this Mr. Elgar should be well satisfied.[148]

---

[145] 26 Oct. 1900 (HWRO 705:445:8446).
[146] 6 Nov. 1900 (copy in Novello archives).
[147] Ibid.   [148] 8 Nov. 1900 (copy in Novello archives).

38, Northampton Street, Birmingham.

Nov 9/00

Dear Mr Littleton

   . . . I certainly feel that I am bound morally to carry out my promise to Elgar, and as you say that this work is a commercial failure it seems to me a matter of very little moment to you whether you promise a royalty or not, as the probability according to your letters is that the cost will not be covered, to *me* it means a great deal after having pledged my word . . .

I am
Sincerely yours
G H Johnstone.

Novellos did not immediately reply.

As if to point the last irony, Cambridge University chose this time to offer Edward an honorary doctorate. It had nothing directly to do with the success or failure of *Gerontius*: the honorary degree had been pushed by Stanford (who was enthusiastic over the *Variations*) since the spring, when he accepted honorary membership in Edward's Worcestershire Philharmonic. At that moment Edward had been busy ciphering 'Satanford' into the Demons' Chorus.

The difficulties he felt over the Cambridge offer were expressed hesitantly even thirty years later to his biographer Basil Maine, who wrote:

The thought of being addressed as Dr. Elgar did not at all please him . . . He was unaffectedly grateful for this sign of recognition. But by accepting the degree it seemed to him that he would be—well, it was difficult to say precisely what—perhaps, in some indirect way, he would be forfeiting some of his natural freedom. Living up to the implications of an academic title would cramp his style.[149]

When the Cambridge offer arrived, Edward said he would refuse it. In response to 'an urgent message from Alice Elgar', Miss Burley hurried over to Craeg Lea:

I went up cautiously and found him sitting at his table with his head in his hands. He said gloomily, 'They've offered me a Doctor's Degree at Cambridge University, but I shan't accept it. I'm just writing a refusal.'
   Miss B. 'Won't accept it, but why not?'
   Elgar. 'It's too late.'
   Miss B. 'Too late, for what?'
   Elgar. 'For everything.'
   Miss B. 'I don't understand it. Why, it's the greatest honour they can offer you . . .'
   Elgar. 'But I can't afford to buy the robes.'
   Miss B. 'You need not buy them. You can hire them. You are not the first impecunious Doctor of Music. Why! I could almost lend you enough for the hiring.'
Then I laughed & said, 'I'll go round & sing & collect money in a shell!'[150]

Later that day Edward wrote to accept. As one of the dates offered was 22

---

[149] *Life*, p. 118.   [150] Quoted in *Letters to Nimrod*, pp. 111–12.

November—St. Cecilia's Day—that was chosen for the ceremony. When Miss Burley gave out the news to her friends in the Worcestershire Philharmonic, the resulting flurry caused Edward to decorate one of his letters to Jaeger with a cartoon sketch of 'Philharmonic ladies going to Cambridge for the event. (great excitement) . . . I feel Gibbonsy, Croftish, Byrdlich & foolish all over.'[151] Yet he defended it to his sister Dot, still looking after their old parents at 10 High Street: 'You must not think this is a 2½d. thing like the Archbp of Canterbury's degree but it's a great thing: it has of late only been given to Joachim, Tschaikowsky, Max Bruch & a few others.'[152]

He and Alice travelled to Cambridge by way of London. There he called on Alfred Littleton over the *Gerontius* royalty, and at last the chairman capitulated—writing afterwards to Johnstone without mentioning Edward's visit:

We must do our best to clear this matter up in regard to Elgar's work, & altho' we never expected to pay any Royalty on it at any time we will now agree to pay a Royalty of 4d. on Vocal Scores & a sum equal to 10 per cent of the marked price on any portion of the Vocal Score which may be published separately, these Royalties not to become payable until all expenses incurred by us in connection with the work have been paid by the sale of copies.[153]

On the degree day at Cambridge, Stanford was not present. He had written to say that he had to be in Leeds on 22 November to consult about the 1901 Festival, of which he was appointed conductor-in-chief in succession to Sullivan. As it happened, Sullivan died that very morning. Stanford's rival for the Leeds conductorship, Frederic Cowen, was to be in Cambridge to receive an honorary doctorate side by side with Edward and Stanford seemed to want to avoid him.

At the Cambridge ceremony the University's Public Orator spoke in Latin. He alluded to *The Light of Life*, the *Bavarian Highlands*, *King Olaf*, the *Sea Pictures*, the *Imperial March*, the *Variations*, and *Gerontius*. Then he concluded:

If from his Malvern and its Hills—the cradle of British arts, the citadel fortified by Caractacus—this disciple of the muses saw the towers of all the temples to religion shining below him in the light of the morning sun, and recalled the beginnings of his own life, he could indeed say with the modest pride of the antique bard:
> 'Self-taught I sing; 'tis Heaven and Heaven alone
> Inspires my song with promise all its own.'[154]

Alice described it all in a letter to old Ann Elgar at High Street:

I was entranced by the Orator—he spoke so splendidly & I can hear now the grand sonorous phrases, the way he described 'Gerontius' was beautiful. I *wish* you cd. have

[151] 14 Nov. 1900 (HWRO 705:445:8448).
[152] 17 Oct. 1900 (HWRO 705:445:4597).
[153] 21 Nov. 1900 (copy in Novello archives).
[154] Translated from the text printed in the *Cambridge University Reporter*, 27 Nov. 1900, p. 265.

seen E.—he looked so perfectly beautiful—really it is the only word, in his robes with a strong light on his face . . . E. had *such* a salvo of applause.

[After the ceremony] E. criticised the Public Orator's Latin much to that great person's astonishment & amusement![155]

Two days later Alfred Rodewald was to play the *Variations* with his Liverpool Orchestral Society. Edward travelled up for the weekend and received another warm welcome. Rodewald made the visit an occasion for rejoicing over the Cambridge doctorate, and in the months to come he raised a subscription amongst friends for a set of doctoral robes made to Edward's measure by Hall's of Cambridge.

Edward recognized the beginning of a great friendship. He confided to Rodewald his hopes and fears for the Symphony he wanted to write. Rodewald responded instantly with a generous offer of support during the composition— an offer to be repeated by letter a few days later: 'Think about my proposition regarding your Symphony. Don't be proud but sensible, il faut vivre, mais il faut entendre votre symphonie et belle musique.'[156] Yet Rodewald's offer only showed how much more than money was still lacking for the completion of Edward's Symphony. Back at Craeg Lea, he wrote to Jaeger at the end of November: 'It's raining piteously here & all is dull except the heart of E. E. which beats time to the most marvellous music—unwritten alas! & ever to be so.'[157]

If ever the Symphony was written, would it still revolve round Gordon of Khartoum? Near the time of Rodewald's offer, Edward wrote to Sir Walter Parratt at Windsor about the chance of setting a poem which seemed to bring forward the Gordon débâcle to the current British humiliations in South Africa. It was Kipling's Diamond Jubilee poem 'Recessional':

> The tumult and the shouting dies—
>   The captains and the kings depart—
> Still stands Thine ancient sacrifice,
>   A humble and a contrite heart.
>
> . . . .
> Far-call'd our navies melt away—
>   On dune and headland sinks the fire—
> Lo, all our pomp of yesterday
>   Is one with Nineveh and Tyre!

Parratt responded quickly to Edward's idea of setting this poem: 'Please do it, and soon—This is the psychological moment!—the end of the century, and we hope the end of the War.'[158]

[155] 24 Nov. 1900 (HWRO 705:445:4643).
[156] 29 Nov. 1900 (HWRO 705:445:2731).
[157] 29 Nov. 1900 (HWRO 705:445:8449).
[158] 2 Dec. 1900 (HWRO 705:445:1967).

One attraction of such a setting was described by Edward thus:

I like to look on the composer's vocation as the old troubadours or bards did. In those days it was no disgrace to a man to be turned on to step in front of an army and inspire the people with a song.[159]

But again his song-writing was overtaken by the orchestral impulse for abstract music which had been flowing strongly in him since he had wanted to substitute the orchestral Scenes from English History for *Caractacus* at Leeds. Now that orchestral impulse took a new turn:

We are a nation with great military proclivities, and I did not see why the ordinary quick march should not be treated on a large scale in the way that the waltz, the old-fashioned slow march, and even the polka have been treated by the great composers; yet all marches on the symphonic scale are so slow that people can't march to them. I have some of the soldier instinct in me . . .[160]

On the first day of the new year 1901 he sketched a 'Quick March' in A minor.[161] A busy introduction led to a primary theme swinging through ebullient triplets over a descent of steps:

The Trio melody supplied the descending steps with a series of answers rising from below:

Two days later he sketched another 'Quick March'.[162] This one began in E flat—only to drop grandly into a D major tonic with its primary theme made from another descent of steps. Then came a Trio melody founded again on descending steps:

This melody was more reflective than any he had written since *Gerontius*. Perhaps it sounded his own 'Recessional'. But old or new, it seemed then and later (as he himself expressed it) 'a tune that comes once in a lifetime'.

---

[159] *The Strand Magazine*, May 1904, pp. 543–4.   [160] Ibid.

More might be made of such a tune than merely the Trio for a Quick March. Could it become the vehicle of longed-for Symphony writing? It almost seemed so, as Edward interrupted a letter to Jaeger on 12 January to exclaim: 'In haste & joyful (Gosh! man I've got a tune in my head) . . .'[163] On the 20th Alice was writing to Jaeger: 'I think there cd. be no *nobler* music than the Symphony. I *long* for it to be finished & have to exist on scraps. Do write and hurry him, it always does *some* good.'[164]

Jaeger did. But Edward could not be drawn. He was happily revising the old Overture *Froissart*, which Novellos were to engrave and print at last in the full score, and he twitted Jaeger:

Why not start an Elgar society for the furtherance of the *master's works*. I'm very well but working hard. Oh! my string Sextett.—& I have to write rot & *can* do better things.[165]

When Jaeger duly enthused over the prospective sextet, Edward answered:

No time for kompoliments—that Sextett will not be ready for years—trade, my boy, trade before everything—every damn thing.[166]

The letter finished with a fragment of music *scherzando* labelled 'Cockayne'.

*   *   *

Its history went back to the autumn—to the weeks after *Gerontius*. Then he had sought some self-insight in rereading *Piers Plowman*, which had been quoted at the head of the *Musical Times* article and referred to again in the newspaper reports of the Cambridge degree ceremony.

In the poem the weary wanderer on the Malvern Hills lies down beside a stream, falls asleep, and dreams. His is not, at the outset, the dream of Gerontius. It is an earth-vision, with the whole Severn Valley below the Hills seen as a 'a faire felde ful of folke'.[167] Amongst them are musicians—'Japers and janglers, Judas-children':

> And somme murthes to make · as mynstralles conneth,
> And geten gold with here glee · giltles, I leve.
> Ac iapers & ianglers · Iudas chylderen,
> Feynen hem fantasies · and foles hem maketh.[168]

The dream-vision extends to London, where flourish the seven deadly sins: pride, lechery, envy, wrath, avarice, and the rest. Then Piers Plowman arises—a 'People's Christ' of the countryside 'who tills for all'. But when Piers

[161] BL Add. MS 47903 fos. 7–8.   [162] BL Add. MS 47903 fos. 2–3.
[163] Elgar Birthplace.   [164] HWRO 705:445:8824.
[165] 24 Jan. 1901 (HWRO 705:445:8483).   [166] 3 Feb. 1901 (HWRO 705:445:8486).
[167] Prologue, line 17. Elgar's copy of Skeat's seventh edition (OUP, 1893) is dated in his hand 'Nov.12.97' (now in the possession of Raymond Monk).
[168] Prologue, 33–6.

wins God's pardon for those whose works match their faith, his triumph is denied by a priest who stands for the Church's monopoly of pure faith. Their quarrel penetrates the dream and the sleeper wakens to an empty and darkening world:

> And I Þorw here wordes a-woke · and waited aboute,
> And seighe Þe sonne in Þe south · sitte Þat tyme,
> Metelees and monelees · on Maluerne hulles,
> Musyng on Þis meteles . . .
> Ac I haue no sauoure in songewarie · for I se it ofte faille . . .[169]

The old poet was also a punster: his 'metelees' ('meat-less') was close to the word 'meteles' in the next line, meaning 'dream'. His 'songewarie' stood likewise for the observation of dreams. Therein he mirrored Edward's mood at the end of the *Gerontius* year. For Christmas Edward had sent Jaeger a copy of *Sea Pictures*, and wrote on the last day of the old year and the old century:

I have given up sending cards, so sent you the score of the miserable Mal de Mer—I forget what quotation I put on—from Piers Plowman's Vision I expect—that's my Bible, a marvellous book.[170]

There had been an enquiry from the Philharmonic Society in London stiffly offering to produce a new orchestral work. Edward wrote a letter of careful gloom to sound Jaeger's response:

Seriously, my dear friend, look at the position: e.g.—I'm asked to write something for the Phil:—well I've practically got a Concert overture ready: the P. won't pay *anything*.
Now look at this.

To copying parts	12.0.0.
Rehearsal ex[pense]	3.0.0
Do. & Concert	6.0.0.
net loss	21.0.0.

Now what's the good of it? Nobody else will perform the thing—if I take it to your firm they might print the strings but the result wd. be the same.

No thank you: I really cannot afford it and am at the end of my financial tether. Don't go and tell anyone but I *must* earn money somehow—I *will not* go back to teaching & I think I must try some trade—coal agency or houses . . .[171]

He had given up his teaching for the most part, but the writing of even a *Gerontius* did not replace the emolument.

Again Jaeger's encouragement arrived. And then the presence of this most faithful friend in London might seem in itself to suggest a character for the new music—as Edward wrote at the start of another season with the Worcestershire Philharmonic:

169 Passus 7, 139–42, 148.   170 HWRO 705:445:8453.
171 26 Oct. 1900 (HWRO 705:445:8446).

I do want to see you or somebody as knows suffin—I am bored to death with commonplace ass-music down here—the bucolics are all right when they don't attempt more than eat, drink & sleep but beyond those things they fail . . .

Don't say anything about the prospective overture yet—I call it 'Cockayne' & it's cheerful and Londony—'stout and steaky'.[172]

As he explained to Hans Richter, 'Cockaigne is the old, humourous (classical) name for London & from it we get the term Cockney.'[173] A Cockney overture for London could tweak the superior Philharmonic nose while at the same time it expressed his own reaction to the entire world of county societies and provincial festivals right up to Birmingham. But the name held other implications to sift through post-*Gerontius* hours and days. He noted the literary definitions:

Cockaigne
('the land of all delights: so taken in mockerie'. Florio)
'In Cockaigne is mēt & drink
Withvte care, bow [i.e., anxiety] and swinke.' (Land of Cokaygne. Early English Poem)
  1. An imaginary country of idleness & luxury;
  2. The land of Cockneys: London & its suburbs. (Obsolete except in historical use or in literary or humorous allusion.)
Vast variety of spelling.
Usually associated with *Cockney*—but the *con*nection, if real, is remote.[174]

The subject of London, he recalled,

. . . was first suggested to me one dark day in the Guildhall: looking at the memorials of the city's great past & knowing well the history of its unending charity, I seemed to hear far away in the dim roof a theme, an echo of some noble melody:

175

The 'noble melody' echoed here was a new variant of the old descending steps. The small notes of approach, busy rhythm, and near-sequential movement all revived elements of the 'tune from Broadheath'. Yet it was not a child who stood in the Guildhall listening: the melody, when it came now, sounded the C major tonality through which the defeated Caractacus had been marched in chains to Rome, through which the dead Gerontius heard the choirs of a heaven where his own judgment would exclude him. Still if Gerontius could be purged, and Caractacus pardoned, then Edward might find for himself the 'unending charity' which seemed borne upon the dim echoes of London.

The descending shape of the 'Guildhall' melody began to generate variants for 'Cockayne'. One would make a primary theme for the new Overture:

[172] 4 Nov. 1900 (Elgar Birthplace).   [173] 7 Apr. 1901 (Richter family).
[174] MS at the Elgar Birthplace.
[175] 'The question of programme music' (undated MS fragment at the Elgar Birthplace).

(Analysing the music years later, Donald Tovey found 'a magnificent Cockney accent in that pause on the high C'.[176])

A development of the primary theme sounded like a military band:

That in turn suggested a Salvation Army group playing their hymn out of tune. The bass could be thumped out on F accompanied by tambourine, while the wind began tentatively in G flat:

When the bass players obliged by raising their part a semitone, they would only incite the upper wind to restart their hymn tune a semitone higher again. It was a precise parody of the 'Angel of the Agony' formula in *Gerontius*. By 14 November he was parading to Jaeger:

I've written some tunes which will 'make a methody swear'—the whole quotation is too good for you—(Devon hunting song ancient adapted) (I believe its Somerset after all)

> 'Oh! Elgar's work's a d—able work
> the warmest work o' the year,
> A work to tweak a teetotaller's beak
> And make a methody swear.'
> (This is Cokayne!)[177]

The Cambridge degree had followed, and Rodewald's offer for the Symphony. Through much of January 1901 he had sought a way to the Symphony through the great tune sketched for the D major Quick March; but it was no use. Going back to 'Cockayne', he began privately to test the wind of opinion among friends and critics. Edward Baughan responded: 'With best wishes, especially for that "Cockayne" Overture (I don't like the title). Why not "In London Town"?'[178] Edward adopted the suggestion as a subtitle. The original title he kept, but respelt it *Cockaigne*.

[176] *Some English Symphonists* (OUP, 1941), p. 33.
[177] HWRO 705:445:8448.   [178] 5 Feb. 1901 (HWRO 705:445:2823).

In London in February he was ready to show the sketch to publishers. After the trouble over the *Gerontius* royalty, he did not offer it to Novellos, but went to Schotts, who were always asking for something to follow the earlier music they had published. Schotts did not offer suitable terms: so he took it to Booseys, who did.

The Boosey firm was run by two brothers, Arthur and George. Arthur was just Edward's age. But George, nine years younger, became his special friend. George Boosey was a sporting bachelor: he was a good shot and a first-class cricketer, exemplifying the 'county' virtues that Edward admired. Moreover, he had attended Malvern College, where he had mixed with sons of local gentry, including the 'cricketing' Fosters of Malvern[179] George's breezy geniality offered instant contrast to the denizens of Novellos.

Back at Craeg Lea, Edward laid out the structure for *Cockaigne*. The Overture opened in C major directly on its primary theme. A sequence of thirds

(a cheeky reminder of the 'Enigma' B figure) led to the 'Guildhall' melody. Here was the idea which had begun all this music: it was set now as an emblem in the midst of an exposition made entirely from its progeny.

The tonality shifted to E flat for a second subject which was close to the 'Guildhall':

Sequentially extended and varied to some length, it suggested a point of view—a point of hearing—for the rest: it was described by Edward as a 'lovers' theme.[180] Then came a 'Guildhall' diminutive

this was to sound for the critic Ernest Newman the whistle of 'the perky, self-confident, unabashable London street boy . . . just as Wagner obtained the theme of his Nuremberg apprentices out of the Master-singers.'[181] Edward himself noted:

[179] Conversation with Arthur Boosey's son, Leslie Boosey, 1976.
[180] 26 Oct. 1901 to Henry Ettling (Elgar Birthplace).
[181] *Elgar* (John Lane The Bodley Head, 1922), p. 157.

Of course the process is Wagner's but it interests me to say that the idea of 'diminution', to mark youth, came to me from Delibes' 'Sylvia' which I knew before 'The Meistersinger'. I made a note for future use that the broad 'Bacchus' theme (minims) was reduced to crotchets for Tarsichon & the *'Jeunes filles'* of her escort.[182]

Formal development soon brought the 'Band' figure extending anticipation through a brilliant crescendo, and then the full force swung round the corner *grandioso*. The joyful noise faded gradually toward the Salvation Army's out-of-tune hymn. That in turn led to a solemn variant

to suggest the dim spaces of a large church.

Thus Edward ingeniously adapted sonata development to his own skill with variations. Where the development in *Froissart* had functioned as little more than a delaying tactic before recapitulation, the *Cockaigne* development carried on the exposition's task of introducing new variants. All the development's variants had come out of the Overture's primary material; the second subject ('the lovers') functioned as presenter of the experiences. Yet all of them together had derived from the emblematic 'Guildhall'. This interaction of all the Overture's ideas rather finessed the sonata-contrast of primary and secondary themes.

Recapitulation brought back the Band, now in the tonic C major, to play a telescoped version of the opening theme and the cheeky sequence in thirds. A restatement of the 'lovers' music also found C major. The coda led through a crescendo of the 'streetboy' figure to the 'Guildhall'—resounding at last *Nobilmente* with the orchestra joined by full organ. A quick reference to the opening theme and the Overture was done.

In total length *Cockaigne* was very close to *Froissart*—359 bars against 340.[183] But *Cockaigne* had far greater unity, smoother opening out of its material, incomparable boldness of rhythm and harmony, an ingenious and idiosyncratic sonata structure. So it stood as a milestone on his path toward the Symphony: the new Overture was an emphatic success in drawing on his own private impressions to fill a respected abstract form.

There was no doubt of his wish to avoid the implications of a private 'programme' when he wrote to Joseph Bennett about doing an analysis for the Philharmonic première:

You know I don't believe in symphonic poems at all but I don't mind a 'Name' for a piece. The work is in regular form somewhat extended: in the 'working-out part' [i.e., development] sundry contrapuntal japes occur, and certainly a military band passes, but

---

[182] Draft letter of late 1932 to D. Millar Craig (Elgar Birthplace).
[183] The timings of Elgar's gramophone recordings of 1933 are 13 min. 25 sec. for *Cockaigne*, 12 min. 27 sec. for *Froissart*.

there's nothing to give the astucious mystagogue (that's you as analyst!) much trouble in describing.[184]

So he repeated his plea for the *Variations*: ideas of a private 'programme' might have generated the music, but once it moved toward 'regular form' he craved attention only for the abstract achievement.

Yet *Cockaigne* was not the Symphony. When he finished the score on 24 March he inscribed its final page with the line from *Piers Plowman*: 'Metelees & monelees on Malverne hulles.' He wrote to Richter to enquire if the great man might think of conducting the new Overture, and unhappiness came out again: 'The work is not tragic at all—but extremely cheerful like a miserable unsuccessful man ought to write . . .'[185]

Two songs written against the composition of *Cokaigne* showed the darker face. The words of Clifton Bingham's *Come, Gentle Night* looked backward through *Gerontius* to the *Sea Pictures*:

> Come, holy night!
> Long is the day, and ceaseless is the fight;
> Around us bid thy quiet shadows creep,
> And rock us in thy sombre arms to sleep!

Then just as he was finishing *Cockaigne* an elderly resident of Birmingham, Frank Fortey, sent him some verses by the Polish poet Krasinski; and at once he set them in spite of Fortey's amateur translation. The title was *Always and Everywhere*:

> But say, when soft the grasses o'er me wave,
> That God is kind to hide me in the grave:
> For both my life and thine I did enslave,
> Always and Everywhere.

His own interest in words and letters would have reminded him that the initials of the repeating 'Always and Everywhere' were those of Alice and Edward.

He decided to perform as much as possible of *Gerontius* in the Worcestershire Philharmonic's Spring concerts. They might manage a good deal of it if the Demons' Chorus was cut out. Novellos were only too eager to send the score and parts for rehearsing: they were not in demand anywhere else then. Jaeger had written to Dora Penny: 'I am still trying hard to get Gerontius performed in London, but it is almost hopeless. I still hope Wood will do it . . .'.[186]

Henry Wood agreed to put the Prelude and Angel's Farewell into his Ash Wednesday Concert at Queen's Hall if the choral part of the Farewell could be

---

[184] J. A. Westrup, *Sharps and Flats* (OUP, 1940), p. 94.
[185] 1 Apr. 1901 (Richter family).
[186] 27 Dec. 1900 (quoted in Powell, *Edward Elgar: memories of a variation*, p. 33).

removed. Edward made the arrangement, and used it himself when he conducted a concert with Frederic Cowen's orchestra in Bradford on 16 February. The programme also included the revised *Froissart*, and *Sea Pictures* sung by Muriel Foster, a young mezzo-soprano who had just completed her studies with Anna Williams at the Royal College of Music. Miss Foster combined beauty of voice and of presence with a sensitive intelligence quite new in the interpretation of Edward's music. She proved to be the first of a group of young singers who would come forward as Elgar specialists during the next few years.

From Bradford he went to London for Henry Wood's Ash Wednesday Concert on the 20th. Jaeger reported to Dora Penny:

Your dear Doctor E. E. is in town & this morning we went together to Queen's Hall to hear Wood conduct the Gerontius Prelude & Angel's Farewell . . . Wood conducted it with loving care, spent 1½ hours on it & the result was a performance which completely put Richter's into the shade.[187]

It was the beginning of a real interest on the part of Henry Wood, who now planned to conduct the *Variations* in May. Wood's interest attracted the attention of foreign conductors, Edouard Colonne and Felix Weingartner— who promised to conduct the *Variations* and the new Overture *Cockaigne* in Germany.

And there was attention from a quarter still more august. Queen Victoria had died in January: she was hardly in her grave before the air was full of plans for the new Coronation. On 12 March Sir Walter Parratt wrote in his capacity as Master of the King's Music, enclosing verses for a Coronation Ode: 'These words have been sent me by Arthur Benson, son of the Archbishop [of Canterbury], a genuine poet and known to the Royalties. Should it take your fancy—he would be honoured & I most grateful if you could set it to immortal music.'[188] Benson was a schoolmaster at Eton. His skilled and deeply felt poems—at a time when the Laureateship was in the hands of the untalented Alfred Austin—had made him an increasingly familiar figure at Windsor. His Ode was in short lines which would lend themselves to singing. The idea was to have it ready for the Coronation ceremony in June 1902. Edward's mind leapt ahead. When he sent Jaeger an heraldic design his friend twitted him: 'Are you designing *your 'arms'*, gules & all? Oh you sly young Sir Knight!'[189]

There was also an enquiry from Windsor about a Coronation March. This, in the opinion of Novellos (when he told them about it months later) should have been taken as a Royal Command.[190] Lacking experience, Edward asked the advice of Frank Schuster. Schuster was away on one of his continental holidays,

[187] 18 Feb. 1901 (ibid., p. 34).

[188] HWRO 705:445:1969. The invitation to set Benson's verses had previously been extended to Stanford, who refused (it was said) on account of publishers' meanness. See David Newsome, *On the Edge of Paradise* (John Murray, 1980), p. 104.

[189] 13 Mar. 1901 (HWRO 705:445:8524).

[190] 7 Nov. 1901 (Novello archives).

but he was only too eager to be involved, as he wrote from Dresden on 28 March:

Anything you do—or think of doing—will *always* interest me—and I shall do what I can to further your interests, the moment I get back and *go amongst people again*. It can only be done by talking—and when I run against the right person—as I hope to do—I shall *talk* no end! *Of course* you are the man to do it—and no other![191]

For the royal purpose his sketch of the March with the great Trio tune was ready to hand. When Dora Penny came to Craeg Lea in early May, Edward called out from the study:

'Child, come up here. I've got a tune that will knock 'em—knock 'em flat,' and he played the Military March No. 1 in D . . .

E. E. came in to dinner that evening in a bright red golf blazer with brass buttons, over his evening shirt.

'I say, you *are* smart,' I remarked admiringly.

'Well, if W. M. B. wears a pink coat at dinner why shouldn't I wear this?'

'He only wears a pink coat on state occasions when he has grand people to dinner, like you; when I'm there he wears an old Ledbury Hunt coat which I like better; it's quieter!'

'Well, I'm not quiet. Far from it.'

I pretended to look under the table.

'It's no use looking. You won't see satin knee-breeches and silk stockings!'[192]

He seemed to be thinking already of Court dress.

Dora Penny's visit coincided with the Worcestershire Philharmonic concert on 9 May which included the almost complete *Gerontius*. It gave great satisfaction, as Edward wrote to the editor of *The Musical Times*:

You know in certain circles—connected alas! with the futile Birmingham Chorus—a great point has been made about the supposed difficulty—our chorus sang *well*—in tune throughout & with real devotional feeling—the work made a *profound sensation* really & was a noteworthy event.[193]

In drawing the editor's attention to his own success, Edward also remembered the smaller success of a lesser composer at the same concert. After *Gerontius*, the second half had concluded with a *Romance and Bolero* by John Austin, the Worcestershire Philharmonic leader who was also Edward's redoubtable helper whenever scores and parts arrived for proof-reading. When Austin's work came on, Edward made him conduct, while he himself retired to the orchestra to lead the second violins. It was a graceful and popular gesture, of which he said nothing to the *Musical Times* editor, only asking attention for Austin's composition.

That was of a piece with his dedication of the *Cockaigne* Overture 'to my many friends The Members of British Orchestras'. On 20 June Edward

---

[191] HWRO 705:445:6966.

[192] Powell, *Edward Elgar: memories of a variation*, pp. 35–6

[193] 18 May 1901 (BL Egerton MS 3090 fo. 60–1).

conducted the première of *Cockaigne* to close the Philharmonic season in London. 'Great glorious success', wrote Alice in the diary, and all the papers echoed her. The critic of *The Referee* wrote:

Technically, the score is a masterpiece of contrapuntal science, and the orchestration is so rich and full as to suggest comparison with the Overture to 'Die Meistersinger'. Better than all its science, however, is its powerful expression of healthy & exuberant life. It is music that does one good to hear—invigorating, humanising, uplifting.[194]

Joseph Bennett in *The Daily Telegraph* gave his encouragement a personal note:

There is no suggestion here of a city of dreadful night . . . It may be urged that the composition of street scenes is not activity on the highest plane of music. Granted, but we cannot have everything all at once. Mr. Elgar is yet in an early stage of his career as a creative musician, and he must be allowed some freedom . . . Some day the composer will give us his symphonies, but on the way to that achievement he is in no more hurry than was Brahms. The pace is for himself to determine.[195]

In the months to come, when the Symphony still failed to progress, the 'City of Dreadful Night' idea returned to suggest an opposing 'Cockaigne No. 2'.

* * *

After *Cockaigne* there was nothing. The question of Coronation Marches had grown complex when it appeared that Mackenzie, Edward German, and several others were already at work. Then the Musicians' Company announced a competition for the best Coronation March: 'We must now look for a masterpiece,' Edward wrote to Frank Schuster on 6 June,[196] and promptly retired from the fray.

Finding no work of his own to do, he busied himself with work for a friend. The Gloucester Cathedral organist Herbert Brewer had a cantata called *Emmaus* in the Three Choirs programme for September 1901; he had finished the vocal score, but then a misunderstanding with the librettist caused such delay that there was no time left to orchestrate his work amid all his duties as chief conductor for the Festival at Gloucester. When Brewer contemplated withdrawing his music, Edward did the orchestration for him in ten days, and wrote to Jaeger: 'I . . . hope it will relieve his mind & perhaps his wife's—oh these wives of musicians—what they go through & *suffer*—my heart bleeds for them sometimes—we MEN can buck up & fight; but the others—'.[197]

Jaeger himself had begun to be seriously ill. He had so much trouble with his nose and throat that Edward insisted on taking him to Dr Greville Macdonald and paying the celebrated doctor's fees. Dr Macdonald ordered an immediate operation to stop the onset of consumption, followed by further cautery. When

[194] 23 June 1901.   [195] 22 June 1901.
[196] HWRO 705:445:6937.   [197] 26 June 1901 (HWRO 705:445:8499).

Edward and Alice went up to Birchwood for the summer, it was arranged for Jaeger to spend his holiday at Craeg Lea to try to recruit his strength between treatments.

With him came Julius Buths, the German conductor. Buths had successfully introduced the *Variations* at Düsseldorf in February, and was planning to perform *Gerontius* in December with a view to its inclusion in the Lower Rhine Festival of 1902. He was also undertaking a translation of the Newman text, which Novellos were to engrave in a separate German edition of the vocal score.

Edward faced a summer at Birchwood for the first time without a major project in the final stage of completion. He filled July with cycling. Increasing skill invited longer and longer rides through surrounding countryside. Alice had not really taken to it, and Dora Penny lived too far away to be generally available. But sometimes there was a little party of mostly female friends and neighbours to ride with him. Of these Miss Burley was the most eager companion:

There cannot have been a lane within twenty miles of Malvern that we did not ultimately find. We cycled to Upton, to Tewkesbury, to Hereford, to the Vale of Evesham, to Birtsmorton where Cardinal Wolsey is said to have fallen asleep and come under the fatal shadow of the Ragged Stone, to the lovely villages on the west side of the Hills—everywhere.

Much of Edward's music is closely connected with the places we visited for, as we rode, he would often become silent and I knew that some new melody or, more probably, some new piece of orchestral texture, had occurred to him. Unlike most composers he carried no notebooks in those days[198] but seemed able to register and remember his musical ideas even in the middle of a conversation. On one of these occasions I offered to stop talking in order that he might the better concentrate. 'No,' he said dreamily, 'I like your vain bibble-babble.'

. . . Sometimes we would talk, sometimes we would pedal along in silence. He was very difficult and one never quite knew what would be the mood of the afternoon. There were times when he was gay and hopeful especially at the beginning of the cycling season when the fresh green of the trees seemed an invitation to take longer and longer rides and he was thankful for release from the winter confinement indoors.

I found that he was particularly touched by birdsong and that he loved and knew all the little creatures that darted in and out of the hedges. The only occasion when he was distressed by anything he heard was one hot afternoon when some village children in squeaking shoes came scrunching along a rough road past the gate on which we were resting. The unpleasant noise worked on his nerves to such an extent that I almost expected an explosion of rage.

. . . Often when we had ridden for some time in silence he would say, 'Stop and get off, I want to talk.' And against the spectacular background of, perhaps, Castlemorton Common and the hills he would discuss with me at enormous length the works on which he was engaged and would sometimes explain with pride the ingenuity of his technical devices—with the result that we would sometimes argue as to their merit since I

[198] He did on many occasions carry pocket-sized scraps of music paper for noting ideas.

could not see that ingenuity, for its own sake and inaudible to the listener, had much value.[199]

In the provincial world round Malvern, Miss Burley offered a self-consciously intellectual alternative to Alice's ministry (as Dora Penny offered a self-consciously youthful alternative). Yet she failed to understand how creative eyes and ears must be always dissatisfied with the *status quo*:

> Once when we were walking up a hill, a tramp approached, sat on a felled tree and, taking some bread and cheese from his pocket, began a meal. 'That man is happy,' said Edward, 'How I envy him!' This seemed to me rather naive since the only thing the tramp had which Edward had not was an acceptance of life as it came.[200]

His attractiveness to women was never in doubt: his aquiline features, his dark insecurities, and glinting appeals for quick reassurance spoke louder than any words. Several of these women looked eagerly for a real return of their own romantic feelings. In later years a number of different legends were to be circulated about the possibility of an illegitimate child—always, suggestively, another daughter. Several women laid claim. Investigation has revealed no evidence beyond a degree of emotional volatility in each of the women.

When any woman sought to inspire Edward by her presence, Alice as often as not left her to it. If Alice was sometimes anxious, she never let Edward see that her anxiety was for anything but his perfect comfort. For in her strongest moments Alice understood more clearly than all the others that sex would never be so important to the man who projects himself in creative things as to the man for whom sex is a chief outlet. Edward had once written to Jaeger:

> Oh these boys: if only they knew that a woman's not worth a damn who won't put up with everything except ineptitude & crime.[201]

Alice recognized his need for privacy and a secure background before all things. She knew that in the long run Edward's music must be more important to him than any woman; that every other woman would make more demands than she; and that ultimately he would not stand for it.

So the marriage with Alice provided the indispensable basis on which his works were founded. His first work of any size had been written in the year after the marriage; his last big work would be written in the year before her death but almost fifteen years before his own. During the final silent years all the women who had been romantically interested were still alive and several of them unattached: but no music would emerge, nor happiness either, with any of these friends from his creative years.

The only woman who might have posed a threat was Carice, who was Edward's own flesh. There Alice acted with decision. She arranged everything to make their daughter unbrilliant. She sent her to a school (it was Miss Burley's) close enough to have enabled Carice to be a day-pupil; but Carice

---

[199] *Edward Elgar: the record of a friendship*, pp. 145–6.
[200] Ibid., p. 149.   [201] 21 Aug. 1898 (HWRO 705:445:8316).

must be a boarder for fear of disturbing her father's immortal concentration. She dressed Carice frumpishly in eternal brown stockings, surmounted with an outmoded lavishness certain to arouse the spite of other girls—a red plush cape, a black velvet neckband with a gold cross to advertise her Catholicism to the Protestant community. Edward veered between ignoring his daughter and rallying her as a small masculine assistant with gardening tasks addressed to 'my man'. The result for Carice was a lifelong uncertainty of manner which turned the most obvious statement into a hesitant question—covering a considerable intellect, musical gifts, and sprightly humour under the heaviest cloak she could bear.

When cycling at Birchwood was limited by minor injuries in late July, he was at the end of his resources. In mid-August he finished and orchestrated the two Marches—and thus resigned the claims of the great Trio tune for either the Coronation or the Symphony. He sent the Marches to Booseys with the title 'Pomp and Circumstance'. The quotation was from Shakespeare's *Othello*, and its context suggested a parting:

> Farewell the neighing steed, and the shrill trump
> The spirit-stirring drum, the ear-piercing fife,
> The royal banner, and all quality,
> Pride, pomp and circumstance of glorious war!

He dedicated the Marches to Alfred Rodewald and Granville Bantock, and arranged for the premières to be played by Rodewald's orchestra in Liverpool—as if to insist that their music had no necessary connection with London.

As the summer at Birchwood drew to its harvest fullness and beyond, all the prospects over the valleys below could mock his inactivity. He wrote to Miss Burley, holidaying with her mother, sister, and several young nieces at a cottage on the coast of Cardiganshire at Llangranog.

I had a very melancholy letter from Edward who was in one of his black moods of depression and was unable to do anything. I thought he was needing a change so I asked him to come down to this remote place. We could find him a bedroom in the village and come to meals with us . . .

He had never stayed at the seaside in his life and knew nothing about it but he was very happy. We rigged up a bathing suit for him out of an old pair of pyjamas and he did not mind when my little niece told him that he looked like a monkey! We were in and out of the water all day. The children went off in the boats after lobster pots, and they often got shrimps in the bay. There were mussels and winkles on the rocks. Edward was quite unrecognisable—He shouted with glee and played about like a little boy . . .

We often waded out at low tide to the little island of Ynys Lochtryn[sic]. There were a few sheep on it, but the views of the beautiful coast were lovely . . . In the evenings a little company of men used to collect on the little quay in front of the pub. They used to talk & smoke and then someone would hum a note and they would all sing a hymn or song in four part harmony. It was so natural and beautiful. One day as we looked across the bay we saw a party of folk on the hillside & wondered what they were doing. Presently we heard them singing . . .[202]

Edward himself recalled it:

In Cardiganshire, I thought of writing a brilliant piece for string orchestra. On the cliff, between blue sea and blue sky, thinking out my theme, there came up to me the sound of singing. The songs were too far away to reach me distinctly, but one point common to all was impressed upon me, and led me to think, perhaps wrongly, that it was a real Welsh

idiom—I mean the fall of a third—  Fitting the need of the

moment I made

It sounded almost an echo of the great Trio tune in *Pomp and Circumstance* No. 1. He stayed five days before returning to Alice and Carice at Birchwood.

One afternoon as August was drawing to its close, two young visitors appeared. George Alder, a brother of Edward's former violin pupil Mary Beatrice Alder, was now studying the horn at the Royal Academy of Music. He brought with him a seventeen-year-old composition student named Arnold Bax, who found himself 'perturbed and delighted' at the prospect of meeting Dr Edward Elgar.

We set out on a sultry afternoon, our heads in a cloud of gnats which we tried to disperse by energetic smoking, and as we approached the unpretentious but charming cottage I almost regretted my temerity in coming. My tongue and throat were dry and my heart a-flutter with nervousness, which was part allayed and part aggravated when we were told by a maid that Mr. Elgar was at present out somewhere in the woods. But he would be back at tea-time or soon after, the meanwhile would we sit in the garden where the mistress would join us at once?

The composer's wife, a pleasant-looking fair-haired lady, with—it struck me—rather an anxious manner, welcomed us very kindly in her gentle, slightly hesitant voice. Almost at once she began to speak enthusiastically and a little extravagantly about her wonderful husband and his work.

She spoke of her Edward's early struggles for recognition and referred to the rudeness of well-known literary men to less-known musicians, relating how Elgar in his young days had set to music a lyric by that notorious tough, Andrew Lang . . . Our hostess was continuing, 'On the other hand Dr. Richard Garnett was quite charming about "Where Corals Lie",' when I became aware of footsteps behind me. 'Oh, here he is!' cried Mrs. Elgar, and I rose and turned with suddenly thudding heart to be introduced to the great man.

Hatless, dressed in rough tweeds and riding boots, his appearance was rather that of a retired army officer turned gentleman farmer than an eminent and almost morbidly

---

[202] Quoted in *Letters to Nimrod*, pp. 140–2. The name of the island is actually Ynys Lochtyn.
[203] Written in Jan. 1905 for the first performance programme of the *Introduction and Allegro*.

highly strung artist. One almost expected him to sling a gun from his back and drop a brace of pheasants to the ground.

Refusing tea and sinking to a chair he lay back, his thin legs sprawling straight out before him, whilst he filled and lit a huge briar, his rather closely set eyes meanwhile blinking absently at us. He was not a big man but such was the dominance of his personality that I always had the impression that he was twice as large as life.

That afternoon he was very pleasant and even communicative in his rumbling voice, yet there was ever a faint sense of detachment, a hint—very slight—of hauteur and reserve. He was still sore over the 'Gerontius' fiasco at Birmingham in the previous autumn, and enlarged interestingly upon the subject. 'The fact is,' he said, neither the choir nor Richter knew the score.' 'But I thought the critics said . . .' I started to interpose. 'Critics!' snapped the composer with ferocity. 'My dear boy, what do the critics know about anything?'

. . . Knocking out his pipe, he suggested that we might like to have a glance at a huge kite he had recently constructed. We duly appreciated the lines of this mighty toy, though as there was no wind its excellencies could not be practically demonstrated, and were then led into a small wood adjoining the garden where we found Elgar's little daughter sitting on a swing. 'Showing rather more leg than I care about, young woman!' remarked her father crisply. Thus admonished, the child dutifully slipped to the ground and I paused to say a few words to her whilst the composer passed on with Alder. The latter told me as we were returning to Malvern that during my short absence Elgar had asked him what were my musical ambitions. On being told that I intended to devote myself to composition Elgar had made no comment beyond a grimly muttered, 'God help him!'[204]

Across Europe in Bayreuth Edward's name was on the lips of Henry Wood. He was besieged by the Irish writer George Moore, trying to persuade him to write music for a new play, *Grania and Diarmid*. Wood protested that he was no composer, and gave Moore a letter to Edward. Moore had further designs beyond the play, as he wrote:

. . . I shall be pleased if my drama inspires you to write an opera. It will be produced in Dublin on the 21st of October by [F. R.]Benson's Shakespearean Company—it is based on a legend and the legend is the great heroic legend of Ireland, Diarmuid and Grania. I do not know if I or Mr. W. B. Yeats, my collaborator in this play, will be able to find time to versify it . . . For the moment my concern is to get you to write me a few pages of music . . . I don't want music for the first or second acts but I do want music for the death of Diarmuid. A moment comes when words can go no further and then I should like music to take up the emotion and to carry it on.[205]

The notion of a funeral march touched Edward so deeply at this moment that he agreed to do it before reading the play, and despite the fact that the Irish Literary Theatre Company could pay nothing.[206] When the text arrived, it offered equal identification with hero and heroine. Grania is betrothed to the

---

[204] *Farewell, My Youth* (Longmans, Green, 1943), pp. 29–31.

[205] Thursday [22 or 29 Sept. 1901] (HWRO 705:445:2266).

[206] Moore's letter of 7 Sept. 1901 (HWRO 705:445:2259) speculates that Benson the producer might pay something.

leader Finn, and the wedding tables are laid. Beside them the Druidess Laban sits spinning. When Finn's troop enter, one of their number is the beautiful youth Diarmid. He and Grania are fatally attracted: both break their vows to Finn and go together. Then Grania tires of the marriage. The spinner breaks thread after thread until she has no more flax. Diarmid, weary of 'this crooked road of morrows', goes to his death by accident or design at the point of Finn's spear, and they carry him out.

Edward planned a big, slow march with the principal subject in the Aeolian mode to give an atmosphere of legend. The A minor tonality and triplet figure

sounded slow, hollow echoes of *Pomp and Circumstance* No. 2. A second theme reminisced over something like 'Praise to the Holiest':

The texture was rich yet open, the scoring restrained. Yet it dwarfed the fourteen or sixteen players likely to occupy the pit at the Gaiety Theatre in Dublin. And the March extended over six and a half minutes. Clearly it was an expression which demanded painting on a larger canvas than that suggested by the practical limitations.

Then Moore asked for a few added fragments—a horn call and a tiny song for Laban to sing at her spinning wheel. After the première he sent his gratitude:

I am therefore more anxious than before that you should write an opera on the subject. Whatever the merits of the piece may be, one merit it certainly has: it has inspired some of the most lovely music. Every time I hear the Dead March I hear something new in it . . . If you can write the whole opera in the same inspiration I cannot but think that you will write an opera that will live.[207]

But Edward had promised, a year earlier, to write a cantata for the Norwich Festival of 1902: he had earmarked his old project of setting Jean Ingelow's '*The High Tide on the Coast of Lincolnshire*', and it was time to begin thinking seriously about it.

For the Gloucester Festival they were the guests of the Bakers at Hasfield Court. Edward involved the three boys of the house in an elaborate charade based on the Royalists and Roundheads of Scott's *Redgauntlet*. The boys must all be Royalists, of course, with Edward himself as their secret ally, the disreputable captain Nanty Ewart. They built a fort, sailed the moat, and found

[207] 23 Oct. 1901 (HWRO 705:445:2331).

Roundheads where they could—principally in the boys' Aunt and Uncle Townshend and Frederic Cowen, also part of the Festival house party. The eldest Baker son, William, recalled:

We boys were in the garden waiting for Nanty to come out. He had waved to us from his bedroom window so I went indoors to meet him. In the hall I saw Mrs. Elgar, a packet of letters in her hand, the afternoon post having just come. She was going to take them upstairs but as she saw Nanty coming down, she waited and held the letters out to him.

'But I'm going out now,' he said, taking them from her and, with a very Nantyish oath, he threw them down on the floor and they scattered in all directions.

'Oh, Edward, that *was* naughty!'

I picked the letters up and gave them to her. Remarking quietly 'These must be answered *at once*,' she held them out to him. With a shout of ribald laughter he took them from her and went straight back, upstairs, without another word. I went out and told the boys that Nanty couldn't come just yet. We did not see him until tea time.[208]

It was a rare sight of the maternal supervision Alice wielded over his career. She was unconcerned about his games until they conflicted with the necessities.

The week's Festival performances included the *Gerontius* Prelude and Angel's Farewell, and *Cockaigne* at the Wednesday Evening Secular Concert. On Thursday evening came Brewer's *Emmaus*. The Elgars were able to hear several performances from high up in the Cathedral Triforium. Far away from any audience, it became a favourite Festival place for Edward, who assembled several chairs and stretched his gaunt length over the lot.

A month later he appeared at the Leeds Festival to conduct the *Variations*. During the week there they were surrounded by friends—Dr Buck, the Kilburns, Henry Embleton, Steuart-Powell, Frank Schuster, the critics Arthur Johnstone, Alfred Kalisch, and Herbert Thompson (son-in-law of the Festival secretary Frederick Spark), the orchestral manager Vert who had sent the *Variations* to Richter in the first place.

At Leeds Alice revived an acquaintance thirty-five years in the past. As a girl she had gone to Brussels for piano lessons with a well-known teacher named Kufferath. There she had made the acquaintance of the teacher's daughter Antonia, a singing pupil of Julius Stockhausen, an intimate of Clara Schumann, Brahms, and Joachim: Antonia had become a chosen interpreter of their music. Later she married Edward Speyer, whose musical associations went back to childhood memories of Mendelssohn and Spohr. Now, through his wife's old friend, Speyer came to know another composer: 'At our first meeting Elgar was particularly interested to hear of our close friendship with Brahms, for whose works he entertained the greatest admiration.'[209]

There were other introductions at Leeds. Schuster came up with Harry V. Higgins of the Covent Garden Grand Opera Syndicate. Higgins was planning a Gala Coronation concert at the Royal Opera House to coincide with the august

[208] Quoted in Powell, *Edward Elgar: memories of a variation*, p. 102.
[209] Edward Speyer, *My Life and Friends* (Cobden-Sanderson, 1937), p. 174.

event taking place then. He asked whether Edward might write a new work for the occasion, and Benson's 'Coronation Ode' was mentioned. Edward countered that he might revise *Caractacus*—which had not been revived at Leeds—and promised to send Higgins the score.

G. H. Johnstone came up and asked whether Edward might be prepared to write the major work for the Birmingham Festival in 1903. Despite the disastrous première of *Gerontius*, its significance had gone home. Johnstone gave assurances about the direction of the new Birmingham Chorus, and Richter would still be in general charge. Edward mentioned the subject of 'The Apostles'—for whose preparation there would now be ample time.

Through the Leeds Festival week it became apparent to many people that Edward Elgar might achieve a really big public career. Novellos were concerned at having lost so much of his latest music to Booseys, and from Berners Street Jaeger wrote on 16 October:

Never mind those 'pompous & circumstantial' Marches, (which I see are down for next Tuesday [their first London performance, at one of Henry Wood's Promenade Concerts]. Hurrah!, won't we have some fun!)

When are you going to send us your Norwich Cantata? Now, you will send *that* to us wont you? You *ought* to anyhow, seeing what we are doing for Gerontius here & in Germany; & we shall pay you just as well as Boosey & Full Scores *wont* be any difficulty in future, I guess. I have had another long talk about you to Messrs. A[lfred] & A[ugustus] L[ittleton] & you need fear no worries in future. They have had an eye-opener over the Leeds Festival & I'm sure they'll meet you in *every way* in future. You know when they *have* taken to a man they'll do *any mortal thing* for him as in the case of Sullivan & Stainer.

Only *understand* each other a little better & all will go like a House on fire & you *cant* deal with a better firm. Think over it & believe me, there's a dear. Our Editors (!!!) won't edit *you*! never fear *that*!

Send your Irish Play music also *please*.[210]

Edward responded:

I note all you say & *when ready* will certainly come to 1, B[erners] St.—*say so*.

The Irish play music I am having *copied* out for Wood [to conduct the London première in January 1902] as I may (silentium) utilise it for an opera later.[211]

He and Alice went to Liverpool for Rodewald's première of the two *Pomp and Circumstance* Marches. The success of the first March especially was frantic. But it was nothing to the London première under Henry Wood (which Edward missed through a confusion of telegrams). To the end of his life Wood recalled the effect of the No. 1 March:

The people simply rose and yelled. I had to play it again—with the same result; in fact, they refused to let me go on with the programme. After considerable delay, while the audience roared its applause, I went off and fetched Harry Dearth who was to sing *Hiawatha's Vision* (Coleridge-Taylor); but they would not listen. Merely to restore order, I played the march a third time. And that, I may say, was the one and only time in the history of the Promenade concerts that an orchestral item was accorded a double encore.[212]

Yet all the success and prospect of success only turned Edward in on himself. Staying with Alice through these days at Rodewald's house in Liverpool, they were joined for dinner one evening by a young man who called himself 'Ernest Newman' and had recently become a full-time musical critic. Newman was a keen observer, and more than half a century later he wrote:

I have the clearest memories, visual and mental, of the Elgar of that first period of significant change. He gave me even then the impression of an exceptionally nervous, self-divided and secretly unhappy man; in the light of all we came to know of him in later life I can see now that he was at that time rather bewildered and nervous at the half-realisation that his days of spiritual privacy—always so dear to him—were probably coming to an end; while no doubt gratified by his rapidly growing fame he was in his heart of hearts afraid of the future.

I remember distinctly a dinner at Rodewald's at which Mrs. Elgar tactfully steered the conversation away from the topic of suicide that had suddenly arisen; she whispered to me that Edward was always talking of making an end of himself.[213]

More than a year after *Gerontius* there was no definite progress toward another major work. Birmingham and Norwich and Covent Garden were nebulous; so was the Symphony. When Richter conducted a sumptuous *Cockaigne* in Manchester on 24 October it brought its composer only discontent. Edward wrote to thank Richter:

My own overture was most exhiliarating & I was glad indeed to hear it under your sympathetic and most masterly direction—but, it has taught me that I am not satisfied with my music & must do, or rather try to do, something better & nobler.

I hope the symphony I am trying to write will answer to these higher ideals & if I find I am more satisfied with it than my present compositions I shall *hope* to be allowed to dedicate it to my honoured friend Hans Richter: but I have much to do to it yet.[214]

With three big commissions in prospect, he still felt such uncertainty that he allowed himself to be tempted into yet another short-term project. The pianist Fanny Davies had made an insistent request for a piano piece, and she was a first-class wheedler:

I am *so* disappointed if you can't let me have just a wee 'little Elgar' for my recital on Dec: 2nd?? I wanted to have from 'Purcell to Elgar', I thought it would be such a delicious idea! I can't have a really adequate nice English group without an Elgar in it, now can I? Won't you write me a Study or an Impromptu—I could learn it *very* quickly if I had it—& the Concert is not till December 2nd

—I won't make up my programme for another week in the hope that you might feel you would be able to send me a thing after all————?[215]

[212] *My Life of Music*, p. 154.   [213] *The Sunday Times*, 30 Oct. 1955.
[214] 25 Oct. 1901 (Richter family).   [215] 6 Nov. 1901 (Mrs Anthony Bernard).

It raised the spectre of his childhood piano-playing and all the associations with his father's shop. But the spirit in adversity might seek its way back to formative things hitherto ignored.

He started with two figures, one moving upward in sequential chords, the other moving down in sequential figurations. Both raised faint ghosts of a piano-tuner's keyboard rhetoric.[216] And both resembled the sketching process of his piano extemporizing as it was to be preserved in gramophone recordings of later years. Out of the two figures came numerous busy developments, and one finely sustained melody that might suit the piano less well.

When it came to assembling the piece, his own piano-playing habits emerged again. Instead of using the striking long melody as a primary theme, he raked together the more pianistic material for a brilliant C major opening: it was as if he was still using the piano in his study to sort out and shape random thoughts toward a goal that was not yet clear. The big lyric melody emerged only as a middle section in E flat (thus repeating the key-structure of *Cockaigne*). Then the piece returned to opening ideas, incorporating the big tune in a virtuoso coda. He marked it *Allegro (Concert Solo)*; and recalling Schumann's example, added the subtitle 'Concerto without orchestra'. He gave it his opus number 41, and sent it to Fanny Davies for her recital.

At the recital on 2 December 1901 the piece won much applause, but the critics were less enthusiastic. Fuller-Maitland complained in *The Times*: 'After a very rhapsodical prelude in C, a marked rhythmic figure is carried on pretty regularly, but beyond this it is difficult to detect at first much organic connection between one part and another.'[217] Organic connection there was: but it was masked by a disparity of finish between the contrasting sections.

He had offered the *Allegro* to Novellos for 40 guineas and royalty. Jaeger wrote about that and the *Grania* music on 6 December:

I have had another talk with Mr Alfred L. We want that Funeral March of yours, please. *Score of course* to be printed. What price do you put on it? Say it & we'll say whether it is beyond us.
    And the P[iano] F[orte] piece? Do you still stick to £42 & the Royalty, after the unkind way the mighty London Press has cut it up? We don't mind *them* any more than you do in this matter, but £42 & Royalty *is* high for us: I cannot imagine Schotts paying that, though of course I take your word for it. Express your views, will you?[218]

When Schotts also declined the piece at Edward's price, he expressed his views by quietly putting the manuscript away. And he replied to Jaeger's query with a recursion to the physical complaint which had plagued him a dozen years earlier at the time of his marriage:

. . . My eyes. The doctor wants me to give up as much music-writing as possible! so do the publishers—and the critix—and the public—and the other composers. & so does
                                                                        Ed: Elgar [219]

---

[216] This suggestion I owe to Sir Clifford Curzon.   [217] 3 Dec. 1901.
[218] Elgar Birthplace.   [219] 9 Dec. 1901 (HWRO 705:445:8516).

Through all his writing of big works in the 1890s and the pressure of composing *Gerontius* he had made no mention of the eye trouble. Now it reappeared—just as it had appeared in 1889—after a long spell of frustration at being unable to follow his music to the goal for which he longed. The dark tower might hold the Symphony, or oblivion.

8

Apostolic Successions

ᴵᴺᵛᴏᴶᴄᴶᴴᴵ

EDWARD now had a national reputation. Each success raised higher the standard by which any new work would be judged. Yet the Symphony remained nebulous, while all his big works developed out of programmes and stories and incidents. At the age of 44, with half of creative life behind him and the world knocking at his door, achievement still depended on finding some outside impulse to give large form to his music.

When some new sketch books arrived from Novellos at his request, one of his first entries began from a specific memory. On a page headed 'Ynys Lochtyn' he set down a number of ideas developing the fall of a minor 3rd which had reached him through the summer atmosphere at Llangranog. The ideas were good, as he wrote to Jaeger on 17 November 1901: 'I want a sketch book bound in *Human skin* to write some of the things I am doing now.'[1] Into what project might they fit?

The promised Birmingham Festival commission came in a letter from G. H. Johnstone written on 1 December; once again Johnstone offered himself as negotiator for the publication of a new Festival work. Meanwhile Ernest Newman was urging precisely the opposite course in an article on Edward's art:

He has been held back by the ideals of the musical generation now on the verge of extinction. Look at the bulk of his Festival work alone, for example, and you might be forgiven the opinion that he is not the leader of the new school but the solitary noteworthy survivor of the broken army of the old . . .

I say nothing of the reasons, other than artistic, for Dr. Elgar writing so many cantatas and oratorios. If the road to success lies through the bog of the Festivals, a composer must needs take that path; but he cannot do so without wasting a good deal of his time and strength as an artist, and without a quantity of undesirable substance clinging to him for some time after . . .

There can, of course, be no question that the *Gerontius* is not only the finest work ever produced at an English Festival, but the finest work Dr. Elgar has written. It is, indeed, the real Elgar, the Elgar that has been made by his heredity, his reading, his reflection upon life, the Elgar one knows in the flesh and the spirit—which is what we cannot say throughout the Elgar of any of the previous cantatas, even *King Olaf*. But the *Gerontius* is so great because Dr. Elgar has had the rare good fortune to come into contact, at the very height of his powers as a musician and as a thinker, with a poem peculiarly fitted to

---

[1] HWRO 705:445:8514.

stimulate and become part of him. I very much doubt whether he will be so fortunate again.[2]

Edward should abandon choral writing altogether, Ernest Newman felt, for his true medium in the orchestra.

The critic's estimate of *Gerontius* was vociferously confirmed abroad when Edward and Alice went to Düsseldorf for Buths's German production of the work on 19 December 1901. Jaeger went with them to represent Novellos and to report the event for *The Musical Times*:

Dr. Elgar was enthusiastically called upon the platform after each part—a rare honour in Düsseldorf—and heartily cheered by the huge audience of some 2,500 people . . . At the conclusion the chorus presented him with a huge laurel wreath and the orchestra greeted him with the cacophonous fanfare called a 'Tusch' . . .

Throughout his stay in the city, Dr. Elgar was received with the utmost kindness, and with real admiration for his genius. To hear him being addressed as 'Verehrter Meister' by the many musicians who had come from near and far to hear his work was a new and pleasant experience.[3]

Edward seized the chance to make his point with Novellos:

As to the performance, it completely bore out my idea of the work: the Chorus was very fine & had only commenced work on Nov. 11—this disproves the idea fostered in Birmingham that my work is *too difficult*. The personnel of the Chorus here is largely amateur & in no way, except in intelligence & the fact that they have a capable conductor, can they (or it) be considered superior to any good English Choral Society.[4]

But at Berners Street the battle of *Gerontius* was already won. Henry Wood, who now conducted the Sheffield Festival, had offered to alter the 1902 Sheffield programme to include *Gerontius*—but only on condition that he had a printed full score to mark and to keep. If a busy conductor took time to prepare a large new work, he must hope to give it more than a single performance: and then his own marked score was essential to avoid duplication of effort. Novellos capitulated and prepared to engrave the *Gerontius* full score. In fact, the directors were considering a suggestion from Jaeger that they should offer Dr Elgar a blanket contract to publish everything he should write in future.

As for Jaeger himself, the prospect darkened. The throat treatments had no success.

Craeg Lea, Malvern.

Jan 13, 1902.

My dear Jaeger:

. . . I'm awfully distressed to hear you are again ill & terribly disappointed that our operation was not, finally[,] successful. I wish you all the good things & only wish I could put this right for you; do tell me what can be done & how & when? I think hourly about you & worry about it until I'm sick . . .

---

[2] *The Speaker*, 21 Dec. 1901, p. 231.   [3] *The Musical Times*, 1 February 1902.
[4] 21 December 1901 (Novello archives).

As to your long letter: my things are successful among musicians but the public don't buy them . . . My music does not arrange well for the piano & consequently is of no commercial value. If I had a free mind I shd. like to write my chamber music, & symphony &c.&c., on all of which forms of art Providence has laid the curse of poverty. Bless it! So I don't see how any publisher cd. be persuaded to endow an *artistic* writer: why should they? If I write any stuff at all somebody is bound to publish it—& they all say it's no good—they are proud to do such stuff 'cos it's in all the high-class programmes &c.&c. But Providence, as I've often told you, is against all art so there's a satisfactory end.

Yours ever

Ed. E.

I shd. be glad to consider anything you have to say of course & thank you a 1000 times for thinking of it.[5]

It was as if the prospect of a blanket contract for the future gave fear instead of hope. Again his music went backward. He looked out the *Pie Jesu* written for old William Allen's funeral in 1887, re-arranged the music as an *Ave Verum*, and sent it to Novellos as his 'Op. 2'. Then on 14 January he worked at two instrumental sketches made years earlier for a 'Children's Suite'.[6] The first, labelled 'Sorrowful', began with a G minor figure that seemed now to echo the 'Enigma':

The second of the two pieces was in the old answering key of G major. He completed the two, orchestrated them, and gave them the title of Charles Lamb's essay 'Dream Children'. One passage was quoted above the music:

And while I stood gazing, both the children gradually grew fainter to my view, receding, and still receding till nothing at last but two mournful features were seen in the uttermost distance, which, without speech, strangely impressed upon me the effects of speech: 'We are not of Alice, nor of thee, nor are we children at all . . . We are nothing; less than nothing, and dreams. *We are only what might have been.*

The dreams were brought close to home by the name of Alice.

In London for Henry Wood's rehearsal of the *Grania and Diarmid* music on 17 January, Edward found himself approached by a twenty-four-year-old violinist in the Queen's Hall Orchestra who introduced himself as William Henry Reed:

I was so thrilled by the music, and by what was to my ear the newness of the orchestral sound, that I left my seat among the first violins and followed him out through the

⁵ Elgar Birthplace.

⁶ BL Add. MS 47903 fo. 66. The music was identified as old work in a letter of 6 June 1905 to F.G. Edwards (BL Egerton MS 3090 fo. 85–6).

curtains until I caught him up half way up the stairs. Breathlessly I begged him to excuse me for thrusting myself forward, but I was anxious to know whether he gave lessons in harmony, counterpoint, etc.

His answer was characteristic: 'My dear boy I don't know anything about those things.'

. . . It was soon very evident that Elgar was not annoyed by my temerity in running after him that day, for afterwards, whenever he came to conduct, he never failed to speak to me on his way to or from the conductor's desk, always finding something friendly and encouraging to say.[7]

Reed was a youngster then, and as English as roast beef. Yet he bore a physical resemblance to Jaeger—slight of stature, a shock of hair and moustache, and an air almost of worship to reassure Edward at every moment. If Jaeger was truly doomed to disappear, here might be another such encourager.

*   *   *

Ivor Atkins was keen to do *Gerontius* at the Worcester Festival in September 1902. The Cathedral clergy at first objected, and finally made it a condition that the text be purged of any references contrary to Church of England doctrine. Edward secured the grudging consent of Cardinal Newman's executors at the Birmingham Oratory. He was present when Atkins announced the projected performance to a meeting of Festival Stewards on 5 February. Two days later, when Atkins came to dine at Craeg Lea, Edward showed him the sketch of a new *Pomp and Circumstance* March headed 'I. A. Atkins gewidmet'.[8]

The new March was laid aside when further news came of the Covent Garden project. Harry Higgins had looked at *Caractacus*, and did not see in it the makings of an opera. But he was very keen that Edward should set Arthur Benson's 'Coronation Ode' on a big scale for the Coronation Gala Concert at the Opera House in June. Benson agreed to expand the poem to meet this scheme. Edward asked for Henry Coward's Sheffield Choir to sing it. The publication of such a 'popular' work should be offered to Booseys. Higgins then secured the approval of the King for the whole plan.

Already there was an idea for the music. At a concert that included the *Pomp and Circumstance* Marches on 26 November Edward had sat with Clara Butt. She afterwards reminded him: '*While* listening to the tune in Pomp & Circumstance [No. 1] I asked you to write something like it for me & after a little talk & persuasion on my part you said "You shall have that one my dear." '[9] He thought of a triumphant final section for the *Coronation Ode* to employ Clara Butt as soloist with the full chorus and orchestra.

[7] *Elgar as I Knew Him*, pp. 21–2.   [8] MS in possession of Wulstan Atkins.
[9] Undated letter at the Elgar Birthplace, written to protest the allegation that the Trio tune had been fitted to words 'at the suggestion of King Edward'. I can find no evidence that Edward VII heard *Pomp and Circumstance* No. 1 before 5 Feb. 1902, when he attended a concert by the Royal Amateur Orchestral Society. He had not then met Elgar.

He wrote to Arthur Benson to ask whether words might be written to fit the *Pomp and Circumstance* Trio tune and Benson had replied: 'I will try & write you . . . a finale on the lines you indicate—though the metre is a hard one—if you could string together a few nonsense words just to show me how you would wish them to run, I would construct it, following the air closely.'[10] Benson himself was a keen amateur musician, and he proved a skilled and willing collaborator, ready to entertain any hint as to subject or metre. Within a week he sent verses for a Finale beginning 'Land of hope and glory . . .'[11] Other sections followed quickly through constant correspondence between poet and composer.

By the middle of February 1902 Edward began to write the music. The opening section was invocatory: 'Crown the King with life!'—a prayer for the man coming to the throne in his sixtieth year. Edward made the tonic key E flat major: he wanted to add a military band to the orchestra and organ. E flat major was also the tonality of many recent and relevant thoughts—the opening of *Pomp and Circumstance* No. 1, the centre of *Cockaigne* and the piano *Allegro*, the big slow melody of 'Nimrod' in the *Variations*, the goal of *Caractacus* in the culminating 'Britain' motive.

The primary figure of the *Coronation Ode*,

seemed to translate the 'Enigma' theme to the newer tonic. A second figure, purely instrumental, sounded a new variant of descending steps turning upward only at the end:

The sudden drops through every third step and the appoggiaturas over bar-lines gave it a quality of 'Recessional'. A third figure shaped the march-triplets of *Pomp and Circumstance* No. 2 and *Grania and Diarmid* to a sequence that also moved downward:

From these ideas he shaped a big orchestral introduction and an imposing

[10] 3 Dec. 1901 (HWRO 705:445:3295).   [11] 10 Dec. 1901 (HWRO 705:445:3296).

choral movement with parts for four solo singers. The conclusion was set to a hushed *Pomp and Circumstance* No. 1 Trio:

> All that hearts can pray,
> All that lips can sing,
> God shall hear to-day;—
> God shall save the King!

Benson's next section made a sharp reminder of the South African War:

> Britain, ask of thyself, and see that thy sons be strong,
> Strong to arise and go, if ever the war-trump peal . . .

It was not all 'pomp and circumstance' either, as the poet drew his listeners within range of horrors:

> Under the drifting smoke, and the scream of the flying shell,
> When the hillside hisses with death,—and never a foe in sight.

But these unseen foes were put to flight by Edward's music, strutting first in triplet declamation and then settling down to a good square tune in C minor for bass soloist and men's chorus, with episodes in G minor and A flat. The melody covered the long lines of verse, but its elaborate formula was hard to vary. The result was close to light opera. (Edward German's reminiscent *Merrie England*, being finished at that moment, contained its patriotic baritone solo with chorus 'The Yeoman of England'.)

The light-opera atmosphere sounded strangely again through a duet and quartet to set the achievements of art side by side with a pure spirit:

> Fiery secrets, winged by art,
> Light the lonely listening soul . . .

The lines ought to have touched Edward's best music; yet working through March 1902 under pressure to complete his work, there was little time to wait for subtler ideas.

A short prayer for 'Peace, gentle peace' introduced the Finale, 'Land of hope and glory'. This was built entirely of material from the opening chorus, beginning with *Pomp and Circumstance* and leading back to a full reprise at the end. It was quickly achieved, immensely impressive, ideally suited to the occasion.

And yet there was something of 'Recessional' again in the subdominant patterning of key relations through the last half of the *Ode*. Beginning with C major in the duet, the music moved through F major ('Peace, gentle peace'), B flat (opening of 'Land of hope and glory') to E flat for the final return. So it matched two lines near the centre of the final chorus:

> Tho' thy way be darkened, still in splendour drest,
> As the star that trembles o'er the liquid West.

It might stand for England's position in the west of Europe—or for Edward's home countryside in the west of England. If his music found a wide public then, it was because his own apprehensions in the air all around those years were moving just ahead of the general mood.

The vocal score of the *Coronation Ode* was finished on the last day of March 1902. Then it struck Arthur Benson that there was nothing referring to the Queen. It was a serious omission, and on 2 April he sent a short lyric built on Queen Alexandra's Danish origin:

> Daughter of ancient Kings,
> Mother of Kings to be,
> Gift that the bright wind bore on his sparkling wings,
> Over the Northern Sea![12]

Edward set the verses in close-moving four-part harmonies under an E flat melody rising up at the end to C minor: its disarming simplicity raised Benson's capable verses to a level of genius.

Boosey's paid Edward £100 for the score, with a 5*d*. royalty on every copy sold. And when they saw 'Land of Hope and Glory', their experience with ballads told them that it would make a hit. At their request, author and composer revised it as a solo song for separate publication. Clara Butt was to introduce it at a 'Coronation Concert' in the Albert Hall a week before the *Ode* at Covent Garden.

Writing the *Coronation Ode* had displaced other projects. Edward decided he could not respond to the blandishments of George Moore over an Irish opera. He cobbled the small fragments from *Grania and Diarmid* into a single piece of 'Incidental Music', joined it to the Funeral March and the tiny song, and sold the lot to Novellos for £100. Other opera proposals from Laurence Binyon and Laurence Housman fared no better. One way and another, none of the stories seemed right for him. His own idea of setting Maurice Hewlett's novel of chivalric romance, *The Forest Lovers*, met the same fate because he was unable to work with the owner of the dramatic rights.[13]

All this time the Norwich Festival conductor Randegger had been pressing for news of the cantata *The High Tide* which Edward had promised for their Festival in the autumn. Before the end of 1901 Randegger had wanted to know the solo voices required so as to engage soloists: Edward specified soprano and possibly bass. In February 1902 Randegger asked for a final title and approximate length. But little work had been done. Before the end of April—only six months before the Norwich Festival—Edward had to tell Randegger and Novellos that the cantata could not be written: his eyes were troubling him again.

He spent the month of April orchestrating the *Coronation Ode* and a new

[12] HWRO 705:445:3306.
[13] Burley, *Edward Elgar: the record of a friendship*, pp. 112–13.

arrangement of the National Anthem which Novellos had requested as the only possible *riposte* to Booseys' *coup* of the *Ode*. He agreed with Harry Higgins to introduce the new arrangement at the Covent Garden State Concert just before the *Ode*. It made the ideal opening for a concert attended by the entire Court, foreign ambassadors, and their legations.

The demand and the prices for seats were both unprecedented. No complimentary places could be found even for Alice or Arthur Benson, and Edward purchased two stalls at twenty guineas each.[14] Twenty guineas were the wages of a top grade residential household servant for a year then. There was no shortage of money in the Elgar household when the occasion was imperious.

On 10 May he conducted the Worcestershire Philharmonic in a miscellaneous concert including Parry's *The Lotos-Eaters* and *Reigen*, a choral work by the Mainz conductor Fritz Volbach, who had been greatly stirred by the German production of *Gerontius* in December. Now a second German performance of *Gerontius* was due in Düsseldorf at the Lower Rhine Festival.

Edward and Alice crossed on the night of 16 May. Several friends went over—Jaeger to represent Novellos again, Rodewald, Henry Wood, the critics Alfred Kalisch and Arthur Johnstone. The Festival opened with a splendid B minor Mass. Next morning Edward found Richard Strauss overrunning his rehearsal time for the Liszt *Faust Symphony*, which was to share the programme with *Gerontius*. 'E. very angry,' noted Alice: 'at last hurried unsatisfactory rehearsal [under Buths]. E. called up after each part.'

The performance was a huge success. Many years later Henry Wood wrote:

The astounding impression the performance made on everyone has remained in my memory ever since. It is quite impossible to describe the ovation dear Elgar received; he was recalled twenty times after the end of the first part.[15]

Alice went to the diary in triumph: 'Great Fest Essen followed. R. Strauss made beautiful speech about Meister Elgar, no end of toasts. In bed at 3.' Edward wrote to Alfred Littleton:

For me, I understand, the thing was a triumph! but I feel rather dazed at the success & will think of it when six months more hard work have rather dulled the memory of these wonderful days . . . Richard Strauss, who never speechifies if he can help it, made a really noble oration over Gerontius—I wish you could have heard it—& it was worth some years of anguish—now I trust over—to hear him call me Meister![16]

But Strauss's remarks had a sting in the tail. *The Times* correspondent reported:

A short speech was made quite unexpectedly by Herr Richard Strauss, in which he deplored that England had hitherto not taken her proper place among musical nations, because of the want of Fortschrittsmänner—that is, of men who represent the forward movements in art at any given epoch—ever since the period of England's musical

---

[14] HWRO 705:445:2125.   [15] *My Life of Music*, p. 249.
[16] Friday 21 May 1902 (Novello archives).

grandeur in the Middle Ages. The creation, however, of a work like *The Dream of Gerontius*, he added, showed that the gap had been filled and that a day of reciprocity in music between England and the rest of Europe was dawning. 'I raise my glass to the welfare and success of the first English progressivist, Meister Edward Elgar, and of the young progressivist school of English composers.'[17]

Strauss's short history of English music had ignored Parry, Stanford, and all the rest of Edward's older colleagues. It was the sort of publicity that did more good abroad than at home. An English composer of a younger generation, C. W. Orr, recalled:

The phrase caused some fluttering in the academic dovecots over here, where Elgar was by no means *persona grata*, originality of any sort being held suspect at our schools and colleges in those days, but it had the effect of stirring up the torpid imagination of the average musician and amateur, who felt that Strauss, while he might be the stormy petrel of musical Europe, was nevertheless someone to be reckoned with . . .[18]

After a fortnight's holiday with Rodewald in Eisenach (the birthplace of Bach) and Dresden, Edward and Alice came back to London on 4 June to find Richard Strauss already arrived for a concert of his own works. After the concert the two composers supped together. The music of Strauss made its sumptuous impression, as Edward wrote to Nicholas Kilburn: 'Strauss is absolutely great—wonderful and terrifying but somewhat cynical—his music I mean. *He* is a real clever good man.'[19] None the less, as Edward confided to Jaeger on the eve of attending a Mozart performance at Covent Garden:

Tomorrow (Deo gratias) I am to hear 'Don Giovanni' (after a fashion) this even Strauss† cannot shake in the depths of the heart & soul of your friend

Edward Elgar

†The Wagner shaker.[20]

Then he plunged into preparing the *Coronation Ode* for its State performance on 30 June. On the 12th he went to Sheffield to rehearse the Choir: they received him as a conquering hero. In London two days later he rehearsed the orchestra at Covent Garden. The military band rehearsed their part. He took Melba, Kirby Lunn, Ben Davies, and David Ffrangçon-Davies through the solo music. The Welsh baritone Ffrangçon-Davies made an immense impression: he was soon to become an intimate friend and another chosen interpreter.

Back in Malvern, Edward sought his own countryside for the little time he had before the final combined rehearsal and the performance. One day after another found him cycling farther into the deep peace of country lanes in early summer. On 21 June he set out for a long day, heading towards Bredon. A

[17] 23 May 1902.   [18] 'Elgar and the Public', *The Musical Times*, 1 Jan. 1931, pp. 17–18.
[19] 7 June 1902.   [20] 18 July 1902 (HWRO 705:445:8545).

series of coincidences that morning sent his mind back over thirty years to when he had worked as a boy in William Allen's law office. He would remember this morning for the rest of his life, recalling it for his biographer Basil Maine who wrote:

At a cross-roads he was intending to turn to the left but was surprised by the sudden appearance of a cart-load of hay drawn by three horses, and was compelled to take the right-hand turning. Farther on, he arrived at another cross-roads and was about to take the left-hand turning when another cart-load of hay, again drawn by three horses, forced him to the right. He cycled on and slowly began to realise he was on the road that led to the village where Allen was buried. Then it was that he resolved to visit the grave. He arrived at the grave-yard and there found the sexton's wife.

'Can you help me to find the grave of William Allen?' he asked.

For a moment, the women, with a scared look upon her face, stood dumb and still. Then she said: 'Funny you should be askin'. Why the stone o'that grave fell down and broke this very mornin'. We were wonderin' if there were any relations or friends we could write to about it.'[21]

He thought no more of it. Writing to Jaeger next day he felt only the perversity of having to leave the summer countryside for his grand occasion:

I *hate* coming to town—shall miss the hay making I fear. Had 50 miles ride yesterday amongst the Avon country—Shakespeare &c.&c. Oh! so lovely but solus 'cos I can't find anybody here foolheaded enough to eat bread & cheese & drink beer—they've all got livers & apparently live in the country 'cos they can't afford to be swells in a town. Oh! lor.[22]

On 24 June, the day before he was to go to London, Miss Burley joined his cycling. They went into the country west of the Malvern Hills and found themselves in the hamlet of Stretton Grandison. It was quieter than ever, Miss Burley recalled, with the Coronation just two days away:

People were already going to London in great numbers.[23] It had been a hot day and after our usual examination of the church we managed to get some tea at the inn . . .

Edward was already very full of the affair, the importance of the occasion and of the Court Dress he was to wear, while conducting the Ode. On an impulse I said as we ate our simple tea, 'Does it strike you that the King is going to have an extremely trying time these next few days?'[24]

We had hardly finished tea and were preparing to go when the landlady rushed in saying 'Oh! Sir. The Coronation is put off, the King is ill and is to have an operation at once. The news is just in at the Post Office.'[25]

Next day he wrote to Jaeger:

Don't, for heaven's sake, *sympathise* with me—I don't care a tinker's damn! It gives me three blessed sunny days in my own country (for which I thank God or the Devil) instead

---

[21] *Life*, pp. 115–16.   [22] 22 June 1902 (HWRO 705:445:8542).
[23] Quoted in *Letters to Nimrod*, p. 165.
[24] *Edward Elgar: the record of a friendship*, p. 158.
[25] Quoted in *Letters to Nimrod*, p. 165.

of stewing in town. *My* own interest in the thing ceased, as usual, when I had finished the M.S.—since when I have been thinking mighty things![26]

Was Jaeger to understand by that the project for Birmingham—or an honour more immediate?

The King slowly recovered. Edward's name did not appear in the Coronation Honours List—though Parry received a baronetcy and Stanford a knighthood. Vernon Blackburn commented in *The Pall Mall Gazette*:

Sir C. Villiers Stanford probably gets what many people would have expected him to get. We are sure that we have never hesitated to praise him for his many excellent qualities, but some may have doubted if Stanford has ever done a really *big* work? . . . We should approve with more complete whole-heartedness if we did not rather miss one name upon whose inclusion some had very confidently counted—Edward Elgar.[27]

* * *

Edward turned his thoughts to the new oratorio due for Birmingham in fifteen months' time. 'The Apostles' was a subject to attract him in many ways. First there was its wide public appeal: its source in the New Testament could touch Catholic and Protestant alike. If he consulted the Anglican librettist of *The Light of Life*, Edward Capel-Cure, he might avoid the doctrinal trouble which beset the *Gerontius* text in its acceptance for the Worcester Festival.

The biblical subject offered Edward a chance to shape his own libretto. With advice from Capel-Cure, he might make it for himself—forming speeches for characters and even whole scenes of psychological interplay by assembling quotations almost line by line from diverse Old and New Testament sources. Thus the libretto could grow with its music, and the whole result would be as intensely the expression of a single spirit as the operas of Wagner.

Then there was Edward's private identification with the story itself. Of all biblical subjects, the Apostles of Christ offered the clearest story of human inspiration from its beginning. It was the story of a promise. So it would open to Edward's music the Christian allegory of a creative life. And so it might engage his own art to a faith in something beyond itself.

The process of defining religious faith through personal art was arduous and long. But here again the subject of 'The Apostles' tempted him. It was literally a story without an end, for where was the end of Christian Apostlehood? It was a continuous laying on of hands. Thus the story itself made a gigantic, unending sequence—a sequence to guide his music through many years, perhaps through all the remaining years of creative life. It must make more than a single oratorio. He saw in it the material for a trilogy—a grand construction to match Wagner's *Ring des Nibelungen*, on a theme sacred to all the Western world.

A subject that was really endless could have no clear hero. So it could

[26] 25 June 1902 (HWRO 705:445:8543).   [27] 30 June 1902.

become the goal of this immense project to discover what the end of life should be. If that invoked Gerontius again, it invoked still more insistently the 'Enigma' formula: beginning with a question, it would consult the 'friends pictured within' the first Christian story. Then, if this ultimate Enigma was solved, the solution would produce its own hero—some immanence of the creative spirit as unforeseeable at this project's beginning as the triumphant 'E.D.U.' was unforeseen in that first dispirited piano extemporizing at Forli three and a half years earlier. It was soon after the earlier extemporizing, indeed, that his close thinking about the Apostles had found an opening in Judas.

Any hero emerging from the end of an 'Apostles' trilogy would have to equal or overtake the power of an earthly Christ, wherein the Apostles had their beginning. Edward remembered the 'Civitas Dei' assemblage made for him by Minnie Baker in 1894. The whole matter might end with nothing less than the Last Judgment. Here again the darker side of inspiration offered something definite. St. Augustine had opened the final vision with the horrors of Judgment and Apocalypse. So the story of Judas could anticipate the coming of Antichrist. Edward entertained the notion of actually naming the final work of the trilogy after that dark eminence.[28]

Yet the whole undertaking—whatever the attraction of its beginning, however splendidly it might be prolonged—was bound in the end to exact the heaviest price. For it set at wager the whole of Edward's creative future; and this was a wager which his art seemingly could not win. If he succeeded in defining an ideal faith by his induction, then his own life would have achieved its fulfilment and would need nothing more from his art. If he failed and no hero emerged, than what conceivable subject could remain to invite any further music?

'I am now plotting GIGANTIC WORX,' he wrote to Ivor Atkins on 2 July 1902.[29] He discussed publication with G. H. Johnstone of the Birmingham Festival, and Johnstone wrote on his behalf to Alfred Littleton to sound Novello's interest. Littleton asked for terms. Johnstone replied by invoking comparison with the greatest oratorio première in Birmingham's history—Mendelssohn's *Elijah* in 1846:

I have no doubt that Elgar's new work will be a most important one & should prove a great success. The subject is a very good one, & I hope for the Birm Festival that it may prove a second 'Elijah'. I am very anxious to do all I can for Elgar, & my proposition is that you should buy the work for a Thousand pounds with a small royalty on each copy sold, & I feel convinced that it will be a most profitable speculation.[30]

Littleton protested vigorously.

Edward meanwhile made a pilgrimage to Bayreuth. He went in the genial

[28] BL Add. MS 47906.				[29] MS in possession of Wulstan Atkins.
[30] 3 July 1902 (Novello archives).

company of Archibald Ramsden, a piano dealer who had joined with the publisher Joseph Williams to found the 'U. B. Quiet' Club for musical friends (the name signified that no 'shop' was to be talked). At Bayreuth they heard *Der fliegende Holländer*, *Parsifal*, and the first three operas of the *Ring*.

He returned on 29 July 1902, and two days later Alice noted: 'Began to be very busy collecting material.' Johnstone wrote from Birmingham:

I am in negotiation with Novellos as to the publishing of 'The Apostles' & they ask to have some idea of the work . . . How are you getting on with it & how soon do you think you can finish the Scoring?[31]

Edward replied:

After the holidays I will have some M.S. for you or the publishers to see.[32]

They did not go to Birchwood that year. Instead Edward took Alice and Carice to stay a fortnight with Rodewald in his holiday cottage at Saughall, near Chester. The Bantocks were there, with Ernest Newman and his wife. No other children were present, but Carice made great friends with Rodewald's cat Sam while the adults were occupied. At the end of it Edward cycled all the way back to Malvern, staying overnight at Shrewsbury, while the others made their way by train.

He did not produce any *Apostles* manuscript for Johnstone, but in late August 1902[33] he set two poems by A. C. Benson as solo songs to meet a request from Booseys. The publishers were looking for a success to follow *Land of Hope and Glory*. But the new songs were far from that. *In the Dawn* celebrated a perfect love: the desolation of parting was looked at, 'But to have loved her sets my soul Among the stars.' The second song, *Speak, Music*, made the clearest statement of why Edward sought his own music:

> Song, take thy parable,
> Whisper that all is well,
> Say that there tarrieth
> Something more true than death,
> Waiting to smile for me; bright and blest.

The two songs were soon completed. Booseys paid 25 guineas each for them and a royalty. They were marked as his Op. 41—the number given originally to the piano *Allegro* for Fanny Davies.

These songs, setting side by side his greatest debts to the love and encouragement of his mother, were the last music he completed in her lifetime. She had grown more and more frail, and after her eightieth birthday earlier in the year she went rapidly downhill. One day in late August he cycled over to Worcester with Miss Burley: 'I waited for him at a tea shop while he paid her

---

[31] 1 Aug. 1902 (Novello archives).

[32] Quoted by Johnstone in his letter of 6 Aug. 1902 to Alfred Littleton (Novello archives).

[33] It has been erroneously supposed that the two songs were written in 1901. Alice Elgar's diary makes the composition dates quite clear.

what was to be his last visit. When he joined me he was so deeply moved that he could hardly speak, for he knew that she was dying.'[34] She died quietly on 1 September 1902, and was buried three days later beside the two sons who had preceded her by more than thirty years.

For the Worcester Festival in the following week, Alice had found them a house in College Green, facing the Cathedral. It was now called Castle House, and when Edward entered it the past came up before his eyes. Castle House was where the Davisons had lived forty years ago—the house where he had extemporized in childhood for the ladies and gentlemen of 'old-world state'. Now his music had a central place in the Worcester Festival. *Sursum Corda* was down for the Sunday Opening Service. The new National Anthem arrangement inaugurated the Festival performances on Tuesday morning. *Cockaigne* and three *Sea Pictures* shared the Wednesday Evening concert with Strauss's *Death and Transfiguration*. On Thursday came *Gerontius*. His mind went back over the years of reading shared with his mother, and at Castle House he sketched a musical portrait of Shakespeare's Falstaff—an *Allegro* in C minor.[35]

The *Gerontius* performance on Thursday morning, conducted by Edward in mourning black, made a deep impression. Granville Bantock wrote to Ernest Newman (whom he still addressed by his original name, William Roberts):

Never have I experienced such an impression before, as I did on hearing 'Gerontius' this morning in the Cathedral. If Elgar never writes another note of music, I will still say that he is a giant, & overtops us all. His music moved me profoundly. Believe me, my dear Will, although Elgar & I look at music through widely different spectacles, his 'Gerontius' is beyond all criticism or cavil. It is a great great work, & the man, who wrote it, is a Master, and a Leader. We were all deeply affected, and gave way to our feelings. While Elgar was conducting, the tears were running down his cheeks. I want to hear nothing better. I have felt as if transfixed by a spike from the crown of my head to my feet. Once on hearing Parsifal at Bayreuth, when the dead swan is brought on, & today, at the words 'Novissima hora'.[36]

The title role was sung by John Coates: his dark ringing tenor gave *Gerontius* a new urgent drama. The part of the Angel was memorably given by Muriel Foster—who had sung the *Sea Pictures* the night before. The performance had been marred only by the Cathedral clergy's insistence on deleting Roman Catholicisms from the text—leaving the singers to get over the lapses as best they might. The *Pall Mall* critic Vernon Blackburn reported this *jeu d'esprit*:

A Roman Catholic priest of known wit, who was present this morning at the performance, suggested that instead of omitting the words 'In Purgatory', the difficulty might have been better solved by simply putting 'Fried souls'.[37]

But *Gerontius* had drawn the largest audience of the *Festival*. Despite heavy rain, *The Westminster Gazette* reported:

[34] *Edward Elgar: the record of a friendship*, p. 161.
[35] Sketchbook II, p. 13.   [36] 11 Sept. 1902.
[37] *The Pall Mall Gazette*, 11 Sept. 1902.

Queens of Sheba in great numbers, their regalia a best hat and a netted bag of sandwiches and octavo scores, came this morning to hear the wonders of King Elgar's wisdom. The Cathedral is said to seat three thousand . . . persons, and there were some who had to stand. By popular acclaim, Mr. Elgar has certainly been crowned.[38]

After the performance Alice gave a luncheon at Castle House. One of the guests was Henry Walford Davies, a young composer whose short oratorio *The Temple* was included in the Festival at Edward's instance. Each of the three Cathedral organists received a copy of the newly printed full score of *Gerontius* as a present from the publishers, with an inscription from Edward. The Speyers were also there, returning home that afternoon as they did not care to hear anything after the unique experience of *Gerontius*. Edward dedicated the new song *Speak, Music* to Mrs Speyer—and thus enrolled himself again in the tradition of Schumann and Brahms. Alice wrote in the diary: '*Crowds* of people came all the aftn. & to tea—& in evening . . . A most wonderful day to have had in one's life. D.G.'

Other wonderful days followed. On 14 September Edward conducted a concert of his own music with Rodewald's orchestra in Liverpool. His host was more delightful than ever. One evening, in company with Ernest Newman, he took Edward across to a pub much frequented after concerts, opened the door to a private room, and in awful secrecy inducted him into a 'Skip the Pavement' Society—named from the need to get there before closing-time after a concert. 'S.T.P.' proved a great source of fun—scrolls of admission, Monteverdean badges, and a 'Curse Book' of rules and transgressions, kept for a time by Edward himself.

Then came rehearsals for the Sheffield Festival. Sheffield was giving not only *Gerontius* but the *Coronation Ode*—in what would now be its première performance. The *Gerontius* on 2 October was a huge success: just as at Worcester, the demand for seats was greater than for any other Festival event. Again Muriel Foster sang the Angel. And on the afternoon of the same day she sang in the *Coronation Ode* under Edward's direction. The other soloists were John Coates, Ffrangçon-Davies, and a young soprano who made a deep impression, Agnes Nicholls. Alice noted: 'Magnificent performance of the Ode—immense enthusiasm.' The newspapers agreed, though some of the critics felt the inspiration unequal, and many did not like the *Land of Hope and Glory* setting. But in that opinion they were soon to be overborne.

At Sheffield the Elgars met several friends of Frank Schuster's—the Duke of Norfolk, Charles Stuart-Wortley (MP for Sheffield), and his wife. Mrs Stuart-Wortley was a daughter of the painter Millais, a brilliant and deeply sympathetic woman with a fine understanding of artists. The Catholic Duke of Norfolk suggested to Mrs Stuart-Wortley that London should hear *Gerontius* in the new and as yet unconsecrated Westminster Cathedral; Mrs Stuart-Wortley promptly passed on the suggestion to those who could implement it.

[38] C. W. J., 12 Sept. 1902. The actual attendance figure was 3130.

On 7 October Edward conducted the *Coronation Ode* at Bristol.[39] Again it scored a big success. On the 26th he conducted the first London performance at the Queen's Hall. W. H. Reed witnessed the scene from his place in the first violins:

At the close the enthusiasm was such that Elgar was brought five times to the platform; then a voice from the gallery was heard: 'Let's have the last part again.' Quiet was only restored when Robert Newman (the promoter of the Queen's Hall concerts of the day) came forward to express the composer's gratitude for the splendid reception of his work, and to beg the audience to allow the programme to proceed, at the same time stating that on the following Sunday afternoon, the 9th [November], which happened to be the king's birthday, the *Ode* would again be performed under the composer's direction.[40]

And so it was—with the same enthusiasm. He conducted it again with Embleton's Choral Union at Leeds, and when he returned to London to take the whole of a Saturday Concert for the ailing Henry Wood on 22 November, the *Ode* found its place in yet another programme; the other works Edward conducted that day were the Grieg Concerto (played by the composer's friend Arthur de Greef), Humperdinck's tone poem *Dornröschen*, and the Love Scene from Richard Strauss's *Feuersnot*.

Growing celebrity brought invitations. One came from Frederick Spark to write a work for the Leeds Festival in October 1904.[41] Edward accepted in general terms, without specifying the type of work he would write.[42] Spark pressed him for a definition.[43] At last Edward hinted that he might produce the long-deferred Symphony.

There were several smaller projects. Fifteen months earlier he had had a visit from the Revd Charles Vincent Gorton, a Canon of Manchester Cathedral and Rector of Morecambe. Canon Gorton had founded a competition festival at Morecambe, and he had offered Edward £100 for a new choral work. Edward had liked Canon Gorton and the idea of his festival; and at last he agreed to write a short part-song to verses by the Manx poet Thomas Edward Brown, *Weary Wind of the West*.

Then he chose five translations from *The Greek Anthology* to make a set of part-songs for male voices. At the centre of the little cycle was a picture of natural inspiration to touch some of Edward's deepest memories:

> After many a dusty mile,
> Wanderer, linger here awhile;
> Stretch your limbs in this long grass;
> Through these pines a wind shall pass
>
> . . . .

---

[39] As an example of Elgar's conducting fees at this time, letters to Vert (who acted as his concert agent) show that he got 20 guineas for the Bristol engagement, which also included *Cockaigne* and a new orchestration of *The Pipes of Pan*, sung by Andrew Black. (Ibbs & Tillett archives.)

[40] *Elgar*, p. 69.

[41] 27 Oct. 1902 (HWRO 705:445:5976).   [42] 28 Oct. 1902 (HWRO 705:445:5975).

[43] 30 Oct. and 18 Nov. 1902 (HWRO 705:445:5977, 5978).

> While the shepherd on the hill,
> Near a fountain warbling still,
> Modulates, when noon is mute,
> Summer songs along his flute . . .

In memory of the *Coronation Ode* invitation, he dedicated the *Greek Anthology* part-songs to Sir Walter Parratt at Windsor.

The fame of his music was reaching farther and farther. At Edward Speyer's suggestion, Richard Strauss produced *Cockaigne* in Berlin. Victor Herbert introduced it in Pittsburgh. And the distinguished old American conductor Theodore Thomas, looking toward his own performances of *Gerontius* in Chicago and Cincinnati, gave his considered view of Dr Elgar's achievements to the American interviewer:

'How do you regard Elgar as a composer, Mr. Thomas?'

'There is not a composer now prominent who is so well equipped, so able as he! Not one in all Europe!' was the positive answer.

'Greater than Richard Strauss?'

'Strauss is a specialist, and as such may be regarded as standing by himself, but Elgar has abilities that make him the superior as an orchestral writer of any man the world knows now, or ever has known for that matter . . .'

'And do you consider him equally eminent from a creative viewpoint?'

' . . . Take his "Dream of Gerontius", for instance. Its orchestral score is tremendous. As a choral work I consider it the greatest the last century has produced—I except none, Brahms' "Requiem" nor any other modern work of similar character.'[44]

Foreign acclaim returned to London when the Brahms disciple Fritz Steinbach brought the Duke of Saxe-Meiningen's private orchestra to St. James's Hall for a brilliant series of concerts sponsored by Edward Speyer. In the same programme with Brahms's Third Symphony Steinbach included, at Speyer's special request, the *'Enigma' Variations*. The distinguished conductor himself caught the enthusiasm, and Edward's music gained another powerful friend in Germany.

For the Meiningen Concerts the Elgars stayed at Frank Schuster's house in Westminster. There they were surrounded by an inner circle of Schuster's friends—the Stuart-Wortleys, Lady Charles Beresford, and the art critic Claude Phillips. With him Edward discussed the possibility of a 'pantomime-ballet' on the subject of Boccaccio's *Decameron*—treated, as Phillips noted, 'from [a] higher point of view . . . plague tableaux tragic and comic—grand finale of all these things combined!'[45]

Then Schuster took Edward's name into more exalted circles. After the Meiningen Concerts he wrote of dining with the Duke and Duchess of Connaught:

They were *both most enthusiastic* over the 'Coronation Ode'—& beyond that knew little

---

[44] 'Chicago, October 5, 1902', *The Musical Courier*.

[45] 2 Dec. 1902 (HWRO 705:445:1977).

about—or by—you. I am happy to say that in a chat of at least $\frac{1}{2}$ an hour's duration I was able to improve the *Duchess'* mind considerably on that subject! Also I want to tell you what *Mrs. Leslie* (the lady you thought interesting) overheard someone say to Pad[erewski] a propos of the 'Ode'—'Who is Elgar & where did he study?' he asked. 'Was he at any conservatoire?'

'No'—said Paddy.

'But who was his master then?'

'Le bon Dieu,' answered our friend.[46]

Early in December Edward was taken by Sir Walter Parratt to Kensington Palace, where Princess Henry of Battenberg had expressed a particular desire to meet him. Yet the shadows were there still: a week later he noted a sketch for *Cockaigne* No. 2—'The city of dreadful night'.[47]

Plans for *The Apostles* were going forward slowly. On 22 October G. H. Johnstone had written to Alfred Littleton at Novellos:

In reference to 'The Apostles' I had a long talk with Elgar the other day. He is getting on very well with it and he himself thinks that it will be much finer even than 'Gerontius'.

His present publishers [Boosey] are pressing him very much to sell the work to them but I would much rather the publication be in your hands.

I still think you would do well to accept my first offer, as I am quite sure it is going to be a great success, and would be a really good speculation.[48]

Again Littleton asked to see some of the new work. But all Edward could send was a rough description of the trilogy, to show:

I. the schooling [of the Apostles]
II. the earthly result
III. the result of it all in the next world . . . Last Judgment & the next world as in Revelations:
each work to be complete in itself—the one bearing on the other . . .[49]

Next day Johnstone wrote to Edward again:

I received this morning a letter from Booseys, asking me what I had done in reference to the publication of 'The Apostles'.

Before I reply to them I shall be glad to know when I shall be able to show either them or Novellos some part of the work, the time is quickly coming when it ought to be in the hands of the printers and I should like to get it completed, before the end of this year.[50]

But once again, with less than a year in hand before the première, there was nothing to show. And this time the very plot outlines were still vague—to say nothing of any libretto. Thinking it over in such a light, the planning of a trilogy was a very large matter to settle in the time remaining before its Part I must be

[46] 26 Nov. 1902 (HWRO 705:445:6896).   [47] Sketchbook II, p. 28.
[48] Novello archives.   [49] 28 Oct. 1902 (Novello archives).
[50] 29 Oct. 1902 (Elgar Birthplace).

produced. So Edward decided to make just a single work after all, in which several leading Apostle figures would be developed and contrasted. It meant abandoning the final picture of the Last Judgment, and with it the chance for some final emergent heroic grandeur. The subject was now more open-ended than ever: it might be taken as far as Antioch, where the converts were first named Christians.

The best hero for this action would be Peter—the 'rock' of Christ's Church. Peter would stand as the man made strong by inspiration and able to pass that inspiration to others. His strength would contrast with the self-doubt of Judas. And between those extremes should be some case of gradual revelation—a sinner converted. Conversion was the central theme of the Apostles: that was precisely the moral of the old schoolmaster at Littleton House who had described the Apostles as 'poor men, young men . . . perhaps before the descent of the Holy Ghost not cleverer than some of you here.'

Conversion stories in the New Testament were legion. What Edward needed was some guide through the maze of biblical material (as the biblical quotation in St. Augustine's *Civitas Dei* would have guided his work on 'The Last Judgment'). He found that guide in the favourite Longfellow, whose long poem 'The Divine Tragedy' was a versified account based on Scriptures of Christ's work on earth. 'The Divine Tragedy' contained many stories and characters, but only a single important conversion—that of Mary Magdalene.[51]

The theme of feminine conversion could make an instant appeal to Edward. All his own closest experiences of conversion had come through women—his mother, Miss Walsh, Alice. Now his youngest sister Dot, born a Catholic, having lived out their mother's life in the old family rooms above the shop, wanted to take the veil of a Dominican nun. Still another example of feminine conversion had met him again three months ago in the figure of Kundry in the performance of *Parsifal* at Bayreuth.

Yet Mary Magdalene was not one of the twelve disciples. Her conversion was not an act of the Apostles: it went back to Christ himself. Making Mary Magdalene one of his chief characters would slow the development of the main action by demanding a bigger role for the earthly Christ. This difficulty seemed small enough at the outset, weighed against the appeal of Longfellow's suggestion.

Longfellow had made his own additions to the Mary Magdalene story. He placed his heroine in a tower at Magdala, ruminating over a highly-coloured past and recalling the sight of Christ coming over the water in the fishermen's boat. At that point the poet had inserted the quite separate story of Peter trying to walk on the water: the sight of the doubting man saved by Christ then made the dramatic vehicle for the woman's conversion. Yet such a vehicle effectively hid the psychological process of Mary Magdalene's inner development to

---

[51] Elgar's annotated libretto for *The Apostles* shows several references to Longfellow's poem. Opposite 'In the Tower of Magdala' he wrote: 'See "The Divine Tragedy" (The First Passover, IX) Longfellow.' (BL ADD. MS 47904B fo. 7^v.)

conversion: and that process toward conversion should make the thematic centre for Edward's Apostles-story.

The lacuna seemed hardly to trouble him as he found other hints in Longfellow's poem. In fact it suggested almost everything he needed up to Christ's Ascension. One scene in the poem showed Christ reading in the synagogue from the Book of Isaiah:

> The Spirit of the Lord God is upon me.
> He hath anointed me to preach good tidings
> Unto the poor, to heal the broken-hearted;
> To comfort those that mourn, and to throw open
> The prison doors of captives . . .

It inspired Edward's opening: he had only to turn up the original passage in the Old Testament to find the words for his choral Prologue.

Longfellow also showed the Transfiguration[52], with Christ praying on a mountain top. That would make the figure of Christ too central for a work about the Apostles. But it suggested the earlier occasion when Christ ascended a mountain to pray all night—a withdrawal, an exit. Here was a beginning for Edward's action. He gave the description of it not to Christ but to a watching Angel Gabriel.

What could follow this scene at night was already suggested in Longfellow's account of Peter walking on the water. After the storm on the lake subsided, the poet made Mary Magdalene say:

> . . . The wind was hushed,
> And the great sun came up above the hills,
> And the swift-flying vapours hid themselves
> In caverns among the rocks!

It was the Eastern sunrise: following directly on the night of Christ's prayer, it could invite Edward's highest response.

One scene in Longfellow's poem showed Christ blessing the children. Another returned to the harvest atmosphere of the opening, with Christ walking 'In the cornfields' among his new Apostles: it could recall the origin of the Catholic 'Sanctus fortis' as a Rogationtide chant to be sung in procession through the fields. On 3 November he wrote to ask Capel-Cure:

Will you sketch out for me a scene in which 'Jesus went about to all the villages *Thro the fields of corn*—Contention of the Apostles for precedence: Peter, *speaking* dignified, *John* also speaking, softly & graciously, Judas *roughly*, the others as chorus

Then with it the lesson of the children. Here the general chorus (S.A.T.B.) can come in—This wd. make my quiet sort of pastoral scene & the chronology of the children episodes in the gospel seems so vague that no violence will be done by putting all these incidents together? Jesus must not speak more than necessary—the Apostles must stand out as the *living* characters . . .

[52] Elgar explained why he did not set the Transfiguration in a letter of 1 June 1903 to Canon Gorton (*Letters*, p. 119).

Will you tell me if any of the enclosed fragments do any *violence* to your feelings—in the selection of words I mean. No one can object to the Angel looking on during our Lord's all night prayer?

The Judas scene I like but please suggest any words . . .

I hope you like the idea of the Prologue—the Isaiah words read by our Saviour.

Capel-Cure compiled a scene to show Christ and the Apostles 'By the wayside'. To spark individual reactions in the different characters, he introduced Christ's Beatitudes. And thus he touched a memory in Edward older than the schoolmaster's comparison of the Apostles with the boys at Littleton House. The small picture given to the schoolboy of 1868 by Fr. Waterworth contained on its *verso* the legend: 'Blessed are the pure in heart, for they shall see God.'

Once again Edward's musical ideas had been emerging independent of words. The Ynys Lochtyn page in his 1901 sketchbook, filled with musical fragments from the summer week in Wales, now provided several motives for *The Apostles*. First came a sequential figure incorporating the falling thirds:

This was now marked 'I': he thought of using it to invoke the great sunrise. Then its opening sixth was written separately and marked 'II', and the following sequence of thirds written separately and marked 'III'—starting again the process of fragmentation. On the verso of this sketchbook page was a wide-ranging idea marked 'IV':

This became a motive of 'Christian Fellowship'. Still another Ynys Lochtyn figure was now to stand for 'Christ's Prayer':

Thus the first *Apostles* music had been called forth not in response to any words or themes of the Apostles' story, but in circumstances quite unconnected. Wagner would never have done that. The cross-referencing powers of Wagnerian Leitmotiv—always a strong aid to dramatic development—would have been impossible until the composer had worked out the complete shape and symbolism of his story: it meant in every case a libretto virtually finished before a note of the music was written.

For Edward the musical abstraction—whenever, wherever, however it came—was first. Literary and dramatic thinking could come after, as a useful means for developing the musical thoughts which already existed. This appeal to the music first showed the fundamental difference between Edward and Wagner. It showed why, despite a lifelong fascination with Wagner's motivic schemes and rich orchestration, the genius of Brahms continued to speak to him. It was why he found it so difficult to accept any proposal for writing an opera. It was why he still wanted above all things to write the Symphony.

On 5 November he went to Leeds for a Choral Union rehearsal and performance of the *Coronation Ode*. He took his sketchbook, and during the days in Leeds he set down some music to follow Peter's walking on the water, when Mary Magdalene would sing: 'The wind ceaseth and they worship Him.' Then he added an idea for a choral Epilogue, 'Turn ye to the stronghold, ye prisoners of hope'. It was not the first time the hint of hope as bondage had appeared in his music.[53]

Back at Craeg Lea, there was another letter from Johnstone. He had yielded to Alfred Littleton's protest over the £1000 price for *The Apostles*—but only to the extent of specifying that Novellos should pay Edward £500 on completion, another £500 when 10,000 copies of the vocal score had been sold, and a royalty after that. Littleton had replied that his firm must see something of the new work before they could agree. Johnstone informed Edward: 'They would like to see the "Libretto" and they want to know how many soloists you are writing for, if the work is likely to be more or less difficult than "Gerontius"—also if you can let them know about how long the work will take in performance.'[54]

Malvern

Nov 25.02

Dear Mr. Johnstone:

I only arrived home late last night & found your letter.

As to the Apostles—the work will be as difficult as Gerontius—except the Demons' Chorus which is exceptional of course—the Chorus work is difficult in the same mystic sort of way as the other work.

It will take, roughly, two hours.

It falls naturally into the two parts.

I The Calling of the Apostles to the Ascension.

II The spread of the Gospel until the climax at Antioch.

  ? The soloists are Peter — Baritone

                John   Tenor

                Judas   Bass

                Sopran Mary

                Contralto Mary Magdalene

& for recits & reflectual passages.

---

[53] Sketchbook II. In a draft libretto (BL Add. MS47906 fo. 12) 'Turn ye' is noted for insertion as an Epilogue.

[54] 22 Nov. 1902 (Elgar Birthplace).

I send this in great haste as I think you want it

> Kindest regards
> Yours very truly
> Edward Elgar

I shd. of course like to know what arrangements are to be made before anything is signed. Of course the full score *must* be published.[55]

He sent no libretto, for that was in flux with the music.

He wanted a musical figure to represent the Apostles generally. He ransacked Gregorian chants, and asked the help of the German publisher Schott: their London office promised to send all the 'old Latin music' they had.[56] None of it was right. Then in Liverpool for another Rodewald concert at the end of November, he and Alice went to church on the Sunday morning with Adrian Mignot, the Catholic president of Rodewald's orchestra, and were entertained to luncheon afterwards at Mignot's house. What happened there was recalled later for Edward's biographer Basil Maine, who wrote: 'He was sitting at a writing-table and aimlessly put his hand upon the page of a book lying there. He looked and saw that his hand was resting upon a Gregorian chant. This, he saw immediately, was the theme that had been eluding him.'[57]

It was a portion of the Gradual 'Constitues eos', celebrating (as Edward felt it) 'the power promised to the Apostles and their successors'.[58] The melodic outline bore a faint resemblance to the first 'Ynys Lochtyn' sketch, which was now to introduce the 'Light and Life' of the great sunrise.

He also found the Antiphon 'O sacrum convivium' for use in the Peter section of *The Apostles* and copied it into his book.[59] And when he began connected work on the music, he worked not at the beginning but at the Peter section:[60] this might provide a climax for the whole design. Edward later described his musical thinking in general terms for W. H. Reed, who wrote: 'He rarely started anything at the beginning. He worked at a theme and brought it perhaps to a climax; for then, as he has said to me, he knew to what he was leading.'[61]

On 5 December the Elgars were in London to attend the first English performance of Richard Strauss's *Ein Heldenleben*. Strauss had been brought to conduct his work in London by Edgar Speyer (a cousin of Edward Speyer at Ridgehurst). Edgar Speyer now headed the Queen's Hall syndicate. He and his

[55] Novello archives.   [56] 20 Oct. 1902 from Charles Volkert (HWRO 705:445:6766).
[57] *Life*, p. 116.   [58] BL Add. MS 47904B fos. 208–9.
[59] Sketchbook II, pp. 43ᵛ and 44, dated 1 Dec. 1902.
[60] Elgar's letters of 28 July 1903 to Alfred Littleton (Novello archives) and 1 July 1903 to Jaeger (Elgar Birthplace) say that the Peter section was 'written first'. This music was largely held over to be incorporated in *The Kingdom* in 1906.
[61] *Elgar as I Knew Him*, p. 129.

wife, the well-known violinist Leonora von Stosch, entertained Strauss at their house in Grosvenor Street, where the Elgars joined the company to dinner. Next day they attended the rehearsal and performance of *Ein Heldenleben*, and afterwards met Strauss again at the other Speyer's house, as Edward wrote to Jaeger:

We rushed off to Ridgehurst immediately: Strauss tore himself out of the crowd & said to me
   'Freund, sind Sie Zufrieden?'
   Ja! gewiss![62]

Strauss's aggressively defined *Heldenleben* made a trenchant contrast to the enigmatic heroism of Edward's *Apostles*. But the contrast had its effect. Edward played *Apostles* ideas constantly on the piano at Ridgehurst until (as Edward Speyer described the scene later for his young friend Adrian Boult):

Elgar was improvising one day when he played a progression of three chords

and promptly jumped up and wrote it down. It became a theme that is heard many times in *The Apostles*, and bears the label 'Christ the Man of Sorrows'.[63]

In fact those chords synthesized the slowly rising figure of 'Christ's Prayer' from the Ynys Lochtyn sketches.

He needed some words to lead up to the great sunrise. As he had drawn on the Bible, it struck him that dawn in Jerusalem might be illuminated from other Jewish sources. He enquired in Malvern for the Talmud and Mishnah, and there found words for a Morning Psalm to be sung in the temple. He wrote to the critic Alfred Kalisch, who recommended him to consult Rabbi Francis Cohen in London.[64] Cohen suggested adding words from Psalm 92, 'It is a good thing to give thanks unto the Lord', and said that it might be accompanied with one of the traditional Hebrew melodies which Ernst Pauer had edited for the publisher Augener. The whole Psalm could be prefaced with a flourish blown on the ancient Shofar, or ram's horn. The Shofar sounded the interval of a rising sixth—the interval which had opened the first Ynys Lochtyn sketch.

Another letter had come from G. H. Johnstone at the beginning of December 1902:

Novellos are anxious for me to get from you the 'Libretto' of the new work 'The Apostles'.

  [62] 10 Dec. 1902 (HWRO 705:445:8553).
  [63] Sir Adrian Boult, Note on Elgar's Piano-Playing, for *Elgar on Record* (EMI RLS 713, 1974).
  [64] Kalisch's letter recommending Cohen is dated 8 Dec. 1902 (HWRO 705:445:3873). Cohen sent his recommendations in Jan. 1903 (HWRO 705:445:3872).

If you could let me have a copy of it I shall be very much obliged, as I am anxious to get the matter of publishing settled as soon as possible.

I hope that the work is progressing very favourably. We ought to begin to print very soon.[65]

Edward could delay no longer. It meant deciding on a definite shape for the libretto without full guidance from his slowly evolving music.

The three stories of Judas, Mary Magdalene, and Peter ought to follow dramatically in that order—the doubter, the convert, the strong man. But Christ's presence on earth, needed for Mary Magdalene, would disappear with Judas. And thus chronology demanded the sacrifice of the ideal design for moving from Judas-darkness gradually toward the light.

He found several suggestions for the Mary Magdalene section in the libretto of an older oratorio, Sir Julius Benedict's *St. Peter*, produced at the Birmingham Festival of 1870.[66] There the storm on the waters began with a solo contralto singing words from St. Matthew:

And Jesus constrained His disciples to get into a ship, and to go before Him unto the other side . . .

Benedict had depicted the storm in an orchestral interlude, and then his chorus sang from Psalm 42: 'Deep calleth unto deep, at the voice of the storm and the tempest.' Edward took the St. Matthew words to open his own scene. His Mary Magdalene was also a contralto, and during his orchestral storm she would sing 'Deep calleth unto deep . . .'. Other suggestions from Benedict's libretto went into later parts of the scene.

He had already made up his mind to follow Longfellow's lead, and to insert Peter's walking on the water at the point of Mary Magdalene's conversion. He wrote to Canon Gorton of this: 'I thought the manifestation of supernatural power on the lake a "sufficient" incident of this (Manifesting) class & the work must go . . .'[67] But go where?

Longfellow had identified Mary Magdalene with the woman who had ended by washing Christ's feet with her tears. Edward followed this too. It meant that his Mary Magdalene would have to show the same rueful character from beginning to end: whatever action swirled round her, this heroine would remain a fixed point. Any 'development' could only come in relieving episodes.

So there must be more episodes. Mary Magdalene's journey in search of Christ was another opportunity for revealing character development. But into it Edward decided to insert the story of Peter recognizing Christ at Caesarea Philippi (also in Longfellow, but unconnected with Mary Magdalene). Thus once again the drop-curtain of a Peter incident would descend to conceal a vital revelation of the central character—the converted. It was a further sign of

[65] 3 Dec. 1902 (Elgar Birthplace).
[66] Elgar's marked copy of the libretto by Joseph Bennett for Benedict's *St. Peter* is bound in BL Add. MS 47906 (fos. 107–12).
[67] 1 June 1903 (*Letters*, p. 119).

Edward's own uncertainty over the process of conversion which should stand at
the focal point of his entire work.

After Mary Magdalene would come Judas. Here again Longfellow offered
material. Amongst many suggestive things, one stood out. As Judas
approached Christ in Gethsemane with soldiers, the poet paraphrased St. John
to make Peter ask:

> What torches glare and glisten
> Upon the swords and armour of these men?

It invited a cyclic return to the night atmosphere of Christ's prayer at the
beginning of the entire action.

After the Crucifixion the poet showed 'The Two Maries' at the sepulchre
very early in the morning, when the Angel asks (after the words of St. Luke):

> Why do you seek the living among the dead?
> He is no longer here: he is arisen!

That was practically the end of Longfellow's 'Divine Tragedy'.

The biblical Acts of the Apostles, however, took up the story just before the
Ascension. Here was the ideal guide to Part II of Edward's plan, where Peter
would emerge as leader. The Acts added several interesting stories which
might come in. One held a special fascination—Simon Magus, the alchemist of
Gitta and lover of Helen of Tyre, trying to buy divine healing power with his
gold: thus Simony. The money recalled Judas, and Edward noted:
'Judas→Simon Magus→Antichrist—*The same spirit*.'[68] Was there a private
echo in the swingeing financial negotiations over publishing what he wanted to
make his own highest inspiration?

But any final formation of Part II must wait. In December 1902 he just raked
together libretto material for Part I and sent it in to Novellos. The first response
came just before Christmas from Jaeger: 'You have set yourself no small task!
Some of the "Situations" should give you superb chances for inspired music.'[69]
Alfred Littleton saw it less clearly when he wrote to Johnstone: 'I could really
gather little or nothing from the libretto as it was in such a fragmentary state.'[70]

Christmas at Craeg Lea meanwhile was festive. Edward Speyer and his wife
sent a set of Wagner's literary works, and when another friend sent the whole
of the *Encyclopaedia Britannica* in its own revolving bookcase, Edward's joy
was complete. He wrote to Jaeger on 21 December:

Dear Augustus:
   This is the shortest day—so I set forth on the longest letter I ever wrote (to you)—a
regular Yule-loggy puddingy, Brandy-saucious letter . . .
   I have had Xmas presents—all Wagner's prose works (translated) 8 vols & & & & & &
the Encyc. Brit. & the bookcase !!!!a present—(£42).

   [68] BL Add. MS 47906 fo. 17.   [69] 22 Dec. 1902 (HWRO 705:445:8567).
   [70] 2 Jan. 1903 (Novello archives).

Behold in me a learning prig: *prig* mark you—I know the height of Arrarat (But don't know how to spell) & all sorts of japes.

Look 'ere: *I'm learned* now & no base Nimrodkim (Hebraic plural) shall look down on me: is not my learning vast, in 35 volumes & in a revolving bookcase—my head revolves too with delight.

I can tell you who was Aaron's mother-in-law's first cousin's bootblack & Infinitesimal Calculus &c.&c. I charge 6d to enter the study now.

I say. I have a lively fine specimen of a Vanessa—pish! I shd. say to one unwise a peacock butterfly who is helping the Apostles & lives in this study. I feed him—no, drink him on sugar & water & he lives in a Chrysanthemum—it's all lovely & Japanesey & pastoral—I'm sure the beast is a familiar spirit—Angel Gabriel or Simon Magus, or Helen of Tyre or somebody: just fancy sitting in this study surrounded with flowers & a *live* butterfly at Xmas—this music's going to be good I can tell you.

Much love to you all (I must read up Love in the Ency:)

A merry Christmas to all at Curzon Rd (limited to No. 37)

Your austere and learned friend (34 vols & a bookcase)

Paracelsus Elgar.

(with a pain in his stomach)

Mince pizon.[71]

The original Paracelsus had been an alchemist, like Simon Magus. When Dora Penny (who also received a letter about the *Encyclopaedia*) asked afterwards what had happened to it, Carice replied: 'Mother had to send it away quite soon. Father would turn it round so!'[72]

*   *   *

Before the end of December 1902 formal composition of *The Apostles* vocal score had begun. The choral Prologue opened a new tonic key for Edward's music—A flat major. It was his farthest journey into a tonic of flats for a major work. After the exploration of G minor and major through the 1890s, the tide had set in *Caractacus* to three flats. It seemed to ebb for a moment in the D minor and major of *Gerontius*. Then the middle sections of *Cockaigne* and the piano *Allegro* as well as the whole frame of the *Coronation Ode* reasserted three flats. Now with the supreme effort of this largest work of his life, the tonic key moved farther into flats and darkness.

The opening motive had the easy flowing rhythm of Gregorian chant. Its first notes suggested a kinship with 'Constitues eos'.

[71] HWRO 705:445:8555.

[72] Powell, *Edward Elgar: memories of a variation*, p. 53.

In the Analysis Jaeger was to write, this motive was called 'The Spirit of the Lord'. The chorus entered in unadorned octaves:

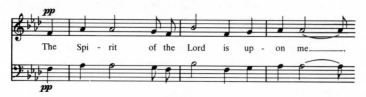

As everywhere else in this complex work, a deep impression was made in Edward's simplest, most disarming music.

At the words 'He hath anointed me' the three rising chords of 'Christ the Man of Sorrows' strode quietly into the orchestral foreground; and immediately, at 'to preach the Gospel', came a figure to signify 'the Gospel':

The opening choral octaves developed in plangent four-part harmonies to signify 'Christ's Mission' at the words 'He hath sent me to heal the broken-hearted', with the 'Gospel' motive sounding softly round it. At 'recovering the sight of the blind' came a quotation from *The Light of Life*. It all drew to a climax surrounded with a new falling motive which Jaeger was to call 'Preachers':

A middle section in the Prologue introduced two more motives. First came a vocal extension of 'Christ the Man of Sorrows' to make a figure of 'Comfort'. Then the music rose to a new idea in 3/4 metre, 'The Church':

The change of metre instantly set off this music from the rest—as it would continue to do at each later appearance in the score. The 'Church' figure was set to words which invoked again the faith of Edward's mother:

For as the earth bringeth forth her bud,
And as the garden causeth the things that are in it to spring forth;
So the Lord God will cause righteousness and praise to spring forth before all the nations.

It made the second climax. Then a recapitulation of 'The Spirit of the Lord' closed the Prologue.

In a playing time of less than six minutes the music had traced a curve of reflective aspiration through widely scattered ideas. Some were developments, others independent. All would play important roles in the music to come. Yet they had mingled so mellifluously through the Prologue that their individualities might hardly yet be taken in. It was 'music in the air . . . music all around', whose unobtrusive beginnings could give later appearances of the themes indefinable familiarity. Like the 'Enigma' theme, this music made an opening which could be drawn upon again and again, rearranged and developed through the personalities to follow.

On the last day of 1902 Johnstone wrote to Alfred Littleton:

I have heard from Dr. Elgar this morning that you have returned the "Libretto" to him and he tells me now that he has a portion of the work ready for the printers.

I shall be glad to know if you are prepared to accept my offer of November 22nd, as I have received this morning an offer equal to yours, from another publishing house, which offer I am holding over until I hear from you.[73]

At last Littleton capitulated. Novellos would pay £500 on delivery of the manuscript full score, another £500 after the sale of 10,000 copies, and a royalty after that. These princely terms were due entirely to Johnstone's tact and strength in negotiating. Edward could not have achieved such terms for himself, as he showed all too clearly in proposing a meeting with Littleton to arrange details: 'I do not think there is much to settle & I trust we shall be able in future to let things "publish themselves", as it were, without our going into commercial discussions.'[74] Perhaps it was another hint that he might accept the publisher's idea for a blanket contract.

Through the first days of 1903 he worked happily at the opening scene. A tenor recitative introduced a new figure which was to appear at other recitatives (as in *The Light of Life*). The scene opened 'In the Mountain.—Night'. A remote oboe development of 'The Spirit of the Lord' evoked the hot Eastern darkness. From it emerged the slow rising 'Prayer of Christ'. The watching Angel Gabriel was accompanied with the Angel figure from *Gerontius*. 'Constitues eos' sounded among the softly interweaving motives without any clear development, suggesting hours and hours of night stillness. At the end came a new descent to represent 'Christ's loneliness':

It almost echoed the opening notes of *Gerontius*.

[73] 31 Dec. 1902 (Novello archives).   [74] 11 Jan. 1903 (Novello archives).

At last the Shofar sounded. Its upward sixth found accompaniment in Edward's old descending fifths and fourths. The watchers on the roof of the temple greeted the dawn, and the singers within sang their Psalm to the Hebrew melody. When the Shofar sounded again, it went to the Ynys Lochtyn theme now signifying 'Light and Life'. Gradually the ensemble grew until it resolved into descending steps. Then the sun burst over the world, its rays descending and descending in majestic slow syncopating scales over the music of 'Christ's prayer'. It was the most overwhelming page of natural scene-painting in all Edward's music, and the most imposing invocation.

The question of what music could follow that invocation was pondered through a week at Ridgehurst with the Edward Speyers in mid-January 1903. He played the scene as far as it went for his host and hostess, who were deeply impressed. Back at Craeg Lea on the 20th, he prepared this much in vocal score for the printer, and next morning Alice posted it to Novellos.

Still he could not decide how to follow the great sunrise. So he began to work instead at the Mary Magdalene music. On 23 January he sketched a contrapuntal variant of descending steps for Mary Magdalene's 'Anguished Prayer':[75]

Now Novellos were pressing him for a final version of the whole libretto. They were anxious to get it set up in type, especially as Julius Buths was considering whether to make an immediate German translation for printing underneath the English in the score. When Buths had translated *Gerontius* after its first performance, Novellos had to print a separate edition. *The Apostles* should have an equal appeal in Germany. Edward wrote on 23 January: 'I will send the libretto, as complete as possible, at the end of next week. In making the fair copy (8vo. [vocal] score) I am constantly making slight alterations, but the "Spirit" of the whole work & the principal passages would remain.'[76]

He asked Capel-Cure to come up from his vicarage in Dorset to help settle the remainder of the words. He came on the 27th, and they spent the entire afternoon working at the text. The vicar made many suggestions: in particular, as he remembered it, 'the portions . . . which refer to Judas were largely the outcome of these.'[77] As with the Blind Man in *The Light of Life*, Judas would

---

[75] Sketchbook VII, p. 16.　　　　　　　　　[76] Novello archives.
[77] 'Rev. E. Capel Cure: his memories of Elgar', cutting from a Bournemouth newspaper c. 1938. There is an impression in the Capel-Cure and Fitton families that Elgar and Capel-Cure disagreed over Judas.

stand outside the temple, excluded from the singing within. But theologian and composer found so much to do that during the visit they only got through Part I of Edward's plan, ending at the Ascension.

'E. & A. worried over "Apostles",' wrote Alice on 30 January. He went out cycling, covered sixteen miles, and later that day wrote 'beautiful MM Scene music'. But he was not quite happy over it. And there was still nothing to follow the big sunrise. Alice used a new typewriting machine acquired in the autumn to make a fair copy of the libretto for Part I. They posted it to Novellos on 2 February.

There was further correspondence with the Leeds Festival secretary Spark. He had sent a formal offer of 50 guineas for the production of Dr Elgar's Symphony. Compared with the *Apostles* payments these were paltry terms, and Edward demurred. Spark then asked what terms would be acceptable. But with the central question of musical development so heavily clouding even *The Apostles*, Edward decided that on the whole another choral work would be safer: he promised to do his 'very best' for Leeds.[78]

From other directions came cheering news. The Dutch conductor Willem Mengelberg was organizing a Richard Strauss Festival in London, and at Strauss's own request invited Edward to suggest a work of his own for inclusion side by side with the music of his new friend. Edward chose *Grania and Diarmid*. In the same post came his first really big royalty cheque from Booseys, reflecting the *Land of Hope and Glory* success: 'Gott sei Dank', Alice wrote; 'a true blessing after such a long struggle.'

And that day, 3 February 1903, marked a happy reunion with the Worcestershire Philharmonic as they began rehearsing the *Bavarian Highlands* songs. In the previous autumn, after unsatisfactory rehearsals for yet another modern German choral work (Wilhelm Berger's *Totentanz*), Edward had refused to conduct the concert in December and offered his resignation. Now he was prevailed upon to reconsider, and at the rehearsal on 3 February Alice noted: 'Fine attendance—good practice. People *delighted* with singing Bavarians. Many speeches, much devotion & enthusiasm for E.'

Next day he was busy with a chorus to follow *The Apostles* sunrise and conclude Scene I. Uncertainty still hovered, and on 5 February he wrote to Jaeger to hurry the vocal score proofs of the opening portion of music already sent in:

. . . I want something badly to encourage me, as my letters are dreary and the weather is too cold for me to go and sit in the marsh with my beloved wild creatures to get heartened up and general inspiration.'[79]

Longdon Marsh was a strange place he had found cycling under the Malvern Hills south to Birtsmorton near Cardinal Wolsey's 'Ragged Stone'—a flatness

---

[78] Correspondence of Dec. 1902, Jan. and Feb. 1903 (HWRO 705:445:5979–81).
[79] Novello archives.

of eerie silence and willows repeating their endless spiky shapes along the banks of a stream.

The mild winter continued to invite cycling, and the rest of his energy through the next days was poured into a big chorus to finish Scene I. It did not grow out of the preceding sunrise: perhaps nothing could do that. Instead there was almost a break in the texture, before the tenor entered with his recitative figure to describe the choosing of the Apostles. The descending figures of 'Fellowship' and 'Preachers' introduced the chorus:

> The Lord hath chosen them to stand before Him, to serve Him.
> He hath chosen the weak to confound the mighty;
> He will direct their work in truth.

The first sentence was set in big choral octaves to the melody of 'Constitues eos'. The second line opened a vigorous canon developed from the 'Preachers' motive. Beside the last line 'He will direct their work in truth', Edward noted in his own copy of the libretto:

> This of course really means 'He will reward them according to their deserts.'[80]

If it meant that, then Christ's truth would not be opened equally to all men: the individualities of the separate Apostles were about to emerge.

The chorus sang 'Behold! God exalteth by His power' to more octaves and 'The Spirit of the Lord' music. 'The meek will He guide in Judgment' was sensitively set to 'Christ's mission'. 'We are the servants of the Lord' brought back 'Constitues eos'. And all was framed with the 'Gospel' motive sounding softly through the orchestra. But it remained an assemblage: there was no vital development because the goal of this drama had not been defined.

The lack of close musical development could be forgotten as the Apostles emerged one by one out of the ensemble. Peter led off with a new variant of 'Constitues eos' to describe the 'Apostles' Faith':

Thou wilt shew us the path ⎯ of life ⎯

John made another variant to sing a preliminary Beatitude:

> O blessed are they which love Thee,
> for they shall rejoice in Thy peace: and shall be filled with the Law.

But before he could finish the last significant remark, Judas cut in with his own conclusion:

> We shall eat of the riches of the Gentiles,
> and in their glory shall we boast ourselves.

[80] TS in possession of A. Rosenthal.

He was accompanied by a new 'Earthly Kingdom' motive—a descent of steps to echo the doubting Jews in *The Light of Life* and the old Gods in *King Olaf.*

The chorus recapitulated 'He hath chosen them', extending the music through masterful interweavings of many motives as the Angel Gabriel floated above with a reminiscence of the scene's noctural beginning. Jaeger was to describe it all as

. . . a perfect maze of leitmotives, of highly ingenious combinations of different rhythms, and very Elgaresque spinning of long melody-threads out of the filaments distributed broadcast over voices and instruments. With consummate art the different themes are dovetailed, linked together and combined . . . The combination of three different rhythms (3/2, 6/2, and 4/2) on pages 46 to 48 is an interesting *tour de force.*[81]

It was a towering polyptych, with every individuality immobilized.

Edward turned that immobility to stunning advantage when at the moment of coda he introduced Christ, majestically singing his first words in the work:

> Behold, I send you forth.
> He that receiveth you, receiveth Me;
>   and he that receiveth Me, receiveth Him that sent Me.

The Angel recalled again the nocturnal music to sing 'Look down from heaven, O God, and behold, and visit this vine;' the chorus sang a *pianissimo* 'Amen'; 'Constitues eos' sounded once more in the orchestra, and the scene was done. On 10 February 1903 the remaining vocal score manuscript was posted to the publishers.

He turned to the pastoral 'By the Wayside', which was to follow as Scene II. Despite the submission of the libretto to the publishers, the text was still in flux when Dora Penny came to spend the afternoon and evening of 12 February at Craeg Lea. She found him interspersing Christ's Beatitudes with private comments for the Apostles. Dora, as a clergyman's daughter, took an immediate interest.

The study seemed to be full of Bibles. He had a Bible open on the table in front of him and there seemed to be a Bible on every chair and even one on the floor.

'Goodness!' I said. 'What a collection of Bibles! What have you got there besides the Authorized and Revised Versions?'

'I don't know; they've been lent to me. I say, d'you know that the Bible is a most wonderfully interesting book?'

'Yes,' I said, 'I know it is.'

'What do you know about it? Oh, I forgot, perhaps you *do* know something about it. Anyway, I've been reading a lot of it lately and have been quite absorbed.'

He appeared to be looking out texts and I offered to help.

'I want something that will fit in here'—pointing to a line [where Christ sang 'Blessed are they which do hunger and thirst after righteousness; for they shall be filled.']

[81] *The Apostles: Analytical and Descriptive Notes* (Novello, 1903) pp. 18–19.

I thought for a moment and fortunately something suitable occurred to me and I quoted it. ['Mercy and truth are met together: righteousness and peace have kissed each other.']

'You don't mean to tell me that comes in the Bible? Show it to me.'

I found the eighty-fifth Psalm in one of the Bibles and laid it before him.

'Well! That's extraodinary! It's just what I want here.'[82]

When the text was settled, final shaping of the music followed swiftly. 'By the Wayside' began with a pastoral development of 'Preachers':

Christ opened the Beatitudes with a melody developed from the general shape of Peter's 'Faith' motive:

Again the music of simplicity held the greatest power, making a quiet centre in the action for Christ. The comments of the Apostles were set in brackets in the libretto to indicate private thoughts. Mary, John, and Peter sang '(He setteth the poor on high from affliction;)' while Judas plunged to his own conclusion: '(He poureth contempt upon princes.)' It was the conservative view, and Edward noted in his own copy of the libretto: 'Judas here sees a great deal further than the others.'

Each of Christ's Beatitudes returned to rising, aspiring music. Each was followed by more or less descending private comments from the others. Almost all wove an unbroken texture of softly flowing lyric. The exception was always Judas. After 'Blessed are the merciful', Mary, John, and Peter sang '(He that hath mercy on the poor, happy is he;)' while Judas interrupted them to look at the question from underneath: '(The poor is hated even of his own neighbour:

[82] *Edward Elgar: memories of a variation*, pp. 39–40, where these and other events are attributed to Dec. 1901. I can find no independent evidence of specific work on *The Apostles* text or music before July 1902. Alice Elgar's diary cites no visit from Dora Penny between June 1902 and February 1903. But Dora, in reconstructing later the chronology of her visits to the Elgars, overlooked the visit of 12–13 February 1903 (described in Alice Elgar's diary) when Elgar had just begun work on 'By the Wayside'.

Dora Penny also attached to her *Apostles* libretto visit a scene in which Alice, typing the libretto, mistakenly put £-signs for question-marks through a quotation from St. John's Gospel (which does not now appear to figure anywhere in the libretto). This would have been impossible before 31 Oct. 1902, the date the typewriter was acquired. Here Dora Penny may have been describing what she had been told rather than what she saw for herself. For on 12 Dec. 1902 Elgar had written to Edward Speyer: 'Alice has been very busy typing (that's rather a horrible verb) my libretto and has only distributed one "£" amongst all the twelve apostles!'

the rich hath many friends.)' '(Draw out thy soul to the hungry,)' pleaded the chorus, and John sought rising music to add, '(And satisfy the afflicted soul;)' but Judas reached down to the irony of self-interest: '(Then shall thy light rise in obscurity.)'

One Beatitude went back to Fr. Waterworth's little engraving thirty-five years earlier: 'Blessed are the pure in heart, for they shall see God.' The scrutiny of that enigma by the characters now before him found the trenchant comments of opera:

Mary: (Thou art of purer eyes than to behold evil.)
John [taking to himself an extravagantly rising figure]: (Blessed are the undefiled.)
Peter [already showing the man who finds it difficult to be certain]: (Who can say, I have made my heart clean?)
Judas: (The stars are not pure in His sight.)

And the chorus replied with a disarming simple progression taken from one of Edward's earliest sketchbooks to set '(How much less man?)'

At the end all the Apostles joined together in a Beatitude of their own while Christ watched silently:

Blessed are they which have been sorrowful for all Thy scourges,
    for they shall rejoice with Thee, when they have seen all Thy glory,
    and shall be glad for ever.

It was just what Edward's Judas might have said if he could have seen Christ's glory proved to him. On 20 February the short scene was finished and posted to Novellos.

Three days later Edward rehearsed the Worcester Festival Choral Society in the *Coronation Ode*, and on 24 February conducted a performance which shared the programme with Richard Strauss's *Wanderers Sturmlied*. But already he was into the next scene of *The Apostles*, the conversion of Mary Magdalene.

A tenor recitative described the Apostles entering the ship while Christ went up into the mountain to pray. Then 'In the Tower of Magdala' a B minor *Allegro* opened yet another figure of descending steps. It led to the motive of Mary Magdalene's 'Anguished prayer': the repeating notes

dragged at the *Allegro* pace. The heroine sang a new motive of 'Forgiveness':

It was a fresh instance of upward leaps unable to interrupt inevitably descending steps. There followed a long chromatic descent which might have described Gerontius's Angel of the Agony softened by feminine remorse.

A 6/8 variant of Mary Magdalene's opening B minor theme evoked a reminiscence of the heroine's gay past in 'Feasting':

This music was soon to quicken to a 'Fantasy' chorus of memories: 'Let us fill ourselves with costly wine and ointments.' Yet Edward broke the linkage of solo and chorus to repeat the slow chromatic descent and 'Anguished Prayer' for his heroine to sing 'My tears run down like a river day and night.'

The break epitomized the basic difficulty with this entire Mary Magdalene story. There was nothing to develop in the life of a character who must maintain her sorrow to wash Christ's feet with her tears at the end. No relieving development must overstrain this fragile thread of sorrow. When at last the 'Fantasy' chorus did appear, Edward had the heroine sing over it a figure in which (as he himself described it) 'the insistent A's hold one down to the unwished thoughts of past vanities while the chromatic harmonies shew a feeling of struggle to shake off the memory . . .'[83] All this complication merely elaborated what it could not develop.

After a time a light wind rustled through the strings to prepare the storm on the lake. Mary Magdalene went sorrowfully on until it came up. The storm was orchestral, and Edward's orchestration was prodigious. But again the storm must not develop at the expense of the unifying mood. The storm abated suddenly when Christ sang 'Be of good cheer'; it moved by fits and starts as Peter tried to walk on the water; and when Christ stretched forth his hand, it disappeared.

'The wind ceaseth and they worship Him,' sang Mary Magdalene. Sorrow returned. And her resolve to seek Christ—which ought to have made a tremendous moment—was framed with a negative Psalm, 'Thou hast not forsaken them that seek Thee.' The wish to celebrate was there: 'My soul followeth hard after Thee,' she sang, and attempted a climax at 'Thy right hand upholdeth me.' But the dragging pace remained the only unity.

Peter's recognition of Christ at Caesarea Philippi stirred the mood again. It opened with the tenor recalling music from 'By the wayside'. Christ's questions were asked and answered—first by a puzzled chorus singing the *Enigma* theme, and then triumphantly by Peter. That led to a grand choral entry on the motive of 'Christ's Prayer':

---

[83] 22 Aug. 1903 to Jaeger (HWRO 705:445:8606).

Proclaim unto them that dwell on the earth, and unto every nation, and kindred, and tongue, the everlasting Gospel.

But instead of real development to fulfil the pent-up sense of something accomplished at last, this too was curtailed for fear of destroying the fragile unity. Soon Mary Magdalene was back, recalling her old sorrowful motives as she went to find Christ.

Here came still another insertion. Edward designed a short solo of comfort by the soprano Mary. He set it off with a further idea from Longfellow: when Mary Magdalene came into the house where Christ was staying, the poet included an ironic comment from the host based on words from St. Luke:

This man, if he were a prophet, would have known who and what manner of woman this is that toucheth him: for she is a sinner.

Edward reassigned the words to a chorus of women—'always hardest on their own sex,' he observed:[84] one sketch of the music was labelled 'Women mocking (brutes!)'.[85] It moved *più mosso* to an accompaniment of chattering descent.

With each slow return after animating interruption, the grip of the drama loosened. As Mary Magdalene reached the goal of her sorrow in the tears to wash Christ's feet, a coda reassembled 'Forgiveness', 'Christ the Man of Sorrows', 'Anguished Prayer', and the 'Gospel' motive, dropping into their centre the B minor figure of descent which had opened the whole section—now translated to D major. But the motives at the end of this long portrait remained more scattered than all the separate motives at the end of Scene I.

The only hints of development remained in the Peter episodes. The man of small faith was strangely upheld in this music, while the woman who should have lit the central lamp for *The Apostles* found little. Her conversion revealed nothing to its composer. His own vision at the centre of his biggest work remained immobilized somewhere between the Gerontius who had found his Purgatory and the Judas who would come. As if to emphasize the void, his music for *The Apostles* had travelled from the A flat major of the Prologue to the 'diabolical' tritone to D.

There was no time to consider. When the end of the Mary Magdalene section was reached on 23 March, he was only half way through Part I—perhaps one quarter through the whole plan submitted to Birmingham and Novellos. Already there was over an hour's music. He outlined the rest of the plan for the editor of *The Musical Times*. After Mary Magdalene's conversion would follow Judas, the Crucifixion, and the Ascension set to a 'mystic chorus' of words combined from the Psalms and St. John's Gospel to end Part I. 'Part II opens with the first gathering of the Apostles, the descent of the Holy Ghost,

[84] 17 July 1903 to Canon Gorton (*Letters*, p. 122).
[85] BL Add. MS 47905.

and the exhortation of Peter. The troubles and trials go on until we arrive at Antioch.'[86]

There had been more correspondence with Leeds. When secretary Spark pressed him to define his proposed choral work so as to plan the Leeds Festival programme of 1904, Edward said he could not think of any other subject in the midst of *The Apostles*. Spark cleverly replied by offering 100 guineas for the production of the Symphony. Feeling by this time cornered, Edward at last accepted.[87] How the production of the Symphony at Leeds would square with the promise of its dedication to Richter in Manchester remained to be worked out.

Work on *The Apostles* was running into a variety of spring engagements. Rehearsals were going steadily on for the Worcestershire Philharmonic concert in May to include the *Bavarian Highlands*. Several days had been spent at Hanley rehearsing and conducting *Gerontius* there. Then there was an interim concert in the Worcestershire Philharmonic series by the Adolf Brodsky Quartet from Manchester. On 25 March the Brodskys played a programme which included Mozart's *Sinfonia concertante* for violin and viola, with Edward supplying the orchestral part at the piano. There were meetings with Troyte Griffith and other friends keen to form a Concert Club to bring such groups as the Brodsky Quartet to Malvern. Lady Mary Lygon's brother Lord Beauchamp (now returned from Australia) accepted the Presidency, and Edward agreed to become Vice-President.

Earlier in March *Gerontius* had come again under the baton of Richter—who with his new Hallé chorus-master R. H. Wilson was this time on his mettle. But Edward could not attend Richter's performance in Manchester because that night he himself was rehearsing *Gerontius* at Hanley with the North Staffordshire Choral Society. The performance was a great success, and Edward wrote afterwards to the chorus secretary: 'I place the chorus in the highest rank, and I thank the members for giving me the opportunity of hearing a performance of my work almost flawless. I do not praise choruses indiscriminately.'[88]

The remainder of March 1903 saw a profusion of *Gerontius* concerts. The Wolverhampton Choral Society did it under Bantock, Cowen conducted it at Liverpool and Sinclair in Birmingham: Sinclair had also put it down for the Hereford Festival in September, but there was clerical opposition in the Cathedral.[89] The first American performances were reported from Chicago under Theodore Thomas and from New York under Frank Damrosch. Another German performance was given at Danzig. Still others were

---

[86] *The Musical Times*, 1 Apr. 1903, pp. 228–9. The editor's interview with Elgar had taken place on 14 Mar.

[87] Correspondence of Mar. 1903 (HWRO 705:445:5985–6).

[88] Quoted in R. W. Ship, *A History of the North Staffordshire District Choral Society* (Wood Mitchell & Co., [1909]), pp. 18–19.

[89] Mentioned in a letter of 23 Mar. 1903 from Elgar to Alfred Littleton (Novello archives).

scheduled for Middlesborough and Bristol during April. And plans were maturing for the first London performance at the new unconsecrated Westminster Cathedral in June: Edward asked the North Staffordshire Chorus to sing there.

In Birmingham the new Orchestral Committee chairman Charles Beale had begun negotiating with soloists for *The Apostles*. Some of the correspondence devolved upon Edward. Ffrangçon-Davies, first considered for the role of Judas, was soon chosen to sing Christ. Madame Albani—in her mid-fifties still the leading light among oratorio sopranos—was asked for the dual role of the Angel Gabriel and Mary. Muriel Foster had long been chosen for Mary Magdalene, John Coates would sing the narrator and John, Andrew Black Peter. The American bass David Bispham was suggested for Judas, but his schedule was too full (so it was said) to attend all the rehearsals.[90] It was worrying, for Edward tried to design each of the roles round an individual voice—as he had used the friendly personalities of 1898 to suggest their Variations. Writing to Jaeger to complain of delayed proofs for the Mary Magdalene scene, he explained:

I wanted, if possible *Sc.III*, to go thro' with Miss Foster—now its not ready—so our trying of it must wait—that's all: I want the part to suit her exactly & not to have to put any alteration *after* printing.[91]

The question of who was to sing Judas was still unsettled when Edward began formal work on that music. But here his own identity gave him the grip he could not find in Mary Magdalene's conversion. An orchestral Introduction reviewed the night-motives from the opening scene; the 3/4 'Church' figure again stood apart to suggest that all was moving as ordained; then the dark music returned. It was the second evocation of night in the work as a whole.

To begin the vocal music of the Judas scene, the tenor introduced Peter again. After a choral setting of 'I will smite the shepherd', Peter sang 'Be it far from Thee' to an echo of the 'Enigma' figure. Then came his facile promise of loyalty: a few minutes later he would deny Christ.

First the music made way for its new chief character. The men of the chorus took over the narrative and the colour darkened. As they impersonated the Chief Priests and Pharisees, the orchestra sounded a new rising motive of 'Temptation': the notes of the minor triad climbed slowly upward while chromatic harmonies dragged below them. Judas entered to ask in a near monotone 'What are ye willing to give me?', and the money was told out in a pianissimo descending triplets to be scored for woodwind, triangle, glockenspiel, harp, and organ.

---

[90] On 18 June 1903 Elgar wrote to Alfred Littleton: 'I imagine—quite out of my own head—that Bispham's unfortunate domestic affairs may have influenced the committee in the matter of engagement.' (Novello archives.)

[91] 22 Apr. 1903 (Novello archives).

Now the chorus of men impersonated the soldiers going with Judas to arrest Christ. Following Longfellow's lead, Edward found St. John's description of them moving 'with lanterns, and torches, and weapons': he made them chant the words in the softest monotone over a tiny march.

Before the marchers arrived at their goal he placed the first big statement for Judas—again in brackets to indicate the interior monologue of silent thought:

> (Let Him make speed, and hasten His work, that we may see it;
> He shall bear the glory, and shall sit and rule upon His throne,
> the great King,—the Lord of the whole earth.)

The words were to be sung 'rhapsodically', '*con entusiasmo*' over a glint of 'Constitues eos', a sequence of 'Apostles' Faith', the descending march of the 'Earthly Kingdom'. The vocal line distorted 'Apostles' Faith' as Judas sang, 'Whomsoever I shall kiss, that same is He.' At 'Hail, Master!' the orchestra pointedly recalled the 'Wayside' theme.

The Betrayal was over in a moment, and they were leading Christ back to the High Priest 'with torches, and lanterns, and weapons' as softly as they had come. Peter's three denials moved through half-remembered motives, each adding its subtle comment if the listener should be quick enough to catch it. Then a lingering description of Peter's bitter weeping brought the women of the chorus to make a soft polyphony of the rising thirds in 'Temptation'. The moment for Judas's development had come.

The new music was shown to several friends. One was Lee Williams, the former organist of Gloucester Cathedral (whose part-song Edward included in the Worcestershire Philharmonic programme to be given at the end of the month). After a sight of *The Apostles* Lee Williams wrote on 9 April:

Dear boy, it is *really* wonderful what musical ideas you have got in your head, or under your waistcoat, or wherever you keep them about your person; how I wish I could find the fountain of inspiration that comes so easily to you. I write & write sometimes, only to find on reviewing my efforts candidly, & sternly, that some other chap has said the same thing so much better than myself! But this disheartening discovery does not in the least prevent my profound admiration for those, like yourself, who tell us something *really* new each time you write.[92]

Yet with only six months left before the première in Birmingham, what still remained to be done was staggering. From his study in Craeg Lea, Edward wrote to Ernest Newman:

I am sadly tired out & this vast view from my window makes me feel too small to work: I used to feel that I 'expanded' when I looked out over it all—now I seem to shrink and shrivel . . .[93]

The first part of the Judas music had already gone to Novellos, but on 20 April he asked Jaeger to send back the manuscript. He wanted to recast it.

[92] HWRO 705:445:3579.   [93] 14 Apr. 1903.

Then a new suggestion for the 'Judas' soloist came from Johnstone in Birmingham: 'Do you think that Van Rooy could do the part to your satisfaction? It would be a good name to get on the programme and I think that his English would be alright.'[94] Anton Van Rooy's would be a very good name. The Dutch baritone was one of the great Wagner singers at Bayreuth, a celebrity of the International Season at Covent Garden, a friend of the Speyers at Ridgehurst. It was a prospect to tempt inspiration.

Yet again there were interruptions. On 22 April Edward had to fulfil a long-standing engagement to conduct *Gerontius* with Nicholas Kilburn's choir at Middlesborough. On the 27th and 28th he rehearsed and conducted the Worcestershire Philharmonic concert: beside the *Bavarian Highlands* were Liszt's *Orpheus*, a cantata by Humperdinck, and two extracts from a *Russian Suite* by Granville Bantock. On the 29th he conducted the *Coronation Ode* for Lady Mary Lygon's Madresfield Musical Competition in Malvern. On the 30th he and Alice went to Morecambe for *The Banner of St. George* and the first performance of the specially written part-song *Weary Wind of the West*, which he conducted before an audience of 6,000. The Morecambe people welcomed Edward as their most distinguished guest, and Canon Gorton was a superb host who could also discuss *The Apostles* on the highest level of Anglican sympathies.

In Gorton's house Edward saw 'my ideal picture of the Lonely Christ as I have *tried* (and tried hard) to realise . . . the Character'.[95] The photograph of a painting by the Russian artist Kramskoi showed Christ in the desert seated on a rock, leaning forward with hands clasped convulsively in his lap, staring at the ground before him while a landscape of utter desolation stretched back to the horizon.

Afterwards he wrote a letter of fulsome thanks to his host at Morecambe:

I cannot well express what I feel as to the immense influence your Festival must exert in spreading the love of music: it is rather a shock to find Brahms' Part songs appreciated and among the daily fare of a district apparently unknown to the sleepy London press: people who talk of the spread of music in England and the increasing love of it, rarely seem to know where the growth of the art is really strong and properly fostered: some day the press will awake to the fact, already known abroad and to some few of us in England, that the living centre of music in Great Britain is not London, but somewhere further North.

It had the tone of public utterance, and the letter was duly printed in the Morecambe Festival brochure. Copied by *The Musical Times*, it started a lively resentment among several London critics who felt that they and their city deserved better at Dr Elgar's hands.

When Edward and Alice got back to Craeg Lea on 5 May, they found the

[94] 24 Apr. 1903 (Novello archives).
[95] Quoted in Marjorie Ffrangçon-Davies, *David Ffrangçon-Davies*, p. 39. The history of how Elgar acquired his own copy of this picture is contained in HWRO 705:445 6790 and 6792.

proofs of the Mary Magdalene scene at last. Muriel Foster came down on the 9th to stay the weekend and go through the music. But then Novellos wrote to say that the Birmingham authorities were 'very anxious' to have some *Apostles* music for the Chorus to begin rehearsing.[96]

Five months before the première, the great scheme was still less than half complete. For the second half not even a libretto had been sent in. The need for curtailment was manifest. He decided to make one division after the Mary Magdalene section. To mark the division, and perhaps to give her portrait a more definite frame, he would insert there a choral setting of 'Turn ye to the stronghold'. That should also go some way to placating the Birmingham Chorus in their demand for material to rehearse. If the new chorus finished Part I, Judas and the Ascension could then form Part II, Peter's emergence Part III: the Peter section would have to be drastically shortened to emerge at all in the time remaining.

Replying to Novello's letter, Edward eased into the subject of curtailment by complaining of the difficulty over finding a singer for Judas:

Craeg Lea, Wells Road, Malvern.

May 14:1903

Private concerning the *Judas soloist*
Dear Sirs:

I am sorry I could not reply at once—I have had a chill.

I had quite hoped that the 1st. pt of the Apostles cd. have been ready by now: but, owing to the difficulty in arranging about a suitable soloist for 'Judas' I have been holding the M.S. back as it may be nec[essar]y to recast the 'order' of some of it.

Will you ask your printers to go on with the *chorus* parts as far as p.108 only in the vocal score [the end of Mary Magdalene's music]: I hope to send a chorus to *follow* this on Saturday next.

The scene beginning on p.109 [i.e., Judas] will have to be repaged & the rehearsal numbers altered.—the engraving of this portion (voc. sc.) can be proceeded with with dummy paging & *no* rehearsal figures.   Pt. I will now end with the Chorus which I mentioned. I am sorry for all this delay but I think now all will go on smoothly.

Pt II will end at the Ascension & Pt III (short) will conclude the work.

For concert performance a pause can be made after Pt I.

In great haste
Yours faithfully
Edwd. Elgar[97]

Unless he compared the account of the project already published in *The Musical Times*, a reader of this letter would hardly realize that the altered plan covered little more than half the original proposal.

'Turn ye to the stronghold' was quickly written to finish what was now Part I. The music was diatonic and repetitious in a way unparalleled elsewhere in *The Apostles*. A soft opening statement moved the chorus slowly up and down the D major triad. It was echoed by a quartet of the soloists. The chorus re-entered

---

[96] 12 May 1903 (HWRO 705:445:8239).   [97] Novello archives.

cantabile with a comfortable, old-fashioned fugato—before the quartet and then the chorus returned to their D major triads. A tiny development led to the near repetition of the whole formula. It was all *tranquillo* and *largamente*—except for the fact of repeating over and over as if an invocation 'Turn ye to the stronghold, ye prisoners of hope.' Those prisoners could only be the Apostles—unless they were the composer and his own hopes. Where then was the stronghold? Just before the end he inserted another extra Beatitude from Ecclesiasticus: 'Blessed is he who is not fallen from his hope in the Lord.' At the very end the chorus sang the two words 'Turn ye' as if they were 'Amen'.

Then he went back to Judas. Gradually the dark character revealed more and more. Edward confided to Canon Gorton:

To my mind *Judas'* crime & sin was *despair*; not only the betrayal which was done for a worldly purpose. In these days, when every 'modern' person seems to think 'suicide' is the actual way out of everything my plan, if explained, may do some good.[98]

The first beneficiary could be Edward himself.

His plan was to show Judas finally as the outsider of outsiders—standing alone while the singers inside the temple sang another Psalm. The music of this Psalm played ironic variation on the Morning Psalm sung before the first sunrise near the beginning. The last line of music in that Morning Psalm made a basis for opening the music of this one; but now the temple singers took as their text the bloodthirstiest of all the Psalms. Underneath it a bass figure marched in soft syncopation.

> O Lord God, to Whom vengeance belongeth, lift up Thyself.
> Thou Judge of the earth, render a reward to the proud.
> Lord, how long shall the wicked triumph?

Overhearing the opening words of the Psalm, Judas sang over them 'My punishment is greater than I can bear': the musical figure leapt upward to fall back in descending steps, echoing the Angel of the Agony. As the Psalm moved relentlessly forward, Judas approached the Priests and flung down the silver to try to reverse the action. But the Psalm thundered on behind him:

> Lord, how long shall the wicked triumph?
> Yet they say, The Lord shall not see.

The Psalm-singing was not actually continuous behind Judas. The power of his self-projection through these moments in Edward's music controlled his own hearing: when his thoughts became tumultuous, the Psalm seemed to fall silent; but when he could hear it again, the Psalm had passed to later verses of crueller suggestion. In contrast to the weakness of Mary Magdalene, here was development with a vengeance, music and drama hand in glove.

Judas began a series of connected thoughts: and instantly his music commanded the compelling recitative of *Gerontius*. 'Whither shall I go from

[98] 17 July 1903 (*Letters*, p. 121).

Thy Spirit? Or whither shall I flee from Thy presence?' found a figure of mirror-reversal—trying to climb in steps only to fall suddenly back—while the accompaniment descended in slow sequences of empty fifths. 'If I say, Peradventure the darkness shall cover me, then shall my night be turned to day—' and the music recalled with deadly accuracy the sunrise figure from the Morning Psalm.

The temple singers found another Beatitude in their Psalm:

> Blessed is the man whom Thou chastenest,
>   That thou mayest give him rest from the days of adversity . . .

Again Judas's thought went home:

> 'Rest from the days of adversity.'—
> Never man spake like this Man; He satisfied the longing soul . . .

The music just recalled Christ's opening Beatitude from 'By the Wayside' before the temple singers finished their verse:

> . . . until the pit be digged for the wicked.

It brought Judas's closest approach to the 'Angel of the Agony' music, but plunging *Allegro molto* down the scale to its tonic fulfilment:

Writing to Canon Gorton, Edward said that Judas had now turned blasphemous.[99] But the words (from the *Book of Wisdom*) were too close to the fears which moved his own art:

Our name shall be forgotten in time, and no man have our work in remembrance, and our life shall pass away as the trace of a cloud, and shall be dispersed as a mist, that is driven away with the beams of the sun, and overcome with the heat thereof.

It was the summer atmosphere which had made a culminating vision in *King Olaf* seven years ago:

> As torrents in summer,
> Half dried in their channels,
> Suddenly rise, though the
> Sky is still cloudless,
> For rain has been falling
> Far off at their fountains:
>
> So hearts that are fainting
> Grow full to o'erflowing . . .

Now the parched heart would expire from its own illumination.

[99] Sunday 5 July 1903 (*Letters*, p. 120).

The Psalm had reached 'The Lord knoweth the thoughts of man' when Judas's reflections were interrupted with cries of 'Crucify Him!' building up in a sequence like the orchestral sequence in *Gerontius* before the Soul's vision of God. But the climactic chord this time was only a single note to be sounded by all the lowest instruments in unison, for its vision was only of the self. 'Mine end is come—the measure of my covetousness,' sang Judas to the figure which tried to climb, only to fall back. But just then the singers finished their Psalm:

> He shall bring upon them their own iniquity.

It was the wage told out to this prisoner of hope.

After the drama of Judas came the Crucifixion and Ascension. Thinking to divert attention from Christ to the Apostles, Edward hardly showed the Crucifixion. A tiny scene at Golgotha opened with the words 'Eli, Eli, lama sabachthani?' printed above a line of purely instrumental music which reached its climax in the same chord which revealed the Angel of the Agony in *Gerontius*. But it avoided looking directly at the *débâcle*, as *Caractacus* had avoided representing the crucial battle for Britain, as Edward's music would have avoided going with the Soul of Gerontius into the ultimate presence unless Jaeger had intervened.

Mary and John sang a succession of scattered motives. Here musical fragmentation served the drama as it had never served Mary Magdalene. Yet the tonality had moved back to E flat major—the key of the sunrise and the choosing of Apostles. The A flat tonality of the Prologue might be reserved for recapitulation with Peter's emergence in Part III.

The Resurrection was passed over as the music moved into another little scene 'At the Sepulchre' very early in the morning. Following the suggestion of Longfellow's 'The Two Maries', it was set almost entirely for women. The men in the temple merely recalled their morning Psalm and the sunrise music. This second sunrise was telescoped into a dozen bars—a potent reminder of all that had passed since the first morning.

To prepare 'The Ascension', an 'Alleluia' sounded a simple figure in a semichorus of Angels. It recalled the Angel's 'Alleluia' in *Gerontius* (and it had a similar Gregorian origin): but this 'Alleluia' took the brightness of major tonality, repeated over and over by the highest voices. The sopranos of the full chorus sang 'Why seek ye the living among the dead? to a vigorous figure. But this time the fugal invitation was refused, and the question was answered in lightly echoing phrases: 'He is not here, but is risen,' set against a soft radiant innocence rising and falling through the accompaniment. Then the heavenly Alleluias returned to lead to the Ascension.

But now it was June. The London performance of *Gerontius* at Westminster Cathedral was upon them. It shared the week with the festival of Richard Strauss's music at Queen's Hall. One report hinted that Strauss himself might conduct the Westminster *Gerontius*.[100] But the North Staffordshire Choir wanted Edward, whom they knew.

He and Alice went up to London on 4 June. They stayed with Frank Schuster, and went that evening to hear *Don Quixote*. 'Cd. not like it at all,' wrote Alice in the diary; but three years later, when Edward conducted a series of orchestral concerts, he wanted to include *Don Quixote*.[101] In June 1903 Strauss's superbly realized variations of eccentric knight and squire made a telling contrast to *The Apostles* still so far from completion. Next day Alice noted: 'E. tired & worried over "Apostles" music.'

There was no time to think, for it was a day of *Gerontius* rehearsal. The title role was to be sung by the German tenor Ludwig Wüllner, who had made such an impression at the Düsseldorf production that Edward had written: 'we never had a singer in England with so much brain.'[102] Wüllner had sung the role in English at Richter's March performance in Manchester, when there had been criticism of his pronunciation. But Edward had the man he wanted: he brought Wüllner to Schuster's house to go through the part and stay to lunch. With them came the tenor's young cellist friend Paul Grümmer, who repeated the request of Brodsky's cellist Carl Fuchs for a cello concerto.[103] In the afternoon Edward rehearsed the other soloists, Muriel Foster and Ffrangçon-Davies, with the Philharmonic Orchestra in the Cathedral.

Next day, 6 June, the North Staffordshire Choir travelled down by special train starting at 3.30 a.m., arriving in London in time for breakfast and a full morning rehearsal. The performance was that afternoon. When the doors opened, the new Cathedral was seen by many of the press for the first time:

The shell is finished, the bells are hung, and the glowing mosaics of the chapels spread broadlier every day.[104]

The great building filled slowly, but noiselessly. As usual, the lower-priced seats were taken up first, but the majority of those who had purchased the higher-priced chairs—£5/5s. each—came in late, so late indeed that the commencement of the performance was delayed for a quarter of an hour . . . The crowd not only filled the nave, chapels, and side-aisles, but quite a large audience had seating accommodation in the apse behind the orchestra. As soon as Dr. Elgar was seen advancing to the front to take his place on the rostrum the occupants of seats in the apse at the rear of the chorus commenced to clap. This Dr. Elgar immediately repressed, and thus this proved to be the only demonstration throughout the proceedings.[105]

Alice wrote of the performance: '*Most beautiful*. Glorious afternoon, & vast audience—E. conducted most splendidly. Chorus very beautiful, Wüllner finer than anyone A. thought, & Angel beautiful & Ffrangçon Davies magnificent.'

[100] *The Staffordshire Sentinel*, 8 June 1903.

[101] *'A Future for English Music' and other lectures*, p. 211.

[102] 21 Dec. 1901 to Novellos (Novello archives).

[103] On 1 July 1906 Grümmer wrote to remind Elgar of this meeting and to ask again for the concerto (HWRO 705:445:3794).

[104] C. J. O. in *The Outlook*, 13 June 1903.   [105] *The Staffordshire Sentinel*, 8 June 1903.

A brilliant group of friends and acquaintances gathered—the Duke of Norfolk, the Earl of Shaftesbury and his sister Lady Maud Warrender, the tenor Gervase Elwes and his wife, Claude Phillips, the Stuart-Wortleys.

Through the next days at Schuster's, Edward continued 'tired & worried'. He went over *Apostles* music with Madame Albani, and met Van Rooy to show him the Judas part. But when Van Rooy saw the renewed criticisms of Wüllner's English in the Westminster *Gerontius* he decided not to risk it. So, with the *Apostles* première now four months away, the role of Judas was still vacant. The new Part II was not finished, Part III only sketched.

They returned to Malvern in a bad state. Alice had fallen in Schuster's dining-room and torn a rib muscle. Edward's troubles were less physical. 'Wretched day,' Alice wrote on 11 June: 'E. very badsley & dreadfully worried &c.' Carice's confirmation took place that day but neither parent attended, leaving all the arrangements to Miss Burley.

On 14 June he was working again at 'The Ascension', and over the next ten days it shaped to an enormous ensemble. From this point the story followed the opening chapter in the Acts of the Apostles. In its words Edward's Apostles asked 'Lord, wilt Thou at this time restore the kingdom to Israel?' He explained to Canon Gorton:

I have kept in view the weakness & vacillation of the Apostles—'We trusted': 'Lord wilt Thou at this time!' etc. The strength & firmness come later.[106]

So the recapitulation to the strength which had opened *The Apostles* was again put off.

The last utterance of Christ, after the Resurrection, was to be his longest. When he sang, 'It is not for you to know the times or the seasons,' the 'Prayer' motive began climbing. At 'Go ye therefore and teach all nations', the 'Gospel' and the 'Preachers' motives achieved a wonderful combining in Wagnerian polyphony. Then the strings would rise up and up, set off with a descending counterpoint: as the contralto described Christ taken into a cloud, the upper strings were to sail slowly away toward another high 'Alleluia'. So unrolled the panorama on which 'The Ascension' would be mounted.

Several days found him cycling back to the vast quiet of Longdon Marsh as he confided later to W. H. Reed, who wrote:

He told me . . . he had to go there more than once to think out those climaxes in the Ascension; for they had to be so built up each time that they never reached such a pitch of intensity as at the last and greatest climax . . .[107]

On 19 June and on the 20th he rode to the Marsh with Canon Claughton's son Alban, who recalled:

I remember we went down there and we passed a field with a gate on the road, and he'd say:
'Look here, do you mind going forward for about a hundred yards or so and just

---

[106] 17 July 1903 to Canon Gorton (*Letters*, p. 121).   [107] *Elgar as I Knew Him*, pp. 99–100.

waiting for me?' And he'd get over the gate and be in the field by himself twenty minutes or half an hour . . .

Then he'd catch me up and we'd go off and have tea somewhere.[108]

On the evening of the 20th Alice noted: 'E. raser tired—Went on with his wonderful chorus.'

'Give us one heart, and one way,' sang the men's chorus on earth, while the high voices resounded their soft 'Alleluias' above. The ensemble built gradually through tightly woven motives toward a big climax, where the Apostles' theme of 'Faith' reappeared in grand ensemble to words from Psalm 22, 'All the ends of the world shall remember . . .'

The 'mystic chorus' followed to the words from Psalm 14 and St. John's Gospel:

> 'I have done Thy commandment,
>   I have finished the work which Thou gavest Me to do;
>   I laid down My life for the sheep.'

Then the problem of development was mysteriously solved. One day Edward found himself in the church at Queenhill, near Longdon Marsh. He recalled it later for his biographer Basil Maine, who wrote:

He was wandering round a church and saw a notice inviting people to take a pamphlet and put a penny into the box. Unthinkingly he gave a penny, took a pamphlet and in it discovered the words

'What are these wounds in Thine Hands?'

—and these corresponded exactly with the idea which had been forming in his mind . . .[109]

The question, from the Book of Zechariah, drew the answer:

> 'Those with which I was wounded in the house of my friends.'

It was a strange catechism for the composer of the *Variations* dedicated to 'my friends pictured within'. He set the question and answer to a chant-like figure softly marked with descending fourths. It was taken up by the women of the full chorus to recall the Crucifixion; whereupon the semichorus sounded their 'Alleluias' again.

The last recapitulation was at hand. 'Give us one heart, and one way' recommenced the upward climbing steps to reach 'The kingdom is the Lord's', sung to the 'Spirit of the Lord' theme *stringendo*. The 'Apostles' Faith' music sounded *tutta forza* as the chorus sang 'All the ends of the world shall remember,' four soloists 'They shall declare His righteousness,' and the semichorus 'From henceforth shall the Son of Man be seated at the right hand of the power of God.' One stroke remained: in the midst of the final 'Alleluias' there appeared the motive of 'Christ's Peace' from *Gerontius*. So ended the

---

[108] BBC broadcast *c.* 1965.   [109] Maine, *Life*, p. 116.

biggest choral ensemble of Edward's life. Its polyphonic mastery almost answered the Apostles' great sunrise with a real vindication of the plan.

The end of Part II was in sight as Edward and Alice went up to London on 24 June for the first performance of the *Coronation Ode* in the presence of the King and Queen. It was part of a concert organized by Lady Maud Warrender in aid of the Union Jack Club. At Frank Schuster's house the Elgars met Earl Howe, the President of the Birmingham Festival and one of the guarantors for the Union Jack concert. The Countess Howe invited them to a brilliant dinner before the concert on the following evening. Edward took his hostess in to a table which included Sir George and Lady Maud Warrender, the Shaftesburys, the Portuguese ambassador Soveral, and Prince Francis of Teck.

The concert after dinner was a long programme with the *Coronation Ode* at the end. 'Most brilliant spectacle,' Alice wrote:

In the interval E. was presented to the King who spoke to him quite a long time & very touchingly told him how he liked his music & in his illness used to have some of his favourite pieces played to him once & sometimes more than once a day—& how it soothed him very much, then the Prince of Wales came up & introduced himself & sd. how disappointed they were not to hear the Ode last year & how looking forward to it now.

Lady Maud Warrender recalled:

[Mrs.] Elgar was with me in my box, which was next to the King's. I noticed that he was fast asleep during the Coronation Ode. I was so anxious that [Mrs.] Elgar should not observe this that I made her change her place, and succeeded in distracting her attention. And as H.M. did wake up when 'Land of Hope and Glory' blazed forth, all was well.[110]

It did not help *The Apostles*. Realizing 'The Ascension' to finish Part II involved many revisions of earlier parts in the score, some of which had to go back to the publishers even after they were engraved. The result was delay all along the line in preparing choral parts for beginning rehearsals at Birmingham. Remembering *Gerontius* three years earlier, Jaeger took alarm:

1, Berners Street, London, W.

24/vi 1903

My dear E.

Here are some more proofs.

May I make a polite suggestion? Time is flitting fast & we should have got Part I of the Apostles to B'ham long ago. But the corrections are so *fearful* & so *upsetting*, (what looks simple enough on paper often takes a devilish long time to do) that I get heart-broken over the delays & the worries. Whatever you do—although as you know, we are most anxious to do *all* you wish—*dont* make more alterations *just now* than are *absolutely imperative*: the *expense* is a *detail*—it is the *delay* which I so deplore greatly. It

[110] *My first sixty years* (Cassell, 1933) p. 188.

seems to me as if we should *never* get done with altering the plates! *After* we have printed the 25 advance copies of the Score for the Soloists, conductors &c you can overhaul the work & *alter ad lib*, I need not say; at present it seems to me fatal to the proper *rehearsing* of the work.

Dont be cross with me; I write only to save if possible, a repetition of the 1900 collapse, due to the 'want of sufficient Rehearsing'.

> . . . Ever yours
> A. J. Jaeger[111]

Face to face with the ghost of the *Gerontius* fiasco, it was Edward himself who collapsed. He gave up the unequal struggle to finish *The Apostles*: it must stop at the Ascension, with all the chief characters but one immobilized between their 'weakness and vacillation' and half-comprehending wonder. Two days after the receipt of Jaeger's letter, it was Alice who bore the message:

June 27th 1903. A. to Novello to say it must be only 1 & 2 parts of the Apostles music. Saw Mr. Clayton—*very* nice & sd. it wd. be quite as well & E. *not* to worry—Back to lunch & then home [to Malvern].

Next morning Edward wrote to assure Alfred Littleton that only a recurrence of his eye trouble had stopped *The Apostles*:

Craeg Lea, Wells Road, Malvern.

> Sunday June 28:1903

My dear Littleton:

I had hoped to have called at Berners St: I have been seeing my London doctor & my eyes are again in trouble—he forbids more work: now I propose to the Birmingham people that they produce Pts I & II of the Apostles—this portion is complete in itself & may well stand alone. My wife went immediately to Berners St & saw Mr. Clayton explaining matters: I trust you will not mind the alteration of plan: of course all financial arrangements are at an end & must be reconstructed as *you* please entirely.

. . . The concluding portion of the work—(Pt III to round it off)—much of which was written first—you can have anytime later.

I am so vexed & sorry to be troublesome but fates work their own way & I think you may not be disappointed after all with the portion proposed to be done.

I am glad on acct. of the supreme difficulty of the vocalists—I am now able to propose [Andrew] Black for Judas, & [Kennerley] Rumford [Clara Butt's husband] for Peter: the part of Peter is not *big* until Part III when he is left 'in charge' & head of the Church—so I think it is more satisfactory all round.

> Kindest regards
> Yrs sincy
> Edward Elgar[112]

His private opinion of the soloist arrangements was reserved for Jaeger:

Now I have to put up with a ——— caste of English . . . The only way out of the singer business was to omit 'Peter'—who shines forth in Pt III & let Black do Judas—It's

---

[111] HWRO 705:445:8656.   [112] Novello archives.

pitiful but there are really no *good* singers beyond three & one (Bispham) we can't have![113]

When Jaeger wrote to ask whether it would not be better to postpone the entire performance until health returned, Edward rejoined:

I'm not ill! & it is of no use to postpone the work as I shall *never* get English vocalists—a complete caste that is—to do my work . . .

Now don't go pouring out your dear old heart to me, 'cos I'm not worth it. I want to do things well but I want power! I *can't* think why you preach to me about my career (once more). My career, to me, is this:—

I live on a pound a week.

I receive applications for autographs.

I receive applications for subscriptions.

I don't read about myself but I invariably get sent to me anything people think I don't like to see!

And I try to write music & don't like that either!

A damned fine '*career*'—wot ye well![114]

On Monday 29 June, while Edward golfed, Alice went to explain the position to Birmingham: 'Mr. Johnstone met A. & took her to lunch at the Grand Hotel. Very nice & soon resigned to only having parts 1 & 2.' Next day Johnstone wrote Edward a careful letter.

I was very glad to hear from Mrs. Elgar yesterday that 'The Apostles' is now in the hands of Novellos, completed, as far as you intend it to be for production at the Festival. I understand that your original intention of writing a third part has now been dropped, owing to the state of your health but I feel quite certain from what Mrs. Elgar told me that the work will seem complete in itself, as you have written it, and that the Grand Finale which she described to me at the end of the second part will be a fine finish to the work. It is well that the public should simply understand that the work is practically a complete one, and no mention should be made of the fact that the third part was originally intended to be written but had to be abandoned. I think this is absolutely necessary for your own sake as well as for the Festival.

. . . I am sorry to hear that you have not been well but trust that it is only temporary.[115]

Had Johnstone gathered from Alice that Part III was abandoned, or was this intended as an acute hint? Writing to Novellos the following day he seemed to make the hint more explicit:

You are probably aware that Dr. Elgar has completed the work of 'The Apostles' in the form in which it is to be presented at the Festival. He has I think wisely abandoned the idea of a third part which would probably have made the work too long and have been an anti-climax.[116]

As it stood, the subject was framed with night-and-morning sequences from the life and death of Christ. Within these was an inner frame of pastoral episodes—'By the Wayside' and 'At the Sepulchre'. In the centre two large

[113] 1 July 1903 (HWRO 705:445:8587).   [114] 1 July 1903 (Elgar Birthplace).
[115] 30 June 1903 (Elgar Birthplace).   [116] 1 July 1903 (Novello archives).

portraits faced in opposite directions—the woman of sorrowful conversion, the man self-betrayed.

*   *   *

Edward began *The Apostles* full score with an inscription:

> 'To what a heaven the earth might grow
> If fear beneath the earth were laid,
> If *hope* failed not, nor *love* decayed.'

The lines from William Morris's 'The Earthly Paradise' made a poignant reminder of Birchwood, where they had spent recent summers. But Edward had now decided to leave the Malvern district altogether. His excuse, as he wrote to Jaeger, was the rumour of new building opposite Craeg Lea 'which will spoil our heavenly view, so it will be of no use to think of Birchwood.'[117]

Local friends could not understand his desire to leave Malvern. But *The Apostles* had drained him deeply—and so perhaps had drained suggestive places like Longdon and Queenhill within easy cycling distance of Craeg Lea. And though Edward's stamina was excellent for a man of forty-six, there were places where cycling could be easier than round the hills of Malvern.

Alfred Rodewald invited them to spend July at a cottage named Minafon near Bettws-y-Coed in North Wales. Edward took *The Apostles* score and his bicycle. But Rodewald had bought a motor car, and the novelty of flying through the country with noise and speed kept most of the party agog. While the others went out in the car, Edward stayed at the cottage orchestrating.

He wanted Jaeger to write an Analysis of *The Apostles* as he had done for *Gerontius*, and Rodewald invited Jaeger to join the party at Minafon. Jaeger sent his regrets to Edward:

Much heavy, weary work keeps me here & though I think I *should* have a chat with you about the Analysis I fear I must alone flounder along through the Labyrinth of your creation.

I say, last Sunday for the *first* time (!) I went for the thing; I spent about 4 hours on it &, oh Lor!! you made me sweat with 'Entoosm' as dear old Hans R[ichter] calls it. Why, you *wretch*, this is even more wonderful than Gerontius, & I shall have to buy a Revised, up to date Lexicon for my supply of adjectives. Really, there is nothing in music like this & I am frightened out of my wits when I think of that analysis which nevertheless I am *burning* to do . . .

As regards your [orchestrated portion of the first] scene just to hand, it looks wonderful, *wunderbar*. That opening! & 'that there' Temple stuff with Shofar, antique cymbals, colour most gorgeous & new, effects most astounding & bewildering, organ! &c&c. Oh! my poor analysis.[118]

Edward wrote to Alfred Littleton to ask for Jaeger's release for a fortnight, and he came on 11 July.

---

[117] 29 Aug. 1903 (HWRO 705:445:8607).   [118] 8 July 1903 (Elgar Birthplace).

The days at Minafon sped by—discussing points for analysis, walking, swimming, cycling, motoring. Over them all rejoiced the tall Jove-like presence of their genial and sensitive host—who beguiled Edward from any disappointment over failing to finish *The Apostles* by celebrating the triumphant scoring. Through the orchestrated pages of the wonderful sunrise, the choosing of the Apostles, the Beatitudes spoken and heard 'By the Wayside', it could seem to Edward that Rodewald of all his friends stood guardian over the happy and achieved things of creative life: 'I used to pass him every sheet as I finished it at Bettws & heard his criticisms & altered passages to please him—God bless him.'[119]

In Birmingham rehearsals with the Festival Chorus were beginning. Prospects were bright. The new chorus-master, R. H. Wilson, had been Richter's chorus-master for the Hallé Concerts and had shown what he could do in preparing the fine Hallé *Gerontius* in March. On 10 July Wilson rehearsed *The Apostles* Part I—all that Novellos could yet supply. The Birmingham Chorus went through it with fervour, 'as tho' they realized that there was something to be "wiped off the slate".'[120]

The chorus music for Part II was soon expected from the printers, and on 18 July Wilson came to Minafon to go through Part II with Edward. Alice chronicled the afternoon:

Mr. Wilson, Richter's Choir Master, came about 12.30. After lunch E. played his 'Apostles' music to him, Rodé & Jaeger in the room & A. some of the time. A. felt & Mr. Wilson sd. it was an aftn. that wd. be remembered & I think, written about—Most marvellous music. Each listener deeply affected & intensely moved & excited.

For the final week at Minafon Carice was allowed to come up. Again she was the only child, but she was almost thirteen and well trained not disturb the special atmosphere surrounding her father. He found several intervals in his work to take her on walks through the beautiful hills and villages.

On the last day of July they all returned to Craeg Lea—to more scoring, and a steady stream of small and large alterations and improvements through the work. Canon Gorton came down for two nights to discuss the writing of an 'Interpretation' of *The Apostles* from an Anglican viewpoint, which Edward persuaded Novellos to print. He wrote to Alfred Littleton:

I really think a sort of article wd. do good for many reasons: I have 'made' the libretto myself & there are some people (Inter alia the gentleman who made mischief over The Light of Life &, later, over Gerontius at Worcester) who are only too anxious to pull it to pieces and say I have treated it solely from the R. Catholic point of view . . . Canon Gorton writes clearly & beautifully . . .[121]

On 17 August he finished *The Apostles* full score. The tremendous concentration of its labour had not been interrupted by further eye trouble.

[119] 11 Nov 1903 to Jaeger (HWRO 705:445:8629).
[120] 10 July 1903 to Elgar (HWRO 705:445:3561).
[121] Friday [31 July 1903] (Novello archives).

The centre of the Three Choirs Festival at Hereford that September was a performance of *Gerontius* in the Cathedral with the text intact—a victory over conservative clergy. Among Edward's many visitors at the Festival was Stanford, 'who came & talked (contra) Strauss to me—rather on the brain I think'.[122] Alice ran a house party which included many friends—Rodewald, Bantock, Ernest Newman, Frank Schuster. Schuster valued keenly his moments alone with Edward. After the Festival he wrote: 'In the matter of friendship we are, and I thank God for it, on terms of equality, but try as I may I can never forget that *you* have written a "Gerontius" and *I* have only listened to it!—the gap *awes* me.'[123]

Not everyone said things so directly. To lighten the atmosphere, Edward put on a cap Schuster had left behind and posed in it when *The Sketch* came to make a 'Photographic Interview' at Craeg Lea in the following week. He wrote: 'I have just been photographed in your cap! there's a conversion for you, if all goes well you will see "it" in Sketch some day.'[124]

The dominating note of the composer's depicted outdoor life seemed to be golf—so much so that after *The Sketch* photographs appeared, *Punch* had a go:

The other day, when playing over the Malvern Links with Sir Charles Stanford, Mr. Elgar gave a wonderful exhibition of his power as a driver. Slicing his tee shot at the short hole over the railway, Mr. Elgar managed to land his ball in a passing motor-car, which was not stopped until it had gone half a mile, thus surpassing all Mr. Blackwell's records.[125]

A week after *The Sketch* visit, he was interviewed again by the Birmingham journalist R. J. Buckley. Asked about *The Apostles*, Edward insisted once more that he did not write to order. Then he added a comment calculated to set at rest any lingering rumour about rushed or uncompleted work:

'I have been thinking it out since boyhood, and have been selecting the words for years, many years. I am my own librettist; some day I will give you my ideas on the relationship between librettist and composer.'

Like many busy men of active brain, Dr. Elgar relegates an infinity of things to the shadowy morrow.[126]

Having carefully made sure of Richter's blessing, Edward determined to conduct *The Apostles* première himself: it would have been difficult for anyone else to learn the vast work in what would amount to no time at all. But reports of the Birmingham Chorus rehearsals were favourable. On 21 September

[122] Undated letter, *c*.9 Sept. 1903, to Jaeger (HWRO 705:445:8590).
[123] Saturday [11 Sept. 1903] (HWRO 705:445:6898).
[124] 16 Sept. 1903 (HWRO 705:445:6990). *The Sketch* 'Photographic Interview' appeared on 7 Oct.
[125] 21 Oct. 1903.
[126] 'Dr. Elgar at Home', interview of 24 Sept. later drawn upon for Buckley's book, *Sir Edward Elgar* (John Lane The Bodley Head, 1905).

Edward himself took a rehearsal; again on the 29th, when he expressed pleasure at the result.

In London at the beginning of October he rehearsed the six soloists. Then came two rehearsals of the orchestra in Manchester, since most of the Festival orchestra were Hallé players. Through it all he worked ceaselessly with John Austin correcting orchestral parts as they came from the printers and from a succession of three Novello copyists: on 3 October they were at it for nine hours, and he wrote to Frank Schuster next day, 'I am *dead* with fatigue . . .'[127]

But all was ready for the combined rehearsal in Birmingham on 9 October: it was a great success. The faithful band of friends assembled at Birmingham included the Speyers, Stuart-Wortleys, Lord Northampton, Jaeger, Rodewald, Kalisch, Dora Penny, and Schuster—who motored them in his new car to spend the weekend with Lord and Lady Windsor at Hewell Grange.

On the performance morning, 14 October, Alice noted: 'Everyone in Artist room thrilled with excitement.' Every place in the Town Hall had long ago been sold. Even with many extra seats fitted in, the demand was such that a ballot had to be instituted, in which more than 700 people were disappointed. The authorities even thought of mounting a second performance at the end of the Festival week, but they were hindered by the system of issuing serial tickets for the week's entire programme. As it was, the whole Festival was focused on this performance.

The Birmingham Daily Mail began its account with the distinguished audience:

The President, Earl Howe and Countess Howe, arrived about a quarter past eleven, and they were followed soon after by the members of their house party at 'West Grove', the beautiful Lady Cynthia Graham, attired in a dove grey dress with large feather ruffle and black hat, who arrived in the city last evening, being escorted by Mr. Schuster. A large party came again from Highbury, Mrs. Chamberlain, who was wearing a red costume with black hat, being accompanied by the Misses Chamberlain and several other ladies, and Mr. Neville Chamberlain . . . So fashionable was the assembly this morning that comparatively few of the visitors arrived on foot.

. . . When Dr. Elgar took his seat shortly before half-past eleven every inch of the interior of the vast building was occupied by an audience which will undoubtedly rank as one of the most brilliant of the many distinguished companies seen within the walls of the historic hall. A conspicuous figure in the front row of the great gallery was Madam Clara Butt, who had as her companion Mrs. Elgar. There was also present an unusually large contingent of London pressmen . . .[128]

Alice wrote: 'Wonderful performance & wonderful impression on audience, the quiet & silence at end of 1st part, the highest tribute. Last chorus overwhelming.'

The Worcester Herald reported:

At the conclusion of the work the audience remained for a few moments as if

[127] 4 Oct. 1903 (HWRO 705:445:6107).   [128] 15 Oct. 1903.

spell-bound, and unwilling to mar the devotional effect of such a masterpiece by applause. It was, however, only for a few moments, and then the enthusiasm was not to be restrained . . .[129]

'The verdict could not be mistaken,' reported the *Mail*:

It conveyed a complete, a triumphant success for Dr. Elgar, who was three times recalled, the audience, band, and chorus giving vent to their feelings by a perfect storm of applause and cheering.[130]

In the Artists' room Richter was determined as all the rest on a grand event. The chorus-master Wilson remembered his old chief's expatiation:

'This is the greatest work since Beethoven's Mass In D.'
    And when I said—knowing his intimate connection with Wagner and his works—I said 'excepting Wagner?' he replied almost savagely:
    'I except *no* man.'[131]

'E. happy,' wrote Alice in the diary: 'Everyone saying Marvellous. Quite a reception at Hotel later. Dined with the Littletons. Wonderful day.'
    The critics thought so too. Old Joseph Bennett wrote in *The Daily Telegraph*:

The occasion may be described as in some respects unique. It was so in my own personal experience, for through all the years in which I have known the Birmingham Festival it has never happened that the whole musical world, not only in this country but also abroad, has gathered more or less closely around the production of an Englishman. It is a good omen that at last a man of our own race and nation has come to the extreme front, and drawn to himself the wondering admiration of all who profess and call themselves musicians and lovers of the art.
    There is something impressive in the position now occupied by Elgar. He is not an intriguer. He does not compass heaven and earth making his proselytes to believe in his own powers, neither does he trim his sails to catch the varying breezes of popular opinion. Having something to say in the fashion which appears to him best, he says it straight out, and leaves the issue to the fates. Yet, though sturdily independent, courting nobody, he now occupies the position of a man with whom most people are determined to be pleased. There must be something in him much more than common to bring about this result.[132]

One factor discouraged any final judgement on the music. It was the unfinished state of *The Apostles*, which was discussed openly by all the critics. Alfred Kalisch wrote in *The World*:

In one respect *The Apostles*, as far as we have gone, is unique in the history of music. The first two parts deal with the Ministry, the Passion, and the Ascension, subjects which have inspired composers in all ages. But they have never treated them merely as a prelude to a later drama.

---

[129] 17 Oct. 1903.   [130] 15 Oct. 1903.
[131] 4 June 1931 to Elgar (HWRO 705:445:4908).
[132] 15 Oct. 1903. The following four quotations are of the same date.

Fuller-Maitland wrote in *The Times*:

It is abundantly evident that a great deal of the first two parts of *The Apostles* is but an index to that which will occur in the third part . . . In one sense the book, or its arrangement, is a little disappointing, as matters at present stand, since it presents rather a series of more or less detached pictures or scenes which are intended ultimately to lead to a definite end.

For Joseph Bennett the detachment arose out of a system of thematic motives which in the end seemed to defeat purely musical development:

It may be true that the method here so conspicuously exemplified helps the composer in the accomplishment of his work by reducing it to the level of mere deftness, but it is possible to buy ease as well as gold at too high a price. When the exigencies of his method demand orchestral assertion of the gospel theme, the preachers motive, and so on, the vocal part, if any, must conform to the conditions thus imposed, and so loses its freedom.

Thematic motives in Wagner were kept short to encourage combining and musical development. Many of Edward's *Apostles* motives were longer and more elaborate. The result was a crowded effect which puzzled the *Daily News* critic Baughan. Unlike Bennett, Baughan upheld the system of thematic motives in modern music drama:

But I must admit that it is apt to be illuminative on paper more than in hearing . . . unless the invention of themes is distinctive. They must have unmistakable character; they must be inspirations, or else they conjure up no definite ideas, and their constant recurrence either passes unnoticed in the swirl of the orchestral current, or it seems to produce a feeling of irritation. One has the feeling of chasing shadows instead of realities, of watching faint outlines which should develop clearly but do not . . .

Taking 'The Apostles' as it stands, its shifting pictures convey an impression of illusiveness. The central idea is not clear . . . Dr. Elgar has invented some characteristic themes, and, especially, some characteristic harmonies, which seem to give connection to the music, but as a whole his specific musical invention has not been equal to his imagination.

Canon Gorton replied to these criticisms in *The Manchester Guardian*:

The work *is* incomplete. We were bid recognise that fact from the outset. But so also is the Gospel of St. Luke, which Dr. Elgar most closely follows, incomplete, though it too ends with the Ascension. St. Luke felt that it was incomplete, and therefore wrote the Book of the Acts; but that in turn is incomplete . . . His subject is the incompleted Kingdom.

The third part of 'The Apostles' we wait for.[133]

That did not answer the critic of *The Athenaeum*, who seemed to see a flaw in the way the whole subject was cast:

[133] 'Elgar's "Apostles" from a Churchman's Point of View'.

In the final chorus attention is again drawn to the Man of Sorrows. Thus . . . we pass from the greater to the less, from the divine to the human.[134]

* * *
* *
*

When he gave up finishing Part III of *The Apostles* for the 1903 Birmingham Festival, Edward had told Alfred Littleton that the financial arrangements would have to be reconstructed. On the day after the performance Littleton made his proposal: as the finished work consisted of two out of three parts, Novellos would pay two-thirds of all the sums originally agreed. Edward emerged from this meeting, Alice noted, 'very depressed re. financial arrangements'.

He turned to G. H. Johnstone, pointing out that the length of *The Apostles* as it stood was over the two-hour estimate submitted for the original plan in December 1902. Johnstone undertook to get the original sums reinstated for the present score, and wrote afterwards to confirm this:

I want you to feel quite safe about the £500 being paid to you [as the initial instalment on completion], but if they do not pay it all at present, please remember that *I* am responsible to you for the amount & will find you the balance until they *have* paid it.[135]

Alice took up the fight in a different direction. She marched up to Frank Schuster, with all his powerful friends at the Opera House and elsewhere. Alice appeared to Schuster 'very tiny indeed, and had a quiet, intimate way of speaking. This caused her to come up close to anyone to whom she had anything important to say' (as he recalled for Adrian Boult):

'You know the way dear Alice used to come up to one and confide in one's tummy? Well, one day she said to my tummy:
"Frank, dear, we are always going to [Hereford] Festivals or [Birmingham] Festivals and so on. Don't you think we might have an Elgar Festival some time?"
'My tummy reported what she had said and I went off to see Harry Higgins . . .'[136]

Higgins, with the knowledge of the *Coronation Ode* success (even if not in the end premièred at Covent Garden), was interested. He suggested a three-day Festival entirely devoted to Edward's music during the coming winter. The proposal was passed to Novellos, where it added its effect to Johnstone's remonstrances over *The Apostles*. Alfred Littleton did the gentlemanly thing and made up the £500. A further £500 was to become due when 10,000 copies had been sold: it looked as if that might be within a year or two.

Edward's response was immediate. When even Novellos acknowledged the

---

[134] 15 Oct. 1903.   [135] 22 Oct. 1903 (Elgar Birthplace).
[136] Sir Adrian Boult, *My Own Trumpet* (Hamish Hamilton, 1973) p. 70.

promise of the subject in its first instalment, then its promise to himself was re-extended.

Malvern

Oct 28: 03

My dear Littleton—

Very many thanks for your letter & the enclosure (formal receipt in the envelope). It's very kind of you & your directors to take the large view of the matter & this makes it possible for me to proceed with my work comfortably: I knew you wd. like to know this.

My ideas now revert to my colossal scheme of years ago—but I may not live to do it.

I. The Apostles (which you have)

II. A continuation as talked over with you &

III. The Church of God (or Civitas Dei)! . . . I have the IIIrd. part libretto done in one shape.

(It was the selection which Minnie Baker had arranged for him in 1894.)

Now as to performances [at an Elgar Festival]. I have telegraphed to Mr. Higgins asking him to see you today & I hope the interview has been accomplished. If necy & if I get a wire before twelve I could come up to town tomorrow.

> Kindest regards
> Yours v sincy
> Edward Elgar[137]

The Covent Garden scheme coincided with another proposal from Richter's current London agent, Alfred Schulz-Curtius (who had also managed the Westminster Cathedral *Gerontius*). This was that Richter should conduct the first London performance of *The Apostles* in March with the entire Hallé Chorus brought to Queen's Hall for the occasion. In the end Schulz-Curtius met with Harry Higgins to combine their proposals in a three-day Festival uniting the advantages—summarized by the Novello secretary in a letter to Edward:

The artistic consideration—which caused you to favour Richter & Queen's Hall
The prestige—which caused you to like the Covent Garden cycle.[138]

The presence of Richter as director of the Elgar Festival would constitute Covent Garden for those three days as Edward's Bayreuth. If ever there was an invitation to go on with his *Ring* of oratorios here it was.

The second oratorio could not possibly be written in the time before March 1904, when the Festival was scheduled. But Edward's thoughts went to the final 'Miscellaneous' concert planned for the third night at Covent Garden. There was the place for the work to show his gratitude and devotion to Richter. The ideal tribute would be the Symphony which he had promised to dedicate to Richter. It had been discussed between them again at Birmingham; then it was

---

[137] Novello archives.   [138] 29 Oct. 1903 (Novello archives).

glanced at in a letter describing the Elgars' motor journey with Frank Schuster to Hewell Grange:

. . . All the incidents are being worked into the Symphony in Eb dedicated to Hans Richter

by his friend
Edward Elgar.[139]

Though the Covent Garden Festival was still a secret, the Symphony was beginning to be rumoured in the press as Dr Elgar's next project. *The Worcester Daily Times* reported:

It is understood that before he completes the oratorio he will finish scoring his first symphony. This may be heard in the spring at Manchester from Dr. Richter's orchestra, or it may be postponed until the [summer] season, and be produced at Queen's Hall. Rumour asserts that the Philharmonic would like to produce it, but they do not offer substantial inducements as a rule. There is another story that it may be secured for the Leeds Festival.[140]

If the Symphony was for Richter, then its première was not for Leeds. Edward made the best he could of a final withdrawal from Leeds by telling the Festival secretary Spark that the Symphony might somehow interfere with the production of a big choral work for Leeds at some future time. Spark's patience was wearing thin: he begged for at least a short choral work for the Festival in October 1904, if only to save face with the press.[141] But even that was not to be forthcoming.

On 30 October Dora Penny came to stay the night at Craeg Lea. They had 'a glorious evening of music'.[142]

One of the things that always fascinated me in the Craeg Lea study was a collection of books bound in green linen which were piled on a shelf of a tall 'what-not' close to the piano. As soon as I could, after arriving on a visit (making sure that wandering around looking at things did not disturb), I used to gravitate towards that shelf and turn over the green books. He wrote, generally in blue pencil on the cover, the name of what was inside. I remember seeing one fat book with 'Symphony' on it, and I said:

'Are you writing a Symphony? How perfectly splendid! Couldn't you possibly play some of it?'

'Possibly.'[143]

One idea for the E flat Symphony had started a particularly promising

[139] Sunday [11 Oct. 1903] (Richter family).
[140] Undated cutting at the Elgar Birthplace. Similar reports appeared elsewhere toward the end of Oct. 1903.
[141] 31 Oct. 1903 (HWRO 705:445:5988).
[142] Powell, *Edward Elgar: memories of a variation*, p. 57.
[143] Ibid., p. 38.

development. It began on a half-sheet of music paper in pen and ink—a rise of crotchet steps followed by a fall of quavers:

The pace was later marked ♩ = 72. Seventy-two beats to a minute is the normal human adult pulse rate: ♩ = 72 is known to musicians as 'walking pace'. Edward's sketch opened with four crotchet steps walking upward to meet an overlapping quaver descent. The measured rise to a quicker fall could suggest that this music had come through walking up hill and down dale—or pedalling a bicycle through those places.

When the whole figure repeated sequentially in the ink sketch, the bare notes began to acquire harmonic definition. The crotchets became first-inversion triads mounting up in parallel—and then falling through variegated chromatics.

With harmonies moving about in this way, they might be supported by a pedal-point in the bass—a fixture under the movement. And that fixture, looking back at the three-flat signature, could be C. The idea gave pause: after adding the tied notes of a single C sounding the bass clef, Edward laid down his pen. He did not use it again in the sketch.

How to develop it further? W. H. Reed recalled many later occasions in developing an idea when 'with his characteristic restlessness, he had started a new hare.'[145] He picked up a pencil—thus keeping the chance to erase—and

---

[144] '*Rough Sketches for Symphony no. 2 given to Alice Stuart Wortley*' (bound volume of MSS at the Elgar Birthplace) fo. 45.
[145] *Elgar as I Knew Him*, p. 26.

under the still unharmonised first statement sketched a new bass-figure round
the pedal C. The pattern of dotted rhythm gradually smoothing in downward
progress made an effective counterfoil to the steady march above. He marked
the new idea 'Good', and repeated it in the same C-position beneath each
sequential statement.

The contrast of a sequence moving through tonalities in the treble with a
fixed figure in the bass invoked again a dialogue that was all-pervasive in the
musical thinking of Edward's generation, when composers everywhere were
pushing tonal relations farther and farther to expand the range of their
expression. A line from one of Shakespeare's sonnets came to his mind: ' . . .
Art made tongue-tied by authority'. (It was from Sonnet 66, beginning 'Tir'd
with all these, for restful death I cry.') Atop a new full sheet of music paper he
wrote 'Art made tongue-tied by authority', and developed the ensemble in
short score. After 8 bars the ensemble was inverted, and the C-centred *idée fixe*
came out on top.

Still this figure obsessed him with its tiny drama of rhythmic contrast
smoothing to equality. Repeated over and over, it would give an irresistible
propulsion to any music it inhabited. So the 'new hare' began to outstrip its
progenitor.

One help for developing a musical shape, as Edward confided to Reed, was
the thought of some particular instrument: 'He told me that if he thought of any
instrument's tone quality he would naturally invent music proper to it.'[146] The
bass register of this figure's origin, its rhythmic definition and punctuating
repetition all suggested the tuba: and it was when considering the tuba's timbre
and sonority, as he said to Reed, that this development began to show itself:

The rhythm made a framework on which endless figures could be shaped.
Where they might lead remained yet to be seen.

The Elgars were planning to go to Italy for the winter: the change of scene and
warmer climate might bring the Symphony to its flowering. Closer to home, the
proposed absence gave an opportunity long sought to resign the conductorship
of the Worcestershire Philharmonic in a final and friendly way: there would be
no time in future for taking the many rehearsals.

Before departing to Italy, however, he was to go to Liverpool for another
concert with Rodewald's orchestra. On 3 November Rodewald sent a cheery
card looking forward to the concert: he had had 'a bad attack of Flue' but said
he was mending.[147] Soon afterward the influenza brought complications, and at

[146] Ibid., p. 130.   [147] HWRO 705:445:2745.

last gradual paralysis. Rodewald sank into unconsciousness, and the reports grew worse and worse. On 9 November Edward could stand it no longer and caught the morning train to Liverpool—only to find that Rodewald had died just before he arrived at midday. He wrote to Jaeger: 'I broke down & went out—*and it was night* to me.' They were the words which had headed the Judas sketch sent to Jaeger four years earlier. 'What I did, God knows—I know I walked for miles in strange ways . . .'[148]

Returning to Malvern, he spent the next days walking and walking. A letter came from Ernest Newman:

Why are men like this taken from us when so many vermin are left to encumber the earth? I get sadder & sadder as I think of what poor helpless insects we are. What is the use of our befooling ourselves in the fictitious world of art when a brutal hand can descend on us like this, in a moment, and crush a score of us at once?[149]

Newman was an agnostic. Many years later he was to measure the extent of Edward's agreement with him:

I often discussed philosophical—*not* religious—questions with him, and I can vouch for it that on these he invariably turned a purely philosophic mind. Where all matters of this kind were concerned he was a dual being, coolly rational when the matter under discussion was purely intellectual, religious when it was artistic in the first and intellectual only in the second place. His Catholicism seemed to me to be in large part the product of the impact on him from boyhood onwards of all the magnificent art that Christian emotion has called into being throughout so many centuries. And so, when put to a sharply realistic test his religion, I think, was apt to give him scant support.[150]

Rodewald had been the first close friend of Edward's maturity to disappear without warning, almost from one moment to the next. When Edward arranged with Alfred Littleton in October to extend his *Apostles* music so largely, he had added 'but I may not live to do it.' And within a fortnight this splendid Apostle of his music was struck down at an age less than his own. Weeks after Rodewald's death Edward wrote to Walford Davies:

I have had the most severe shock that had ever happened to me & I bear it ill I fear
The man was a *good* man—one of the best—& he has gone & the blank can never be filled & I resent this in some way I cannot define.[151]

It was as if he had always suspected that the good things must be always in the past: this was the suggestion of the nostalgia that was in his music. Most recently he had given the words to Judas: 'Our life is short and tedious, and in the death of a man there is no remedy.'

They were not due to start to Italy for another week, but on 17 November Alice persuaded him to go up to London for distraction during the intervening days. Passing through Worcester on the journey, he wrote to Ivor Atkins: 'I stole an hour between trains to say good-bye. I've been into the Cathedral,

[148] 11 Nov. 1903 (HWRO 705:445:8629).
[150] *The Sunday Times*, 6 Nov. 1955.
[149] 8 Nov. 1903 (Elgar Birthplace).
[151] 3 Dec. 1903 (HWRO 705:445:2178).

which I have known since I was four, and said farewell. I wanted to see you. I am sad at heart and feel I shall never return . . .'

In London he settled details of the Covent Garden Elgar Festival with Harry Higgins and Schuster (who would act as guarantor). *Gerontius* would occupy the first evening, *The Apostles* the second, and on the final night a miscellaneous concert comprising *Froissart*, a selection from *Caractacus* (thus exposing part of it at last to a Covent Garden audience), the *Variations*, *Sea Pictures*, the two *Pomp and Circumstance* Marches, and the new work. He also discussed with Higgins the idea of writing a ballet for Covent Garden later—not *The Decameron* now, but Rabelais's *Gargantua and Pantagruel*.

On 19 November Edward and Frank Schuster went overnight to Manchester for Richter's third *Gerontius*—a resounding success. Afterwards Adolf Brodsky came to the Grand Hotel to offer an appointment as Professor of Instrumentation and Composition not only in his own Manchester College of Music but in the City University as well. The stipend was £400 per annum and 'all personal freedom': a staff assistant would do the drudgery, as Brodsky wrote afterwards: 'As little we would expect Franz Liszt to be a piano teacher, as little we are expecting you to be a "teacher" in the common sense of the word. It is your personality we want to secure. Your name would give glory to the Institutions & attract, I am sure, all the talent of the country.'[152] Despite his total lack of academic background, it was a flattering offer. Edward promised to think it over during the winter in Italy.

* * *

They left London on 21 November. Before the end of the month they were in Bordighera, looking for villas to rent: but the season was advanced and many had been taken. Finally they found an acceptable house at Alassio, with two old servants easily engaged. They went in on 11 December: Alice's first task was to supervise the moving of all the furniture so that Edward could have his study upstairs, as at Craeg Lea. He bought a quantity of fine rag scoring paper made in Turin, and together they hired a piano.[153] The weather turned to rain and wind, but Edward wrote to Jaeger:

What matters the Mediterranean being rough & grey? What matters rain in torrents? Who cares for gales—*Tramontana*? we have such meals! such wine![154]

On 21 December Miss Burley arrived with Carice for the Christmas holidays:

From the balcony of the villa one had a good view of the Mediterranean and I remember

[152] 5 Apr. 1904 (HWRO 705:445:2402).

[153] In her reminiscences Miss Burley said that the piano was hired in Genoa when she and Carice arrived. Alice Elgar's diary places the hiring and arrival of the piano ten days earlier at Alassio, and makes no mention of any second piano hired at Genoa. Miss Burley's memory years later may have confused her own experience with what she had been told.

[154] 13 Dec. 1903 (HWRO 705:445:8632).

Edward's joy at discovering that the fishermen were still drawing in their nets in the precise manner described by Virgil.[155]

More than fishing nets were being drawn through those waters. A few days before Miss Burley's arrival a ship had sailed in with casks of wine, which had been unloaded by the simple expedient of throwing them into the water and drawing them ashore with ropes.[156] On the 23rd Edward took Miss Burley down to the pier, made her interview the customs inspector and find out where this wine was being sold:

I learnt that it was from Sardinia (Vino di Sardegno) and that it was delivered to a little trattoria in a side street.

It was clear that if my reputation [as a linguist] was to be maintained I must get some of this wine, so I found my way to the trattoria and went in. (I did not take Edward as he tended to embarrass one's shopping by making a running commentary on the proceedings in English: 'Tell him he's a squinting pirate,' he would say while one bargained.)

The principal room of the trattoria was very large and dark with a low ceiling and long tables. It looked a good setting for any imaginable wickedness. However the proprietor sold me a large fiasco of the wine, no doubt at about five times the price paid by the habitués, but even so ridiculously cheaply, and I carried it back to the villa in triumph. Edward greeted me with shouts of joy and would drink nothing else.[157]

The windy weather turned cold, and soon this affected Edward's spirits. On the last day of 1903 Alice wrote:

'Still cold & grey & windy—E. & A. much depressed at these conditions & wondering if they will not pack up & go home. E. feeling no inspiration for writing.

When he played the sequential idea for the E flat Symphony to Miss Burley, she showed no enthusiasm: 'Like many of his tunes, it did not strike me as being particularly good, a two-bar phrase that was repeated sequentially *ad nauseam*.'[158]

On 3 January 1904 he wrote to Jaeger:

This visit has been, is, artistically a complete *failure* & I can do nothing: we have been *perished* with cold, rain & gales—five fine days have we had & three of those were perforce spent in the train. The Symphony will not be written in this sunny(?) land . . .
I am trying to finish a Concert overture for Covent Garden instead of the Sym . . .[159]

So again the Symphony eluded him. The time remaining before the Covent Garden Festival in March was now very short—even to write what he would describe as a 'Fantasia Overture'.

---

[155] *Edward Elgar: the record of a friendship*, p. 169.
[156] Elgar noted the ship's arrival on 16 Dec. (notebook at the Elgar Birthplace). Again Miss Burley's reminiscence was to place herself at the centre of a scene which occurred before she arrived.
[157] *Edward Elgar: the record of a friendship*, pp. 169–70.
[158] Ibid., p. 168.   [159] HWRO 705:445:8676.

At last there was some sun. In the afternoon of 3 January they went an outing to an old church at Moglio. Miss Burley recalled it as

. . . a dear, dirty little place spectacularly placed on the hillside. The minute houses were built beside and below the level of the steep *salitas* with the result that the returning inhabitants would walk rapidly down hill and suddenly disappear into their homes like rabbits into a burrow. Edward loved this and never tired of seeing them disappear.

'There one really could roll home,' someone said.

'Mòglio, Moglio, roglio, roglio,' said Edward with a pleasing access of fatuity.[160]

Headmistress superiority might find pleasure in pointing to fatuity, but this little jape for Carice's amusement started an idea for the Overture. In later years W. H. Reed noted

. . . how a sentence, however frivolous, or perhaps only a word, would run in his head almost *ad nauseam*, just as a fragment of a tune will do . . . I learnt from [Carice] afterwards that he kept repeating this ridiculous name until at last he actually put it into his music:

On 7 January Alice wrote to Frank Schuster:

I hope you like the idea of the Overture 'From the South'. I feel sure you wd. think what is written already is simply fascinating.

Do you not think the King will come to the Festival? That really wd. be a great point for it, how shd. he be asked? You will say, 'there, again, sheer audacity'—but the Fest. is altogether so wonderful (thanks to you) to think of that nothing seems impossible.

. . . E. & Carice & our friend who brought her out are gone for one of the usual wild afternoon walks, up precipitous mule tracks, 'Salitas', frightfully steep & wild but leading to *lovely* spots & views . . .[162]

Two days later Alice went with them on another excursion: 'Lovely day. To Andora by train after lunch—Walked to . . . Roman bridge, & then up to San Giovanni Battista . . .' The little church stood near a group of pine trees, particularly remembered by Miss Burley (who was soon to return to England for the opening of the new school term):

We had climbed one of the *salitas* and came suddenly upon a little chapel by a group of pine trees. Classical in style, like a temple, it was falling into ruin and the sudden impact of its beauty silenced us.

'It really only needs a shepherd with his pipe to make the picture complete,' I said.

At that moment to our amazement a shepherd did in fact appear from behind the chapel. He was dressed in a sheepskin, and unconcernedly drove his flock along the path and out of sight.[163]

[160] *Edward Elgar: the record of a friendship*, p. 170.
[161] *Elgar as I Knew Him*, pp. 33–5.   [162] Elgar Birthplace.
[163] *Edward Elgar: the record of a friendship*, p. 171.

For Edward it could revive pine scented memories of Broadheath, of Spetchley, of Fontainebleau. The 'shepherd with his flock and his home-made music'[164] suggested a crotchet rise and quaver fall that echoed the 'walking' figure of the Symphony sketches:

This was joined with 'Moglio', and the two figures telescoped together would call

The opening of Tennyson's poem 'The Daisy' came into his mind:

> . . . what hours were thine and mine,
> In lands of palm and southern pine . . .

To reach this enchanted spot they had trod the steps of Romans, as Edward wrote: 'the massive bridge and road still useful, and to a reflective mind awe-inspiring.'[166] Again Tennyson's poem said it:

> What Roman strength Turbia show'd
> In ruin by the mountain road . . .

Edward developed his 'Shepherd' music 'to paint the relentless and domineering *onward force* of the ancient day and give a sound-picture of the strife and wars ("the drums and tramplings" [Sir Thomas Browne]) of a later time':[167]

Suddenly the entire Overture was in focus as a record of this afternoon's experience—'streams, flowers, hills; the distant snow mountains in one direction and the blue Mediterranean in the other.'[168]

In a flash it all came to me—the conflict of the armies on that very spot long ago, where now I stood—the contrast of the ruin and the shepherd—and then, all of a sudden, I

---

[164] Quoted from a letter written by Elgar about 22 Feb. 1904 to Percy Pitt and Alfred Kalisch, in *Analytical and Descriptive Notes* for the Covent Garden concert of 16 Mar. 1904.

[165] The words appear thus pencilled over the clarinet figure at cue 11 in Elgar's MS full score (Royal Academy of Music).

[166] Quoted in Pitt and Kalisch, *Analytical Notes.*

[167] Ibid.

[168] Ibid.

came back to reality. In that time I had composed the overture—the rest was merely writing it down.[169]

But writing it down also meant evolving the structure to display all these ideas and hold them together. Two days later he was deep in it.

As an opening theme he used the idea written in Sinclair's Visitors' Book in the summer of 1899 for 'Dan triumphant (after a fight)':

The tonality was E flat major—the key of his unfinished Symphony. This wide-ranging figure in the E flat tonality might recall Strauss's *Heldenleben*—if Edward's idea had not been written down in 1899, before ever he heard Strauss's work. Still he was using the idea only now, in the light of an acquaintance with Strauss's music which had culminated in the big Strauss Festival at Queen's Hall in June 1903. So this music for his own London Festival went toward a Straussian opulence of themes and orchestral tuttis and solos of insistent lengths rare in his earlier music.

The opening 'heroic' theme of the Overture was immediately given a counter-melody to produce a first hint of 'Moglio':

A second counter-melody appeared:

Then the descending steps of the first counter-melody generated a new figure of descending and ascending steps:

The descending steps shaped a *Nobilmente* to reflect the great sunrise in *The Apostles*:

[169] Interview with Miller Ular, *Chicago Sunday Examiner*, 7 Apr. 1907.

All this was nothing more than a first subject group, and already there were more than a hundred bars.

The 'shepherd' entered, with 'Moglio', telescoping to 'Fanny Moglio'. This was not the Overture's second subject, but only an episode in the midst of exposition. So it was like the 'Guildhall' melody in *Cockaigne*—a hint, between primary and secondary subjects, of where all this music had begun.

At last came the second subject in a new key and new metre:

The descending steps in the middle (echoing first-subject ideas) were extended with falling fourths to echo 'Moglio' once more:

It completed an exposition of 235 bars.

The Overture's middle section also recalled *Cockaigne* by continuing to bring in new variants. The 'shepherd' and 'Moglio' figures gradually worked up to introduce the '*Grandioso*' development of the 'Romans'. The 'Romans' in turn produced a huge suspension of falling fifths, spanning the interval of 'Moglio' over and over again:

—a sequence answered and repeated in tremendous *tuttis* through 65 Bars.

Another sequential figure mixed the 'heroic' primary theme with 'Romans' *con fuoco* through a further hundred bars. Then still another new idea appeared from out of the exposition's F major second subject:

In the Analysis for the first performance, Edward's friends Percy Pitt and Alfred Kalisch stated that this melody was '. . . founded on a canto popolare, [though] the composer tells us that he does not know who wrote the tune and that he has not copied it note for note.'[170] Later he admitted to Ernest Newman that it was entirely his own.[171] The melody would sing first through the solo viola: thus it invited memories of Berlioz's *Harold in Italy* and perhaps of Strauss's *Don Quixote*.

A *pianissimo* augmentation of the opening 'heroic' theme breathed across the romantic distance to hint at recapitulation. Clarinet and horn echoed the 'canto popolare'; 'Moglio' floated through the violins; and the lonely viola sang a last lingering phrase to close a middle section of 348 bars. The whole effect, as with the middle section of *Cockaigne*, was of variations. So there was a real problem in turning that forward-moving progress round to recapitulation. In *Cockaigne* he had managed it by brilliant telescopings and recombinings. But that needed time and thought, and time for *In the South* was running short.

On 21 January came a letter asking Edward to dine with the King and the Prince of Wales at Marlborough House on 3 February: so it seemed that Alice's second suggestion to Frank Schuster had also borne fruit. The dinner would precede a 'Smoking Concert' in which it was hoped that Dr Elgar would conduct his *Pomp and Circumstance* March. Alice telegraphed Edward's acceptance. It meant leaving Italy in just over a week. In those final days at Alassio he finished his short-score sketch for the entire Overture.

The last *tranquillo* note of the 'canto popolare' was interrupted with a full recapitulating tutti to re-sound the opening. The music followed the pattern of the exposition with hardly a change to re-present almost literally a great part of the first-subject material. To this he attached all the second-subject material similarly reproduced from the exposition—now in the tonic E flat major but otherwise much as it was. At length the figure of long descending steps from the middle exposition re-entered—but changed and softened, as if recalling far places of the sunny spirit. Below it the Overture's earlier themes climbed toward a brilliant end.

The whole work extended to nearly 900 bars—almost three times the score length of *Froissart* and *Cockaigne*—and would occupy twenty minutes in

---

[170] *Analytical and Descriptive Notes*, p. 33.   [171] *Elgar*, p. 173.

performance. Its length expressed partly nostalgia for a vision of the south only glimpsed through the cold weather of the actual visit, partly the wish for the Symphony still unachieved. When Jaeger offered an initial comparison with the tone poems of Richard Strauss in an analysis he would write, Edward demurred:

I do not think I should put that about Strauss at the beginning—not necessary—. S. puts music in a very low position when he suggests it must hang on some commonplace absurdity for its very life.[172]

The large expression of Edward's Overture had been achieved inside a month. Before they left Alassio on 30 January he was beginning the orchestral score.

Edward, Alice, and Carice reached London on 1 February 1904 and went to stay with Frank Schuster. Next evening Edward rehearsed *Pomp and Circumstance* No. 1 with the Royal Amateur Orchestral Society, and on the evening of the 3rd went to dine at Marlborough House. With the King and the Prince of Wales were Prince Christian, Prince Louis of Battenberg, Lord Shaftesbury, the Marquis of Soveral, and Lord Howe (who was placed next to Edward). Parry and Mackenzie were also present, each to conduct a work. Alice summarized Edward's description for the diary: '. . . the King talked music to E. & took him out 1st after dinner.' At the concert *Pomp and Circumstance* was encored by the royal audience.

The first section of *In the South* score, covering most of the exposition, had been left with Jaeger that morning. Calling at Novellos two days later, Edward saw some of it already copied out for the engraver. That afternoon Alice took Carice back to Malvern to open the house, air it and warm it for Edward's arrival next day. At Craeg Lea he settled instantly to work: 'E. wrote splendid "Romans" in his Overture,' noted Alice on 8 February.

The Covent Garden Festival, now barely a month away, had begun to focus national attention on Edward. The Birmingham journalist Robert Buckley was at work on a biography. On 11 February the first of a series of magazine interviewers arrived at Craeg Lea. He was Rudolph de Cordova for *The Strand Magazine*. He skilfully manoeuvred Edward through the early days and developing experiences. Coming to recent times, Edward confided some of the changes of plan attending *The Apostles* and announced his intention of returning to the immense trilogy scheme:

'There will, therefore, be two other oratorios.'
    This definite pronouncement of Dr. Elgar's cannot fail to evoke the warmest anticipations on the part of the music-loving world.[173]

In response to a question near the end about his method of composing,

[172] 13 Aug. 1904 (HWRO 705:445:8698).
[173] *The Strand Magazine*, May 1904, p. 543.

Edward's mind went first of all to the current task of orchestration and then backward to the source of it all:

How and when do I do my music? I can tell you very easily. I come into my study at nine o'clock in the morning and I work till quarter to one. I don't do any inventing then, for that comes anywhere and everywhere. It may be when I am walking, golfing, or cycling, or the ideas may come in the evening, and then I sit up until any hour in order to get them down. The morning is devoted to revising and orchestration, of which I have as much to do as I can manage. As soon as lunch is over I go out for exercise and return about four or later, after which I sometimes do two hours' work before dinner.[174]

Next day the 'Romans' section of the Overture went off to Novellos in score, and three days after that he sent the *con fuoco* music leading into the 'canto popolare'. Jaeger responded on the 16th: 'Thanks for Score "From the South". The Copyist (Dodd) is hard at it, but the Engravers (Geidel) cannot do much until they see *more "copy"*. So kindly let us have the Remainder as fast as you can . . .'[175] On the 17 and 18 February he scored the rest of the 'canto popolare' and on into the recapitulation through a bad headache. Several hours on the 18th had to be given over to another journalist, Pilkington of *The Graphic*, who took his own photographs as well as interviewing Edward. None the less late that afternoon another parcel of score went off.

Three days later, sitting late at his task on the evening of 21 February, he finished the Overture. It was now entitled *In the South*. He signed it, made Alice and Carice put their initials, and added the signature of 'Fanny Moglio'. The music was dedicated to Frank Schuster, for the Covent Garden Festival was in many ways his achievement.

On 25 February Richter conducted a preparatory performance of *The Apostles* in Manchester, but Edward felt too exhausted to go. He was hoping to see Richter at Craeg Lea to go through the score of *In the South* before its rehearsal at Manchester. It all depended on the speedy return of printed sheets from Leipzig, where much of Novello's engraving was done.

Malvern

Feb 28.1904

My dear Jaeger: (the hunter)

. . . The rehearsal of the new overture is in Manchester Wedy March 9th in the morning: so on Tuesday the parts corrected must be there.

Richter will be *here* on Friday [4 March]—can I have the score back by that time?

Mr. Dodd could send me all the wind & score here by Wednesday morning *next* & I could correct them & the remainder of the strings as they come in.

I think it looks smoothe enough to work it so, nicht wahr? Wire me tomorrow (Monday) if the score & parts (wind & any strings ready) can be *here* on Wednesday because I must arrange with Mr Austin to come & help me as usual . . .

Yours ever

Jaeg*ee* (the hunted)[176]

---

[174] Ibid., p. 544.   [175] HWRO 705:445:8710.   [176] Novello archives.

Novello & Co.
1, Berners Street, London, W.

Feb 29 1904

My dear Elgar

Your letter has taken my breath away! I wired at once to Dodd, & here is his reply. Moreover I consulted Oppenheimer who had this morning a letter from Leipzig saying that the completion of the proofs of the String parts would be dispatched tomorrow, Tuesday. This means that they won't reach us till Thursday, which means that they *can't* be read, returned to Leipzig, corrected & printed & dispatched to reach Manchester on the *8th*! so after much calculating & pondering we have come to the conclusion that the *only* thing to be done was to print *copies* for the Band from the *un*corrected plates; Supposing they dispatch these parts (sufficient for the Hallé orchestra) on Wednesday, they would reach us on Friday, when all our Readers & copyists (available) would devote themselves to checking the parts with the Score & copying the corrections into the *duplicate* String parts. You will not have to see them at all; there is no time, if they have to be with Richter on Tuesday or Wednesday. You really waited too long with your Coda. (I *don't* mean to suggest of course that You could have been any quicker!) Things have been driven so close, that it is *physically impossible* to do your behests.

. . . Yours, ever, & this time very much disturbed over things

AJJgr[177]

And he added later:

I shall insist on a *close time* for your next work, in which to produce all, Score & parts in peace instead of in pieces. This present method is Bedlam, to be sure![178]

Edward himself had to conduct the Overture's première since Richter would now have no chance to study it. As for the correcting, he telegraphed for the score and all the orchestral parts to come to Craeg Lea, where he and John Austin could go through them: then he would carry the whole lot to Manchester for the first rehearsal on 9 March. But Austin was ill and could not come. So the entire task of reading and playing through each of the engraved and manuscript parts fell upon Edward. With the help of Alice and Carice, he worked through 5, 6, 7, and 8 March up to the moment of leaving for Manchester.

On the morning of the 9th the Hallé Orchestra were eager and waiting, and their reading of the new work gave everyone present the excitement of hearing the music for the first time. Then another gentle admonition reached him from Jaeger, who wrote:

I had quite a *long* talk to Richter yesterday. Well, I heard amazing things . . . You *will* have festive times next week! Don't let 'em spoil you 'dear innocent guileless Child', as dear old Hans calls you in his fatherly, loving way.[179]

[177] HWRO 705:445:8715.
[178] 4 Mar. 1904 (HWRO 705:445:8718).
[179] 8 Mar. 1904 (HWRO 705:445:8720).

Grand Hotel, Manchester,

March 9:1904

Dear Mr Jaeger:
    The time has now come when I think all familiarity between us should cease: the
position I now hold—greatly owing to your exertions & friendship—warrants me in
throwing you over.

Yours truly

Now, you old moss, read the other side.

Dearie Moss:
    What an old frump you are! whenever anything of mine is to be done you beg me not
to be conceited & not to forget my old friends &c. You are an *old PIG* & deserve some
such letter as the unfinished one on the other side [–] anyhow you always seem to be
expecting it: be assured you won't receive anything of the kind from me. There is no fear
that I shall forget anyone.
    That Overture is *good* & the Roman section absolutely *knocking over*. They read it
like angels & the thing *goes* with tremendous energy & life. *Fanny Moglio* figures largely
all through to Carice's intense amusement . . .
    Don't bother me about conceit again—I haven't any except that I always resent any
familiarity from outsiders & I *do* stand up for the *dignity* of our art—not profession

Every deary Moss
with love
Yours—
Edward.

I hope the letter on the other side, if only you had the luck to read it first gave you a very
proper & deserved fit.[180]

From Manchester he went directly to Frank Schuster's house in Westminster, where Alice joined him. Newspaper reporters had begun hanging about
Schuster's front door in Old Queen Street, and Edward took to slipping in and
out by the garden wicket opening on to Birdcage Walk. When it was announced
that the King and Queen would attend the opening concert of the Elgar
Festival, the attention of the press redoubled. *The Sunday Times* wrote:

Unusual interest attaches to the Elgar Festival which is to be held at Covent Garden
to-morrow and the two succeeding evenings, for it is the first tribute of the kind that has
ever been paid to an English composer during his lifetime, and in its *locale* and
patronage is an indication that our upper classes are no longer disdainful of any
independent movement in native music . . . That in the first place special honour should
be paid to Dr. Elgar is right and fitting, for he is at once the foremost and the most
individual of the younger generation of our composers, and he has compelled even the
Continental critics to admit that English music is deserving of serious attention.[181]

The Referee (on the same day) said it more trenchantly:

It is not too much to say that the Elgar Festival scheme is unique in the history of British

---

[180] Elgar Birthplace.   [181] 13 Mar. 1904.

music. A living composer honoured in his own country by three days performances of his works at Covent Garden Theatre is positively startling.

On that Sunday, the day before the Festival opening, Edward remained in bed all the morning with a 'dreadful headache'. In the afternoon 'E. rested & got better—still precarious till evening—then came Frank's beautiful dinner party . . .'. Alice described the scene in a letter to old W. H. Elgar in Worcester:

To our surprise we found the dining room decorated with E.E. (initials) & the names of his works in flowers on the walls quite lovely. Afterwards there was a reception, everyone you can think of. Lord Howe took me in to dinner & told me the King is *so* interested in the Festival. E. took in Lady Maud [Warrender] & had Lady Radnor on his other side.[182]

It was a deeper personal challenge than anything yet faced by the piano-tuner's son who had once upon a time extemporized his own music in the house of old-world state. Henry Wood, seated on the other side of Lady Maud at Schuster's table, witnessed its effect:

Elgar, however, was in one of his very silent and stand-offish moods. In fact, his manner was so noticeable that Lady Maud Warrender, who was sitting next to me, drew my attention to it.

'What's the matter with Elgar tonight?' she whispered. 'He seems far away from us all. I suppose it is because he doesn't like this sort of homage.'

We all felt rather uncomfortable when Schuster rose and asked us to drink the health of his illustrious friend.[183]

'Frank proposed E.'s health in the most touching way with his heart in his voice,' wrote Alice.

We obeyed, and when we resumed our seats we naturally expected Elgar to make a suitable reply. Instead he went on talking to [Lady Radnor] and probably had no idea his health had been drunk at all.[184]

Next morning he and Schuster donned Court dress and went to the King's levee. Alice wrote to W. H. Elgar:

He looked quite beautiful in the charming velvet suit &c. Lord Howe presented him, and the King gave him a nice smile . . . Now we are going to the Rehearsal & this evening is 'Gerontius'. Afterwards a supper here. Prince Francis of Teck & 3 ambassadors & others are coming.[185]

At Covent Garden all was in readiness:

With the lower boxes removed, and rows of stalls occupying their place, the auditorium was much larger than usual, while the stage [was] completely filled from front to back with the orchestra and singers, nearly four hundred strong . . .[186]

[182] HWRO 705:445:4598.   [183] *My life of music*, p. 179.
[184] Ibid.   [185] HWRO 705:445:4598.
[186] *The Pall Mall Gazette*, 15 Mar. 1904.

The partition which separates the floor of the house from the orchestra was taken down, and a space in front of the stage was railed off with banks of flowers. There was a blue canopy over the stage, and there were Venetian masts and festoons of roses in which glowed electric lights. The back of the stage was arranged to represent a baronial hall.[187]

In the evening the press reported the arrival of the royal party for the opening concert:

The King looked in the best of health. Her Majesty was beautiful and stately in white, veiled with black net embroidered in silver, and a knot of pink flowers in her bodice. Princess Victoria wore black and silver, and Princess Ch[ristian] was in pink crêpe.[188]

Joseph Bennett wrote in *The Daily Telegraph*:

The auditorium was filled with one of those mighty and distinguished gatherings which only London can show. And all this in honour of an English composer who, a few years since, was an unknown provincial musician.[189]

Other papers followed suit. Only *The Times Literary Supplement* wondered whether, as an answer to early neglect, the pendulum had swung a little too far in the other direction.[190]

The King and Queen came again on the second evening to hear *The Apostles*, though it was noticed that the audience was not quite so large. On the final evening the Queen came alone, but the audience again included Lord Howe (seated with the Duchess of Marlborough in the stalls), Lady Maud Warrender, Lady Cynthia Graham, Prince Francis of Teck, Mrs Ronalds, Lord Howard de Walden. Richter conducted the first half, comprising *Froissart*, portions of *Caractacus*, and the *Variations*. Then Edward came on to conduct *In the South*:

On his appearance at the conductor's desk after the interval, Dr. Elgar was accorded an enthusiastic greeting. Then in his forceful, energetic manner the composer raised his bâton and the new concert overture struck at once the note of the joy of living in the midst of 'blue Ionian weather'. Although obviously nervous, Dr. Elgar immediately showed himself master of his orchestra, and his alert, vigorous method, with the strict beat and the characteristic wrist movement, kept his executants in a firm grip.[191]

There was some criticism of length and diffuseness in the new Overture, but Vernon Blackburn wrote in *The Pall Mall Gazette*:

Elgar, as usual, is completely original. He goes to Italy, and without any *parti pris* he quietly and meditatively observes the shining atmosphere, the lean brown soil, the olive trees, crooked, yet lovely in their nakedness, the Tuscan hills, the edges of the sea near such a port, say, as Genoa; and all this has been transmuted by a musical alchemy into the purest expression of an utterly sincere art.[192]

Alice noted: 'The evening was most beautiful—Immense enthusiasm, Orch.

[187] *The Manchester Guardian*, 15 Mar. 1904.   [188] *The Daily Mail*, 15 Mar. 1904.
[189] 15 Mar. 1904.   [190] 18 Mar. 1904.
[191] Unidentified cutting at the Elgar Birthplace.   [192] 17 Mar. 1904.

magnificent, Queen sent for E.   E. changed at C. Garden & then went to Lord Howe's, He took me in to supper. Most delightful evening, we stayed till the last.'

* * *

After the Elgar Festival the parties went on:

March 18. E. & A. & F[rank Schuster] dined at Lord Northampton's. E. took in Lady Jekyll. A. Balfour there . . .

Edward seized the chance to talk to the Prime Minister about musical copyright. He had joined William Boosey's 'Musical Defence League' to protest the growing piracy in song printing. As Boosey pointed out:

Popular songs only required two or three pages of paper, and they could be photographed or litho'ed in any old shed or barn which happened to be handy. They could then be retailed to an army of street hawkers for distribution. This in fact was what was done, and in 1902 popular songs were sold by thousands, both openly on the London streets and everywhere throughout the provinces.

. . . Some wag hit upon a further device to draw attention to our position . . . Mr. Balfour . . . had just published a little treatise on Free and Fair Trade. It was in a paper cover, was very brief, and was published at 1s. Suddenly a pirated copy of this little work appeared on the streets, retailed at one penny. It contained a note by the editor on the front page, stating that the work educationally was of such value to the masses that it had been found necessary to bring out a penny edition . . .[193]

In due course *Land of Hope and Glory* had fallen victim to this treatment. Edward asked whether Balfour would dare bring the composer's case before the committee then drafting new copyright legislation, and Balfour told him he could not 'as he would not be taken seriously'.[194] Then the Prime Minister ventured a question of his own: would Dr Elgar be disposed to accept an honour if one were offered?

Edward hinted at this a day or two later when he was interviewed by E. A. Baughan for *The Daily News*: '. . . but matters were not in train for a definite announcement . . .' The rest of the interview turned on musical copyright.

'While the world of fashion, as well as the middle classes—the real supporters of music—were honouring the art at Covent Garden, at the other end of the town our legislators were heaping indignities on it by whittling down the Musical Copyright Bill by inserting clauses which will make it quite inoperative.

'Why, of all the arts, should music be supposed to have no rights?' asked the composer, knocking out his pipe with savage emphasis. '. . . A pirate may photographically reduce the score of an important work which has taken perhaps a couple of years to write and sell it at an absurd price, and there is no practical remedy . . . The Committee on the Copyright Bill is evidently determined that the composer

---

[193] *Fifty years of music* (Ernest Benn, 1931), pp. 112–13, 116–17.
[194] Elgar's recollection of the exchange, as told to the Leicester family on 2 June. 1914 (written down by Philip Leicester).

shall be a hermit. I am practically a hermit and don't mind, but, once more, where are the younger men to come in?'

Appropriately enough, Dr. Elgar is at present engaged in writing music for chorus and orchestra to O'Shaughnessy's ode . . . 'We are the music makers'.[195]

It was the first announcement of Edward's intention to set this poem. O'Shaughnessy's verses might not be immortal poetry, but in the week after the great London Festival surveying all his music from *Froissart* to *In the South*, the presence of the King and Queen and the conversation with the Prime Minister, the child who had listened by Severn side sought his reflection still:

> We are the music makers,
>> And we are the dreamers of dreams,
>
> . . . .
>
> World-losers and world-forsakers,
>> On whom the pale moon gleams:
> Yet we are the movers and shakers
>> Of the world for ever, it seems.

In Leeds on 22 March to arrange performances of *Gerontius* and *The Apostles* with the Choral Union, he showed 'The Music Makers' to Henry Embleton. Embleton said at once he would like to commission Edward's setting on the biggest choral scale.[196]

Back at home, flattering attention continued from every side. Proofs of the new biography were brought by the author, Robert Buckley. Then Edward found himself elected to The Athenaeum under the Club's Rule 11, providing for special invitation to 'persons of distinguished eminence in science, literature or the arts, or for their public service . . .': he had been proposed by Parry and seconded by Stanford.

Brodsky wrote from Manchester to remind him of the offered professorship and to beg his acceptance. But Edward concluded at last that he could not do it. He had already made a private memorandum of his reasons:

It is a mistake to put a composer at the head of a Music school: it gives a false standard—a poet should not be a schoolmaster—pupils are too apt to imitate & think their professor's works finite—Like a crowd of piping bullfinches listening to their master (a whistling tailor) they learn the same tune—& the one who cannot learn to pipe in exact imitation is the discarded one—he has to make his own way & sing his own song. Let us hope that this is the proverbial bird that eventually catches the worm![197]

Another London performance of *Gerontius* was conducted by Felix Weingartner with the title role sung for the first time by the Catholic Gervase Elwes. Further performances of *The Apostles* were given in London, Birmingham, York, and at the Lower Rhine Festival this year under Steinbach

---

[195] *The Daily News*, 25 Mar. 1904.   [196] HWRO 705:445:3189.
[197] Entry dated 'Feb 1904' in a notebook headed 'Alassio' (Elgar Birthplace).

at Cologne—where Edward and Alice graced a festive occasion in May. Jaeger was there, German hospitality flowed, and Steinbach conducted a surpassingly brilliant Festival. In addition to the success of *The Apostles* there was a Bach *Brandenburg Concerto* performance which impressed everyone who heard it as the very finest conceivable.

Canon Gorton's Morecambe Festival again rejoiced in Edward's presence. Two of the *Greek Anthology* part-songs were among the choral test pieces, and there was a surprise, as he wrote to Alfred Littleton: 'I wish you cd. have heard some of the choirs singing—impromptu—in the moonlight on the sands at Morecambe "Feasting I watch".'[198] In the summer he was to go again to Morecambe to address a meeting for conductors: his own ideal, he said, was Richter.

Ever since Booseys had lost *The Apostles*, Arthur Boosey had been trying to persuade Edward to opera. When Edward said he might be interested in the subject of Cleopatra, Boosey approached Arthur Benson. In the previous autumn, when Edward had set some verses of 'Adrian Ross' only to find that the copyright of the words was unavailable, Benson had again turned the trick of providing words for music already finished: the result, *Speak, My Heart*, had done well enough for all concerned. Now Boosey promised Benson that operatic collaboration with Elgar 'would be highly *lucrative*, & that I can well believe'.[199] But Benson was suffering from private depression and was full of doubt: he tried to pass Boosey's offer to Maurice Baring, but Baring would have none of it.[200] So the question hung fire.

At Novellos, Alfred Littleton was trying again to arrange an exclusive contract to publish everything Edward should write for the future. Now he offered a flat royalty of 25 per cent. Edward hesitated, he said, because of commitments to Boosey. Littleton offered to except from the contract any such commitments, as he wrote on 14 April 1904:

It is understood that this arrangement leaves you free to publish with Messrs. Boosey your four Military Marches, a second Overture in connection with your Cockaigne Overture already published by them—an Opera in case Messrs. Boosey should provide you with a libretto by Mr. A. C. Benson which you agree to accept—and a Cycle of Songs to be written for Mr. Plunket Greene.[201]

At last Edward promised to accept, but it was several weeks before he wrote the final committing letter.

Craeg Lea, Wells Road, Malvern.

June 7:1904

My dear Littleton:

In reference to your very kind letters of April 14th & the propositions contained in

---

[198] 3 Sept. 1904 (Novello archives).
[199] Diary, 8 Jan. 1904 (Magdalene College, Cambridge).
[200] The course of the negotiations can be traced in Benson's diary entries for 30 Oct. and 28 Nov. 1903, 8 Jan. and 20 Mar. 1904.
[201] Copy in Novello archives.

them I write now to accept your offer. That is to say in future for five years certain & after that time so long as we may wish—the engagement, that is, to be terminated by six months notice on either side, I send you everything I write (except the things already promised to Messrs. Boosey . . .) & you pay me a royalty of one fourth of the marked price & a sum 'down' for new works . . .

> With kindest rgds
> Believe me
> Yrs v sincerely
> Edward Elgar[202]

On 22 June he was at Durham to receive another honorary degree. That morning an official letter arrived at Craeg Lea from the Prime Minister. When Edward returned in the evening: 'A. told E. of letter, he sd. with such a light in his face "has it come" but then thought it wd only be about copyright. Then he opened the letter & found H.M. was going to make him a Knight.' In the diary next day, Alice returned to the private baby-diction of years ago: 'E. & A. vesy peased & hugged one anosser vesy often.' Edward cycled over to his sister Pollie's house at Stoke to tell the Graftons and old W. H. Elgar (who was now staying with them). Pollie's eldest daughter May, now twenty-four and an enthusiastic photographer, made a camera portrait of the father and the son at the historic moment.

Then it was arranged for May Grafton to join the Elgar household as Edward's secretary. May was very much her mother's daughter: her intelligent beauty and open-hearted charm made her an asset wherever she went. And her companionship would also be wanted when Carice (who was now almost fourteen and in her last term at The Mount) came at last to live with her parents.

On 24 June the knighthood was published in the Birthday Honours. That afternoon Alice gave a tea-party. Miss Burley brought Carice and her friend Cissie Cuthbert.

'I expect', said Cissie, one of those tactful little girls who seem early marked out for a career of domestic diplomacy, 'that we'd better call her "Lady Elgar" as many times as possible.'

The tea party, attended [also] by Ivor Atkins . . . was a happy one. It was impossible not to feel that Edward had fully deserved the honour, impossible not to sympathize with Alice's pride that her husband had achieved, in addition to a musical success, a rank which must command the respect of those who had looked down on him. But in spite of Cissie's excellent resolve we somehow felt shy of the title, and it was Ivor who at last said:

'Well, someone has to take the plunge. May I pass you a cake, *Lady* Elgar?'[203]

But Carice had already summed up her mother's feeling on the instant of hearing the news: 'I am so glad for Mother's sake that Father has been knighted. You see—it puts her back where she was.'[204]

---

[202] Novello archives.
[203] Burley, *Edward Elgar: the record of a friendship*, p. 173.   [204] Ibid., p. 174.

Through the spring Edward had not settled to any further work. Before the Covent Garden Festival they had decided to leave Craeg Lea and move to Hereford. A number of friends there, led by Sinclair and the Dean of the Cathedral, assured them of a warm welcome. The gentler hills round the western city offered a new countryside, quieter and more remote, easier for cycling. It was the county of his mother's birth and childhood.

Alice had found a commodious house called Plas Gwyn a mile east of the city. Like Forli and Craeg Lea, it was neither a city house nor a country house, but stood midway between with access to both. The owner of Plas Gwyn, Edwyn Gurney, was the son of a former mayor of Hereford: altogether it offered a more consequential social outlook than Craeg Lea.

One by one they told their friends. Miss Burley wrote:

The first I heard of the move was when on one of our cycle rides Edward suddenly said:
'They've found a house for me at Hereford.'

This was a locution he often used when he wished to disclaim responsibility for any step he had taken . . . The increase of their means, and of course of Edward's fame, made a larger house than Craeg Lea desirable but there were plenty of large houses available in Malvern . . .[205]

Miss Burley was very unhappy. One day he suggested their cycling over to Hereford to see the house:

Despite protestations that the move would make no difference to our friendship we knew that a period in our relationship was ending and we were more than usually silent. We took the road along the east side of the hills, a road of great beauty with the wooded heights on the right side and the drop to the Severn on the left. At Little Malvern we dismounted and entered the Catholic Church (in the graveyard of which Edward and Alice were both ultimately to rest). Then, refreshed by the silence, we resumed the now very familiar ride through Ledbury, with its black and white houses, and Stoke Edith into Hereford.

Plas Gwyn . . . was a square house placed at the corner made by a side turning called Vineyard Road. The little property was what house agents describe as matured and well-wooded, and the whole suburb attractive and spacious with charming glimpses of the Wye Valley through the trees on the opposite side of the road from Plas Gwyn . . . There were attractive features, the verandah running round two sides and partly covered with climbing plants, the relatively large rooms and the garden . . . Yet to me it looked, with its two plastered sides towards the road and its two brick sides toward the garden (a fairly common Midland arrangement) the sort of house that might be chosen by a prosperous and aesthetically not very exacting merchant rather than a suitable home for a sensitive and highly strung artist.[206]

The move to Hereford took place on the last day of June. May Grafton arrived and began to help with the task of settling in. They were still far from finished when Edward and Alice went to London on 4 July for the knighthood next day. Edward tried his new Court suit at the tailor's, and that evening he

[205] Ibid., p. 174.   [206] Ibid., pp. 174-5.

attended a meeting at the Queen's Hall to protest against the new Musical Copyright Bill. Under the chairmanship of the Duke of Argyll, a number of composers spoke.

Then, amid a veritable roar of applause, Sir Edward Elgar came forward at the Duke's call; and, when the cheering had died away, said very slowly, and in broken accents that could scarcely be heard even in the stillness:

'My life, ladies and gentlemen, has, as you know, been a self-made one. London called me from my country home, and *you* have made me what I am. But you call other composers from their homes to you, and you allow the law to deprive them of their livelihood. That is all I have to say'—and he retired quietly to his chair. It was a strangely moving little speech.[207]

Next morning came the knighthood:

E. dessed & looked *vesy* booful. A. helped him & buckled his sword. Then Frank arrived & drove to Buckingham Palace with E. & then met A. & went shopping with her. Returned & found E. arrived. The King smiled charmingly & said 'Very pleased to see you here Sir Edward.' Then we three lunched at Pagani's . . . E. & A. home late . . .

On 14 July they went to London again to dine with the Prince and Princess of Wales at Marlborough House. Edward stayed several days at Frank Schuster's country house The Hut near Maidenhead, while Alice returned to Hereford to battle with domesticities. She and May Grafton spent many hours arranging the books of Edward's library on shelves in his new study—the biggest room on the ground floor—only to find that when the master returned the arrangement was not approved. So Alice had to spend another day rearranging the books mostly outside the study in the corridor while May Grafton took Edward out for a long cycle ride through the new countryside.

Jaeger wrote enthusiastically anticipating all the new music to be written at Plas Gwyn. But the new house and changed conditions seemed only to weigh Edward down with expectations he might not fulfil. On 15 August, after several weeks at Plas Gwyn, he wrote to Hans Richter:

Work has not yet commenced here and I sometimes wonder if I shall ever invent any more music; we hope, however, that in time I may again take a pen in hand with the old zest, but at present too many worries seem to oppress me and nothing comes, musically, to relieve them.

Six days later, on the 21st, he wrote in the sketchbook an idea beginning:

It seemed to expand Mary Magdalene's

[207] *The Pall Mall Gazette*, 5 July 1904.

He developed the idea through a dozen bars of short score. On the last line of the page came a telescoped version marked 'end', and followed by the final words of Hamlet, 'the rest is silence.'[208]

In London for the Gloucester Festival rehearsals at the end of August, he was introduced by Jaeger to an American admirer who had attended the Festival several years earlier. He was Samuel Sanford, professor of piano at Yale University, a fine pianist, a man of culture and wealth. Sanford insisted on purchasing a new Steinway upright piano for the study in Hereford. He suggested that if Edward should visit the United States he would find a brilliant reception.

During Edward's absence, Dora Penny arrived for her first visit to Plas Gwyn:

I found the Lady and Carice in possession—Sir Edward was in London—and I went all over the house and was shown everything and found old friends in new places.

'The Indian furniture has positively come into its own here,' I said. 'Doesn't it look nice?'

The study was a fine large room on the ground floor. It had a bow window looking out on the veranda and another window which let in the morning sunshine. When I first went into it I could not help thinking of the tiny study at Forli, and I spoke my thoughts aloud, adding, as we stood there arm in arm: 'Isn't it glorious to feel how he has got on, and how people all over the world are beginning to understand and appreciate?' . . .

'I think *great* music can be written here, dear Dora, don't you?'

. . . After breakfast next morning I heard the Lady in the study opening and shutting windows, and then she came out and shut the door behind her.

'Do go into the study, dear Dora, there's such a surprise in there.'

. . . I opened the door and went in. I was greeted with a burst of music! But what curious music it was, and very difficult to describe. It was rather like a harp, and the sound rose and fell in arpeggios of intervals of thirds—minor or diminished. It was very strange and rather eerie—in an empty room. I walked forward and saw that one of the windows which looked on the veranda was only partly open and a framework with vertical strings was fixed in the opening. I wondered if it was an Æolian Harp—I had never seen or heard one. The Lady had come in after me and was now beside me.

'Edward loves it. He thinks it is so soothing!'[209]

It could echo again the wind in the reeds at the bank of the distant river. He had made it himself for this window. But none of his earlier study windows had been fitted with devices to catch that music for him.

. . . The following evening His Excellency came home. My diary says: 'He was very tired but thoroughly cheerful. He brought back a huge box of sweets for her Ladyship from Professor Sanford.'

It was still hot and lovely next day and E. E. and I rode about the lanes on our bicycles. He took me to Holm Lacy and, leaving our cycles under some trees, we went down to

---

[208] BL Add. MS 47904A fo. 87.
[209] Powell, *Edward Elgar: memories of a variation*, pp. 63–4.

the river Wye. It was lovely there. E. E. was in great form that day, talking and laughing about all sorts of things and here, at last, was my longed-for opportunity.

'Why did you give Mr. Jaeger such a grand and noble tune?'[210]

Bit by bit she extracted the story of Edward's depression five years ago, and of Jaeger's heart-warming comparison with Beethoven. But Edward made her promise not to tell this, and then changed the subject. Next day he wrote to Frank Schuster: 'I . . . see nothing in the future but a black stone wall against which I am longing to dash my head . . .'[211]

* * *

For the Gloucester Festival they stayed with the Bakers at Hasfield Court. The Dean of Gloucester had vowed that *Gerontius* would never be sung in his Cathedral while he had power to prevent it. Accordingly, only the Prelude and Angel's Farewell figured in a miscellaneous concert on the Tuesday evening. Edward conducted, but they left before Stanford conducted his *Te Deum* in the second half. On the Wednesday morning Sinclair played an Organ Concerto by the former Gloucester Cathedral organist Charles Harford Lloyd, for which Edward had anonymously supplied a third-movement cadenza as a gesture of friendship.[212] The main Elgar works at Gloucester that year were *In the South* and *The Apostles*.

After *The Apostles* performance, Charles Beale came up with new enthusiasm over the prospects for Birmingham in 1906, as Alice noted:

. . . He said E. must say whether he wd have 2 or 3 performances at next Festival all being well. He sd. to me, 'He is far too great a man for us to dictate anyth[in]g.'

Edward wrote to Alfred Littleton:

. . . I saw Mr. Beale of Birmingham: if the new oratorio is ready *they expect it*—they will give two performances, preceded by the Apostles—*3* performances, I mean programmes, to me alone! enough to satisfy a moderate ambition. I am working away at the oratorio but have nothing to shew you yet; & I sometimes feel overweighted.[213]

Novellos were taking his reputation so seriously that they were engraving the full scores of all his early choral works, up to then only in manuscript—*The Black Knight*, *The Light of Life*, *King Olaf*, *The Banner of St. George*, and *Caractacus* to join *Gerontius* and *The Apostles*. But the only music he finished that autumn was the *Pomp and Circumstance* March built round the trio tune written early in 1902 as a tribute to Ivor Atkins. Booseys' terms for the new March were £50 and a royalty on all arrangements sold.

On 5 October Edward conducted *In the South* at the Leeds Festival. The production of the new work at Covent Garden had caused distress in Leeds.

---

[210] Ibid., pp. 65, 110.
[212] MS in possession of Oliver Neighbour.
[211] 3 Sept. 1904 (HWRO 705:445:6908).
[213] 15 Oct. 1904 (Novello archives).

Several journals at the time thought they saw a connection between the Covent Garden première and Dr Elgar's recent withdrawal from the Leeds commission. Stanford, as Festival conductor, had written to one of the committee:

The *Musical Times* announces an Elgar Festival in March at Covent Garden Theatre, Richter, Manchester Band . . . and a *new orchestral work*, which the *Daily News* of today labels the 'Symphony announced for Leeds.' This is a very big slap in the eye for the Festival.[214]

The Leeds committee drafted their 1904 programme at first to exclude any work by Elgar. But Stanford was impartially fair: he had joined with Hubert Parry in pressing for the new Overture's inclusion:

Fair play, old chap, and a man's artistic work ought to rank independently of his personality. If it had not been that Hans von Bülow had taken this view of Wagner, the Bayreuth theatre would not be standing now.[215]

At Leeds Edward's performance of *In the South* shared its programme with the Brahms Violin Concerto, in which a tremendous impression was made by the twenty-nine-year-old Austrian violinist Fritz Kreisler.

Next day Edward received still another honorary doctorate at the inaugural ceremony for a new University of Leeds. Large groups of notables were enrolled that day as graduates of the new institution. In the musicians' contingent were Stanford, Parry, Mackenzie, Charles Wood, and Walford Davies. There was talk that Edward might be made Professor of Music at Leeds.

That evening he gave dinner to Josef Holbrooke, a young composer enjoying a première performance of his own in the Leeds Festival but complaining that he had no encouragement: it was a familiar problem, and Edward's sympathy gained him an acquaintance that was to prove embarrassing and troublesome more than once in the future. He and Alice remained in Leeds for a Wagner programme on the Friday morning, but they left before two Stanford performances that evening—a new Violin Concerto for Kreisler and *Songs of the Sea* given their première by Plunket Greene.

A week later Edward found himself in the academic world again when he went to Birmingham for prize-giving at the Midland Institute. The Institute provided further education for hundreds of young people who could not attend university: Edward's sympathy with their cause had been seized on by Granville Bantock (head of the Institute's School of Music), who made him visiting examiner. The main speaker of the evening was Sir Oliver Lodge, Principal of the new Birmingham University. After a vote of thanks from the Lord Mayor, Edward seconded the vote in a considerable speech of his own to define a new place for music in the intellectual world:

---

[214] 28 Nov. 1903 (quoted in Harry Plunket Greene, *Charles Villiers Stanford*, p. 154).
[215] 20 Jan. 1904 (ibid., p. 155).

If we musicians are going rather a little ahead and putting a little intellectuality into our music, and trying to be considered reasonable beings, instead of merely puppets to amuse you, the lords of science and literature must not be angry with us. The sun does not chide the morning star because it ushers in the dawn.[216]

It was the symbolism which had opened *The Apostles*.

There were many who wanted Edward to occupy a chair of music in the new Birmingham University. One of the keenest was Charles Harding, a retired solicitor who was vice-president of the Birmingham Festival and a life governor of the University. The University's Dean of the Faculty of Arts and Professor of German was Harding's son-in-law Hermann Fiedler—already known to Edward through his brother Max Fiedler, a conductor who had performed several Elgar works in Germany.

Mr Harding approached his old friend Richard Peyton, a wealthy business-man with a life-long devotion to music and a connection with the Birmingham Festival going back to the first performance of Mendelssohn's *Elijah* in 1846. Harding suggested that Peyton offer to endow Birmingham University with a chair of music on condition that Sir Edward Elgar should occupy it. Peyton had some doubts, which he communicated to Hermann Fiedler on 7 November:

I understood you to say that the appointment in question was of Dr Elgar's own seeking, that he had declined an offer from Manchester & had been in communication with Leeds. That he should desire the appointment is of course favourable as indicating a willingness to perform the duties but I should be glad if I could be informed whether the authorities of the University have been able to satisfy themselves that in regard to questions of temperament, decision, punctuality & the exercise of control over others the appointment would be a desirable one. I have not the pleasure of the acquaintance of Dr. Elgar & therefore speak now only in a general sense & under the impression that his unquestionable genius in regard to musical composition and the high reputation he has obtained as a composer would not be the only qualifications to consider in making the appointment.[217]

It was a shrewd summing up. Edward found himself being tempted by his friends in Birmingham to reverse his judgement that a composer should not head a school of music. Late in the year of his Covent Garden Festival, his knighthood, his move to a larger house, with little new music to answer these events, perhaps the experience of public speaking at the Copyright meeting in London, at Morecambe, and now in Birmingham could after all be turned to some account for the future.

Yet he hesitated. On 12 November Hermann Fiedler came to Plas Gwyn to add personal persuasion, and reported to Charles Harding a 'satisfactory

[216] Quoted in *The Birmingham Post*, 13 Oct. 1904.
[217] HWRO 705:445:3356.

interview. He will *write* early next week.'[218] But when Edward wrote it was a refusal.[219] Harding's next move was to seek out George Sinclair, the Hereford Cathedral organist now also conducting the Birmingham Festival Choral Society. Harding reported to Fiedler:

I had a short conversation with Sinclair who said that the reason for Dr. E's letter to you refusing was that he had the impression from you, that he would have to deliver something like two lectures weekly, which both he *and his wife* thought would spoil his life for original work.[220]

On 17 November Granville Bantock was sent over to lunch at Plas Gwyn with Alfred Hayes, secretary of the Midland Institute and Bantock's predecessor as its teacher of music. They talked to Edward through the afternoon. Bantock stayed the night. Next morning he telegraphed to Fiedler: 'Very favourable. Bringing letter two o clock today.'[221]

Edward's letter proposed several conditions:

1. That I should not be expected to reside in Birmingham.
2. That I do not deliver more than six lectures or addresses in the first year.
3. That if the development of the musical activity in Birmingham should lead to the creation of a Faculty of Music, a lecturer shall be appointed for the tutorial work.
4. That a full & cordial concurrence in the above proposals comes from the donor & the officers of the University.
5. That the post be not advertised, even pro forma.[222]

It amounted almost to a demand that he be let off even the lightest teaching duties at the earliest moment. Then he said he must consult a friend. This was Alfred Littleton at Novellos, and Edward made a special journey to London on 21 November to see him. Littleton advised acceptance, but with the added provision that he might resign altogether after three years.[223]

Peyton, the prospective donor, had misgivings. He wrote to Oliver Lodge:

In consequence of what I have heard from a private source I believe Dr Elgar would prefer to continue to reside at Hereford & would not return to Malvern which is so much more accessible from Birmingham, and I think we ought to be able to feel assured *that he would not take up the work in any temporary or tentative fashion* . . .[224]

But Fiedler and Harding had the bit between their teeth. On 24 November, when Sinclair conducted *Gerontius* with the Birmingham Festival Choral Society, Harding entertained the Elgars at his home, and next morning took Edward to meet Richard Peyton. The meeting was a success on both sides, and Peyton wrote officially to the Chancellor of the University Joseph Chamberlain:

[218] HWRO 705:445:3364.
[219] 14 Nov. 1904 (HWRO 705:445:3244).
[220] 17 Nov. 1904 (HWRO 705:445:3365).
[221] 18 Nov. 1904 (HWRO 705:445:3367).
[222] 18 Nov. 1904 (HWRO 705:445:3245).
[223] 20 and 22 Nov. 1904 (HWRO 705:445:3246–7).
[224] 20 Nov. 1904 (copy: HWRO 705:445:3357).

The proposal I have to make arises from the fact that there is at the present time a special opportunity of offering an appointment to a chair of music in the university to one of the most eminent of English musicians whether of the past or the present time; and the offer which I have the honour and pleasure to make is to contribute a sum of ten thousand pounds for the endowment of such a chair, the only condition being that it should in the first instance be offered to and accepted by Sir Edward Elgar, Mus.Doc., LL.D.[225]

So the complex bargain was struck. The new Professor was to have an annual salary of £400.

That autumn Frank Schuster paid his first visit to Plas Gwyn. Afterwards he sent a gift for the garden to cast its shadow in many directions. Edward's letter of thanks looked at some of them:

Plas Gwyn, Hereford.

Nov 4: 1904

My dear Frank:

. . . I was very *sweetly angry* with you over that Sundial but I love having it & am really happy; coming from you makes it perfect: it is very human,—it *lies* all the morning and tells the truth, repentantly, all the afternoon: the latitude does not suit it;—I mean the actual latitude, not the unveracious latitude it allows itself in mistelling the morning hours. Also it is feminine & only beams & smiles in sunshine; it seems to frown when there is a lack of *gold in the air!*† These be philosophisings.

. . . So much love

Yours ever
Edward

†Alice resents this sentence rigidly.[226]

At the end of November he travelled to the continent with Schuster for performances of *In the South* (at Cologne under Steinbach) and *The Apostles* (at Mainz under Volbach and at Rotterdam). On the evening of the Rotterdam *Apostles* Weingartner was conducting *In the South* in Berlin. Alice remained at home with Carice, who recalled the effect of her father's absences on her mother:

One could hardly get an answer about anything, and until his telegram came saying he was safely in London—or wherever it was—she was terribly on edge, and seemed to think that one was very unfeeling if one did not appear to be worried about the journey. If she ever allowed herself to throw off this terrible worrying, she could be the most delightful companion, but unfortunately this did not very often happen.[227]

A series of telegrams kept Alice apprised of his movements, and on

[225] Quoted in '*A Future for English Music*' *and other lectures*, p. 10.
[226] HWRO 705:445:6917.
[227] Quoted in P. M. Young, *Alice Elgar: Enigma of a Victorian Lady* (Dennis Dobson, 1978), p. 149.

9 December she travelled to London to meet his return at Frank Schuster's house: 'Such joy to have E. safes—& F. Very happy evening.'

The conductor at Mainz, Fritz Volbach, had come with them: he had been invited by Stanford to conduct his own *Reigen* at the Royal College of Music. On the evening of 11 December Schuster gave a dinner party with a reception to follow. The guest-list apparently omitted Stanford, but included Edward and Alice, Volbach and several German friends, the Stuart-Wortleys, Claude Phillips, Sidney Colvin and his wife (the intimates of many artists and writers of the recent past including especially Robert Louis Stevenson), and a fifteen-year-old Westminster schoolboy whom Schuster had befriended through a family connection. The boy was keenly musical. His name was Adrian Boult.

When the great evening came, I was shown straight into the dining room where I found besides our host and Edward Elgar, Generalmusikdirektor Fritz Volbach of Mainz . . .

Elgar had just received the engraved score of *The [Apostles]*, and was delighted with the beautiful work that Novello's experts had put into it. We were sitting together on the long seat which stretched round three sides of the Old Queen Street music room, looking at the score and its accurate alignments, when our host, who had been digging in Miss Edith Clegg's case and discovered a song by Elgar, came over and said:

'Edward, you must come and meet Miss Clegg, who is going to sing *After*.'

The great man got up, walked over and shook hands with the remark, 'Well, you have spoiled my evening for me.'[228]

The boy was shocked at this 'absurd' behaviour a moment after Sir Edward Elgar had been sitting beside him admiring the big new score. But the words alone from this half-forgotten song of 1895 could make their cruel comment on the ambition which had not yet been able to bring *The Apostles* to any completion, or the Symphony either:

> A little while for scheming
> Love's unperfected schemes;
> A little time for golden dreams
> Then no more any dreaming.

He pulled himself together for a visit by Richard Strauss to conduct his own music in Birmingham on 20 December. At a big supper afterwards in the Clef Club he made an elaborate speech of welcome to Strauss which was reported in all the papers. The Professorship at Birmingham University had just been announced.

A few days later came an 'odious' letter from Stanford, as Edward confided to Alfred Littleton: 'Many *disagreeables* arise from a certain quarter over my new appointment which seems to have caused bitter irritation . . .'[229] The

---

[228] *My Own Trumpet* (Hamish Hamilton, 1973), p. 64. Boult thought it was *The Kingdom* they looked at, but the incident is clear in Lady Elgar's diary for Dec. 1904, and Sir Adrian agreed.
[229] 29 Dec. 1904 (Novello archives). The Stanford letter had arrived two days earlier.

irritation could have arisen from several sources. Despite Stanford's gestures of friendship at The Athenaeum, at Leeds, and elsewhere, Edward had quite pointedly avoided him—most recently at Gloucester and again at Leeds.

Then there was Stanford's jealously guarded academic power at the Royal College of Music and Cambridge and elsewhere. The addition of his conductorship at the Leeds Festival should have given him some influence in the choice of a professor for the new University of Leeds. Leeds had extended its invitation to Edward, and had been refused. And in the wake of that refusal came the announcement from Birmingham, where Stanford had no influence. To such a man in such a mood, it could look very much like a snub.

On top of that, Edward had gone out of his way to give Richard Strauss a fulsome welcome in Birmingham. No one had forgotten Strauss's widely advertised remarks about 'Meister Elgar' as 'the first English progressivist' when he had mentioned no other living composer in England. Stanford had widely advertised his own dislike of Strauss's music for 'the general want of refinement in his idiom'.[230] Eighteen months earlier Jaeger had reported: 'A young friend of mine has just been here after an interview with Stanford re modern music. He says C.V.S. *foamed at the mouth* almost dismissing R. Strauss' & other moderns' music. Poor disappointed man!'[231] Now Elgar went out of his way to praise the German in a way he had never praised any British music in public.

To Stanford's 'odious' letter Edward sent a 'most gentle & courteous reply', according to Alice. But it marked the end of a relationship made always uneasy by the conflict of temperaments. On the one side, Edward's dizzying success despite his lack of academic connections could tempt jealousy even in natures more patient than Stanford's. On the other side, Stanford's energetic advice and advocacy could always strike a man of Edward's insecurity as tinged with patronage.[232]

*   *   *

The new London Symphony Orchestra wanted Edward to conduct one of their concerts. Many of the board and the players were acquaintances or friends, for the Orchestra had been founded largely by Queen's Hall players. (The precipitating cause of defection was Henry Wood's imposition of a ban on the old system of sending deputies to rehearsals or concerts whenever more lucrative engagement offered; but the members had for some time wanted more opportunities to play under a wider variety of conductors.[233]) Chief

[230] Harry Plunket Greene, *Charles Villiers Stanford*, p. 99.

[231] 20 May 1903 (HWRO 705:445:8652).

[232] Basil Maine's MS notes of Elgar's conversation on 12 July 1932 recall 'S's "bad, green teeth".' (Royal College of Music).

[233] Conversation in Aug. 1981 with Francis Bradley, who had this information from his father, Adolf Borsdorf, one of the original board members of the London Symphony Orchestra from its foundation.

conductorship of the new Orchestra had been accepted by Richter, but the players themselves made the policies. They asked Edward for a new piece to play at the all-Elgar concert he was to conduct in March 1905. Jaeger enthused over the prospect:

I'll hope you can write the Symphony orchestra a short new work. Why not a *brilliant* quick *String* Scherzo, or something for those fine strings *only*? a real bring down the House *torrent* of a thing such as Bach could write (Remember that *Cologne* Brandenburger Concerto!) a five minutes work would do it! It wouldn't take you away from your *big* work for long. You might even write a *modern Fugue* for Strings, or *Strings & Organ*! That would sell like Cakes.[234]

Edward replied that he could not settle to anything. He only allowed it to be announced that he was at work on the pantomime-ballet of Rabelais Gargantua and Pantagruel. But as 1904 came to its end without any real progress, the triumphs in the first half of the year and the depression which followed could find their echoes farther in the past.

Through these years of producing his biggest works for the triennial Birmingham Festival, other phases of creative life seemed also to find a three-year cycle. After the *Gerontius* première in 1900 had come depression sharper and deeper than the disaster of the performance could account for; the *Apostles* première three years later gave no cause for depression, yet within a month the death of Rodewald at an age younger than Edward's own had deepened depression with a mortality that came as close to home as ever the symbolism of *Gerontius* had come. Out of each experience had come the desire to write a symphony. Out of each had emerged instead a concert overture evoking a far-away place. *Cockaigne* in 1901 had shown a happy place—with all its darker aspects relegated to a possible 'Cockaigne No. 2'; *In the South* in 1904 travelled farther to express wider moods. Each Overture had achieved a brilliant London première. After *Cockaigne* in 1901 had come the invitation to write the *Coronation Ode*; after *In the South* in 1904 had come the knighthood. Each was followed by a summer of black depression—each relieved by the composition of *Pomp and Circumstance* Marches.

After the first Marches in 1901 he had gone for the holiday in Wales, and there sketched music at Ynys Lochtyn. Some of that music had been used in *The Apostles*, but there remained the 'Welsh' idea of a falling third:

Three years later this idea came strongly back to him:

The sketch was forgotten until a short time ago, when it was brought to my mind by

[234] 28 Oct. 1904 (HWRO 705:445:8739). The Third *Brandenburg Concerto* had been conducted by Fritz Steinbach at the Lower Rhine Festival in May 1904.

hearing, far down our own Valley of the Wye, a song similar to those so pleasantly heard on Ynys Lochtyn. The singer of the Wye unknowingly reminded me of my sketch.[235]

Three years ago, as now, Jaeger had encouraged him to write. But now mortality came ominously closer. Three years ago Jaeger had come to the fresh air of Malvern to rest between Dr Macdonald's attempts to halt his consumption. Late in 1904 he had had a further sentence: he must go for the winter to Davos in Switzerland, the never-never land of incurables. At the beginning of the new year 1905, practically on the eve of his departure, Jaeger wrote again—with apprehension less for himself than for his friend:

37 Curzon Road, Muswell Hill

8/1/5

My dear Edward,

I hope to be able to leave on Wednesday—shall be away for *months*. So I just scribble a line (I am *fearfully* busy clearing up affairs & getting ready) to say Goodbye & Auf Wiedersehen under Happier circumstances. I am relieved to say that the firm are behaving liberally to me, though in *any* case this is a great worry & trouble, I need not say.

I worry also over your *muse*, for I fear greatly we shall get less & less out of you. This is the danger of success artistic & social! (especially social, of course). I grieve over it, & so do all those who most sincerely love & admire you. We know you *must live*, but England *Ruins* all *artists*

With kindest regards to Lady Elgar

much devoted love
Ever yours
A. J. Jaeger[236]

After Jaeger's departure, Edward implemented his suggestion for the London Symphony Orchestra piece. On 26 January 1905 he wrote to Jaeger at Davos:

Dear Moss: I'm doing that string thing in time for the Sym:orch: concert. Intro: & Allegro—no *working-out* part but a devil of a fugue instead. G major & the sd. divvel in G minor . . . with all sorts of japes & counterpoint.'[237]

The old keys had made one more triennial return. Three years earlier at the beginning of 1902 he had completed the reminiscent contrasts of *Dream Children* in the memory-laden tonalities of G minor and G major. Three years before that had come the '*Enigma' Variations*; three years earlier still the G major Organ Sonata with its Finale in G minor. Three years before that he had finished *The Black Knight*—which had its beginning just three years earlier in the first summer of his marriage to Alice.

The Introduction in the new work began with a new version of Edward's old falling fourths, sounding through the entire orchestra of strings doubly divided with a solo quartet to sit in front. Then came the primary subject:

---

[235] Note written for the première programme.
[236] HWRO 705:445:8751.   [237] 26 Jan. 1905 (Elgar Birthplace).

and a counter-melody:

After it the 'Welsh' melody entered quietly in the three flats of E flat major, to sound a rhythmic development from the primary shape—an echo that was a private reminiscence.

. . . And so my gaudery became touched with romance. The tune may therefore be called, as is the melody in the overture *In the South*, a *canto popolare* . . .[238]

Like the 'canto popolare', this also was first sung by the solo viola. Gradually it was taken up the whole ensemble, echoed by the solo quartet to end the Introduction in G minor.

The *Allegro* recommenced the primary theme in G major double time. The second subject developed the 'Welsh' theme in semiquavers:

[238] Note written for the première programme.

As with *Cockaigne* and *In the South*, this exposition was again a series of developing variants from a thematic source in the middle. A distant echo of the 'Welsh' theme just punctuated the end of the exposition.

The fugue-in-place-of-development took the variants further. The fugue subject in G minor derived from a figure of semiquavers at the exposition climax; counter-subjects were similarly derived from exposition material; and all worked to a fine climax. But the *Allegro* primary subject never appeared, and the secondary subject only when the fugue wound down toward recapitulation.

Returning to G major, the recapitulation tersely redistributed exposition material through solo quartet and orchestra. Then it was capped with a coda made by the 'Welsh' melody—unheard in its full form since the Introduction. It resounded now through all the ensemble, singing triumphantly toward a quick ten-bar summary and a final chord *pizzicato*.

Thus the very form of this *Introduction and Allegro* enacted Edward's description of taking his music from the air all around. The real theme, first caught in the air at Ynys Lochtyn, appeared casually in a secondary passage near the beginning; then it was overtaken with variants and counter-melodies vigorously developed with hardly a reference to their secret source—until, at the end, the source was revealed in compelling return to what had appeared at first only as if by chance. In the *'Enigma' Variations* the discovery of final significance had come through the ministry of 'friends pictured within': now it was engendered within himself.

Here was the dialogue of experience above and below the level of consciousness—the dialogue of reality and the dream through memory. The melody half heard in Wales had gone into Edward's memory, and had only been recalled to consciousness by a mysterious repeating of experience three years later. He described the process in general terms:

An idea comes to me, perhaps when walking. On return I write it down. Weeks or months after I may take it up and write out the movement of which it had become the germ . . . But the piece has gradually shaped itself in my mind in the meantime, and the actual writing is thus a small matter.[239]

So at last, at the age of forty-seven, Edward's outward-moving excursion through his own countryside of the mind found its way home. It was his mother's recapitulation of life through the hours and seasons of symbolic cycle. If this abstract drama of the past returning to embrace present experience could be extended through diverse movements, it might unlock his Symphony.

The realization of a Symphony on such terms could remove the need to pursue his synthesis through the specific symbolism of his mother's religion. Yet now he was publicly committed to exploring further the enormous epicycles of *The Apostles* for Birmingham. And before that he would lecture in Birmingham as a university professor. The implied requirement of defining his

[239] *The Music Student* (Aug. 1916), pp. 347–8.

ideal creative formula in public could not have come at a more awkward moment.

It was Edward's pride now to live by his own music, for that was proof of the music's power. Since his knighthood especially, he felt that life should show a broader style. But Plas Gwyn entailed heavier expense than the smaller houses in Malvern, and he no longer taught or conducted in a regular way. In the previous summer he had been forced to ask Novellos for an advance of £100 on *Apostles* payments; at the beginning of 1905 he asked for another £100.

Early in February he was given an honorary degree at Oxford. It was arranged by Parry as Professor of Music. Parry brought down the entire London Symphony Orchestra for a celebratory concert in the Oxford Town Hall to include his own *Blest Pair of Sirens* and the *'Enigma' Variations*, each conducted by its composer. Edward's meeting with the new Orchestra was a success on both sides. During the rehearsal he was approached by the London Symphony manager and asked to conduct a short tour with the Orchestra in the North during the coming autumn. The offer of 26 guineas a day and railway expenses was too high to refuse.

A week later came a letter from Professor Sanford inviting Edward and Alice to spend a month with him in the United States. There would be an honorary degree at Yale University if they could attend the Commencement ceremony at the end of June. But the real purpose was elsewhere: several offers were in the air for Edward to conduct in America at very high fees, and the preliminary visit, Sanford wrote, 'might be the way to crystallise the various schemes for your conducting.'[240] Edward accepted the invitation for July, dedicated the *Introduction and Allegro* to Sanford, and then asked Novellos to negotiate any American offers: 'I will not go for less than Weingartner who has £2500 (not dollars) for sixteen concerts: they can either take me or leave me.'[241]

After the last sheets of the *Introduction and Allegro* were posted to the printer on 13 February, he had no occupation. Golf was impossible as the links at Hereford were too far away. Cold, grey, windy weather prevented much cycling. He wrote to Richter once more: '. . . the Symphony does not come . . .'[242]

Thinking about the *Apostles* continuation which must be started, he compiled from the Bible the text for a 'Hymn of Faith' which made a rubric on the larger project. Early in the assemblage he introduced the definition from the Epistle to the Hebrews:

> Faith, the substance of things hoped for,
> the evidence of things not seen.

[240] Quoted in Elgar's letter of 20 Feb. 1905 to Alfred Littleton (Novello archives).
[241] 27 Mar. 1905 to Alfred Littleton (Novello archives).
[242] 19 Feb. 1905 (Richter family).

At its centre, Edward's 'Hymn' looked at the dark side:

Behold, the days shall come when the way of truth shall be hidden, and the land shall be barren of faith.

(II Esdras 5,1)

There followed a verse from the Ninety-fourth Psalm (which had framed the final guilt of Judas):

> Unless the Lord had been my help,
> my soul had almost dwelt in silence . . .

And near the end:

> For Thou, O God, hast proved us;
> Thou has tried us as silver is tried.

(Psalm 66, 10)

Silver in *The Apostles* had been the wage of betrayal. But if this 'Hymn' made its comment on the entire *Apostles* project, Edward did not attempt to set its words to his own music. Instead he showed it to Ivor Atkins, who had been asked to write a work for the Worcester Festival that year, and Atkins decided to set it.

Having entered a second art, Edward toyed with a third. His portrait was being painted at Plas Gwyn by a London artist, Talbot Hughes. Alice wrote to Jaeger at Davos:

. . . It fired E. with a great desire to paint & he has bought a box of oil colours, & paints strange symbolical pictures a la Böcklin, & Segantini & Blake! He certainly has a power of representing a scene from his imagination, & one that he has done of a river with sombre trees and a boat crossing is very suggestive.[243]

On 1 March Edward took himself up to Birmingham to look over the scene of his first University lecture in a fortnight's time, and 'returned mis'. Two days after that they left Plas Gwyn for a round of engagements to include the inaugural lecture and two concerts with the London Symphony Orchestra at Queen's Hall: the first was to include not only the *Introduction and Allegro* première, but the first performance of *Pomp and Circumstance* No. 3. In London they stayed with Frank Schuster. The parties began on the evening of their arrival with a big dinner: there was Lady Charles Beresford, the Stuart-Wortleys, Claude Phillips, Lord Northampton, the Sidney Colvins, John Singer Sargent (whom Schuster had first introduced several years earlier), and new acquaintances: the American writer Howard Sturgis, the actress Constance Collier, the stage designer Percy Anderson, and Burne-Jones's son Philip (who had inherited his father's baronetcy and brushes but only a little of his fragile genius).

Edward's reaction to the London limelight was immediate, and sharper than his nervous bad manners at Schuster's Covent Garden Festival dinner a year

[243] 1 Mar. 1905 (HWRO 705:445:8841).

earlier and again in December when Edith Clegg had offered to sing *After*. Now he had violent headaches: on 4 March after the solo quartet had rehearsed the *Introduction and Allegro* in Schuster's music room, he was '*very* porsley. In evening so poorly we had a Dr. Sinclair about 10.15—A. slept on a sofa & tried to dokker E. up.'

Yet he got through his rehearsals and also several parties in the following days. After luncheon on the 8 March he went to Queen's Hall for the first London Symphony concert—accompanied by Alice 'with bag of restoratives for E.' She wrote to Jaeger:

The Orch. is gorgeous & play splendidly for E. *Grania* was more than beautiful & poetic, the Variations delightful & the new String piece quite fascinating. Many people think it the finest thing he has written, the 4t. comes in with so beautiful an effect, the peroration toward the end *is* fine. The new March is *thrilling*—the most pacific friends were ready to fight! The critics, some of them, of course were frightened at it, but happily the audiences judge for themselves.[244]

In fact the virtuoso string writing of the *Introduction and Allegro* had not been brought to much brilliance with only two rehearsals. It was to be several months before repeated London Symphony performances would show the work's true quality.

The inaugural lecture at Birmingham was practically upon him. It was bravely entitled 'A Future for English Music'. He drafted and redrafted, arranged and rearranged a number of persistent thoughts. The sheets of notes, drafts, crossings-out, and marginal additions accumulated. Their organization was still not clear on the day before the lecture, when a shorthand writer came to Old Queen Street to help him get it into a form which could be typewritten for reading out.

Next day Frank Schuster accompanied them to Birmingham, where they were met by Richard Peyton and taken to tea with a large gathering of University officials before proceeding to the Midland Institute. *The Manchester Guardian* reported:

The large hall of the Midland Institute was crowded, and the staff of the University was on the platform. During the interval of waiting, selections from Sir Edward Elgar's works were played on the organ . . . The new professor, who wore the robes of a Doctor of Music of Cambridge, was loudly cheered as he entered with Sir O. Lodge, the Principal of the University, whose remarks were commendably brief and to the point . . .

Edward rose to speak:

I fear that in this opening address there may be rather a strong personal note . . . As I am

---

[244] 28 Mar. 1905 (HWRO 705:445:8839).

speaking from my own experience—experience dearly bought—I fear that it cannot be otherwise.[245]

Through this lecture and three others for the autumn—'Composers', 'Executants', and 'Critics'—he would speak his mind in a mood familiar to all his friends. One was to write: 'He had an unhappy knack of doing this at times; even in an ordinary short speech he would let a word or a phrase drop which had the effect of annoying someone and which, if he had stopped to consider it, he might have left unsaid.'[246] The habit, even in a prepared lecture, could be aggravated by a sense of uncertainty. And sure enough, almost at the outset he implied scorn for two groups of people very prominent in his audience— academic musicians and professional lecturers. But these were not his only targets:

The history of music from the time of Purcell onwards is well known, and it would be merely a tiresome repetition of the ordinary commercial lecture to go over the two centuries preceding 1880. Some of us who in that year were young and taking an active part in music—a really active service such as playing in orchestras—felt that something at last was going to be done in the way of composition by the English school . . .

I am not going to criticise these works in detail: it is not necessary. Happily for us some still live and give their quota of joy and satisfaction: the greater portion are dead and forgotten and only exist as warnings to the student of the twentieth century . . .

Dr. Richard Strauss, in a vivid speech made at Düsseldorf three or four years back, threw a brilliant illumination on this somewhat darkened picture. We all knew, although we dared not say so in so many words, what he then told us: that Arne was somewhat less than Handel, that Sterndale Bennett was somewhat less than Mendelssohn, and that some Englishmen of later day were not quite so distinguished as Brahms.

That last could well have been a hit at Stanford. Stanford had completed his studies in Germany through the interest of Joachim, the great friend of Brahms—whose style Stanford's music was often thought to imitate. But Strauss in his Düsseldorf speech had toasted only 'Meister Elgar'. Everyone in the room knew that.

It is saddening to those who hoped so much from these early days, to find that after all that had been written, and all the endeavour to excite enthusiasm for English music—'big' music—to find that we had inherited an art which has no hold on the affections of our own people, and is held in no respect abroad.

. . . I find that [English music] is commonplace as a whole. Critics frequently say of a man that it is to his credit that he is never vulgar. But it is possible for him—in an artistic sense only, be it understood—to be much worse: he can be commonplace. Vulgarity in the course of time may be refined. Vulgarity often goes with inventiveness . . . But the commonplace mind can never be anything but commonplace, and no amount of education, no polish of a University, can eradicate the stain from the low type of mind which is the English commonplace.

[245] This and the following quotations from Elgar's Birmingham lectures are taken from the edition by Percy M. Young, *'A Future for English Music' and other lectures*.
[246] W. H. Reed, *Elgar*, p. 89.

For all the social and artistic snobbery which had dogged him from his provincial childhood, revenge could not resist its opportunity. In defending vulgarity, moreover, he could implicitly defend the over-compensation for private insecurity which was showing itself through his own music in ever more aggressive writing for the brass instruments.

He named a single exception—'Sir Hubert Parry, the head of our art in this country . . .' But then he named a special category of work to stand for what he found most offensive:

Twenty, twenty-five years ago, some of the Rhapsodies of Liszt became very popular. I think every Englishman since has called some work a Rhapsody. Could anything be more inconceivably inept? To rhapsodise is one thing Englishmen cannot do . . . This, you will say, is a trivial incident. So it is, but nevertheless it points a moral showing how the Englishman always prefers to imitate.

He named no names. But everyone in the room knew also of Stanford's *Irish Rhapsodies* as among his most successful works.

The audience were growing restive. It was not only with embarrassment. Through the lecturer's lack of organization and consequent repetition, each of his points extended to ten times its needed length. Rosa Burley remembered it as:

. . . one of the most embarrassing failures to which it has ever been my misfortune to listen. The opening was greeted with the respectful attention which Edward's eminence deserved but as the evening wore on and point after point missed its mark feet were shuffled, a cross-fire of coughs set in and one gradually realized that the day was hopelessly lost.[247]

At last he approached the main point of his title.

But still something moves, and the day is coming, if the younger generation are true to themselves, are strong, if they cease from imitation and draw their inspiration from their own land.

Yet in the end he could only fall back upon his own private vision:

There are many possible futures. But the one I want to see coming into being is something that shall grow out of our own soil, something broad, noble, chivalrous, healthy and above all, an out-of-door sort of spirit.

It all had little enough to do with formal teaching.

Now what part is the University of Birmingham going to play in this advance? Do not be disappointed—I cannot say . . . It is impossible to forecast any results unless the material is found to work upon.

A library was the prime necessity: and the formation of the University Music Library was to be his chief work as Professor in the months to come. Birmingham must also have a rich and continuous musical life to attract good

---

[247] *Edward Elgar: the record of a friendship*, p. 183.

students. Here he referred with approval to the Festival Choral Society (whose present conductor, Sinclair, was in the audience), and the attempts of William Stockley to establish permanent orchestral concerts in the city. Old Stockley was also present, and Edward's very affectionate reference showed that he attached no blame over the première of *Gerontius*.

I see here many of my fellow musicians and I am complimented by their friendly interests in to-day's proceedings. I have wearied you by insisting on the indefinite nature of music in this University. One thing is *not* indefinite and that is my goodwill and I will say love of my fellow musicians. I hold very definite views as to our art, but I do not intend to use my position here in any other way than for the great, broad, open, and let us say, peaceful advancement of the best. I therefore invite the co-operation of all Birmingham musicians to this end: let us sink small differences and hold in view the good of our beloved art only.

Whether the new Professor himself had done that was the question asked in the press reports of the lecture next day. There was a good deal of predictable reaction from critical allies of the academies, and also from a press element eager to build up any occasion for 'copy'. Altogether the reports of what he said, Edward wrote to Professor Fiedler, 'are absolutely beside the point & futile to a degree more than ordinary even in musical criticism.'[248] Yet there had been a positive response from *Musical Opinion*:

It was only to be expected that Sir Edward Elgar's speech at Birmingham would arouse a deal of opposition. I hear of thunderbolts ready to be launched from South Kensington [the Royal College of Music, where Parry was Director and Stanford Professor of Composition] and Tenterden Street [the Royal Academy of Music, where Mackenzie was Principal]. It is generally held that a composer in the position of Sir Edward ought not to have made public such opinions. Many men in sympathy with what he said agree with this . . .

Personally, I think that Sir Edward was quite justified in speaking out to his audience on this matter . . . None of these composers has developed as he should have developed, and I am firmly convinced that the dull life of professor has killed his talents. The future of British music rests on no far fetched idea of looking to the literature of our country as an inspiration or to any real attempt to embody the spirit of the nation, but simply on our composers not being compelled to eke out their existence by undertaking the onerous duties of heads of teaching institutions. And it is in the nature of irony that Sir Edward Elgar himself is doing his best to dig his own grave at Birmingham. Hitherto, whatever the difficulty has been, he has been able to live without thus enchaining his gifts: and he should be thankful that this has been so.[249]

* * *

There was no music in prospect, but a season crowded with concerts and social engagements. Returning to London with Frank Schuster on 17 March 1905,

[248] 19 Apr. 1905 (quoted in '*A Future for English Music*' *and other lectures*, p. 69).
[249] Apr. 1905, pp. 498–9.

they attended a banquet to celebrate the hundredth birthday of the great singing teacher Manuel Garcia, the brother of Malibran and Pauline Viardot. The next night Edward dined at Buckingham Palace. The day after that came his second concert with the London Symphony Orchestra, in which he was to conduct Schumann's Second Symphony as well as his own music. Again he had a violent headache:

March 19.   E. dreffuly badly. A. stayed to nurse him up. Went from bed to Queen's Hall, A. with bag of restoratives—Conducted splendidly. Back straight to bed . . .

March 20.   E. feeling so ill. A. sent for Dr. Ashe—He sent us home [to Hereford]—E. raser better on arriving . . .

He wrote to Mr Peyton in Birmingham: 'I am afraid my illness is not from overwork.'[250]

He did not feel well enough to go to Leeds on 21 March to conduct *Gerontius* with Embleton's Choral Union, but made the effort to go to Hanley a week later to direct *The Apostles*. In April came a visit of nearly a week to Leeds—to conduct the Choral Union's second *Gerontius* in less than a month, and *The Apostles*. At Morecambe in May an audience of 5,000 gathered to hear him conduct *King Olaf*. And these were not a tithe of the Elgar performances being given by choral societies all over the country and abroad. In Canada a singing group had named itself 'The Elgar Choir'.

Back at Plas Gwyn at the end of a cold spring, there was at last some cycling weather. His odd days at home were spent in domesticities—arranging a fountain in the garden to reproduce the old fountain at Broadheath, making a rockery and houses for Carice's new pets—two rabbits, of which the white angora was named Peter. A pair of doves nested in a shed at the back of the house: the shed was promptly named The Ark. Yet the headaches recurred, sometimes leaving him prostrate for a day at a time. For now he faced the trip to America.

On 6 June 1905, the day before beginning his journey, Edward took himself over to Worcester for the afternoon: 'Returned terribly depressed.' He wrote to Jaeger (who was back from Davos but not cured):

I have no news of myself as I have for ever lost interest in that Person—he ceased to exist on a certain day when his friends interfered & insisted on his ———
  It is very sad.[251]

It was almost six years to the day from the première of the *'Enigma' Variations* dedicated 'to my friends pictured within'.

On 7 June he went to London, where Alice joined him. Two days later at

[250] 3 Apr. 1905. (quoted in *'A Future for English Music' and other lectures*, p. 68).
[251] 6 June 1905 (HWRO 705:445:8746).

Dover they waited on the pier for two hours in the rain before the tender took them out to the *Deutschland*:

After Cherbourg they let us have a Cabin de Luxe. Changed all our things. E. rather badsley, headache &c—Dreadful vibration & noise. E. very mis.

The six-day voyage was relieved with interesting shipboard acquaintances including Mrs Julia Worthington, a charming friend of Professor Sanford's with a palatial flat at The Wyoming in New York City and a big house up the Hudson at Irvington.

They arrived in New York on 15 June in hot, steamy weather. Sanford was there to meet them. Next day he accompanied them on the seventy-mile journey to New Haven, where they were to stay in his grand house with its well-trained servants and gold dinner-service. The oppressive weather was combatted with long motor drives along the Connecticut coast to Sanford's summer home at New London: he was a genial, resourceful host, nicknamed by Edward 'Gaffer'. But for much of the time Edward felt ill, as he wrote to Alfred Littleton:

This climate has been *awful* & I have been really well only about two clear days—don't tell anyone this! but it has really been quite seriously trying & the doctor continually hovering about me: the intense humidity is intolerable, the heat actually not very great but we sit in pyjamas bathed in perspiration from morning till night.[252]

He managed some of the social engagements arranged for him, including the Commencement on 28 June—for which he was presented with his Yale Doctoral robes and mortarboard. Yale's Professor of Music, Horatio Parker, played *Pomp and Circumstance* No. 1 on the organ—for perhaps the first of all the countless times that music was to accompany American graduations.

The last day of June brought Lawrence Maxwell Jr, President of the biennial Cincinnati May Festival, to interview Edward in New Haven. At the previous Cincinnati Festival in 1904, the late Theodore Thomas had conducted a large number of Elgar works. Now Thomas's successor Frank van der Stücken (who had met the Elgars at Cologne when Steinbach conducted *The Apostles*) was keen that Edward himself should conduct at Cincinnati for the next Festival in May 1906. But Van der Stücken had found the negotiations with Novellos staggering, as he wrote to Alfred Littleton on 21 June:

After many delays and discussions, I received a cablegram from Cincinnati, authorizing me to engage Sir Elgar's services for the sum of *one thousand pounds*. I never believed that it would come to this, for *never* has such a price been paid for a composer before this.[253]

Back came Novello's answer:

. . . The sum mentioned by Mr. Littleton, after consultation with Sir Edward, was Fifteen hundred pounds . . .[254]

[252] 8 July 1905 (Novello archives).   [253] Novello archives.
[254] 24 June 1905 (Novello archives).

Edward wrote to Alfred Littleton: 'My *feelings* are dead against coming here again but my pocket gapes aloud.'[255] So when Mr Maxwell arrived in New Haven, they struck the bargain: in May 1906 Edward should go to Cincinnati.

There was a two-day visit to Boston in suffocating heat to meet the American composer George Chadwick. Chadwick conducted the Worcester (Massachusetts) Festival, whose directors had also expressed interest in Edward's presence if they could have the première of a new work. But from this nothing emerged. The American visit finished with a torrid stay on the twelfth floor of a New York hotel, relieved by an excursion to Mrs Worthington's house at Irvington-on-Hudson, dinner at Delmonico's, iced drinks, and electric fans. One American newspaper interviewer caught Edward at the end of his tether on the subject of American music:

'Your national anthem is even worse than England's . . . There is "Yankee Doodle", which has words that are stark idiocy, while the music would set the teeth of a buzz-saw on edge.'

At last it was time to sail. On 11 July, accompanied by Sanford and several Yale acquaintances, they embarked for Liverpool.

Back at Plas Gwyn on 17 July, both were thankful for quiet home landscapes and the gentle English summer. Through the rest of July and August Edward alternated cycling with sessions of helping Ivor Atkins orchestrate the *Hymn of Faith*. But when Jaeger wrote to ask if he might do an Analysis for the continuation of *The Apostles*, Edward replied:

I know nothing about Apostles pt. 2. or any analysis: if it is ever finished I imagined you might take on the analysis *if* properly recompensed: my life now is one incessant answering of letters & music is fading away.[256]

Jaeger himself was in a bad way. More surgery had failed to halt his consumption, and he would have to go back to Switzerland in the autumn. Before that, Edward asked him to join their house party for the Worcester Festival.

In mid-August came the first stirrings of music for six months. He was looking at Matthew Arnold's 'Empedocles on Etna'. The three scenes represented a man's life. Early in the morning in a cool forest region (recalling the morning forest near the Severn in *Caractacus*) a young musician Callicles sings of the beauty of earth. He invokes one of the symbols of Edward's Broadheath:

> . . . The sun
> Is shining on the brilliant mountain crests,
> And on the highest pines . . .

[255] 29 June 1905 (Novello archives).   [256] 6 Aug. 1905 (HWRO 705:445:8749).

In the second scene, at the boundary between the forest and the bare mountain slopes at noon, the mature Empedocles hears the young man's singing and reflects:

> The Gods laugh in their sleeve
> To watch man doubt and fear,
> Who knows not what to believe
> Since he sees nothing clear,
> And dares stamp nothing false where he finds nothing sure.

Finally, on the rim of the volcano at evening, the ageing Empedocles resolves his suicide:

> Before the soul lose all her solemn joys,
> And awe be dead, and hope impossible,
> And the soul's deep eternal night come on—
> Receive me, hide me, quench me, take me home!

They were close to the words of Gerontius himself. But the setting did not progress.

What Edward finished that autumn was a gentler passage—a part-song to verses from Coventry Patmore's 'The River', entitled *Evening Scene*:

> The aspen leaflets scarcely stir:
>   The river seems to think;
> Athwart the dusk, broad primroses
>   Look coldly from the brink,
> Where, listening to the freshet's noise,
>   The quiet cattle drink.
>
> The bees boom past, the white moths rise
>   Like spirits from the ground;
> The gray flies hum their weary tune,
>   A distant, dream-like sound;
> And far, far off to the slumb'rous eve,
>   Bayeth an old guard-hound.

He set the verses in D minor—his first use of the tonality in a separate work since *Gerontius*—and dedicated it to the memory of the recently dead and much beloved manager of the Morecambe Festival, R. G. Howson. Pointing the symbolism of the river, Edward inscribed the end of the score with the name 'Rotherwas', a quiet place along the Wye close to Plas Gwyn. It was just three years since he had written his last part-songs in *Weary Wind of the West* for Morecambe and the Songs from the *Greek Anthology*.

For the Worcester Festival, as three years ago, Alice had again taken Castle House in College Green and filled it with Edward's friends and acquaintances: Canon Gorton, Jaeger, Frank Schuster, Henry Ettling, Prof. Sanford (in England again), Mrs Worthington, Frank van der Stücken from Cincinnati. Young Havergal Brian from Hanley came to the Festival as Edward's guest and had generous hospitality: 'He introduced me to all his friends—found me rooms, and when I reached them I found tickets for all performances, including an invitation to lunch and the banquet when he was made Freeman of the City of Worcester.'[257]

The Freedom of the City had been arranged by Hubert Leicester, who was now the Mayor of Worcester: he always said his chief reason for accepting the Mayoralty was to be able to give Edward the Freedom. Hubert was the city's first Catholic mayor, and on the opening Sunday of the Festival he made the Corporation come to St. George's Church in company with Edward, Alice, Carice, May Grafton, and Madame Albani, who was again to sing in *The Apostles*.

The Festival music began on the Tuesday with *Gerontius* (which displaced *Elijah* from its traditional position). Before the performance came the ceremony of presenting the Freedom of the City—on a scroll encased in a silver casket which Hubert himself had designed to show the Arms of the City and vignettes of well-loved places. Edward's speech of thanks recalled the fond boyhood they had shared; then he paid tribute to the influence of his mother, who had died just three years ago: 'Many of the things which my mother told to me I have tried to carry out in my music.'

As they emerged from the Guildhall in brilliant sunshine, the pavements toward the Cathedral were lined with people, including several of the orchestra (now recruited more and more among London players). One of them was the young violinist W. H. Reed who had first spoken to Edward three years earlier about becoming a composer. Now he saw

. . . the procession making its way from the guildhall to the cathedral with the mayor, the high sheriff and all the aldermen in their civic robes and Elgar walking solemnly in their midst, clothed in a strange gown which puzzled most of the onlookers. Upon inquiry this turned out to be the Yale University gown and hood which Elgar hastened to wear on the very first occasion that a Doctor of Music's robes were needed at any of his public engagements.

Remaining in the memory is another thing in connection with this procession and civic honour, and that is that Elgar turned as he passed a certain house in the High Street on his way to the cathedral and saluted an old gentleman whose face could just be seen looking out of an upper window. It was his father, who was watching the honour being paid to his son by the city of his birth. Being very old and feeble, he was unable to leave his room; but what must his feelings have been on looking out of that window and seeing before his very eyes the fulfilment of his wildest dreams![258]

It was two days before W. H. Elgar's eighty-fourth birthday. The window from which he looked was in the family rooms above the old shop, whose business was now in the hands of his younger son, Frank.

In the Cathedral Ivor Atkins conducted *Gerontius* before a huge audience. Then Hubert Leicester gave the Mayor's luncheon for the new Freeman. In the

---

[257] Quoted in Reginald Nettel, *Ordeal by music* (OUP, 1945), p. 37.
[258] *Elgar*, p. 87.

afternoon came the *Hymn of Faith*. On Wednesday evening Edward conducted the *Introduction and Allegro*, and on Thursday morning *The Apostles*: 'Beautiful performance,' Alice wrote: 'Wonderful hush & awe—Many weeping.' Madame Albani, Muriel Foster, and John Coates sang their original roles. Plunket Greene sang Judas, and a young local baritone, William Higley, made such an impression in the role of Peter that Edward thought of giving him the role in the next portion of *The Apostles* for Birmingham, where Peter was to emerge as a grand figure.

Frank Schuster had brought an invitation for Edward from Lady Charles Beresford, whose husband had recently been made Commander-in-chief of the Mediterranean Fleet. The invitation was to join a party in the HMS *Surpise* to cruise with the Fleet in the Mediterranean for a fortnight as guests of the Navy. Edward dithered. Finally on the Festival Friday morning he decided he could not go. Schuster regretfully left for London to prepare his own journey. As soon as Frank had left, Edward decided he would go after all:

A. went out & got things for E & a box & a suit case—had his initials painted on box & packed. May[Grafton] fetched things from Herefd. E. went 2.45 train—A. in cab with him—E. much touched by all the week & success of works & love shown him. Suddenly all the house party appeared to see him off.

Jaeger wrote afterwards to Alice:

What impressions will crowd upon his wonderful Brain as he nears Athens & beholds from afar the Acropolis which stands—even in ruins—for all that is greatest & noblest in art & History. The journey will stimulate him & his creative faculty tremendously, I have no doubt & he will return strong in health and brimful of ideas. Only!—What about 'Apostles Part III' for B'ham next year!? I begin to fear me that we shall hope in vain to see our soaring expectations fulfilled.[259]

Edward and Frank went by train to Brindisi, and crossed in a small ship by Corfu to Patras. At Athens they were met by Lady Maud Warrender's husband Sir George (in command of the *Carnarvon*) and Henry Harris. They all went aboard the *Surprise*, where they found Lady Charles Beresford, the authoress Mrs Craigie (better known under her *nom de plume* John Oliver Hobbes), and Lady Maud Warrender. Their expeditions ashore were recalled by Lady Maud:

Lady Charlie and Mrs. Craigie had a somewhat flamboyant taste in clothes and floating veils. Their appearance on the quay was such an astonishment to the Greeks that they would be surrounded by a mob, and Frankie Schuster, Bogie Harris and I found it less disconcerting to land at another time![260]

Edward kept his own diary to record the sights and sounds of the voyage that followed.

Friday (22nd Septr) Fleet began to prepare to move about 6—Sailed off one by one

[259] 20 Sep. 1905 (HWRO 705:445:8792).　　　　[260] *My first sixty years*, p. 82.

about 7 . . . At eleven anchor raised & we started, taking a long long look for the last time at Athens & the Parthenon.

Then began a lovely voyage: round Cape Colonna—thro' Doro Channel to Lemnos—during the day we passed the Fleet in singles and twos—much firing at targets &c. The most gorgeous sunset we have yet seen . . .

Saturday (Sep 23rd) Arrived off Lemnos about six o'clock . . . Fleet arrived, very grand sight, solemn procession & *grand* noise anchoring. Admiral [Lord Charles Beresford] to lunch—in afternoon Frank & I ashore—all Turks—poor dried up little village, quite eastern—dogs about: we walked thro the village & out to open country—heard a pretty shepherd boy playing on a pipe [—] quite beautiful—gorgeous sunset [—] back to ship with the Warrenders . . . Decisive news that the fleet *nor* Lady Charlie may not go near Constantinople on acct of tension. Frank decided that I must go—so to bed.

Sunday morning. Surprise started at six a.m. to take us to Tchanak, which is the limit under the circumstances . . .

There they found an Austrian steamer going to Constantinople.

[Monday, 25th September 1905] Glorious sunrise & the minarets of Stamboul began to come thro the mist—wonderful. wonderful . . . Endless dreariness over passports, but at last free, & drove to Perah Palace Hotel—thro' goods, Turks, donkeys, pack trains, and dogs & dogs.

. . . Drove up to the higher part for views: in the sunset the Bosphorus & Stamboul were insanely beautiful . . .

Sep 26th Tuesday: No headache. up at 8.30 glorious weather—providence is kinder to Moslems than to Xtians . . .

Sep 27th Wednesday. at 10 drove to the Seraglio—special permit . . . We all sat round the gorgeous room [—] many Turkish servants—one superior—first, one carried round a stand, with a raised centre, on which stood a glass jar of rose leaf jam—tumblers of iced water surrounded it—you took a teaspoon & ate the jam, & drank the water—this left a beautiful, delicate taste of roses all down your throat & round your mouth . . .

Then on this glorious evening—sun & wave & sky—up the Bosphorus (Oxford)—as far as Teraphia . . . Dressed, & bathed & rested & then to dine with the Ambassador . . . Lady M. sang many songs & Frank accompanied beautifully—then we had 'In Haven' & 'Where corals lie' which Lady M. sang well & I accompanied [—] walked back with Frank & Mr. Harris along this wonderful shore.

To bed at twelve. Very quiet only the lapping of the waves just under the window—but the steamers calling made it impossible to sleep about 5.30.

Next day they boarded a Mauritius steamer for Smyrna, where they rejoined the *Surprise*:

Saturday (30th). Rose early—glorious day. Frank, Lady M, & I ashore—went to the bazaar—much finer sight than Constinople. Colour movement & camels—100's—led by a donkey through the bazaar. (This was my first touch with Asia & I was quite overcome—the endless camels made the scene more *real* than Stamboul—the extraordinary colour & movement, light & shade were intoxicating [.] Lady M. & Frank bought heavenly jam (cherry & *roseleaves*) & camel-bells, rugs & coverlets.) Lady M. gave me a silver camel lamp in remembrance of my first eastern camel . . .

Sunday (Oct 1) rose late. Very, very hot & sirocco blowing—Peculiar feeling of intense heat & wind . . . early lunch & then (at 2 o.c.) ashore & drove to the Mosque of dancing dervishes . . . Small mosque. Received in great style. Music by five or six people very strange & some of it quite beautiful—incessant drums & cymbals (small) thro' the quick movements.

The accumulated impressions shaped some music of his own, extemporized on the ship's piano and written down with the title *In Smyrna*. The subject was ideally suited to Edward's piano playing: it produced keyboard writing of idiosyncratic delicacy—a tremulous pastel of distant shades and shapes and rhythms opening in the old G major once more and moving gradually to G minor—never to return.

As the ship made its way back to Patras, the weather broke. Lady Maud Warrender recalled:

H.M.S. *Surprise* had a most uncomfortable way of behaving in a rough sea, quite unlike any other craft I have ever been in, a sort of corkscrew motion which, good sailor though I am, completely defeated me. Lady Charlie was the only one who did not succumb. She even managed to sit on a surging music stool and play 'The Ride of the Valkyries' and the 'Fire Music' at the height of the storm, when everyone else was prone and utterly miserable.

At Patras, where we were to leave the yacht, there was a big sea running in the harbour. Unless we caught the little steamer to Fiume which ran only once a week, it meant staying in a very bad hotel in a very dull place. Mrs. Craigie and myself hated the idea of this, so we made up our minds to make a dash in a small boat for the steamer, leaving 'Frankie Schu' and Edward Elgar quivering on the quay, not daring to face the risk of getting alongside in a horrible sea.[261]

The ladies got off successfully, the men had two days at Patras before another ship took them back to Brindisi, whence a train carried Edward to Paris and England:

Thursday Oct. 12 arrived Dover at four oc—Train at once to Charing X—arrd. Langham at 6.30 rested.
    Went to East End & tried to get Eastern food—dried sprats &c—then back to this dreary civilization.

Friday Oct. 13 midday to Norwich—Stayed with Mr. Oddin Taylor, rehearsal of 'Apostles' [for the Norwich Festival]—and the last music I had heard was the Dervishes in Smyrna!

One of the artists scheduled to appear at the Norwich Festival was Fritz Kreisler. He was clearly anticipating another meeting with Sir Edward Elgar when he gave a provocative statement to the press:

If you want to know whom I consider to be the greatest living composer, I say without

---

[261] Ibid., pp. 83–4.

hesitation, Elgar. Russia, Scandinavia, my own Fatherland, or any other nation can produce nothing like him. I say this to please no one; it is my own conviction. Elgar will overshadow everybody. He is on a different level. I place him on an equal footing with my idols, Beethoven and Brahms. He is of the same aristocratic family. His invention, his orchestration, his harmony, his grandeur, it is wonderful. And it is all pure, unaffected music. I wish Elgar would write something for the violin. He could do so, and it would be certainly something effective.[262]

It was a striking invitation, and it tempted Edward away from the *Apostles* continuation due for Birmingham in twelve months' time. Back at Plas Gwyn on 22 October (the day before leaving for the Norwich Festival) he sketched several extended ideas for a Violin Concerto.[263] One theme was marked 'Allegro Orch':

B minor was a favourable key for a violin concerto (as Saint-Saëns had shown).[264] Edward himself had used the tonality most recently for Mary Magdalene in *The Apostles*: her

was in fact a near anagram of the Violin Concerto opening theme. This theme found its reply in a figure whose main notes traced the old descent of fourths

The sketch showed a '2nd theme' in the old tonality of G major:

At Norwich they enjoyed meeting Kreisler again, and sat through Stanford's *Te Deum* in order to hear him play the Bach E major Concerto. Edward conducted the *Introduction and Allegro* and *The Apostles*—in which W. H. Reed had to leave his place in the violins and take over the organ at short

---

[262] *The Hereford Times*, 7 Oct. 1905.

[263] Pp. 32–5 in Sketchbook IV, later removed and given to Kreisler, are now in the Library of Congress, Washington, DC.

[264] Elgar had heard the Saint-Saëns Concerto in B minor at Queen's Hall during the London Musical Festival of 1899, played by Eugene Ysaÿe with the orchestral accompaniment played on the piano by Raoul Pugno. Both performers signed his programme (now at the Elgar Birthplace).

notice owing to the illness of the regular player. Afterwards Edward wrote in gratitude to ask Hans Richter to listen to a short piece of Reed's which his London Symphony Orchestra would play through at one of their rehearsals. Richter unhesitatingly consented.

From the personality and presence of Richter emerged another thematic idea—as if it was an extra 'Enigma' variation. Again the new idea was instrumental, though it might have had a hidden source is Gerontius's

And I hold in ve - ne - ra - tion, For the love of Him a - lone,—

Without necessarily bringing that music back to his full consciousness, Edward now gave it greater boldness and greater concision in a sketch labelled 'Hans himself!':

Back at home, he was waited on by a deputation from the Hereford City Council asking him to be their Mayor. Alice noted: 'They spoke very nicely & touchingly—offering him the highest honour in their power.' She wrote to Hubert Leicester (who had just been re-elected Mayor of Worcester):

I only *trust & hope* he *will* accept it as I feel *sure* it wd. be the very best thing for him, he wd. *like* it so much & I feel convinced it wd. be the best chance for the next part of the 'Apostles' as coming in contact about things with men, always interests him & does him good.[266]

Edward consulted with Hubert, but decided he must not be tempted.

He did take the chair at the Worcester Guildhall on 4 November for a meeting to organize a Worcestershire Orchestral Society. The Worcestershire Philharmonic had failed and disbanded at the beginning of the year, and something was wanted to take its place. Lord Beauchamp accepted the presidency of the new organization, Edward became a vice-president, and Isabel Fitton honorary secretary.

His name and fame were also in request on larger horizons. Before the end of the year Edward received a circular letter signed by the artist William Rothenstein about deteriorating relations between England and Germany—whose growing navy had been largely the cause of the Mediterranean Fleet manoeuvres which Edward had seen in September:

---

[265] Pasted into Sketchbook II, p. 46. Elgar told Ivor Atkins of the invention, and Atkins responded on 27 Oct. 1905: 'I must hear the "Hans" theme . . .'

[266] 31 Oct. 1905.

Dear Sir,

A movement was set on foot some months ago in Germany with the object of doing everything possible to allay the feeling of mistrust which has lately grown up between England and Germany. To this end the most distinguished representatives of art, literature and science prepared a letter which they have signed and which they desire to print in the English press, with the hope that a letter written in a similarly friendly spirit by their British confrères may be published simultaneously.

I enclose a copy of the English reply which has been prepared to be printed with the signatures of forty of the most distinguished men of letters and science, musicians and artists, in this country, among which I sincerely hope you will allow your name to appear . . .

It is hoped that such a protest of mutual esteem may prove of real service in helping to bring about a better understanding between the two countries.

Faithfully yours
W. Rothenstein[267]

Edward enrolled his name among the distinguished forty.

*   .*   *
*   *
*

In snatched hours he had been working at the new *Apostles*. It would open on the Peter material sketched for *The Apostles* three years earlier. Guided by the narrative in the Acts of the Apostles, the new work could include the reassembly of the disciples under Peter and the choosing of Matthias to replace Judas (Acts 1), the descent of inspiration in 'tongues like as of fire' with Peter's first sermon and the baptism of his hearers (Acts 2), his miracle of healing the lame man (Acts 3), the arrest of Peter and John and their release (Acts 4). Part I of the new work would finish with the Lord's Prayer, whose setting had been nearly completed in 1902. Musical ideas had also been evolved for the Pentecostal scene (Acts 2),[268] and perhaps for a good deal more of what would become Part I of the new work.

Part II might take in several of the later stories in the Acts: Stephen the first martyr (Acts 6–7), Simon the magician and his bribe of Gold (Acts 8), the conversion of Saul to Paul (Acts 9), Cornelius the Roman centurion as the first Gentile convert (Acts 10), leading to the mass conversion at Antioch (Acts 13).[269] Thus the second oratorio would cover just the material proposed for Part II of the original *Apostles* in late 1902.

Much of the new music would be built on motives from the first *Apostles*.

---

[267] 20 Dec. 1905 (Elgar Birthplace).

[268] On 26 Oct. 1903, after *The Apostles* première, Edward Capel-Cure wrote to Elgar: 'I am a little puzzled about Part III [to come]. Pentecost we have already discussed & I know how magnificently you will develop its musical possibilities—but after that, what next?' (HWRO 705:445:3575).

[269] Elgar's libretto sketches for these incidents have now been divided between BL Add. MSS 47905B and 47906.

There were some new themes as well, and these he played on 29 October for Alfred Littleton at a house Littleton had recently taken in Hereford within walking distance of Plas Gwyn. Then further work was interrupted by the need to prepare the next lecture announced at Birmingham for 1 November. He wrote to Frank Schuster:

All else must wait as I am killed with the University—Oh! how I wish when we (*I*, I mean *not we*) had passed Hotel Kroecker [in Constantinople] the bombs *had* gone off & killed no one but your poor distracted

E.E.[270]

The lecture on 'English Composers' was as vaguely organized as its predecessor, and as blunt. 'Our art has no hold on the affections of our own people,' he quoted from the inaugural lecture, 'and is held in no respect abroad.' He criticized the younger composers' symphonic poems for what he called their 'abhorrence of form . . . It matters not if it is in classical "form", but use proportion.' To explain 'proportion', he quoted from Charles Kingsley:

The utterance must be the expression, the outward and visible autotype, of the spirit which animates it . . . It is in vain to attempt the setting of spiritual discords to physical music. The mere practical patience and self-restraint requisite to work out rhythm when fixed on, will be wanting; nay, the fitting rhythm will never be found, the subject itself being arhythmic . . .

The question of a larger rhythm generated from within was the fundamental problem of his own *Apostles* music. And once again he reached the end of his lecture subject without any clear generalization:

I would like to sketch an education necessary for a musician in case he should develop (or dwindle) into a composer, but I fear it would only be an enumeration of suggested helpful surroundings.

Again there was controversy. Stanford, having kept silence after the Inaugural Lecture, now fired off a letter to *The Times* protesting against the claim that English music was held in no respect abroad. *Musical Opinion*, on the other hand, agreed that there was little native music of quality: it only thought that the whole matter had been given undue concern because of the solitary success of Sir Edward Elgar.

He had barely a week at home to work on *Apostles* sketches before the next lecture. As his forthcoming tour with the London Symphony would include a concert in Birmingham with the Brahms Third Symphony in the programme, he was glad enough to interrupt the more contentious general subjects with a straightforward analysis illustrated by his own piano playing:

The form of the [Third] Symphony is strictly orthodox, and it is a piece of absolute music. There is no clue to what is meant but, as Sir Hubert Parry said, it is a piece of music which calls up certain sets of emotions in each individual hearer. That is the height of music . . .'

[270] 29 Oct. 1905 (HWRO 705:445:6921).

Then he took up what was for him the Third Symphony's most interesting feature:

A curious 'motto' theme: introducing the first subject, running through the whole first movement and knitting it together . . . The movements are 'related' to each other in the following way. The 'motto' theme reappears in the last movement twelve bars from the end—the first subject proper of the first movement reappears in the last nine bars from the end.

It was close to the plan of his own *Introduction and Allegro* (also in the London Symphony programme for Birmingham).

One member of the audience that afternoon was entranced. He was a twenty-year-old composition pupil of Bantock named Julius Harrison. More than half a century later he recalled it: 'I can still see and hear him, his eyelids batting rapidly with that nervous tension and energy which all those who knew him will remember: his whole mind concentrated on a subject so near to his own ideals in music.'[271] Yet even such an analysis could draw acrimony from the press. Ernest Newman (who had left *The Birmingham Post* to return to *The Manchester Guardian*) took violent and lengthy issue with the inference that absolute music is superior to programme music. He cited Edward's own works with programmatic titles, and asked: 'Why does Sir Edward choose to work in a medium that his judgment condemns?'[272]

Edward wrote privately to Newman offering to talk it out when the London Symphony tour reached Manchester. When Newman said he could not manage a meeting there, Edward invited him to Plas Gwyn: that too proved difficult to arrange. After forming a high opinion of the younger man's abilities, Edward could only deplore Newman's pugnacity.

He rehearsed the London Symphony for their tour, and next morning went to Paddington for the train to Cheltenham, where the first concert would take place. The Orchestra was already in the train when he arrived, as W. H. Reed recalled:

Elgar was in his fur coat and, his luggage being collected and duly placed upon a hand-barrow, he was in the act of striding off up the platform to the reserved carriages, when the porter let go his luggage and ran after him saying:

'I wouldn't go up there if I were you, sir; there is some sort of a bloomin' band or something, going away, up there.'

'That's all right,' said Elgar as he continued his progress, '*I'm one of them.*'[273]

Alice came down to Cheltenham by a later train, and after the concert she took him back to Plas Gwyn for the weekend before he started north. Altogether he had had fewer than twenty days at home since early September.

On Monday 13 November they went to Birmingham, where he was joined by John Cousins, a valet engaged at Norwich to accompany him on the tour. The

[271] Elgar: a personal reminiscence (BBC broadcast, 27 Oct. 1960).
[272] 9 Nov. 1905.   [273] *Elgar*, pp. 88–9.

programme that night comprised the *Marriage of Figaro* Overture, *Sea Pictures* sung by Edna Thornton, *In the South*, the *Introduction and Allegro*, two *Slavonic Dances* by Dvořák, and the Brahms Third Symphony. *The Birmingham Post* reported:

We have to record a fine—even a great performance of the Symphony, though we thought the Andante taken rather too slowly—its 'peace and contentment' seemingly too greatly insisted upon. The first and last movements were magnificently played . . .[274]

The programme was more or less repeated elsewhere, sometimes with the substitution of the Schumann Second Symphony for the Brahms Third.

The next concert took place the following evening in Liverpool, and the night after that they were in Manchester. There Edward found himself pursued to his hotel by a journalist called Kenyon, who wrote under the name of 'Gerald Cumberland':

He walked towards me with head slightly bent, his tall spare frame attracting attention by its aristocratic aloofness. He looked what very few eminent men do look—distinguished. It was not his face, though that was fine; it was not his head, though that was well poised: it was something I know not what—a way of holding himself, a manner of walking as though even here he had dreams and visions. He entered into conversation with an air somewhat cold and distant; but in a minute the frostiness had disappeared, and he talked with the hesitating, interrupted eagerness of a nervous man who has something important to say . . . My personality, swept from its normal path, was aware of nothing but a pair of eager eyes that probed, of an astonishing brain that questioned . . . With Elgar the thought strikes one that within that brain are being born unimaginable thoughts, that in *there* is hurry and nervous bustle—sensations being rapidly garnered, to be fused together some day and delivered to the world in sound.[275]

Of the concert that evening, Ernest Newman wrote about the Brahms Third:

Now and then last night Elgar threw the whole force of his nervous temperament into the music, filling it with an electric quality that is of course inapplicable to the work as a whole, and the cessation of which left us rather colder than we otherwise need have been.[276]

Sheffield, Glasgow, Edinburgh, Newcastle, and Bradford—he 'got frightfully sick of the tour' before it finished on 21 November.

There was another week before the next Birmingham lecture in which to pick up the threads of his *Apostles*. The arrest of Peter and John would recall the earlier arrest in Gethsemane 'with lanterns, and torches, and weapons'. He decided to strengthen the parallel by introducing after the new Arrest a big section to set the theme of suffering against the coming of another night.

[274] 14 Nov. 1905.
[275] 'First Impressions: No.1—Sir Edward Elgar', in *The Musical World*, 15 Dec. 1905.
[276] *The Manchester Guardian*, 16 Nov. 1905.

Therein his music would reflect the wonderful Eastern sunsets seen in the Mediterranean in September. He resolved that this new section should take the form of a solo for the Virgin Mary. Thus it might transmute the evening suicide of Empedocles to a mother's vision of her son's triumph.

The ideal medium for an evocation of sunset brilliance and then dusk would be the orchestra. The problem was to juxtapose it with the singing voice so that the musical design would not compete with the drama. It was the problem which had presented growing difficulty in *The Apostles*.

He arrived home at last on 23 November with a cold—tired, depressed, and worried—to find Dora Penny encamped for a visit of several days. He barricaded himself in the study. Alice mounted guard. She had already warned her guest that the time was not propitious:

'Dear Dora, it *is* nice to see you, but H.E. is *very* busy and I am afraid you'll have a very dull visit!'

I felt for a moment that she would have been thankful if I had said I would go straight back home there and then. But I didn't say it. Instead I said:

'That's all right! I expect there are heaps of things I can do for you, and if he is busy I can keep you company.' . . .

'Well, make haste and come down, dear Dora. Tea will be ready soon. You remember that when H.E. is busy like this I *never* have a bell rung for meals and we are all as quiet as possible!' (More hints!)

I had been with them, of course, at other times when he had had what I irreverently called 'a composing fit', but this one promised to be the best—or worst—that I had experienced so far.

The Lady and I had a cosy tea by the fire and had much to say. During tea she filled a thermos flask with tea and put it on a tray that was all set ready, and I noticed that there were eatables in covered muffin-dishes. I sprang up to open the door—I knew it would be useless to offer to take the tray—and she put it down on an oak chest outside the study door. She came back and we finished our tea. 'If H.E. opens the study door he'll see the tray and take it in.'

But no door was opened and no tea was taken in. There was just silence.

I had brought some needlework and I fetched it, and we talked about what I could do for her next day; and so the time slipped away and we went up to dress for dinner.

Dinner was just coming in and the Lady was in the hall when the study door opened and E.E. appeared.

'Where's dinner?' he said rather roughly

'It's here now, dear darling—we are just going in.'

He looked up and saw me on the stairs:

'Hullo, you here? I'm busy.'

'Yes,' I said, 'so I hear. I'm keeping the Lady company.'

We went in and had our dinner. He never spoke. When he was not looking at his plate he looked straight in front of him with rather a tense expression. He was very pale and looked tired and drawn. Half-way through dessert he pushed his chair back, hit my hand, which happened to be on the table, quite sharply, and left the room. He banged the study door and turned the key. For an instant I thought, 'That's to keep me out!' I looked at the Lady inquiringly.

'He always locks himself in now that the study is downstairs,' she said, 'he feels safer! . . . Oh, *dear* Dora, look at your *poor* hand! That *was* naughty of Edward, really!'

'It looks much worse than it is,' I said, rubbing it, 'I expect it will be all right soon.'

So we went back to our drawing-room fire, and then began one of the most remarkable evenings I have ever spent . . . Coffee came in, and the thermos on the tray was emptied and washed and filled with coffee. I had mine good and strong: I wanted to keep awake. We sat on, talking, reading, working, and when 10.30 came the Lady said:

'Oughtn't you to go to bed, dear Dora? I'm sure you're tired.' But I said please mightn't I stay up with her, I should so much like to. So she said no more about that, for which I was thankful. Presently she said:

'Don't you think it would be nice to make ourselves some tea?' I went with her to the kitchen, and there was a tea-tray all put ready and a large plate of sandwiches covered over, and plates of cake and biscuits. 'It isn't the first time this has happened,' I thought, and carried the tray of eatables into the drawing-room. We had two sandwiches each and took all the rest to the tray on the oak chest.

While we were drinking our tea we heard the piano at last! The piano in the study was an upright and it stood against the wall with its back to the drawing-room fire-place, and the sound seemed to come down the chimney. As the house was all quiet we heard quite well and we just sat and listened, and forgot the time. It was really most wonderful hearing the scene as it grew, phrase by phrase: once a reminder of something in *The Apostles*—the Lady and I looked at one another—and then it was all new again.

I don't know how long he went on playing, but silence came at length and we both realized that it must be very late and that we were greatly in need of another brew of tea. I went out and made it this time, and the hall clock struck half-past one as I passed it. He was playing again when I came back with the tray, but we had not finished a first cup when the music stopped. We heard his key turn, and the Lady got up and opened the drawing-room door.

'Hullo! You still up? and Dorabella too? and tea! Oh, my giddy aunt! This is good!'

I went and fetched in the other tray and we had a grand meal . . .

After we had drunk up all the tea and eaten all the sandwiches and most of the cake and biscuits we went into the study.

'Come and turn over, Dorabella, will you?' and he showed me where part of it was on the back of another page and that sort of thing. Then he played the whole of that evening's work, and more, straight through, and we recognised passages we had heard down the chimney. I saw the words, 'The sun goeth down; Thou makest darkness, and it is night . . .'

It was well after 2.30 when I made a move to go to bed.

'I think we'll all go now, dear Dora'—the Lady got up—'I must just go and put things straight in the drawing-room.'

Seeing me make for the door, E. E. called out: 'Oh, do stop and talk to me, Dorabella, I haven't heard half the news yet.'

'Yes, *do* stay, dear Dora, and talk to him; I *promise* not to carry any heavy trays!'

He opened the door for her and I remember the sound of her quick little steps going across the hall.

'Fancy your staying up all that time—why ever did you?'

'I just loved it.'

'You do look charming in that frock. When I saw you on the stairs—'

'I wished I'd brought something quieter and more ordinary, but you see I never bring—'

'Don't you dare to bring any dingy, smoky frocks when you come to stay with me, because I won't stand it—and you only looked at me twice during dinner!'

'Twice, was it? Well, I was terrified! I simply daren't look at you for fear of putting you off your stroke or something!'

'At first I hoped you wouldn't and then, as dinner went on, I hoped you would. Finally I went away; you'd won, and that was why I hit your hand so hard. Did it hurt? I meant it to!' He picked up my hand and inspected it.

'It stung a bit at first, but there isn't much of a mark left. Here's the Lady. Do you know that it's nearly three to-morrow morning?'[277]

Forcing success in this way took its toll. Next day 'E. cold & very tired & depressed—dreadfully worried.' Alice took Dora Penny out of the house to pay a round of social calls. The day after that Dora left. In the evening Alice summoned the doctor: 'Long talk over E. & his work. *Much* worried.'

26th November . . . Much worried—Fate of 'Apostles' for Festival trembling in the balance.

On 28 November, as he tried to plan his next lecture for Birmingham, Henry Wood arrived with his wife for their first visit to Plas Gwyn: Mrs Wood wanted to run through the soprano solos in *King Olaf* for a forthcoming performance. They stayed the night and left late the next morning. 'E. frantically busy with Lecture, A. & May frantically writing it out till time to start. Worked all the way in train. Peyton's carriage met us . . .'

The subject was 'English Executants'. He opened it with the question of education, and cited a recent book by Ffrangçon-Davies, *The Singing of the Future*, for which he himself had written a Preface:

In circles of lesser value the modern ballad, with its unanalysable inanities, is still accepted as a recognisable form of art, but our better singers—our real interpreters and our teachers—have long ceased to affront their own intelligence by presenting the rubbish demanded by the uneducated for their pleasant degradation.[278]

The matter of education cast its influence equally on choral singing. An English chorus might be trained to be

. . . a machine as regular as a steam engine and perhaps as explosive. Every detail, every syllable is studied, but often there is something wanting—that something is *an understanding* of the subject sung about . . .

This sort of thing would be impossible in a good German Chorus: on the whole, they have superior education, accustomed to hear classics in their theatres in every town and to become acquainted familiarly with subjects and personalities . . .

[277] *Edward Elgar: memories of a variation*, pp. 69–73.
[278] David Ffrangçon-Davies, *The Singing of the Future*, with a Preface by Sir Edward Elgar (John Lane The Bodley Head, 1904).

There was no equivalent available in England. Hence the total lack of English singers of opera. In fact his own experience of London play-going suggested that

. . . We have no drama in England. That is to say we have no real dramatic (stage) art; we have in the whole ranks of the theatrical profession enough good actors to properly cast one play—and *no more* . . .

The want of drama carried over into English instrumental playing. He praised some of the pianists—Fanny Davies and Leonard Borwick in particular—and the English orchestral profession as a whole. Yet student orchestras at the colleges lacked the sense of drama; and that lack, he felt, had been inculcated into them:

This is the fault of studying under an uninspiring conductor. Everything should be done to make the players appreciate to the full the nuances and general 'sweep' which we admire in our adult orchestras; it is easier to restrain too great an ardour than to excite a dulled—not dull—player—dulled by his tuition—to the same feeling.

And so to the dearth of English conductors. He recalled the 'pedantic mechanics' of thirty and forty years earlier. There were several composer-conductors of present eminence—he mentioned Stanford and Cowen—but these had not the breadth of sympathy or experience on the rostrum needed for the full repertory.[279]

It would seem that we have, up to the present, had room for only one conductor, i.e., a conductor who does nothing but conduct—himself a giant—Henry J. Wood.

Then Edward proposed what Arthur Nikisch was beginning at the Conservatory in Leipzig—a class for conductors:

It is all very well to say 'Conductors are born not made'—but have we ever seriously attempted to make them? or to assist them seriously in acquiring practical knowledge . . . If musical degrees are ever granted in this University, conducting will become one of the main points to be considered and form an important part in the examination; facilities for practice will be provided.

It was an unprecedented idea in England, and it was to prove years ahead of its time. Not until 1919 did the first English class for conductors emerge at the Royal College of Music under Adrian Boult.

Again there was reaction in the press, which found it easy to ignore the praising parts of the lecture and concentrate sensationally on the blame:

Sir Edward Elgar is perhaps just a trifle too outspoken. Lecturing recently at Birmingham University on the subject of 'English Musical Executants', he deplored the existence of 'too many brainless singers' and declared that even the influence of the

---

[279] Sir Adrian Boult conveyed to me his vivid memories of both Stanford and Cowen 'beating their way' through works like the Brahms *Requiem* with no intent beyond keeping the big forces more or less together. Stanford would wave his stick with one hand while consulting his watch with the other 'as if he wanted his dinner'.

teaching in our colleges had been unable to raise the standard all round. Mr. Wood, he says, is our one and only conductor . . . An eminent organist has made perhaps the most decided answer to all that, and he does it so: 'My opinion of such language is simply this: the statements are unworthy of a professional chair.'[280]

On 1 December Robert Buckley came to Plas Gwyn to secure an interview 're storm raised by misleading quotations from E.'s lecture', as Alice noted. Edward wrote to Frank Schuster, 'Poor Alice is terrified at the thought that I shall be asassinated.'[281] Buckley reported Sir Edward 'genial, cheery, at peace with all the world'—and discovering the joys of chemistry in a laboratory which he had fitted out in the basement.[282]

Four days later he was busy with the next lecture, and the morning of 6 December saw the entire household again toiling at the assemblage up to the moment of leaving for Birmingham. The subject was 'Critics', in which he included audiences. He defined the critic's function: to give a composer, after he has learned all he can from friends and teachers, 'the final polish'.

He quoted at length from Hanslick on Brahms. He praised Ebenezer Prout for bravery in retracting his own earlier criticism of *Parsifal*. He cited the prose of Joseph Bennett; the judgement of the late *Manchester Guardian* critic Arthur Johnstone; Baughan, Kalisch, and several others:

The lively and unstable pen of Bernard Shaw fluttered over the page of the Star in 1890, and subsequently in the world to our great amusement: naturally so fascinating a writer produced a host of imitators and gave a new life to musical criticism which, to say truth, it needed in some quarters. But in Bernard Shaw's writing there was always a substratum of practical matter, or (to put it chemically) the volatile and pellucid fluid held in solution—matter which was precipitated into obvious solid fact by the introduction of the reader's own common sense. His imitators give us the watery fluid without the useful precipitate.

. . . I speak with admiration of a critic gone from among us, Mr. Ernest Newman—a champion of the new school and an author of repute . . . It is true we do not agree on several points, but I have never found difference of opinion make daily intercourse with real strong men less possible.

So again he offered the olive branch.

He would avoid coming too close to 'the shady side of musical criticism'. He made the single exception of an attack in *The Cornhill Magazine* on the music of Sir Arthur Sullivan immediately after Sullivan's death, citing it as 'that foul unforgettable episode'. He did not name the writer. Many people present knew it was Fuller-Maitland, the friend of Stanford; and everyone knew that Stanford had succeeded Sullivan as conductor of the Leeds Festival.

Some of the press reaction this time was a little more muted, but as *The Musical News* wrote: 'Sir Edward Elgar must really be beginning to feel sorry that he accepted the Birmingham professorship, seeing that now he cannot

---

[280] *Musical Opinion*, 1 Jan. 1906.   [281] 2 Dec. 1905 (HWRO 705:445:6920).
[282] *The Birmingham Gazette and Express*, 2 Dec. 1905.

open his mouth apparently without finding himself embroiled in some more or less lively controversy.'[283]

A week later came the final lecture of the first series—'A Retrospect' over the year of the professorship. Near the beginning Edward sounded a rueful note:

It seemed possible to me that I might in some way help those engaged in their studies and pledged to an artistic life, and perhaps clear the way a little here and there, and perhaps to set some sort of ideal . . .

In fact he had saved for this final lecture of the year his own answer to the cult of programme music and the symphonic poem:

I still look upon music which exists without any poetic or literary basis as the true foundation of our art. As four-part harmony remains still the foundation of Choral writing, as the Cantilena of the Violin is the real essence of the genius of the instrument, so is absolute music the real staple of our art. No arguments I have yet read have altered this view.

. . . I hold that the Symphony without a programme is the highest development of art. Views to the contrary are, we shall often find, held by those to whom the joy of music came late in life or who would deny to musicians that peculiar gift, which is their own, a musical ear, or an ear for music. I use, as you notice, a very old-fashioned expression, but we all know what it conveys: a love of music for its own sake.

It seems to me that because the greatest genius of our days, Richard Strauss, recognises the Symphonic Poem as a fit vehicle for his splendid achievements, some writers are inclined to be positive that the symphony is dead. Perhaps the form IS somewhat battered by the ill-usage of some of its admirers, although some modern Symphonies still testify to its vitality; but when the looked-for genius comes, it may be absolutely revived.

In noticing my remark made incidentally in the Lecture on Brahms' Symphony, some one has said 'my views are astounding seeing what things I have written myself' . . . I HAVE written Overtures with titles more or less poetic or suggestive: am I then so narrow as to admire my own music because I have written it myself? . . . When I see one of my own works by the side of, say, the Fifth Symphony, I feel like a tinker may do when surveying the Forth Bridge . . .

Next day Alice wrote to Jaeger at Davos:

Yesterday was perfectly delightful, the audience did thoroughly enjoy it & were in a suppressed state of clapping all the time. The reports give *no idea* of what his Lectures are.

Now he is turning to Music again which is a great joy.[284]

But it was not to the long-desired Symphony that he could turn. As he took up the new *Apostles* in earnest, his headaches returned.

On 23 December the Manchester journalist Kenyon came to Plas Gwyn to write a longer article. He recalled:

---

[283] 9 Dec. 1905.   [284] 14 Dec. 1905 (HWRO: 705:445:8831).

It was Christmas week, and within ten minutes of my arrival Lady Elgar was giving me hot dishes, wine and her views on the political situation. The country was in the throes of a General Election . . . Radicals were the Unspeakable People. There was not one, I gathered, in Hereford. They appeared to infest Lancashire, and some had been heard of in Wales . . .

After lunch Elgar took me a quick walk along the riverbank. For the first half-hour I found him rather reserved and non-committal, and I soon recognised that if I were to succeed in obtaining his views on any matter of interest I must rigidly abstain from asking direct questions.

Kenyon worked round to the subject of musical criticism. Edward confessed to general disappointment.

'There are exceptions, of course,' I ventured. 'Newman, for example.'

'No; Ernest Newman is not altogether an exception. He is an unbeliever, and therefore cannot understand religious music—music that is at once reverential, mystical and devout.'

'"Devout"?' whispered I to myself. Aloud I said:

'A man's reason, I think, may reject a religion, though his emotional nature may be susceptible to its slightest appeal. Besides, Newman has a most profound admiration for your *The Dream of Gerontius.*'

Elgar was silent for a few minutes. Then, with an air of detachment and with great inconsequence, he said:

'Baughan, of *The Daily News*, cannot hum a melody correctly in tune. He looks at music from the point of view of a man of letters. So does Newman, fine musician though he is. Newman advocates programme music.'

And Edward reviewed the argument. Kenyon went in hot pursuit:

'Do you, for that reason, declare that Strauss regards music from the literary man's point of view—Strauss, who, of all living musicians, is the greatest?'

He paused for a few minutes, and it seemed to me that our pace quickened as we left the bank of the river and made for a pathway across a meadow. But he would not take up the argument; stammering a little, he said:

'Richard Strauss is a very great man—a fine fellow.'

A more sensitive interlocutor might have realized that these were not the questions Sir Edward found it easiest to discuss at this moment, and taken the very polite hint of a snub. But Kenyon only concluded 'that either his mind was wandering or that he could think of no reply to my objection'.

A little later, on our way home, we discussed the younger generation of composers, and I found him very appreciative of the work done by his juniors. He particularly mentioned Havergal Brian . . .

'[But] you must not, as many people appear to do, imagine that I am a musician and nothing else. I am many things; I find time for many things. Do not picture me always bending over manuscript paper and writing down notes; months pass at frequent intervals when I write nothing at all. At present I am making a study of chemistry.'[285]

---

[285] 'Gerald Cumberland', *Set Down in Malice* (Grant Richards, 1919), pp. 80–5.

On the last day of 1905 Edward wrote to Walford Davies:

You ask of me: well I am the same depressed (musically) being & the same very much alive (chemically & every other 'ally) mortal; keen for everything except my avocation, which I feel is not my vocation by a long tract of desert: I am working away & some of the themes are not bad.[286]

*   *   *

Plas Gwyn, Hereford.

Jan 4 1906

My dear Littleton:
     By this post I send the first scrap of the new work—title to be considered: it should be 'The Kingdom of God'. This portion is only the introduction but the rest shd follow soon: this portion must of course end a page.

Yrs ever
Ed: Elgar[287]

The opening page of the manuscript, written three years ago for the first Apostles, was now marked 'Intro for The Kingdom of God'—with a query. It began with the Gregorian antiphon 'O sacrum convivium' which Edward had saved for his Peter music. Peter entered quietly recalling the words of Christ: 'Where two or three are gathered together in My Name, there am I in the midst of them.' The other soloists—representing the Virgin Mary, Mary Magdalene, and John—replied: 'These are the words of the Lord Jesus . . .' But then there was a better text in Acts 20: deleting 'These are', Edward substituted the single word 'Remember'.[288] In the year after the Introduction and Allegro with its 'Welsh' melody forgotten and recalled, remembrance was to make the entry also into his new work.

     When Alfred Littleton saw this first portion of manuscript, he did not quite like the suggested title The Kingdom of God. Thinking it over, Edward decided to make a new musical beginning altogether. He planned a major orchestral Prelude to open with a survey of motives from Peter's past (almost as the three-year-old idea for an orchestral work about Shakespeare's Falstaff might have traced that character's inner history). Such an orchestral character study had a particular point now: it offered a middle ground between the ideal of symphonic abstraction made public in his last lecture at Birmingham and his own present necessity of continuing the Apostles music to make a story.

     The new Prelude began Allegro maestoso with a triumphant recalling of the 'Preachers' and 'Gospel' polyphony which had accompanied Christ's final words to the Apostles, 'Go ye therefore and teach all nations . . .' The music's E flat major announced the basic tonality for this entire oratorio. The

[286] HWRO 705:445:2191.
[287] Novello archives.
[288] BL Add. MS 47905A fos. 19–22. These pages have been stamped with rehearsal cue numbers (beginning at '1') for printing.

'Gospel' soon combined with 'Constitues eos'. A motive appeared for Peter himself:

The upper line was made from 'Christ's loneliness'; the semiquavers underneath recalled the chromatically climbing triplets which had accompanied the accusing questions of the Servants when Peter denied Christ. Then the 'Servants' figure itself appeared, interwoven with 'Christ's loneliness' to drive home its quiet point. After that came the music which had set 'And the Lord turned and looked upon Peter, and he went out, and wept bitterly.' The Prelude had telescoped Peter's relevant past in a connection of memories which closely paralleled the 'stream of consciousness' techniques being used in these years by Henry James and younger writers to open a new world of dramatic narrative in the novel.

Now the Prelude turned to the future. The pace broadened for a superb idea. In the Analysis Jaeger wrote, it was called 'New Faith':

NEW FAITH

The upper voice climbed through triads and diatonic sequences whose triplets recalled the Servants' questions. The bass counterpoint soon resolved into descending steps: so the oldest expression of Edward's music still made a basis for this 'New Faith'.

A variant of the descending steps made the bass for another new motive, which Jaeger would call 'Penitence'. Its melody seemed to sound a far-off rhythmic echo of 'Christ's loneliness':

'New Faith' sounded once more, once more 'Penitence' followed—this time rising to an impassioned climax and a pause.

Then came a third new motive, 'Contrition', leading directly into 'O sacrum convivium':

The falling shapes that opened the old antiphon spread toward a final new motive, 'Prayer':

As the bass climbed slowly upward, the treble sequences reached higher and higher to begin their descents. The paradox was emphasized by repeating this 'Prayer' through a big climax.

The multiplicity of themes and motives had been assembled in a masterful sweep. When set out in Edward's brilliant orchestration, this Prelude would make an arresting piece. Yet it had found no return to the driving energy of its opening: through its long *diminuendo*, each momentary climax made one more ruminative echo.

'E. very happy with his introduction,' Alice wrote on 9 January 1906. The London critic Alfred Kalisch arrived for a two-day visit. Ivor Atkins joined them, and there was much playing through the new music. On 12 January the Prelude was posted to Novellos.

When Kalisch returned to London, he took Carice and May Grafton to stay with the Littletons at their London house in Lancaster Gate. Carice had been quite unobtrusive at Plas Gwyn in her own little room with its small iron bedstead at the top of the house near the servants' quarters, well away from the study: there was a separate entrance to these quarters on the opposite side of the house. Asked many years afterwards whether she had found it easy 'living with father, a great composer', she responded bravely:

No, I wouldn't say it was always easy. After all, being the only child, I was somewhat sacrificed to the moods and needs of the moment. But still there was always this atmosphere of gaiety and fun about it. The only time I think that the household routine was at all disturbed was when he was writing the really serious works like *The Apostles* and *The Kingdom*, which took it out of him so tremendously that there were late nights and times when he shut himself up in his study and was not to be disturbed.[289]

Carice had been much at home since the move to Hereford had entailed her leaving The Mount. If the dinner with Dora Penny in November was an indication, then Plas Gwyn during Edward's difficult time of composing was no place for Carice. After her ten days with the Littletons it was arranged for her to stay in Malvern with the Acworths and their young daughter Rosamund. And when she returned to Plas Gwyn, Carice was to attend every day at the Girls' High School in Hereford.

On 13 January Edward and Alice entertained John Stanhope Arkwright, the young Conservative MP for Hereford. In the evening they all went to Arkwright's political meeting just before polling day in Hereford for the General Election. Arkwright held his seat. But in the election as a whole,

---

[289] BBC broadcast 'The Fifteenth Variation', 1957.

extending through the country over three weeks, the Conservatives suffered a staggering defeat. Balfour himself lost his seat: the former Prime Minister discerned in the result 'a faint echo of the same movement which has produced massacres in St. Petersburg, riots in Vienna, and Socialist processions in Berlin'. Political observers were less fascinated by the size of the Liberal majority than by the distinct new Parliamentary presence of a 'Labour' Party—all men of working-class origins. It opened another crack in the foundation of old-fashioned aristocratic ideals. And to that extent, it could focus sensitive conservative thought on the very themes of retrospection already resounding through Edward's music for the most private of creative reasons.

The Fourteenth of January found him in his study, writing more or less all day at the first scene of the new oratorio. It was laid 'In the Upper Room' of the Apostles' private gatherings. Soloists and chorus joined in a simple hymn, expressing what was now instinct in Peter's orchestral conscience. Jaeger was to call this choral figure 'Aspiration', but it embodied still more descending shapes:

Peter emerged, framed between 'Church' and 'Gospel' motives, to sing a soft greeting, 'Peace be multiplied unto you.' It was linked to the portion of vocal score sent in to the publishers before the Prelude. The others added their reflections, all laden with motives from the first *Apostles*. At last the 'Aspiration' of the choral opening shrank to

The goal of all this reminiscence was the re-pointing of 'Aspiration' to a vigorous new choral opening, 'O praise the name of the Lord'. The basses began it, and the other voices duly entered in contrasting keys: but it was only an appearance of fugal exposition, against a background filled in by the solo quartet. Soon it too dwindled to scattered recollection—until the downward shapes made a soft contrapuntal 'Amen'. The energy of hope was gone into the past because the future was not clear.

It was all done in two days and sent off to Novellos on 16 January. Next morning 'E. not well & depressed. Turning against his work . . . E. to Ross at 2 [p.m. train]—thinking of walking back but only walked about there & home by train—very depressed—cd. hardly touch anything . . .'

When he had wondered how to finish the opening scene in *The Apostles* three years ago, he had gone forward to work at a later phase. Now he finished off the three-year-old setting of The Lord's Prayer, for the end of Part I in the new work. It was to show the Apostles' inspiration reaching all the people—who would then speak together as with a single voice. Edward's setting approached the accents of speech in chromatic, unharmonized octaves:

Strangely it sounded again the figure of chilling change which had opened *Gerontius*. He posted the setting to Novellos on 18 January—as if to commit himself to finishing the music which must intervene.

Then the shape of melodic descent seemed to generate an ending for Scene I. On the day he sent off the 'Lord's Prayer' setting he was at work on another figure:

It echoed an old idea from the Organ Sonata Finale of 1895. Yet now it promised an excitement so far missing from the new oratorio. Over the next two days he sketched the big chorus to end the scene, capping the choice of the new disciple Matthias to replace Judas.

Then he was able to go back and complete the intervening music for the choosing of Matthias. The morning of 20 January found 'E. muss better—working very hard & keenly.' After the 'Amen' at the end of the earlier portion sent in, Peter introduced the motive of 'Pentecost':

His address to the assembled company began with more recollection: he shaped old motives in yet another finely woven sequence—'Christ the Man of

Sorrows', 'Judas', 'Fellowship', 'Constitues eos', the 'Beatitudes'. When he recalled the Resurrection, the orchestra re-sounded the *Apostles*' 'Alleluia'

in a big climax. But still it was all in the past, and the climax dwindled. At last the chorus of men converted the opening notes of 'Alleluia' to a soft hymn of supplication:

The women joined in developing this figure to describe the casting of lots.

When Alice returned from errands and calls in Hereford that afternoon 'Found E. hard at work & he played at once the wonderful passage 'Casting lots'. He spent no time over dinner & wrote a little more . . .'

As the lot fell upon Matthias, John led the others in reviving a strophe of 'The Lord hath chosen you' from the opening scene of *The Apostles*. That moved directly toward the exhortation Edward had assembled from diverse bibilical sources:

O ye priests!
  Seemeth it but a small thing that God hath separated you to bring you near to Himself, to stand before the congregation to minister unto them?

So, when motives and themes and symbols of the earlier *Apostles* were all reassembled and the twelve disciples reconstituted, it virtually asked the question: What was it they were there to do?

The exhortation thundered again: 'O ye priests!' But then came only a further gloss, beginning softly now:

For it is not ye that speak, but the Spirit of your Father
  Which speaketh in you:
The Lord hath chosen you . . .

The thrice-familiar words this time revived the ensemble of motives which had begun the orchestral Prelude to this new work, as the chorus sang 'Ye are the messengers of the Lord'. But once again the energy evaporated in reflection. At the end the chorus whispered:

O ye priests!
  This commandment is for you.

But what was the commandment? At the end of this big opening scene, the enigma was as dark as ever. He had found in the *Introduction and Allegro* the power that reminiscence could give his music. Yet that was only half of the Apostles' story. The other half was the application of their memories to the world of present action. But the busy present held nothing for Edward's ideal of reminiscence making its revelation to the future. Again and again his libretto sought to stimulate the music by going to the New Testament Revelation in preference to the historical Gospels. Peter would emerge as a noble visionary. But what then? When his vision was taken into the world of action and passed from the original Apostles to newer recruits, its original force would attenuate. Perhaps the Apostles had already shown him all they could.

January 29 1906 . . . E. very porsley, altered & corrected Scene of choosing Matthias. A. . . . posted it. Dr. East came [from Malvern] by 5. & stayed till 8–10 train . . . E. not well & thinking cd. not finish his work. Dr. East talked & advised him. Very bad night . . .

January 30. E. still vesy porsley—Not out—not able for anythg. & very depressed. A. dreadfully worried . . .

At the end of the month, with only a single scene finished, there was still a wide stretch of story to traverse before he reached even the Lord's Prayer. And that was only the half-way point in the new work. The second half, to culminate somehow at Antioch, had hardly been planned: there were only some disorganized sketches over a vast tract of New Testament material. How was it all to be done in time for Birmingham, with the eight-week American trip obliterating April and May? Beyond that stood the spectre of a third oratorio to draw forth from the Last Judgement some gigantic eminence to answer the earthly Christ. Edward decided he must give up altogether—now. But he reckoned without Alice.

Carice was to remember behind her mother's small stature and soft voice 'the most indomitable will . . . I think you might almost call her ruthless where my father was concerned . . . Everything had to give way to what was right for him . . .'[290] On 1 February Alice went to see Alfred Littleton at his house in Hereford. If the Birmingham commission were to be resigned, this greatest work of Edward's life would stand a self-confessed failure. Yet if its burden could be somehow reduced, there might still be a chance: for if there was at least something to go on with at Birmingham, then they might hope for the future to produce more. Alice's proposal was to get Birmingham to accept half the new work—the 'Peter' material up to The Lord's Prayer, for which Edward had many sketches. Littleton was sympathetic. Everything depended upon the Orchestral Committee in Birmingham—and then upon Edward himself.

Through the first days of February she nursed him. His headaches were relieved with cycling and with chemistry. On the 10th he sent a birthday telegram to Lord Charles Beresford. The congratulations had an extra point.

[290] BBC broadcast, *The Fifteenth Variation*, 1957.

That day saw the launch of the *Dreadnought*, the dream-ship of Beresford's rival Admiral Fisher: its combination of speed and range and armour rendered obsolete every battleship in the world—including the entire German fighting navy. Later that day 'E. bad headache & saying he must give up his work. A. dreadfully worried.'

The press began to get wind of it:

A report has been gaining ground to the effect that the health of Sir Edward Elgar is such as to cause his closest friends anxiety. Of a highly nervous disposition, and a constitution which can hardly be called robust, Sir Edward is devoting himself with his accustomed self-absorption to the composition of the last part of 'The Apostles'. The strain on the nervous forces of the man must be enormous, and it is therefore a matter for universal regret to hear it suggested that a nervous breakdown is likely to cause even a temporary cessation of his activities.[291]

Alice sent for their physician in Hereford, Dr Collens:

February 12. E. still badsley & saying he must give up his work. A. to go & say so. Dr. Sinclair came up & then Dr. Collens, & then E. said A. might say he hoped to do half. A. dreadfully worried but Dr. Collens said he thought E. wd. be better & able for it.

February 13. A. & May [Grafton] to Birmingm. at 9.42 . . . Met Mr. Johnstone at Station. Long talk. Very nice about it all & ready to accept a morning's part instead of a whole day. A. & May walked about, ordered large R[evised] V[ersion] Bible for E . . .

The final word at Birmingham rested with Charles Beale, the Chairman of the Orchestral Committee. A week later he signified the Committee's acceptance of the half-scheme: the half-achievement was to be kept strictly secret. Edward wrote to Alfred Littleton: 'I wd. prefer *not* to do it but it seems the only way to make things pleasant for everybody & so I suppose it must be done.'[292]

Alice had meanwhile been carefully at work. She persuaded Edward to invite the young baritone William Higley to come and sing the new 'Peter' music:

February 16. E. a little better but head still bad—Dr. Collens came. Mr. Higley came at 9.30 & sang some of the new Apostle music very finely—quite carried away by it & thought it wd. be wonderful . . .

February 17. E. better & busy. A. & May to Malvern for Concert Club . . . 9pin [Troyte Griffith] returned with A. & May. We walked up from station. Found E. very engrossed in writing.

February 18. E. all well . . . Played some new Apostle music to Troyte—much impressed.

The second scene was his own insertion into the story as related in the Acts. He had made a short pastoral for Mary and Mary Magdalene coming to the temple to offer alms—'the First Fruits'—on the morning of Pentecost. 'At the

[291] *The Staffordshire Sentinel*, undated cutting at the Elgar Birthplace.
[292] 21 Feb. 1906 (Novello archives).

beautiful gate' of the temple they would see the lame man whom Peter was later to heal by his miracle.

The music was in G major, but it continued the quadruple metre which had dominated both the Prelude and the big opening scene. The new section was devoted entirely to a quiet duet which recalled the little scene 'At the Sepulchre' in *The Apostles*. The dawn-motives sounded yet again, now interwoven with the figure of 'Pentecost'.

After a few days' break in London with Frank Schuster and his friends, Edward turned to the next scene:

March 3. E. beginning to look at his work again—but not much in the mood.

Scene III returned the story to the Acts. Its music (sketched for the earlier *Apostles*) maintained the 4/4 metre, and its first section kept the atmosphere of soft contemplation as well. It opened in B minor—the tonality which had opened the third scene (showing Mary Magdalene) in *The Apostles*.

The setting was again 'the Upper Room' of Scene I, where the Apostles now gathered later on the day of Pentecost. The men sang softly to the 'Spirit of the Lord' music in A major:

> When the great Lord will,
>   we shall be filled with the Spirit of understanding.

The women replied through an A flat motive of 'Supplication' (faintly echoing the 'Pity' of Scene II and Mary Magdalene's 'Anguished Prayer' from *The Apostles*).

At length the upper voices sang 'Come from the four winds, O Spirit.'

> And suddenly there came from heaven
>   a sound as of the rushing of a mighty wind:

the solo alto entered over an orchestral *Allegro* recalling the Storm on the Lake in *The Apostles*. And like the *Apostles* storm, the orchestral utterance was fragmented again so as not to overwhelm the solo voice. But when 'Tongues parting asunder, like as of fire' came down in the old descending steps, the men's chorus of Apostles broke into a resistless march drawn out of 'Constitues eos':

It moved majestically through sequence after sequence until the women climaxed it with 'The Lord put forth His hand, and touched their mouth,' set to the 'Pentecost' motive. It was not developed, but development was just round the corner.

The solo contralto moved the scene to 'Solomon's Porch' in the outer temple, where the multitude gathered to hear of the miracle. Here was the centre of this entire oratorio and its subject. Against eccentric chromatic leaps of a 'Bewilderment' figure, the basses began a seeming fugal exposition reminiscent of the Demons in *Gerontius*. The women shouted 'What meaneth this?' and everyone made different leaping and chromatic variants. The whole ensemble exactly revived the doubting Jews in *The Light of Life*: yet Edward's music was woven now with a mastery of which he could only have dreamt ten years ago.

'Bewilderment' combined with the swift descending 'Tongues', and suddenly from them sprang the 'Fear' motive from *Gerontius*:

Peter sang the 'Constitues eos' marching figure, and the chorus divided: altos and basses cried 'These men are full of new wine,' sopranos and tenors sang, 'They are truly full of power.'

There was an interruption—a call of distress from Miss Burley in Malvern. In the time since the Elgars had left the neighbourhood, her school had suffered 'a disastrous series of epidemics' (brought on, it was whispered, by conditions in the kitchen presided over by Miss Burley's mother). Now Miss Burley lost her old resilience, and she appealed for Edward's counsel. On 9 March he went over to Malvern 'to see if he cd. advise Miss Burley—Home by 5 [pm] train & in fairly good spirits'. Next day he wanted to work but was prevented by a headache. On the 11th 'E better & . . . wrote magnificently.'

Peter was about to answer the question 'What meaneth this?' He prepared himself with a brief interior recollection of Christ's words, 'I have prayed for thee, that thy faith fail not . . .' to the music of 'Christ's Prayer'. Then the Peter-motive sounded *fortissimo*, and he rose to the address:

Ye men of Judea, and all ye that dwell at Jerusalem, be this known unto you, and give ear unto my words.

He sang a broad and steady recitative in flats, supported by 'Church', 'Peter', and 'Pentecost' motives. But when he came to the centre of his statement (quoting the words of the Prophet Joel) the music changed to C major as the 'New Faith' theme—unheard since the Prelude—sounded over and over again at the psychological moment:

> It shall come to pass in the last days, saith God,
> I will pour forth of My Spirit upon all flesh:
>> And your sons and your daughters shall prophesy,
>> and your young men shall see visions,
>> and your old men shall dream dreams,
>> and it shall be that whosoever shall call on
>>> the name of the Lord shall be saved.

The setting of the later lines was to be revised right up to the time of printing in vocal score.[293] In the final version, at 'Your young men shall see visions', the far distances of it called forth a cadence that dropped suddenly to A flat. At 'Your old men shall dream dreams', Edward could not resist a finely diffused reference to *Gerontius*. For the concluding words, the 'New Faith' music rolled forward once again.

His libretto deleted several verses from Peter's sermon in the Acts referring in a fiery way to the Last Judgement. If that third oratorio was to be written, Peter's reference would have made keen anticipation. But Edward's Peter remained the noble visionary of an earthly paradise.

Next day, 12 March, Henry Embleton journeyed down from Leeds just to hear some of the latest finished work: 'E. played to him & he thought the new music the most beautiful he ever heard.' After Embleton left, William Higley came to sing the Peter part. Alice and Carice and May sat in the study to hear it—the girls with awe, Alice with a sense of what she could not have shared with anyone in that room.

Then a letter came from Jaeger at Davos. He had seen the earlier chorus 'O ye priests' in proof, and could not enthuse. Alice intercepted his letter and sent a firm reply:

I think yr. surroundings &c must have depressed you when you wrote it.

It is curious that that Chorus did not fire you, it works up all who have heard it to a great pitch of excitement—& I think you might have given E. some credit for his really fine literary taste & poetic feeling in his selection of words. If you cannot feel the Sacerdotalism of any *Church*, there is the eternal priestdom of Elect Souls in all ages, who have stood above the lower minds & dragged them up; to those who believe by religion, & to others by art, literature & pure & noble character & aspirations; so instead of 'Matthias' meaning nothing to us, it is the type of everything wh. can still infuse heroism, self-sacrifice & great thoughts into all who are not dead to such things.

Wait to judge of the new work, & especially to *remark* to anyone on it, till you have heard E. play it—all those who have, & all those have been real musicians, think it the most original & greatest thing he has done. He is very very busy & has much to think of, & so soon to start for America so *please* not remark on anything I have said. He is better I am thankful to say but the strain of the work is very great for him & makes him very easily worried. He wd. send his especial love I know, did he know I was writing.[294]

---

[293] A sketch for the final revision between cues 98 and 99, sent to the publishers when the vocal score was in proof, is in possession of the writer.

[294] 17 Mar. 1906 (HWRO 705:445:8851).

Peter reminded his now fascinated congregation of the life of Christ—the 'Beatitudes' music sounded from *The Apostles* 'By the wayside' with the rising 'Adoration' motive from 'The Ascension'—and of Christ's death. He rubbed in the contrast of the human Crucifixion, as the orchestra sounded the music of 'Eli, Eli, lama sabachthani' from 'Golgotha' in *The Apostles*; the people ruefully recalled their boast 'His blood be on us, and on our children'; and Mary revived her music of grief at the Cross. 'Men and brethren,' sang the chorus (repeating the words of Peter, but set now to a figure anticipating 'The Lord's Prayer') 'what shall we do?'

The music had moved through many keys; but when 'New Faith' sounded softly now in D major, it set the tonality for a final ensemble. D major was the tonality which had closed Scene III of *The Apostles*. Yet where 'Turn ye to the stronghold' had been added in afterthought, this Finale would rise from long lines of preparation. Peter began it with his answer—set in the quickening 3/4 metre which had been almost withheld from the work until this moment:

> Repent,—and be baptized every one of you,
> in the Name of Jesus Christ;
> for to you is the promise, and to your children,
> and to all that are afar off,
> even as many as the Lord our God shall call unto Him.

Each of these phrases was set to memorable music. 'In the Name of Jesus Christ' climbed in strong diatonic steps to culminate in a falling seventh that measured the distance traversed from similar figures sung by the Angel of the Agony and Judas. 'For to you is the promise, and to your children' sounded melodic symmetries round the opening notes of the D major scale. 'And to all that are afar off' rose a sudden octave over plagal G harmonies; then the harmonies slipped toward A major so as to make another 'return' through the tonic of Edward's finest reminiscence.[295]

The whole of Peter's statement was repeated by the chorus, who sang it over to themselves through line after magnificent line. They sang again in octaves, only at the very end finding harmony; but the massed repetition moved against the richest orchestral counterpoint to make (as Jaeger was to write in his Analysis) 'a *Tutti* of astounding sonority'. Once more Peter sang 'In the Name of Jesus Christ,' and the chorus recalled the 'Supplication' motive from near the beginning of the scene twenty minutes earlier.

Now a *stringendo* brought the entire ensemble to join in sounding a 2/2 descent of 'Pentecost' over and over again to the words:

> The First Fruits of His creatures, of His own will,
> God brought us forth by the word of Truth.

---

[295] The music and its orchestration had been in Elgar's mind since at least 17 Feb., when he wrote to consult the London Symphony Orchestra's principal horn player Adolf Borsdorf about a proposed scoring for the passage (MS in possession of Borsdorf's son).

'Pour on us the Spirit of grace' quietly recapitulated the music of Peter's 'Repent, and be baptized.' But when 'In the Name of Jesus Christ' came this time, it brought the 'New Faith' music resounding through the orchestra: and on the instant the whole vocal force answered with the Apostles' triumphant 'Alleluia'. This sudden, inevitable coda in the midst of what had seemed a continuing episode was a stroke of sheer genius: for the conversion was complete. And inspiration had descended on the great project yet once more.

When Edward posted the last parcel of Scene III to Novellos on 27 March, the work as now planned was more than half finished. There would be time to complete the final scenes after returning from the United States. He was to sail in nine days' time. He had decided at last—after keeping her waiting through all the winter—that Alice should accompany him. While she prepared and packed, Edward bicycled.

On 3 April the Birmingham chorus-master R. H. Wilson came to Plas Gwyn to go through the finished portion of the new work before commencing his rehearsals. He was deeply impressed. Then the conversation, Alice noted, inevitably recurred to 'the day at Minafon when he heard the Apostles for the 1st time & thoughts of dear A.E.R.[odewald].' On that day another of the friends died quite suddenly—Billy Baker's wife Mary, the hostess of Hasfield Court. The following afternoon Edward made the short journey over to Worcester to bid his aged father goodbye for the voyage.

On the morning of 6 April they were seen off from Hereford station by the two girls—'May very tearful, C. a little pale but quite composed—though I know her heart was very full . . .' They boarded the *Celtic* at Liverpool for the nine-day voyage, which was calm and easy until the morning of landing at New York. Then Edward awoke with a headache at the preparations for landing. The ship crept up the Hudson River in thick fog to her pier. The city was '*overpoweringly* hot and close', and there was confusion at the customs examination. Professor Sanford—'Gaffer'—met them, but he was not alone. Alice wrote: 'Reporters & photographers were there waiting. Gaffer persuaded E. to submit to be taken.' Leaving his servant to collect the luggage, Sanford drove Edward about in his car and then settled his guests at West 52nd Street for dinner with Mrs Worthington and the Damroshes. Afterwards Edward wrote to Alfred Littleton: 'It is, quite privately, talked of having a three day festival in New York next winter & they want the first [American] performance of the new work . . .'[296]

The following evening they boarded the overnight train for Cincinnati. As they rolled through western Pennsylvania and Ohio next morning, they encountered the extremes of American climate in April:

Most uninteresting country. Snowy early & cold & had to ring & have windows shut.

[296] 18 Apr. 1906 (Novello archives).

Clearings & little wooden houses & hideous towns. Arrd. all safely D. G., about 1. Van der Stucken & Mr. L. Maxwell met us, crowds of reporters & photographers.

The Cincinnati Post reported:

A tall, spare figure in ice-cream clothes, with drawn brows and a worried expression, that's Sir Edward Elgar, the famous English composer, who reached Cincinnati Tuesday to conduct his own works at the May Festival. Lady Elgar, small, plump, with laughing blue eyes and the prettiest manners imaginable, accompanied him.

. . . When the train pulled in at 1 p.m. the first to be exposed to the Cincinnati atmosphere were two English travelling rugs, gorgeously striped and barred, three English looking 'boxes' of leather, liberally decorated with labels, and a small woven fibre handsatchel—the personal property of Lady Elgar.

Following the luggage came the English-looking owners.

Sir Edward Elgar is the typical Englishman, silent, reserved and unsocial—until after dinner. He pulled at his drooping moustache nervously and waved the interviewers away with a thin, but firm, hand.[297]

One of the reporters saw Edward look toward the waiting automobile of Mr J. G. Schmidlapp, a member of the Festival Committee:

Mr. Schmidlapp's closed motor waited the company near the depot entrance, and drawn up beside it were the photographers, seeing which the English conductor gave a hollow groan and murmured 'Again.' Very genially the American Van der Stucken faced the recalcitrant ones towards the cameras. 'What's one more or less?' he remarked, as the artists gave frantic high signs to a porter to depart from the field of vision. The visitors accepted the situation with rather ill grace and as speedily as possible entered the closed car accompanied by Mr. Maxwell and were driven to the Country Club, where apartments have been reserved for them.[298]

One of the photographs showed Edward in the act of plunging toward the door of the car as if nothing mattered but to get inside.

His trials were not over. At the Country Club another reporter was eager to interview him on a variety of topics. Next morning:

Mr. Maxwell & Mr. Schmidlapp called at 8.30! E. not quite dessed . . . Then people called & called—Mrs. Taft, Mrs. Fleischmann, Mrs. Holmes, Mrs. Longworth &c &c &c.

The presence of Sir Edward Elgar had made an enormous attraction for the Cincinnati May Festival: all seats had been sold and the boxes auctioned at millionaires' prices.

Then rehearsal. Chorus knew their parts but very unpoetical & not as if they understood the words. E. *very* kind & good—back late & *very* difficult to get any food for him.

April 19. E. began to orchestrate his new work—Very hot. Mrs. Maxwell called & took A. out for drive, very hot & dusty & hideous place.

[297] 17 Apr. 1906.   [298] 17 Apr. 1906.

On the title page of the new score Edward set down lines from a Canadian poet:

I would write
  'A Music that seems never to have known
  Dismay, nor haste, nor wrong—'

(Bliss Carman)

in Cincinnati
April 1906
E.E.[299]

As the days went on the hazards at the Country Club increased:

Very hot & close—Girls & ladies come & play Bridge in the afternoon & shriek & make a wretched noise.

April 20. Very hot & close . . . A party here today so E. & A. going over to Schmidlapps for the night to escape the noise & disturbance . . .

April 21 . . . Parish Priest called, E. argued hotly with him—Walked across to Club about 10.15. E. back to his Orchestn. *so* dood. Very hot, A. found cooler white dress. The evening was most unpleasant, noisy rowdy sounding people, man whistling loudly, most exasperating, E. cd. not write, noisy party dining. We went down & E. expressed himself forcibly about the noise &c, we were disgusted. Ordered dinner in our room & went to bed early. E. much disturbed by noises. About 1 - 2 men came up laughing loudly & talking & throwing boots onto floor &c—disgusting . . .

April 22. E. woke with bad headache. A. made him tea early, he lay quiet & improved a little. Out of the question to attempt to move. House quiet, many motors & golfers on the Links. E. hard at work orchestrating.

The days went on through rehearsals and luncheons and dinners. Then Mrs Worthington arrived for the Festival.

April 28. E. had a *great* day at his Orchestration, worked for some hours. Had a little turn with A. late but it was *so* hot & close—& exhausting. E. & A. & Mrs. Worthington in auto to Rehearsal. [Introduction and] Allegro first & E. had hard work—they did not seem to care to play it. Then he was much tried by Sopr[ano, Corinne Rider-Kelsey in *The Apostles* rehearsal] & Witherspoon not knowing their parts & A. felt he was irritated . . . E. quite annoyed with Rehearsal. We came straight back in *clouds* of dust & lightning going on . . .

Next evening they dressed for dinner with the Maxwells. It was the house-warming for a 'gothic room' which Maxwell had built on to his house, complete with carved furniture, stained glass windows, and an organ. They found a large party: 'Crowd & *war* of voices—all the people we knew & introduced to heaps more. Organ opened—E. played on it—& was most sweet to everybody.'

On 1 May, the opening day of the Festival, a telegram came from Edward's brother Frank in Worcester: W. H. Elgar had died very quietly two days earlier. Soon there was a letter from Frank Schuster:

[299] Bodleian MS Mus b.32, Oxford University.

My heart goes out across the seas to you . . . I know that you are too sensible to fret & repine—that you had feared yourself for this sorrow for a long time past—that your uppermost sentiment will be gratitude for the long & fairly active life granted your father—and yet I know too that you will not be able to stifle altogether the regret of being absent when your dear one was taken . . . But there is no *sting* in such a death, and it is perhaps as well that the long farewell was spoken when you thought it only an 'auf wiedersehn'—for what could you have done in the way of devotion & attention at the last that you had not given freely & fully all your life through?[300]

Alice noted: 'We stayed some time talking & musing on & thro—Decided not to go to the Opening Concert & had a quiet day & evening.'

On the night of 2 May came *The Apostles*. John Coates and David Ffrangçon-Davies had come to Cincinnati to sing with the Americans. Alice noted: 'Fine performance. Ffrangçon & Coates splendid & giving the true keynote. E. conducted splendidly, & the impression was profound—Great audience.' At the end of each scene the audience insisted on applauding. Immediately afterwards Edward found himself being interviewed about it:

'Why', asked Sir Edward, somewhat irritated, 'why ruin the dignity of the performance by applause between the episodes?'

'But aren't you satisfied with Cincinnati's enthusiastic greeting?'

'I know nothing about it,' replied the composer. 'I walk to the desk, take up my baton, and set about my work. I am merely an artist, and as an artist I'm satisfied with nothing on earth!'[301]

The next afternoon he conducted *In the South*. On Saturday afternoon came the *Introduction and Allegro*, and *Gerontius* followed in the evening with eight hundred extra audience allowed to stand. The day after was the last in Cincinnati:

May 6. E. with the Schmidlapp party & Mrs. Worthington to lunch at The Pillars—a 'Kentucky Lunch'—very noisy, negroes singing &c, but no national songs . . .

Two days later they were back in New York, where Edward did more orchestration through the hot weather. Their final ten days in America were filled with elaborate lunches and dinners. One was at the mansion of the millionaire Andrew Carnegie:

. . . Fine large house but *very* dull—very commonplace dinner & ugly dinner table. He spoke very excitedly & effusively about E.—he took A. in, it was a party to E. *Very* dull altogether. We rejoiced to escape—

But not before Carnegie had extracted a half-promise that Edward should attend his new Carnegie Institute at Pittsburgh to accept an honorary degree when it opened in the spring of 1907—if the date would fit with the New York 'Elgar Festival' planned by the Damrosches. On 18 May they boarded the *Celtic* for Liverpool and home.

[300] 4 May 1906 (HWRO 705:445:6956).
[301] Telegraphed by the New York correspondent of *The Evening Standard*.

It was a dull voyage, and the ship was pursued by fog all the way across the Atlantic and into the harbour at Liverpool. There the sun came out:

Lovely afternoon—Country looked beautiful—Caught the 3.5 train, crossing the ferry to Birkenhead—Late in the afternoon, the train stopped at every station after Shrewsbury, the Church bells sounded so sweet & lovely, sounding across fields as we stopped at the little country stations. E. & A. loved it all—Found all well [at Plas Gwyn].

Among the accumulation of post there was unhappy news of Miss Burley. Her troubles at The Mount had forced her to close the school, and she had thereupon left England altogether to take a private position in Portugal: 'The wrench was the hardest I have ever had to face. The work of the school, the lovely Malvern countryside, the growth of Edward's fame—interests which had seemed to be part of the very texture of my life—all had to be torn out and left.'[302] Clearly more than school epidemics had been involved in Miss Burley's wrenching departure. It was not quite two years since Edward had taken himself away from her environs with the move to Hereford.

The interruption of the American trip was measured when Edward tried to return to his music. Alice noted: 'Each day E. so unwell—& quite unfit for his work.' His mind was elsewhere. On 29 May he wrote to Hubert Leicester about a possible by-election in Worcester: 'Do they want a Unionist Candidate? . . . If they have no man ready, in case of an election, do you think they wd. have me? Don't say a word of this publicly: I should like it but am not sure how the land lies.'

Next day he had a 'drefful headache'. His forty-ninth birthday on 2 June was spent 'in bed nearly all day'. Four days after that he was 'dreadfully worried as he cd. not get on with his work'. Alice wrote to welcome Jaeger home from Switzerland for the summer, but it turned to warning:

Do not trouble to write but rest *all* you can. I trust E. will be well again in a day or two but it is so hindering for him & he gets *so worried*—so that we must not hurry him or any thing—I do trust all the work will be finished before long & then he can recuperate. Meanwhile I prevent anythg. possible from worrying him.[303]

Charles Beale had written from Birmingham to say that his committee would now like to announce the title of the new work. Edward wrote again to Alfred Littleton, and they settled on *The Kingdom*. But all Edward could send to the publishers on 8 June was a portion of full score begun in America.

Two days later he started a heavy cold. Dr Collens had already recommended a holiday at New Radnor in the mountains near the Welsh border, and Alice found a house there to be taken for a fortnight. There he went down with a fresh cold, and Alice thought she saw a white spot in his

302 *Edward Elgar: the record of a friendship*, p. 185.
303 6 June 1906 (HWRO 705:445:8849).

throat. A local doctor, hastily summoned, said a gargle would take it away, and from that moment Edward improved.

June 20. E. very porseley all morning but began a little composing. After lunch thought he wd. like to go on with it . . .

June 21. E. certainly better. Worked all the morning. After lunch went for a ride but returned before long hot & tired . . . E. for another ride after tea & came in better—After dinner played his new introduction—D.G.

This Introduction opened Scene IV with a short pastoral, whose D major answered the D minor of the 'Judas' Introduction to *The Apostles* Scene IV. New descending steps were developed from the beginning of Scene II, where the women had seen the lame man. At the end its descending shape turned to 'Pentecost'.

A contralto recitative described Peter's converts after their baptism. At the mention of breaking bread, the music went quietly to F major to sound a new figure which would return in the Finale. Then, against the 'Pity' motive, Peter and John were placed 'At the Beautiful Gate' of the temple, where the lame man asked for alms. Peter invoked a series of motives in a powerful recitative, and healed the man 'In the Name of Jesus Christ'. 'Pity' accelerated in downward strokes *Allegro* as the chorus exclaimed: 'This is he which sat for alms.' 'Bewilderment' merged with the 'Temple' themes from near the beginning of *The Apostles* as the man walked into the temple to praise God.

Yet all the old Temple-motives were mere scene-painting at what might have been an action of high drama. It was the only incident in the whole of *The Kingdom* to show its hero's power in practical terms. A decade ago, in *The Light of Life*, Edward had made half his oratorio out of just such a healing. Now Peter's healing recitative and the swift chorus to follow were over in a moment. *The Apostles* 'Turn ye to the stronghold' was echoed in a short duet for Peter and John,' Turn ye again': it just carried the healing incident to an acceptable length.

Novellos were beginning to worry about printing the final music in time for rehearsals at Birmingham. On 25 June Edward wrote to reassure Alfred Littleton:

This illness had quite upset my plans for everything. I now see my way D.V. to let you have the final M.S. of this portion in July. You can print in August & the Chorus can quite well learn the remainder in September: so all will be well. I very much regret causing any inconvenience but I find I cannot 'work up' my sketches when I'm ill.[304]

Minutes after writing that letter came another small disaster:

Going out in to the garden after breakfast E. slipped on wet stones with his deck shoes on & bark[ed] his shoulder & knee. Dreadful to see him in such pain—Decided to return that afternoon—E. A. & C. came by 3.15 train . . . the Dr. said nothing was broken D.G.—dreadfully miserable—cd. not move in bed, the shock so upsetting to him.

[304] Novello archives.

Massage over the next days eased the pain, and by 30 June he was beginning to write again. But when Jaeger sent to ask when they could have the final version of the concluding Lord's Prayer setting, Alice replied: 'He is sorry for the delay, but please have a little more patience & all being well, the new work will be about his best and in time convincing even to pagans, but you must *please* not worry or hurry him in any way.'[305]

The whole of Scene IV was entitled 'The Sign of Healing'. But what really occupied Edward, as in *The Light of Life* ten years earlier, was the punishment which must follow on working miracles in the world. The result of Peter's miracle was to bring his arrest by the priests and Sadducees. The passing reference in Acts 4 to 'eventide' was opened to insert in Edward's music a monument to suffering. This was the big sunset piece to be sung by the grieving, exulting mother—the longest solo in the entire *Apostles* music. He had assembled the text literally line by line from the most diverse Old and New Testament sources to say precisely what he wanted. It was the rough working out of this most private expression which had cost him such anguish at the time of Dora Penny's visit in the previous November.

First a marching figure of descending steps echoed the march of Judas and his rabble in *The Apostles*. The contralto described the placing of Peter and John in ward 'until the morn' (the 'Watching' and 'Dawn' figures sounded from the earlier work). But at the word 'eventide', the descending thirds of 'Prayer' made a new figure of 'Night'

A solo violin gave out fragments of two old Hebrew hymns (found in the same collection which had produced *The Apostles* Morning Hymn). Each of the old melodies made its appeal to Edward's style—the 'Hymn of Weeping' a slow syncopation, the 'Hymn of Parting' still another figure of slowly descending steps.

Mary's quiet opening was a meditation on 'Parting' and the earlier march:

The   sun   go·eth   down__;

Edward resisted the temptation to sound a lower note on the word 'down'. The rising fifth at the end of the phrase looked forward to the brilliance to come: so again the needs of his music overtook the significance of the words he had chosen. Mary revealed her darkening landscape in phrases that leapt upward and sank slowly back to reminisce with the lonely violin over the Hebrew

[305] 1 July 1906 (HWRO 705:445:8846).

hymns: 'I commune with mine own heart, and meditate on Thee, in the night watches.'

She foresaw the rejoicing 'when His glory shall be revealed': and a series of sunset climaxes flashed and thundered about the orchestra behind the high voice. The singing of Agnes Nicholls, who was to create the role at Birmingham, could soar effortlessly over scoring that would rival the magnificence of the orchestral Prelude to this oratorio, as she sang: 'The Gospel of the Kingdom shall be preached in the whole world . . .' But that triumphant phrase from St. Matthew was followed in Edward's text by another from Revelation—'. . . the Kingdom and the patience . . .'. After all that had been felt and thought and written and even performed, the time was still not yet.

The brilliance which had waxed and waned turned to dusk. The violin ruminated over its memories with a lingering beauty which might someday find a place in the Violin Concerto. And the final soprano phrase answered her opening descent in mirror image:

As a whole this big set piece answered the orchestral dawn near the beginning of *The Apostles*. In the end it mattered little what words the voice had sung. The desolation of a lonely presence against the darkening horizon had revealed Edward's orchestral genius more masterfully than ever.

After a three-day visit to Frank Schuster, he settled to the last scene of *The Kingdom*. It returned to 'The Upper Room' of Scenes I and III. The chorus revived their hymn of 'Aspiration' from the opening of Scene I to other words as familiar:

> The voice of joy is in the dwelling of the righteous:
> The stone which the builders rejected is become the head of the corner.

As John described their release, the whole chorus and orchestra joined to thunder the final new music of the work:

It seemed to echo the triumphant solution at the end of the 'Enigma':

Such profound triumph followed a little strangely on a single night's imprisonment. But the imprisoning incident had been too small for the anguish of Edward's hurt, too small for the joy of his release.

His libretto was arranged to conclude with a final communion for the Apostles—a last sacrifice of present engagement to past and future hope. At its centre came The Lord's Prayer, sounding its *Gerontius* echoes in choral speech—rising to a climax in the 'Spirit of the Lord' music from the beginning of *The Apostles*, sinking back in descending steps 'for ever and ever'. John and Peter turned those steps to the descent of 'Prayer' to bring the chorus to their final words: 'Thou, O Lord, art our Father, our Redeemer, and we are Thine.' 'New Faith' sounded its sequence again in E flat major, and *The Kingdom* was finished.

Compared to the tremendous events of *The Apostles*, almost nothing had happened. As there was less challenge, so there was less chromaticism. The relative security of diatonic expression made a kind of musical unity. And even more than *The Apostles*, *The Kingdom* was bound in almost continuous quadruple metre. Even more than its predecessor, it made a continuous mixture of solo and choral declamation. Its 100-minute length—shorter than *The Apostles* by twenty minutes—had immobilized the action.

July 23 1906. E: really finished the composing of his beautiful work—Most thankful . . .

In London to discuss the Analysis with Jaeger, Edward lunched with Professor Sanford and two great singers—Enrico Caruso and the French baritone Victor Maurel, who had included *The Pipes of Pan* in a recent Bechstein Hall recital. Back at Plas Gwyn on 28 July, he took up the orchestration of *The Kingdom*. In ten days he scored 68 new pages. Every other post brought fresh proofs of vocal score to correct, and Jaeger sent suggestions for several musical emendations to the new work which Edward gladly adopted. Each day found him in the study pouring out the work.

Near the end of August, with the full score close to completion, Jaeger brought his Analysis to Plas Gwyn. He was very ill. There was talk of his having to retire from Novellos, though he was only forty-six. He could not attend the Hereford Festival in September because the doctor had ordered him to the seaside before leaving for Switzerland once more. He stayed several days now while Edward heavily revised the Analysis.[306]

[306] In his letter of 16 Oct. 1907 to the Novello company secretary, Henry Clayton, Jaeger expressed doubt about accepting the firm's payment for the Analysis 'especially after Elgar knocking it about in the way he did.' (Novello archives.)

Alice's house party for the Hereford Festival that year included the Kilburns. Edward conducted only the *Introduction and Allegro* at the Festival, leaving *Gerontius* and *The Apostles* to Sinclair while he himself sat at home working through orchestral parts for *The Kingdom* with John Austin and Nicholas Kilburn, who recalled:

Mr. Austin and I found ourselves along with Sir Edward hard at work, they each with a violin and I at the piano . . . With the piano score it was interesting to watch how Sir Edward 'spotted' the printer's mistakes, and how dextrously he and Mr. Austin played the various parts, including, of course, those of the transposing instruments, sometimes octaves higher or lower than the printed notes.[307]

On 19 and 20 September he rehearsed the Birmingham Chorus in *The Kingdom* and *The Apostles* (which would be sung the night before *The Kingdom* première). On the 25th he rehearsed the orchestra in Manchester. Alice wrote to Jaeger of this first orchestral hearing: 'Gorgeous seas of sound. The band broke now & again into irresistible applause.'[308]

After the rehearsal Edward caught the night train to Aberdeen, where the distinguished musicologist and professor Charles Sanford Terry had invited him to join a large convocation including Sir Oliver Lodge and Andrew Carnegie to receive honorary degrees at the celebration of the University of Aberdeen's Quatercentenary in the presence of the King and Queen.

He returned to Birmingham late on the 28th, and the following day took the combined chorus and orchestra rehearsal. He began with the opening scene of *The Apostles*, and then went right through *The Kingdom*. During the short luncheon break he did not leave the hall. In the afternoon they had the rest of *The Apostles*, and then Alice took him home for the weekend to rest. They were back in Birmingham on Tuesday for *The Apostles* that night. Alice noted: 'Very fine performance.'

Next morning was *The Kingdom*. *The Birmingham Daily Post* reported:

The Town Hall never presented a more crowded aspect at a Festival than it did this morning. There was not a corner in the great hall that was not occupied, right from the organ loft to the extreme end of the great gallery. It testified to the popularity of the great English composer, the man of the hour, who has instituted a new art form in oratorio.[309]

The audience included the new Festival President Lord Plymouth, the Duchess of Marlborough, Mrs Joseph Chamberlain, Austen Chamberlain, and many friends. Alice wrote:

Enormous audience. Some standing room (guinea each) allotted—a special train from London. E. looked most booful & conducted splendidly—no haste & much dignity.

The Chorus saw something else:

[307] 'Elgar—A Personal Note', in *The Music Student* (Aug. 1916), p. 362.
[308] 27 Sept. 1906 (HWRO 705:445:8844).
[309] 4 Oct. 1906.

Sir Edward Elgar's emotions were so stirred by his own wonderful work that, according to the observation of the choristers, tears were streaming down his face several times during the oratorio.[310]

The accumulated strain of the last months was beginning to find its release. Frank Schuster realized something of it when he wrote afterwards: 'You have been entrusted with one of the greatest messages ever sent to mankind. Do not falter, do not fail, but God willing, go on and deliver it to the end.'[311]

Many people, both friends and strangers, agreed. Old Joseph Bennett of *The Daily Telegraph* was amazed, as he had been amazed before:

I confess that had my opinion been asked of 'Gerontius' or 'The Apostles' before either was tested publicly, I should have declared that such music stood no chance of acceptance by our good English folk, by the nation which, in art matters, and especially in this art, loves the simple and direct. But Sir Edward Elgar, by a happy stroke of fortune, captured his countrymen just as they had become in sufficient measure sensible of the modern spirit . . . The two enormous audiences [for *The Apostles* and *The Kingdom*], the patient attention given to the music, and the response elicited by every *tour de force*—a response none the less obvious to experienced listeners for being soundless—these things are not to be explained away by flippant tongues . . .[312]

Like many other critics, Vernon Blackburn in *The Pall Mall Gazette* was ecstatic: 'I believe that in the history of art [*The Kingdom*] will rank definitely with the "Matthew Passion" of Bach.'[313]

But Joseph Bennett had already seen the question still underlying the entire project when he reviewed the vocal score, published before the première:

In 'The Kingdom', the kingdom is still to desire, and on almost the last page believers are praying 'Thy Kingdom come.'[314]

Baughan, in *The Daily News*, felt that the overall shape was wrong:

The filling of the disciples with the Holy Spirit [in Scene III] is, no doubt, the dramatic, or even theatrical, climax of the oratorio, but the composer's intention has been to deal with the teaching of the Apostles, and therefore the oratorio could not end at what seems a natural conclusion. But having that intention Sir Edward Elgar ought surely to have planned his music on different lines.

It remained for Ernest Newman, now returned to Birmingham as critic of *The Daily Post*, to slate the whole endeavour as plan and result:

The general level of inspiration is, in my opinion, below that of 'Gerontius' or 'The Apostles'. Some of the choral portions are so obvious in sentiment that one can hardly believe they came from the delicately spiritual brain that conceived 'Gerontius'; but much of the choral writing is highly effective, and the laying out of the parts for the voices is as expert as ever. A good deal of the music must frankly be called dull in itself, but Elgar is now so consummate a master of effect, particularly of orchestral and choral

[310] *The Birmingham Mail*, 4 Oct. 1906. [311] 6 Oct. 1906 (HWRO 705:445:6958).
[312] 3 Oct. 1906. [313] *The Daily Telegraph*, 1 Oct. 1906.
[314] 4 Oct. 1906.

effect, that he can often almost persuade us against our own judgment that the actual tissue of the music is better than it really is.[315]

It was a judgement on the whole notion of fitting his music to a programme of story and event and libretto for the purpose of maturing that music. The experience of writing the oratorios had developed further Edward's skill in making intricate variations. But it had shown him nothing about taking those variations toward any vital recapitulation—toward any sense of returning to find a future more richly stored than the past.

He had begun this immense project without knowledge of what its end would be—with an implied wager that his sacred subject would light the way to some overwhelming, definitive recapitulation. Indeed it was the deepest purpose of the *Apostles* project to find that recapitulation—for the reassurance of Edward's private faith. But his wish to believe held the very question which had begun the project. And thus the whole quest was circular. Circularity served musical argument, but not theological argument in pursuit of revelation. So the subject of his choice and long consideration from boyhood, the libretto of his own making, ended by defeating his musical impulse.

Near the end of the year, just before Christmas, he wrote ruefully to recall a jape with Ivor Atkins in which he himself had figured as 'Reynart the Foxe':

Times is awry . . . I know that, owing to the ripeness of the year that Xtians wax gross in their food & religion, of whilk twain the first concerns you not whilst the second may hold you fast to tinkle silver into collecting plates with warily selected tunes—such is the Xtian way.

. . . I am fairly well in my own self—but the eyes! I cannot spring upon my quarry with the old certainty—& one who has pulled down harts of grease despises field mice, so I wait for amended vision.[316]

[315] 4 Oct. 1906.   [316] 22 Dec. 1906.

9

The Looked-for Genius

WHERE now was Edward's path forward? A few months after the première of *The Kingdom* the entire trilogy plan was denounced by Ernest Newman:

He has seen fit to fasten upon his own back the burden of an unwieldy, impossible scheme for three oratorios on the subject of the founding of the Church; and until that scheme is done with, and Elgar seeks inspiration in a subject of another type, the most sanguine of us cannot expect much from him in the way of fresh or really vital music.

The wisest thing for him to do now is to abandon the idea of a third oratorio on the subject and turn his mind to other themes. These may bring him new inspiration and a new idiom; at present he is simply riding post-haste along the road that leads to nowhere.[1]

This was couched unmistakably as an exhortation to the composer. Edward always said he did not read criticism of his works. Yet Newman's article was collected by Alice and duly mounted in the press cutting book with the rest. And Edward's own thoughts had been as clearly occupied with the new direction his music must take when he said in a Birmingham University lecture given just before beginning formal composition of *The Kingdom*:

. . . Some writers are inclined to be positive that the Symphony is dead . . . but when the looked-for genius comes, it may be absolutely revived.

Gradually it would emerge that the work he must now face was his own long-awaited Symphony. Without the failure of the *Apostles* project, desperation might never have brought him there.

The process of the revelation was slow. One signpost was seen in the new series of lectures beginning at Birmingham less than a month after *The Kingdom* première. He planned this series to examine the orchestral scores of great composers, beginning with a general lecture on 'Orchestration':

The modern orchestra is capable of an unending variety of *shades of tone*, not only in succession, but in combination. We see that a whole world of new harmonies is at the disposal of a composer for orchestra: harmonies which may sound execrable and impossible on the piano but which may give the greatest pleasure when scored for instruments. To give a single instance—The dissonant notes of a chord may be merely suggested by the soft-toned instruments, or they may be thundered out by the

[1] *The Birmingham Daily Post*, 22 Mar. 1907.

heavy-toned ones while the principal notes of the chord—the backbone—is [sic] merely suggested. Every shade of even dynamic force is possible for every note of every chord, and this simultaneously. We have only touched on the fringes of the possibilities of modern harmony . . .

Orchestration can best be learnt or assimilated by listening to orchestras. The opportunities in this country are comparatively few . . .

And he turned to practicality and finance: 'We want larger halls and municipal orchestras: the loss must be cheerfully borne at first.' But the final conclusion was as personal as it was unhappy:

The writer of successful high-class orchestral music can look forward to the usual comforts of the work-house &, I fear, nothing more. I speak advisedly.

It might be heard as an excuse for not yet, even by the time of his fiftieth year, having produced his own essay in the symphonic form he had so praised in that room a year ago.

A week later came a lecture on Mozart's G minor Symphony—the symphony which had been the model for his own professional study almost thirty years earlier. On the appointed day, 8 November, the weather was frightful:

Pouring in torrents—E. wd. not let A. come—Went to Birmingham by his dear souse . . . Small room, more people than cd. well find place—no light for piano &c&c&c—dreadfully depressed.

He began the lecture:

Turning from a modern score to this small, attenuated orchestra, a pitiful array of instruments, we may wonder how it is possible that a great art-work could be evolved from such sorry materials . . . We have to marvel that with such a selection of instruments, variety and contrast can be found sufficient to hold the attention for thirty minutes.

And he gave a close analysis blending insight with affection.

Looking at the scoring, some mechanical limitations were obvious. Mozart used only the open notes on the horns of his day—'and great ingenuity is shown in "dodging" . . . to get a useful note'. Inevitably moments came in the music when no available horn note would fit the chord. If it was a climax, the climax must suffer. And he cited a case:

Can anyone say that Mozart would have omitted the upper note of the phrase if he had an instrument which could play it? No.

The whole question of 'improving' the orchestration of the old masters might turn on this very simple point.

The press (having been more or less silent about the previous 'Orchestration' lecture as their columns were full of local politics at that moment) fell on this. 'Sir Edward Elgar', it was solemnly reported, 'expressed himself in favour of a revision of the works of the old masters.'

Depression gathered again into Edward's eyes with an irritation of the lids.

In London for a Royal Society of Musicians concert with the London Symphony Orchestra, he consulted the doctors again. The prescription was rest. He seized the opportunity, and wrote to both Lodge and Fiedler saying he must give up the lectures: he offered to pay back the stipend and resign. Lodge replied that he was not to think of resigning but only not to worry and get well. The next lecture, already announced for 22 November, was 'postponed'.

November 20. A. a little less worried about E. He happier feeling clear of Birmingham Lectures—quite good spirits . . .

November 21. E. (& A.) to Mr. W. Lang—oculist. D. G. he said eyes were excellent, & insisted on change & warm climate . . .

It was exactly what Alice wanted to hear. Before the Mozart lecture she had written to Mrs Kilburn: 'I want to get him away to Italy when Carice's holidays begin. I think he needs a change so much, if we can only manage it.'[2] Thus once more, as three years ago after *The Apostles*, the depression that followed a Birmingham oratorio première would be answered with a sojourn in the south. The Italian plan was quickly made:

November 23 . . . E. & Frank [Schuster] to ship office in morning & reported cabins; returned after lunch & engaged passages in Orient [Line] Orontes 28 Dec—to Naples.

Alice went back to Plas Gwyn with a heavy cold, leaving Edward to divert himself in town with Schuster.

When he returned to Plas Gwyn on 27 November, Alice noted 'Eyes looked pretty well.' At home they soon worsened. He was advised to try the waters at Llandrindod Wells, despite the season. On 5 December they made the railway journey in rain and wind:

Very wretched journey—changing[—]wet platform &c. Llandrindod looked deserted & wretched. A. went out in pouring rain to find Pump Hotel. Shops shut. Cabs put away for summer . . .

There was nothing to do but attend the cure. It was a vision of the underworld, as Edward wrote to Sidney Colvin in London:

I was one of four or five others: we met like ghouls in the pumproom at 7.30 a.m. in the dark: mysterious & strange: hooded & cloaked we quaffed smoking brine & sulphur & walked thro' dim-lit woods, sometimes in snow . . . The squalid pumproom is built of *blue & yellow* bricks, placed alternately; the latter lend themselves to inscription and are scored over, by youthful Wales, with uneconomic remarks on the private matters of their friends &, presumably, enemies also.[3]

For the trip to Italy, Edward asked Alfred Littleton to send £250 from his royalty account (now receiving 25 per cent from sales of all his music published since the agreement). They sailed after Christmas and arrived in Italy on

[2] Wednesday [7 Nov. 1906] (HWRO: 705:445:BA8089).
[3] 26 Dec. 1906 (HWRO 705:445:3418).

6 January 1907. For the first week they were in Naples—walking, seeing the treasures of the Museum and of Pompeii in the company of Canon Gorton, who was serving as Chaplain of the English Church in Capri as he tried to recover from illness. As in Alassio three years earlier, they were pursued by cold winds.

Then Alice persuaded Edward to go to Capri—the inspiration of her old poem 'Love alone' which had begun the *Sea Pictures*. There the warmer air transformed him overnight. Alice wrote to Mrs Kilburn on 15 January:

This morning E. went over to the Barber's to have his hair cut, the Barber was always playing the Mandolin, so E. took up a Violin & they performed a Duet, then a Guitarist arrived & they performed a Trio brilliantly to a delighted audience! E. so gay & amused.[4]

Later that day he pencilled part of a trio 'for the Barbers' in his sketchbook.[5]

From Capri they went for a final ten days to Rome. It was Edward's first sight of the Eternal City, and he was overwhelmed. They were taken over the Vatican by the young maestro at the Sistine Chapel, Dom Lorenzo Perosi: he had effected sweeping reforms in the music of the church, was a composer in his own right, and a deep admirer of Edward's music. They also met the leading figure of secular music in Rome, Giovanni Sgambati—a pupil of Liszt, founder of the 'Liceo musicale' at St. Cecilia Academy, and a composer with a European reputation. Sgambati and his wife gave a reception for the Elgars which was widely reported in the Roman press. Edward basked in the attention: he thought of returning to Rome for a longer stay in the following winter, and they looked at several apartments before turning homeward on 23 February.

The cloud on the immediate horizon was the prospect of his third visit to America. Back at Plas Gwyn, Alice wrote on 27 February to Mrs Stuart-Wortley (whose Christian name was also Alice):

My dearest namesake

We arrived last evening & my first lines from home are to you. So many thanks for your dear letter & invitation, it is *horrid* not to be able to come, but what is still more horrid is the reason, that the dear E. has to start on Saturday, 2 March, for U. S. A. in the Carmania, & as I have been away so long, he says I must stay at home now . . .

I trust he will be home in April & I am thankful to say he is taking a nice good man [John Cousins of Norwich] to look after him which is the one consolation to me.

I cannot *tell* you how much better he seems, I only wish you cd. see him—& I wish you cd. have seen him in Rome, he was *utterly* blissful . . .

Yr loving
C. A. Elgar[6]

After just four days at home, Edward sailed from Liverpool. The voyage was beset with gales and snowstorms, which delayed the ship's arrival in New York.

---

[4] 15 Jan 1907 (HWRO 705:445:BA8089).
[5] Sketchbook VIII.   [6] Elgar Birthplace.

On 19 March he conducted Frank Damrosch's Oratorio Society in *The Apostles* at Carnegie Hall: it was his first professional appearance in New York. A week later he conducted the same forces in the American première of *The Kingdom*. Reviews were generally favourable and enthusiasm was great—so great that he was invited to lead a prayer meeting. He recalled it later for an interviewer, who wrote:

It was at the time when Strauss's *Salome* was to be given, and as Elgar was at the time conducting his oratorios there, it was taken for granted that his sympathies would be with the good people who made it a matter of public prayer that New York might be saved from the calamity of a performance of the dreadful work. Elgar is, however, an old friend of Strauss, and he did *not* 'lead the prayers'.[7]

He went to Chicago for a concert with the Theodore Thomas Orchestra—the *Variations*, *In the South*, and *Pomp and Circumstance* No. 1. To the Chicagoans he cut a strange figure on the rostrum: 'His beat is the long straight beat of the British bandmaster and it is frequently indefinite and unsteady. He stands with feet well apart and with body noticeably bent at the hips, and there is little of grace or flexibility about his leading.'[8]

He found himself interviewed again. The question was programme music, and though he had just conducted *In the South* his answer was uncompromising:

'I will never write "programme music", never limit my musical ideas by forcing them to express what they cannot express.'[9]

Yet there was the old admiration for Richard Strauss:

'Strauss is the greatest genius of the age,' he affirmed with decision, 'and his later works I like best of all, much distinguished opinion to the contrary notwithstanding. His "Don Juan" is the greatest masterpiece of the present, and his "Heldenleben" and "Zarathustra" I find almost as inspiring.'[10]

From Chicago he went to Pittsburgh for the honorary degree at the Carnegie Institute and two concerts with the orchestra there. The local press reported:

With the exception of the applause given Sir Edward Elgar, little attention was paid to the music. The whole company was busy seeing who was there. It was the most unrivalled exhibition of local wealth that Pittsburgh has ever seen and the best feature of the whole evening's performance was the crowd of pretty girls and debutantes gowned in luxurious elegance. It took the audience nearly ten minutes to return to its seats after the warning bell rang following the intermission, and Sir Edward Elgar had to stand waiting his turn until they were all settled in their places before he could begin.[11]

At the new Carnegie Institute he was duly made an honorary graduate in

---

[7] Percy Scholes in *The Music Student*, Aug. 1916, p. 348.
[8] *Chicago Daily Tribune*, 6 Apr. 1907.
[9] Interview with Miller Ular, *Chicago Sunday Examiner*, 7 Apr. 1907.
[10] *Chicago Inter-Ocean*, 7 Apr. 1907.
[11] Unidentified Pittsburgh newspaper (cutting at the Elgar Birthplace).

company with Guglielmo Marconi, W. T. Stead, Sir William Preece, and the astronomer Sir Robert Ball. But here was another annoyance. The millionaire Carnegie had seemed to offer such lavish expenses for Pittsburgh that Edward had accepted a low fee of $1500 for his New York Oratorio performances. When it came to Carnegie's paying up, however, the story was different. Edward wrote to Alfred Littleton:

I have much to say about the disgraceful way old Carnegie treated his 'guests'—& the old *fibber* put me in quite a false position with Damrosch & the secretary of the Oratorio [Society]. He told them he was paying me *600£* ($3,000) to come. *Now* he refuses to pay any of his 'guests' more than 16£ (lowest possible fare!) each way & they are all at much expense![12]

It meant there would be small proceeds from this American journey, and it was to be several years before Edward could be induced to go there again. He was glad enough to return to New York, where he saw much of Mrs Worthington—and to sail home in the *Campania*, reaching Liverpool on 27 April. Soon he had to ask Littleton for 'a *sort of* idea as to how my royalty account is going on. It will depend upon that—now old Carnegie seems hopeless—how we proceed about our Italian winter arrangements.'[13]

*   *   *

In June 1907 Edward faced his fiftieth birthday. For a reflective man, the fiftieth birthday should mark a point of self-evaluation: by then the whole career must have revealed much of its final shape and proportion. Often in the past depression had come at birthdays and holiday seasons—anniversaries of time vanished never to return. A month before the fiftieth birthday Edward's mind went back to the music of his own past.

From fragments of anthems sketched long ago for the choir of St. George's Church in Worcester, he culled an *Ave Maria* and an *Ave Maris Stella*. They could join the *Ave Verum* he had arranged in 1902 out of the funeral music for old William Allen. He sent them to Novellos with a note to Alfred Littleton: 'They are tender little plants so treat them kindly whatever is their fate.'[14] They duly appeared with the *Ave Verum* to complete his published 'Op. 2'.

The fiftieth birthday, 2 June, fell on Sunday. While Alice was at evening church with Carice and May Grafton, Edward wrote a small part-song. The words were by Arthur Macquarie; they were not distinguished, but they said precisely what he wanted now.

> Like the rosy northern glow
> Flushing on a moonless night,
> Where the world is level snow,
>   So thy light.

[12] 25 Apr. 1907 (Novello archives).   [13] Wednesday [15 May 1907] (Novello archives).
[14] 24 May 1907 (Novello archives).

In my time of outer gloom
Thou didst come, a tender lure;
Thou, when life was but a tomb,
    Beamedst pure.

. . . .

Oh glow on and brighter glow,
Let me ever gaze on thee,
Lest I lose warm hope and so
    Cease to be.

His setting was an *Allegretto* of four-part harmonies which might have suggested a hymn but for the varied phrase lengths. When Alice returned he showed her the manuscript completed with a dedication to C. A. E.—'wh. made A. feel very unworthy & deeply deeply touched.' It was entitled simply *Love*. They sent it to Novellos to join *O Happy Eyes* (the first part-song he ever wrote to Alice's words) as Op. 18 No. 2.

On 7 June he finished another reminiscent work—a fourth *Pomp and Circumstance* March in the old G major tonality, with a great plagal trio melody climbing up and down the triad of C. It was the most joyous of the series—appearing almost three years after its immediate predecessor, six years after the first two.

Not all his reminiscences were happy. When *The Kingdom* was performed in King's College Chapel, Cambridge, Edward took the chance to renew acquaintance with Arthur Benson (who had recently come to Cambridge from Eton). After the performance, Benson wrote in his diary:

He told me his eyes were overstrained & he cd. do no work—then he said simply that it was no sort of pleasure to him to hear *The Kingdom*, because it was so far behind what he had dreamed of—it only caused him shame & sorrow . . . He seemed all strung on wires, & confessed that he petitioned for a seat close to the door, that he might rush out if overcome . . .[15]

On returning home he found the newly engraved full score of *The Kingdom*. Novellos had made a splendid job of it in the printing works at the back of their distinguished new office and showroom premises in Wardour Street. It was a noble folio, and the expert engravers had laid it out so lavishly that the new full score appeared to be almost as imposing as *The Apostles*.

Then a second parcel arrived, from a more remote past. It was a painted iron chest from W. H. Elgar's sea-faring ancestors at Dover. The chest had been inherited by Frank Elgar with the shop in High Street, Worcester, and Frank had sent it to Edward as a birthday gift:

June 23 . . . Very busy at the old Chest. Colours came out wonderfully. E. writing out his Children's Music.

It was almost as though the family chest opened again the old play of

---

[15] 11 June 1907 (MS at Magdalene College, Cambridge).

childhood—to show the woodland glade intersected by the stream, with old age waiting on the opposite shore to be lured back to 'our fairyland'. Quickly in the fiftieth birthday summer the idea took shape: he would recast this oldest music of all in the full orchestral dress of modern experience.

It marked a crossroad in his creative life. Faced with a crisis of self-confidence, the creative spirit might return to scrutinize the basis of its earliest teaching and learning. But Edward had had no real composition teachers: he had only his own earliest music of childhood. Now he was reaching back to the identifiable beginning of creative life in these little tunes—to test whether they could yet make their appeal to the audiences who had applauded his later music through half the civilized world. If they could, then the instrumental, orchestral origins of his music might even now sustain the creative effort of a Symphony. He gave the mature orchestration of the children's play music a significant title: *The Wand of Youth*.

He began to finish and join and round off the fragments with the lightest hand, to preserve their innocence: 'Occasionally an obviously commonplace phrase has been polished but on the whole the little pieces remain as originally planned.'[16] He divided them into two Suites. The first contained a busy Overture, a graceful Serenade, the Minuet 'a la Handel' that once introduced the Two Old People, the 'Sun Dance' which had later awakened them, 'Fairy Pipers' passing in their boat to charm them asleep, 'Fairies and Giants' built largely on the 'tune from Broadheath'. But just before the end a 'Slumber Scene' preserved another memory: the ground bass in the music consisted of three open-string notes of the old double-bass so that the part could be played by one of the children. Edward's music set the three notes in ascending order toward the little piece's tonic of G major. The middle tune in the 'Slumber Scene' was another figure of descending steps:

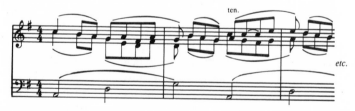

Suddenly in the midst of working on the children's music came a most extraordinary discovery:

June 27. E. much music. Playing great beautiful tune.

It was the occurrence of such a tune which had raised his hopes for a Symphony in 1901. What he played now was a variant of the descending 'Slumber Scene' figure in *The Apostles*' opening key of A flat major—accompanied by a counterpoint:

---

[16] MS notes for *The Wand of Youth* Suite No. 2, 15 July 1908.

Even the specific figure of descent in the upper voice was not new. W. H. Reed noticed later that the opening notes and the harmonic implications were almost precisely those which had closed the *'Enigma' Variations* coda in 1899:

[I] called Elgar's attention to this strange fact on one occasion, pointing out that it was not just a similar passage such as one often meets in comparing the works of any composer with one another, but an *exact* repetition of the theme subconsciously written [eight] years previously, the relative note-values, the intervals *and the accompanying harmony* being identical. The whole theme is transposed into A flat major for the Symphony. Elgar confessed that he was at a loss to account for it, being quite unaware that the repetition existed until it was pointed out to him.[17]

It had returned in company with an idea still older. For the bass counterpoint of the new invention reproduced exactly the bass figure in the scene 'And King Olaf Heard the Cry' at the moment when the hero remembered his mother:

This bass was the bass in the 'great beautiful tune' of 1907—tonality, notes, rhythm, and harmonies through bar after bar.

[17] *Elgar*, p. 157.

So the inspiration of June 1907 was really a bringing together of two ideas which had existed separately in his mind's music over many years. And thus it was the paradigm of mature self-expression: separate experiences from the past drawn together in new and renewing ensemble—the power of being touched and transfigured by the Wand of Youth.

The new ensemble of tune and counterpoint was a pure expression of the diatonic in music—the musical 'old-world state'. It sang its long phrases as if the chromatic threats which had risen steadily through the past half-century did not exist. Yet much of the significance might remain below the conscious level of the composer's thinking: so his 'great beautiful tune' carried implications that needed long and careful working out.

Early in June 1907 he went to Birmingham for the University's Degree Day. On the afternoon before the ceremony he addressed the inaugural meeting of a University Musical Society, of which he had accepted the presidency. The new Society had been presented with a 'semi-grand' piano by Richard Peyton. Edward tried it over at the meeting, and then welcomed the piano with gentle deprecation:

Some day the instrument might be looked on as a curiosity, for mechanical players are multiplying in such vast numbers that the human piano-player will probably disappear. Ever since the piano came into vogue there have been mechanical players, but of the human variety, and there is one point upon which we may congratulate ourselves, viz., that music from an instrument of which you [have] merely to turn the handle, so to speak, is almost invariably better than that produced by a pianist who simply strums on the keys.[18]

Behind the Peyton Professor of Music, the piano-tuner's son still watched.

That night he and Alice dined with the Peytons: among the guests was Neville Chamberlain. Next day, at the huge convocation in the Town Hall, an honorary MA was conferred upon Edward by the Vice-Chancellor (none other than Charles Beale, the chairman of the Birmingham Festival Orchestral Committee)—to a rout of student cheers, penny whistles, and paper darts. There was no doubt about the Peyton Professor's popularity with the undergraduates.

But there was serious work afoot in Birmingham. He was attempting, with Peyton, Beale, G. H. Johnstone, Bantock, and others, to form a scheme to bring regular orchestral concerts to Birmingham—and therefore to the University—under the conductorship of such men as Richter, Henry Wood, and Max Fiedler. It was a question of finding financial support. Edward despaired of it, but in the end they managed to plan a series beginning in the autumn.

He cycled over to spend three days with his sister Pollie and her family at

---

[18] *Birmingham Daily Mail*, 6 July 1907.

Stoke Prior, and they revived further memories for *The Wand of Youth*. Back at Plas Gwyn, work on the Suite was varied with chemistry in The Ark. This was the former dovecot, converted to its new purpose after the earlier chemistry in the basement had sent smells through the house. 'Here', as W. H. Reed observed, 'Elgar would retire and ease the burden of his destiny as a composer by pretending to be a chemist.'[19] One adventure took on the quality of fable:

He made a phosphoric concoction which, when dry, would 'go off' by spontaneous combustion. The amusement was to smear it on a piece of blotting paper and then wait breathlessly for the catastrophe. One day he made too much paste; and, when his music called him and he wanted to go back to the house, he clapped the whole of it into a gallipot, covered it up, and dumped it into the water-butt, thinking it would be safe there.

Just as he was getting on famously, writing in horn and trumpet parts, and mapping out wood-wind, a sudden and unexpected crash, as of all the percussion in all the orchestras on earth, shook the room, followed by the 'rushing mighty sound' he had already anticipated in *The Kingdom*. The water-butt had blown up: the hoops were rent: the staves flew in all directions; and the liberated water went down the drive in a solid wall.

Silence reigned for a few seconds. Then all the dogs in Herefordshire gave tongue; and all the doors and windows opened. After a moment's thought, Edward lit his pipe and strolled down to the gate, *andante tranquillo*, as if nothing had happened and the ruined water-butt and the demolished flower-beds were pre-historic features of the landscape. A neighbour, peeping out of his gate, called out:

'Did you hear that noise, sir: it sounded like an explosion?'

'Yes,' said Sir Edward, 'I heard it: what was it?'

The neighbour shook his head; and the incident was closed.[20]

On 16 July there was a visit from the Oxford musician Ernest Walker, who was writing programme notes for the Leeds Festival in the autumn. Having failed again to get an Elgar première, Leeds had asked Edward to conduct *The Kingdom*. But Edward's thoughts were anywhere but on his unfinished trilogy. Dr Walker recalled the visit for his own biographer, who wrote:

In the afternoon the composer with Lady Elgar and their daughter took him a motor drive along the banks of the Wye. For tea they were back in the drawing-room, where rows of Elgar's works bound in white [buckram], were prominent on the bookshelves, and presented an air of 'pomp and circumstance'. And on the staircase this was emphasized by enormous laurel wreaths that had been presented at German concerts, with golden letters on big coloured ribbons.

A fresh-air saunter round the garden between tea and dinner was welcome and Elgar explained, with nervous gestures, his experiments on Luther Burbank lines; the netted twigs and the strange fruit he had cultivated: a hybrid between an apple and a pear. He spoke of his experiments in chemistry which the day before had caused an explosion . . .

After dinner Walker was taken to the study, a room that held a valuable heraldic

[19] *Elgar as I Knew Him*, p. 38.   [20] *Elgar as I Knew Him*, p. 39.

library, works on chemistry, and no sign of music except one upright piano: Elgar said if he had been able to live a completely independent life he would have liked to be an M.P. Innocently Walker asked for which party: 'Conservative, of course' Elgar snapped out.

Again a silence and again Walker tried to get conversation on to *The Kingdom*, but Elgar went to the piano and played from his own 'Wand of Youth', stopping now and again to say he would like to incorporate this or that musical idea in a bigger work.[21]

Only at the end of the visit Edward said barely enough about *The Kingdom* to enable Walker to write his note.

On the last day of July he wrote to Hans Richter: 'My eyes are not allowed to work much so I devote myself to small & uninteresting things: I think my work in the world is over.'[22] Yet then there was a beginning of something else. It seemed to come again from the childhood dream of Severn side:

August 2. Fine morning. E. wrote *lovely* river piece. You cd. hear the wind in the rushes by the water . . .'

But then the 'river piece' was laid aside to finish orchestrating *The Wand of Youth* Suite No. 1.

Henry Wood had just conducted the première of *Pomp and Circumstance* No. 4 to a packed house of wild enthusiasm. He wanted the première of *The Wand of Youth* Suite as well, and Edward agreed. And when at the end of the year this earliest music came to its London hearing, all the critics noted the child's fatherhood of the man. *The Westminster Gazette* wrote of

. . . how early the characteristic Elgarian idiom seems to have manifested itself, some of the phrases and melodies of the suite being remarkably similar to those which have long since become familiar with the composer's later works—and also how little of it is in any marked degree suggestive of other composers or in any marked degree derivative. Elgar was apparently himself from the beginning.[23]

In the Gloucester Festival that year they had *The Apostles* and *The Kingdom* side by side. The two oratorios drew huge audiences, and Alice thought the performances 'splendid'. But after the Festival Edward wrote to the publisher of O'Shaughnessy's poems for formal permission to set 'The Music Makers'—whose vision was utterly secular:

> We are the music makers,
>   And we are the dreamers of dreams,
> Wandering by lone sea-breakers,
>   And sitting by desolate streams.

At the Cardiff Festival to conduct *The Kingdom* in late September, he heard Parry direct a new choral work to a poem of his own entitled *A Vision of Life*. Like 'The Music Makers', Parry's poem dealt with entirely human aspiration:

---

[21] Margaret Deneke, *Ernest Walker* (OUP, 1951), pp. 110–11.
[22] Richter family.   [23] 16 Dec. 1907.

> The heavens are full of visions,
> The air is full of voices,
> And we are faint with longing
> To hear the message clearly.

Edward wrote to Jaeger: 'I say! that "Vision" of Parry is *fine stuff* & the poem is literature.'[24] At the Leeds Festival in October, where Edward next conducted *The Kingdom*, there was another choral work of secular prophecy. It was a setting of Whitman's 'Toward the Unknown Region' by the thirty-five-year-old Ralph Vaughan Williams, a pupil of Parry and Stanford of whom great things were predicted. In that final decade of the hundred years' Pax Britannica, Edward found himself not the only musician seeking to pierce the veil.

He conducted yet another triumphant performance of *The Kingdom* at Hanley. But when he and Alice went to Birmingham on 16 October for the first of the new orchestral concerts there, Hans Richter met him with a 'fervent appeal to E. to finish the Sinfonie—dear noble old man.' Could the 'great beautiful tune' of the fiftieth birthday summer be made a basis?

But then he seemed to avoid it. Ten days later, on the morning of a Brodsky Quartet concert at Malvern, he began a String quartet. It was interrupted only when the time came to start for London and Italy, where they were to spend the winter.

He arranged for a series of Birmingham University lectures to be given by others—among them Walford Davies and Ernest Newman. In Birmingham to introduce the first of these lectures, Edward explained that his eyes were 'not of much use to him', but he 'trusted the ailment was a passing one'. Then the Elgars let Plas Gwyn for several months and started for Italy.

Passing through London on 2 November, they saw the Vedrenne-Barker production of *The Devil's Disciple* by Bernard Shaw. ' . . . the thing lacks conviction,' Edward wrote: 'Shaw is very *amateurish* in many ways.'[25] Shaw reacted to Edward's presence on the scene in different terms a month later when he answered an impresario's enquiry for an opera libretto:

I wonder whether Elgar would turn his hand to opera? I have always played a little with the idea of writing a libretto, but though I have had several offers nothing has come of it.

When one is past fifty and is several years in arrears with one's own natural work, the chances of beginning a new job are rather slender.[26]

The chances applied equally to Edward himself.

'The serious work waits for Rome,' he had written to Canon Gorton in

---

[24] Tuesday [8 Oct. 1907] (HWRO 705:445:8780).
[25] 4 Nov. 1907 to Troyte Griffith, who had designed sets for the production (*Letters*, p. 177).
[26] 4 Dec. 1907 to Col. Mapleson; printed in *The St. James's Gazette*, 13 Dec. 1907.

September.[27] What that work was *not* to be was hinted in his first Roman greeting to Schuster:

My dear Franky: Here is my Mecca & I love it *all*—Note the fact that I am pagan not Xtian at present.[28]

And on 3 December he wrote to Alfred Littleton 'definitely & finally to give up the idea' of a third oratorio on the Apostles.[29] The same day he began to write out the first movement of his Symphony in short score.

The Symphony opened directly on the great tune, stepping down and then upward *Molto maestoso* through its phrases of unequal length:

Edward described it for Ernest Newman:

. . . the opening theme is intended to be simple &, in intention, noble & elevating . . . the sort of ideal *call* (in the sense of persuasion, not coercion or command) & something above everyday & sordid things . . .[30]

It stood at the head of his Symphony like the motto themes in Brahms's Third Symphony and Tchaikovsky's Fourth.

At a double bar the key changed to D minor, and there rose up an *Allegro* which opposed the motto theme in every way:

Where the motto theme moved *maestoso*, this was *appassionato*. Where the motto descended in steps, this rose through a triad. Where the motto measured long unequal phrases, this rhythm insistently repeated. Where the motto was diatonic, this was aggressively chromatic. And what signified deliberate attack more clearly than everything else was that the *Allegro*'s D minor was a tritone from the motto's A flat—the opposite, most distant point on the circle of tonalities—the 'diabolus in musica'.

---

[27] 16 Sept. 1907 (*Letters*, p. 190, where it is misdated 1909: Elgar's '7' and '9' were often indistinguishable).

[28] Undated, *c.* early Nov. 1907 (HWRO 705:445:7057).

[29] Novello archives.

[30] 4 Nov. 1908.

So Edward's Symphony embodied again the central theme of *The Apostles*—an ideal set in the diatonic music of an older world, and then put to the extreme test of a chromatic future. The Apostles' 'motto' had been the figure of Christ—whose music was diatonic when it referred to the heavenly ideal, and chromatic only for worldly matters. Afterwards Mary Magdalene and Judas had followed Christ with chromatics. And the tonal territory of *The Apostles* also lay between the A flat major Prologue and the D minor uprising of Judas—precisely the tritone of the Symphony. Only the Symphony's D minor 'Judas' now rose up directly on the heels of its master. The challenge not finally met in *The Apostles* stood at the head of the Symphony with peremptory insistence.

By 1907 the need to test traditional diatonic ideals with chromatic questions was very clear. Through Edward's lifetime chromaticism had made steady inroads upon the diatonic music of Europe. Now men only a generation younger than himself contemplated the destruction of tonality altogether. Such a destruction would take with it half a millennium's inheritance of understanding between composer and audience. The nostalgia of Edward's music passionately sided with that understanding.

The problem of music reflected the problem of the world that produced it. After nearly a century's *Pax Britannica*, imperial aspiration in Germany was mounting the gravest challenge to Britain since the era that ended at Waterloo. And in these last years, Edward's music had reached a breadth of public appeal unknown to any composer in England since Handel. The music which made such an appeal was clearly saying something the people profoundly wanted to hear. What they heard in *The Apostles* and *The Kingdom*, what they were to hear in the Symphony, was the need to hold fast to the old ideal—together with the equal need to test that ideal by the sharpest of future tests.

In the world of politics and nations, fears had risen steadily since the turn of the century. During the autumn of 1907 Alfred Littleton had been asking and asking Edward to write a 'marching song for soldiers'. It was to be commissioned by the Worshipful Company of Musicians, of which Littleton was a member. The Company's wish was part of a tide of nationalism rising in England as the German challenge rose.

Edward protested that such a song would never be used; yet Littleton continued to send sets of verses which might be appropriate. It was least of all the sort of composition Edward thought he wanted to undertake at this moment, as he complained to Frank Schuster:

Yes: I am trying to write music—but the bitterness is that it pays not at all & I must write & arrange what my soul loathes to permit me to write what *you* like & I like. So I curse the power that gave me gifts & loathe them now & ever. I told you a year ago I could see no future: now I see it & am a changed man & a *dour* creature.[31]

If the future lay in marching songs, even that prospect might turn to account

---

[31] Undated, probably early Nov. 1907 (HWRO 705:445:7057).

for the Symphony. On another page of the short score Symphony sketch, he wrote out a 1/2 strutting march for what he described as a 'Scherzo (so called)':

The short-breathed descent through the triad and the little sequences of self-important rhetoric made another spiritual opposite to the 'great beautiful' tune of the 'motto'. Then came a figure of semiquavers—also containing triadic shapes and punctuated at the end with quaver triads to recall the D minor *Allegro* opening:

The same notes were repointed and greatly lengthened to form the theme of a big *Adagio*.

How all this might be related to the motto theme and its musical opposition was considered over the next days, as Edward explored the sights and sounds of Rome. He was still suffering from headaches, and his eyes troubled him. The ink of the short-score sketch yielded to pencil, the pencil yielded to questions.

He decided to write some part-songs after all. Aside from the matter of Littleton's *Marching Song*, several requests coming in just then seemed to touch on the past and future questioning that dominated the Symphony sketches. The editor of *The Musical Times* had asked for a setting of a simple hymn beginning 'How calmly the evening once more is descending'. Edward wrote music which he described as 'homely but *"felt"*' when he sent off the part-song on 2 December.[32]

Three days later Alice produced words for a *Christmas Greeting* which they might together send back to Sinclair in Hereford for his Cathedral choristers' holiday concert. Edward set this poem for high voices (plus optional tenor and bass parts) with two violins and piano—reviving the ensemble of *The Snow* and *Fly, Singing Bird* written to Alice's words a dozen years earlier. The old keys of G minor and major were used once more. In the midst, as the pipers 'wandered far', came a quotation from the Pastoral Symphony in Handel's *Messiah*. But the poem's final stanza touched Edward's secret fear of anniversaries:

---

[32] BL Egerton MS 3090 fo. 88.

> Our England sleeps in shroud of snow,
>> Bells, sadly sweet, knell life's swift flight,
> And tears, unbid, are wont to flow,
>> As 'Noel! Noel!' sounds across the night.

On 8 December the little work was finished and posted with an affectionate note to Sinclair.

Then at last he agreed to write the *Marching Song* for Littleton. The words, by a Captain de Courcy Stretton, hurried on with a confident impulse:

> Thousands, thousands of marching feet,
>> All through the land, all through the land,
> Gunners and Sappers, Horse and Foot,
>> A mighty band, a mighty band.
>
> . . . .
> What's in the wind now, what's toward?
>> Who cares a bit, who cares a bit?
> Marching orders, we're on the way
>> To settle it, to settle it.

Edward's tune was grand, with suspensions and sequences more adapted to professional musicians than to marching soldiery.

But W. G. McNaught (who worked at Novellos in several capacities) suggested a poem for a more subtle marching song.[33] *The Reveille*, by the American writer Bret Harte, was a dialogue between the drum of compulsion and the philosophic conscience:

> 'Let me of my heart take counsel:
>> War is not of life the sum;
> Who shall stay and reap the harvest
>> When the autumn days shall come?'
>> But the drum
>> Echoed, 'Come!
> Death shall reap the braver harvest,' said the solemn-sounding drum.

Edward set *The Reveille* for men. Its tonic was the D minor of the Symphony's *Allegro*. It opened with a soft bass figure more of rhythm than of tune. The only fragments of noble melody came when the drum spoke—a semitone away in D flat major. It dramatized as sharply as the tritones of the Symphony an intimate battle. In slow crescendo the argument developed to the drum's 'Better there in death united,' where 'death' sounded a *fortissimo* A major chord for the doubly divided tenors against A flats and D flats for the basses. Yet what must win the day at the end of this *Reveille* was the relentless ancient rhythm of the drum, culminating in a slow fanfare at 'My chosen, chosen people, come!' That final climax was followed only by a single line of total surrender:

> For the great heart of the nation, throbbing, answered:
>> 'Lord, we come!'

[33] McNaught's letters of 16 Dec. 1907 and 1 Jan. 1908 (HWRO 705:445:8250-1).

The Reveille was finished on 26 December 1907 and posted to Novellos the following day.

The year's end found Edward at work on four further part-songs for mixed voices to poems of his own choosing. These poems also contributed their themes to his thinking over the Symphony. The first was from Tennyson's 'The Lotos-Eaters':

> There is sweet music here that softer falls
> Than petals from blown roses on the grass,
> Or night-dews on still waters between walls
> Of shadowy granite, in a gleaming pass;
> Music that gentlier on the spirit lies,
> Than tir'd eyelids upon tir'd eyes;

(It was an acute personal reference for Edward amid these vistas in the south.)

> Music that brings sweet sleep down from the blissful skies.
> Here are cool mosses deep,
> And thro' the moss the ivies creep,
> And in the stream the long-leaved flowers weep,
> And from the craggy ledge the poppy hangs in sleep.

Edward's setting quietly continued the harmonic conflict of his recent music. The opening lines were sung first by the men in the old G major of the *Christmas Greeting*; then the women sang the same words in the A flat major of the Symphony motto. So it repeated the semitone juxtaposition of *The Reveille*, but with a difference that was almost total. Here the lyric flowed mellifluously within a dynamic range from *p* to *pppp*. Rhythm almost ceased, with phrases beginning in mid-bar, syncopations, suspensions, bars in 5/4 and 10/4. In this warm and shadowy land the two tonalities sounded only the pitch displacement of far echoes—to end in women's voices moving down the A flat triad as the men replied with triads of G on the final significant word

The other three part-songs of the group were all marked with a three-flat tonality, and all explored semitone contrasts. *Deep in my Soul* set Byron's verses of love's eloquence turning inward:

> Deep in my soul that tender secret dwells,
>   Lonely and lost to light for evermore,
> Save when to thine my heart responsive swells,
>   Then trembles into silence as before.
>
> There, in its centre, a sepulchral lamp
>   Burns the slow flame, eternal—but unseen;
> Which not the darkness of Despair can damp,
>   Though vain its ray as it had never been.

The opening music wandered toward E flat. 'Save when to thine my heart responsive swells' brought a *forte* climax in the tritone A major, before 'it trembles into silence as before' through the wandering keys of the opening. As the sepulchral lamp was lit, the music found a momentary centre in E flat minor. Then the opening returned to flare its unsteady flame against the catacombs of doubt.

O Wild West Wind set the final section of Shelley's *Ode*:

> O Wild West Wind!
> Make me thy lyre, even as the forest is:
> What if my leaves are falling like its own . . .
> Drive my dead thoughts over the universe
> Like withered leaves to quicken a new birth! . . .
> Be through my lips to unawakened earth
> The trumpet of a prophecy! O, wind,
> If Winter comes, can Spring be far behind?

Edward's setting was marked *Nobilmente* 'with the great animation but without hurry'. Its big tune rose out of the E flat triad through grand plagal gestures to reach an octave. It was altogether more diatonic than its predecessors, slipping into enharmonic keys briefly at 'a new birth' and the foresight of Spring.

The final song was entitled *Owls (An Epitaph)*, to a poem by Edward himself. It was his first mature setting of his own lines. The opening image answered Shelley's 'West Wind':

> What is that? Nothing:
> The leaves must fall, and falling, rustle;
>   That is all:
>   They are dead
>   As they fall,—
> Dead at the foot of the tree;
> All that can be is said.
> What is it? Nothing.

He wrote to Jaeger: 'It is only a fantasy & means nothing. It is in [a] wood at night evidently & the recurring "*Nothing*" is only an *owlish* sound.'[34] Yet the

---

[34] 26 Apr. 1908 (HWRO 705:445:8784). The subtitle 'In a wood' was written on the MS but later deleted.

three stanzas moved from the sound of leaves to the cry of a wounded creature, and then to an unmistakably human funeral.

The *Owls* setting was the most chromatic and fragmented of the four. A triadic rising question was answered by the single word 'Nothing' set to a falling semitone harmonized in minor thirds. A wisp of softly chanted slow march led to 'All that can be is said' in bare octaves. The second stanza rose enharmonically to E major, the third slipped chromatically back. The music was dated 31 December 1907, and its end was *pppp mesto*. The final

was unresolved.

In setting his own lines as 'An Epitaph' to those of Tennyson, Byron, and Shelley, Edward made a private distinction. The dedications of the first three part-songs were all to musical friends—Canon Gorton, Mrs Worthington, McNaught. *Owls* bore a dedication 'To my friend Pietro d'Alba'. White Peter was Carice's angora rabbit back at Hereford amid the end-of-year snows. So this was a Christmas Greeting of another kind.

On the penultimate evening of the old year 1907 the Elgars had a visit from Arthur Benson, wintering in Rome with a friend:

We found them in a big comfortable salon—a good deal of china about which E. had bought. E. in dress clothes—we were not—came eagerly out to fetch us in . . . He looked well, with his pale face, high-bridged nose, quick movements. But he said his eyes were weak & he cd. only work an hour a day. With all his pleasantness & some savoir faire, one feels instinctively that he is socially always a little *uneasy*—he has got none of [Walter] Parratt's courtesy or Parry's geniality . . .

Lady E. *very* kind but without charm, & wholly conventional, though pathetically anxious to be *au courant* with a situation. Elgar's daughter about 16, a quiet, obedient, silent, contented sort of girl, interested in Rome. Also a niece . . . They are here for some months—it is a quiet street, & the rooms have a fine view.

He took me into his sanctum, a little room opening off the salon, where there was a piano, & big sheets of music paper, being scored.

'An hour a day,' he said—'it isn't much, but it just keeps one going—& then it is a pleasure to write on this paper—really artistic music paper from Milan.'

. . . Elgar showed me over his writing-table a photograph of angels playing instruments from S. Gregorio. He pointed out the trombonists, & said 'I like that—that is the instrument I blow myself—& look at those upturned faces—that helps me while I work, more than you would think.'

. . . The kindness of the party won my heart. E. lighted us down the long stairs. The worst thing about him is the limp shake of his thin hand, wh. gives a feeling of great want of stamina.[35]

---

[35] 30 Dec. 1907 (MS Diary at Magdalene College, Cambridge).

In the first days of 1908 the household in the Via Gregoriana was sadly disturbed. May Grafton was called back to England by the sudden illness of her father, and within a few days he was dead. Will Grafton had been only a little older than Edward, and a warm friend from before the time of his marriage to Pollie in 1879. It was the first mature family death in Edward's own generation.

After May's departure, he began a heavy cold which turned to influenza. Before he recovered Alice went down with it. On 30 January he wrote to Alfred Littleton: 'I think this has settled any idea of music which has waned frightfully during the last month.'[36] Their troubles were much in the mind of Arthur Benson when he returned to London and visited Frank Schuster:

I like Schuster, he is so harmless, kind, innocent & refined. We had tea & a long talk mostly about Elgar. I learn that Lady E. is *not* well off, some £300 a year only, & that E. has to make money. Just now he is in a mood of hopeless depression, hating Rome, unable to work or to do anything whatever.[37]

Yet they stayed on. Through the Mediterranean spring Edward explored Rome in long walks (called 'murks') with Carice. Alice, when she began to recover, saw to it that they answered their invitations in Roman society both native and expatriate. They attended parties of Bonapartes and Sermonetas and Polignacs. They studied Italian and French at the Berlitz school, and Carice took singing lessons until the sounds disturbed her father's concentration. They went to Puccini's *Tosca* and *Madam Butterfly* at the opera. Sgambati proposed that Edward conduct an all-Elgar concert in a series which already included personal appearances by Debussy, Sibelius, and Strauss. But the hall was a cavernous place full of echo, and Edward declined.

They entertained Lord Northampton and also the Brodskys (whose Quartet were playing a series of chamber-music concerts in Rome). Mrs Worthington arrived, and at her parties they met the poet Alice Meynell, and Abbot Gasquet, one of the most famous English Catholics. But Rome offered no inspiration to Edward's faith, as he would describe it later for Hubert Leicester and his family:

If you have any religious feeling whatever, don't go to Rome—everything money— clergy gorgeous & grasping . . . 'Special music' (bombardon & side drum & Gounod's Ave Maria). Present Pope a good holy simple man but knows nothing . . . should be replaced by permanent Commission with secretary who could be dismissed.[38]

All this time music was nowhere. He wrote excuses to Jaeger:

I cannot afford to get a *quiet* studio where I might have worked & my whole winter has been wasted for the want of a few more pounds: it seems odd that any rapscallion of a painter can find a place for his 'genius' to work in when a poor devil like me who after all *has* done something shd. find himself in a hell of noise & no possible escape! I resent it

---

[36] Novello archives.   [37] 21 Feb. 1908 (MS Diary).
[38] Conversation at the Leicesters' home in Worcester, 17 June 1908, as noted by Philip Leicester.

bitterly but can do nothing. It is just the same now at Hereford, noise has developed in the neighbourhood—I dodged it doing the Kingdom at great expense by going to Wales, but I can't do it again: my lovely works do not pay the rent of a studio![39]

Yet those works had paid for the family's entire six months in Italy. On 8 May the Elgars sailed for home, and eight days later reached Tilbury Docks.

The arrival home only emphasized his empty-handedness in respect of the Symphony. On 2 June, his fifty-first birthday, Edward sketched a song 'In Memory of a Seer'. He began cycling, but came in depressed on account of the motor cars that were multiplying even through the quiet lanes of Hereford-shire. He had headaches. But then, in the summer landscapes of home, the Symphony slowly came.

It was not achieved all at once. Gradually the news began to reach close friends. On 8 June Frank Schuster wrote to Alice: 'So glad to know you settled down once more & Edward at work! This is splendid & *is* it true about the Symphony?'[40] A week later Alice wrote to Jaeger: 'E. sends his love . . . & he wants to say to you "the Sym. is A I"—it is gorgeous—steeped in beauty.'[41] Through the latter part of June he was playing some of its music for friends.

June 29 . . . E wrote all day, possessed with his Symphony . . .

June 30. Lovely day. E . . . for lovely ride late. He thought he had never felt or seen such lovely atmosphere in country & sights. Warm & still, Lovely roses. Happy over his work—quite lovely, wonderful light sky, light till 11 & later.

*   *   *

The Symphony's opening A flat 'motto' melody offered its world of diatonic serenity. Then the opposing D minor *Allegro* rose up to present its primary subject in two phases:

It was as if all the pent-up energy of trying to write the Symphony over so many years had finally burst its bonds: but the imprisoned thing, instead of finding joy in its release, was deeply marked with the struggle of bitter confusion.

[39] 26 Apr. 1908 (HWRO 705:445:8784).   [40] HWRO 705:445:6959.
[41] 15 June 1908 (HWRO 705:445:8856).

The *Allegro* energy began to cool, and the primary subject was surmounted with an echo of the motto's stepwise descent

The exposition continued with a series of developing variants:

 (which echoed Mary Magdalene's 'Anguished

Prayer' in *The Apostles* ),

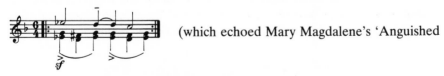 (from Ib)

(from Ia)

and a soft synthesis

which Edward described as 'sad & delicate'.[42]

Thus the Symphony first movement gradually approached its second subject, which proved to be another synthesis, in augmentation—a quiet refuge from the *Allegro* world:

The whole progress from the Symphony's beginning toward this second subject encapsulated the process by which Edward's Apostles were led into their ambiguous Kingdom.

[42] 23 Nov. 1908 to Ernest Newman.

Then the Symphony exposition did what his Apostles had hardly done: it moved back from its quiet centre to face the strife. At the climax of a long crescendo came the 'sad & delicate' synthesis (which had opened the way for the second subject) now blared *fff tutta forza* in brazen augmentation. At this revelation of opposing force within the same music, the exposition fell back as if stunned. And into the breach stepped the motto, quietly asserting its separation from all this conflict. The exposition was closed.

As Edward's exposition had once again included much development his formal development section again proceeded by further variation. The exposition's little surmounting figure over the primary theme was extended:

Another variant made an arabesque of descent:

A third opened wide chromatic intervals to sound (in the composer's words) a 'restless, enquiring & exploring'[43] figure:

This 'exploring' led to another big crisis. But its pressure no more resolved the conflict than the exposition's pressure had done.

The music disintegrated to fragments which Edward asked to 'be played [from four bars afer **29** to **31**] in a veiled, mysterious way—a sort of echo.'[44] The disintegration suggested that further search through these perspectives would be fruitless. And the motto stole upon the ear—but now in the tritone D tonality which had opened all the *Allegro* conflict. It regarded the *Allegro* conflict as serenely as Edward's Christ had regarded his Apostles contending among themselves and with the world in which they must bear his ideal.

Recapitulation turned the extended surmounting figure which had opened the development to counter-melody for the returning primary subject:

[43] 4 Nov. 1908 to Ernest Newman.
[44] 14 Dec. 1908 to Hans Richter.

Edward wrote of this: 'As [the treble] gives a feeling of A minor & [the bass] of G maj:—I have a *nice sub-acid feeling* when they come together at 32.'[45] The recapitulation soon found the four flats of motto tonality and in that tonality it followed the pattern of exposition rhetoric for a long distance.

On 15 July Edward wrote to Walford Davies, 'I . . . am covering sheets of paper to no good end.'[46] That day found him scrutinizing again the oldest of all his music as he wrote an account of *The Wand of Youth* Suite No.2 for the Worcester Festival programme: he had offered the première to Ivor Atkins in lieu of the large new work for which Atkins had again asked him. Two days later he was in Worcester itself for a Festival Committee meeting.

He lunched with Hubert Leicester and his wife at the house they had built in The Tythyng soon after their marriage. Hubert was now in his third term as Mayor of Worcester. With them at luncheon was his eldest son Philip. Philip was a serious young man of nineteen, and he had been reading Boswell's Life of Dr Johnson. He was so impressed with the presence of a great man now in his own home that he made private notes of all Edward said and did that day. After Edward's trenchant observations on modern Catholic Rome over the luncheon table, there were intimate details to observe and set down:

Fond of milk pudding—in old dish—will descend to tapioca—2 servings only—cigarette—no cigar till evening . . .

Up to cemetery . . . greatly affected at W[ill] G[rafton]'s grave—home past Loretto Villa[, where he had lived with Will and Pollie at the beginning of their marriage in 1879].

Feels present position keenly—been looking forward to retiring & coming back to old friends & old life—'now I come home & find poor W. gone . . . Hubert, how nice it would be to start fresh.'

That afternoon he went down past the old music shop to the Cathedral, and returned late to Hereford.

[45] 23 Nov. 1908 to Ernest Newman.   [46] *Letters*, p. 183.

Once more the wand of youth had been waved over creative problems. Two days later the Symphony was progressing again—interspersed with cycling through the hot weather, as he wrote to Troyte Griffith:

Do come over: I am writing heavenly music (!) & it will do you good to hear it . . .
   I said today 'Most prayers are inverted imprecations'—are they?
   It was very hot & riding we felt like ghosts of hippogriffs grazing charred grass upon the meads of hell.[47]

First-movement recapitulation had retraced the exposition until

blared again through all the brass. Then at last the central  emerged as if to answer the *Owls'* unresolved 'Nothing'. It repeated in a pattern of sequences rising elaborately up to descend through C♯, B, A, G♯—the opening figure of the motto.

Suddenly the G♯ became A♭. Insistence faded, and the motto just emerged. Edward wrote of this:

I have employed the *last desks* of the strings to get a soft diffused sound: the listener need not be bothered to know *where* it comes from—the effect is of course widely different from that obtained from the *first desk soli*: in the latter case you perceive what is there—in the former you don't perceive that something is not there—which is what I want.[48]

The motto went toward repetition as if in triumph. But it had been joined by the surmounting figure—which at last was left alone to begin a coda. This coda sounded 'a sort of echo'.[49] Scattered fragments alternated with the descending arabesque—which shaped a wistful echo of the motto. The movement ended with the 'sad & delicate' variant sounding still a semitone from the motto's A flat tonic:

It could stand for the mirages of promise and hope which had shaped Edward's Symphony so far.

Through the first movement, the motto ideal had been attacked by chromatics and insistent rhythm. Now these two threats were to be scrutinized one at a

---

[47] 19 July 1908 (*Letters*, pp. 184–5).   [48] 14 Dec. 1908 to Hans Richter.   [49] Ibid.

time. First came the rhythmic question: it occupied the second movement, a 1/2 *Allegro molto* (in place of a Scherzo). A short opening figure turned the first movement's  round a blind alley:

Over it raced the headlong semiquavers set down in Rome:

The goal was the self-important little march also in the Roman sketch:

Prodigious rhetoric of syncopation, orchestration, and dynamic contrast were brought to bear on this stiff figure without any developing result. The rushing semiquavers returned with redoubled fury and explosive punctuation expiring in sterility.

The tonality slipped to B flat major for an alternative to insistence and strut:

This pastoral ensemble suggested the 'river piece' started in the previous summer, in which Alice heard the wind in the rushes. Afterwards W. H. Reed remembered this music rehearsed in too matter-of-fact a way to satisfy Edward, who rapped his stick on the conductor's desk:

'Don't play it like that: play it like'—then he hesistated, and added under his breath, before he could stop himself—'like something we hear down by the river.'[50]

Again his music, set within this *Allegro molto*, irresistibly suggested some retreat from the rough world of action—some inward turning to a woodland glade of private happiness.

The pastoral tune was softly interrupted by triplets cascading through chromatics of thirds—which quickly rearranged themselves in a reminiscence of the 'Welsh' falling thirds from the *Introduction and Allegro*:

Beneath them the *Allegro molto* semiquavers rushed back through the lower strings. The semiquavers were then overlaid with the short-breathed march: played simultaneously, these quick figures were doubly short-breathed, like the Demons in *Gerontius*.

In a lightning take-over, the 'river' tune overwhelmed the whole ensemble *fortissimo* and brought it back through steadier rhythm to B flat major. A moment later the 'river' was its quiet self again, flowing through woodwind, harp, and violins. The 'Welsh' falling thirds re-sounded, and then all the strings joined to extend the 'river' melody in a series of diatonic sequences rising up over a slow striding bass that recalled the motto:

---

[50] *Elgar as I Knew Him*, pp. 140–1.

The memory of the motto dealt a mortal blow to the forces of rhythm and hot
energy. The rushing semiquavers muttered *pianissimo* like the Demons being
left behind. The muttering slowed to quaver triplets, to quavers, to crotchets:
and what had once been furious semiquavers moved through a slow procession
of quiet approach, with the motto steps sounding high above. The monomania
of speed sank out of hearing as the 'river' had risen gradually through it. The
lonely child who had listened long ago by Severn side was ready to write what
he could now clearly hear.

With the crossing of that river on a single bridging sound, the rushing
semiquavers of the *Allegro molto* became 'note for note the theme of the
Adagio'.[51] The notes were extended and repointed as boldly as the word
settings in the recent Roman part-songs:

Where the *Allegro molto* had sent its headlong rage up to A, the *Adagio*
paused on the penultimate G♯—the A♭ of the motto. This note opened
another semitone shift of harmony—to B flat minor—with a keening
wistfulness of first violins

The shadow passed, and the music swung slowly back to D major. But the real
subject of this movement was revealed. Now that rhythmic insistence had been
left behind, there remained the matter of how to set within a style of nostalgia
and reminiscence the chromatics which threatened the very basis of traditional
diatonic expression. Through lingering chromatic descents, the atmosphere
went suddenly cold: woodwind, horns, and trombones called softly to one
another in dim three-note chromatic figures which took the opening *Adagio*
notes into the world of *Owls*. The horns found the minor second of the *Owls*
'Nothing'.

Out of these mysteries came another big diatonic melody—a second subject
for the *Adagio*. It was accompanied by a further descent of steps over a slow

marching bass to recall the motto. As Edward wrote, 'the germ of this is of course in the introduction':[52]

Again the melody found effortless extension. The movement as a whole was gaining a profound eloquence which would be quickly recognized by Hans Richter when he rehearsed the Symphony for its first London performance: 'Ah! this is a *real* Adagio—such an Adagio as Beethove' would 'ave writ.'[53]

Again the diatonic song sank through chromatics. But therein this *Adagio* was finding the very definition of nostalgia: a vision of the past brought to focus by the implied contrast of what came after. Without the sense of later contrast, the retrospect would not exist.

The discovery was celebrated at the climax with a secret revival. One despondent day in 1904, midway between *The Apostles* and *The Kingdom*, Edward had set down in his sketchbook a melody which he was to inscribe with Hamlet's final words, 'The rest is silence':

Molto espressivo e sostenuto

52 Ibid.
53 Quoted in W. H. Reed, *Elgar*, p. 97.

Then it stood as a hymn to what might have been. Now the descending steps could be heard as emergent from the Symphony's motto.

So this older idea found its place at the penultimate point in his Symphony's evolution, where it could sing of disputes fallen silent, of goals envisoned. It held up the mirror to Edward himself in a way which was keenly to strike a younger composer. Speaking out of personal acquaintance, Arthur Bliss said half a century later:

Elgar was a man, I consider, of inspired musical personality. He wrote because he was compelled to do so . . . I think he was a very sensitive, highly imaginative, often harassed human being, and whenever I hear the Slow movement from the First Symphony I see the man: especially do I see a clear-cut image of him in the final bars.[54]

The Symphony up to this point had almost retraced the Apostles' story as far as Edward had taken it. Here again had been a beginning 'motto' ideal redolent of a diatonic past which stood untouched before a series of chromatic attacks through the first movement. Here in the second movement was the insistent march of purpose opposing the ideal—as Judas had marched the soldiers through Gethsemane. There followed the Symphony's *Adagio*—as Peter's great slow statements had been the most memorable achievements in *The Kingdom*. But there the *Apostles* project had stuck, for Edward found no way to blend his own retrospective ideal with the pragmatic, disputatious Apostles of later biblical history. Trying to force that issue through *The Kingdom* in 1906 had brought him to the verge of nervous breakdown.

The Symphony would succeed where *The Apostles* had failed for one reason: the Symphony traced not some other story, but precisely his own. He wrote to Ernest Newman:

As to the 'intention': I have no tangible poetic or other basis: I feel that unless a man sets out to depict or illustrate some definite thing, all music—absolute music I think it is called—must be (even if he does not know it himself) a reflex, or picture, or elucidation of his own life, or, at the least, the music is necessarily coloured by the life. The listener may like to know this much & identify his own life's experiences with the music as he hears it unfold: I have had a wide experience of life but . . . as to the phases of pride, despair, anger, peace & the thousand & one things that occur between the first page & the last, I prefer the listener to draw what he can from the sounds he hears.[55]

The Symphony Finale opened *Lento* with *tremolando* Ds sounding in the strings. The 'exploring' figure rose up from the first *Allegro* to begin the Symphony's fourth consecutive movement in either duple or quadruple metre. Now the 'exploring' led to a new variant of the motto—closest of all in notes, most distant in its sinister spirit:

[54] BBC broadcast, 'The Fifteenth Variation', 1957.
[55] 4 Nov. 1908.

A solo clarinet recalled the rising triad from the first-movement *Allegro*, and then coupled it to an echo of the motto descending steps: so the first-movement opposites were now joined in a single figure

Edward called it to the special attention of Ernest Newman, who was preparing the programme note for the first performance: 'Could you note the Clar[ine]t entry "*romantico*" two bars after $\boxed{108}$ . . .'[56] The motto itself sounded for the first time since the opening movement—far away in the last-desk strings. A fuller ensemble brought the '*romantico*' from A flat to the tritone D minor.

Instantly there rose up another 2/2 sonata *Allegro* in D minor. The old first-movement drama seemed about to be re-enacted. But there was a telling change in the new melodic shape:

In bars 1 and 3 the C♯–G sounded a tritone: yet this was now coupled in melody with the diatonic steps of the motto (bars 5–6, 7–8). So this primary Finale theme forged the link between the future and the past.

The '*romantico*' rose *con passione* to a variant which made the Finale's second subject:

Despite the triplets and key-changing seventh harmonies, the melody was rhythmically secure and unambiguously diatonic. Edward wrote of it: 'No

[56] 23 Nov. 1908.

connection was intended between the march-like theme in the finale & the opening [motto] theme.'[57] But he had confided to Jaeger over the connection of motives in *Gerontius*: 'I really do it without thought—intuitively, I mean.'[58] And so there was indeed a shining innocence about this second subject of the Symphony Finale.

Yet intuition was not a complete answer to the Symphony's conflict. The sinister D minor motto march of the Finale introduction strutted in double-quick time through an expanding series of chromatic sequences. This most intimate enemy rising up in mid-Finale might have suggested an Antichrist for the third oratorio. And the Symphony was about to return an answer neither the crucified Christ nor the nostalgic Peter had returned in Edward's *Apostles*.

Sonata development opened headlong with a reprise of the primary Finale subject. The sinister march instantly sought its bargain: it mounted a staggering technical display—close canon, chromatic inversion, contrapuntal sophistication. The primary subject was drawn in, and even the innocent second subject

felt the pull as horns and trumpets called

through the gathering ensemble.

Thereupon the tempter made his final demonstration: he lifted the second subject to place it as his capstone. But the moment this image of innocence was raised to the head of the corner, the structure began to crack. The second subject's triplets fatally interrupted the marching insistence which had driven up such an astonishing height. Still the march tried to force the second subject to a sequence. And with that, the fatal crack split the whole structure to its foundation. A tritone C♭ sounded *sforzando* through all the lowest instruments, and instantly the rest fell silent. Edward wrote later to a young critic of the Symphony: 'You do not see that the fierce quasi-military themes are dismissed with scant courtesy; critics invariably seem to see that a theme grows, but it appears to be a difficulty to grasp the fact that the coarser themes are well quashed!'[59]

The quashing at last was of mood rather than theme. With the 'quasi-military' threats removed, what had been the menacing strut of the D minor motto march became a wondrous melody in the semitone distance of E flat minor—the dominant minor relation of the motto's A flat:

[57] 4 Nov. 1908 to Ernest Newman.
[58] Tuesday [28 Aug. 1900] (HWRO 705:445:8401).
[59] To Neville Cardus (quoted in *The Manchester Guardian Weekly*, 2 Mar. 1934).

On it came, gradually gathering to itself the orchestral ensemble through page after wonderful page, filling the entire second half of the Finale development with a consoling strength in utter contrast to the demon energy which preceded it.

Recapitulation reviewed the struggle. Then a coda brought the '*romantico*' rising up through the A flat triad. The descending motto steps replied in simultaneous augmentation and diminution. String arpeggios descended, the upper wind added triplets, lower instruments syncopated the quadruple metre: so the apostles of every character were assembled for the judgement, each voice chattering its separate individuality.

Now at last the motto re-entered its own tonic A flat major. The re-entry began in a few winds and last-desk strings, with the varied voices of the rest swirling round it. Gradually every instrument joined to enrich what emerged for Edward as 'the conquering (subduing) idea'.[60] He wrote of the Symphony to Walford Davies: 'There is no programme beyond a wide experience of human life with a great charity (love) and a *massive* hope in the future.'[61]

The Symphony's music had not minimized the cost of achieving such a hope. In this also it reflected its composer's experience. For his entry after a ten-year struggle into this ultimate kingdom of musical abstraction was achieved at the cost of sacrificing all future hope in programmes of faith outside himself. He wrote of the Symphony to Mrs Stuart-Wortley: 'I *am* really alone in this music!'[62]

The Symphony's creative hope in the future was none the less something to which the majority of Edward's countrymen could respond in 1908—even though Germany was laying down ships to outreach the *Dreadnought*. But there again the music's loneliness as well as its hope anticipated the future now barely half a dozen years distant.

*   *   *

All through the summer of 1908 he had worked tirelessly on the Symphony despite interruptions. At the end of July he had to go to Birmingham for a long interview about University degrees (he had given no professorial lectures for nearly two years). Yet he finished the second movement *Allegro molto* in time to take it with the first movement to Novellos on his way through London on 10 August. He and Alice were travelling to an Elgar concert in Ostend, where he received a hero's welcome from the musicians of France and Belgium headed by Vincent d'Indy. Returning through London, he went to see Jaeger, now practically confined to his house in Muswell Hill. In the previous autumn, Edward had insisted on ceding all his own royalties in Jaeger's Elgar-Analyses to his friend. Now he asked Jaeger to join their house party for the Worcester

[60] 4 Nov. 1908 to Ernest Newman.
[61] 13 Nov. 1908 (*Letters*, p. 187).
[62] Monday [31 Jan. 1910] (Elgar Birthplace).

Festival in September, but the journey was beyond him. On 18 August, back at Plas Gwyn, 'E. feeling his way to his Symphony again.' By the 23rd the *Adagio* had been finished and posted to the publishers.

Perhaps it was the assurance of seeing his way to the end of the Symphony—of finding his own solution at last to the central problem of his creative life—that gave another resolution: on 29 August he wrote his formal resignation as Peyton Professor of Music in the University of Birmingham. In the original agreement he had stipulated that he might resign after three years. Writing to old Mr Peyton, Edward tried to soften the blow:

Owing to a longstanding promise the dedication of this, my first symphony, is to our friend Hans Richter,—otherwise I should have asked your acceptance of the dedication: however I hope I may be able to write another large work & I shall then ask you to allow your name to be upon it.[63]

Peyton was sad for the University, as he wrote to the Secretary: 'The actual result & virtual waste of time has, I need not say, been a great disappointment to me.'[64] Edward must privately have agreed. In his sense of release, he took the Symphony Finale score to work on through a visit to Frank Schuster at The Hut.[65] Schuster wrote to Alice at Plas Gwyn:

I have not seen him in such health of body and mind for years! I think it is in great part due to that superb Symphony, the composition of which *should* make any man radiant beyond measure—only it didn't follow that it would have this effect on *Edward*, and it is as glorious to behold him as it is to listen to it![66]

For the Worcester Festival, Alice had taken the King's School House, where their house party that year included many friends. They had some difficulty in converting a school to a domestic residence equipped with servants for the week. It was described later for W. H. Reed, who wrote:

Upon the arrival of the guests, and when they were being shown to their various rooms, it was discovered that none of them was furnished with a bell or other means of communication with the domestic staff.

Sir Edward at once saw his chance and took it. He set off to the local toyshop and bought up everything that would produce an individual sound. Then he drew up a scheme for the various bedrooms, as follows:

No.1. Rattle
No.2. Tin trumpet
No.3. Squeaker
No.4. Toy drum
No.5. Penny whistle
    etc.

[63] 17 Nov. 1908 (HWRO 705:445:3243).
[64] 28 Aug. 1909 to G. H. Morley (copy in Peyton's hand: HWRO 705:445:3413).
[65] In Schuster's copy of the Symphony miniature score, p. 122 (the second page of the Finale) is inscribed: 'This was written at the Hut (Orchard room) 1st September 1908. Ed: Elgar.' (In possession of Sir Adrian Boult.)
[66] Monday [31 Aug. 1908] (HWRO 705:445:6962).

and hung up the list in the kitchen. He told me it was worthwhile waking up early in the morning to hear Mrs Worthington appeal plaintively on the penny whistle from her door, or Frankie Schuster blow a fanfare on the tin trumpet from his, or three or four of them together set up a din which only his trained ear could disentangle.[67]

A late arrival, before *Gerontius* on the Tuesday evening, was Lady Maud Warrender. After the performance Edward met another brilliant woman who was to bring him many friendships through his later life. This was Mary Anderson, who had enjoyed a brilliant career on the American stage in the last years of the old century, before retiring to marry a wealthy American, Antonio de Navarro. De Navarro wanted only to lead the life of an English country gentleman, and they had settled in the picture-book Worcestershire village of Broadway. The de Navarros were Catholic, and after the Worcester *Gerontius* they commiserated with Edward over the mutilation of the text on which the Cathedral clergy had again insisted.

Next morning Edward conducted *The Kingdom*: at The Lord's Prayer the entire audience silently rose. But for this, as for *Gerontius*, Ernest Newman reported 'a regrettable superfluity of empty seats'. There were no empty seats at the Secular Concert that night for the première of *The Wand of Youth* Suite No. 2. Alice recorded a 'tremendous ovation to E.'

A week after the Festival Mary Anderson de Navarro came to Plas Gwyn with her twelve-year-old-son 'Toty'. She recalled:

There was no far off conventionality in that party. He was a hearty host, and greatly pleased with an excellent Friday dish he had evolved . . .

While we were having coffee, I asked him what was his favourite smoke.

'Cigars, good cigars—but I am too poor to smoke them, so I have taken to cheap cigarettes.' Handing me the box: 'Have one, my darling.'

I was astonished before the joke flashed on me—the cigarettes were called 'My Darling'. He loved jokes and twinkled over them.[68]

But the point about the cigars was not forgotten. Soon a package arrived from Antonio de Navarro with a note:

Dear Sir Edward,

I hear that my wife sent some spanish pimentos to your good wife. As we always 'do things together', may I approach you with a small box of cigars. They should be bay, and *not* tobacco leaves, I know, & send also to you my apologies in consequence. But will you take the will with the weed? & believe me to be

Yours very sincerely
A. F. de Navarro.[69]

After the last note was set to the Symphony score on 25 September, reaction followed swiftly:

[67] *Elgar as I knew Him*, p. 33.

[68] *A Few More Memories* (Hutchinson, 1936), pp. 204–5. The Friday dish, as J. M. ('Toty') de Navarro recalled, was curried vegetables.

[69] 9 Oct. 1908 (Mrs J. M. de Navarro).

September 26. E. very badsley. In bed all day—Dr. Collens. Down in evening.

At the end of the month he went to London, taking the Finale score to Novellos, but returned 'rather out of heart with world in general'.

Within the week he had finished a new song for Lady Maud Warrender. The words of *Pleading* (by Arthur Salmon, who had sent a book of verses in hopes that Edward might find music for some) showed the opposite face to the Symphony's nostalgic triumph:

> Will you come homeward from the hills of Dreamland,
>   Home in the dusk, and speak to me again?
> Tell me the stories that I am forgetting,
>   Quicken my hopes, and recompense my pain?
>
> Will you come homeward from the hills of Dreamland?
>   I have grown weary, though I wait you yet;
> Watching the fallen leaf, the faith grown fainter,
>   The memory smoulder'd to a dull regret.
>
> Shall the remembrance die in dim forgetting—
>   All the fond light that glorified my way?
> Will you come homeward from the hills of Dreamland,
>   Home in the dusk, and turn my night to day?

Here was another sign of the loneliness and insecurity never far from the surface. It made a powerful appeal: Edward's quick and restless sensitivity—especially now he was a great man—was bound to attract vaunting imagination in others. Edward courted it. He wrote to Mrs Stuart-Wortley at Christmas: 'Alice & Carice & a friend have gone off to a far church in the car—I am worshipping several things by the fire—memories mostly of the New world geographically & emotionally—that Symphony *is* a new world isn't it? Do say "yes".'[70]

His own Alice kept her counsel. She had seen through a longer perspective than any others the development of Edward's personality and his genius. She was in the best place to realize that his growing nostalgia was volatile. She understood how necessary it was for him to find in his memories a richness that time alone could put there. It was a quality which had been noticed by Miss Burley when they lived at Craeg Lea:

In the old days innocent visitors to Malvern complimented him on the beauty of his surroundings and said how easy it must be to compose music there. This always provoked him to a sort of dull rage; and indeed his warmest appreciation of Malvern, as of every other place in which he lived, was not expressed until after he had left it.[71]

Again and again his fancy embroidered some past time in rich colours—colours unsubstantiated by any diary references for that day or week or month or year.

---

[70] 25 Dec. 1908 (Elgar Birthplace).
[71] *Edward Elgar: the record of a friendship*, pp. 186–7.

No matter. To invest the past with splendour was vital, if his music was to go forward in the retrospect and prospect which had just given his Symphony its 'massive hope in the future'.

On 12 October Edward took himself alone to London. He stayed at Queen Anne's Mansions (near Frank Schuster's house overlooking St. James's Park) where service flats could be hired for a few days or a week at a time. Alice joined him there for the first London performance of *The Wand of Youth* Suite No. 2. Back in Hereford, he fell ill again until more Symphony proofs arrived.

He had agreed to publish five hundred copies of the Symphony in miniature score before the first performance. It meant dispensing with any chance of making alterations afterwards. A piano arrangement by Sigfrid Karg-Elert was also in process. He wrote the dedication, as he had promised

> To Hans Richter, Mus.Doc.
> True artist & true friend.

On 6 November Richter arrived at Plas Gwyn to go through the new music, in preparation for the première with the Hallé Orchestra in Manchester. It was his first Elgar première since *Gerontius*, and neither man was taking any chances.

Three weeks later came a private rehearsal of the Symphony at Queen's Hall with the London Symphony Orchestra (who were to give the London première four days after Manchester). During this visit Edward consulted yet another doctor over the swings of physical and mental health that had been overtaking him with more regular violence:

November 25. E. to Dr. Bertrand Dawson. A. saw him at end. Nice seeming cultured man—understanding temperaments. Trust advice will do muss good. All sound. D. G. Lunched early & caught 1.40 train home—E. pleased with Ark.

Even chemistry seemed to be about to yield publishable results, as Edward wrote to his friends of a new 'Elgar Sulphuretted Hydrogen Machine' which he hoped to patent. Chemistry occupied him constantly up to the eve of the Symphony's première in Manchester.

December 2. E. busy & seeming well but very porsley towards evening. Rather fearing for next day—

December 3 . . . E. & A. left 10.45 [for Manchester]. E. so porsley, doubtful if going till last minute. Foggy & train a little late. E. hurried to rehearsal . . . Fine playing.   E. sat in Hall but then stood up by Dr. Richter & showed what he wanted & did splendid good. A. sat with Mr. Wilson [the Hallé chorus-master] at rehearsal. Dr. Sinclair there. E. better. D.G.   Dined with Frank . . .

The scene at the première that night was described by Robert Buckley for his paper:

Manchester was shrouded in fog which penetrated the Free Trade Hall and made the lights burn blue. According to the authorities it had the further effect of reducing the attendance . . . The galleries as well as the floor had many empty benches . . . The band scarcely seemed at home with the music; the strings were rough, and other departments failed to exhibit the perfection we expect from an orchestra of such reputation.[72]

But the adverse conditions did not long delay the response of the audience present. *The Musical Times* reported:

It was most interesting to watch the somewhat formal appreciation which followed the first movement ripen into enthusiasm as the work progressed. There could be no doubt that it was the slow movement which struck home and called forth the most sympathetic response . . .[73]

It was the custom in those days to applaud after every movement. Alice wrote: 'After 3rd movement E. had to go up on platform & whole Orch. & nos. of audience stood up—wonderful scene. Also at end.'

The critics spoke next day with one voice in recognizing a great work. *The Daily Mail*, under the heading 'The Musical Event of the Year', went right to the point:

It is quite plain that here we have perhaps the finest masterpiece of its type that ever came from the pen of an English composer.'[74]

Baughan in *The Daily News* saw the importance of the personal achievement:

The new symphony is certainly a work of high endeavour and extraordinary accomplishment. It bears out a contention I have often made, that Elgar's genius lies in the direction of orchestral music rather than towards choral composition. In the symphony there is a completeness of utterance which we do not find in either 'The Apostles' or 'The Kingdom.'

That completeness was seen by Samuel Langford in *The Manchester Guardian* as a brilliant synthesis of abstract symphony with the programmatic symphonic poem:

. . . To Elgar, as to every thoughtful musician, the conflict which has raged in music for the past eighty years between the rhapsodic and the architectural schools or between programme music and absolute music has presented a difficulty. He has steadfastly preached on one side of the question and practised on the other. And he has excused his practise as a kind of amateurism. What I have done, he said plainly, is to the masterpieces of Beethoven and Brahms as nothing. But now, at last, he has tackled the problem in earnest; he has put away his self-accusation of dilletantism and has written a Symphony, but without forsaking his old rhapsodic style at all. And in so doing he has shown the true solution of the problem. He has refertilised the symphonic form by infusing into it the best ideas that could be gathered from the practice of the writers of symphonic poems. And he has proved that these ideas can be used, and that they give the symphonic form a new kind of unity, both poetic and technical.

[72] *The Birmingham Gazette and Express*, 4 Dec. 1908.
[73] 1 Jan. 1909.
[74] 4 Dec. 1908. The following four quotations are of the same date.

Ernest Newman, in *The Birmingham Daily Post*, saw the implications of the Symphony for Edward himself:

It is quite clear that Elgar has found a new vein . . . Now and again there is a suggestion of the highly-strung mood of *Gerontius*, and once or twice the more open-air Elgar, the Elgar of *Cockaigne* and the *Pomp and Circumstance* Marches, peeps out. But, as a whole, the Symphony is genuinely new Elgar, new both in feeling, in idiom, and in workmanship.

For a symphony a composer needs, if not before all other things, at any rate in addition to all other things, long breath and lasting power. It is the form that tries the composer's endurance to the utmost. If he is only a miniaturist, with only the capacity for short sprints, as it were, the symphony inevitably finds him out.

In his previous orchestral works Elgar has undoubtedly given us the impression of being a delicate and imaginative miniaturist. The 'Enigma' variations are all miniatures, refined and poetic little jewels, each in its own setting. In the 'Cockaigne' and 'In the South' overtures, though the picture is meant to be more continuous, the method of structure is still that of the miniature painter. Each section is expressive and complete in itself, but does not blend very intimately with the other sections.

The power of continuous and connected thinking that is shown in the Symphony is therefore all the more desirable. The work is unusually long, lasting to-night something like fifty-five minutes, yet rarely do we feel that the thread of it has even for a moment broken. It was worth Elgar's while to wait thus long before he produced his first symphony.

The work has profited by all his varied experience of life and by thirty years of continuous exercise of his imagination and his technique. Both on the inventive and the formal side the work is very original . . . The finale is also strong with that tempered philosophical strength that makes the first movement so truly remarkable. Wherever this quality comes uppermost it suggests Gerontius with a difference—a Gerontius who instead of dying has continued to live and is all the better for the agony of spirit he has been through . . . It is a work not merely of English but of European significance.

The Star explored the implications in a comparison with Strauss's *Heldenleben*:

In both we see a personality struggling against opposing elements; in both the central figure arouses sympathy by the union of indomitable purpose with purely human despondencies. But here the likeness ends. The Symphony is more spiritual than the tone-poem. We seem to see the conflicts and the final higher development from within in Elgar's work: from without in the other . . . Both seem to speak of love: but in 'Heldenleben' it is a human love, in the Symphony it suggests the amor intellectualis Dei . . .

The whole work should put new hope into all who wish British music well, and in its peculiar mood there seems to be something we have so long looked for in vain—the really British note.

From Manchester Edward and Alice went to London. There on 6 December Richter opened the London Symphony rehearsal with words which were to remain in the memory of W. H. Reed, seated in the first violins:

'Gentlemen,' he said, 'let us now rehearse the greatest symphony of modern times, written by the greatest modern composer,' and he added, *'and not only in this country.'*[75]

Frank Schuster was in the stalls with Gabriel Fauré, and that night he gave a dinner party for the two composers.

An hour before the concert on the following evening, the pavements outside Queen's Hall held a scene of wonder in the eyes of the *Daily Telegraph* critic: 'A spectacle, unhappily not usual in London, was that of a crowd too large for the space in Queen's Hall attempting to enter that building . . . bent upon hearing the first performance in London of Elgar's Symphony. Many an enthusiast was turned empty away.'[76] Jaeger was there, having come from Muswell Hill in a motor cab to be present at what must be his last Elgar première. He wrote:

I never in all my experience saw the like. The Hall was *packed*; any amount of musicians. I saw Parry, Stanford, E. German, F. Corder, E. Faning, P. Pitt, E. Kreuz, etc. The atmosphere was electric . . . After the first movement E. E. was called out; again, several times, after the third, and then came the great moment. After that superb Coda (Finale) the audience seemed to rise at E. when he appeared. I *never* heard such frantic applause after any novelty nor such shouting. Five times he had to appear before they were pacified. People stood up and even *on* their seats to get a view.[77]

The critic of *The Clarion* saw

. . . a tall, thin man with a slight stoop, black hair turning grey; a thin, ascetic face, yet the face of a fighter if it were not for the earnest eyes and brow of the mystic; a resolute chin, the lower half of the face showing indomitable energy, yet it is the face of a student. Elgar's bearing is aloof and dignified, yet also distinctly diffident. Altogether a curious and lofty personality, intensely English, yet with a touch of something strange and apart.[78]

The critical praise of Manchester was amplified. *The Morning Post* wrote of the Symphony as ' . . . a masterpiece such as no other British hand has yet produced. The work increases the author's fame; to the world it goes forth as a masterpiece and our incontestable claim to be regarded as a creative musical nation.'[79] After a round of London parties, Edward and Alice returned to Plas Gwyn on 11 December. There the late afternoon sun shone against the towers of Hereford, as Alice recorded their arrival:

Found all well D.G. & sunshine in our hearts. Glorious sunshine all the afternoon & wonderful sunset, seemed a good omen. The sunset, I pray, presaged the 3rd Pt. of 'Apostles', it was like the Civitas Dei pinnacles, towers set in gold.

But Edward found no joy in the approach of another holiday season. When

[75] *Elgar*, p. 97.
[76] 8 Dec. 1908.
[77] Dec. 1908 to Dora Penny (quoted in Powell, *Edward Elgar: memories of a variation*, pp. 79–80, where Corder's initial is printed as 'J.').
[78] 18 Dec. 1908.
[79] 8 Dec. 1908.

Frank Schuster's sister sent a small piano for use in The Ark, his letter of thanks written on Christmas day concluded:

I am still disappointed that you have not heard the Symphony: it is making a very wild career & I receive heaps of letters from persons known & unknown telling me how it uplifts them: I wish it uplifted me—I have just paid rent, Land Tax, Income Tax & a variety of other things due to-day & there are children yapping at the door, 'Christians awake! salute the yappy morn'. I saluted it about seven o'clock, quite dark, made a fire in the ark & mused on the future of a bad cold in my head & how far a carol could get out of the key & still be a carol: resolve me this last.[80]

The same day he wrote to Mrs Stuart-Wortley: 'I am dulling in the house with a cold, which is depressing for me & *I mean to make it so* for all around me—that's a nice Christmas feeling.'[81]

Yet the demand for the Symphony at Queen's Hall had caused the London Symphony Orchestra directors to announce 'an extra concert for the purpose of repeating the Elgar Symphony'. Thus Richter had conducted it again on Saturday 19 December. Jaeger wrote to Dora Penny: 'I went on Saturday too & took my wife & we enjoyed the work immensely. The House was packed again, & Busby, the managing Director of the Symph: Orch: Co. told me the day was a *record* for them in the way of selling tickets . . .'[82]

The American première took place in New York in early January 1909, conducted by Walter Damrosch—with such success that it had to be repeated three weeks later. Other performances followed in Chicago, Boston, and Toronto. On the continent of Europe it was heard in Vienna under Bruckner's pupil Ferdinand Löwe, in St. Petersburg under Alexander Siloti, in Leipzig at a Gewandhaus Concert conducted by Nikisch. Nikisch was so impressed that he issued a statement to the press:

I consider Elgar's symphony a masterpiece of the first order, one that will soon be ranked on the same basis with the great symphonic models—Beethoven and Brahms . . . When Brahms produced his first symphony it was called 'Beethoven's tenth', because it followed on the lines of the nine great masterpieces of Beethoven. I will therefore call Elgar's symphony 'the fifth of Brahms'. I hope to introduce it to Berlin, with my Philharmonic Orchestra there, next October . . .[83]

At home the Symphony had received its third London performance on New Year's Day 1909. This time the Queen's Hall Orchestra was conducted by Edward. Again the hall was packed. The composer's reading was contrasted to Richter's:

[Richter] fulfils so completely the average musician's ideal of a fine performance, an ideal that is largely exhausted if a fool can steadily tap the time without colliding, so to speak, with the conductor's bâton. Under Elgar the music was *sung*, and with infinite pathos.[84]

[80] HWRO 705:445:6837.　　　　[81] Elgar Birthplace.
[82] Quoted in Powell, *Edward Elgar: memories of a variation*, p. 80.
[83] Quoted in *The Musical Times*, 1 June 1909.
[84] J. H. G. B. in *The Musical Standard*, 9 Jan. 1909.

Six days later Edward conducted the Symphony at Queen's Hall yet again, as the centrepiece in all-Elgar concert. The London reporter for *The Birmingham Daily Post* telegraphed to his paper:

Music just now is for Londoners summed up in the name of Elgar. His new symphony is on every lip; nay, it has found its way to every heart. Only a few pedants stand peevishly aloof. Everyone else is metaphorically on his knees before the composer and his new masterpiece. One performance a week seems a necessity for London's comfort, and a long vista of Elgar is opening before the eyes of our concert-goers.[85]

Fuller-Maitland allowed himself to wonder acidly in *The Times* whether supply had created demand, 'for to announce frequent performances is a sure way of arousing public interest'.[86] Baughan in *The Daily News* had answered such carping in advance when he wrote after the first London performance: 'Those composers who complain of neglect, and imagine that Elgar's vogue is due to all kinds of causes but the music itself, may rest assured that when they can write compositions which can move an audience they also will be popular.'[87]

The Symphony was even put down for a series of free concerts sponsored by Harrod's store to celebrate the sixtieth anniversary of their establishment. This extravagance soon reached Mr Punch, who anticipated the Spring Sales with a piece entitled 'St. Cecilia at the Sales: or, The New Handmaid of Commerce'. Here the popularity of the new Symphony took it from shop to shop under the speculative batons of leading West-End actors and music-hall entertainers:

Messrs Torrey and Dems, of the great Emporium on Campden Hill, announce a monster musical entertainment to be held under the dome of their new buildings on May Day in celebration of the 25th anniversary of their association with Kensington. The proprietors, with an enterprise which does them infinite credit, have planned Sir EDWARD ELGAR's Symphony in the forefront of their programme. This epoch-making work, we may note, will be conducted for the first time by Mr. GEORGE ALEXANDER, and the sermon will, of course, be preached by Dr. TORREY, who comes from America for the purpose . . .

Messrs. Bark and Bark, the well-known Kensington outfitters, propose to commemorate their jubilee, which falls on the anniversary of the battle of Waterloo, by a grand orchestral concert, at which Sir EDWARD ELGAR's Symphony will be conducted for the first time by Mr. LEWIS WALLER . . .

The game was good and *Punch* made further play with the musical patriotism inspired by the Symphony, drawing on a complex political background. The increasing German naval menace had caused the First Sea Lord, Admiral Fisher, to take many ships from Lord Charles Beresford's Channel Fleet to form a new 'Home Fleet' under his own command stationed at Scapa Flow, opposite the German entrance to the Baltic. Beresford had publicly denounced the scheme; forced to resign his command, he demanded a public enquiry, which was to sit in April 1909. Meanwhile Fisher assured the nation that with the 'Home Fleet' England might rest easy. One of *Punch*'s Symphony

---

[85] 8 Jan. 1909.   [86] 9 Jan. 1909.   [87] 8 Dec. 1908.

performances, therefore, was sponsored by Messrs Dormy and Mendoza, at whose concert

> . . . Sir EDWARD ELGAR's Symphony will be conducted for the first time in the Pyjama Saloon by Sir John Fisher. The great sailor has also kindly promised to sing his favourite appeal to the nation, *Dormi pure*, together with several German *Wiegenlieder* of a most deliciously narcotic and tranquillising character.[88]

Someone on the *Punch* staff may have known of Edward's friendship with Beresford. Three days after the magazine appeared, when Edward dined at the Stuart-Wortleys' house in Cheyne Walk, Beresford told them all 'a terrible tale of naval unpreparedness'.

*   *   *

*   *

*

While the Symphony advanced from success to success, its composer descended into 'a grey time'. 'Very depressed,' Alice noted, 'not caring for music or Ark.' Then came the financial account of Symphony performances from Novellos. At the time of sending in the Symphony in the autumn, Edward had demanded a composer's fee for every performance. The firm had never in their history charged such a fee. In the end they agreed, and this account now yielded five to ten guineas from each concert where the Symphony was played. But no composer's fee had been charged to the Queen's Hall manager Robert Newman in respect of the Symphony performance conducted there by Edward himself on New Year's Day. He objected immediately. His letter drew a careful response from Novello's secretary, Henry Clayton:

> I am awfully sorry that you are not satisfied with the arrangements which we made with Mr. Newman for the production of your Symphony at Queen's Hall under your direction. I can assure you that both Littleton and I have a good deal of anxiety about performances of this work. Our object has been to get as much as possible in the shape of performing fees for your benefit, and at the same time, as I put it to Littleton, not to kill the goose for the sake of the eggs of gold. And I can assure you it needs very delicate handling. For we know of several cases where the idea of performing the work has been abandoned, where we quoted £15.15.0 or £10.10.0—*fees which included our charge for the loan of the music* . . . Therefore when we knew that Newman was going to pay you Twenty-five Guineas for conducting, & us Five Guineas for the loan of the music, we felt that the total Thirty Guineas was about as far as we could ask them to go.[89]

Yet it reached Edward's ears that both Debussy and Sibelius were conducting concerts of their own works in Queen's Hall that season at fees substantially higher than his own.

[88] *Punch*, 24 Mar. 1909. The magazine appeared on the day Beresford hauled down his flag at Portsmouth. The large department stores of Derry & Toms and Barkers were in Kensington High Street. Messrs Dormy were makers of pyjamas. 'Dormi pur' is the beginning of a celebrated ensemble in Flotow's *Marta* in the Italian translation by which it was best known in England.

[89] 1 Feb. 1909 (Novello archives).

Plas Gwyn, Hereford.

Feb 2. 1909

Dear Clayton:

All right & thanks for your letter: I am sure you do your best for me but the incredible meanness of Queen's Hall annoys me, more than I should reasonably let it do so, after the exuberant protestation of Edgar Speyer [the head of the Queen's Hall financial syndicate] & the rest . . . They pay *any* foreigner 4, 5, 6, 7 or even 8 times the amount given to me & lose largely over the visitor because they say it's good for art. It annoys me that the money they really make out of me is spent on other people.

This is all by the way but it relieves my feelings

Kind rgrds
Yours sincerely
Edward Elgar [90]

Grey weather and a heavy chest cold depressed him further. Two visits to Llandrindod Wells did no good. Then their American friend Mrs Worthington invited them to spend several weeks of April and May at a villa she had taken near Florence. Alice sent Edward to London and Paris to divert himself, while she made arrangements to let Plas Gwyn for their absence. The upheaval involved the departure of Edward's niece May Grafton: 'You see,' Alice wrote to Mrs Kilburn, 'she came when Carice was so young & now she is grown up there is really nothing to do . . .' (This in case Mrs Kilburn could hear of any employment for May.) On 15 April May Grafton returned to her mother's house at Stoke Prior, while Alice and Carice caught the train for London. Crossing to Paris, they found Edward and Mrs Worthington, and on the 20th they all took the night train to Italy.

Eighteen hours later they were ensconced in Careggi at Villa Silli. The house was of antique origin. Alice wrote to Mrs Stuart-Wortley (who knew the place intimately) of

glorious weather, the world bathed in sunshine, the air scented with flowers & resounding with nightingales—. It is a very nice spacious Villa, the hall in Roman days was the Atrium & in later ages, it was one of the Medici Villas; the great Medici Villa where Lorenzo died is close by. I trust you will hear E.'s impressions, tonally, some day—some days we hear them already.[91]

A new sketchbook begun at Careggi was labelled 'Opera in three acts'. Into this he wrote many musical ideas, but no hint of a subject. Numerous offers of opera libretti had come. Laurence Housman and George Moore had been assiduous in trying to find subjects to tempt him. In 1908 Moore had suggested *The Lake* by Lord Howard de Walden. Edward objected that it had a similarity to *Tannhäuser*. Moore reminded him that when they had discussed making *Grania and Diarmid* into an opera, Edward had objected that it was similar to

[90] Novello archives.
[91] 4 May 1909 (transcript by Clare Stuart-Wortley now at the Elgar Birthplace).

Siegfried; then he said that at this rate Edward could get to the end of his life without finding a subject.[92] Again at Careggi nothing came of opera writing.

Since finishing the Symphony, however, his thoughts had returned to the E flat Symphony sketches of 1903 and to the Violin Concerto sketched just before meeting Kreisler in 1905. The old *King Olaf* motive of 'heroic beauty'

 had seemed to stand behind the Concerto

primary theme of 1905:

 Now at Careggi came a variant for solo violin:

And there were ideas for a slow movement; but they did not draw together.[94]

When his thoughts for the first Symphony had failed to come together in Rome at the end of 1907, he had turned to writing part-songs with words which might shape his thoughts. Now again at Careggi he wrote part-songs. One was a little picture of Tuscany called *The Angelus*, framed in the *idée fixe* of a falling second. The words, adapted from Tuscan dialect, as he told Mrs Stuart-Wortley, 'are of the place & not far from your own monastery on the Fiesole road of which C[harles] B. S.W. has memories also. Anyway it wd. give me the greatest pleasure to put your belovèd name on it if you both allow it.'[95]

The other part-song written at Careggi was more imposing. Its words were translated by Dante Gabriel Rossetti from the medieval Italian poet Cavalcanti. Yet they had their significance for Edward in 1909—the year that might have seen the conclusion of the *Apostles* trilogy, but saw instead the Symphony with its 'massive hope in the future' heard by many of its audience in a nationalistic way:

> Dishevelled and in tears, go, song of mine,
>> To break the hardness of the heart of man:
>>> Say how his life began
>> From dust, and in that dust doth sink supine:
>>> Yet, say, the unerring spirit of grief shall guide
>>> His soul, being purified,
>> To seek its Maker at the heavenly shrine.

---

[92] 14 July 1908 (HWRO 705:445:2272).

[93] Sketch headed *'written later in Italy'* (Library of Congress).

[94] The short-score manuscript of the *Andante* written later in London was inscribed: 'Jan & Feb 1910 Careggi–Q. Anne's Mansions SW.' (Elgar Birthplace).

[95] 23 June 1909 (HWRO 705:445:7827).

The key was B minor, and the opening repeated the Violin Concerto opening—but with the big rising interval significantly reduced:

In the same way, the part-song's answering phrase echoed Mary Magdalene's B minor

from *The Apostles*:

A middle section developed the melody in D major for the words 'Yet, say, the unerring spirit of grief shall guide his soul . . .'. At 'purified' a big sequence assimilated the song's opening figure so exactly to the descending steps of the 'Liebestod' that it must have been deliberate:

Yet again there was an Elgarian precursor—in *The Apostles*:

After a sustained climax 'at the heavenly shrine', the music fell back gradually to a soft reprise: 'Go, song of mine, To break the hardness of the heart of man.'

On 15 May Alfred Littleton arrived in Florence with his daughter. The old man had lost his wife in the autumn after a long and happy marriage, and he had been hit very hard. In these years he and Edward had drawn closer, until their friendship now realized the hope Jaeger had once entertained: 'You would like him very dearly if only you *knew* him.'[96]

Three days later, as Edward and Alfred Littleton shared the wonderful Florentine weather, news came that Jaeger's long suffering had reached its close in London. Edward wrote to the widow: ' . . . I cannot realise that the end

[96] 5 Feb. 1904 (HWRO 705:445:8711).

is come & I am overwhelmed for the loss of my dearest & truest friend.'[97] *Go, Song of Mine* might have written the apostrophe to that friendship. Too late to inscribe it to Jaeger, Edward dedicated it now to Alfred Littleton.

Yet it had been a fine visit—so fine that the Elgars toyed for an instant with the idea of buying a villa there: one which they liked, and which was for sale, was reputed to have been designed by Michelangelo. When Mrs Worthington's lease of Villa Silli came to an end, they parted with real regret. 'Pippa' (as her friends called her after the heroine of Browning's poem) left for America. Edward and Alice turned north to Pisa, Bologna, and a final Italian week in Venice.

The heat and noise of early summer crowds there were a shock after the quieter delights behind them. But one morning as Edward sat in the Piazza San Marco, his ear was caught by 'some itinerant musicians who seemed to take a grave satisfaction in the broken accent' of a rhythmic phrase in their music.[98] He noted it down—a crazy pattern falling across every bar-line. Then he walked into San Marco itself, where the golden gloom suggested something totally opposite for his music. These two impressions were to be carried side by side into his ideas for the Second Symphony.

The Italian impressions were to be framed with older reminiscence, as Alice wrote to Mrs Stuart-Wortley on 5 June: 'We are turning homewards on Monday via Bavaria—& its memories.'[99] Lermoos, Innsbruck, a wagon through high passes full of Alpine flowers—and so to Garmisch, as Carice remembered:

We paid a solemn call on Richard Strauss and his wife, my mother being their interpreter. We were amused when Strauss wanted to show my father some score or paper of interest, he called out to Frau Strauss who produced a bunch of keys from her underskirt, & duly locked it all up again when they had finished.[100]

That afternoon, between Alpine thunderstorms, they went far up the passes to meadows full of flowers, and saw cows going home with their bells ringing under the heavy evening skies. They lingered until 14 June. Two days later they were back in London. Alice and Carice went to stay at the Langham, while Edward took himself off separately to stay with Frank Schuster in Old Queen Street.

Taking a leaf from the book of his host, perhaps, Edward felt the need of a valet. Interviewing several candidates, he and Alice selected Arsène Jaulnay, an agreeable young Frenchman who was also a considerable athlete. Jaulnay had won the amateur lightweight boxing championship of France, and when

---

[97] 21 May 1909 (in possession of Richard Brookes).
[98] MS notes by Charles Sanford Terry, based on his conversations with Elgar about the composition of the Second Symphony (Athenaeum Club).
[99] Transcript by Clare Stuart-Wortley (Elgar Birthplace).
[100] Quoted in Percy M. Young, *Alice Elgar*, p. 164.

they all returned to Hereford he ran successfully in a local marathon. Moreover he was a great success in his adopted profession as a gentleman's gentleman, and quickly settled down in the household at Plas Gwyn.

At home Edward felt unwell again. He turned down another opera proposal from Laurence Binyon. But he did complete a little *Elegy* for strings, requested by Alfred Littleton for a Musicians' Company memorial to their Warden. The opening of the music transmuted the falling seconds of *The Angelus* to a meditative sequence, out of which the *Elegy* slowly unfolded. The end was inscribed 'Mordiford Bridge'—an ancient structure down the road from Plas Gwyn. Years later he recalled:

Most of my 'sketches',—that is to say the reduction of the original thoughts to writing, have been made in the open air. I fished the Wye round about Mordiford & completed many pencil memoranda of compositions on the old bridge, of which I have vivid & affectionate memories.[101]

The summer found his spirits improving in an old combination: 'July 1, 1909. E. getting ready for cycle excursions & looking up sketches.' He went so far as to hint to Novellos that Part III of the *Apostles* trilogy might after all emerge.[102] But a letter had come from the Secretary of the Philharmonic Society, Francesco Berger, to ask about producing the Violin Concerto.[103] Ever since *Cockaigne* Berger had been trying to get another Elgar première for the Philharmonic; and Edward had agreed to open their new season in November by conducting a concert of his music. Late August found him 'possessed with his music for the Violin Concerto'.

The Hereford Festival intervened. Sinclair conducted *The Apostles*. Edward shared the Wednesday Evening Secular programme with Bantock and Frederick Delius (who directed his *Dance Rhapsody*). The première of *Go, Song of Mine* was sung to a deeply appreciative audience. But the centre of his week was his own performance of the Symphony, of which Ernest Newman recalled 'the tenseness, the ardour, and the sense of living through a series of highly charged experiences'.[104]

He went to Liverpool on 24 September for the first gathering of a Musical League for composers, of which he had accepted the Presidency. No music of his own was played, for the object of the League was to promote lesser-known composers. Ethel Smyth conducted in a bright blue kimono, and Vaughan Williams produced a cantata on Rossetti's *Willowwood*. At a festive Sunday luncheon in the Town Hall, Edward made 'a witty and sympathetic speech' to the assembled company—who also included Percy Grainger, Havergal Brian, and Arnold Bax. Bax was to recall the result of this meeting:

In the following year I was invited for the first time to send in a work for performance at a

---

101   16 Jan. 1931 to G. H. Jack (MS in possession of Keith Harvey).
102   15 and 17 July 1909 to Henry Clayton (Novello archives).
103   6 July 1909 (HWRO 705:445:2878).
104   *The Birmingham Daily Post*, 6 Oct. 1909.

Queen's Hall Promenade Concert, and when I went to Sir Henry Wood for a preliminary run through he told me (to my intense pride) that it was none other than Elgar who had recommended him to take up my work. It seems that he never forgot my visit to Birchwood (I think his days there counted as the happiest in his tormented life, and he kept a special regard for anyone who had seen him in those surroundings).[105]

In October came the Birmingham Festival—the Festival which might have seen the completion of the *Apostles* trilogy. Richter was to conduct *Gerontius* and the Symphony at what would be his last Festival. Edward anticipated this meeting, as he had anticipated the meeting with Kreisler in 1905 with the sketch for a Violin Concerto: now he worked at a Second Symphony. The material came from the E flat Symphony sketches of 1903, recently looked at again in the perspectives of Italy.

On 5 October he went to Birmingham to hear Richter conduct the Symphony that was. Next morning they had *Gerontius*. In the audience Edward met a ghost from the past—a young ghost. It was the beautiful Muriel Foster, happily married since the last Birmingham Festival in 1906, and retired from the concert platform at the wish of her husband. Edward gave her a page from his sketchbook which he inscribed:

First sketch for Intro: to Section IV of 'The Kingdom', which section was written for Mrs. Goetz & was, apparently, not good enough for her as she sings it no more!

leaving
Edward Elgar
to his fate.
(which is sad)

Oct 6. 1909[106]

There was no continuation for any creative project then. After Birmingham he conducted *The Kingdom* at Southport, and that began a 'dreary procession' of autumn performances.[107] A northern tour with the London Symphony Orchestra was to pay him a hundred guineas. It visited Newcastle, York, Hull, Doncaster, Middlesborough, Nottingham, and Leeds. A critic in Nottingham drew this profile:

Sir Edward Elgar . . . is a man of nerves. As he raises the baton, which he holds between his thumb and first two fingers, as one might take hold of a pen, he seems to quiver with excitement. The tension of his nerves communicates itself to his muscular system, and he becomes quite angular and stiff—indeed, almost awkward—in his movements. The elbow of his right arm is bent and rigid, and when he makes a downward or horizontal movement with his baton his shoulders and body follow the direction. As one might expect from these manifestations of a febrile temperament, his orchestral renderings are marked by waves of emotion.[108]

At the end of this tour of 'the coallied North' (as he wrote to Alice

[105] *Farewell, My Youth*, pp. 31–2.
[106] MS in possession of L. A. Foster.
[107] 11 Oct. 1909 to Frank Schuster (HWRO 705:445:7003).
[108] *The Nottingham Daily Express*, 30 Oct. 1909.

Stuart-Wortley), he was taken by Henry Embleton to view engine works: 'I saw locomotives building & torpedoes & other nasty maleficious things.'[109]

Meanwhile another orchestral tour had been organized round Edward's Symphony by young Thomas Beecham. With the indefatigable energy that was to mark his long career, Beecham had formed his own orchestra and was touring it entirely among towns where he could give the Elgar Symphony its first local performance. In little over a month they played it in Cardiff, Exeter, Torquay, Bournemouth, Southport, Reading, Bradford, Cheltenham, Malvern, Burton, St. Helen's, Chester, Wigan, Lancaster, Bolton, Kendal, Harrogate, Belfast, Dublin, Preston, Hanley, and Cambridge. But as the tour went on, it was noticed that the Symphony was shorter and shorter. The last *reductio* came at Hanley. Havergal Brian wrote in dudgeon to *The Musical Times*:

The first movement was cut down one half: part of the 'exposition' and the whole of the 'development' were cut out, and some minutes were sacrificed in the succeeding movements. Those who know the Symphony will be astonished to hear that the actual time occupied in its performance was only thirty-eight minutes! It was an insult to the composer and also those responsible for the concert.[110]

An official protest was entered, and Beecham addressed his orchestra: 'the composer of this immortal masterpiece' had communicated his desire that it be played complete or not at all. 'Therefore,' said Beecham, 'we shall play it complete—with all the repeats.' There was not a single repeat marked in the score. Counting the Beecham concerts, the Symphony had received eighty-two performances in the year since its production. It was said to be a record in musical history.

*   *   *

On 11 November Edward conducted the opening concert of the season for the Philharmonic Society. The secretary, Francesco Berger, sent the conductor's fee of 35 guineas together with a suggestion:

We all hope you will consent to honour us again by taking a Concert in our next season, and making that the occasion for introducing some new Work. Will you let us have your Violin Concerto? . . . We open on the 10 November.[111]

Edward marked this letter 'Yes praps!'

Back at Plas Gwyn he had a heavy cold. He was just able to meet a request from the Master of the King's Music, Sir Walter Parratt, for an anthem to be sung in the mausoleum at Frogmore on the anniversary of Queen Victoria's death in January.[112] For the words he returned to Cardinal Newman:

[109] 7 Nov. 1909 (HWRO 705:445:7669).   [110] 1 Feb. 1910.
[111] 30 Nov. 1909 (HWRO 705:445:2860).
[112] Parratt's letter approving Elgar's choice of words is dated 10 Nov. 1909 (HWRO 705:445:1963).

> They are at rest.
> We may not stir the heaven of their repose
> By rude invoking voice, or prayer addrest
>   In waywardness to those
> Who in the mountain grots of Eden lie,
> And hear the four-fold river as it murmurs by.

The reference to Eden touched death with an innocence that could bring it close to his own private borderland, the river of childhood. His setting of this Elegy started with a final reduction of the Violin Concerto opening figure in B minor: reduced already from its big upward leap for *Go, Song of Mine*, now it was reduced to a single semitone up and down:

At the end of each stanza came the ghosts of descending step and falling fourth:

It was a striking phrase, and it would reappear in his music before long.

As if in response, on 27 November 'E. had most touching letter from the dear Prof. Sanford. The Prof. thought he had not long to live & sd. most loving farewell.' Sanford was sixty—eight years older than Edward, some months younger than Alice. When Edward tried to write a note of condolence to Frank Schuster on the death of another friend, the condolence turned on himself:

Here we are in gloom—Alice is nursing her cold in bed & I am creeping about under the Doctor's hands—cough &c.—I begin to believe in Doctors for he says I must be worrying over something as there's no reason it (the cold &c.) shd go on; the poor man thinks it's nerves over composition—when it's only heartbreak for something or somebody else . . .

It *is* deadly dull here & enough to drive one to despair—and the world is so nice & waiting for me—if I cd. only get to it.[113]

When the Stuart-Wortleys' friend Sir Gilbert Parker sent a volume of his poems, Edward saw a means for utilizing the mood. From Parker's poems he decided to make a cycle of solo songs with orchestra. Therein, a year after the achievement of the Symphony, he might explore the depths of his consciousness as he had explored them after the achievement of the '*Enigma*'

---

[113] 25 Nov. 1909 (HWRO 705:445:7005).

Variations ten years earlier in *Sea Pictures*. Now the solo voice against orchestra practically invoked the ensemble of the hoped-for Violin Concerto.

He did not begin the new cycle at its beginning, but wrote three songs for its middle and end. The song he marked 'No. 3' took its title from the opening words:

> Oh, soft was the song in my soul, and soft beyond thought were thy lips,
> And thou wert mine own, and Eden re-conquered was mine . . .

The *Allegro* melody moved persistently downward—even at the climax

Love went farther into the past in the next song of the projected cycle, which he numbered '5'.

> Once in another land,
> Ages ago,
> You were a queen, and I,
> I loved you so:
> Where was it that we loved—
> Ah, do you know?
>
> Was it some golden star
> Hot with romance? . . .

Edward's music was marked *fantastico*. Each of its first two stanzas opened with the same figure:

It suggested the symmetry of the primary figure for the Violin Concerto. Over 'Hot with romance' came a high counterpoint

like the Concerto second subject

> But you were a queen, and I
> Fought for you then:

brought back for a moment the world of *Froissart*. But then it went on:

> How did you honour me—
> More than all men!
> Kissed me upon the lips;
> Kiss me again.

The final song for the Parker Cycle was entitled *Twilight*:

> Adieu!—and the sun goes a-wearily down,
> The mist creeps up o'er the sleepy town,
> The white sails bend to the shuddering mere,
> And the reapers have reaped, and the night is here.

An *idée fixe* of descending steps took the opening of the Symphony's 'motto' to the B minor of the looked-for Violin Concerto. So, despite the words, there was promise in Edward's music.

Near the end of the year, nostalgia went back into the distant past when he opened a Christmas present and found the scores of the Beethoven String Quartets. They had been sent by Edward Speyer and his wife, to whom he wrote:

Thanks to you & Tonia a thousand times: the books are a delightful possession & I renew my growth in reading some of the old dear things I played when a boy,—when the world of music was opening & one learnt fresh *great* works every week—Haydn, Mozart and Beethoven. Nothing in later life can be even a shadow of those 'learning' days: now, when one knows *all* the music and all the mechanism of composition, the old mysterious glamour is gone & the feeling of *entering*—shy, but welcomed—into the world of the immortals & wandering in these vast woods—(so it seemed to me) with their clear pasture spaces & sunlight (always there, though sometimes hidden), is a holy feeling & a sensation never to come again, unless our passage into the next world shall be a greater & fuller experience of the same warm, loving & *growing* trust—this I doubt.

The child's fairyland beyond time was there still, and it glimmered through the second movement of Beethoven's Op. 53 No. 3 where C minor goes suddenly to C major. Edward added to his letter:

Was there ever anything so wonderful as

This is the real Christmas feeling—apart from the world of puddings & beef.[114]

And his song-writing turned from the frustrated love of the Parker cycle to 'A Child Asleep', by Elizabeth Barrett Browning (whose 'Sabbath Morning at

[114] 15 Dec. 1909.

Sea' had been at the centre of *Sea Pictures*). This poem raised another hymn to sleep and dreams:

> Vision unto vision calleth,
> While the young child dreameth on.
> Fair, O dreamer, thee befalleth
> With the glory thou hast won!
> Darker wert thou in the garden, yestermorn, by summer sun.

The opening notes of his music linked descending steps

with the shape of melody which had evoked the murmuring of death's river in *They are at Rest*. There was also an echo of the Beethoven Quartet phrase for

the last words of the stanza

He inscribed the song to the baby son of Muriel Foster (whose first birthday would fall on the day after Christmas) 'for his mother's singing'.

On 21 December 1909, the day he finished *A Child Asleep*, there was news of the death of another friend: it was the young and beautiful wife of Henry Wood, Olga Ourousoff, whose soprano singing had meant much in recent performances of such works as *King Olaf*. The approach of another Christmas, with its atmosphere of past happiness and future question, seemed only to show something missing from his own present. That day he wrote to Alice Stuart-Wortley on a sheet of blue paper acquired in Italy:

This is what Petrarch and the rest wrote on! I would write a sonnet to you but it would not rhyme & if it did, it would not be good enough for you otherwise. Anyhow I can *think* sonnets to you & America [i.e. Mrs Worthington] which probably had better not be scanned.

Alice is busy amending everyone's conscience to the 25th, which is quarter-day & a variety of disagreeable things also including Noël.[115]

His own Alice emerged in this description as a mother instructing her wayward child.

Suddenly the vision of the lover left alone with his memories was intolerable. He left the unfinished Parker songs, and the day after writing to Alice Stuart-Wortley began a new cycle of four solo songs to his own words. He called them 'paraphrases' of folk-songs from Eastern Europe—lest their tumultuous emotions appear too close to home. The words were by 'my confidant & adviser Pietro d'Alba'.[116] The closeness of Peter Rabbit made a

---

[115] HWRO 705:445:4109.
[116] 3 May 1910 to Frances Colvin (HWRO 705:445:3421).

measure of Edward's loneliness then. For these were to be songs of passion.

The first was entitled *The Torch*. 'My heart is aflame,' it said, 'As the beacon that lights thee to me.'

Come, O my love!
      Come, fly to me;
All my soul
      Cries out for thee:
Haste to thy home,—
      I long for thee,
      Faint for thee,
      Worship thee
Only,—but come!

Twice more the cry was renewed, each time set to a rising triad:

Come, O my love!— Come, fly to me____.

But between those cries came two visions as desolate as the vision of *Owls* (which he had dedicated to Pietro d'Alba before the Symphony two years ago) set to minor steps:

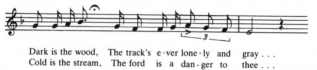

Dark is the wood, The track's e-ver lone-ly and gray . . .
Cold is the stream, The ford is a dan-ger to   thee . . .

The two figures in the song seemed to develop opposing possibilities from the

Violin Concerto second subject

On 23 December, as he worked at the song with its image of the cold and dangerous stream, the Wye at Hereford overflowed its banks and flooded all the meadows in front of Plas Gwyn.

Close to this time he paid another reminiscent visit to Hubert Leicester and his family in Worcester. Hubert's son Philip recorded:

Hair now very grey, but otherwise unchanged in appearances—eyes open & shut rapidly & continuously . . . hands small, slender & nervous—right hand shakes nervously as it rests on table—complexion rather dark, almost olive colour, strong well-shaped moustache—nose almost hooked—teeth large, strong, white & show when he smiles.

Some remark suggested next world.

(F[athe]r) 'Ah, Ted, you need not be frightened to go there.'

(EE) 'No No—I'm not at all frightened—only curious.' E.E. said he was always working, working, working away at something or other—

(Fr) 'Ah, Ted, you were never idle.'

(EE) 'Yes, yes, Hubert, but none of it is of any use. Here have I been, working all my life & done nothing of any practical use to anybody. My time has been absolutely wasted. I often think how many a man can go out with a £5 note & do more real good in no time than I ever did in my whole life.'

Says any amount of his work is thrown away as useless. Loses all interest in a thing as soon as finished, & goes on to something else . . . (Speaks rapidly, sharply & distinctly. Not very loud but clear.)

E.E. said he was very unhappy. F[athe]r said he (Fr) was the happiest man in the city . . .[117]

At the end of 1909 came another death—that of Basil Nevinson, following a long illness and depression, at the age of fifty-seven. It broke the circle of 'friends pictured within' the *'Enigma' Variations* a second time within the year. As Alice closed her diary, she tried to sum it up:

A strange year—Glorious Symphony phenomenal Success—Lovely time at Careggi—but Influenza in Spring . . . & loss of *many* friends. Jaeger in Spring—Basil Nevinson in Dec—Olga Wood in Dec . . .

D.G. for all Mercies—& a truce from *some* anxieties—& being yet spared to one another & some dear friends—God grant 1910 have happiness for E. in store.

Edward had taken himself to London again. Alice joined him in time for a big New Year's party at the home of the eminent solicitor Sir George Lewis: 'Dinner to 67—evening receptn. & theatricals *most* amusing, Crackers, Songs, Caps on—E. pink & blue headdress!' Over the weekend they attended a Royal Academy private view, went to theatres, met friends who diverted him. Alice began to look at possible flats to rent in London, but found nothing.

Edward finished the visit by arranging to produce the three songs from the unfinished Parker cycle as they stood for a Jaeger Memorial Concert in late January to benefit the widow and children. As an added inducement, the beautiful Muriel Foster would come out of retirement especially to sing the songs. Even a partial Elgar première would be a great draw for the concert and the fund. Yet he could never have agreed to perform only the fragment of a work if its progress still involved him closely. It was almost as if he had taken the occasion of Jaeger's memorial to dismiss those songs of nostalgia and disillusion. But then news came that Professor Sanford had died in New York: at a memorial concert Walter Damrosch was to conduct Edward's Symphony.

Back at Plas Gwyn, he began to orchestrate the Parker Songs for the Jaeger concert. And with that came another impulse toward the Concerto:

January 11 1910 . . . E. very busy with Concerto & Bassoon piece . . .

The Bassoon piece was a promise to Edwin James, principal bassoonist of the London Symphony and the new Chairman of the Orchestra. It was entitled *Romance*, and its opening figures for both orchestra and solo instrument

[117] MS note of 15 Feb. 1910.

became anagrams of the orchestral and solo openings already planned for the Violin Concerto:

Below each of them descended once more the steps which had underlaid the motto theme of the Symphony and so much else. Thus the Bassoon *Romance* scrutinized one principal idea for the Violin Concerto as *The Torch* had scrutinized the Concerto's other idea.

At last his own sense of loneliness and loss drove him to grasp the opportunity held out by the two original themes of 1905. In the Concerto first movement they would function as first and second subjects in the usual way. But he did not object when, after the Concerto was finished, Ernest Newman sent him an analysis for *The Musical Times* in which the second subject was described as 'the perfect feminine counterpart and complement to the masculine themes in the first subject-group'.[118] He changed much of Newman's article, but left that definition unaltered. The Concerto's goal was to find the dialogue to link these opposites.

Such a dialogue was not at Plas Gwyn then. He must get away to work on the Concerto. It was an opportune moment. Alice was just then enjoying a little triumph of her own. She had written words for the Trio tune of *Pomp and*

---

[118] 'Elgar's Violin Concerto', in *The Musical Times* (1 Oct. 1910), pp. 631–4. When Newman submitted it before publication, Elgar in a letter of 18 Sept. 1910 called the analysis 'splendid'.

Circumstance No. 4, calling the new song *The Kingsway* after the road recently opened in London. The new song was accepted for immediate publication by Booseys, with a première by Clara Butt at the Alexandra Palace. On 14 January 1910 two of Alice's stanzas appeared in *The Daily Mail*: 'A. *was* peased . . . C. Alice E[lgar] today, not Lady E.' But by then Edward had departed to London, where he took a flat in Queen Anne's Mansions to pursue his Violin Concerto alone.

* * *

Once again his music began in the middle, with a slow movement *Andante*. The primary subject was an old theme:

These short, regular phrases of diatonic expression recalled the innocence of *Dream Children*, the introduction to Part II of *Gerontius*, *The Wand of Youth*. Again the bass descended in steps. It would be an orchestral statement. The violin responded with a counter-melody, extending the orchestra's fact in solo imagination.

A second subject rose over a bass of descending fourths:

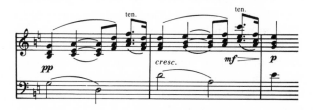

The longer range of this sequential figure, with its modal harmonies, seemed to hint at blending the original 'masculine' and 'feminine' subjects for the Concerto's first movement. To begin a development section, the solo violin linked something very like the original 'feminine' figure to an inversion of the 'masculine':

The orchestra responded with a clear reminiscence of the original 'masculine' theme, now marked *Nobilmente*:

On 20 January, after less than a week at Queen Anne's Mansions, he was ready to try the new music with a violinist: he asked the beautiful Leonora von Stosch, an accomplished pupil of Ysaÿe and the wife of Sir Edgar Speyer of the Queen's Hall syndicate. On the 22nd Alice joined him in London. And two days after that, on the day of the Jaeger Memorial Concert, Edward sketched the continuation of the *Nobilmente* development.[119]

Sequences through solo violin and orchestra rose gradually to a climax of towering passion in the entire orchestra

That passion sounded only once. Like the climax of 'Nimrod', it faded in a single bar. The ardour which had risen gradually through exposition and development had been instantly hushed, and then in half a dozen bars the solo violin restored the music to something like its first innocence as a mother would tuck up a child disturbed by its dreams. It might be an image of Alice's relationship to Edward: she had joined him in London just before the climax was written. The *Nobilmente* which had thus been silenced, Edward told Charles Sanford Terry, could be carved on his own gravestone.

Yet if this climax and its result had occupied the centre of the movement, they were not the movement's goal. For all was set in the perspective of full recapitulation. The end of the recapitulation etherealized the development's passionate climax in a violin solo climbing higher and higher—until the orchestra slipped under it the innocent first subject to begin a coda. Here at last the two *Andante* themes were set side by side in miniature like two Dream Children. Listening to these final phrases in a gramophone recording made more than twenty years later, Edward said to a much younger friend listening with him: 'This is where two souls merge and melt into one another.'[120]

What did it portend, this innocence that went *Nobilmente* in quest of experience and then, having glimpsed a single climax, allowed itself to be led back to etherealized reminiscence? Had his music entered the lists of experience once again to win a victory only for what might have been? If the

---

[119] Short-score sketch p. 7a, containing the music between the present cues 55 and 56 (Elgar Birthplace).
[120] Vera Hockman, 'Elgar and Poetry' (unpublished TS, 1940).

Concerto was really to move on from the lonely hope descried in the Symphony, this innocence still haunting the gates of his creative life was a ghost to give pause indeed.

At the Jaeger Memorial Concert on 24 January Muriel Foster sang the Parker songs so well that *Oh Soft was the Song* had to be repeated, and Hans Richter conducted the *Variations*. Alice wrote: 'He turned to the Orch. spreading out his arms as if to draw every sound & made the Nimrod gorgeous.' Afterwards Edward and Alice went to supper with the Edgar Speyers, where the *Andante* was played again: 'All loved it much, it was repeated . . .' Alice was staying on for a performance of the Symphony under the eminent Russian conductor Vasily Safonov on 31 January. She spent the intervening time looking at more London flats.

On 26 January they dined with the Landon Ronalds. Ronald was thirty-six, a keen-eyed conductor who had seen his chance and taken it. When a 'New Symphony Orchestra' was formed and then abandoned by Thomas Beecham, Ronald had made it his own: he gathered a list of patrons that began with Princess Henry of Battenberg, proceeded through Lady Maud Warrender and Frank Schuster, and ultimately included Sir Edward and Lady Elgar. Seeing the triumph of the Elgar Symphony, it was Ronald who had conducted the Harrod's performances of March 1909, Ronald who conducted it twice again at Sunday Concerts, and Ronald who gave its first Italian performance in Rome on 6 January 1910 with great success. His creed was simple: 'I wish any and every one to share my enthusiasm for what I consider the best in the art.'[121] That naïve ambition could strike a responsive chord in Edward's memory of his own young days.

Four days later they were entertained to luncheon by Mrs Marie Joshua. She was the sister-in-law of Sir George Lewis, at whose New Year's party they had met. Mrs Joshua was a German who had lived her life in the belief that music was a German monopoly. But at Richter's urging she had heard Edward's Symphony, and had become such an instant convert that when she happened to see Edward himself at an art gallery private view she 'waded straight up to him' and asked him to lunch.[122] In a letter confirming the engagement, she had written on the day before the Jaeger Concert:

When they applaud to-morrow, will they all know what they applaud? Will they realize that it has taken Nature all the centuries from Purcell to now to express what is finest in their Nation with musical Genius which is yours. Dr. Richter with his unerring 'classicity' made me understand this soon, very soon, & I feel grateful. Grateful too I shall feel to hear that you & Lady Elgar can be with us next Sunday.[123]

[121] *Variations on a Personal Theme* (Hodder and Stoughton, 1922) p. 172.
[122] Recalled by Mrs Joshua's granddaughter, Mrs Winifred Christie (conversation with the writer, 1975).
[123] 23 Jan. 1910 (HWRO 705:445:1941).

At Mrs Joshua's house Edward basked in attention from Richter and Percy Pitt. After they left, 'E played the Andante from Vio. Concerto. Mrs. J. wept & was most devoted & sympathetic.'

He and Alice saw Beerbohm Tree's stage portrayal of Beethoven. The thought of lonely genius haunted him as he wrote to Alice Stuart-Wortley about attending Safonov's performance of the Symphony: ' . . . I will look at you from a distance; I shall sit *alone*, if I go, which is doubtful—I *am* really alone in this music!'[124] He did attend the Safonov concert, sitting apart from both Alices. On the morning after it Alice Elgar went back to Plas Gwyn, having found no suitable residence in London. Edward stayed on.

On 6 February 1910 the Stuart-Wortleys came to Schuster's house to dine, and Edward again played his *Andante* with Lady Speyer. Charles Stuart-Wortley was so impressed that he asked for the loan of the music to play it for himself. Edward made him a present of the short-score sketch. But it was to Alice Stuart-Wortley that he confessed his own despondency next day:

I promised to tell you of my London visit—I do not think it has been a success: it is too lonely & I cannot see how we are to 'take' a place big enough for us all: you shall hear of any plans but I think a decent obscurity in the country is all I can attain to—there is really no 'place' for me here as I do not conduct or in fact do anything & I am made to feel in many ways I am not wanted. I suppose I shall still pay an occasional visit to conduct a new thing—if any new things are ever finished.

I am not sure about that Andante & shall put it away for a long time before I decide its fate. I am glad you liked it.[125]

If this was aimed at drawing encouragement, it found its mark. Alice Stuart-Wortley's daughter wrote afterwards:

I remember Mother saying to me in consternation that Edward was threatening not to go on with the Violin Concerto, how dreadful his leaving it wd. be . . . I am certain that Mother, with the moral support of my father and myself, said to him, 'Edward, you *must* go on with it.'[126]

Late in the afternoon of the day he had written to Alice Stuart-Wortley came another idea for the Concerto—conceived, Edward said, 'in dejection':

It seemed important enough to inscribe the sketch with the hour of its invention: 'Feb 7 1910 6 30 p.m.'[127] Ultimately this idea was to find its place

---

124 Monday [31 Jan. 1910] (Elgar birthplace).
125 7 Feb. 1910 (HWRO 705:445:4043).
126 16 Mar. 1936 to Carice Elgar Blake (Elgar Birthplace).
127 Elgar Birthplace.

between the first-movement primary and secondary subjects: building a bridge between them, it would open a way to the entire first movement—and so to the Concerto itself.

Before it arrived at the second subject, this idea conceived first 'in dejection' went on to find a remarkable climax:

It developed the climax from *O Soft was the Song*:

and E - den re - con - quered was mine_____.

Opposite the new climax in the Concerto short-score sketch he wrote:

> This is going to be good!
> "When Love and Faith meet
>   There will be Light"
> Feby 1910. Queen Anne's Mansions.[128]

The faith outside himself that might stand up to a searching wind of self-doubt reminded him of the little spring flower he had known from childhood, called the windflower: he gave its name to the connecting theme that stood so in his music. But then the symbol went deeper.

Ever since coming to Christian-name terms with the Stuart-Wortleys, the coincidence of the two Alices had irritated him, and he had sought another name for the second Alice. Now he had it: always afterwards Alice Stuart-Wortley was to be called his 'Windflower'.[129] In later years he wrote a description under the title 'The Vernal Anemones: a beautiful native':

The pleasant legend which couples the tears of Venus with the anemone is not one that need try the receptive imagination very high, for in its simple, graceful beauty the flower may well have had a celestial origin.

The little group of anemones commonly called windflowers are happily named, too, for when the east wind rasps over the ground in March and April they merely turn their

---

[128] This inscription was written between 7 Feb. (when the idea had been invented 'in dejection') and 9 Feb. (when he left Queen Anne's Mansions).

[129] Alice Stuart-Wortley's daughter Clare recalled: 'I believe he ultimately called her Windflower after the themes—not the themes after her.' (Reminiscence at the Elgar Birthplace.)

backs and bow before the squall. They are buffeted and blown, as one may think almost to destruction; but their anchors hold, and the slender-looking stems bend but do not break. And when the rain clouds drive up the petals shut tight into a tiny tent, as country folk tell one, to shelter the little person inside.

Our native windflower, *Anemone nemorosa*, is often overlooked by gardeners . . . Who that has read it can forget Farrer's story of his finding the blue wood-anemone, which, like many another, he had pursued all his life as a will-o'-the-wisp? It was in Cornwall, and doubtingly he had plunged into the wood at twilight in search of the phantom flower . . .[130]

Cornwall was the Lyonesse of Tristan and Isolde, and the site of Arthur's castle at Tintagel. And it was the favourite retreat of Alice Stuart-Wortley, whose daughter Clare remembered: '[We] loved the place, and usually went there at Easter or Whitsuntide—sometimes in summer too. Mrs. Stuart-Wortley had long wished Sir Edward to see it, thinking the majestic cliffs and sea would appeal to him.'[131]

Instead he had to return to Plas Gwyn, where his own Alice was expecting him. There he picked up the last of his four 'Paraphrases of Eastern Folk-Songs' (the intervening two were never to be done). This was entitled *The River*. His words[132] might have found their inspiration when the Wye flooded the fields round Hereford at Christmas, for the verses bore the date 1909 and carried an explanatory quotation:

The river was in full flood and, had it remained so another twenty-four hours, would undoubtedly have overwhelmed the enemy; but it sank far below its normal level more rapidly than it had risen three days before.

Edward's *River* appeared first as a powerful mother:

> River, mother of fighting men,
> (Rustula!)
> Sternest barrier of our land,
> (Rustula!)
> From thy bosom we drew life:—
> Ancient, honoured, mighty, grand!—
> Rustula!

But then it showed deceit in an image of younger love:

> Like a girl before her lover,
> (Rustula!)
> How thou falteredst,—like a slave;—
> (Rustula!)

---

[130] *The Daily Telegraph*, 28 Apr. 1924. *Nemorosa* means 'of the wood'.

[131] MS reminiscence (Elgar Birthplace).

[132] A MS draft of the poem in Elgar's hand with many alterations is in possession of Miss Margaret Elgar.

> Sank and fainted, low and lower,
> When thy mission was to save.
> Coward, traitress, shameless!
>   Rustula!
>
> On thy narrowed, niggard strand,
>   (Rustula!)
> Despairing,—now the tyrant's hand
>   (Rustula!)
> Grips the last remnant of our land,—
> Wounded and alone I stand,
> Tricked, derided, impotent!
>   Rustula!

The verses bore a mysterious place-name 'LEYRISCH–TURASP': it was Edward's anagram of a Germanized Peter Rabbit—'PETRUS HAS[E] LYRIC'.

The music opened *Allegro con fuoco* in a descent of B minor steps—followed instantly by a powerful echo of the 'Windflower' climax in the Violin Concerto *Allegro*:

In the song that phrase introduced the opening image of the river as mother: it did not return when the image changed to the girl before her lover or at the final impotence. He finished the song on 19 February 1910 amid storms of wind and rain which that night blew down trees and tangled telephone wires. Two days later he took himself back to London to stay with Frank Schuster.

Alice came up on 24 February to continue flat-hunting: she stayed with Maud Warrender's sister Lady Margaret Levett. On the 28th she found a splendid flat at 58, New Cavendish Street: it had belonged to the late Librarian at the House of Lords. Edward could not see it that day, as he was dining with the Prince of Wales at Marlborough House in the evening. But he saw it next day, was impressed with its big library of books, and they took it for three months.

He stayed with Frank Schuster at The Hut, while Alice shot home to prepare Plas Gwyn for tenants once more and to collect things they would need at the flat. The work there exhausted her sixty-one-year-old energies. Yet when she returned to London five days later, it was she and the valet Jaulnay (with Carice in the background) who coped with endless details of inventories and servants while Edward contemplated the great library. He wrote to Schuster's elder sister Adela—who had invited him to visit her at Torquay:

I cannot travel just now as a period of unexampled domesticity has set in in this tiny flat but it cannot last for ever. Dear Frank has been here today & has blessed & approved of everything including the overwhelming library of books—magnificent really & unused. How ignorant we are & feel when we see what specialists get together . . .

I have no news of anything artistic—such things seem to be dropping out of my small weary life—I am like a sham book in a library, only a name & useless for wit or wisdom—one in a row of useless things & even my binding is shabby![133]

He plunged into London theatre—sometimes with Alice, more often with Frank or the Stuart-Wortleys. Maud Warrender took him to see Beecham's production of Strauss's *Elektra* at Covent Garden. Alice noted: 'E. went & enjoyed his evening. Much impressed with E[*lektra*] but kept on saying The pity of it! the pity of it . . .' One day he took Carice to Greenwich. Another day they all went to the zoo. They attended a cinematograph. They dined at the House of Commons with the Hereford MP Arkwright. Alice Stuart-Wortley came to tea on the day before she was to leave with her family for Tintagel. Afterwards Edward sent a note to follow her:

Dear Windflower (nemorosa)

. . . I hope you will have a not too cold journey & a serene time at the sea—which is mine by birth, adoption & heritage.

Love to you all
Yrs.
E.[134]

Then Frank Schuster proposed again the visit to his sister at Torquay—to be followed by a motor tour of Cornwall. Edward accepted. His own Alice, having fitted out another home for him, was left behind once more. Schuster was so overjoyed at the prospect of having Edward to himself that he planned their route wide of Tintagel, as Edward wrote to the 'Windflower' there:

Tomorrow F & I commence a motor tour 'round' Cornwall—avoiding Tintagel it seems but I am not *director* or dictator or Heaven knows what wd. happen to you or to Tintagel.[135]

But the lady at Tintagel made her own representations, as her daughter remembered:

It was Mrs Wortley who after much persuasion induced Sir Edward and their mutual great friend Mr. Schuster to visit Tintagel, saying that if only Sir Edward once saw it, he would write something so wonderful! Sir Edward, highly amused, used to say that she would have to be responsible for anything, however dreadful, that he might compose as a result of the visit![136]

[133] 11 Mar. 1910 (HWRO 705:445:6840).
[134] Monday [21 Mar. 1910] (HWRO 705:445:4106).
[135] 21 Mar. 1910 (HWRO 705:445:4121).   [136] MS reminiscence.

Sunday 3 April 1910 found the two travellers driving with Schuster's chauffeur over the moors through snow. But they arrived safely at Tintagel in the afternoon, as Clare Stuart-Wortley wrote:

We all walked down to the sea in the 'Cove', below the Castle ruins; and saw it all in very bad weather, at its most stern and forbidding; we three Wortleys loudly bewailing that Tintagel should so badly greet so great a guest. Sir Edward said very little, but did not complain.[137]

Next day Charles Stuart-Wortley left for London. The remaining four went to the little harbour village of Boscastle for the afternoon. Clare recalled:

On the drive home some evening sunshine was enjoyed; and the party walked (a regular custom with Mrs. Wortley) the steep parts of the road down into, and up out of, the Rocky Valley. It was then . . . that the austere yet lyrical beauty of the Tintagel country really showed itself to Sir Edward at last.[138]

On Tuesday the travellers left for Land's End—where Edward 'walked out as far as possible & waved to Pippa' in New York.[139] He sent a postcard of Land's End to Clare Stuart-Wortley with the message: 'I restrained myself from jumping in with great difficulty—a very fine coast,' and added two climactic

chords from the 'Windflower' theme

After returning to London he wrote to Adela Schuster of the whole motor tour in Cornwall: 'Oh! it was a confusion of interest & makes a background for present sorrow. I find many business things here which are wearying & headaching things & no music possible.'[141]

He had returned to the flat in New Cavendish Street on 11 April. His reaction had been immediate in the music he played then. Dora Penny was there—still attractive, still unmarried at thirty-five, and on the eve of a trip to Italy. Instead of showing her the Concerto, he played a Slow movement for his Second Symphony—a long treading 4/4, the sound of a funeral march. Dora wrote: 'The whole movement seemed to me to tell a tragic story of anxiety and sorrow, fears and hopes. [It] stayed in my head all the way to Italy.'[142]

After dinner that evening he went with Alice to a reception for Richard Strauss given by Sir Edgar Speyer. Sir Edgar approached Edward on behalf of Queen's Hall: they were planning another 'London Musical Festival' for May 1911, and hoped to bring off at the same time an International Musical

---

[137] Diary of the motor tour (Elgar Birthplace).
[138] MS reminiscence.   [139] Diary of the motor tour.
[140] Postmarked from Penzance, 6 Apr. 1910 (Elgar Birthplace).
[141] 14 Apr. 1910 (HWRO 705:445:6841).
[142] 'Some of Elgar's Greater Works in the Making' (TS in possession of the writer).

Congress to draw leading musicians from all across Europe. Could Edward give them a new major work for première then? If the Philharmonic had the Violin Concerto, what about the Second Symphony?

Two days later Edward returned from a matinée with Frank to find

Sir E. Speyer's letter with cheque—fruit of a speculation he made for E. E. felt of course he cd. not accept it & returned it next day with a *beautiful* letter.

April 14 . . . E. with Frank went to see Sir E. Speyer at 6 & had financial talk. Sir E. Speyer sd. E. was the 1st who ever refused fruits of a speculation. Very human kind touching episode—Dear of Sir Edgar & of course perfect of Edoo—

With the Symphony negotiation in the background there could be no thought of any other course. But in the end he settled with Sir Edgar for the new Symphony's première at the London Festival in a year's time. He said to the Stuart-Wortleys when they returned from Cornwall: 'I shall dedicate the Symphony to Edward the Seventh so that dear kind man will have my best music.'[143]

On 20 April he wrote to Alice Stuart-Wortley: 'I am now ablaze with work & *writing hard*; you *should* come & see (& hear it!).'[144] Two days later, when Claude Phillips came to tea, 'E. played Concerto & Symphony no. 2.' But on the day after that Alice Stuart-Wortley was at New Cavendish Street, and it was the Concerto First movement which came uppermost again:

April 24 . . . E. very thrilled with his Concerto & working at it.

April 26. E. very intent writing.

Through these days, everything seemed to depend on the proximity of Alice Stuart-Wortley. He wrote to her on the morning of 27 April about attending an afternoon concert at Queen's Hall and a Private View at the Royal Academy:

My dear Windflower

I fear the cold wind has been too much for you.

Please come to-day & tea here after—or tea only if you cannot come to the Concert . . .

I have *no* tickets for the Private View RA only for the dinner—do let me go with you to the P.View if you can manage. I will wear a nice hat. Tell me this afternoon.

You *must* come either to the concert and tea—or the latter only . . .

I have been working hard at the Windflower themes but all stands still until you come & approve!

Love

Edward[145]

---

[143] Clare Stuart-Wortley recalled the remark more clearly than its occasion. In later years she thought it might have been made on 3 Mar. 1911 before they all attended the theatre. But its tone, together with Elgar's later remarks about having planned the Symphony and especially the Slow movement before King Edward's death, make it almost certain to have been made during the lifetime of the King. The moment of renewed interest in the Symphony right after Tintagel seems to me the likely time for this remark.

[144] HWRO 705:445:7816.   [145] HWRO 705:445:7853.

His own presence was requested in other directions. That evening he attended a dinner in Carleton House Terrace given by Sir Gilbert Parker to entertain the High Commissioner for Australia. The guests included A. J. Balfour, the Dukes of Argyll and Marlborough, Lords Cromer and Curzon and Charles Beresford, Austen Chamberlain, Sir George Lewis, Sir Edward Poynter PRA, Sir Hubert von Herkomer, and Edmund Gosse. Then the three mornings 28–30 April were booked for him to go to the Aeolian Hall to edit the mechanical rolls made of the entire Symphony for performance on the 'Orchestrelle', a player-piano that reproduced orchestral music in the home. A feature was that the composer himself could determine the playing speeds.

After the first morning with the 'Orchestrelle', he was committed to his own Alice for a tea-party in New Cavendish Street. In the midst of it he took himself off to write to the other Alice—who had now arranged a ticket for him to go with her to the Royal Academy Private View:

My dear Windflower

I am so tired & there are people here & I am very *sad* indeed—a most interesting dinner last night at the Parkers'—you will see it in the Mg.Post. I *do* like to be amongst brains & my wretched music takes me amongst them so seldom: so *you* must come tomorrow to the Orchestrelle Bond St.—any time 10.30—to 1.0.   I cd. give you lunch somewhere if you like & R.A. after . . .

It is so dreary to-day & the tunes stick & are not Windflowerish—at present

Your

E.[146]

So the second 'Orchestrelle' day was spent together. His own Alice noted: 'E. lunched with S. Wortleys & then to private view at Academy. A. went too but cd. not find him—mis.' On the final 'Orchestrelle' morning, 'A. & C. allowed to go & hear.' But that night Edward attended the Royal Academy Dinner without her: he made the acquaintance of Rudyard Kipling.

Next morning, 1 May, he had a headache. While Alice and Carice fulfilled an engagement made for all three to lunch with Mrs Joshua, Edward met Alice Stuart-Wortley to give her an ink sketch of the 'feminine' second subject of the Concerto. This had become a second 'Windflower' theme in the first movement which now was all before him.

The Concerto opened with the 'masculine' primary figure

an answering descent

[146] HWRO 705:445:7862.

and a variant: together they made a first subject group. Then came the first 'Windflower' idea, written originally 'in dejection' but leading its music to the extraordinary climax 'where Love and Faith meet'. And this in turn led to the Concerto's 'feminine' second subject, the other 'Windflower' theme:

Its music blossomed to a variant which seemed to distil the earlier passionate invitation in *The Torch*

to a sequence:

These sequences were met by the 'masculine' descent with a resolute energy

that poured its strength into the rest of this exposition.

At a return of the original primary figure *Come I^a*, the solo violin made its first entry in the Concerto with the answering phrase invented at Careggi:

Following old Concerto models, the entry of the solo instrument inaugurated a second exposition. It was not a repeat but an extension, where the previous material of orchestra 'fact' would be translated by the violin to solo 'fantasy'. Orchestra and solo were to retain these roles throughout the Concerto.

As violin and orchestra conversed, there emerged an ease of exposition that was new in an Elgarian *Allegro*. After the entry of the 'feminine' second subject now, the violin turned minor rise to exquisite major triad:

The idea had been in the original Concerto sketch of 1905: now it found its place. The second exposition culminated in a solo that reminisced over the original 'feminine' form with new softness.

In development, the violin took the 'masculine' figure to more distant places:

And when that music was taken up by the orchestra, the violin over it began a long *animando* of semiquavers, octaves, double stops in every virtuosity—culminating in a *fortissimo* bar of quadruple-stopped invitation. The entire development thus far was of 'masculine' primary material. The orchestra thundered the first 'Windflower' *con passione*, driving up the biggest climax of the whole *Allegro*.

To it at last rose the 'feminine' second subject, but with a harshness unlike any other appearance of this music through the entire Concerto—*Maestoso* in the minor, rising to *fortissimo*, and raked with 'masculine' descents:

In response to the great 'masculine' invitation, the 'feminine' presence had turned its back and refused. The 'masculine' descent hurled itself *strepitoso* round the triad

like Empedocles at the rim of the volcano or Judas overwhelmed with his despair. Down it plunged and down again through huge orchestral turbulence—while the violin stood by in silence, as the hero of reflection might quietly regard waves of passion raised in the apostles of his fantasy by a religion of his own love.

Thus the storm at the centre of the Violin Concerto first movement showed a strange and tragic meeting. As soon as the 'masculine' fantasies had developed to clear inviting, 'the perfect feminine counterpart and complement' majestically turned its back. And this denial at the very climax of development was final. When the storm abated, the primary 'masculine' figure stood alone to open the recapitulation.

Now the violin's answering phrase traced a sequence of descending steps:

At last the second subject 'feminine' rise itself turned downward

(an eventuality which had been suggested in the exposition). On the verso of the page containing this music in short-score sketch Edward wrote 'Anemone nemorosa', and then quoted three lines from 'The Sovereign Poet' by William Watson:

> The undelivered tidings in his breast
>   Suffer him not to rest.
> He sees afar the immemorable throng.[147]

In a coda the 'masculine' primary emerged in what Edward described as 'an "affliction" of the opening phrase'.[148]

So this opening *Allegro*—to be recognized universally as one of its composer's most intimate statements—could relate a private history. For as Edward had developed his 'feminine' second subject, the ghost of Helen

[147] p. 25ᵛ (Elgar Birthplace).   [148] 30 May 1910 to W. H. Reed (*Letters*, p. 196).

Weaver seemed to rise again. She had broken his engagement and his love in the days of his creative beginnings. Then he had journeyed to remote landscapes alone. And three weeks ago, on returning from Alice Stuart-Wortley at Tintagel, his impulse before his own Alice and Dora Penny was to go to the piano and play a funeral march. If the Concerto's opening *Allegro* showed the past, it was to lead to the *Andante* which found its present climax in the mothering of the older Alice.

Yet it was just here that the artist in Edward triumphed. The Concerto did not advertise his private experience by making programme music. Instead it used the experience to drive his music to a powerful abstract shape. This use of private experience might not reach any full consciousness; and there would be no need for an audience to know Edward, to read his particular history, or to recognize a love theme. It was only clear that desperation played as strong a role in this *Allegro* as in the opening *Allegro* of the Symphony, and that desperation was borne of a deep desire thwarted—a desire that remained *Nobilmente*. The interaction of melodic, harmonic, and rhythmic figures in the Concerto told all these things, and would tell them as clearly if the figures were merely 'A' and 'B' or '1' and '2'.

Behind it all stood a man strong enough to seek out the consistency that resolved his private strengths and weaknesses into motives, wise and generous enough to turn those motives to music whose very abstraction from the personal specific could touch and console and illuminate those less strong than himself—those who suffered but could not formulate. It was the *Nobilmente* that made all the greatest art, and as long as he kept it his art would find its way.

In the midst of the Concerto came a small death and a large one. On 3 May he had word from Plas Gwyn that Peter Rabbit was no more. He wrote to another friend made in London, Mrs Sidney Colvin—a sympathetic woman of seventy, whom Robert Louis Stevenson had regularly addressed as 'Madonna' and 'Dearest Mother': 'It is terrible to think how many human beings could be spared out of our little life's circle so much easier than my confidant & adviser Pietro d'Alba.'[149]

Three days later there was a rumour that King Edward VII was mortally ill. Alice wrote:

May 6 . . . Terrible news only too true—So difficult to believe the King was really dangerously ill—So terribly sudden. 1000s round Buckingham Palace. E. walked round. Maud Warrender to tea. She thought the illness was very dangerous—Oh! our own King . . . E. A. & C. to Covent Garden for 2nd Act of Siegfried—They told E. the worst news might come any moment & the performance be stopped. Ghastly feeling to watch for it. At 1.30 A. heard the dreadful words called out in S[tree]t and again later—

May 7.　Intense feeling of sadness. E. feeling desolate. A. & C. to find dress for C.

---

[149] 3 May 1910 (HWRO 705:445:1669).

Nos. & nos. of people in black already—Papers with beautiful articles—& people saying King Edward VII's loss the greatest calamity wh. cd. befall the country.

Edward wrote to Lord Howe offering to finish a funeral march: it might have been the Slow movement for the new Symphony which he had played a month ago to Dora Penny. But the regretful answer was that there would be no time for the band to rehearse new music for the funeral, even supposing it could be finished quickly.[150] Whatever reflections the King's death and the passing of his era might have for Edward's music then, they would have to wait for future exploring. On 8 May (his twenty-first wedding anniversary) he wrote to Frank Schuster in a confusion of emotions:

My dear Frank
These times are too cruel & gloomy—it is awful to be here now—that dear sweet-tempered King-Man was always so 'pleasant' to me.
I have the Concerto well in hand & have played(?) it thro' on the P.F. & it's *good*! awfully emotional! too emotional but I love it: 1st movement finished & the IIIrd well on—these *are* times for composition.
. . . Alice Wortley came to tea today & had a dose of Concerto which beseemingly she liketh well. Also Lady M[aud] yesterday to a similar meal & *corrective*. We are dismally gay—walk like ghosts & eat like ghouls. Oh! it is terribly sad.

Your
Edwd[151]

On 12 May he tried the first movement with Lady Speyer. Some of the solo passages were not yet fixed, and he needed an expert player to help him toward their final embodiment. But Lady Speyer was not now the right person: her large house was always full of guests interested in her husband's Queen's Hall syndicate, and her own very feminine beauty might no longer be such a clear aid to working out the implications of this music.

Then suddenly Edward saw the person he wanted. It was Billy Reed, just graduating to leadership of the London Symphony Orchestra and physically almost a younger image of the lost Jaeger. The day he ran into Reed was in fact nearly the anniversary of Jaeger's death. Here was another small, self-effacing man, unquestioningly devoted to his music. Such a man posed no sort of threat—in contrast to some of the women who attracted Edward.

The young man himself was staggered. His memory chronicled the experience minutely:

Naturally very much flattered that he should remember my existence, what was my astonishment when, meeting him one day in Regent Street, he stopped me to know whether I had any spare time, and if so could I come up to see him at a flat in New Cavendish Street where he was then living. He was sketching out something for the fiddle, and wanted to settle, in his own mind, some question of bowing and certain intricacies in the passage-work. As can easily be imagined, I leapt at his suggestion.
. . . When I arrived at the New Cavendish Street flat on my first visit, [Saturday 28

---

[150] 11 May 1910 (HWRO 705:445:3421).   [151] HWRO 705:445:7019.

May] about ten o'clock, I found E. striding about with a lot of loose sheets of music paper, arranging them in different parts of the room. Some were already pinned on the backs of chairs, or stuck up on the mantelpiece ready for me to play.

The idea was clearly to get a view of the whole pattern in the sequence of sheets placed round the room—with the chance to change, add to, or take away from it.

After my introduction to Lady Elgar, we started work without losing a moment. What we played was a sketchy version of the Violin Concerto. He had got the main ideas written out, and, as he put it, 'japed them up' to make a coherent piece . . .

   This morning's work was a unique experience to me: it gave me a very intimate view of Elgar as a composer . . . He was always very diffident; but he knew when anything he had written gave him pleasure. There was no false modesty about his joy in hearing the solo violin boldly entering in the first movement (Fig. 9) with the *concluding* half of the principal subject instead of with the beginning, as if answering a question instead of stating a fact.[152]

The novelty of this idea so pleased his fancy that I had to play that unusual opening many times with him, he thundering out the first two bars on the piano as if issuing a challenge to the solo violin to come in and see what *he* could make for it.[153]

Afterwards Edward wrote to Reed:

I do not know how to thank you sufficiently for your kind help & most beautiful playing on Saturday; I am anxious to send the first movement to the printers & am only waiting for the final 'flourish' . . . At 44 the passage is built on an 'affliction' of the opening phrase. I think it is possible to make a good, festive noise but I am not sure about the bowing: whether it wd be best detached or slurred (*dug out*) in twos. Any wisdom you may have to spare will be thankfully received by your very grateful friend

Edward Elgar.[154]

Reed came again the following day. And the day after that, 1 June, the first movement went to Novellos in the violin and piano score. Orchestration was yet to come.

   The short lease of the flat in New Cavendish Street was at an end. On 2 June—his fifty-third birthday—Alice packed at the flat while Edward journeyed to Lincoln to rehearse *Gerontius*. He went alone again for the performance a week later, but was joined there by Alice Stuart-Wortley. From Lincoln he went to The Hut—where he had already sent a warning to Schuster: 'I want to *end* that Concerto but do not see my way very clearly to the end—so you had best invite its stepmother to the Hut too. Do.'[155]

---

[152] *Elgar as I Knew Him*, pp. 22–24.
[153] 'Elgar and his Violin Concerto', in *The Listener*, 10 Nov. 1937, p. 1042.
[154] Monday 30 June 1910 (*Letters*, p. 196). Details of passages at issue in the first-movement sessions were given by Reed in 'Elgar's Violin Concerto', *Music and Letters*, vol.XVI no.1 (Jan. 1935), pp. 31–2.
[155] Friday [27 May 1910] (HWRO 705:445:7055).

Alice Stuart-Wortley was not the only visitor during those June days by the Thames. W. H. Reed remembered:

It was not very long before I received an urgent summons to go there. The slow movement and the first movement of the concerto were almost finished; and the Coda was ready. Could I, therefore, come and play them with him? I went on the following Sunday [12 June].

I can see it now as it looked that spring morning when I first arrived. It was a sweet riverside house . . . Across the lawn, and almost screened by trees, was the studio ['the Orchard Room'], away from the house and approached by stones placed in the grass about a pace apart. It had a rather barn-like exterior, but inside, it was a home for most of the curios, Chinese ornaments, rare and extraordinary objects which Frankie had collected and brought home from all parts of the world. I remember particularly well a stuffed lizard, or a member of that genus—a fine specimen, though perhaps rather large for a lizard. It was suspended from the ceiling in such a way as to be swayed by every gust of wind coming in through door or window, and I always felt that it swung round to have a good look at us when we played the slow movement . . .[156]

Several slow-movement solo passages received special attention. One was the violin run up through softest high notes to bring the entry of the second subject:

We tried the last few notes in harmonics, the sound of which pleased him so much that that method of interpretation was instantly adopted, and he at once wrote 'armonico' over it.

We had great fun over the 'ad lib' passage four bars before **51** [at the end of slow movement exposition] with the sudden leap of a twelfth up the G string. The first time we tried it this way instead of going to a more reasonable position on the D string, the effect so electrified him that I remember he called out 'good for you' when I landed safely on that E with a real explosive sforzando.

Much work had also to be done with the passages at **52** and **54** shaping these arabesques, and deciding whether they should be demisemiquavers or whether the groups should be written as broken triplets . . .[157]

Through the days at The Hut, Edward seemed to move inside an invisible sphere hardly penetrated even by his host or Alice Stuart-Wortley. W. H. Reed recalled:

They all seemed a very happy party, each going his own way and meeting on the raised verandah for meals. Sir Edward spent most of his time at work in the studio, where the others wisely left him alone unless invited to come and hear some of the concerto if I happened to be staying there.

When we were tired of playing, or if Sir Edward wanted to go out in the air for a change, the fiddle was laid in its case and we went off together, strolling about the river-bank, watching the small fish in the water and enjoying the quiet beauty of the

---

[156] *Elgar as I Knew Him*, pp. 26–7.
[157] 'Elgar's Violin Concerto', *Music and Letters*, vol.XVI, no. 1, p. 34.

place. During these walks he took me more and more into his confidence, until to my great joy I found myself gradually becoming one of his intimate friends.[158]

During that visit to Schuster's country house the main lines of the Concerto Finale emerged. If the first movement *Allegro* had evoked a private past and the *Andante* hinted at a central experience in the present, this Finale might now look at a future. It was cast as still another sonata structure in 4/4 metre—suggesting the third stage in an order of events. The Finale (like the Finale of the Symphony) returned to the special form of its first-movement *Allegro*: here it was an *Allegro* of double exposition.

The F♯ and G which had opened the first movement now oscillated rapidly through the orchestra, raising a chill wind. The violin swirled in continuous turns up the B minor triad. The orchestra sounded upward fourths, but in sequences that descended. The violin interrupted with insistent gestures upward. Against them came a new orchestral descent that was far steeper

On its appearance, the violin's aspiration turned rapidly downward. That was the Finale introduction.

Continuing violin semiquavers led to the primary Finale theme—a *vivace* whose triple shape reversed the first movement's 'masculine' descent

The sequence of this rising primary theme still descended, and the vigour of its beginning slipped away. Soon the plunging figure from the Finale introduction made a new sequence:

The plunging sequence appeared between this movement's first and second subjects—just where the 'Windflower' theme written 'in dejection' had appeared in the first *Allegro*. But where the 'Windflower' idea had gone forward in syncopation, the steady crotchets of the new sequence sounded resignation. The violin drooped through a little repeating figure

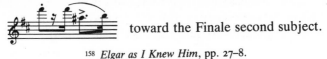

toward the Finale second subject.

[158] *Elgar as I Knew Him*, pp. 27–8.

The second subject was a ghost of the 'Windflower'

The violin's aspiring seventh only emphasized the descending steps in the orchestra below. However the violin embroidered it, the result was the same; and at the end of this first Finale exposition the orchestra had taken the violin's drooping figure for its own.

In the second exposition the violin pursued its fantasies with more frenetic intent—often in echoes of first-movement passage work. Through another restless introduction over descending steps, the violin semiquavers ran up and down, up and down, until they drew savage imitation from the orchestra—a rare instance of irony in Edward's music. The violin sounded the Finale primary subject *con forza* in triple stops—all by itself. Repeated by the orchestra, the vigour fell away through one section after another of the upper strings until the solo violin was left alone once more. When it paused, the whole orchestra sounded the plunging sequence that moved toward the second-subject echo of the 'Windflower'. Through the second exposition, a gulf had opened between solo and ensemble as the horizon darkened.

Returning to Plas Gwyn, he wrote to Alice Stuart-Wortley on 19 June: ' . . . it took me a long time to "find myself" here but the work goes on and the pathetic portion is really fixed.'[159] This 'pathetic portion' of the Finale was the only part of the Concerto to be 'fixed' in the presence of his own Alice since the climax of the *Andante*. And the new section of the Finale was in fact the same *Andante* music recalled. It replaced Finale development by harking back to the *Andante*'s *Nobilmente* figure. From there it went forward again to the great cry

which in the *Andante* had been instantly hushed. Now that cry sounded again and again, as if by sheer force of repeating it could strike through the veil. But a fury of exposition semiquavers swept dismissively through the orchestral strings: above the turbulence the solo violin sounded the Finale first subject alone. Solo and tutti met only in the plunging sequence. When the violin essayed one echo of its old aspiring fantasy, it was instantly overborne by the orchestra sounding the plunging sequence once more.

What should have been Finale recapitulation went backward again. The music returned to first-movement themes, weaving them into a Cadenza of solo fantasy. Yet as this Cadenza would play a large role in the Concerto's structure, it could not leave out the orchestra. Edward devised the Cadenza 'on a novel plan I think—*accompanied* very softly by a few inst[rument]s.'[160] There would

[159] HWRO 705:445:7854.   [160] 29 June 1910 to Frank Schuster (HWRO 705:445:6660).

be horns, timpani, and strings *pizzicato tremolando*—'to be "thrummed" with the soft part of three or four fingers across the strings.'[161] '. . . *Rustle* 3 or even 4 fingers *flatly* (not hooked) over the strings & let the sound be sustained, soft & harmonious . . .'[162] 'The sound of distant Aeolian harp flutters under and over the solo.'[163] These phases of the Concerto were the only parts of the music to be finished at Plas Gwyn, where the summer air still sang through his own Aeolian harp in the open study window.

Against that sound, as he wrote, the solo violin 'sadly *thinks over* the 1st movemt.'[164] It began *Lento*, with the old 'masculine' figure in the thrumming orchestra, answered by the violin with the first 'Windflower' theme. Wandering arpeggios reached toward the topmost registers of hearing, returned to echo the first 'Windflower' and then the second. '. . . I have . . . brought in the real inspired themes from the 1st. movemt.', he wrote to Alice Stuart-Wortley, '& the music sings of memories & hope.'[165] Yet the hope was with the memories, in the past. As the whole final prospect filled with nostalgia, it held him in its thrall. He wrote again to Alice Stuart-Wortley on 23 June: '. . . I am appalled at the last movement & cannot get on:—it is growing so large—too large I fear & I have headaches (here); Mr Reed comes to us next *Thursday* to play it through . . . & we shall judge the finale & condemn it, if you like.'[166]

Reed recalled his visit to Plas Gwyn on 30 June 1910 as vividly as the previous meetings:

On arriving at the house I found the studio in the state I had become accustomed to at the London flat: music-paper all over the room, scraps at any vantage point, many different versions of the same thing with the different bowings to be tried for each.

At once we plunged into it. Passages were tried in different ways: the notes were regrouped or the phrasing altered. The Cadenza was in pieces; but soon the parts took shape and were knit together to become an integral part of the concerto. Ivor Atkins . . . came and played the piano accompaniment, while Sir Edward strode about the room, listening and rubbing his hands excitedly.[167]

After the appearance of the second 'Windflower', the thrumming orchestra dropped away altogether as the violin traced slow figures of descent—ending at last in the gentlest little rise, as if just to look up from some far distance of reminiscing over a most private past. Reed remembered:

The lento between **105** and **106** nearly moved him to tears as he repeated it again and yet again, dwelling on certain notes and marking them 'tenuto', 'espress', 'animato', or 'molto accel.' as he realised step by step exactly what he sought to express.[168]

To end this Cadenza recapitulation, the orchestral strings would take their bows again to sound *ppp* the *Andante*'s *Nobilmente*: to it came the 'feminine'

[161] Note in the printed full score, p. 88.
[162] 15 Mar. 1911 to Nicholas Kilburn.
[163] 18 Sept. 1910 to Ernest Newman.
[164] 29 June 1910 to Frank Schuster (HWRO 705:445:6660).
[165] 16 June 1910 (HWRO 705:445:7851).
[166] HWRO 705:445:7854.   [167] *Elgar as I Knew Him*, pp. 28–9.
[168] 'Elgar's Violin Concerto', *Music and Letters* vol.XVI, no. 1, p. 35.

and 'masculine' figures of the first movement. A coda brought to that *Nobilmente* the second and first subjects of the Finale.

So the symmetry was complete. The private past evoked in the Concerto first movement was also the private future evoked in the Finale. As the masculine hero stood alone at the end of the first *Allegro*, so he might stand alone again in the future. For his own Alice was past sixty now, and the end might find Edward as lonely in old age as the young man who had dreamt of Helen Weaver in Leipzig.

Here once again the 'friends pictured within' had led Edward to his music. But now half the friends were dead. In the Concerto first movement, the living relationship with Alice Stuart-Wortley had been haunted by the ghost of Helen Weaver. In the second movement behind his own Alice stood the ghost of his mother. In the third, round the generous friendship of Billy Reed hovered the spirit of Jaeger. So the Concerto's heroism was older and newer than the lonely heroism of the Symphony. As he finished the Concerto he wrote to Frank Schuster:

The world has changed a little since I saw you I think—it is difficult to say how but it's either larger or smaller or something.

This Concerto is *full* of romantic feeling—I should have been a philanthropist if I had been a rich man—I *know* the feeling is human & right—vainglory![169]

On 1 July Edward showed the Concerto short score to Fritz Kreisler in London, and the great violinist was delighted: 'he said at one passage "I will shake Queen's Hall!".' The Queen's Hall Orchestra, the London Symphony, Landon Ronald's New Symphony Orchestra, and the Philharmonic Society were all fighting hard for the Concerto première and offering high fees. The London Symphony promised £125 for the first performance and 50 guineas for a second performance immediately afterwards. But the Philharmonic won the day by offering the composer £100 for each of two performances with Kreisler on 10 and 30 November.

Orchestral scoring was interrupted by conducting engagements at a festival in York. Alice went with him, and Alice Stuart-Wortley was there. Afterwards she came home with the Elgars to Plas Gwyn. W. H. Reed joined them there on 26 July to play through the entire Concerto. So for a day and a night Plas Gwyn held the three living souls who stood behind the three movements.

For the ghosts who stood behind these living presences, he found a passage from Lesage's *Gil Blas*, where the student deciphers the inscription on a poet's tomb.[170] And though the Concerto was dedicated to Kreisler, he set the tomb inscription opposite the opening page of music (the place in the oratorios where he had caused to be printed 'A.M.D.G.') with the poet's name replaced by five dots:

[169] 29 June 1910 (HWRO 705:445:6660).
[170] Quoted at the beginning of William Ernest Henley's *Echoes*, where Elgar probably found it.

'AQUÍ ESTÁ ENCERRADA EL ALMA DE . . . . .'
('Herein is enshrined the soul of . . . . .').

Those dots might stand for numbers of names, or for one. At one moment he wrote to ask Antonio de Navarro: 'If I want it to refer to the soul of a feminine shd. it be—de la . . .?'[171] But then he left it undefined, as he wrote to Nicholas Kilburn:

Here, or more emphatically *In here* is enshrined or (simply) enclosed—*buried* is perhaps too definite—*the soul of* . . .? the final 'de' leaves it indefinite as to sex or rather gender
  Now guess.[172]

That night, after the others had retired, he sat up late talking to Reed. He reminisced over early struggles at home and in London. He talked of cycling recently with Sinclair, who had begged for his company and then outdistanced Edward up every hill to each new prospect, waiting only for the older man to catch him up before plunging on again. At the age of fifty-three, one could not command the energies of earlier years, and Edward's cycling in recent summers had declined: when Sinclair and his assistant Percy Hull came in to hear the Concerto on 14 July, Edward had told them that he and Alice had decided to give up Plas Gwyn and Hereford altogether.

He also told Reed that evening how Alice helped his music:

'You know, Billy'—I was Billy by this time—'my wife is a wonderful woman. I play phrases and tunes to her because she always likes to see what progress I have been making. Well, she nods her head and says nothing, or just "Oh, Edward!"—but I know whether she approves or not, and I always feel that there is something wrong with it if she doesn't. She never expresses her disapproval, as she feels she is not sufficiently competent to judge of the workings of the musical mind; but, a few nights before you came . . . I played some of the music I had written that day, and she nodded her head appreciatively, except over one passage, at which she sat up, rather grimly, I thought. However, I went to bed leaving it as it was; but I got up as soon as it was light and went down to look over what I had written. I found it as I had left it, except that there was a little piece of paper, pinned over the offending bars, on which was written, "All of it is beautiful and just right, except this ending. Don't you think, dear Edward, that this end is just a little . . .?" Well, Billy, I scrapped that end. Not a word was ever said about it; but I rewrote it; and as I heard no more I knew that it was approved.'[173]

The Concerto orchestration was finished early in August. Then Edward and Alice paid a little round of reminiscent visits in the summer countryside of Worcestershire—to the Whinfields at Severn Grange, to one of Mrs Hyde's parties in Foregate Street, to the wonderful Cotswold house of the de Navarros at Broadway. Mme de Navarro wrote:

---

[171] 9 Oct. 1910 (de Navarro family).   [172] 5 Nov. 1910.
[173] *Elgar as I Knew Him*, pp. 22–3. The passage referred to was probably the violin figures at the very end of the Finale, which were altered to their present form when the music was already in proof. Reed quotes the original version in 'Elgar's Violin Concerto', *Music and Letters* vol. xvi, no. 1, p. 36.

He enjoyed the meals out of doors, especially the dinners in the old courtyard, seeing the stars slowly coming out, the moon rising from behind the hills . . . One day there happened to be many callers, and it was comic to see him bolting up the hills to escape. He came to Mass each day in our chapel, and after breakfast would spend the morning by the bathing pool with a football which he tried to kick across it. When I taunted him with 'You can't do it,' he said 'I will. I bet a Sonata I'll get it across.' But he never did. Even so, the Sonata was not forthcoming. He knew all about newts, water-boatmen and the like, and played about the pond like a boy.[174]

Later in the month he worked at an anthem setting of Psalm 48. Its music echoed the Concerto too closely. The opening instrumental motive

followed the outline of the Concerto's first-movement primary subject. Another figure, prominently repeated near the beginning and end of the anthem, traced again the Concerto's Finale primary subject note for note:

Edward put the anthem on one side, and it waited eighteen months before reaching the publishers.

Kreisler was to be at the Gloucester Festival to play Bach. So Edward arranged to rehearse the Concerto with him there and to give a private violin and piano performance at the Cookery School, which Alice had taken for their house party that year. The rehearsal took place on Saturday 3 September. Mme de Navarro motored over to listen to the rehearsal in an upstairs room of the house. The only other listener was Robin Legge, who had succeeded Joseph Bennett as chief critic of *The Daily Telegraph*. He wrote:

We duly seated ourselves upon the canonical sofa with a box of cigarettes between us. Elgar took his appointed place upon one of the old-fashioned round piano stools which swing round according to the wish of the occupant. Fritz Kreisler stood, splendidly dignified as always, fiddle in hand. The corrected proofs were in their appointed place on the desk of the piano, and the atmosphere was of the most genial order . . . though interruptions were more or less frequent.

At one moment Kreisler would edge Elgar off the piano stool while he himself took the composer's place, and, with a deprecatory remark: 'Na, na, Edward,' and an appeal to us of the audience, would attempt to show the composer his idea of the right and proper manner in which the music should go. Of course, Elgar would lift him from the piano stool laughingly and say: 'Na, na, Fritz, this is how *I* want it played.'[175]

Downstairs Dora Penny had arrived with her hostess for the Festival, Miss Amy Danks, to help Alice prepare a large tea party:

[174] *A Few More Memories*, pp. 205, 206.   [175] Ibid., p. 208.

Unusual noises were coming from somewhere upstairs and the Lady told me that Herr Kreisler and E. E. were going through the Violin Concerto behind locked doors! Having done all I could for the moment downstairs I went to look for Amy.

'Look here, I'm not going to miss all this. What about you?'

So we both slipped away upstairs and sat on the top step outside the door. It *was* interesting! Kreisler was trying bit after bit—not playing it properly, of course—but he was getting the composer's meaning and ideas. They did not speak one another's language very well and it was difficult at times. Kreisler became worried and anxious now and again and then at last he understood and raced off with it joyously. Loud applause from the piano. At last we tore ourselves away. Tea was coming in and the Lady might want me . . .

Tea was finished at last and soon afterwards Amy and I left.

'Come to-morrow evening at about 9 o'clock, dear Dora, will you? and bring Miss Danks.' Then the Lady added mysteriously: 'There's going to be something *quite interesting!*'[176]

Schuster had made the suggestion that Billy Reed, who was at Gloucester to lead the Festival orchestra, should be given the chance of a private performance before Kreisler's, and so it was arranged for the Sunday evening. Twice they went through the Concerto to prepare it, as Reed recalled:

I must confess I had some inward qualms. I knew every note of the concerto, and exactly how he liked it played: every nuance, every shade of expression; yet I felt a little overwhelmed at being asked to play the solo part at what would actually be the very first performance before an audience. It was one of those facts that you cannot annihilate by just calling it private.

When the time arrived I went over to the house and found the guests assembled. Nearly all the prominent musicians engaged at the Festival were there: the three Festival conductors, Sinclair, Atkins, and Brewer; the past organists of Gloucester Cathedral, Harford Lloyd and Lee Williams (known as the 'Father of the Three Choirs'); some of the musical critics, and the house-party. The room was full; and all the lights were turned out except for some device arranged by Frank Schuster for lighting the piano and the violin stand.[177]

Dora Penny and her friend arrived:

. . . we wormed our way to where the Lady was dispensing coffee.

'Dear Dora,' she said, 'you and Miss Danks must please pack yourselves away in a very small space, there are many more coming than I expected and I'm *very* doubtful about the chairs!'

So we wedged ourselves into one of the two window-seats and were pretty close to the piano . . . It was a very warm evening; every door and window was open. People sat about all over the place, on the arms of chairs, up the stairs, and on the floor.

'No one minds if I play in my shirt-sleeves, I suppose,' said E. E., taking off his coat. 'You know I can't play this stuff.' Then, to Mr. Atkins: 'You come and play the treble and I'll play the bass.'

Just before they began E. E. said in a low voice to Mr. Reed: 'You won't leave me alone in the tuttis, will you?'

[176] *Edward Elgar: memories of a variation*, pp. 91–2.   [177] *Elgar as I Knew Him*, pp. 30–1.

I was half-afraid that E. E. might indulge in his curious habit of 'singing' while he was playing. It was an odd noise: it seemed to be a kind of filling in of parts that he had not fingers enough to play. It was really more like grunting than singing. He actually did do it at the start, but whether it was that one became so absorbed, or whether he stopped doing it as anxiety lessened, I do not know. Nothing, however, spoiled the beauty of the performance.[178]

On the Tuesday evening Ralph Vaughan Williams directed the first performance of his *Fantasia on a Theme of Tallis* before Edward conducted the first complete performance of *Gerontius* to be given in Gloucester Cathedral. On Thursday Kreisler gave his private performance of the Concerto, and the house was filled again. Afterwards Robin Legge wrote:

Frankly, I have never seen a keener enthusiasm in one musician for the work of another than Kreisler showed for Elgar's Concerto . . . I believe that Elgar has succeeded in a very high degree in revivifying the once moribund concerto form, and I believe that that will be the universal verdict on Nov. 10 . . . Nov. 10, then, is likely to prove to be a date of rare historic importance in modern British music, for we shall obtain then the reply to the question so often asked—Is this the long-awaited master-work, the fourth violin Concerto in the literature of music?[179]

Between the Gloucester Festival and the Concerto première Edward felt restless. They were not going to the Leeds Festival that year, he wrote to Alice Stuart-Wortley:

. . . as I am not asked: my popularity shews, in dismal relief, the unpopularity of someone else!

They propose to ruin the Variations, to travesty (the accompaniments to) the Sea Pictures & conventionalize Go song of mine. The festival has steadily gone down in interest & is now a dull affair of only Kapellmeister interest.[180]

It was the final Leeds Festival under Stanford, and Edward's judgement was wrong. Stanford's performance of the *Variations* drew praise from every quarter, and Vaughan Williams produced his *Sea Symphony*—a work which showed the dimensions of a creative spirit perhaps to rival Edward's own.

While the Festival was going on at Leeds, Edward found himself working at another *Pomp and Circumstance* March—just three years after the last one. Then the first orchestral score proofs of the Concerto arrived—later than they should have been to permit comfortable printing of the study score being prepared for sale at the time of the première. But Sanford Terry arrived coincidentally, and he helped with the proof-correcting. One day Edward showed him sketches for the E flat Symphony.

The orchestral parts of the Concerto were also delayed by a printer's misunderstanding. By the time Edward had them he was forced to correct them at terrible speed. Troyte Griffith was at Plas Gwyn on Sunday 23 October:

---

[178] *Edward Elgar: memories of a variation*, pp. 92–3.
[179] *The Daily Telegraph*, 26 Sept. 1910.
[180] 31 July 1910. (HWRO 705:445:7859).

When the parts came from the printers, he played through every note of every part on the piano with Mr. Austin playing the violin or viola, ruthlessly stopping in the middle of a bar, never wasting a second. They played the violin concerto one Sunday at Hereford from 10 o'clock in the morning to 11 o'clock at night only stopping for meals. Elgar objurgating Novellos whenever he found a mistake.[181]

Two days later he wrote to Alice Stuart-Wortley: 'I have been working too hard over this absurd printing muddle & have a slight headache—& have also been making a little progress with Symphony No. 2 & am sitting at my table weaving strange & wonderful memories into very poor music I fear.'[182]

Then it was time for the Concerto première. On 5 November he wrote to the Windflower again: 'The piano arrgt was published yesterday—how I detest its' being made public.'[183] He and his own Alice went to stay with Frank Schuster in London. On the morning of the 9th came the first rehearsal. The press was full of it. *The Evening News* reported that Kreisler 'has even gone so far as to say that it is the greatest violin concerto produced since Beethoven's. Already the concerto has been set down for a very large number of performances, both in this country and abroad.'[184]

Kreisler himself had volubly answered the questions of an American interviewer in London:

'In your opinion, does it rank with the Brahms and Beethoven?'
'Yes; we have not yet had a romantic concerto of this value.' . . .
'Taking into consideration the newer developments of musical art, how does this work stand?'
'In a way, quite outside; although from a player's point of view it is perhaps the most difficult of all concertos for endurance, and it is the first to have all the intricacies of modern scoring. Elgar regards it as one of his finest works. He tells me he has used many youthful themes and that for emotional force it surpasses anything he has yet written . . .'[185]

Only the critic of *The Court Journal* was a little bemused by all the exposure before the first performance in public:

I say in public advisedly, since, according to various accounts, each of which vies with the other in the exuberance of the admiration and praise lavished by the writers upon the music, it has already received many private baptisms . . . 'Ravishing dialogue—one of the loveliest snatches Elgar has ever written—most arresting force—exquisite gentleness of appeal—brilliant enough to satisfy the most ardent virtuoso—splendid fire and energy.' Sir Edward Elgar will find it hard to live up to such a comprehensive eulogy.[186]

A second rehearsal on the morning of 10 November, and the première was upon them. That evening nothing could dampen the general ardour, as Alice

[181] MS reminiscences (Elgar Birthplace).   [182] 25 Oct. 1910 (HWRO 705:445:7856).
[183] HWRO 705:445:4071.   [184] 9 Nov. 1910.
[185] Hadden Squire in *The Christian Science Monitor*, Boston, 19 Nov. 1910.
[186] 9 Nov. 1910.

wrote: 'Poured in desperate torrents. Crowd *enormous*. Excitement intense
. . .' *The Daily Mail* reported:

With rapturous applause such as might have greeted the victor of Trafalgar the great
company gathered in the Queen's Hall last night acclaimed Sir Edward Elgar and the
triumph of his new concerto . . .

When Sir Edward Elgar, who was conducting the whole concert, walked, a tall,
military figure, baton in hand, to the conductor's desk to direct the National Anthem
and the preliminary item [Sterndale Bennett's *Naiades Overture*] he faced an audience
as great and expectant, even if not so fashionable, as the premiere of a musical comedy
might have attracted.

When after the preliminary numbers Herr Fritz Kreisler, the soloist of the concerto,
stepped, violin in hand, upon the platform with the composer a tense silence fell upon
the crowded hall, and was not broken until at the end of the first of the three movements
of the concerto there burst out a hurricane of proud recognition of an English
masterpiece.

The storm of applause hushed itself for the assured delight of the second movement,
to break out again when that was ended, and then to still itself once more not to miss a
note of the concluding and culminating phase of the composition.

Then, when the end came, the huge audience went wild with pride and delight. For a
quarter of an hour they called and recalled the man who had achieved a triumph not only
for himself but for England, and hailed him with wonder and submission as master and
hero.[187]

Back at Schuster's house, supper was laid for forty. One of the guests was
young Adrian Boult:

We sat at three separate tables, filling the big music room, and the menu at each table
was headed with a theme from each of the three movements. I heard Elgar say to Claude
Phillips, the great art critic, 'Well Claude, did you think that was a work of art?'[188]

In the audience at Queen's Hall had been many opinions. *The Star* reported:

. . . here one would be aware of a violinist who said, 'It is not a Concerto at all— it is just
a Rhapsody;' there of another, devoutly: 'Thank Heaven! At last a real Concerto, not a
Symphony with Violin Obbligato.' Some said that to write music so purely abstract was a
step backwards: others that it was program music from first to last, and, if so, it was a pity
the program had not been told us. Some prophesied that every violinist would play it,
because it is so effective; others that nobody would touch it, because the solo part is so
ungrateful. 'What a falling off!' said one critic. 'Elgar stands higher than ever after this,'
said another . . .

When people talk like this about a work, two things are certain: first, that it is a work
of unusual power, and second, that it has qualities of newness and unexpectedness—if
such a word be permitted—which rudely move people out of their grooves.[189]

The unexpectedness, for Baughan in *The Daily News*, resolved itself into the

---

[187] 11 Nov. 1910.
[188] *My Own Trumpet*, pp. 18–19.
[189] 11 Nov. 1910. Next quotation is the same date.

apprehension that the central insight was not at the end of the Concerto but in the quiet middle movement:

The obscure psychological warfare of the first movement gives place to the serene dream of the Andante, a dream touched by passion and full of intricate sensitiveness and strange imaginings. The finale, in which a very beautiful cadenza for the solo violin echoes the dreaminess of the Andante, only with more certainty and definiteness . . . is also meant to be full of force and emotional fire. The movement is not strong enough in this respect, however. There is a want of climax . . .

The difficulty, according to Ernest Newman, was that of writing new music in old forms:

Here and there in the finale, as in that of Elgar's symphony, we feel that the thematic material is not so weighty nor the tissue so closely woven as in the earlier movements. The fault lies not with the composer, but with the form. One realises more and more each year that composers ought now to abandon the three-movement form and cast their symphonic works in one continuous movement . . . The symphonic form of the future must surely be more free, more improvisatory, as it were. In the extraordinarily beautiful and impressive cadenza in the present concerto Elgar has shown us the lines on which the new music could safely run . . .[190]

Among younger spirits everywhere across Europe there was now a deliberate questioning of all forms laden with associations from the past. There were musicians who attacked the basis of tonality itself. There were painters who shredded the subject-matter of their pictures with revolutionary analytic formulae calculated to divide the spectator ruthlessly from any way of looking that could be familiar to him. Novelists attacked the novel's traditional broad social contract with a 'stream of consciousness' technique to focus exclusively on one lonely viewpoint, to explore illusions which divided the individual against himself.

Yet Ernest Newman was clear about the richness of reflection in the Elgar Concerto:

Human feeling so nervous and subtle as this had never before spoken in English orchestral or choral music. That this was the secret of Elgar's hold upon the public was shown later in two ways—negatively by the modified success of the two later oratorios, in which people felt at times that the broader human feeling had been narrowed down to merely theological feeling, and positively by the glad leap the musical public gave towards the [First] Symphony, in which it heard Elgar again speaking eloquently to it of matters that concern the emotional life of each one of us . . .[191]

Now the Violin Concerto had carried Edward's apprehension of that life a step beyond his Symphony—to where the end was no longer a massive hope in the future, but instead an understanding that the best was perhaps with them at that moment, perhaps already in the past. If this was the later fruit of his

---

[190] *The Birmingham Post*, 11 Nov. 1910.
[191] *The Nation*, 16 Nov. 1910.

experience, there was nothing to say that was less relevant to 1910 than the Symphony had been to 1908.

Younger spirits in London were none the less beginning to question Edward's influence. Six days after the Concerto première, *Vanity Fair* published a piece by Francis Toye, a critic in his twenties, entitled 'VELGARITY':

We must all hope that Sir Edward Elgar has a considerable sense of humour, otherwise it is to be feared that he is likely to be overwhelmed by the torrents of snobbery, advertisement, and flattery that now accompany the production of his every new work. The Symphony is the finest piece of music ever written; so is the Concerto and ever shall be, till the advent of a new masterpiece turns the present duality of perfection into a Trinity not unworthy the metaphysical analysis of the compiler of the Athanasian Creed.

. . . Just as an hysterial invasion-scare is the worst possible preparation for a contest with the Germans, so an ignorant, exaggerated, hysterical appreciation of Elgar is the worst possible preparation for a proper recognition and a sympathetic criticism of the music of British composers.

. . . My impression of the Concerto was one of great length, extreme technical ability, wonderful beauty in places, and a passionate addiction on the part of the composer to the indication *nobilmente*, which I dislike very much.

. . . When next Sir Edward Elgar produces something in London I trust that a few more people will remember that, 'master and hero' though he may be, the time is not yet come for—his deification. They might remember too that this unbridled enthusiasm is bound to produce a reaction sooner or later, and that the cause of Elgar is best served by a total abstention from 'velgarity' . . .

*   *   *
*   *
*

The new Symphony was following so hard on the heels of the Violin Concerto as to hint at some further revelation. The '1st Sketch of Symphony No 2' was a descending sequence labelled by Alice 'Ghost'.[192] But this was not where the new Symphony should begin. It would begin in the midst of life—with a *fortissimo* E flat major which sought the affirmation of Beethoven's 'Eroica' opening in the same key:

Allegro vivace e nobilmente

But straight against those affirmative triads came plunging descent

[192] Elgar Birthplace.

It almost retraced the Concerto's central

—the climax which had linked that music's *Nobilmente* with its dreams. The new Symphony was taking its departure from this point.

The dotted rhythm at the end of the new plunging figure bred a series of derivatives. The first fell in steps through sequences

the last (from the sketchbook started at Careggi) rose in fifths through sequences

So the opening E flat major affirmation gave gradual ground to formidable chromatic presence. Ideas for a slow movement to follow suggested a dirge.

He played this much of the music when Sanford Terry was at Plas Gwyn in early October 1910:

> I remember vividly only the jubilant opening theme of the 1st movement and the funeral-march-like subject of the slow movement . . . I remember that in October it was in his mind to use in close context the present opening subject of the [Rondo] & slow movement, and he explained that they represented the contrast between the interior of St. Mark's at Venice, & the sunlit & lively Piazza outside.[193]

Much of October 1910 had gone in preparing the Violin Concerto for its first performance. November was full of conducting engagements. And Ivor Atkins was eager for Edward's collaboration in a new performing edition of Bach's *St. Matthew Passion* to be used in the Worcester Festival of 1911. The idea was that Edward should deal with the music and Atkins revise the translated text. Through the winter he was to come for repeated consultations.

It was the last week in November before Edward could settle to the Symphony again. He had bought a date-stamper (in imitation of Richter, who always used one), and with it he stamped many Symphony sketches made over the next two months and more.[194] The result was to give a unique insight into his way of preparing a big abstract composition. On 25 and 26 November he worked at an *Impetuoso* climax for the first-movement recapitulation. He also made a rough sketch for the end of the whole Symphony, where upper strings would rise over the figure played again and again for Miss Burley at Alassio, to meet the plunging descent in a final fading prospect:

[193] MS notes on the composition of the Second Symphony (Athenaeum Club).
[194] These are now divided between the Stuart-Wortley papers at the Elgar Birthplace and the Sanford Terry material at The Athenaeum.

This figure cast its retrospective light across his entire career: one germ of it had been in the violin *Reminiscences* of 1877

another was in the child's 'tune from Broadheath'.

He went to London on 26 November to address the Institute of Journalists Dinner, pleading for more cheerful music to engage and gratify the people. He stayed at Queen Anne's Mansions to continue work on the Symphony. The next day, 27 November, found him writing out the *Impetuoso* in a related key to climax the exposition. He also made a bridge passage to lead from recapitulation into coda. And that day he sketched a development section based on the 'Ghost' figure which had come first of all—a skeleton of the Symphony's opening plunging descent sounded over tritones:

It measured an infinitude of distance travelled from the music's first joyous affirmation.

This emotional polarity within a single utterance recalled the opposition of 'motto' and first *Allegro* subject which had launched Edward's First Symphony. Such polarity was widespread in the music of Europe then. It informed the symphonic poems and operas of Richard Strauss and the gigantic symphonies of Mahler. The same thinking was at work in the new study of psychoanalysis in Vienna and elsewhere. In the body politic, it was the ghost that raised *ententes* at a distance as it alienated neighbours nearer home.

On 30 November came the very successful second performance of the Violin Concerto. Alice joined him in London, and they continued at Queen Anne's Mansions through the first week in December. Plas Gwyn was lent to the Conservative MP Arkwright for his campaign in another general election. Arkwright held his seat. But the Liberals remained at the end in control of the Commons. They were laying the foundations of welfare-state socialism—amid parliamentary opposition of a bitterness never seen through the whole of Edward's lifetime. This election moved the country closer to a constitutional crisis over the power of the Lords. The 'old-world state' could appear afflicted with an enemy bred inside itself, and the growing tension within the country would find its reflection in Edward's Symphony.

Edward and Alice returned to Plas Gwyn on 7 December. He continued to work at the first movement—for which there was now a beginning and climax for the exposition, an opening for development, beginning and climax for the recapitulation, and an entry to the coda.

He worked in a similar way at the second movement *Larghetto*. To open it

the first movement's  was re-woven in a

close counterpoint:

Then the first movement's opening triad re-sounded in a minor shape to begin the *Larghetto* first subject:

The music found more and more distinct echoes of E flat major as it emerged in a powerful slow march:

The next *Larghetto* music was sketched on 13 December. It was an echo of the close-moving music which opened the movement, re-shaped to a slow extending turn:

He would recall inventing this figure from 'only the vague remembrance' of

[195]

---

[195] The sketch is annotated by Alice Stuart-Wortley: 'Sir Edward Elgar's own writing, done to explain a point raised in conversation.' (Elgar Birthplace).

He suggested to Ernest Newman that the passage might be 'a wistful colloquy between two people'. But this music could reach back farther still: the slow thirds shifting through softly undulating patterns made just such a sound as could stir memories of the wind falling and rising through river-side reeds.

On the day after sketching that music, 14 December, Edward and Alice went to Crefeld in western Germany, where he conducted the First Symphony. It was another triumph, with the distinguished composer and conductor Max Schillings leading the applause at the concert and the toasts at the heavy *Fest Essen* afterwards.

They were back for Christmas. It was to be the last Christmas at Plas Gwyn, and they had a little party: the chief guest was Alice Stuart-Wortley. On Christmas Day he played the new Symphony sketches to her. His own Alice noted: 'A very dear day—E. so much happier than ever before over Fest[ival anniversaries].' On Boxing Day the chief guest had to return to London, and on the 27th 'E. badsley headache went to bed very early.'

Two days later he wrote to Alice Stuart-Wortley: 'I have reached out the sheets of the Symphony & am going on or rather—trying to go on.'[196] He worked that day at a complex texture for the middle of the *Larghetto*. From an old sketchbook page of ideas for 'Cockaigne No.2—The City of Dreadful Night' came a slow chromatic rise through polyphonic confusion, a formidable contrast to the broadly diatonic *Larghetto* expression.

That day and the next he was also shaping the third movement *Rondo* at critical points of structure. One sketch developed the quick figure noted in the Piazza San Marco. Its triple-metre rise and fall now made another echo of the Symphony opening:

It developed in two directions. One was a climax of augmentation

The other was a pastoral for the centre of the movement

On 31 December he travelled up to Liverpool to conduct the Violin

<hr />

[196] 29 Dec. 1910 (HWRO 705:445:4055).

Concerto for Kreisler in its first northern performance. Samuel Langford's review in *The Manchester Guardian* praised the Concerto's 'unique intimacy of expression'. But the critic was worried by the music's constant juxtaposing of short figures. It had been a characteristic of the First Symphony as well, but there the great tune at the beginning and end had bound all together. Langford concluded:

The Concerto triumphs in every way but one, and that is in real scope of melody. In that essential feature of great writing Elgar seems to have lost more ground than he has gained by putting himself to school as he has done in his last works.

He is not, perhaps, so much imprisoned as other composers might be by similar restrictions, for his fancy is so delicate and intricate that he can find marvellous freedom in very little space. And the freedom for him is found as much in the manner of performance as in the writing itself. His conducting allows play for an intimacy and delicacy of fancy as distinct on the one hand from the rigour of classical style as on the other from the piquancies and brilliance of the virtuoso conductor.[197]

Edward's secondary self-taught art of conducting had begun to respond to his composing in a way that would be of vital advantage to the performance of his music, and therefore to the music itself—especially when his own readings later came to be preserved by the gramophone.

But why was melodic inspiration fragmenting in more and more short figures? Was his own fount of melody flowing less freely than in the past? Or was the mounting pressure of life through these post-Edwardian days precluding the sustained melody of former years? Whatever the answer, short figures swarmed more densely than ever in the new Symphony.

The first days of January 1911 found him back at Plas Gwyn beginning the full score. Such was his technical mastery now that many important aspects of form and actual thematic discourse could be left for settlement as the work took its final orchestral shape. At the head of the new score he set an epigraph from a 'Song' by Shelley:

> 'Rarely, rarely comest thou,
> Spirit of Delight!'

He wrote of it:

To get near the mood of the Symphony the whole of Shelley's poem may be read, but the music does not illustrate the whole of the poem, neither does the poem entirely elucidate the music.[198]

My attitude toward the poem, or rather to the 'Spirit of Delight' was an attempt to give the reticent Spirit a hint (with sad enough retrospections) as to what we should like to have![199]

---

[197] 2 Jan. 1911.   [198] 13 Apr. 1911 to Alfred Littleton (Novello archives).
[199] 9 May 1911 to Ernest Newman.

The poem as a whole revealed the same opposition of desire and experience, of dream and actuality.

Rarely, rarely, comest thou,
　　Spirit of Delight!
Wherefore hast thou left me now
　　Many a day and night?
Many a weary night and day
'Tis since thou art fled away.

How shall ever one like me
　　Win thee back again?
With the joyous and the free
　　Thou wilt scoff at pain.
Spirit false! thou hast forgot
All but those who need thee not.

As a lizard with the shade
　　Of a trembling leaf,
Thou with sorrow art dismayed;
　　Even the sighs of grief
Reproach thee, that thou art not near,
And reproach thou wilt not hear.

Let me set my mournful ditty
　　To a merry measure,
Thou wilt never come for pity,
　　Thou wilt come for pleasure,
Pity then will cut away
Those cruel wings, and thou wilt stay.

I love all that thou lovest,
　　Spirit of Delight!
The fresh Earth in new leaves drest,
　　And the starry night;
Autumn evening, and the morn
When the golden mists are born.

I love snow, and all the forms
　　Of the radiant frost:
I love waves, and winds, and storms,
　　Everything almost
Which is Nature's, and may be
Untainted by man's misery.

I love tranquil solitude,
　　And such society
As is quiet, wise and good;
　　Between thee and me
What difference? but thou dost possess
The things I seek, not love them less.

> I love Love—though he has wings,
>   And like light can flee,
> But above all other things,
>   Spirit, I love thee—
> Thou art love and life! O come,
> Make once more my heart thy home.

The Symphony under this epigraph showed so far the outlines of three movements pursuing the same psychology. First came an opposition of diatonic and chromatic expression—an opposition perhaps of past and future. Second came a sorrow of remembrance, and third a manic, despairing search. Of the remarkable answer to come in the Finale there were as yet few dated sketches; but that music turned out to be the oldest of all.

Late on 3 January Sanford Terry arrived for another short visit. He found Edward deep in his work, and they discussed it in detail:

He had no intention . . . he said, of making the new Symphony an organic whole by means of such a connecting-motif as that which opens and ends the Symphony in Ab. He wanted it to be the frank expression of music bubbling from the spring within him.

On January 4 and 5 he spent the greater part of each morning in playing over his sketches, and it was interesting to follow the process on which he worked. In every movement its form, and above all its climax, were very clearly in his mind—indeed, as he has often told me, it is the *climax* which invariably he settles first. But withal there was a great mass of fluctuating material which *might* fit into the work as it developed in his mind to finality—for it had been created in the same 'oven' which had cast them all. Nothing satisfied him until itself and its context seemed, as he said, inevitable. In that particular I remember how he satisfied himself as to the sequence of the second upon the first subject in the first movement.[200]

Sketches dated 5 and 6 January showed this working toward the second subject. From the E flat opening *Allegro vivace e nobilmente* sprang a group of ideas gradually contrasting in longer values. The first hint of the new mood

intersected the primary dotted energy 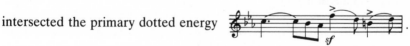.

Before long this interruption of primary rhythm brought a crisis, and the new mood extended in hushed chromatics:

Soon it reached what Edward specified as 'the second principal theme'[201]—a shade of the Symphony's original plunging figure, with a tritone in the midst of its drooping descent:

---

[200] MS notes on the composition of the Second Symphony (Athenaeum Club)
[201] 13 April 1911 to Alfred Littleton (Novello archives).

The contrast with the joyful primary subject was complete.

January 8 1911. E. terrible headache—in bed till afternoon—A. cd. not leave him . . .

January 9 . . . E. still vesy porsley & worried over his musics.

It was not only the dimensions of the problem revealed in the new Symphony. The practical limitation was that he had long ago agreed to go to America in late March to conduct. The Second Symphony, if it was to be ready for its scheduled première at the Queen's Hall in May, would have to be finished before he left—in little more than two months' time. He wrote to Alfred Littleton on 10 January: 'I am all behind with my work & I have grave fears for the 2nd Symphony, but I will decide its fate next week—I have been too cold to do anything.'[202]

On 12 January he made an *animato* linkage for the first movement between the second subject and the *Impetuoso* climax already planned. That *Impetuoso* was in the primary mood: it could recall the primary 'masculine' invitation at the centre of the Violin Concerto first movement. On 14 January he set out the reply—a huge unyielding *Maestoso* statement of the Symphony second subject, recalling the Concerto's *Maestoso* of denying 'feminine' reply. Steely descents again followed the double apparition. Then the music sank through hollow echoes set with drumbeats to a single harp note sounding over and over the end of exposition. On the same day he also sketched a long *animandosi* to lead toward recapitulation. The intervening development was not yet finished.

Then he had to go to London to conduct another performance of the Concerto for Kreisler—who was soon to introduce it in Germany under Mengelberg. At Queen's Hall on 16 January Edward's Concerto was played

[202] Novello archives.

side by side with the Violin Concerto of Beethoven, and each had 'a great reception'.

During the week that followed, Alice and Carice pursued house possibilities all over the home counties, while Edward immured himself in Queen Anne's Mansions with his Symphony. On 21 January he sketched a quiet memory of the joyful opening—to find its place at the centre of the development section as the 'happiness [of] real (remote) peace'[203].

January 23 . . . E. decided to return, all home by 4.45 train. E. looking rather tired but happy over Symphony. Found want of light in London tired his eyes . . .

At Plas Gwyn he continued at white heat. On 26 January his own Alice wrote to Mrs Stuart-Wortley: 'Dear Alice, the Symphony is wonderful. One is led away to regions beyond worlds—He is working *very* hard & I trust will not get knocked up.'[204]

The first-movement development led away to distant regions indeed. It opened with the 'Ghost' descent over tritones sketched in late December— threaded through a soft primary variant in dotted rhythm. Suddenly in the midst of their dialogue arose a new and more formidable ghost—a twisted shape which linked this music to the 'Judgement' motive in *Gerontius*:

The spectre of *Gerontius* came close to Edward's consciousness when he described this figure in the Symphony as 'remote & drawing some one else out of the everyday world'.[205] The entire passage, he wrote, ' . . . might be a love scene in a garden at night when the ghost of some memories comes *through it*;—it makes me shiver . . .'[206]

The strange presence faded to the 'real (remote) peace' music. But then that atmosphere was 'broken in upon & the dream "shattered"' by the *Gerontius* spectre again—set this time to 'the inevitable march of Trombones & Tuba pp.'[207] The march rose to *forte* before it yielded to the first dawning light of recapitulation.

Primary sonorities blazed again, led again to secondary oppositions. Then a quiet-beginning coda gathered the disparate textures in a tremendous chromatic sweep through the whole orchestra up to a single tonic stroke. The climax—achieved in the First Symphony only at the end, in the Violin Concerto

[203] 13 Apr. 1911 to Alfred Littleton (Novello archives).
[204] Transcript by Clare Stuart-Wortley (Elgar Birthplace).
[205] 13 April 1911 to Alfred Littleton (Novello archives).
[206] 29 Jan. 1911 to Ernest Newman.
[207] 13 April 1911 to Alfred Littleton (Novello archives).

at the centre, stood now at the beginning of everything that could follow. He wrote to Alice Stuart-Wortley:

I have recorded last year in the first movement to which I put the last note in the score a moment ago & I must tell you this: I have worked at fever heat & the thing is tremendous in energy.'[208]

His own Alice wrote:

January 28 1911. E. very ardently at work & finished his first movement. Very wonderful & gorgeous. He was hardly over a fortnight scoring & writing this from his sketches.
        For a walk with Mr. Holland who stayed to tea . . .'

Vyvyan Holland was the younger son of Oscar Wilde. The young man's guardians had sent him to a Hereford solicitor who had been a family friend, and he found a warm welcome at Plas Gwyn:

I remember Edward Elgar as a tall man overflowing with energy and nearly always in a hurry . . . If the weather was fine, he would take me for a walk along the banks of the Wye and discuss any subject that might crop up, with youthful enthusiasm.
        He always carried small sheets of music paper about with him, and from time to time he would take one out of his pocket and jot down notes of some theme that had come into his head, humming to himself as he did so. He once told me that he had musical day-dreams in the same way that other people had day-dreams of heroism and adventure, and that he could express almost any thought that came into his head in terms of music.[209]

Next day, 29 January, the Symphony first movement went to Novellos with instructions to begin engraving the score.

On 30 January Edward started scoring the second movement *Larghetto*. 'It is elegiac', he wrote, 'but has nothing to do with any funeral March & is a "reflection" suggested by the poem.'[210] None the less, the 4/4 C minor music was still following Beethoven's 'Eroica', whose 4/4 C minor second movement was marked 'Marcia funebre'. Edward's first subject beginning

echoed the 'Eroica'. The quiet menace of his timpani made another echo. An oboe solo planned later in the movement made a third.
        Yet the slow treading march that unfolded was entirely his own, and one of

[208] HWRO 705:445:7809. This letter was written on 29 Jan. 1911, the day after Alice Elgar's diary records the completion of the Symphony first movement. 29 Jan. was devoted to revising the movement before its posting to the publishers.
        [209] *Time Remembered* (Gollancz, 1966) p. 20.
        [210] 13 Apr. 1911 to Alfred Littleton (Novello archives).

his finest inspirations. To it came the winding thirds of 'wistful colloquy'. And these led in turn to the Larghetto second subject (also written out on 30 January)

From the distant past it re-traced in slowest motion the 'heroic beauty' motive in *King Olaf*; from the earlier discourse of this Symphony, it evoked again the first-movement 'Ghost'. On 1 February Edward took a moment to send another extract from Shelley to Frances Colvin in London:

> 'I do but hide
> Under these notes, like embers, every spark
> Of that which has consumed me.'[211]

After the *Larghetto* second subject he set the chromatic rise from the 'City of Dreadful Night' sketch—gathering slowly through scales and arpeggios in the strings with soft relentless triplets in harps and wind toward a climax. This climax, when it came, showed the essence of Edward's aspiration. It combined

the Violin Concerto's central *Nobilmente*

with the plunging figure which had followed this Symphony's opening 'Delight'

yet the new *Nobilmente* was the simplest of all:

Nobilmente e semplice

It was Edward's best answer to the ghosts of 1910.

In London fresh prospects for a house were offered, and he had asked Alice to go up and look at them. She went on 1 February and returned three days later with news of a house she liked and thought eminently suitable. It was set in large gardens in a quiet corner of Hampstead, only a quarter of an hour from the West End. The house had been designed by the distinguished architect Norman Shaw for a painter, and the unique feature was a huge first-floor studio superbly floored and ceilinged and wainscotted. It would make the grandest study for Edward's music, as she wrote to Alice Stuart-Wortley: 'I still dream of it.'[212]

---

[211] From 'Julian and Maddalo' (HWRO 705:445:3425).
[212] 5 Feb. 1911 (transcript by Clare Stuart-Wortley at the Elgar Birthplace).

By the time Alice returned, he had nearly finished the Symphony second movement. After its *Nobilmente* climax, the *Larghetto* returned to its winding thirds. And that 'wistful colloquy' brought a repetition of the entire discourse from the primary theme forward. Thus the *Larghetto* would take a binary form—exposition followed by immediate recapitulation with no intervening development. It was as if the journey toward the latest *Nobilmente* had gone as far as possible, and there was nothing left but to re-state its significance.

Yet the return did enrich the experience after all. The *Larghetto* primary theme was set now against a multitude of whispering voices: soft brass, harps, and timpani sounded the off-beats, strings and bassoons syncopating semiquavers in slow motion, while a lonely oboe traced arabesques in wandering triplets through the quadruple metre. It was this oboe sound that focused private symbolism behind the music. Edward wrote: 'at $\boxed{79}$ the feminine voice *laments* over the broad manly 1st theme.'[213] So this music followed still the heroic identity born in the First Symphony and matured in the Concerto. Now it had become a subject for lament.

To it directly came the second subject *ppp dolcissimo*, taking the music almost out of hearing. Again the gathering 'City of Dreadful Night' chromatics rose to the utterly diatonic *Nobilmente e semplice*—this time surmounted with a big echo of the Symphony's opening 'Delight'. Again the prospect turned to dusk, and the plunging figure which had followed 'Delight' at the Symphony's opening sounded quietly again: 'may not $\boxed{87}$', Edward asked, 'be like a woman dropping a flower on the man's grave?'[214] It entwined with the thirds of 'wistful colloquy', and at the end the opening slow counterpoint returned in funeral tread to close the movement whose sorrow had twice drawn the response of simple nobility.

He finished the *Larghetto* on 6 February, and the following day took the score to Novellos in London. While there he heard another performance of the Concerto by Kreisler at the Queen's Hall, with Henry Wood conducting; and he went up to Hampstead to see the Norman Shaw house before returning to Hereford on 8 February. Alice wrote to Mrs Stuart-Wortley:

Edward loved the house, the perfectly beautiful one, I mean—He came home rather sad, as there are many difficulties in the way but *if* he really likes it, wh. he does, I feel he must have it, & have a proper room to dream dreams of loveliness in.[215]

The difficulties in the way were financial. Edward and Alice had always rented their houses, and though the income from his music had risen high, so had his expenditure. They had no savings to approach the purchase of a big London property. The hope was in Alice's inheritance from her mother, on

[213] 13 Apr. 1911 to Alfred Littleton (Novello archives).
[214] Ibid.
[215] 11 Feb. 1911 (transcript by Clare Stuart-Wortley at the Elgar Birthplace).

which they had partly lived since their marriage. But the inheritance was tied in a trust.

The morning after his return from London, 9 February, found him busy with the third movement *Rondo*. Here the two subjects pursued opposites of rhythm. The first turned the opening movement's plunging descent to nervous syncopation:

(Through the upper notes flitted the *Gerontius* 'Judgement'  .)

The second subject was obsessed with rhythmic repetition:

Presently this second subject was overlaid with a gentler memory of the

plunging figure  : it seemed to catch a pastoral

echo from *In the South*—the music which had replaced his earlier attempt to write this Symphony. The primary figure returned, but it was also overlaid; and the new overlay soon exploded in another development of the first-movement plunging figure:

(This was one of the ideas written down at the end of December.) As soon as this music approached more sustained melody, it was instantly attacked by the syncopated first subject chattering furiously up and down like the Demons in *Gerontius*. Once again Demon energy was short-lived, and over it came another pastoral strain (also sketched in December):

Yet the demons of this *Rondo* were not to be left behind. When the primary syncopation appeared yet again, it was surmounted with the *Gerontius* 'Judgement' in ghostly outline

—first in upper strings, gradually gathering more and more of the orchestra to itself. The triplets which had shaped Delight in the first movement and lament in the second began to beat relentlessly until they carried the spectre up to the most ferocious climax of the entire Symphony. This music exemplified for Edward himself (as he told a friend) the lines in Tennyson's *Maud* where the hero's frustrated love turns into a fantasy of his own burial:

> Dead, long dead,
> Long dead!
> And my heart is a handful of dust,
> And the wheels go over my head,
> And my bones are shaken with pain,
> For into a shallow grave they are thrust,
> Only a yard beneath the street,
> And the hoofs of the horses beat, beat,
> The hoofs of the horses beat,
> Beat into my scalp and my brain . . .[216]

The growing tension of pursuing his own heroism through a world that showed itself more and more alien to the real old-world values the more loudly it applauded his music had exploded in this *Rondo* with a concentration of all the headaches which had split the composer's brain with ever-increasing relentlessness through these years of his greatest success. Rehearsing the passage years later, Edward seized the private metaphor to explain vividly what he wanted. His explanation stamped itself on the memory of one young player in that orchestra, Bernard Shore:

With a great urgency he would say, in a shaking voice:
'Now, gentlemen, at this point I want you to imagine that my music represents a man in a high fever. Some of you may know that dreadful beating that goes on in the brain—it seems to drive out every coherent thought. This hammering must gradually overwhelm everything. Percussion, you must give me all you are worth! I want you gradually to drown the rest of the orchestra.'[217]

The terrible crescendo recalled only the crescendo he had introduced in *Gerontius* when the Soul sees its God. The crisis now showed the Soul facing its own burial beneath all the things that would oppose and fragment and shatter its vision.

The terrible spectre receded, only to reveal the *Rondo* second and first subjects in unaltered rhythmic opposition. Even now there were wraiths of 'Delight'—before a coda set primary syncopation and secondary repeating rhythms at each other's throats through savage counterpoint to a final ear-splitting stroke. That was the cost of Edward's *Nobilmente* now.

On 16 February he finished the *Rondo*. That day in Hereford there was a rehearsal of the *Romance* he had written for Edwin James, principal bassoonist

---

[216] 'Maud', Part II, v,i. See Maine *Works*, p. 167.
[217] *The Orchestra Speaks* (Longmans, Green, 1938), p. 135.

of the London Symphony Orchestra. Next morning James came to Plas Gwyn in his capacity as the Orchestra's chairman, and asked Edward to become their principal conductor on the retirement of Richter in the spring. Edward agreed, and Alice noted: 'Mr. James talked to A. with deep content of the idea of E. conducting the Concerts in Dr. Richter's place—Touching devotion to E.' So residence in London moved closer. Yet the west of England had been the setting for Edward's entire creative life. Did his acceptance of this notable conductorship in the metropolis suggest that the composer doubted the continuance of his own creative powers?

Up to this moment the Symphony had pursued a psychology of the divided self. Such a drama could find ready understanding in a century whose future was to be scarred by world wars and systematic abasements of the human spirit. Yet as Edward wrote: '. . . the whole of the sorrow is smoothed out & ennobled in the last movement . . .'[218] So this Finale would turn back, to give the new century one of the last answers it might receive from the older world which had gone before.

The obsessive rhythm played over and over at Alassio was now embodied in a long arc of melody for the Finale primary subject:

The repeating rhythm enacted over and over the 'smoothing out' in Edward's description of the movement as a whole. The music's *Moderato e maestoso* moved at the pulse-rate of breathing and walking, ♩=72: it was the pulse of life persisting after the drama of delight, sorrow, and self-questioning. The beginning of its melody had turned the *Gerontius* 'Judgement' toward a diatonic rise.

The rhythmic *idée fixe* rode the entire primary presentation forward to meet the second subject—another theme from the past:

It was 'Hans himself' idea which had now acquired a private poignancy, as Edward had written to Richter on 13 February: 'More than half my musical life goes when you cease to conduct . . .'[219] This second-subject presentation now led through broadly diatonic ways to a *Grandioso* climax, and then to a pendant

[218] 13 Apr. 1911 to Alfred Littleton (Novello archives).
[219] Richter family.

that reflected the Finale second subject in Edward's first big abstract work in several movements, the Organ Sonata of 1895

etc. :

Nobilmente

B flat major slipped into B major for half a dozen bars, and regained its locus to go on to the climax sketched in November for the end of both Finale exposition and recapitulation:

etc.

This showed the Symphony's opening music in the longest perspective of all.

February 21. . . . E. happy over his work D.G. it is really sublime—No one with any feeling cd. hear it without an inward sob—It resumes our human life, delight, regret, farewell, the saddest word & then the strong man's triumph.

February 22 . . . E. wrote 12 pages all but a few bars of full score—writing the wonderful music of the end of the 4th movement.

The full meaning of such an end, however, could only emerge when its diatonic openness had been subjected to a last scrutiny of development.

February 24 . . . E. sailing on with his Symphony & writing the intermediate part. Rather worried over it for a few hours—Then happy over an illumined idea.

In this development the 'Hans himself' second subject was twisted through a chromatic distortion of the plunging figure from the Symphony's first-movement opening. The contest rose and rose until a new figure mounted

suddenly  to a piercing trumpet note

held through the bar. Years later, in a gramophone studio, the London Symphony first trumpet Ernest Hall held his supreme note longer:

Sir Edward asked me at a recording why I held my top B♮ over to the next bar, & I replied that I was so pleased to get the note I didn't like to leave it. His reply was: 'I intended to write it so, but thought it would be too high to hold.'[220]

[220] *Music and Friends: seven decades of letters to Adrian Boult* (Hamish Hamilton, 1979), p. 196.

So it became a tradition which every trumpeter would be proud to inherit.

The significance of this climax was fundamental to the Symphony. Its four rising steps initiated the sequence

It was the simple 'walking' figure of long ago, when it generated both the 'Shepherd' music of *In the South* and the melody which had become the first subject of the present Finale. And this first subject now appeared in ensemble with the crisis call and the 'walking' figure: indeed its presence in the development was invoked by those more primal figures. More than twenty years later, listening to gramophone records of his own performance, Edward told the young conductor John Barbirolli how the 'walking' figure had started the entire movement in his mind.[221] Here at the movement's centre was the fount from which all its music sprang.

The rest of the development ruminated over the discovery. As the ensemble grew quieter, it was overlaid with the Finale second subject and then with a reflection of the *Nobilmente* pendant:

So every idea at last showed itself the child of that parent.

In recapitulation this descending figure made an ensemble with the Finale primary theme. At the end a coda descried the distant opening of the Symphony—floating high on soft woodwind, burnished in rich colours of central brass, fading through the strings to a final dusk where it mingled indistinguishably in the first and last themes of the Finale. Thus the Second Symphony found a cycle of experience both subtler and more ineluctable than the cycle defined in the First.

February 28 1911. This is a day to be marked. E. finished his Symphony. It seems one of his very greatest works, vast in design and supremely beautiful.

This latest big abstract work, to last fifty minutes in performance, had been entirely written and scored from sketches inside two months. By any standard Edward's art had reached its maturity. The fact that the Symphony's ideas reached farther into the past was only superficially a paradox. For the central subject of its music was the life of the past shaping the present. In three months he would be fifty-four.

---

[221] Michael Kennedy's *Barbirolli* (Macgibbon & Kee, 1971), p. 84, reports the conversation of 19 Aug. 1933. Barbirolli identified the figure for Michael Kennedy, who confirms it to me. The actual process is shown in a sketch at the Elgar Birthplace: see above, p. 421.

In the second week of March 1911 Edward and Alice went to Brussels for a Belgian production of the First Symphony. They received a joyous welcome from Eugène Ysaÿe, the great Belgian violinist who conducted symphony concerts in Brussels. Ysaÿe paid Edward the compliment of leading the orchestra he usually conducted through the final Symphony rehearsals and at the concert under Edward's baton. Ysaÿe also proposed to take up the Violin Concerto.[222] His enthusiasm was especially welcome then: Kreisler was already tiring of the Concerto's length and had severely cut the Finale in two recent London performances.

Edward returned to Plas Gwyn with a cold, which worsened so much that he could not sail to America on 18 March. He was to go through eastern Canada and the United States on the first leg of a world tour by the Sheffield Choir. The tour had been organized by Charles Harriss, the English-born organist of Ottawa Cathedral, who had married the widow of a millionaire and so could meet the £60,000 guarantee out of his own pocket. In the event, Edward sailed a week after the Choir, accompanied only by the valet Jaulnay.

He missed the first concert at Montreal, and went on to Toronto to await the Choir there. He looked upon his part of the tour as pure commerce and had made up his mind to a loathsome five weeks, as he wrote to Frances Colvin:

Here in this awful place . . . every nerve shattered by some angularity—vulgarity & general horror . . . I had a dreary crossing—found myself in New York & was looked after *well* & *motherly* by our dear friend [Mrs Worthington]—travelled all night & am now here in ice & snow, brilliant sun & piercing wind & longing for home.[223]

The following evening he conducted *Gerontius* in a hall which the regular Sheffield Choir conductor Henry Coward described as 'crowded far beyond its normal capacity of 3,200, by an audience whose lowest priced ticket was ten shillings.'[224]

He went on with the Choir to Buffalo, but after that had no engagement until Cincinnati. From there he wrote to Alice Stuart-Wortley on 18 April:

It was drearily cold in Toronto & I nearly died of it. *Now* it is summer—not spring—but summer without any leaves or flowers. I rushed from Buffalo to NY. where Pippa & nice Mr. Gray [Novello's American agent] took care of me. Trains full this way so I had to leave on Easter Sunday morng 11 oc & travelled until 7.30 the *next* morning: I loathe and detest every moment of my life here! . . . All I can do is to count the days . . .[225]

That night he conducted *Gerontius* with the best orchestra they encountered on the tour. Its regular conductor was the young Leopold Stokowski, and Edward agreed to arrange Stokowski's début with the London Symphony Orchestra in a year's time.

[222] Ysaÿe did not perform the Concerto at this concert, as is sometimes claimed.
[223] 3 Apr. 1911 (HWRO 705:445:3426).
[224] *Round the World on Wings of Song* (J. W. Northend, 1933), p. 38.
[225] HWRO 705:445:7805.

Three days later they met with a triumphant welcome at Indianapolis, vividly recalled by Henry Coward:

The main streets were decorated with the American and British flags. Across the principal road leading to our hotel a huge motto met our eyes: 'Welcome to the Sheffield Choir' . . . Mrs. Ona Talbot, a public-spirited lady who was chiefly responsible for our visit, gave a supper in honour of Sir Edward Elgar, Dr. Harriss and myself. It proved a very smart function. Speeches were made by Vice-President Fairbanks and Mrs. Talbot . . . An account of the gorgeous gowns and gems of the ladies at Cincinnati had duly reached Indianapolis, and . . . the wealthy section of the citizens had evidently decided to do honour to the concerts . . .[226]

Edward sent his own version to Alice Stuart-Wortley: 'We are met with flags & banners & "processed" all without a smile, entirely *betement*, ludicrous to the last degree but *dead serious*.'[227]

At Chicago there was another *Gerontius* on 24 April with the Theodore Thomas Orchestra (now regularly conducted by Frederick Stock). Chicago was not moved by *Gerontius*, meeting it with 'barely enough hand clapping to bring out Sir Edward Elgar to bow his acknowledgements . . .'[228] But there was excitement enough, for on the day previous a man had been murdered close to their hotel. Coward recalled:

This brought about the order that no lady or ladies were to go out unless they were accompanied by a gentleman . . .

After the performance, Sir Edward, Dr. Harriss and myself were invited by a lady of the smartest set to take supper with the 'upper crust' of the second city of the 'Land of the almighty dollar'. There were perhaps twenty-five select people seated round the table, and my next neighbour whispered to me, 'There is more than twenty million pounds represented here.'[229]

Their arrival at Milwaukee on 26 April was described in a fulsome newspaper account replete with compliments from Dr Harriss on the appearance of the city and its weather. Edward sent the cutting back to Alice with a note: 'Pubd. an hour or two before we arrived! Dr. Harriss is in N. York.'[230] He wrote to Alice Stuart-Wortley:

I am living in the special train—the winds are bitterly cold off this lake—but it will soon be over *soon! soon! soon!*

. . . My mind is a blank in which these people scrawl, or try to, their offensive ideas. (Perhaps!) Pippa is the only bearable one! & I have seen only too little of her . . .

Now, if for *myself*, I would never degrade myself for all the wealth of America to go thro' this awful depth of infamy! They asked me what I wd. take to settle in the States & conduct one of the big orchestras—I said nothing in the world wd. induce me to spend 6 months here—not $10,000,000—this they do not understand—Well—all the rest I

[226] *Round the World on Wings of Song*, pp. 63–5.
[227] 26 Apr. 1911 (Elgar Birthplace).
[228] Cutting from a Chicago newspaper, c.25 Apr. 1911 (Elgar Birthplace).
[229] *Round the World on Wings of Song*, pp. 66–7.
[230] Mounted in Newspaper Cuttings book at the Elgar Birthplace.

reserve until we can talk & then 'I guess' we will not talk of U.S.A. but of Sea pictures and clean things like the Symphony.[231]

A concert two days later at St. Paul brought the end of his commitment. He returned to New York, and on 2 May thankfully went aboard the *Mauretania* for the voyage home.

He summed up his American impressions for Ernest Newman:

I sympathise with the 'good-feeling' U.S. man & woman but they are wholly swamped by the blatant vulgarity of the mediocre crowd. America is getting worse—I see it in four years: it is a curious study—the 'Union' system makes wholly for mediocrity and the orchestras do not & *can* not improve under the system—and we *had* hopes of that land![232]

So ended Edward's experience of the nation whose life was said to give Europe a look at the future. He was never to go there again.

May 9 1911. Lovely day—Sunshine. D.G. Edoo safely home & looking well . . . A. vesy choky with happy joy. E. delighted to be home. Telling of all experiences & detesting U.S.A. except the nice people. Such a joyful day to have him safe.

The diary was full of London engagements, beginning with rehearsals of the new Symphony for its première on 24 May. Negotiations for the house in Hampstead had reached no conclusion, but they had taken a short lease of a house at 75 Gloucester Place. From there he went to conduct the first sectional rehearsals for the Symphony on 17 May. Next day *The Globe* reported:

Something of a record in music publishing has to be chronicled in connection with the new Elgar Symphony, the full score and all the orchestral parts being prepared by Messrs. Novello in about seven weeks, no fewer than 759 plates having been engraved for the work.

When, in preparation for the London Musical Festival, Sir Edward Elgar took his first rehearsal of the symphony with the orchestra this week, it was found that in all the many thousand notes which had been printed there was only one error, a flat sign instead of an accidental in a horn part, a circumstance which has earned for the publishers the composer's grateful thanks. The publishers may take the whole credit for this achievement, as Sir Edward was in Canada and America during the time that the work was being printed, and thus had no opportunity of personally revising the proof-sheets.

Two days later came the first full rehearsal. The Queen's Hall Orchestra, increased to 130 players, had new four-valve trombones to meet the lower Queen's Hall pitch, as Henry Wood noted: 'Elgar, I remember, was delighted with the beautiful *legato* these instruments produced in the finale but I did not care for the horn-like quality of them.'[233] Another rehearsal took place on the 20th, when Alice heard the Symphony for the first time: 'Quite overwhelm-

[231] 26 Apr. 1911.   [232] 17 May 1911.   [233] *My Life of Music*, p. 250.

ing—Wonderful the way one cd. hear E.'s very soul in so many parts—Most most beautiful.'

The London Musical Festival opened on Monday evening 21 May with Henry Wood's performance of *Gerontius*. On Tuesday afternoon Wood conducted the Violin Concerto for Kreisler. Wednesday morning saw a final rehearsal of the Symphony, with the performance in the evening. The rest of the programme was devoted to new works by Bantock and Walford Davies, relieved only by two vocal solos for the Dutch singer Julia Culp. The management's thinking was clearly that the new Elgar Symphony would draw a capacity crowd. But it was the thinnest house of the entire Festival week. The *Academy* critic wondered at the spectacle of a gallery half empty:

It was difficult to repress an inquiry as to the possibility that the number of enthusiasts about the composer's music is not increasing, nay, and it may even be diminishing.[234]

Young Francis Toye could hardly suppress a sneer of satisfaction:

So keen was the interest, indeed, that the more expensive seats at the Queen's Hall at the first production were respectably filled, while not more than one half of the balcony stood empty.[235]

'Sir Edward Elgar conducted with clearness and decision,' reported *The Referee*, 'and the orchestra responded splendidly, and at the close the applause was long and loud.'[236] But there was a hollowness about it which Edward mistrusted. The whole scene stamped itself on the memory of W. H. Reed:

That the composer noticed the coolness of its reception at this first performance was very clear. He was called to the platform several times, but missed that unmistakable note perceived when the audience, even an English audience, is thoroughly roused and worked up, as it was after the Violin Concerto or the First Symphony. He said pathetically . . .
'What is the matter with them, Billy? They sit there like a lot of stuffed pigs.'[237]

The matter was that they were puzzled by the new Symphony—audience and critics alike. Some seized on the word 'Delight' in the Shelley epigraph. 'A BACCHANAL RIOT IN MUSIC', headlined *The Daily Mail*: 'Orgy of colour and joy.'[238] But when other things were noticed, there was less certainty. The uncertainty centred on the close development of opposing things side by side in the music. *St. James's Gazette* wondered

. . . if Elgar's outlook on art was getting too worldly, not for us, but—for himself. At any rate it was difficult to follow the spirit or purpose of the orchestral scramble which formed the third movement. The mood was undoubtedly genial though the gait was often clumsy. It was as though the composer had taken too many instruments along with him in his frolic . . .

Of all the critics, only Ernest Newman was able to take the measure of the

234 3 June 1911.   235 *The Bystander*, 7 June 1911.
236 28 May 1911.   237 *Elgar*, p. 105.

new music. Having written an analysis for *The Musical Times* before he could hear a note of it, he now published a review of the performance under the modest title 'Sir Edward Elgar's Symphony No. 2: a second critique.' Above all he perceived the music's unity in diversity:

In [the *Rondo*] occur passages which seemed to the present writer to approach a state of things only to be described as terrible—terrible in intensity of black import . . . At the end of the symphony a flavour of nobleness and progressive joy do find a place, and if the darker shades do prevail in places it cannot be denied that they are presented with a skill and command that are worthy of the deepest respect and admiration.

    . . . Elgar is always saying something fresh. Because an actor has only one voice it does not follow that he can play only one part . . . So it is, it seems to me, with Elgar. There is not a page either of the symphonies or of the violin concerto that does not reveal him like a photograph; but within this apparently restricted circle of style he manages to say an infinite variety of things, or, if you will have it so, to give an infinite number of fine shades to the highly personal things he has been saying all his life.

    To me the Second Symphony is full of new shades of this kind. And that he is less the slave than the master of his very pronounced style is shown precisely in the way he handles these sequences of his, the largeness of outline he can draw with them, the distance he can see through them in the direction of his goal. I was greatly struck with this towards the end of the *finale*, where the chain of sequences—as long, indeed, as the movement itself—rolls on to its beautiful end like a winding and broken river that at last gathers all its waters together and finds the sea. There is, too, an exquisite mellowness in the graver moods of the Symphony, as well as new strength and freedom in the gladder moments. In the episodical passages, again, as in the cadenza of the violin concerto, I feel that Elgar is on the point of finding a new style, one more flexible and varied . . .[239]

And once more Newman entered his plea that Edward should translate this impulse into music in a single-movement form.

    In the spring of 1911 there were few listeners who penetrated so far as Newman. Anticipating a success to equal the First Symphony, the concert managements had scheduled no fewer than three additional performances of the new Symphony in the remaining weeks of the London season. On 1 June Edward conducted the work with the London Symphony Orchestra for the International Musical Congress at Queen's Hall, but the audience was smaller than for the première. A week later, in the first of two London Symphony concerts to introduce him as principal conductor, the Hall was not a quarter full. Next day he wrote to Alfred Littleton:

No one came to the concert last night so I have told the L.S.O. that I receive no fee of any kind (performing or otherwise) for this concert (8th) or for the concert next week (15th)—most depressing![240]

His regular LSO fee was to be 50 guineas a concert. The audience on 15 June was only a little better.

[238] 25 May 1911. The following quotation is of the same date.
[239] *The Musical Standard*, 27 May 1911.
[240] 9 June 1911 (Novello archives).

If, as Newman felt, the Second Symphony revealed its composer as fully as its predecessor, these sparse audiences clearly said that the public which had followed Edward's music through the decade and more since the *'Enigma' Variations* would not follow where his thinking took him now.

The First Symphony, with its climax of achieved triumph at the end, had touched their deep response. The Violin Concerto, with its climax of private insight in the middle, had half touched them—though with doubts about the long Finale which moved on again. The new Symphony's climax came right at the beginning, with everything afterwards sliding away. It was a prescience that no one in the England of 1911 wanted to recognize. The pattern of the three works together suggested that the world of Edward's experience had gone over its peak: and with that world his art still moved in sympathy.

BOOK IV

FOR THE FALLEN

Death on the Hills

IN the decade since the beginning of the new century, success itself had seemed to divide Edward from his self-doubt. Now the lack of response at the Second Symphony première raised all his old insecurity in an instant. And side by side with personal fear there began to grow a new fear of the future: that he and his music would find less and less understanding in the evolving world.

The Coronation of 1911 was to follow the Second Symphony première by less than a month. Edward made three contributions to it. First, Arthur Benson had pointed out that the *Coronation Ode* of 1902 could be used again with two changes. The chorus to greet the Danish Queen Alexandra would not apply to the young Queen Mary. So Benson wrote a new lyric, 'True Queen of British Hearts and Homes'. It reflected the growing democratic temper:

> Oh kind and wise, the humblest heart
> That beats in all your realms today
> Knows well that it can claim its part
> In all you hope, in all you pray.

Edward set it as a simple hymn. The other change in the *Coronation Ode* was more ominous. It was to omit the hushed prayer which had prefaced the 'Land of Hope and Glory' Finale in 1902:

> Peace, gentle Peace, who, smiling through thy tears,
> Returnest, when the sounds of war are dumb . . .
> Our earth is fain for thee! Return and come.

The Coronation of 1911 would be celebrated without that prayer.

For the Coronation Service in Westminster Abbey, the Abbey organist Sir Frederick Bridge had asked Edward for a March and the Offertory. The Offertory was to be set to words from Psalm 5 beginning 'O hearken Thou unto the voice of my calling'. Edward's setting was filled with soft echoes of the Second Symphony's downward plunging figure.

His *Coronation March* extended the dark reflection. It was to be the Recessional in the Abbey Service. Through rehearsals and performances of the new Symphony, he had finished and scored a huge movement of symphonic proportions. As the occasion was far from a symphonic one, the construction was deliberately loose. Half a dozen separate ideas filled the long exposition:

the first, in a slow triple metre, was transfigured from a sketch for the *Rabelais* ballet. Yet the tone of the music was utterly consistent throughout. Scored for the Coronation orchestra with harps, full percussion, and a big part for the Westminster Abbey organ, Edward's *Coronation March* would sound a dark-hued, angry splendour to tower up over the royal proceedings with a *Nobilmente* that looked away beyond festivity. Together with the subdued *Offertory*, it sounded his own sense of what the new reign was to promise.

The Elgars' invitations arrived for the great ceremony—followed a week later by another official envelope:

June 17 1911. E. was up looking at letters. He suddenly looked up & said 'It is the O.M.' What a thrill of joy—A. cd. see the pleasure in his face—*The* thing he wished for so much—D.G. for such happy moments—Such a dear day followed. E.'s first idea was to give a present to each servant & joy was general. He wrote a beautiful letter of acceptance . . .

He also wrote to Adela Schuster—in the tone of a gifted child exhibiting some special success to a favourite aunt:

. . . of these two letters I am very proud;—please do not reprimand me for this: it is of course a secret until announced but I dare trust you with this and anything & everything belonging to

Edward Elgar. O.M.
Think of that![1]

Three days later the Coronation Honours were published:

June 20. E. A. & C. to Rehearsal at Westminster Abbey—there before 10. Most delightful time. After Music rehearsal, E.'s Offertorium & March, most beautiful, we went into the Nave & sat in front seats & saw rehearsal of procession, peers in robes &c. Some of them & Mary [Lygon] Trefusis saw E. & called out their Congratulations . . .

Did the sight of all this inherited rank and splendour mock even his own Order of Merit? Whatever it was, he suddenly announced that he was not going to the Coronation and he refused to allow Alice to do so either. The refusal astounded their friends. Miss Burley (recently returned from Portugal and now living in London) wrote: 'Alice, whose devotion was proof against almost any humiliation, was really hurt by this prohibition.'[2]

There was disappointment again over their continued attempts to buy the big house in Hampstead. When Edward called at Novellos to ask if future prospects would answer to such an expenditure, he had learned that only a small number of performances of his music were booked—fewest of all of the new Symphony. On receipt of Novello's annual royalty account, he sent a formal notice to terminate the mutual agreement by which the firm had published virtually all his music since 1904. It did not mean that Novellos would

---

[1] 18 June 1911 (HWRO 705:445:6843).
[2] *Edward Elgar: the record of a friendship*, p. 190.

never publish any new Elgar work. But now they would bid against every other interested party for future scores. By the same token, Edward's music was cast on the open market to stand or fall at popular response. If that response was waning, it was the worst move he could have made.

A year's notice was required on either side, but the firm's secretary Henry Clayton sent a reply generously offering to disregard the waiting period and terminate immediately. This took Edward's breath away. Indecision was all too clear in a long letter he wrote to the chairman Alfred Littleton:

I want to tell you at once that I have made no other arrangements and have not contemplated making any. I am not dissatisfied with the firm, although there are some minor points we might have adjusted—not worth considering really apart from the big question, which is as follows.

I have never deceived myself as to my true commercial value & see that everything of mine, as I have often said, dies a natural death;—if you look at the accounts you will see that a new thing of mine 'lasts' about a year & then dies & is buried in the mass of English music; under these inevitable circumstances it seems to me that the royalty system we adopted in 1904 cannot really be satisfactory to either of us. I am now well on in years & have to consider a 'move' & make a new home—under the depressing state of my music I have to reconsider this entirely & shall probably go abroad or to a cottage in the country & leave the musical world entirely.

My reference to a 'sum down' refers to the fact that other publishers have offered me in the past, a substantial sum for a new work: under the present strain this wd. suit me better & there is no reason, that I see, that your firm shd. not do this: only I have *no* work on hand & contemplate no large work in the future—I may *think* of large works but I shall not write them; to write them is labour lost.

. . . I thought that a formal notice was absolutely necessary for the firm as a Ltd. Company:—Also the notice (12 months) which the firm waives if I like, had better be adhered to.[3]

Despite the protest of going abroad or to a cottage, his principal conductorship of the London Symphony Orchestra meant he would live in London. On 17 July he took Rosa Burley to the house in Hampstead. She saw the contradictions:

. . . as we walked through the empty rooms I saw that he meant, if it were financially possible (which I rather doubted), to live there. This was one of the strangest afternoons we ever spent together and I have never known the duality of his character so strongly marked as it was that day.

On the one hand he clearly took a natural pride in the importance of the house with its fine panelling, its long music room, and its great staircase at the head of which Alice would stand to receive her guests. But on the other hand he wanted equally clearly to make me feel that his success meant nothing to him and that there was always some lovely thing in life which had completely eluded him. As we explored the empty house he drew my attention to its beauties, but he also told me that the only part of his life that had ever been happy was the period of struggle at Malvern, and that even now he never

---

[3] 30 June 1911 (Novello archives).

conducted his music without finding that his mind had slipped back to summer days on the Malvern Hills, to Birchwood, or to the drowsy peace of Longdon Marsh.[4]

The final summer at Plas Gwyn was a blank. At the end of July Alice wrote to beg the Stuart-Wortleys to come and stay at Hereford. But they were off to Venice and Bayreuth—where Hans Richter had again invited Edward, and Edward felt again he could not go. In August he went to The Hut for a week—though it was the week in which Carice came of age. She finished a difficult year, culminating in tonsillectomy, but was uncomplaining as ever. Alice chronicled:

August 14 1911. E. at Hut—Carice's 21st birthday—D.G. Well again I hope & grown very helpful & wise & very charming. A very fine lofty character—strong feeling of duty—& full of bright spirits—Sorry not enough scope for those always . . .

They visited the Speyers at Ridgehurst—where Edward discovered the game of billiards, together with the place of the billiard-table in a wealthy and fashionable house. He took himself off alone to the Lakes to see again the places he had seen with Dr Buck close to thirty years earlier. But that did not serve, and in a few days he telegraphed for Alice to come and join him for the rest of the holiday.

He had told Ivor Atkins they were not coming to the Worcester Festival. But then he relented, asking Atkins to make certain of placing the 'OM' directly after his name in the List of Stewards:

Worcester people (save you!) seem to have small notion of the glory of the O.M.—I was marshalled correctly at Court & at the Investiture *above* the G.M.C.G. & G.C.V.O. —(the highest Ld. Beauchamp can go!)—next G.C.B. in fact: such things as K.C.B.s &c are *very cheap* it seems beside O.M.[5]

Alice's father, the old Major-General, had been a KCB.

Once again for the Festival they had Castle House, with its memories of a childhood now nearly half a century in the past. On the opening Sunday 'E. down to river & reeds where the wind sang to him amongst them.' On Tuesday *Go, Song of Mine* prefaced Beethoven's Ninth Symphony. On Wednesday Edward conducted the Second Symphony in the afternoon, and dominated the Secular Concert in the evening with the *Coronation March* and *Sea Pictures* (sung once again by Muriel Foster). On Thursday morning and afternoon they were to have the *St. Matthew Passion* performed for the first time in the new edition—on which Atkins had after all done most of the work. Edward arranged two of the Chorales for brass instruments to be sounded—as at Bayreuth—before each part of the performance. Three trumpeters, four horns, trombones, and tubas climbed to the roof of the Cathedral tower both morning and afternoon to sound each Chorale four times—to north, east, south, and west. Robert Buckley wrote: 'Aloft, out of sight, a hundred and

---

[4] *Edward Elgar: the record of a friendship*, pp. 191–2.   [5] 17 July 1911.

seventy feet in the air, the brass of the orchestra gave forth the old German chorales harmonised by Bach two centuries ago. They rang out over the College Green, those perfect harmonies, over the city, over the river, the meadows, and the hills.'⁶

In the Cathedral, Atkins conducted the *Passion* with distinguished soloists and the violin obbligati played by Fritz Kreisler. That evening Kreisler played Edward's Violin Concerto.

Through the summer Eugène Ysaÿe had been studying the Concerto, of which he was to give four performances with Henry Wood in September and October: one was for the Norwich Festival. Ysaÿe was a generation older than Kreisler, and there were rumours of increasing trouble with his bowing.⁷ Yet he held his place as the greatest lyrical player alive, and Edward looked keenly forward to his interpretation.

At the beginning of September Ysaÿe sent a peremptory demand to Novellos for the orchestral material—to go through the Concerto with the Brussels Orchestra, of which he was conductor. Novellos pointed out that a hire charge would be made. Whereupon Ysaÿe cancelled all the English performances of the work and sent an outraged letter to the firm:

Je vois par la lettre que vous m'écrivez que je me suis trompé, qu'il y a a *Marchandage*, que vous me tenez un langage qui s'éloigne singulièrement de l'idée artistique qui est la seule qui m'ait guidé, moi, pendant le long travail du Concerto. Je ne discute pas les raisons d'interêt que vous faut agir; je me refuse d'entrer dans une discussion pénible à ce sujet, je crois seulement qu'il est peu possible que le Compositeur vous ait invité a m'écrire les misérables considérations dont la lecture fait mal au coeur. Je regrette de devoir vous apprendre que je renonce a exécuter le Concerto d'Elgar; cette résolution m'est dictée par une raison de dignité où les transactions commerciales n'ont rien à voir.⁸

Alfred Littleton wrote to Edward:

The *great artist!* thinks of nothing but collecting his own exorbitant fees and it seems to me that the words 'transactions commerciales' and 'Marchandage' apply much more to him than to anyone else. I should like to show the whole thing up in the public press.⁹

The firm's secretary, Henry Clayton, advised that if they gave way to Ysaÿe, it would be difficult to justify making performing charges to Kreisler or any other violinist who proposed to play the Concerto. But Edward had expressed such eagerness to hear Ysaÿe perform his music that Novellos capitulated and sent the score and parts to Brussels without charge. They specified a hire charge

---

⁶ *The Birmingham Gazette*, 15 Sept. 1911.

⁷ After a Berlin rehearsal of the Concerto in Jan. 1912, Ysaÿe said to Carl Flesch: 'Ah, si j'avais la tranquillité de votre archet!' (Flesch, *Memoirs* [Rockliffe, 1957] p. 81.) Ysaÿe himself described the Berlin performance which followed that rehearsal as 'just a fight'. (Antoine Ysaÿe and Bertram Ratcliffe, *Ysaÿe* [Heinemann, 1947] p. 112.) Yet Kreisler praised this performance highly.

⁸ 13 Sept. 1911 (Novello archives).

⁹ 17 Sept. 1911 (Novello archives).

only for the actual performance, to be paid by the orchestra and not the soloist. Ysaÿe retorted that he would not play the work if any fee was paid by anybody whatever.

In respect of the Norwich Festival performance, Henry Wood gave Edward his assurance that the fee would be paid whatever Ysaÿe thought, and begged him to give way on that basis. So Edward telegraphed to Ysaÿe: 'Pour vous aucunes conditions.' But when Ysaÿe arrived at Norwich he would not hear of the Elgar Concerto, and played the Beethoven instead. Edward conducted *The Kingdom* at Norwich. The diary chronicled no meeting with Ysaÿe.

But they had not heard the last of him. As he could not play the Concerto in England without a fee being paid, in the coming months Ysaÿe used the orchestral material retained from Novellos to give performances in Bremen, Berlin, Königsberg, Vienna, and possibly in Russia (where European copyright could not be enforced). The publishers' demands for the return of the music were ignored. They proposed to sue him in the Belgian courts. At last Edward himself had to write a diplomatic letter which finally brought the Ysaÿe performances to an end.

Then Kreisler, having heard Ysaÿe's Berlin rehearsal, demanded equal right of free performance. At last Edward and the publishers were forced to announce that performing fees for the Concerto would no longer be charged. Then they had the task of explaining why the charges had been made at all, and why the Symphonies were not made free as well.

The autumn of 1911 was meanwhile full of concert engagements for Edward. In October he conducted two concerts at Turin as part of an international exposition: the orchestra of 125 had been drilled in both the *Variations* and the *Introduction and Allegro* by Toscanini (who had toured both works). Then there were the six London Symphony concerts extending through the winter. They demanded the preparation of a wide repertory, including on 16 November the Brahms B flat Piano Concerto with Donald Tovey as a rather diffident soloist, and the Saint-Saëns Cello Concerto with Casals on the 20th. On 4 December Edward asked Alexander Mackenzie to take the baton for a performance of his own *Tam O'Shanter*: it was the first of a series of invitations to British composers whose works were less played than his own.

Headaches and depression surrounded continuing negotiations over the big house in Hampstead. Alice fought a long legal battle to break the trust which held her capital, and at last succeeded. On New Year's Day 1912 she could write: 'Entered E.'s own House—May it be happy & beautiful for him.' Carice evaluated it thus:

It was by no means everybody's house as it would only accommodate a small family such as ours as everything was sacrified to the long stately corridor and the large music room and annexe, a large dining room and large basement, two large bedrooms and three quite small ones, and two even smaller for the staff. The music room was panelled and

Mr. Edward Speyer gave my Father a wonderful present by having the small room off the music room fitted with shelves and cupboards to make him a lovely study and library. Every fitting was of finest workmanship, on the floor of the small entrance hall was a wonderful mosaic and on the door itself a marvellous beaten brass design . . .[10]

Edward named it Severn House. When Dora Penny came and remarked, 'You *are* in clover here,' he answered: 'I don't know about the clover—I've left that behind at Hereford—but Hereford is too far away from London; that's the trouble. Look here, do you see that I've got room for a billiard-table? . . .'[11]

One of the first visitors was Robin Legge, who reported his visit in *The Daily Telegraph* under the title 'Sir E. Elgar, O.M., at Home':

Having settled down in Elgar's fine-proportioned music-room, which has a charming library attached, I plied him with questions as to his work, and a more difficult subject to 'draw' on a matter of this kind I have not often met.

Still, I did obtain the news that the work upon which Elgar is at present engaged is a setting of the 'Ode' [*The Music Makers*] with which the late Arthur O'Shaughnessy opened his beautiful little volume of poems entitled 'Music and Moonlight' . . . I do not think I am committing an indiscretion when I confess my belief that in the composition of the contralto solo in this 'Ode' Elgar steadfastly holds in his mind the lovely quality of voice and the distinction of Miss Muriel Foster . . .[12]

Having reached back to 1905 for the Violin Concerto beginning, and to 1903 for the Second Symphony, he was now returning to the same era for his newest project. Before finally assembling his sketches into a connected work, he negotiated its production at the Birmingham Festival in October. But just as he began to work at it, a commercial project offered to pay for the move to Severn House.

The impresario of the Coliseum Theatre, Oswald Stoll, was to produce an 'Imperial Masque' to celebrate the Indian Coronation of the new King and Queen in 1912. *The Crown of India* was to show India and all her cities assembling with 'John Company' (a personification of the British East India Company) and St. George, to do honour to England. The press announced that £3,000 was to be spent on the production, with costumes and settings by Percy Anderson. The actress Nancy Price was engaged for the role of 'India'. Through pantomime and processions, music was to play a vital role. Stoll worked hard to persuade Edward that this was a project worthy of his muse. The financial reward was great—especially if the composer would rehearse and conduct the run, which would take pride of place in a varied music-hall programme. At length Edward agreed, and began to look through old sketchbooks for usable ideas. W. H. Reed remembered:

He sent me an S O S one day to come and help with some of the orchestration. Such an invitation was always more than welcome; but we had not been at work very long when, amid intense excitement, a billard-table arrived, and was duly installed in another

---

[10] Quoted in Young, *Alice Elgar*, p. 167.
[11] Powell, *Edward Elgar: memories of a variation*, p. 98.   [12] 6 Jan. 1912.

good-sized room leading out of the main studio. After that *The Crown of India* faded out.

A day or two later the telephone bell rang—I answered—Sir Edward's voice speaking—when could I come and play billiards? Now, I don't know whether other people had the same experience as I had with that billiard-table; but I never once succeeded in getting Sir Edward to *finish* a game. We would get as far as 30 or 40, or perhaps even 50; but he always switched off to the study of diatoms—he had added a microscope to his *ripieno* instruments . . .

Personally I much preferred this diversion to the ordinary game of billiards: a game at which I have no skill. But that did not matter; for, as far as I could gather from our interrupted attempts, Sir Edward hadn't very much more himself, although, as in everything else he set himself to do, he was dead serious about it, and would never attempt the most obvious shot until he had thought out exactly where the balls were likely to be, not only for the next shot, but for the one after that. It was too brainy for me—for either of us, I imagine, as I look back on the scene—at any rate, after these exhausting calculations, the obvious shot was missed more often than not; so the rest of the plan did not materialise; and it was my turn to take the cue and make either a *faux pas* or a fluke, after which the calculations were resumed with Einstein-like intensity.[13]

Yet *The Crown of India* went forward. He raided sketchbooks and in one case a published piece—*In Smyrna* of 1905—to create introductions, melodramas to support speeches, songs, interludes, and marches. The Indian elements were replete with exotic rhythms and harmonies. St. George was drawn in a good square tune, and 'John Company' was given a broad *Tempo di menuetto* to identify the established state with happiness and order. Elsewhere amid the pomp and show, violin solos against a quiet ensemble recalled the Violin Concerto not only in texture but more than once in melody as well.

Inevitably *The Crown of India* went farther down the path of the *Coronation March*—simpler melody, smaller development, more repetition. He wrote to Alfred Littleton on 8 January that the masque would last 30 minutes; but when finished and put into rehearsal at the end of February, it turned out to take an hour.

Littleton came to Severn House on 12 January to discuss publishing the new music on the basis of a big sum down. Edward referred him to Stoll's manager at the Coliseum, Bertie Shelton. But Shelton made other arrangements—with the firm of Enoch, who had recently published the Humperdinck music to Max Reinhardt's enormously successful pageant play *The Miracle*. When Henry Clayton gently protested that it was by Edward's own wish that the agreement with Novellos was still in force, Edward apologized that he had quite forgotten:

. . . I will only say that I do not think the Masque is in your line at all & I do not see how your firm could make it a commercial success as others, who exploit this sort of thing, might do.[14]

On top of everything else, as he told Clayton, he was ill again.

[13] *Elgar as I Knew Him*, pp. 49–51.   [14] 25 Jan. 1912 (Novello archives).

He had been ill amost continuously since the move to Severn House. Now there was an alarming new trouble. He had turns of giddiness, which caused loss of balance. More than once he had actually fallen down in the great panelled music room and study. The eminent physician Victor Horsley diagnosed 'gout in the head'—curable with massage and, as soon as possible, with rest.

But there was no rest, as conducting engagements loomed with the Hallé Orchestra in Leeds, the Philharmonic Society, and the continuing London Symphony concerts. At each of those Edward asked a younger composer to direct a work of his own: Hamilton Harty conducted *With the Wild Geese*, Percy Pitt his Symphony in G minor, and Josef Holbrooke *The Raven*.

Then daily rehearsals began for *The Crown of India*. Edward attended the Coliseum every morning except Sundays for a fortnight, and the press were invited to witness the efforts of the latest OM for the music-hall. *The Standard* reported on 1 March:

Unlike most great composers, Sir Edward Elgar plays his own score at the piano, accompanying chorus and solos with extreme care and wonderful patience. He goes over separate bars, repeats special passages, and suggests alterations in phrasing, emphasis, and light and shade with untiring zeal. Then he will suddenly leave his place at the piano and, while a deputy succeeds him at the instrument, beat time with his walking-stick . . .

Sir Edward Elgar would commit himself to no special opinion regarding his first definite contribution to the programme of a big music-hall. 'It is hard work, but it is absorbing, interesting,' he said, during a pause in the proceedings. 'The subject of the Masque is appropriate to this special period in English history, and I have endeavoured to make the music illustrate and illuminate the subject.'

All the while his energy was drained by the strange and secret illness. The actress Nancy Price recalled:

Alas, his bright spirit wore his body to a rind. He gave too generously of himself and made too heavy a demand on his physical strength. I remember once—when he was utterly exhausted during a rehearsal of the 'Crown of India'—he allowed me to take his place and conduct a rehearsal. I have never forgotten the thrill of this! Naturally, little was required of me, as the orchestra knew him and his wishes so well . . . I remember Elgar laughing and saying,

'You will never forget that once, when I was resting on the couch in your dressing-room and sipping hot milk, you conducted one of my works.'[15]

The production opened on 11 March. The scene showed

A Temple typifying the legends and traditions of India. At the back is a view of the Taj Mahal at Agra. In front of it and occupying the entire scene is a semi-circular amphitheatre of white marble, its boundary defined by tiers of steps at the summit of which is a semi-circle of sculptured and fretted seats of marble, for the Twelve Great Cities of India . . . After a musical prelude, the curtain rises on darkness, upon which a

[15] *Into an Hour-Glass* (Museum Press, 1953), p. 213.

faint steel-blue light gradually dawns, warming by degrees to amethyst, which slowly changing to rose, is finally succeeded by a golden glow . . .[16]

So dawned the light of Empire. The Masque proved enormously popular with Coliseum audiences, and through the first fortnight of the run Edward conducted two performances a day—often interspersed with further rehearsals. He wrote to Frances Colvin:

When I write a big serious work e.g. Gerontius we have had to starve & go without fires for twelve months as a reward: this small effort allows me to buy scientific works I have yearned for & I spend my time between the Coliseum & the old bookshops: I have found poor Haydon's Autobiography—the which I have wanted for years—& *all* Jesse's Memoirs (the nicest twaddle possible) & metallurgical works & oh! all sorts of things—also I can more easily help my poor people [his sisters Lucy and Pollie and his brother Frank and their families]—so I don't care what people say about me—the real man is only a very shy student & now I can buy books—Ha!ha! I found a lovely old volume 'Tracts against *Popery*'—I appeased Alice by saying I bought it to prevent other people seeing it—but it wd make a cat laugh. Then I go to the N. Portrait Gallery & can afford lunch—now I cannot eat it. It's all very curious & interesting & the *people* behind the scenes so good & so desperately respectable & so honest & straight-forward—quite a refreshing world after Society—only don't say I said so.

My labour will soon be over & then for the country lanes & the wind sighing in the reeds by Severn side again & God bless the Music Halls![17]

But the end of this turn was otherwise. On 28 March Alice noted: 'E. very uneasy about noise in ear.' It was a recrudescence of the 'gout' which Victor Horsley had diagnosed. Now Edward consulted another eminent physician, Sir Maurice Abbott Anderson, who ordered an immediate rest cure—no writing of any kind or going out of the house for several weeks. The prognosis was deeply disturbing. Mrs Joshua's son-in-law Dr Hugh Davidson said quietly to his family that there were all the symptoms of Menière's Disease—an affliction which could cause progressive deterioration of the middle ear to total deafness.[18] The Elgars' friends kept rigid silence both in public and especially before Edward himself. But he knew, after years of undiagnosed illness and depression, that the threat to his creative life now inhabited a definite shape. There was, however, another possibility—not recognized by the medical profession in patients until many years later: giddiness might be only the symptom of a deep desire to escape.[19]

Almost the whole month of April 1912 was spent 'in cold storage', as Edward said. Gradually he was allowed to walk as far as Hampstead Heath, and at the end of the month to write a few letters. On 27 April he wrote once more to Frances Colvin: 'I took a walk again this p.m. & saw a flock of sheep—bless them—talking a language I have known since I was two years old . . .'[20]

[16] Henry Hamilton, *The Crown of India* (Enoch & Sons, 1912), p. 3.
[17] 14 Mar. 1912 (HWRO 705:445:3435).
[18] Conversation with Dr Davidson's daughter Mrs Winifred Christie, 1976.
[19] Conversation with Dr John Barretto, 1980.   [20] HWRO 705:445:3436.

All this time Severn House was growing more beautiful under Alice's hands. As Edward began to emerge, she instituted regular Sunday afternoons 'At home' quite in the old style of her Victorian girlhood, when friends might count on finding them disengaged. She hoped to attract a salon of 'eminent men' to amuse and distract Edward. Amongst the first guests was another gentleman of the press:

One of the most striking features is the huge music room, where Sir Edward may sometimes be persuaded to sit at the great piano on a Sunday afternoon and improvise for the pleasure of intimate friends who drop in to tea. In the billiard room—billiards is one of Sir Edward's chief recreations—Lady Elgar has arranged a fascinating collection of wonderful trophies presented to her husband at the various musical festivals at which his works have been produced.

Most attractive of all is the Blue Study—Lady Elgar's special pride—where carpet, chair covers and hangings are all a lovely shade of deep blue, blue-bound books abound, and some fine specimens of old blue Bristol glass give a note of glorious colour. Deep blue flowers are chosen, and even the blotting paper in the wide writing pad is a deep shade of blue.[21]

*　*　*

At last in early May he was able to begin formal composition of *The Music Makers*, promised for the Birmingham Festival now just five months away. He was to write of the poem:

The mainspring of O'Shaughnessy's Ode is the sense of progress, of never-ceasing change; it is the duty of the artist to see that this inevitable change is progress. With a deep sense of this trust, I have endeavoured to interpret the Ode as shewing the continuity of art, in spite of those dreamers and singers who dream and sing 'no more'.[22]

It was the paradox in the poem's final stanza:

> Great hail! we cry to the comers
> 　From the dazzling unknown shore;
> Bring us hither your sun and summers,
> 　And renew our world as of yore;
> You shall teach us your song's new numbers,
> 　And things that we dreamed not before:
> Yea, in spite of a dreamer who slumbers,
> 　And a singer who sings no more.

To every Gerontius waiting at his riverbank, then, there should appear the mirror-image of the child lingering still on the opposite shore to catch the faintest stir of promise from the Broadheath side of things.

In interpreting O'Shaughnessy's Ode, I have felt that his 'music makers' must include

---

[21] Cutting from an unidentified paper, 13 May 1912. On 10 May Alice Elgar had noted in the diary: 'Tea party, our first at Severn House.'

[22] 'Introductory note' to *The Music Makers* sent on 14 Aug. 1912 to Ernest Newman, for his guidance in writing programme notes for the first performance.

not only poets and singers but all artists who feel the tremendous responsibility of their
mission to 'renew the world as of yore'.

As I have felt, so I have insisted on this responsibility, therefore the atmosphere of the
music is mainly sad; but there are moments of enthusiasm, and bursts of joy occasionally
approaching frenzy; moods which the creative artist suffers in creating or in
contemplation of the unending influence of his creation.

Yes, suffers:—this is the only word I dare to use; for even the highest ecstacy of
'making' is mixed with the consciousness of the sombre dignity of the eternity of the
artist's responsibility.[23]

That responsibility would have been easier to hymn in the heady days of the
First Symphony success than after the relative failure of the Second. But it was
precisely now, when the realized dreams of position and residence were
mocked by the whispers of a troubled ear, that he most needed the poem's
assurance.

He decided to set the entire Ode without change or deletion. Not since *The
Black Knight*, begun in 1889, had he undertaken a major setting without
altering, abridging, or shaping the text. Had these constructive powers
been exhausted by the efforts of Symphony and Concerto? Or did he see
the Ode's difficult and occasionally embarrassing lines as sharper tests for
his art?

In the years since he first thought of setting the Ode, some melodic ideas had
appeared. But they were short and relatively few. To offset this weakness, he
resolved to enrich the setting with quotations from the whole range of his own
past music. So the best ideas over all his maturity would pass in
procession—perhaps toward some synthesis to replace the lost *Apostles*
synthesis.

At the head of his past themes stood the 'Enigma'. Thirteen years after the
Variations had appeared, the significance of the 'Enigma' was still not
understood—either by the variations' subjects or by the public at large.
Therefore the 'Enigma' music would dominate *The Music Makers* in a way
which ought to explicate the whole matter for everyone with ears to hear the
words to which it would now be set. The primary and secondary 'Enigma'
themes were in fact shadowed in the primary and secondary themes for *The
Music Makers*:

23 Ibid.

Each of the *Music Makers* subjects found a descending reply. Set side by side and all together, these first and second subjects and answers filled a whole page of short score. He sent it to Alice Stuart-Wortley inscribed 'The complete understanding'. The second subject was marked 'molto cantabile *with great joy*'.[24] This page of music would make an orchestral epigraph for the Ode. Next came the first 'Enigma' subject, quoted in full and then intertwined with the *Music Makers* first subject. Edward wrote:

I have used the opening bars of the theme (Enigma) of the Variations because it expressed when written (in 1898) my sense of the loneliness of the artist as described in the first six lines of the Ode, and to me, it still embodies that sense . . .[25]

The chorus entered in quiet hymn-like harmonies over descending steps which summarized the whole panoply of Second Symphony 'Delight' and its plunging answer in half a dozen chords of hushed grandeur:

Edward called it his '"artist" theme'.[26]

' . . . We are the dreamers of dreams' brought the 'Judgement' motive from *The Dream of Gerontius*. 'Wand'ring by lone sea-breakers' invited the 'Sea Slumber Song' from *Sea Pictures*. 'And sitting by desolate streams' gave the 'Enigma' its initial verbal illustration. But the opening stanza ended with the first of the poet's big claims:

> Yet we are the movers and shakers
> Of the world for ever, it seems.

The suddenness was emphasized by the weak ending that served the rhyme. Edward set it in a precipitous crescendo, and he supported the final 'it seems' with another big 'Enigma' quotation. It showed the trouble that was to pursue the entire setting: the assurance he was seeking was beyond the assurance of his own art.

On 23 May Alice wrote: 'E. not very happy over his work—not "lit up" yet.' The following afternoon he conducted the *'Enigma' Variations* at a Memorial

[24] 15 May 1912 (Elgar Birthplace).
[25] Introductory note.
[26] 14 Aug. 1912 to Ernest Newman (copy in Alice Elgar's hand, BL Add: MS 47908 fo. 85).

Concert to benefit family survivors of the musicians who had gone down in the *Titanic*, whose fatal maiden voyage in April epitomized so many broad-based hopes and fears of 1912. Then he wrote to Hubert Leicester (who had carefully kept all the youthful writings for the wind quintet):

> I wonder whether you would lend me the old *Shed*-books: I should very much like to see some of the old things & perhaps to copy some of them . . .[27]

May 30 1912. E. going on with his [*Music Makers*] work—Not very happy over it—Gives A. a pain in her heart for him to be so tensed.

In the second stanza he met the worst of O'Shaughnessy's lines:

> With wonderful deathless ditties
> We build up the world's great cities.

This jingling feminine rhyme he set to the 'artist' theme, roared out by the chorus through orchestral exclamations. 'And out of a fabulous story' found a strain of 'Enigma' reminiscence—converted to quasi-*fugato* for 'We fashion an empire's glory'. Here the words were too much for him. He set the accompaniment with trumpet scraps of the *Marseillaise* and *Rule, Britannia*, and wrote to Ernest Newman: 'You will be interested to see how they go together & the deadly sarcasm of that rush in horns & trombones in the English tune—deliberately commercialising it . . .'[28]

'One man with a dream' brought a sudden hush over the falling second of Gerontius's 'Judgement'—only to give place to truculent counterpoint

> . . . at pleasure,
> Shall go forth and conquer a crown;
> And three with a new song's measure
> Can trample a kingdom down.

The 'artist' theme thundered up to descend *fff* on the final word. The sheer volume was too big for the short melodic figures. But the Birmingham Chorus were waiting for the first printed sheets to begin their rehearsals.

So far there had been little musical development. In the softer third stanza he introduced two variants of the rising second subject. The first variant

sounded in the orchestra against a hushed choral chant calling up the dimmest pages of history. Nineveh emerged in a figure which seemed to cross *The Crown of India* with 'Dies Irae':

---

[27] 29 May 1912.　　[28] See note 26.

pp legatissimo

Babel was overthrown in a mighty counterpoint against the whole-tone scale. Out of the ruins came the second variant

mellifluous choral polyphony prophesied 'To the old of the new world's worth.' Then it entered the one brief commanding development in all this music:

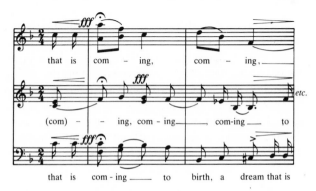

But the arrival was not yet. Without so much as a repetition the music sank to a *pianissimo* return of the opening 'We are the music makers, and we are the dreamers of dreams'. So ended the first third of the setting.

On 3 June (the day after his fifty-fifth birthday), he went to the Temple Church to hear Walford Davies try over the big anthem on the 48th Psalm written nearly two years earlier in the wake of the Violin Concerto. He had finally sent it to Novellos as compensation for *The Crown of India*. The first performance was to take place in July at Westminster Abbey in a service to commemorate the 250th anniversary of the Royal Society.

He turned back to *The Music Makers*, 'happier over it'. The Ode's central three stanzas showed the artist's place in the living world. 'A breath of our inspiration' raised a ghost of the *Variations*' triumphant coda. 'Unearthly, impossible seeming' extended the *Gerontius* falling second in the slowest of all descending steps. Again the words broke in on his music's reverie. 'The soldier, the king, and the peasant Are working together in one': Edward drove the 'artist' theme toward a *grandioso* quotation of the music which approached the final climax in the First Symphony. The climax itself did not follow, but only a choral muttering over 'our dream' which almost echoed the receding Demons.

Stanza 5 brought the first contralto solo for Muriel Foster. 'They had no vision amazing . . .' juxtaposed several figures without developing any. 'But on one man's soul it hath broken, A light that doth not depart . . .' Here at last was a chance for real musical development. Yet Edward could not resist quoting 'Nimrod' through its entire slow melody of nineteen bars to set 'And his look, or a word he hath spoken, Wrought flame in another man's heart.'

Here I have quoted the Nimrod Variation as a tribute to the memory of my friend, A. J. Jaeger: by this I do not mean to convey that his was the only soul on which light had broken or that his was the only word, or look that wrought 'flame in another man's heart'; but I do convey that amongst all the inept writing and wrangling about music his voice was clear, ennobling, sober and sane, and for his help and inspiration I make this acknowledgement.[29]

The grand melody, quoted at full length, dwarfed all the short surrounding *Music Makers* figures—especially when 'Nimrod' found a wonderful coupling with the final descending figure from the Second Symphony in the same key.

That pairing of two big inspirations from the past hurt the setting of stanza 6. The verse enthused again, but Edward could not really believe. At 'therefore to-day is thrilling With a past day's late fulfilling' he sent a new rising variant to meet a ghost of Second Symphony farewell. The goal was another series of attempted climaxes from the earlier Nineveh and Babel music. 'We are the music makers' the chorus fairly screamed: but then came the inevitable falling away 'And we are the dreamers of dreams.'

In the Ode's final three stanzas, the Music Makers were lifted above the world of their ambiguous influence. 'Our souls with high music ringing' brought back the phrases of 'To-day is thrilling'. Straight to it came the 'Enigma', appearing for the last time to set

> O man! it must ever be
> That we dwell, in our dreaming and singing,
> A little apart from ye.

At these words came two memories from the Violin Concerto—the solo fantasy over the 'feminine' Windflower theme, the *Nobilmente* from the central *Andante*. Over it sounded *The Apostles* 'There shall ye seek Him' as a benediction. The soft ensemble made another exquisite moment. Edward marked a long pause.

Stanza 8 began quietly *come prima*: 'For we are afar with the dawning . . .'. Now came the long-awaited climax of the First Symphony: 'Out of the infinite morning Intrepid you hear us cry—' as chorus and orchestra thundered the

---

[29] Introductory note.

great 'motto' tune at last. But having no organic place in this music, the big climax only flashed as spasmodic inspiration and was gone. It was followed by the message of what it was they were all intrepidly crying:

> How, spite of your human scorning,
>   Once more God's future draws nigh,
> And already goes forth the warning
>   That ye of the past must die.

As he had retreated before the prospect of leading the Apostles toward their Last Judgement, so now he retreated again. Instead of finding any synthesizing development, he set these words to another single *stringendo* of the 'artist' theme *fff*; then the whole-tone scale counterpoint slid softly away from the final 'die'.

For the final stanza's idea of renewal in the past, the solo contralto reminisced at length over the *Music Makers* second subject and answer. Then for a coda she went back to the words of stanza 7. At 'Your souls with high music ringing' she found again the brief, lonely climax which had set the third stanza's 'coming'. The descending *Gerontius* shape of 'Unearthly, impossible seeming' set the final lines of the poem:

> Yea, in spite of a dreamer who slumbers,
>   And a singer who sings no more.

To it came a last quotation—Gerontius's 'Novissima hora est'. Then he ended the setting with a hushed recursion to the ambiguous opening:

> We are the music makers,
>   And we are the dreamers of dreams.

The concentrated pressure exerted through a forty-five minute setting for the biggest forces had not galvanized the poem's promised revelation in his own art. The collective presence of the quotations from his earlier works suggested that somewhere behind them all lay some final synthesis for his music—a synthesis which remained unachieved.

Thus the 'Enigma' quotations were especially apt. Yet when Dora Penny came and he played her the music, she understood it so little that she asked whether the theme hidden in the 'Enigma' was not 'Auld Lang Syne'. It was virtually the end of their friendship.[30] He dedicated *The Music Makers* to Nicholas Kilburn—because Kilburn was a choral conductor, and because with his high old-fashioned solemnity he had tried hard to take upon himself the dead Jaeger's role of chief encourager. Kilburn received it as the greatest honour of his life.

On 18 July Edward wrote the last note into the vocal score, and inscribed it with the name of a high place in Hampstead which he would recall as 'my

---

[30] Although this may seem to contradict some published recollections, it was conveyed to me very clearly by Mrs Powell herself in a conversation of July 1960, and it is confirmed by her son.

"spiritual home" for some years'[31]—'Judge's Walk, N.W. 1912'. The judgement seemed to pursue him through the weather itself, as he wrote to Alice Stuart-Wortley next morning:

Yesterday was the usual *awful* day which inevitably occurs when I have completed a work: it has *always* been so: but this time I promised myself 'a day!'—I should be crowned,—it wd. be lovely weather,—I should have open air & sympathy & everything to mark the end of the work—to get away from the *labour* part & dream over it happily. Yes: I was to be crowned—for the first time in my life—But—I sent the last page to the printer. Alice & Carice were away for the day & I wandered alone on to the heath—it was bitterly cold—I wrapped myself in a thick overcoat & sat for two minutes, tears streaming out of my cold eyes and loathed the world,—came back to the house—empty & cold—how I hated having written anything: so I wandered out again & shivered & longed to destroy the work of my hands—all wasted.—& this was to have been the one real day in my artistic life—sympathy and the end of work.

'World losers & world-forsakers for ever & ever'
How true it is.[32]

Through late July he arranged a Suite from the *Crown of India* music and orchestrated *The Torch* and *The River* for Muriel Foster to sing at the Hereford Festival. Then he went a quarter century into the past to orchestrate *The Wind at Dawn*: it had been his first setting of Alice's words, and on 28 July he brought the new score to the breakfast-table and laid it before her.

Every day was dark and wet, every evening they had fires. 'Summer flying away,' Alice wrote on 30 July, and Edward's health was bad. In the midst of orchestrating *The Music Makers* he wrote to Alice Stuart-Wortley on 12 August:

My gout has quieted down but I am suffering from my ear noises &c &c—enough of this.

I have just finished the orchestration of 'the multitudes &c' [stanza 6] so nothing complicated remains . . .[33]

Eight days later he finished the score and inscribed it with the quotation from Tasso which he had set at the end of the *'Enigma' Variations* in 1899 and which he translated: 'I essay much, I hope little, I ask nothing.' He wrote to the Windflower again: ' . . . the end of my work is as dreary as that awful day when I finished the composition & perished with cold on the Heath. I cannot live much longer in this weather & loneliness.'[34]

That day saw the loss of another friend. Canon Gorton had retired to Hereford suffering from partial paralysis. His son wheeled his bath chair out to the bank of the Wye, which was in high flood, left his father for a moment, and somehow chair and occupant plunged down the bank into the swirling waters. The body was not recovered for ten days.

On Tuesday of the Hereford Festival week Edward took Muriel Foster to see

[31] 2 Apr. 1931 to Alice Stuart of Wortley (HWRO 705:445:7732).
[32] 19 July 1912 (HWRO 705:445:7676).
[33] HWRO 705:445:7828.   [34] HWRO 705:445:7673.

the old cottage at Broadheath. On Thursday he conducted *Gerontius* to the largest audience of the Festival week. After Hereford came rehearsals for *The Music Makers* at Birmingham (as he wrote to Alice Stuart-Wortley): 'now I have lost *all* interest in it. I have written out my soul in the Concerto, Sym II & the Ode & you know it & my vitality seems in them now—& I am happy it is so—in these three works I have *shewn* myself.'[35]

In the opening night concert of the Birmingham Festival, 1 October 1912, *The Music Makers* shared pride of place with the first British performance of Sibelius's Fourth Symphony, also conducted by its composer. Alice had ears only for Edward's music:

To performance of Music Makers—Most splendid & impressive. Wonderful effects of Orch[estratio]n & Chorus beautiful rendering. Muriel [Foster] splendid. E. conducted magnificently. Had a great reception—Dear Kilburns there & many friends.

The papers were less than enthusiastic. The poem, it was felt, had little bearing on twentieth-century experience:

Where 'The Music Makers' falls short is in the unreality of the theme. Arthur O'Shaughnessy's ode celebrates the feats of the world's poets in forging to ideals the destinies of their fellow-men . . . But did a single member of the chorus who sang those words, or one person in to-night's audience, really believe them? The fallacy of the poem lies, of course, in the fact that poets have nothing to do with 'teaching humanity' or in the building of empires or cities, but solely with the charming of one's finer senses and the enrichment of one's inner life. Music set to this ode could not therefore be expected to have great strength or sincerity.[36]

Robin Legge in *The Daily Telegraph* noticed the ambiguity of the musical setting:

I do not find on one hearing of 'The Music-makers' that its note is so much of sadness as of unsatisfied yearning. The composer seems to long himself to be convinced that the music makers are what the poet represents them to be; if they are, then surely here is a case of the most glorious optimism . . . Where the poet speaks in general terms, Elgar appears to look at the personal aspect of the matter. The music is often of exquisite beauty, but . . . its very mood is against it—this mood of yearning, alternating with a confident mood of massive power, and finally bringing a return to the prevailing lack of confidence, as if the subject were greater than the composer could translate into terms of music.[37]

A later reviewer did not mince words in dealing with the new score's overweighted quotations from previous works:

Sir Edward Elgar's Birmingham Festival novelty, 'We are the Music Makers', is not a case of new wine tasting like the old; in fact, it is the nips he permits us of the latter that make the latest vintage seem lacking in flavour and bouquet.[38]

---

[35] 29 Aug. 1912 (HWRO 705:445:7806).
[36] Cutting from an unidentified newspaper (Elgar Birthplace).
[37] 2 Oct. 1912.
[38] Cutting from an unidentified Brighton newspaper, 15 Nov. 1912.

Two nights later Edward and Alice went with Carice to the Birmingham performance of Verdi's *Requiem*. The tenor soloist was the young John McCormack, who recalled the scene years later for his biographer:

The quartet consisted of Aino Ackté, soprano; Muriel Foster, mezzo-soprano; John, tenor; and Clarence Whitehill, baritone . . . On paper this quartet seemed to promise an excellent performance. The promise was not fulfilled, and the soprano was to blame. Madame Ackté was . . . completely at sea. It might almost be said that she was still at sea, for her intonation was affected by a bad Channel crossing the day before. For that she was hardly to blame; but she did not know her part. In one place, where the others had to sing changing harmonies under a supposed E flat sustained through three or four bars, the result was more like Stravinski than Verdi . . .

In the artists' room afterwards the singers were looking at each other and saying, with little conviction, 'Well, it wasn't so bad,' when suddenly the door was flung open, and in walked a distinguished, soldierly-looking man with a splendid head and finely carved features. He had a magnificent flowing moustache, and it was bristling with indignation. He went straight up to Muriel Foster, and said in a voice loud enough for all to hear,

'That is the worst performance of Verdi's *Requiem* I ever heard.'

The singers knew it was none of the best, but John resented this criticism. He turned to Whitehill.

'Who the hell is this major-general?' he demanded, 'and what does he mean by rushing in here and giving his opinion unasked?'

'Good heavens,' Whitehill replied, 'that's no general! That's Sir Edward Elgar.'

John did not know whether to be more angry or less. All knew who was to blame for the performance, and he replied sulkily, 'He should not put us all in the same category.'

It soon became evident that he did not. Madame Ackté came forward smiling, and asked Muriel Foster to introduce her to the great composer. Elgar barely acknowledged the introduction. He nodded coldly, and would not shake hands. Ackté burst into tears.

This increased John's resentment. He went on fuming and grumbling, despite Whitehill's attempts to soothe him.

'After all,' said the baritone, 'Sir Edward Elgar ought to know something about music.'

'He may that,' John replied truculently, 'but thank God his music is better than his manners!'

And he nursed a prejudice which lasted until he next met Elgar, nearly [twenty] years afterwards.[39]

Later Edward wrote to Alice Stuart-Wortley: 'I was really ill all last week & you must forgive much to a sick man.'[40]

He conducted the second performance of *The Music Makers* at Brighton in November. Kilburn was doing it with his choir in the north, Ivor Atkins with the Worcester Festival Choral Society, and Frederick Bridge with the Royal Choral Society in London. But there was nowhere the eagerness which had attended the earlier Elgar choral works. When Fritz Steinbach came to tea at

[39] L. A. G. Strong, *John McCormack: the story of a singer* (Methuen, 1941) pp. 104–5.
[40] Monday [7 Oct. 1912] (HWRO 705:445:7109).

Severn House, he talked politely of conducting it at a Lower Rhine Festival; but Novello's pursuit of the matter was met with evasion.

When the critic Alfred Kalisch described Severn House for his paper, he paid special attention to the small study off the music room where Edward's real work was done:

On the floor repose Sir Edward's violin with which he fought his way through the world in his youth, and—incongruously enough—a trombone which he used to play. On the walls are memorials of more brilliant phases of his career—wreaths, all faded, given him on various occasions now historical . . .[41]

But little work was being done there now. When Ivor Atkins offered a commission to complete *The Apostles* trilogy for the Worcester Festival of 1914, Edward refused by saying it would not pay: 'At the present moment the whole world is given over to short things—plays & music suffer most. I am very busy just now with trifles which must be done . . .'[42]

He was extending and orchestrating a piece of the wind quintet music borrowed from Hubert Leicester. He called it *Cantique*, and allowed Landon Ronald to conduct it at one of his Sunday Concerts, as he wrote to Alice Stuart-Wortley: 'It will not be announced as *new* or *first time*—only to see how the public like it.'[43] At the concert on 15 December he found himself disappointed.

There was the new series of London Symphony concerts to prepare, involving the study of works he knew less than well. Several times he asked Alice Stuart-Wortley to help him play through a piano-duet arrangement of the Franck Symphony, before the concert on 24 November. On 9 December an all-Elgar concert drew a big audience. But he was ill and depressed, though the ear seemed quiescent. Carice recalled:

It is really impossible to say that there was anything definite wrong with him; it stemmed from digestive troubles which in their turn sprang from the fact that things were not going well; if something favourable happened digestion was forgotten. In the same way, he would complain of a terrible headache but if one could find something to interest him or if something exciting happened, the headache would be quickly forgotten.[44]

He decided on a more elaborate billiard-table with special lighting, and resolved to pay for it by selling the precious Gagliano violin back to Hill's. Alice was assigned this mission against her will: 'A. felt such a traitor to the thing she loves . . .' But when he and Carice introduced a small dog into Severn House, Alice put her foot down. After ten days' trial the dog was banished as 'not quite trustworthy'. A second animal suffered the same fate; a third ran away.

[41] *The World*, 22 Oct. 1912.   [42] 2 Dec. 1912.
[43] Wednesday [4 Dec. 1912] (Elgar Birthplace).
[44] Quoted in Young, *Alice Elgar*, p. 171.

In the depths of a blank winter, he gave up further London Symphony concerts for the season and made plans to go to Italy by himself:

January 29 1913. E. not well but feeling he must go somewhere—Sir Maurice [Abbott Anderson] came & advised his starting & said he cd. go alone. A. most mis at this idea—Getting things ready for E.

January 30. E. not at all equal to starting. Suggested going by sea & A. to go. Drove to [Thomas] Cook's in P.M. & gave up his ticket. E. very depressed at going to Riviera, said in car, 'If only we cd. go to Naples'. A. said 'Why not?' Naples then. & E. brightened from *terrible* depression & took passage for Naples next day.

They sailed from Tilbury Docks, and on 8 February arrived at Naples. After a few days Edward's depression was deeper than ever, and they went to Rome. When an enquiry arrived there from Hubert Leicester about the *Shed* books still at Severn House, he sent a promise of their return, and then found his mind turning back to St. George's Church where Hubert was still choirmaster: 'I think very much of Worcester—do you do the *old* Holy Week music? I should like to come down to the old spot for that week—but I suppose the memory of the old days is better left to die.'[45]

Yet it would not die. On another sheet of the hotel stationery he copied out a passage from *The Fourth Generation* by the social novelist Walter Besant about

. . . the great & numerous class which has to get thro' life on slender means, & has to consider, *before all things*, the purchasing power of sixpence. This terrible necessity, in its worst form, takes all the joy & happiness out of life. When every day brings its own anxieties about this sixpence, there is left no room for the graces, for culture, for art, for poetry, for anything that is lovely & delightful.

It makes life a continual endurance of fear, as dreadful as continued pain of body. Even when the terror of the morrow be vanished, or is partly removed by an increase of prosperity, the scars and the memory remain, and the habits of mind & body.[46]

Bad weather in Rome added to his troubles, and they decided to go home. They were packed and ready to start when bad news came from New York:

February 22 . . . First post a dear Valentine card from Pippa to E. & before we started another post brought terrible news of her illness—saying hopeless malady of wh. she knew nothing. E. & A. dreadful shock & mourned & wept all the way along.

Back in London they saw a nearly completed portrait of Edward by Sir Philip Burne-Jones, the son of the Victorian artist who had won a baronetcy for his Pre-Raphaelite dreams: Edward stood in profile in the curve of a grand piano, his hands gracefully poised round the lapels of his morning coat. It was to be included in the Royal Academy Summer Exhibition. But when Hubert Leicester proposed later to buy it for the Worcester Guildhall, Edward wrote to Troyte Griffith: 'I *wish* you cd. persuade the Worcr. people *not* to buy that

45 17 Feb. 1913.
46 MS at the Elgar Birthplace. The underlinings are Elgar's.

weak-kneed portrait! *do*. Phil is a good creature but is feeble & makes me stand like himself . . .'[47]

Another cure at Llandrindod Wells produced no improvement in his health. Still another Harley Street physician pronounced that there was no organic disorder. But when a long letter of philosophical encouragement arrived from Nicholas Kilburn, Edward answered:

Well, you talk mysteriously as becomes you & your northern atmosphere. I cannot follow you. I could have done a few years back—but the whole thing (—no matter how one fights & avoids it—) is merely commercial—this is forced into every fibre of me every moment. Not long ago one could occasionally shake this off & forget, but an all-foolish providence takes care that it shall not be *un*remembered for a moment & so—& so—?? You say 'we must look up?' To what? to whom? Why?

'The mind bold
and independent
The purpose free
Must not think
Must not hope'—

Yet it seems sad that the only quotation I can find to fit my life comes from the Demons' chorus! a *fanciful* summing up!![48]

* * *

He had been asked yet again to write a new work for the Leeds Festival, whose next meeting was to take place in October 1913. Stanford had departed as chief conductor, and Edward agreed to share the Leeds Festival conductorship with Nikisch and Hugh Allen. He specified a conducting fee of 50 guineas for each of his two concerts, and 100 guineas as a performing fee for the new work. Leeds agreed without a murmur.

To fulfil the Festival commission, he decided to carry out the *Falstaff* idea sketched a decade earlier—the portrait of chivalry as an ageing jester. So again, as at the beginning of his career, his music would be guided by a story, and again the theme was knightly. Yet the latest hero dissipated his energies until he was rejected by a new monarch. Edward wrote to Ernest Newman:

Falstaff (as programme says) is the name but Shakespeare—the whole of human life—is the theme. A theatre conductor cd easily have given a heavy scherzo & called it Falstaff—but you will see I have made a larger canvas—& over it all runs—even in the tavern—the undercurrent of our failings & sorrows

It traced in faint outline the plunging figure which had pursued the Second

[47] 12 Dec. 1922 (*Letters*, p. 278).   [48] 26 Mar. 1913.   [49] 26 Sept. 1913.

Symphony's 'Spirit of Delight'. Behind it still hovered the ghost of the 'tune from Broadheath' written almost fifty years earlier.

He based his 'symphonic study' on the Falstaff of Shakespeare's *Henry IV* plays (ignoring the buffoon in *The Merry Wives of Windsor*) and on the remarks of two critical authorities—Edward Dowden, the most warmly evocative of Victorian critics; and Maurice Morgann, whose 'Essay on the Dramatic Character of Sir J. Falstaff' (1777) found in the hero a nobility to match his weakness. Edward quoted Morgann to characterize the Falstaff his own music would show—'in a green old age, mellow, frank, gay, easy, corpulent, loose, unprincipled, and luxurious'.

He worked a little at the music in March 1913. Then he spent the first fortnight of April with his sister Pollie and her family at Stoke. Returning on the 16th, he sketched next day a descending figure to retrace the very notes of the 'tune from Broadheath' in an unequal rhythm:

It was to be the centre of an Interlude for Falstaff to rest in Shallow's Gloucestershire orchard. But the scattered ideas did not come together until close to the end of May.

The music opened with an unharmonized descending figure:

Edward wrote: 'This, the chief Falstaff theme, appears in varied *tempi* throughout the work, and knits together the whole musical fabric.'[50] It could function, then, like the motto theme of the First Symphony; but the harmonic implications were far indeed from that '*massive* hope in the future'.

Directly there came a second theme, showing Falstaff the humourist: 'I am not only witty in myself but the cause that wit is in other men.'

---

[50] *The Musical Times*, 1 Sept. 1913. This is the source of other verbal quotations and comments by Elgar unless noted.

Then followed a picture of the young Prince Hal 'in his most courtly and genial mood':

Prince Hal:
(statement)

(reply)

(In common with every other leading theme in *Falstaff*, the Prince Hal 'courtly and genial' theme was to be played through twice. Edward wrote out the *fortissimo* repetition on 26 May.[51]) The Hal figures were more diatonic than any of the Falstaff music so far, though their shapes of statement and reply reflected the reply and statement respectively of Falstaff's 'witty' theme. It was another friendship in music. But where fifteen years earlier in the *'Enigma' Variations* the hero had moved through perceptions of friends to his own definition, now the hero looked backward through his friend as if to a strength he himself no longer possessed.

A third Falstaff theme was inspired by the old hero's response to the Prince: 'Sweet wag, when thou art king . . .'. The music showed Falstaff 'cajoling and persuasive'. It bore the impress of Hal's diatonic expression:

III:
(statement)

But its reply went as easily down

as up

A quick recapitulation rose to a climax rounded off with the opening Falstaff I, as Edward wrote: ' . . . the persuasive Falstaff has triumphed, the dominating Sir John is in the ascendant.' So ended the first of four sections into which the music would divide. Its playing time of three minutes made it the shortest of the four—an introduction.

On 9 June came the long-expected cable from New York announcing the death of Mrs Worthington. That day Edward had to rehearse and conduct a second London performance of *The Music Makers*. Afterwards he sought escape,

[51] MS given to Alice Stuart-Wortley, now at the Elgar Birthplace.

talking *Falstaff* down to The Hut. But while he was there news came of another death—that of Lord Northampton: and the sudden end to one of Frank Schuster's old friendships cast a gloom over that usually sunny household.

Left behind in London, Alice took Carice to a cinema showing a film on the life of Wagner:

June 14 . . . A. & C. to Cinema of Wagner. Hindered by crowd looking at Suffragette Funeral—Cinema wonderful—Very harrowing, creditors interrupting composition &c. Splendid how possessed he was with his works, & no opposition stopped him.

When Edward returned on 17 June she was too excited: 'A. raced to meet E. in Music Room upstairs when he arrived, & slipped & fell on back, (E. tried to break fall). Silly impetuous!' No bones were broken, but such an accident in the mid-sixties was hard, and she was in pain for a long time afterwards. Yet she was on duty next day for the Sunday afternoon 'At home' to receive Henry James, J. L. Garvin, and Arthur Nikisch. Nikisch 'insists The Music Makers must be translated & done in Germany.' But it proved another false dawn.

Five days later came a letter from the London Symphony Orchestra regretfully terminating Edward's principal conductorship, as his concerts had resulted in losses for the Orchestra. Alice noted:

June 23. E. had letter from London S. Orchestra. It all hurt very much. A. most mis knowing it . . .

June 24. E. had Novellos a/c—disappointing—E. vesy depressed & A. *most* mis to see him so . . .

Two days later he was back with *Falstaff*, 'very hard at work'. The second section of the score was a musical amalgam of scenes scattered through the *Henry IV* plays showing life at the Boar's Head Tavern, Eastcheap. The new setting opened with a variant of the 'cajoling' Falstaff III to echo *Cockaigne*:

The 'short, brisk phrases', Edward wrote,

. . . should chatter, blaze, glitter and coruscate; no particular incident is depicted, but the whole passage was suggested by the following paragraph:

'From the coldness, the caution, the convention of his father's court, Prince Henry escapes to the teeming vitality of the London streets and the Tavern where Falstaff is monarch. There, among ostlers and carriers, and drawers, and merchants, and pilgrims, and loud robustious women, he at least has freedom and frolic.' [Dowden].

Instantly the music fell into the hands of the 'honest gentlewomen' of the tavern, lewdly twisting the Eastcheap figures round their own convenience. They, like everything else in the score, had risen from the implications of Falstaff's music. 'This most virtuous company', wrote Edward, 'flits across the fabric to find its full expression later.'

Through the women's figures stole a scrap of Eastcheap music repointed 'to fit the rhythm of Falstaff's

> When Arthur first to court [did come]
> And was a worthy king . . . '[52]

The melodic shape faintly shadowed the 'Judgement' in *Gerontius*. Falstaff's singing rose to an ensemble of 'Eastcheap' music set against the 'cajoling' theme wrenched up and down to raucous, off-the-beat explosions of low brass in isolated chords all too rich. The 'Gentlewomen' shrilled in *sforzando* lurches. Edward wrote to Troyte Griffith: 'Alice is horrified I fear with my honest gentlewomen—of course they must be in—do you think I have overdone them?'[53]

Then came the final Falstaff theme, with all the dotted rhythms smoothed to plethoric breathing. Edward wrote: 'The gargantuan, wide-compassed *fortissimo*, first given in the strings in three octaves, exhibits his boastfulness and colossal mendacity:'

IV:

It was inspired by the line: 'I am a rogue if I were not at half-sword with a dozen of them two hours together.' Thus Falstaff recounted his exploit of snatching the purse at Gadshill before arriving at Eastcheap. 'Eastcheap' figures tiptoed in solo bassoon and double bassoon toward what Edward described as a 'cheerful, out-of-door, ambling theme' in a rhythm only a step from the opening of Falstaff I. After 'muffled calls through the wood', a *scherzando* version of the Prince Hal theme rushed in to snatch 'the twice-stolen booty' from the benighted Falstaff 'mendacious' in double time. The laughter of the tavern 'Gentlewomen' shrilled again, and over quick breathing the 'Arthur' song began once more.

In *fortissimo Allegro molto* the 'honest gentlewomen' ran what Edward

---

[52] Letter of 8 Nov. 1928 to Robert Lorenz. The words are from 2 *Henry IV*, II. iv.
[53] 2 Sept. 1913 (*Letters*, p. 213).

described as a 'scherzo-like course' with a rollicking trio from the 'Arthur' song in the middle. But over it, in a sketch dated 8 July,[54] he set 'the undercurrent of our failings & sorrows' figure. The scherzo recapitulation brought the music close to *Gerontius* again, for Edward wrote: 'Through the heavy atmosphere a strange, nightmare variant of the women's theme floats, and with an augmented version of [the opening theme] Falstaff sinks down [in] heavy sleep . . .'

As Gerontius had dreamed, so now would Falstaff. But this hero's dream was only of the past. Out of Shakespeare's single line 'He was page to the Duke of Norfolk', Edward created 'a dream-picture' that had no equivalent in the plays:

. . . Simple in form and somewhat antiquated in mood, it suggests in its strong contrast to the immediately preceding riot, 'what might have been'.

The last quoted words were those of Lamb's 'Dream Children'. A solo violin moved through wide intervals to trace another figure of descending steps. Its opening trill and semiquavers laundered the 'honest gentlewomen' music:

Thus an old man's vulgarity was transfigured in a dream of the childhood fairyland still glimmering beyond the stream of time. It ended the longest section in Edward's score.

The contract with Novellos had finally terminated after they published *The Music Makers*. Now all Edward's music was in theory available to the highest bidder. Yet he had kept in touch with Novellos over the progress of *Falstaff*, and when it was far enough advanced, Alice was sent as emissary:

July 10. A. to Novello for E. to see Mr. Littleton about Falstaff—Were they going to have it? A. Littleton wanted it. So it was arranged.

July 11. E. happy with his work. A. to Novello with 1st portion . . .

Through that season of working at *Falstaff* he went often to the opera—Debussy's *Pelleas et Mélisande* (which he 'liked much'), Puccini's *Tosca* once more, and especially Beecham's season of Russian opera at Drury Lane. A performance of *Boris Godunov* seen from Lady Maud Warrender's box led to acquaintance with the gigantic bass in the title role, Chaliapin. He duly arrived at Severn House for Sunday afternoon tea with the conductor Emil

[54] Elgar Birthplace.

Cooper, and began to talk of an Elgar opera. Edward wrote to Sidney Colvin: 'Chaliapine has been here & talking over many schemes for a great part for himself—'*Lear*'! for instance: you have seen him I suppose—a splendid *man*—figure & mind—& the finest artist I have ever heard.'[55]

Colvin meanwhile had been trying to interest Edward in opera collaboration with Thomas Hardy, from whose house he wrote:

Max Gate, Dorchester.

Sunday July 20 [1913]

My dear Elgar,

I am at Hardy's, and this is a line to say that my embassy is successful, in so far as I find the old man not only willing but *keen* to co=operate in an opera with you (there's a rotten kind of a pun come without my meaning it).—In a first afternoon's chat over the matter three possible alternatives shape themselves in his mind:—

1) Something founded on the *Trumpet Major*, which has obviously pleasant and picturesque materials for an opera, and has already been cut very successfully into a play for the local company here: this would be rather in the nature of a comic or at least light & whimsical peasant opera, without any very deep or tragic elements.

2) Something founded on the *Return of the Native*: this would be a strong country tragedy, also with very striking & picturesque scenic & dramatic elements.

3) A section, or something founded on a section, of the *Dynasts*, preferably the 100 Days section: this would have to be a new kind of work altogether, bringing on big cosmic forces (choruses, perhaps invisible, of the Pities & the Years) as well as historic personages and events.—

If you feel that you would find anything inspiring in any of these themes, T. H. would much like you to think over them severally and a little later to have the chance of talking them over with you. He is delighted at the idea, and convinced, as we all are, that the combination, if it could come off, would have a huge effect . . .

My love to you all,
Sidney Colvin[56]

Yet this collaboration met the same fate as past operatic proposals. If Edward's music, after Symphony and Concerto, was moving back toward plots and texts, it had not arrived there yet. The unique advantages of *Falstaff* were detailed by Edward for the journalist 'Gerald Cumberland' on 17 July:

I have, I think, enjoyed writing it more than any other music I have ever composed, and perhaps, for that reason, it may prove to be among my best efforts . . .

But it must not be imagined that my orchestral poem is programme music—that it provides a series of incidents with connecting links such as we have, for example, in Richard Strauss's 'Ein Heldenleben' or in the same composer's 'Domestic' Symphony. Nothing has been farther from my intention. All I have striven to do is to paint a musical portrait—or, rather, a sketch portrait.

. . . I have finished all the preliminary sketch-work, and of the actual scoring, only a

[55] 22 July 1913 (HWRO 705:445:3446).
[56] HWRO 708:445:3445. Hardy had previously proposed *A Pair of Blue Eyes* as an opera subject for Elgar.

little remains to be done. I shall say 'good-bye' to it with regret, for the hours I have spent on it have brought me a great deal of happiness.[57]

Again there was the attempt to free even this music from its programme.

The third section of *Falstaff* measured the warrior and the man of peace in less than half the space of the tavern scene. Falstaff awakened with his 'witty' theme—to hear a battle fanfare which turned the 'Arthur' song into a mirror-image of *The Music Makers* 'To-day is thrilling'. The hero's mind went instantly to the 'Gentlewomen' and his own 'cajoling'. At a second, louder fanfare he emerged 'to take up soldiers as he goes'—Moudly, Shadow, Wart, Feeble, and the rest. The fanfare contracted in a rat-bitten march:

It was a new variant of Falstaff's 'mendacious' theme, to evoke the boast: 'I have foundered nine score and odd posts.' He reached the battle with his 'scarecrow army' to a shred of fanfare.

Falstaff 'witty' went into rapscallion battle amid raucous descents, as the 'Eastcheap' music scurried through the brass to emerge in pseudo-*nobilmente fortissimo*. As the scarecrow army disintegrated, there came a clear figure of diatonic fifths descending in the pattern of the 'tune from Broadheath', and Edward wrote:

The march, as we approach the fields and apple-trees, assumes a song-like character:

until we rest in Shallow's orchard.

But 'the undercurrent of our failings & sorrows' showed it to be only another lost Eden. The landscape of Eden rose in a second interlude for small orchestra, weaving 'some sadly-merry pipe and tabor music' with the descending fifths written down after the visit to Stoke in April

Edward called it his 'orchard theme'.

[57] *The Daily Citizen*, 18 July 1913.

Again the dream was shattered by the real world. Prince Hal was proclaimed King in the 'courtly and genial' figure with which Falstaff's fond imagination had endowed him at the outset—and Falstaff rushed away with his 'mendacious' theme and the ragged march, faltering only at approach to the new majesty.

The final section opened *Giusto* in a Coronation March giving firm resolution to the scrambling Falstaff 'witty' and 'Eastcheap' figures:

At the centre of the new pomp and circumstance came a final development of Falstaff's music:

Edward wrote this out on 29 July near the end of a long day's work in the Orchard Room at The Hut, where he had come again for a few days. Alice Stuart-Wortley had been there, but she had left earlier in the day: and Edward sent her the sketch of this music with the inscription: '(Farewell to the Hut) July 1913—written after you left: & now Goodnight etc.'[58]

The new King's Coronation came on, but like everything else in the music it was framed in the wishful thinking of the doomed hero: the approach of Henry V sounded in Falstaff's ears the *scherzando* music of Hal stealing the booty at Gadshill—and here, as there, it was counterpointed in the bass by the descent of 'our failings & sorrows'. Yet now the Hal music sounded its new insistence. When the original Falstaff theme tried to respond in the lower instruments, the rest of the orchestra swept over it in *fortissimo* quavers running the gamut through two complete octaves up and down. Falstaff 'cajoling' and a ghost of 'mendacious' were 'rudely blasted by the furious fanfare' of battle to set the dismissal:

> 'How ill white hairs become a fool and jester—
> I banish thee on pain of death.'

Immediately the royal march is resumed, and dies away: the King has looked upon his ancient friend for the last time.

The fading march took the ancient friend to his final dissolution. Edward wrote of the soft coda: 'in short phrases the decay of the merry-hearted one is

[58] Elgar Birthplace.

shown. The broken man weakens until, with a weird, final attempt at humour, we enter upon the death scene . . .' When the hostess of the Eastcheap tavern related ''A babbled of green fields,' Edward brought back the orchard-theme 'with many changes of harmony, faltering and uncertain'. Yet this Gerontius could not cast off his earthly experience. Edward wrote:

True as ever to human life, Shakespeare makes him cry out even at this moment not only of God, but of sack, and of women; so the terrible, nightmare version of the women's theme darkens (or lightens, who shall say?) the last dim moments.

The last moment of all in Falstaff's life recalled the 'courtly and genial' Prince of earlier times. His music was set now in the same warm retrospect as the closing pages of the Second Symphony. But the golden glow of memory no longer suffused the final bars:

In the distance we hear the veiled sound of a military drum; the King's stern theme is curtly thrown across the picture . . . The Prince, arrived at his kingly dignity, fulfilled the prophecy of Warwick: 'He will cast off his followers, and their memory shall a pattern or a measure live.'

Their memory does live, and the marvellous 'pattern and measure'. Sir John Falstaff with his companions might well have said, as we may well say now,

'We play fools with the time, and the spirits of the wise sit in the clouds and mock us.'

The entire strength of pattern in Edward's *Falstaff* came from its close thematic development—a prodigious skill of clothing consequent ideas in fitting orchestral dress which went beyond even the Second Symphony. He called *Falstaff* a 'symphonic study', thus fulfilling Ernest Newman's prescription for the future of the symphony as a single-movement form. But this close development would fit none of the great inherited forms of his art, whose mastery had been the goal of Edward's creative life.

Falstaff was the story of disintegration to a point where there was nothing left to emerge out of any personal past. In writing it Edward upset the diatonic basis of his own expression. Here for the first time his themes were predominantly chromatic and short-breathed; the 'old-world state' of diatonic melody was reserved for 'what might have been'.

It was the rhetoric of a man whose survival had dragged him unwilling into a disillusioned age. The rhetoric was in places discontinuous, in places repetitious. Edward told several friends that the new work would play for 20 minutes, and it had turned out more than half as long again. The scrutiny of this hero's disintegration had held him in its thrall.

When the score was finished on 5 August 1913, it left less than two months to engrave and print before the first performance at Leeds. Edward decided to supplement the music with his own explanatory essay. He wrote to Alice Stuart-Wortley: ' . . . as the score is behindhand I undertook to furnish the notes (analysis) *myself* . . . no outsider cd. write about Falstaff without the score.'[59] Yet Kalisch and Percy Pitt had done it for *In the South* in 1904 with less time than was now in hand. Was Edward uncertain of his music's effect? Through an August holiday in Wales the essay was completed, and proofs of the score corrected.

At the Gloucester Festival in September he conducted the *Coronation March*, *Gerontius*, and the Second Symphony. Ernest Newman wrote:

I doubt whether any two successive symphonies by any other composer show so great an intellectual growth, so great a deepening of the man's whole nature, and so great a development of musical feeling and technique as are visible between these two symphonies of Elgar. It is because he is now living on a psychological plane higher than that any other English composers have ever touched that this Second Symphony is beyond the popular grasp at the moment.[60]

How then would the popular grasp extend to *Falstaff*?

The London Symphony Orchestra was engaged for the Leeds Festival, and the first *Falstaff* rehearsal took place in London at the small St. Andrew's Hall, Newman Street. Alice wrote:

September 22. Orchestral parts only arrived this very morning from Germany— Orchestra played straight from them—Very few mistakes & they read wonderfully— Falstaff sounded magnificent & wonderful—audience greatly impressed.

Among the rehearsal audience was the *Yorkshire Post* critic Herbert Thompson, who wrote:

The London Symphony Orchestra occupied the area of the hall, the conductor's seat being placed on the platform, and the gallery running round the hall being reserved for a few privileged listeners. The rehearsal commenced at 10 a.m., and Sir Edward Elgar devoted the first half-hour to particular points which he desired emphasised in Beethoven's immortal overture [Leonora No. 3, which was in the first of his Leeds programmes].

Another quarter of an hour was spent trying certain parts of 'Falstaff', which presented exceptional difficulties, the brilliant violin triplet passages in the 'Gadshill' section being one of those selected. Sir Edward Elgar took this quite up to time, and afterwards smilingly asked, 'Any wrong notes?' Another tried-over part was the important theme given to the 'cellos, suggestive of Falstaff's cajoling and persuasive tactics.

After these nibbles at the score, Sir Edward said he would take the work straight through without stopping, but in spite of the wonderful sight-reading of the orchestra, the exceptional difficulties of the part-writing made many stoppages necessary, and it was an hour before Sir Edward shut the score. The work is down again for rehearsal on Thursday morning, when it will be possible to get a better idea of the value of the music . . .[61]

The second rehearsal was attended by Landon Ronald, who was to conduct

[59] 22 Aug. 1913 (HWRO 705:445:7830).
[60] *The Birmingham Daily Post*, 12 Sept. 1913.
[61] *The Yorkshire Post*, 23 Sept. 1913.

the first London performance in November. *Falstaff* was dedicated to Ronald in gratitude for the younger man's staunch championship of his later works, especially the Symphonies. But Edward himself was to conduct *Falstaff* at Leeds—one item in the two enormous programmes he was booked to direct there.

Saturday 27 September found him in Leeds rehearsing the Prologue from Boito's *Mefistofele* through a long hot and humid day. On Monday the 29th he rehearsed *Gerontius* for its first presentation at a Leeds Festival—a palpable sign of Stanford's departure from the Leeds directorate. Next day came the final rehearsal of *Falstaff*. Alice noted:

September 30. E.'s rehearsal was to be at 11—but Nikisch had a cold & E. let him finish first. So E. had only a short time & tired Orchestra. A. drefful nerves about it.

Among the select audience that morning was young Adrian Boult, who had become a devoted follower of Nikisch and an aspiring conductor on his own account. He was enormously impressed to see Edward hand Nikisch the finished printed miniature score of *Falstaff* in advance of the first performance.

Next day Edward opened the Leeds Festival with a morning-and-afternoon programme which included the National Anthem, *Leonora Overture* No. 3, *Gerontius*, Parry's *Ode to Music*, Brahms's *Alto Rhapsody* (with Muriel Foster) and Third Symphony. It drew the heaviest attendance of the entire week. The following evening (in the wake of an afternoon programme which included the première of George Butterworth's *A Shropshire Lad*) came Edward's second concert. *Falstaff* took its place in another immense programme embracing Bantock's new *Dante and Beatrice*, the *Mefistofele* Prologue, and Mozart's G minor Symphony all conducted by Edward, with *The Mystic Trumpeter* under its composer Hamilton Harty, and numerous solo and part songs. Alice wrote:

October 2 . . . E. conducted the Bantock splendidly but it seemed long & dreary. A. had drefful fits of nervs—Then Falstaff. E. rather hurried it & some of the lovely melodies were a little smothered but it made its mark & place. E. changed [clothes], very depressed after . . .

'The audience', noted Richard Capell in *The Daily Mail*, had been 'chary of applause'. The critics found themselves in difficulty. Ernest Newman wrote:

Not only are the Leeds Festival people so overcrowding their programmes that it is a pure impossibility for the critics to hear them throughout, but they have added to our difficulties by placing three novelties on the same day. In addition they imposed on Elgar the strain of conducting three works before his own came on . . . The style of the score shows us in many places quite a new Elgar, and one that the public used to the older Elgar will not assimilate very easily.[62]

[62] *The Birmingham Daily Post*, 3 Oct. 1913.

Many of Newman's colleagues found the same trouble. 'What a queer Falstaff it is!' exclaimed Capell:

As the orchestra crackles and effervesces one wonders how Falstaff found his way to such Walpurgis-Night revelry, how he acquired all the impish agility of a Till Eulenspiegel. It is indeed rather hard to see a point of contact between the Falstaff one knows and Sir Edward Elgar's art, which has shown itself to possess a rapid mercurial wit rather than downright humour . . . The music's volubility of utterance is so impetuous as to be quite a novelty. The pace and the brightness are dazzling for a moment before one feels that after all little consecutive effect has been made and none accumulative.[63]

Young Ferruccio Bonavia wrote of the new score:

The best qualities of 'Falstaff' are dignity and extraordinarily interesting craftsmanship, but for once Falstaff failed to be the cause of wit. Except one or two clever strokes of wit there is none.[64]

Robin Legge strove to do justice in *The Daily Telegraph*, but could not find the music's 'point of view':

I have not yet fathomed the (to me) mystery as to why what I believe is called in theatrical language the 'fat' of the music is applied to other folk and their doings, or to description, and not to the protagonist himself.

Falstaff cajoling and persuasive is wholly delightful, and so is all of what I may call the retrospective music after Falstaff is inexorably swept aside by the King. But he is, indeed, a most complicated person, and most unobvious, and it seems to me unlovable, in spite of his cajolery. I do not think even Elgar has ever written more complicated music, and it is for this very reason that I wish a greater conductor than he had explained his complications last night, for much of the wondrous maze of detail in his score did not become audible at all to me.

Yet in spite of this I heard more than enough to cause me to realise that here again Elgar has given us a masterpiece of music, prodigiously stamped with his own remarkable personality, and I firmly believe that if he will take his courage in both hands and perpetrate a good deal of judicious pruning—say, cut some five or more minutes, so to speak, from his score—his work will be vastly improved, and be infinitely quicker in its appeal.[65]

In fact Edward rescored several passages after the performance, causing Novellos to re-engrave a number of plates: here again his uncertainty showed. But he cut nothing.

It remained for Ernest Newman to sum up after the Leeds Festival what all responsible critics have felt in confronting important premières from that day to this.

Elgar devotes, perhaps, the best part of a year to thinking out a work like his 'Falstaff'. The critic is then expected to tell the world all about it, and to pass some sort of

---

[63] *The Daily Mail*, 3 Oct. 1913.
[64] Unidentified paper (cutting at the Elgar Birthplace).
[65] 3 Oct. 1913.

judgement upon it, at about 10.30 one night, with the knowledge that he has several other things in the programme to discuss—sometimes including another new work, as happened at Leeds—and that if he is much more than half an hour over it all he runs the risk either of his message not getting over the wires in time, or of it being mutilated beyond recognition by a tired and hurried telegraphist at one end and tired and hurried compositors at the other . . . I can remember no new work of importance that has been treated more hurriedly and with less credit to the critics than 'Falstaff' was by every one of us last week.

Then Newman proceeded to correct them all with reflections brought by no more than a single week's thought over the new music and the composer it revealed:

He shows us, by the use of certain leading motifs, the jovial Falstaff, the cajoling Falstaff, the braggart Falstaff, in company with the generous and thoughtless prince; the hardening of the latter's nature under responsibility, the impossibility henceforth of the two characters mixing, the veering round of the moral standard of the world in which they have hitherto moved, and the necessary fracture of the more ill-co-ordinated character of the two against the new and hard reality that has suddenly come into his orbit.

And so this extraordinary critic caught the glint of a light *Falstaff* was playing over them all in that final year of the long *Pax Britannica*:

The subject of the symphonic study is really the mad, pathetic mixture of contrarieties in us all, and the sense of something vast and inscrutable above us, putting an end—a harsh but perhaps bracing end of its own—to all our moral oscillations when the time comes . . .[66]

If the prescience was there, no one seemed disposed to respond to the message—or even to attend it. The London première of *Falstaff* took place in November 1913 in a concert entirely devoted to Edward's music conducted by Landon Ronald. In the audience was the young composer C. W. Orr, who recalled:

None present will easily forget the desolate scene. Here we had a programme conducted by a musician who had prominently identified himself with Elgar's music; a brand-new work from our greatest composer; a first-rate orchestra, and—to mark the occasion—an array of empty benches! The music critic of the *Daily Telegraph* commented caustically on the absence of students and representatives of our academies and schools of music.[67]

Landon Ronald courageously programmed the new work all over again, as he wrote to Alice Stuart-Wortley:

I wonder if you have heard of my cheeky & desperate action! I am repeating the Elgar concert in toto at the Queen's Hall on Friday evening Nov. 28th. I am doing it out of

[66] *The Nation*, Oct. 1913.
[67] 'Elgar and the Public', in *The Musical Times*, 1 Jan. 1931, pp. 17–18.

pure 'cussedness'! No second performance of 'Falstaff' has been arranged [in London], & therefore the work was bound to die almost before it had been heard. Although Monday's concert has cost me an awful lot of money, I feel I simply *must* fight Elgar's battles . . .

Will you, in the kindness of your heart, help me by telling everybody you know about it and making them take tickets. If by any good luck there should be a profit, I should hand it to Elgar . . .[68]

At the second concert the hall was less empty—perhaps because Ronald had persuaded Muriel Foster to sing *Sea Pictures*. Ronald conducted *Falstaff* a third time in a Sunday Concert at the Albert Hall on 14 December. This, according to Alice, drew 'an enormous audience'. But Edward did not delude himself that it was for his newest work. '*Very* depressing time—almost all these days,' Alice had noted in late October. Near the end of the year, it was still so.

* * *

One of Landon Ronald's interests was the gramophone. He had been musical adviser to 'His Master's Voice' since the turn of the century. He had persuaded such singers as Melba and Patti to approach the primitive recording horn, and later had pioneered in making some of the first records of orchestral music. Even in 1913 orchestral recording was a crude process. The acoustic horn could only capture sounds made close to it: with the utmost crowding, forty players might be got in, with lower strings replaced by tubas, wholesale re-orchestration of melodies, and ruthless abridgement where needed to fit the four-minute record. On the face of it, the prospect of recording Elgar was entirely impractical. But Ronald had noted Edward's interest in science and mechanics; and Muriel Foster's brother-in-law, Jeffrey Stephens, was in the Gramophone Company.

On 3 December 1913 Stephens invited Edward to lunch with W. W. Elkin, a publisher whose list included much light music. The result was a contract for two short pieces to be added to the opus number of *Salut d'amour*. Elkin agreed to pay a 'sum down' of 100 guineas and a royalty of 3d. on every copy of each piece sold. And there was an important addition:

Two thirds of net royalties received in respect of mechanical instrument reproduction to be paid to the Composer.[69]

The idea was to record the new pieces before their publication, so as to have records in the shops when the music appeared.

The first of the new pieces, written in December,[70] was called *Carissima*. It revived the innocence of Edward's early melodies:

---

[68] 8 Nov. 1913 (Elgar Birthplace).

[69] Contract signed by Elgar 23 Dec. 1913 (now in the archives of Elkin's successors, Novello & Co.)

[70] 'Composed only last month', according to Robin Legge (*Daily Telegraph*, 11 Jan. 1914).

The little piece was immediately booked for a trial gramophone recording in January 1914: he inscribed it to Mrs Jeffrey Stephens, Muriel Foster's sister.

The other piece, according to the Elkin contract, was to be called 'Soupir d'amour'. Its melodies in fact echoed both *Salut d'amour* and *Carissima*:

The music was quite diatonic. Yet the wide intervals in the second theme especially suggested an ache of longing: one sketch had been headed 'Absence'.[71] He acknowledged a kinship with the Violin Concerto by making it a violin solo and dedicating it to Billy Reed. It was clearly no good as one of the light things for Elkin. Edward arranged its separate publication by Breitkopf & Härtel under the Italian title *Sospiri*.

Many months later he fulfilled the Elkin contract by revising the 'Douce pensée' written in 1882, the year before his engagement to Helen Weaver. The thought was sweeter: but it was more retrospective than ever under its new title, *Rosemary ('That's for remembrance')*.

Thinking further of the Violin Concerto, perhaps, he sent a fragment for a Piano Concerto to Alice Stuart-Wortley, charging her to learn it and play it to him when she came. He would call for its performance thus again and again through the next years: but the sounds did not fire him to go on with it. And when Ivor Atkins asked him again to complete the *Apostles* trilogy for the coming Worcester Festival, Edward again refused:

I have been wofully overwhelmed with busyness—quite a different thing from business—a tortured forced fervidity over necessary trifles—which are more difficult to overcome than the big things—the performance of these last is denied me—the Lord desireth in these his days the pleasant degradation of Man (how foolish wd. be an empty Hell to its maker!) & so—& so—& so—you understand.[72]

Novello's choral adviser W.G. McNaught was more successful when he urged the writing of new part-songs: choral groups all over the country would take them up. For texts Edward turned to the seventeenth-century Welsh poet of religion and the countryside, Henry Vaughan. In Vaughan's *Silex Scintillans*

---

[71] Elgar Birthplace.   [72] 29 Dec. 1913.

he found two isolated stanzas—one of rain, the other sunshine. The rain was in the final stanza of a poem called 'The Shower':

> [Cloud,] if, as thou dost melt, and, with thy train
> Of drops make soft the earth, mine eyes could weep
> O'er my hard heart, that's bound up and asleep.
>   Perhaps at last,
>   Some such showers past,
> My God would give a sunshine after rain.

The other Vaughan setting took its lines from a poem entitled 'Regeneration', to waft him back:

> The unthrift sun shot vital gold,
>   A thousand pieces;
> And heaven its azure did unfold,
>   Chequered with snowy fleeces;
> The air was all in spice,
>   And every bush
> A garland wore: thus fed my eyes,
>   But all the earth lay hush.
>
> Only a little fountain lent
>   Some use for ears,
> And on the dumb shades language spent
>   The music of her tears.

Could it revive the fountain in the old cottage garden whose music was in *The Wand of Youth*? Edward called his setting *The Fountain*, and its diatonic innocence echoed the dream-interludes of *Falstaff*; only several times, without warning, came sudden harmonic shifts.

He dedicated the Vaughan settings to ancient West-Country friends—*The Shower* to a neighbour at Forli before the turn of the century, *The Fountain* to an ageing singer from the Worcester Cathedral Choir who had grown up beside Edward in the old Glee Club. But there were hints closer to his present home in the place-names he wrote at the end of each setting. Both were rustic places north-west of London where one could still catch country sights and sounds. *The Fountain* was marked 'Totteridge', perhaps in memory of a summer's day chronicled by Alice:

June 4 1913 . . . E. & A. for lovely taxi drive to Totteridge. E. fished for creatures. Lovely there, larks singing & water lilies coming out . . .

Of *The Shower* he wrote to Alice Stuart-Wortley: 'I send you this very simple little thing for voices—only at the end I put *Mill Hill* to remind us of our afternoon when it was so cloudy & nice & lovely in the Church yard looking over the vale.'[73]

Then to the Vaughan settings of childhood he added three other part-songs

---

[73] 11 Mar. 1914 (transcript by Clare Stuart-Wortley at the Elgar Birthplace).

of far different shores. The words of all these were adapted from the Russian poets Maikov and Minsky by Mrs Rosa Newmarch. The first was entitled *Death on the Hills*—hills which might be his own Malverns, overlooking Piers Plowman's 'Field Full of Folk':

Why o'er the darkening hill-slopes
  Do dusky shadows creep?
Because the wind blows keenly there,
  Or rain-storms lash and leap?

No wind blows chill upon them,
  Nor are they lashed by rain:
'Tis Death who rides across the hills
  With all his shadowy train.

The old bring up the cortège,
  In front the young folk ride,
And on Death's saddle in a row
  The babes sit side by side.

The young folk lift their voices,
  The old folk plead with Death:
'O let us take the village-road,
  Or by the brook draw breath.

'There let the old drink water,
  There let the young folk play,
And let the little children run
  And pluck the blossoms gay.'

    (*Death speaks.*)

'I must not pass the village
  Nor halt beside the rill,
For there the wives and mothers all
  Their buckets take to fill.

'The wife might see her husband,
  The mother see her son;
So close they'd cling—their claspings
  Could never be undone.'

He set the lines to a D minor *Moderato (quasi alla marcia)*. From descending intervals endlessly repeated, the music escaped only to sound in monotone the cortège of the old, to raise the prayer in single steps barely harmonized. But underneath the prayer Death returned to the inexorable falling intervals, and the sinister ensemble continued to the end. *Death on the Hills* was only a little longer than the Vaughan settings. But Edward recognized a different scope when he finished it on 22 January and sent the sketch to Alice Stuart-Wortley: 'It is one of the biggest things I have done.'

During the writing of *Death on the Hills* he had gone one afternoon to conduct the test recording of *Carissima* for the gramophone. Before the result was known, the Company showed their gratitude by sending one of their latest machines with a selection of recent records to Severn House. Most of the afternoon of 24 January was spent listening to these, and they were played repeatedly for friends and visitors: 'Chaliapine as Boris the most wonderful,' noted Alice.

Those sounds were in Edward's ears as he wrote his next part-song from the Russian. It sharply contrasted *Adagio* and *Allegro con fuoco* in a setting which he himself named *Love's Tempest*:

> Silent lay the sapphire ocean,
>   Till a tempest came to wake
> All its roaring, seething billows
>   That upon earth's ramparts break.
>
> Quiet was my heart within me,
>   Till your image, suddenly
> Rising there, awoke a tumult
>   Wilder than the storm at sea.

January 28 1914. Much milder—A. & C. into town . . . E. very absorbed in his booful new part songs—When A. returned, he said 'Come in & sit down' & then played record just come of 'Carissima'—most lovely & exquisite record.

The last of his Russian part-songs set Mrs Newmarch's version of a ghostly 'Serenade' by Maikov:

> Dreams all too brief,
> Dreams without grief,
> Once they are broken, come not again.
>
> Across the sky the dark clouds sweep,
> And all is dark and drear above;
> The bare trees toss their arms and weep.
> Rest on, and do not wake, dear love,
> Since glad dreams haunt your slumbers deep,
> Why should you scatter them in vain?
>
> Happy is he, when Autumn falls,
> Who feels the dream-kiss of the Spring;
> And happy he in prison walls
> Who dreams of freedom's rescuing;
> But woe to him who vainly calls
> Through sleepless nights for ease from pain!

This Gerontius sought only oblivion. The first three lines were set in monotonous rocking figures of D minor—rising to repeat a fourth higher, sinking back to tonic. Over this intoning opened the longer lines. He marked the score 'Hadley Green', another country place north-west of London.

February 1 . . . Dr. McNaught to lunch to hear new booful part songs—greatly impressed. E. & he had long talk—*nice* man—fine . . .

McNaught suggested the collective title 'Choral Songs' for the Vaughan and Russian settings. He conveyed to Novellos Edward's proposal for a sum down of 125 guineas and an immediate royalty of 25 per cent.

Alfred Littleton was now nearly seventy, and more and more executive decisions were in the hands of his younger brother Augustus—tall, domineering, with a limp and a malacca cane. Augustus Littleton was wintering on the French Riviera when the Elgar part-song proposal reached him. He sent his reply to Henry Clayton:

Le Grand Hotel, Cannes.

Feb 6.14

My dear Henry

I sent you a wire this morning saying must agree Elgar's terms. I don't think we ought to hesitate a moment. The price is high amounting to extortion, but the point is that plenty of other houses would jump at the stuff at the price. We must fight these people until the position becomes absolutely absurd, then we can retire. This is a very different matter to the Indian ballet [*Crown of India*]. Here we are on our own ground and if the Part songs catch on we shall make money, anyway we cannot afford to throw up the sponge just yet. The future must take care of itself, and if Elgar repeats himself, which he won't in his own interest just yet, we must consider each case on its merits.

But you must on no account do the part songs at fourpence [each]. They should be at least sixpence . . . The higher price won't stop the part songs going if they catch on, if people grumble we must quietly let them know that the higher rate is owing to the composer's greed.

Competition gets keener every day, and composers getting scarce and some of those slipping through our fingers. I don't want any more Elgar symphonies or concertos, but am ready to take as many part songs as he can produce even at extortionate rates . . .

Yrs sincerely

Augustus Littleton[74]

Clayton translated Augustus Littleton's calculation into a letter of gracious acceptance for Edward, and the songs were put in the hands of Novello's engravers.

Through the winter and spring he finished several smaller works to answer requests. Some were frankly pot-boilers. There was a thumping song for Clara Butt called *The Chariots of the Lord*, for which Boosey paid a hundred guineas and a royalty:

> Where hands are weak, and hearts are faint,
>   Through conflict sharp and sore;
> Where hearts that murmur no complaint
>   Shrink at the thought of more:
> Then let the power of God be shown,
>   To quell satanic might;
> To rescue those who strive alone,
>   Despondent in the fight.

[74] Novello archives.

A quieter lyric, *Arabian Serenade*, brought a more ordinary royalty. There was a simple marching song, *The Birthright*, for boys—with optional bugles and drums; and three still simpler children's songs for an American publisher. A modest Harvest anthem, *Fear Not, O Land*, was added to Novello's Octavo Series for Church Choirs. A bigger anthem, requested by Sir George Martin for the 200th anniversary of the Sons of the Clergy at St. Paul's Cathedral, was set to Psalm 29: 'Give unto the Lord, O ye mighty . . . the glory due unto His name.' It was impressively scored for orchestra and organ; the ripe sequences served the comfortable Anglican style for such things.

*   *   *

Every serious prospect was more and more dominated by peremptory demands for novelty. In France a critic called Marcel Boulestin pronounced upon the development of English music in an article entitled 'Les Post-Elgariens', proclaiming the folk-song revivals of Percy Grainger and Vaughan Williams as the spirit which would overtake the achievements of Elgar. It was not the first time such things had been said. Ernest Newman replied with vigour: 'I have heard these young lions roar so often that I know their tune by heart . . . At present there is hardly one of them whose intellectual stature brings him up to Elgar's knee.'[75] Alice sent Newman a note of heartfelt thanks: 'You say exactly what I have often wanted said & have again & again felt indignant that no one had the perception or felt impelled, as so many should have done, to say—& to acknowledge greatness while it was yet with them.'[76]

Edward himself addressed the demand for novelty when he gave a presidential speech to the Union of Graduates in Music on 5 May. It was not the new influences in England he feared, but those that came from outside. He began by recalling his 'Future for English Music' lecture at Birmingham in 1904:

I had the pleasure of making a speech about ten years ago in which I dealt with the question of modern harmonies, suggesting what the music of the future might be. I suggested that it would tend in the direction adopted by composers you have had very much before you in the last few months—that the *main development* would be chromatic harmony and strange blending of keys, especially in the matter of orchestration—that it was vulgar to twist our scale, which was good enough for Beethoven and Brahms and Bach, and make it do the monkey tricks I foretold would be forthcoming. I had no belief that the modern ear would become, or desire to become, more refined—everything is designed to brutalise it. The more subtle refinement is not yet with us and can only come by the use of a scale more minutely divided than our own; this would educate the ear to something finer than we have yet heard.

Yet at the end he faced the Graduates in Music with their inheritance from the

[75] *The Birmingham Daily Post*, 16 Feb. 1914.
[76] 19 Feb. 1914.

past: 'I think we may consider ourselves to be conservators of what is the necessary basis of music . . .'[77]

Conserving a necessary basis was beginning to seem as difficult in the life of nations as in music. In Europe Germany had built her big navy, but Great Britain had built a bigger one. At home the social pressures which had brought the Liberals to power and kept them there were fermenting everywhere. Increasing ease of transport and communication were doing their work: as people discovered more and more about their neighbours' lives, there was a new rage for competitive standards. Yet there was also the old insistence that every identity be kept sacrosanct.

The struggle for Home Rule in Ireland rose to a bitterness beyond any memory in living experience. As the Liberal government prepared to pass the Home Rule Bill through Parliament for a third time to override the Lords' veto, opposition mounted to such a pitch that part of the British army was close to mutiny. For the first time in nearly three centuries there was a shadow of civil war. The Unionists demanded a general referendum, and sought the signatures of 'twenty distinguished men' for a pledge:

I . . . do hereby solemnly declare that, if the [Home Rule] Bill is so passed, I shall hold myself justified in taking or supporting any action that may be effective to prevent it being put into operation . . .

They already had the signatures of Lord Milner, the Catholic Lord Lovat, Field Marshal Lord Roberts, Rudyard Kipling, and numerous others when they approached Edward through the former Hereford MP John Arkwright. Alice wrote:

March 2, 1914 . . . Mr. (& Mrs.) Arkwright came up with a stirring manifesto wh. 20 distinguished men were signing protesting against Home Rule & pointing out the great danger—E. asked wd. Lord Lovat undertake to answer letters from the Catholic side. Mr. Arkwright went to investigate & returned saying 'yes'. E. took a little more time to consider, his heart was in it from the first. Lord Milner telephoned later to know his decision, wh. was 'yes'.

Next day the pledge appeared in *The Times* and *The Telegraph* over the twenty names, In very little time it gathered more than a million supporters.

Privately it seemed more than ever a time for counting blessings in the present. Edward wrote to Miss Burley: 'The spring *is* the saddest season of the year if you do not take what is offered to you, and only yearn for the things that are far off . . .'[78] The Elgars' silver wedding anniversary was on 8 May, and a few nights later Alice took him to dine between the acts of a *Meistersinger* performance at Covent Garden.

[77] Quoted in *The Organist and Choirmaster*, 15 June 1914.
[78] 16 Feb. 1914 (quoted in Young, *Alice Elgar*, pp. 172–3).

But his fifty-seventh birthday on 2 June found him travelling alone to Worcester. There he was to take part in giving the Freedom of the City to a distinguished old Worcestershire painter, Benjamin Leader. Addressing the assembled company in the Guildhall, he called Leader 'the poet of the Severn', and the richness of his own past rose up before him: 'Worcester is home to a Worcester man . . . As Worcester people cannot always be at home, they depend in their travels upon the beautiful works of art which Mr. Leader has given them.' Then he made a plea for conserving old Worcester:

There are great changes in the town—improvements, I suppose I ought to say, but we who knew the town 40 years ago do not quite see that every change is an improvement . . . Let us not destroy the half-timbered houses . . . Simple things have the greatest possible influence on the child mind.[79]

A month later Worcester saw him again. He came to stay the night of 8 July at Hubert Leicester's house, The Whitstones. Hubert's son Philip wrote:

After dinner, it being a warm evening & very light, he suggested a walk & so he & I lit pipes & started off up Barbourne.

He knew every house & street. When near Thorneloe he said he would like to call on his sister Mrs Pipe in Barbourne Terrace. I therefore took him to the house.

Mrs. P. was surprised & delighted. She was alone, Charlie P. being out on business. We went in to the tiny place. Everything neat & tasteful tho' on very small scale. E. & Mrs. P. went all over the house (which the Ps had only taken a few months since) while I smoked on the little strip of lawn in front. E. remembered everything—every book, picture & ornament. Just before we left he said 'Where is that dead pheasant?' Sure enough there was an old painting of the Bird over the parlour door.

E. & Mrs. P. on the best of terms. Before we left they said 'good bye' most affectionately in the little parlour, where for a moment I felt 'de trop'. All he could talk of for a few minutes after was her deafness which he thought such a dreadful affliction. It seemed quite to affect him.

Philip knew nothing of the 'Menière's disease'.

Then he took me round St. George's Square, pacing slowly & examining every house & garden minutely. We stopped by the laundry, which he said was over the militia barracks. All the time he talked rapidly, eagerly but in a curiously low voice as if to himself. I just listened, & at times he seemed almost to forget my presence & to be thinking aloud.

He remembered the former occupants of nearly every house & his talk was a constant stream of odd & humourous little anecdotes concerning these citizens of a former generation.

We passed down Thorneloe & along Stephenson Terrace. Here we stood some time, gazing out over Pitchcroft & over to Malvern, the lights of which were faintly visible in the dusk. He turned to me at last & said he was so glad I liked Worcester. Said he loved the old town & the old houses & hoped it might remain unspoilt & unchanged . . .

Then we strolled back to the Whitstones where Aunt Nan & Mother were in the drawing room. We all sat & chatted & smoked. E. sat in the armchair in the corner &

rambled on about all sorts of things. About 10 o'clock Father came in & E. seemed very glad to see him. Then the fun began, & for an hour or more E. rattled out funny tales of the old days in Worcester . . . his eyes rapidly opening & shutting & his pleasure in his memories was evidently intense.

Between the two Worcester visits there had been a big performance of *The Apostles* at Canterbury. It was organized by Henry Embleton. Embleton had also raised again the question of completing the oratorio trilogy, and again Edward declined. Yet Embleton was not so sure: in addition to a very big fee (to come out of his own pocket), he had baited the hook with splendid performances of *The Apostles*—in these days not often heard—first at York Minster and now, under the composer's direction, at Canterbury.

The 270 members of the Leeds Choral Union arrived from the north in two special trains laid on at Embleton's expense. The scene in Canterbury Cathedral on 19 June was described for the local press:

A large stage crowded with indistinct figures, white dresses of the sopranos and contaltos standing out against the sombre masses of the men; in front and below serried ranks of the orchestra, with here and there a flashing gleam of light on some brass instrument. In front the slight solitary figure of the composer-conductor, a gesture, and the chorus rise, their open books giving flicking curves of white, undulating as the singers yield unconsciously to the music's rhythm.[80]

In the wake of a fine performance, Embleton made his descent upon Severn House:

June 30 . . . Mr. Embleton came & had tea & long talk with A. He said beautiful things & made proposals about the 3rd Part of The Apostles in such a beautiful & delicate way like a friend who really loved E. & his work. E. had a talk when he came in & they clasped hands at parting, E. consenting. A feeling of great quiet joy settled down on us.

Three days later Edward and Alice attended a supper given at The Savoy by Harley Granville Barker and J. M. Barrie. Barrie had written some dramatic sketches especially for performance by the distinguished actors present. A cinema camera was there to record this evening of splendour in the face of the world's troubles. The guest list of 150, headed by the Prime Minister, also included the artists Charles Shannon and Charles Ricketts. Ricketts wrote in his diary:

July 4 1914. Savoy Theatre. Barker supper. Quite an amusing evening. All the theatre stars: Shaw, Yeats, Rupert Brooke, Chesterton, Barrie for letters, Asquith, his wife and pretty daughter, countless people in society . . . On arrival we were met by a fierce light; this it seems was a cinema machine, both Shannon and I shaded our eyes and made faces. At my table old Lady Lewis, who is a dear, talked my head off; the pretty Lady Lytton sat on the other side; facing me was the beautiful young actor Godfrey Tearle . . . At the next table sat Shaw, Mrs Patrick Campbell, Lady Mond, an English blonde woman who is an Indian Princess, and the beautiful Rupert Brooke. Chesterton was just

[80] *The Kentish Gazette*, 19 June 1914.

behind me with titled women and Asquith's son . . . Altogether the evening was most entertaining and everybody fizzled a great deal.[81]

A fortnight later all three Elgars were off to a holiday in Scotland. They went to Oban, Iona, and up the Caledonian Canal past Fort William to Inverness—retracing the ground Edward had travelled alone thirty years earlier after the engagement broken with Helen Weaver. Now they went as far as Gairloch in the north-west. Carice wrote:

This entailed a railway journey to Achnasheen and then a thirty mile drive to the Gairloch Hotel. The taxi driver was slightly drunk, and Father and I in the back of the car with our hearts in our mouths sang nonsense songs of which we had built up a good repertoire over the years; this at least distracted us and prevented our staring at the edges of deep precipices or watching with great uneasiness the edge of Loch Maree which came right up to the side of the road with no fence. Mother sat in front, quite happy and imperturbable; she never seemed to sense any danger in a car.

We luckily arrived safely and found the hotel delightful. It had lovely grounds, sands quite near and lovely walks. Time passed all too quickly.[82]

And then with astonishing suddenness the ghosts of disintegration which had haunted Edward's music with closer and closer intent rose up and overwhelmed the nations of Europe.

[81] *Self-Portrait: taken from the letters and journals of Charles Ricketts, R.A.* (Peter Davies, 1939), p. 202.
[82] Quoted in Young, *Alice Elgar*, p. 174.

The Land of Lost Content

IᴺᴗᴖᴗᴍI

Iɴ remote north-western Scotland, the beginning was merely strange. After many telegrams of enquiry to friends, one to the Stuart-Wortleys drew response. From Gairloch Edward replied on 9 August:

I do not know how to thank you for your telegrams: we were nearly mad to get any news & our friends seemed to fail us thinking, no doubt, that we had moved on. We had intended to do so but we are thirty miles from a station; the means of communication, I mean transport, were the motor-charabanc—the hotel motors (two) to Achnasheen & the steamship to Lochalsh. The public vehicles were all comandeered to carry the territorials & were so occupied for two days—gradually all the guests left & we are now the only people left except an elderly gentleman & his wife.

We decided to leave by steamer—but this was also commandeered to pick up reservists &c. from the islands—since that voyage it has ceased to run. In the meantime *both* the hotel motors are hopelessly broken down & the necessary parts must come from Glasgow & now only the road is available so the repairs may not be completed for a month. The last charabanc informed me on Friday that he wd. not return—So we felt that all was over but to our delight he arrived here last night & we hope & trust that tomorrow (Monday) will see us on our way south. We have great difficulty over money as English 5£ notes will not be accepted anywhere.

It has been a weird & affecting time: seeing these dear people bidding good bye but the spirit of the men is splendid—the Seaforths went first—later in the week the mounted Lovat's Scouts rode through—were given a sort of tea meal here by the manageress & rode off in the moonlight by the side of the loch & disappeared into the mountains.

I purposely refrained from any rush or excitement but I am returning to London as soon as possible to offer myself for any service that may be possible—I *wish* I could go to the front but they may find some menial occupation for a worthless person.[1]

At last they got away, and Carice remembered 'all the way to London we were aware of the vast movement of men.'[2]

For Edward, as for so many others that summer, the beginning of war made a break from the old life almost to be welcomed. He wrote to Lady Colvin: 'The times are spacious & we must win through . . . I want you to know that I am going to die a Man if not a musician.'[3] The same feeling permeated an article by

---

[1] HWRO 705:445:7808.
[2] Quoted in Young, *Alice Elgar*, p. 175.
[3] 25 Aug. 1914 (HWRO 705:445:3452).

Ernest Newman entitled 'The War and the Future of Music' which appeared in the September *Musical Times*.

In other fields than the political the war, if it be prolonged, will mean the drawing of a line across the ledger and the commencement of a new account. It is impossible for the Continent to pass through so great a strain as this without a setting free of great funds of dormant emotion, and a turning of old emotions into new channels . . . There will be new horizons to envisage, new hopes and fears and joys and despairs to be sung.

Were we writing about the situation as if it were five hundred years behind us, and so a subject merely for unimpassioned scrutiny of forces and correlation of causes and effects, instead of something blindingly and terrifyingly near to us, we might perhaps say that some such war was necessary for the re-birth of music. For there is no denying that of late music has lacked truly commanding personalities and really vitalising forces. . . . In England Elgar is still the one figure of impressive stature; the men who are almost contemporary with him are not fulfilling their early promise, while in the crowd of younger men it is impossible to distinguish one who has the least chance of making history. Never has there been an epoch of such general musical capacity, but great figures and great ideas are not so plentiful. It is hard to believe that out of the new order of things there will not be born the figures and the ideas we long for.

(If this was unfair to Vaughan Williams, no other young composer in England had yet produced a quantity of music on which real hopes might be built.) At the end, Newman concluded that the musical future lay on the side of internationalism:

The day has gone by when one country can build up a school in ignorance and contempt of what is going on in other countries; it will reject a foreign culture at its peril. We can only hope that the result of the war will not be a perpetuation of old racial hatreds and distrusts, but a new sense of the emotional solidarity of mankind. From that sense alone can the real music of the future be born.

So the nation's most distinguished critic tried to show its leading composer that the European tradition at the basis of his style must triumph in the end over the younger enthusiasm for folk-song. But where would Edward or his world be when this war ended? All the musical festivals, beginning with Worcester, were suspended.

On 17 August he volunteered as a special constable for Hampstead, to take the place of younger men gone to fight. He showed his emblem and baton to W. H. Reed, who saw it

. . . so utterly unlike the baton to which he was accustomed. It was very dreadful to think of a man of his genius and sensitive temperament going on duty at stated times, night and day, with a semi-lethal weapon like that heavy baton, and every one of his family and his close friends prayed devoutly that he would never be called upon to use it.[4]

'I am a fool & look a fool', he wrote to Alice Stuart-Wortley, 'but I am doing what I can!'[5] And to Frank Schuster on 25 August:

[4] *Elgar*, p. 115. [5] 27 Aug. 1914 (HWRO 705:445:4168).

. . . London looks normal; it seems incredible that things shd. go on so well. Alice is well but worried. Carice spends her whole time in *practical* ways—learning nursing &c &c. I am a s. constable & am a 'Staff Inspector'—I am sure others cd. do the work better but none with a better will. I was equipping (serving out 'weapons') & taking receipts & registering my men for hours last night: this morning at six I inspected the whole district—so one does what one can—it's a pity I am too old to be a soldier—I am so active.

Concerning the war I say nothing—the only thing that wrings my heart & soul is the thought of the horses—oh! my beloved animals—the men—and women can go to hell—but my horses;—I walk round & round this room cursing God for allowing dumb brutes to be tortured—let Him kill his human beings but—how CAN HE? Oh, my horses.[6]

There spoke the man born and raised in the deep Victorian peace of the English countryside—and with an overpowering sense that his world was being destroyed. Few indeed saw so far so quickly.

The men and women were reacting in different ways. Jaeger's widow changed her name and her children's to Hunter in fear of anti-German sentiment. Muriel Foster's husband Ludovic Goetz later took his wife's maiden name by deed poll. Frank Schuster kept his name but offered his London house and car and chauffeur to the Belgian ambassador to accommodate the refugees already pouring into every harbour of the south-eastern coast. Sir Edgar Speyer was forced to resign his funding of the Queen's Hall management, and was later made to leave the country for alleged collaboration with the enemy—which turned out to be the sending of food parcels to starving German relatives. In Germany the seventy-year-old Hans Richter renounced all his English honours to the sorrow and indignation of his English friends. Yet the old man's action may have been brought about by political pressure, for shortly before his death in 1916 he was able to get a message through to his son-in-law in London which said:

Give my love to my friends & all the artists who worked with me, when you meet them. They are with me in my waking hours & in my dreams, & my thoughts of them are always good & pleasurable. With thankfulness I think of the hours I spent with them. They were the happiest of my artistic life.[7]

On 6 September Edward wrote a Soldier's Song to a poem entitled 'The Roll Call' by the journalist Harold Begbie. It was sung at the Albert Hall in a patriotic concert organized by Clara Butt in October. But by then it was already clear that much more would be required than the bold retort. Later he withdrew the song.

In Belgium, just across the Channel, Louvain was destroyed. Soon there were raids on the English coast. It seemed that London might be next:

September 21 1914. E. to his Station. Heard of destruction of Rheims Cathedral—much

---

[6] HWRO 705:445:6944.
[7] 15 Nov. 1916 (quoted in Young, *Elgar, O.M.*, p. 176).

upset. Felt must go & see Westminster Abbey were there, so E. & A. went—Saw the Bath Chapel & Banners & Crests . . .

Next day, as if moving by instinct, Edward interrupted his police duties to go alone to Worcestershire. He stayed with his sister Pollie and her family at Stoke Prior for several days. He was 'too oppressed with the war' to linger, and soon he was back in London. Yet a month later he wrote to Ivor Atkins:

I feel as if you in the 'country' were doing something but altho' I am busy from morning till night—the *houses* seem to choke it all off—we are fighting for the *country*. I wish I could *see* it . . . If it is sunshiny just go round to the W[est] end [of Worcester Cathedral] & look over the Valley towards Malvern—bless my beloved country for me—& send me a p.c. saying you have done so.[8]

In November he went down again, as he wrote to Alice Stuart-Wortley on the 13th:

Yesterday I went to Worcester & had the joy of sitting in the old Library in the Cathedral amongst the M.S.S. I have often told you of—the view down the river across to the hills just as the monks saw it & as I have seen it for so many years—it seems so curious, dear, to feel that I played about among the tombs & in the cloisters when I cd scarcely walk & now the Deans & Canons are so polite & shew me everything new—alterations, discoveries &c.&c. It is a sweet old place—especially, to me, the library into which so few go. I will take you in one day.

I . . . hope you will continue to be brave & strong through all these inevitable troubles—it is too dreadful to dwell upon but we must go on—on—on.[9]

Across the Channel a terrible battle had raged for three weeks round Ypres—'the massacre of the innocents', it was called, as the bloodshed on all sides was incalculable. Liège had fallen, Malines, Antwerp, Brussels, and almost the whole of Belgium. The remnant of the Belgian army under their brave King Albert were huddled behind the flooding Yser.

In London a patriotic anthology to be called *King Albert's Book* was organized by Hall Caine with contributions from leading artists, writers, and musicians, the profits to go to the Belgian Fund. When Caine's request arrived, Edward demurred. Caine replied: 'This is not the moment when Elgar should be silent.' Edward recalled reading in *The Observer* a poem by the Belgian writer Emile Cammaerts. Cammaerts was married to the daughter of Marie Brema, who had sung the role of the Angel in the first *Gerontius* at Birmingham. The daughter, Tita Brand Cammaerts, had translated her husband's poem into English. Its title, in memory of all the ruined bell towers of Flanders, was 'Carillon'.

> Sing, Belgians, sing!
> Although our wounds may bleed,
> Although our voices break,
> Louder than the storm, louder than the guns,

[8] 26 Oct. 1914.   [9] HWRO 705:445:7668.

Sing of the pride of our defeats
'Neath this bright Autumn sun,
And sing of the joy of honour
When cowardice might be so sweet.
. . . .
With branches of beech, of flaming beech,
To the sound of the drum,
We'll cover the graves of our children.
. . . .
We'll ask the earth they loved so well
To rock them in her great arms,
To warm them on her mighty breast,
And send them dreams of other fights,
Re-taking Liège, Malines,
Brussels, Louvain and Namur,
And of their triumphant entry, at last,
In Berlin!
. . . .
Sing of the pride of charity
When vengeance would be so sweet.

On 10 November Edward consulted Mme Cammaerts by telephone about the possiblity of setting 'Carillon' either in French or in English. Approval was vociferous. He mentioned the project to Rosa Burley, who wrote:

. . . I ventured to suggest that he should not try to tie himself to the metre of the words, as he would have to do if the piece were treated as a song or a choral item, but that he should provide an illustrative prelude and entr'actes as background music for a recitation of the poem.[10]

Thus the poem could be given with the music in either language. And so it was done.

The bells sounded Edward's old four descending steps over and over, but in a triple metre. The accents, as in bell-ringing, would seem to fall at random. But the descending notes never stopped tolling except when the music itself fell silent for the speaking voice. So they enacted the Allies' persistent intent with a simplicity plain to the most untutored ear.

On the last day of November a contract to publish *Carillon* was signed with William Elkin, who rushed through his presses arrangements for piano, organ, military band, and as many instrumental combinations as might play the new music on a cresting wave of great public emotion. Edward conducted the première on 7 December at the Queen's Hall with Tita Brand Cammaerts reciting her husband's poem. At the end there was uproar. A reporter for the American *Christian Science Monitor* rubbed his eyes in amazement:

Elgar sweeps his listener along with a vehemence which makes even the staid,

[10] *Edward Elgar: the record of a friendship*, pp. 197–8.

respectable folk who go to symphony concerts wish that they might accept the poet's invitation, 'Formons la ronde.' But that says a good deal, for the English middle classes do not usually dance on ruins; they are far more likely to put a fence round them and charge sixpence for admission. Perhaps only an Elgar could achieve such a startling result.[11]

Early in the new year 1915 *Carillon* was recorded for the gramophone with Henry Ainley (who as a young man had acted beside Frank Benson in the Dublin *Grania and Diarmid*). Plans were made to perform it up and down the country. Henry Wood engaged the great French actress Réjane for the recitation. The impresario Percy Harrison announced *Carillon* to climax a programme touring Sir Edward Elgar and the London Symphony Orchestra through cities in the north. The rest of the programme consisted of *Eine kleine Nachtmusik*, Beethoven's *Prometheus* Overture, the *'New World'* Symphony of Dvořák, the Saint-Saëns G minor Concerto with the Belgian pianist Arthur De Greef, and De Greef's own arrangement for piano and orchestra of Belgian folk-songs. Billy Reed, leading the London Symphony violins, recalled to the end of his life the fun they had with this

. . . fantasia on *Melodies Flamandes* (or some title very much like that) . . . In the last movement of the latter work the composer, who was himself the soloist, had a curious way of moving his whole body to the rhythm and of throwing up his hands at the ends of the phrases, so that after a few performances this work came to be renamed by the orchestra and was known as 'The Dancing Flamingo'. De Greef never knew this; but we told Elgar, and he used to bite his lip every night as he opened the score and saw the title, trying to keep a straight face in spite of his own strong sense of humour. This became more difficult at each repetition; and as we approached the last movement his facial expression, which he strove vainly to control, broke down, his hand stole over the lower half of his face, concealing his twitching mouth, and his eyes roved round in our direction, saying quite distinctly, 'You wretches!'[12]

. . . One Sunday [7 March 1915], Sir Edward and I took a taxi-cab and drove out to Queensferry on the shores of the Firth of Forth. We could see across to Rosyth, where several cruisers which had been engaged at the Dogger Bank were being overhauled and having their wounds attended to. We were very thrilled about the submarine nets which we were told stretched right across under the Forth Bridge—and then we drove back and thought how awful it all was, and how much we hoped that it would all be over quickly.[13]

In the opening land battles, soldiers on both sides had been cut down in carnage without parallel in human memory. The opposing armies had dug trenches from which to wage defensive war. As each side tried to outflank the other, the line of trenches grew northward through France and Flanders—until they stood facing each other behind what Winston Churchill described as 'ramparts more than 350 miles long, ceaselessly guarded by millions of men, sustained by thousands of cannon'. Nothing like it had ever been seen. Only the mute witness of tens of thousands of young bodies rotted in the no man's land

---

[11] Jan. 1915.   [12] *Elgar as I Knew Him*, p. 162.   [13] Ibid., p. 55.

between—the 'flower of the youth' to whom each nation should have entrusted its future.

* * *

At the beginning of 1915 Sidney Colvin had sent a suggestion: 'Why don't you do a wonderful Requiem for the slain—something in the spirit of Binyon's For the Fallen . . .'[14] Laurence Binyon had been Colvin's assistant at the British Museum, and was now his successor. He was a considerable poet, and he had tried several times to interest Edward in operatic collaboration. Now the war had lifted Binyon's verse to national recognition: through the later months of 1914 he had published a dozen war poems, collected at the end of the year in a slim volume entitled *The Winnowing-Fan*. Near the end came 'For the Fallen':

> They shall grow not old, as we that are left grow old:
> Age shall not weary them, nor the years condemn.

It was a first serious attempt to reckon up the psychological cost of the war, and it had made an enormous impression.

Edward sketched settings of three of these Binyon poems for chorus and orchestra. He played the sketches to friends, all of whom were deeply moved. Mrs Joshua called them his 'Elegies'. But when he consulted Binyon and Novellos, he discovered that 'For the Fallen' was already being set by a younger composer, Cyril Rootham, for publication by Novellos. To make the situation more awkward, Rootham was a pupil of Stanford and had followed his master to Cambridge—where a perceptible anti-Elgar faction was in existence. Edward wrote to Binyon on 24 March:

I cannot tell you how sad I am about your poems: I feel I cannot proceed with the *set* as [at] first proposed & which I still desire to complete: but I saw Dr. Rootham, who merely wished to thank me for my 'generous attitude' etc. & said very nice things about my offer to withdraw—but his utter disappointment, not expressed but shewn unconsciously, has upset me & I must decide against completing 'For the fallen'—I have battled with the feeling for nearly a week but the sight of the other man comes sadly between me & my music. I know you will be disappointed, but your disappointment is not so great as mine for I love your poem & love & honour you for having conceived it.

I am going into the country & will see if I can make the other settings acceptable without the great climax.[15]

27 March 1915

My dear Sir Edward.

Your words about my poem touch me deeply. My disappointment matters nothing, keen as it is: but think of England, of the English-speaking peoples, in whom the common blood stirs now as it never did before; think of the awful casualty lists that are coming, & the losses in more & more homes; think of the thousands who will be craving to have their grief glorified & lifted up & transformed by an art like yours—and though I

---

[14] 10 Jan. 1915 (HWRO 705:445:3453).   [15] Binyon family.

have little understanding of music, as you know, I understand that craving when words alone seem all too insufficient & inexpressive—think of what you are witholding [sic] from your countrymen & women. Surely it would be wrong to let them lose this help & consolation.

Rootham tells me he purposely planned his work on simple lines so as to be within the compass of small local choral societies: so I cannot see why his should clash, or why both settings should not be published . . .

Yours very sincerely
Laurence Binyon[16]

March 31:1915

My dear Binyon:

Very many thanks for your most kind & sympathetic letter: I quite feel all you say & would give anything that the publication might proceed; but under all the circumstances this is not conceivable.

There is only one publisher for choral music in England: Mr. Rootham was in touch with Novello first—my proposal made his M.S. wastepaper & I could not go on: that is exactly a bald statement of the case: I hope his composition, which on account of its simpler form will go more amongst the people than mine probably would have done, will bring your work with increased point & force to thousands & help them in their trouble—only I would have liked to have been the man to interpret you.

Best regards,
Yours sincly,
Edward Elgar[17]

All their friends quickly came to know of his refusal. One was the musical writer R. A. Streatfeild, another colleague of Binyon's at the British Museum. Streatfeild related to Rootham his own part in meeting Edward's refusal.

This, as you can imagine, was a great blow to Elgar's friends. However I need not linger over our disappointment, except to say that we consulted in the hope of devising some means of rescuing Elgar's work from destruction—for the benefit of ourselves & of the whole English world. We decided that I should have an interview with Mr. Augustus Littleton & should endeavour to persuade him to publish both works. This I did with no very great difficulty . . .[18]

Meanwhile, at Severn House, Sidney Colvin returned to the charge. When he came on 11 April, Alice wrote: 'Sidney overwhelming in his *attack* on E. to go on with L. Binyon's Poems—E. a little moved I thought . . .' Lady Colvin joined the attack. Binyon offered to add an extra stanza especially for the Elgar setting. And when Novellos announced their acceptance of Rootham's score, the basis of Edward's objection seemed to disappear. But his first enthusiasm was dampened, and he retreated to his old plea that the public did not want him. Colvin replied instantly:

April 13 1915

My dear Elgar

I heard with great distress last night that my wife's utmost petition to you on behalf of

<hr>

[16] HWRO 705:445:6350.   [17] Binyon family.   [18] 28 Mar. 1916 (HWRO 705:445:3731).

us all seemed to be in vain and that you persist, now that the original occasion for such a misfortune has passed away, in abandoning the scheme on which we had built such hopes. Well, it is cruelly hard on the fine poet & fine fellow, my friend Binyon, to whom the association with you would have brought just the lift in fame & status which was lacking to him and which, since the war began, he has so splendidly deserved . . .

You put it on the indifference of our race & public to art: but what has the poor British public done now which it had not done a month ago, when you were full of the project and raised all our hearts with the anticipation of a great & worthy expression & commemoration of the emotions of the hour, such as you alone are capable of giving them? Honestly I think you take far too censorious & jaundiced a view of your countrymen: but whatever their shortcomings surely it is you who have changed in the interval & not they?

And surely, granted that the majority of them are not very sensitive to the appeal of art, you cannot in your heart fail to realise that there is a big minority passionately sensitive to it, to whom your work makes all the difference in their lives, & whom—forgive me—you have no right to rob of such a hope as you were holding out to them a month ago. It is mere self-deception to blind yourself to the existence of this minority.

And then as to the majority—even though it may be true that they are not much awake to the touch of art & music, surely they have shown under extreme trial qualities & virtues & heroisms that are the fittest inspiration for art & music—as Binyon's verses have helped to prove, and as your music would prove ten times more if only you would not make all your friends unhappy by abandoning it . . . I cannot bring myself to believe—nor to crush the hopes of the poet by having to tell him—that in the end you will not consent.—

Ever affectionately yours
Sidney Colvin

P.S. The above is not to be answered except by deeds. SC[19]

Thus the man whose life had been honoured with the friendship of Ruskin, Gladstone, Rossetti, Browning, Meredith, George Eliot, Henry James, Robert Louis Stevenson, and Conrad tried to bring Edward home to his own contrarieties. Alice wrote next day: 'E. turned to his beautiful music again, loved it himself. So there is hope. A. to see Frances Colvin who was longing to hear if her & Sidney's entreaties had any effect . . .' They had.

> Now in thy splendour go before us,
> Spirit of England, ardent-eyed . . .

Thus Binyon began the first of his war poems. It gave a collective title for Edward's three settings, *The Spirit of England*. The first was of the poem which headed the book, 'The Fourth of August'. He began the music in the diatonic mood which had been his since the days of *Froissart* with a 'theme associated with the courage and hope of the first poem':[20]

[19] HWRO 705:445:3457.
[20] 15 Apr. 1916 to Ernest Newman.

Such simplicity had two strengths. The public strength, as with *Carillon*, was a clear appeal to every man jack of the nation's audience. The private strength, now when the deepest values were most sorely tried, was a return to the landscapes of his own creative youth:

> Enkindle this dear earth that bore us,
> In the hour of peril purified.

Binyon recalled the more recent past:

> The cares we hugged drop out of vision . . .

At these words the 'Windflower' syncopation of the Violin Concerto became a quiet marching second subject:

Another figure raised an echo of the Second Symphony's final dying 'Delight':

> The seed that's in the Spring's re - turn - ing,

What the nation was fighting for now was what Edward music had fought for all its life.

Only one of Binyon's stanzas stopped that music in the spring of 1915—a stanza condemning the enemy as inhuman. Edward wrote to Ernest Newman: 'I held over that section hoping that some trace of manly spirit would shew itself in the direction of German affairs.'[21] His inability to make out the enemy was widely shared then. People were only just beginning to feel a kind of emotion which had hardly touched Edward's generation before now. This emotion was to be characterized later in the war by Ernest Newman:

[21] 17 June 1917.

For the first time in the lives of many of us we find ourselves indulging in a national hatred and not seeing any reason to be ashamed of it; for the hatred is not so much that of a mere enemy—England has always been able to admire a fine enemy—as that of an immoral something that has become for the first time in history incarnated in a whole nation, the quintessence indeed of all the qualities in man that man, as an individual, contemplates in himself with regret and shame. What makes our anger with Germany so terrible is that, even in the moments when we have been strained almost beyond endurance, it is a cold, steel-like anger,—the anger one feels against some malevolent thing that is not quite human. It is this sense of being opposed to a foe that combines all the skill of science with the cunning of a maniac and the non-morality of a machine that has made the war a crusade for us, and haloed with an unusual pathos and holiness the heads of the young who have died for this land of ours.[22]

In 1915 there was only the shadow of such understanding—and horror at that shadow. Edward left the troublesome stanza, to set the final lines from 'The Fourth of August' with the music he heard clearly. The primary 'theme of courage and hope' was now surmounted by a high solo voice singing the old descending steps:

'It is the simplest of all,' Edward wrote of this first *Spirit of England* setting, 'as I felt the subject shd. be treated so.'[23]

The second of his 'Elegies' set Binyon's poem 'To Women'. Here chromaticism entered to enact (as in *Falstaff*) a torturing of spirit. 'To Women' was set in A flat, a semitone higher than 'The Fourth of August'. An aspiring upward fourth began it, but the phrase fell through minor harmonies—with A flats tolling a soft monotone in the bass. 'Your hearts are lifted up,' sang the high solo voice; but the phrase went inexorably down. Only at 'Your hearts burn upward like a flame' the tolling ceased for a moment. Then it recommenced:

> For you, you too, to battle go,
> Not with the marching drums and cheers
> But in the watch of solitude
> And through the boundless night of fears.

A second subject merely shrank the big opening rise and fall to small intervals accelerating through twisted chromatics in the orchestra:

> Swift, swifter than those hawks of war,
> . . . .
> You are gone before them, you are there!

[22] 'Elgar's "Fourth of August" ', *The Musical Times*, 1 July 1917.
[23] 17 June 1917 to Ernest Newman.

And not a shot comes blind with death
And not a stab of steel is pressed
Home,

but in - vis - ib - ly   it   tore,

(the music traced the figure of dying 'Delight')

And entered first a woman's breast.

The primary slow tolling began again:

Your infinite passion is outpoured
From hearts that are as one high heart,
Withholding naught from doom . . .

(Dying 'Delight' stalked the horizon in descending sequences of slow descending steps)

. . . to bleed,
To bear, to break, but not to fail!

The solo voice leapt up an entire octave to descend again over 'the theme associated with courage and hope in the first poem'. A quiet coda drew dying 'Delight' toward the 'Windflower' syncopation

It was, Edward noted, 'a "premonition" of the climax of No. III'.[24]

'For the Fallen' was the last and longest of the Binyon settings. Its tonic key was again a semitone higher, A minor—as if the music had toiled slowly through its tonal progress up to this barren vantage. Now the bass tolled *Solenne* over two notes—low A and lowest D: through that slow plagal drop all the carefully wrought harmonies of civilization emptied through hollowness. Above it the upward fourth which had set the hope of 'To Women' repeated in impotence:

[24] 15 Apr. 1916 to Ernest Newman.

The music warmed for an instant to a glint of *Caractacus* 'Britons'. Then the tolling returned, and almost casually the chorus came in singing bare octaves that belied their first words:

> With proud thanksgiving, a mother for her children,
> England mourns her dead across the sea.
> Flesh of her flesh they were,

(The Windflower-Delight 'premonition' from the end of 'To Women' rose up to make a second theme now)

>   spirit of her spirit,
>   Fallen in the cause of the free.

The tolling came louder. Then chorus, organ, and orchestra thundered the Windflower syncopation to a huge coronation march:

> Solemn the drums thrill: Death august and royal
> Sings sorrow up into immortal spheres.
> There is music in the midst of desolation

(Death's coronation march turned gently toward Windflower-Delight)

> And a glory that shines upon our tears.

The tolling renewed more loudly.

Suddenly the music became *Allegro (tempo di marcia)*. Edward wrote: 'The whole movement here is a sort of idealised (perhaps) Quick March,—the sort of thing which ran in my mind when the dear lads were swinging past so many, many times . . .'[25] Yet now his music sounded the chromatic twists of *Falstaff*:

Thus the flower of the youth marched through Edward's music to softly wrenched emotion as the chorus chanted in near monotones:

> They went with songs to the battle, they were young . . .

[25] Ibid.

Then came the extra stanza Binyon had written for this setting:

> They fought, they were terrible, nought could tame them,
> Hunger, nor legions, nor shattering cannonade.
> They laughed, they sang their melodies of England,
> They fell open-eyed and unafraid.

In the most celebrated line of the poem, two words were transposed to the more direct 'They shall not grow old . . .'. Here the music entered its only 3/4 metre—a crooning orchestral reminiscence of the *Apostles* 'Fellowship' motive and 'The Spirit of Healing' in *The Kingdom*:

Through it the chorus chanted their bare harmonies:

> Age shall not weary them, nor the years condemn.
> At the going down of the sun . . .

Dying 'Delight' made its simplest descent over soft, strong diatonic harmonies also descending, before the 'Fellowship' music returned to 'We will remember them.'

The last section opened a high solo of final songful development

> Felt as a well-spring that is hidden from sight,

(the tolling began again)

> To the innermost heart of their own land they are known
> As the stars are known to the Night;

And the coronation march of Death welled up to the climax:

> Moving in marches upon the heavenly plain,

(until it linked with the final Windflower-Delight)

> As the stars that are starry in the time of our darkness,
> To the end, to the end, they remain.

A last empty tolling ended the music as softly as it had begun. He inscribed the score 'to the memory of our glorious men, with a special thought for the Worcesters.'

The war, through Binyon's poetry, had given an ultimate subject to Edward's nostalgia. The result was his finest music since the Second Symphony. But though all three Binyon settings were 'sketched and nearly completed'[26] in the spring of 1915, he held his hand over their publication.

*   *   *

To the relief of all his friends, he had resigned from the Hampstead Special Constabulary in February 1915. His winter time health was unreliable. And it was more and more clear, as music and concerts were continuing through the war, that his real contribution lay there. Yet he listened to the persuasion of R. A. Streatfeild to join the Hampstead Volunteer Reserve, which involved drilling and rifle practice. In April he ordered his uniform.

Carice joined the Censorship Department in the War Office. Long ago frustrated from any active interest in music by her mother's prohibition of practice noise in the household of genius, Carice had turned to languages under the brilliant tutelage of Miss Burley. She was to continue in the Censorship throughout the war. Soon she brought home a prescription obtainable from any chemist to meet possible attacks of the latest German weapon, gas. '*Do* get it made up,' Edward wrote to Alice Stuart-Wortley: 'soak your handkerchief & inhale . . .'.[27]

On the night of 31 May 1915 Zeppelins dropped bombs for the first time on London. Alice noted: 'about 6 people killed & some homes smashed—*Brutes.*' British ships were sunk in British waters almost every day: the latest, most outrageous disaster was the *Lusitania*. In France the Allied generals, unable to make progress over German entrenchments, were hurling all their forces in the field with no result: the loss of British lives ran as high as 10,000 in one day.

On the Eastern Front the carnage in Polish territories between Germany and Russia was more terrible. Paderewski organized a 'Polish Relief Fund' in London. Charles Stuart-Wortley served as treasurer, and he persuaded Edward to join the committee. On 13 April Edward had a visit from the Polish composer Emil Młynarski, who was planning with Thomas Beecham a Polish Relief Fund concert at the Queen's Hall in July. He wanted for it an Elgarian tribute to Poland, as *Carillon* had paid tribute to Belgium.

Młynarski hinted at a fantasy on Polish airs, of which he wrote out many for Edward's choice. Alice Stuart-Wortley suggested putting in the opening phrase of Paderewski's *Polish Fantasia*, which was to be played early in the concert.[28] A movement from Młynarski's own Symphony 'Polonia' was also in the

[26] In his letter of 15 Apr. 1916 to Ernest Newman, Elgar described the three settings as 'sketched and nearly completed twelve months ago'.

[27] 27 May 1915 (HWRO 705:445:7814).

[28] Mentioned in Elgar's letter of 2 Oct. 1915 to Alice Stuart-Wortley (HWRO 705:445:7798).

programme, and that gave Edward a title. He set to work on his own *Polonia*, and by 21 April was able to play some of it for the excited Młynarski.

He began with a flaring theme of his own, and then developed each of two Polish themes at length: both of them opened in descending steps. He made a central episode from another figure of descending steps—the opening of Chopin's G minor Nocturne Op. 37 No. 1—in combination with the Paderewski fragment. After recapitulation came a third national tune, 'Poland is not lost'—caught first in soft snatches, connecting gradually in a big melody, growing finally to the grandeur of full orchestra and organ. The entire score of 68 pages was finished on 1 July 1915.

At the Polish concert on 6 July there were elaborately engraved programmes, each tied with a red and white ribbon and containing messages from Paderewski, together with a coloured reproduction of a Polish painting heading a big subscription form. Edward conducted his première and Beecham the rest. Alice wrote: 'Enormous audience, the real *roar* there is for E.—his own roar I always calls it—& recalled again & again.' But no second performance of *Polonia* was booked before October. Perhaps Poland was too remote from British fears and hopes in 1915. And this music lacked the immediacy of recitation which laced the Belgian piece. *Carillon* was going from strength to strength, with a new round of performances planned for the Coliseum in August.

Émile Cammaerts wanted Edward to set another of his Belgian poems, part recitation, part song. Its title was 'Une voix dans le désert', and once again there was an English translation by Tita Brand Cammaerts.

> A hundred yards from the trenches,
> Close to the battle-front,
> There stands a little house,
> Lonely and desolate.

From the half-ruined house a girl's voice sings:

> When the spring comes round again,
> Our cows will greet the day,
> They'll sound their horn triumphant
>
> . . . .
> Sound it so loud and long,
> Until the dead once more shall hear.
>
> . . . .
> Every church will ope its door,—
> Antwerp, Ypres and Nieuport—
> The bells will then be ringing,
> The foe's death-knell be ringing.
>
> . . . .
> And then our graves will flower
> Beneath the peace of God.

When *Polonia* was finished, Edward devised music for this second Cammaerts poem. Beginning in hollow drumbeats, scattered notes grew to phrases, and then to warm melody—which flowered in the song set for mezzo-soprano. At the line about the church doors opening, a climax came in the *Ave Maria* attributed to the early Netherlands composer Arcadelt. Then Edward's music sank back through the song's beginning to the narrator, the disconnected phrases, the drumbeats, silence.

Une voix dans le désert was finished in mid-July 1915. But *Carillon* was already booked for the Coliseum in August; and the new recitation, requiring also a singer, costumes, and scenery, would need costly preparation. The publisher Elkin, having taken *Polonia* in hopes of another *Carillon*, said he would be glad to have *Une voix dans le désert* when there was a definite prospect for its production.

The production did not come until the end of January 1916, at the Shaftesbury Theatre in a triple bill with *Cavalleria Rusticana* and *I Pagliacci*. *The Pall Mall Gazette* described it:

It is night when the curtain rises, showing the battered dwelling, standing alone in the desolate land, with the twinkling of camp fires along the Yser in the distance, and in the foreground the cloaked figure of a man, who soliloquises on the spectacle to Elgar's music. Then he ceases, and a voice is heard coming from the cottage. It is the peasant girl singing a song of hope and trust in anticipation of the day when the war shall be ended . . . The wayfarer stands transfixed as he listens to the girl's brave song, and then, as he comments again on her splendid courage and unconquerable soul, the curtain slowly falls.

It is a somewhat novel idea to combine music, speech, and song in this fashion, in conjunction with a stage setting; and if the whole thing is less effective than might have been anticipated, this is due chiefly perhaps to the excessive restraint which has been exercised by the composer . . . and this is, of course, a pity, since it is only the music which affords any justification for the production at all.[29]

Elkin would print only a piano reduction, carrying a small royalty per copy. The contract, signed on 19 January 1916, called also for a third Cammaerts recitation to Edward's music, *Le drapeau belge*. In each case the composer's main royalties were to come from performances. But after its run at the Shaftesbury Theatre, *Une voix dans le désert* was never revived in a stage production.

As if in distrust of his own facility for producing these war works, Edward wrote an article for *The Westminster Gazette* entitled 'Musical Waterwheels'— 'the devices that enable composers to "carry on"' by repetition:

The waterwheel is as ubiquitous as ever in modern music. The Russians, inspired by the repetitions in their folk-tunes, have reduced it to a simple convention, which consists in repeating every two bars. Debussy caught it from them, and at a certain period of his development two bars of consecutive major thirds were certain to be spun out to four, but that is passed [sic] now, and he has found other waterwheels. The minor lights of the

[29] 31 Jan. 1916.

Schola Cantorum have evolved some rather exasperating specimens. Our own composers, too, are quite familiar with turning a wheel until something happens. Possibly the most striking exception is Delius. My appreciation does not extend to all his works, but I cheerfully admit that he stands the water test astonishingly well.

Then, thinking of two most notable writers of sequences:

The masters, from Bach to Wagner, are all deeply indebted to [musical waterwheels]. They were masters, not because they scorned to use them, but because with them the waterwheel is a mere adjunct to the house and not a pretext for the building. The whole history of music is strewn with the forgotten reputations of those who thought otherwise.[30]

When he could, he got away to the country. Again and again he went to the deep Worcestershire peace of Stoke Prior, to stay with his sister Pollie Grafton and her family at The Elms. Two sons were away with the forces. But there was a new housemaid. And the presence of Mrs Grafton's famous brother made an impression on her to be recalled fifty years later for an acquaintance, who wrote:

During the twelve months that Ellen was there Sir Edward came to stay three or four times . . . He always came alone. Ellen wondered at this, and at the fact that though Pollie, Edward and the Grafton children talked together so often on many subjects neither Lady Elgar nor Carice were ever mentioned. One day Ellen asked Mrs Grafton why Sir Edward's wife never came with him. Mrs Grafton hesitated, frowned, and said, 'She does not care for travel.' And those were the first and last words that Ellen ever heard on the subject.

Sir Edward was always a very welcome guest, his visits eagerly looked forward to. He slept in the boys' room, walked the dogs on the common, enjoyed a glass of beer, and drove to Mass at the little church at the bottom of Rock in the trap with Pollie each Sunday morning. Though he had a keen sense of humour and was a great joker at times, Ellen remembers him as a quiet and reserved man with a gentle voice which he never raised. He still spoke with more than a hint of the rounded accents of his native Worcester and the faint double vowel sounds that mark a midland man . . .

Though there was at least one fiddle and a cello in the house, Ellen never heard Sir Edward play anything but the piano, and that he did often. One day while she was dusting the stairs he began to play. Looking across the hall and through the open door she could see him. She sat quietly on the stair watching him.

To her horror he looked up and saw her. He called, 'Ellen, do you like piano music?'
'Please, Sir Edward, yes.'
'Then come down here and I will play for you.'
So the spellbound little girl sat by his side while the great Sir Edward Elgar played just for her. It was a moment that lived with her. Her eyes sparkled as she told me.

One other event she remembers clearly. It was a warm afternoon in summer. Sir Edward, immaculately dressed as usual in open-necked shirt with a stock at the throat, and Norfolk jacket, sat alone in a garden chair on the lawn. His seat lay beside the path

[30] 13 May 1915.

to the privy: Ellen was too embarrassed to go there with Sir Edward sitting so close. Yet, as she peeped out at him from the kitchen door, he seemed to be asleep. To make sure, she quickly made a cup of tea which she took out to him. He sat with his head back and his eyes shut. His chest rose and fell rhythmically. She put the cup down beside him; he said nothing and did not open his eyes, but she was not entirely *sure* that he was asleep. Boldly she tiptoed to her destination beneath the lilacs. When she returned to the house he still sat as she had left him.

An hour later he came in to her from the garden.

'Ellen, you must think me most rude. You brought me a cup of tea and I never opened my eyes to thank you. You see, I was listening to a most wonderful symphony.'

Ellen was genuinely puzzled. 'But please, Sir Edward, I didn't hear any music.'

He smiled at her: 'Why, my dear, it was the birds.'[31]

After conducting *Carillon* through its summer run at the Coliseum with the Belgian actor Carlo Liten, Edward and Alice went for the last fortnight of August to a beautiful house near Midhurst in Sussex. This was Hookland, the home of the Caulfeild family, to whom Miss Burley was acting as 'finishing governess'. While the Caulfeilds and Miss Burley were away, the Elgars had the freedom of the place, surrounded by the vast quiet Sussex woods and fields ablaze with yellow gorse. Local transport was provided by an old character called Albert West, with a wonderful bright green and yellow victoria and an eccentric pony. He would preface a journey with: 'Rooads do be bad, doan't it?'[32]

Then there was Ship, the Caulfeilds' black and white sheep dog, who raised all Edward's old feeling for animals. One of the villagers said of Ship: 'He's a wonderful *civil* dog.'[33] Edward wrote to Alice Stuart-Wortley: 'I have a large dog friend "Ship" who has adopted me & he takes me for walks in the wildest places—he knows them all & I tell him everything!'[34] Between times at Hookland he worked at orchestrating the Binyon settings. But with Rootham's version of 'For the Fallen' now on the horizon, it was difficult to finish his own.

No sooner back at Severn House than Edward found himself restless to be away again. They arranged to join the Stuart-Wortleys at Walls, near Ravenglass on the coastal side of the Lakes. Edward left by himself on 7 September. That night there was another Zeppelin raid over London, clearly visible from the upper windows of Severn House:

A. & C. watched & watched raid. E. said we were to go down in basement but could not leave this extraordinary sight . . .

September 8. E. at Grasmere—from Ullswater. A. busy clearing up & preparing for Walls. Evening came—all in bed except A. Suddenly Boom, Boom, wonderful deep sound. A. ran to window & then fled out to look through other windows; when C. called her, & from C.'s window strange & awful sight. The sky lit by flying search

[31] Letter of 1966 from Nicholas Hale (who interviewed Ellen) to Alan Webb.
[32] Conversations with the Caulfeild family, 1977.
[33] 10 Sept. 1915 to Rosa Burley (transcript by Mrs E. Burley, in the keeping of Dr Percy Young).
[34] Friday [20 Aug. 1915] (HWRO 705:445:4165).

lights—part of a Zeppelin visible like a gilt bar, & star-like shells bursting more or less near it & the boom of guns sounding!

September 9.   E. at Grasmere.   A. raser (*very*) perturbed at leaving—Telegraphed to E but he had seen nothing in papers, as A. thought he wd. have, & said must come. A. did not like going, felt it was like running away: but as E. said come, prepared.

At Ravenglass she found Edward plunged in country pursuits—taking wasps' nests and building bonfires. Back in London at the end of the month, he soon felt ill and took to his bed.

He recovered in time to conduct the next fortnight of *Carillon* at the Coliseum in early October. The reciter was Edward Speyer's daughter Lalla, married to the Belgian Socialist leader Émile Vandervelde (now minister-in-exile in charge of the large Belgian refugee population in England). Earlier in the year Mme Vandervelde had whipped round the United States and returned with £60,000 and a large quantity of food for the refugees—though America maintained official neutrality. To recite *Carillon* she clothed her statuesque figure in black and stood before a blood-red curtain. Alice wrote: 'Lalla looked picturesque—a type as it were of suffering Belgium.'

There and in France the Allied forces were being decimated. Earlier in 1915 Lord Kitchener had raised a 'New Army' of enthusiastic British volunteers. Now these were going to their deaths by the thousand and ten thousand. In August, though Britain had never in her history had regular conscription, a 'Manifesto on National Service' had appeared in *The Times*. Among the distinguished signatures was Edward's.

After the second Coliseum run of *Carillon*, he was still restless for the country. He took Carice for her holiday to Stratford, where they joined the Stuart-Wortleys for *Troilus and Cressida*. He went on to Stoke, but had to return to conduct a long-booked concert in Bournemouth. It included a *Wand of Youth* Suite. Alice noted between the rehearsal and concert on 23 October: 'E. not liking music at all & saying it was all dead &c. E. rested & then conducted splendidly & evidently music had revived.'

*   *   *

On 9 November 1915 there was a telephone call from the *Daily Telegraph* critic Robin Legge, followed by a letter. Legge wanted Edward to write the music for a fantasy play to be produced in the Kingsway Theatre at Christmas. The producer was the actress Lena Ashwell. Next day she brought the play for Edward to see. It was entitled *The Starlight Express*, an adaptation of Algernon Blackwood's novel *A Prisoner in Fairyland* by Blackwood and the lady who had suggested its staging, Violet Pearn. The authors provided a summary:

The 'wumbled' English family, living for economy's sake in this village among the Jura Mountains, are troubled about many things. Various folk in the village are troubled too. And their 'trouble' is traceable to one main source—that mis-understanding which is due to lack of *sympathy*. The story describes the happy change produced by the magical effect of *sympathy*.

This effect is obtained chiefly by the working of a Secret Society organised by the children—a Star Society. For the children believe that, while their bodies lie asleep, their spirits ('the part Mummie says goes to Heaven') get out and play among the stars. They collect starlight, or, as they call it, star-dust . . . Star-dust is Sympathy!

. . . The great desire of the children, therefore, is to get everybody out of their bodies—out of their little Selves, that is . . . With the help of *Cousin Henry*, they accomplish wonders . . . But Cousin Henry has other helpers too. They are the Sprites—the Figures of his own imaginative childhood. Thinking comes true, he explains, and he has 'thought alive' these Sprites, and brought them with him from his English childhood in the *Starlight Express—a Train of Thought!*[35]

The play script described the Sprites:

When these appear in daylight scenes they have only a slight air of strangeness and mystery to distinguish them from their types of every day. In night scenes they are super-tramp, super-sweep, etc., mysterious swift, singing creatures with their strange musical call not unlike a birdcall. They are blue-shadowy, or faintly luminous. They suggest 'something big and vague at the edges' . . .[36]

These Sprites could make a double appeal to Edward. In their daytime existence, they recalled the enduring fascination of old-time characters like Albert West at Hookland—figures who still abounded in country places then to remind Edward of his own remote countryside. By night they took over the function of the Angels who accompanied and pleaded for the Soul in *Gerontius*. And the entire action of this play came astonishingly close to the fable of childhood and age, inspiration and weariness, inner victory and defeat which Edward had tried to frame in the boyish drama of fifty years ago.

A meeting with the author, Algernon Blackwood, proved a great success. 'Blackwood is an unusual man,' Edward wrote to Sidney Colvin, '& sympathetic to me.'[37] Edward agreed to write the music, and through the next weeks he worked at the *Starlight Express* score with the enthusiasm of a major effort.

The 'presenter' of the play was the chief Sprite—a tramp with a barrel-organ. It was the musician grown old and threadbare entertaining a fickle world, and left with an audience of children. For him Edward wrote a barrel-organ tune that went up and down the triad with a plagal caste in the favourite old key of G minor:

[35] 'Explanation' in the programme of the Kingsway Theatre production, 29 Dec. 1915.

[36] Elgar's copy of the TS, now in the Novello archives. Subsequent quotations from the play are taken from this source, which shows alterations and additions in various hands.

[37] 20 Dec. 1915 (HWRO 705:445:3459).

Before each act the Organ Grinder appeared in front of the curtain. At the beginning he sang:

> O children, open your arms to me,
>   Let your hair fall over my eyes;
> Let me sleep a moment—and then awake
>   In your gardens of sweet surprise!
>   For the grown-up folk
>   Are a wearisome folk,
>   And they laugh [all] my fancies to scorn
>   My fun and my fancies to scorn.

The song's opening phrase leapt up to descend in steps. At the words 'Let me sleep a moment' came the central descent of 'Little Bells' from *The Wand of Youth*—now to become a motive for sleep and starlight:

Then came an intransigent D minor to show the adult world:

(It echoed one of the 'quasi-military themes dismissed with scant courtesy' in

the First Symphony  .)

A middle section in the song was based on a carefree development of 'Little Bells' to show the inspiration of childhood:

The rise of the curtain on the family home turned the steady minor up and down of the barrel-organ through a dotted major triad:

It was a tune for Jane Anne, the gentle elder daughter, aged sixteen. But most of the First Act showed grown-up problems in unaccompanied prose—the frustrated inspiration of Daddy, an unsuccessful author; the domestic tribulations of Mother; the Widow Jequier with her Pension full of residents who could not pay, including old Miss Waghorn, always searching for her long-dead brother.

A Sprite appeared to an eccentric figure:

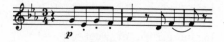

The younger girl and boy appealed for explanations. As Daddy tried to describe inspiration, the barrel-organ's up and down softly contracted to

At the mention of 'Starcaves' Jane Anne's tune developed a passionate dark seriousness:

Solo violin cadenzas raised a dreaming imagination above the world of fact, even as in the Violin Concerto. The arrival of Cousin Henry introduced further night symbolism, and the 'Good night' brought another *Wand of Youth* melody from 'The Fairy Pipers':

    The Organ Grinder's song before Act II was a rollicking waltz with triangle and tambourine describing 'The Blue-Eyes Fairy'—a children's figure who can 'make you forget that you're old'. The night-time scene opened on the edge of a pine forest, to revive Edward's private memories of Broadheath, Spetchley, Fontainebleau. At the edge of the forest the entrance to the Star-cave was almost completely closed with a huge stone. Darkness and dreams were in the ascendant, and Edward's scattered motives from Act I were woven into near-continuous musical texture. The Organ Grinder sang a Curfew Song to a slow, soft echo of the childhood-inspiration theme from Act I: at the end a descending figure of three bells softly echoed *Carillon*. The barrel-organ music developed to a 'tiptoe tune',[38] and the sleeping children were resurrected in dream to stand before the Star-cave.

    Suddenly the Starlight Express appeared, chuffing in to a heavy Sprite figure. One by one the Sprites descended—Sweep, Organ Grinder, Dustman, Lamplighter, Head Gardener, and a 'Laugher' 'who sings trouble into joy'.

---

[38] Elgar's notation in his copy of the play typescript.

They entered the cave, and emerged to scatter stardust on the sleeping villagers. Coming to the ancient Miss Waghorn, the Dustman used his finest of all, kept especially for 'the old and lonely': the 'Starlight' theme linked hands with 'Fairy Pipers', and the old lady rested at last from the search after her dead brother. The Lamplighter made his exit to a reminiscence of the famous words spoken by Sir Edward Grey at the nation's entry into the war: 'Fires are going out all round the world, and I must tend 'em.'

The Organ Grinder's song before the final Act developed the barrel-organ music into another ghost of *Gerontius* 'Judgement'—only now it made a good bracing tune:

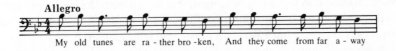

In the middle of the song the *Gerontius* ghost came closer, in *sostenuto* phrases of woodwind, strings, and harp:

> Now I am a Constellation,
> Free from every earthly care . . .

Act III juxtaposed the day and night-time worlds. In the morning Madame Jequier entered rejoicing because Cousin Henry had secretly paid all the debts of her Pension: 'Then the bon Dieu is an Engleeshman.' It was a tribute that would come home to every member of the audience with a friend or relation with the British Army in France. Daddy entered 'on the very edge' of his great idea, speaking through the 'Starlight' music:

The source of our life is hid with Beauty very, very far away. Our real continuous life is spiritual. The bodily life uses what it can bring over from this enormous underrunning sea of universal consciousness where we are all together, splendid, free, untamed; where thinking is creation, and where we see and know each other face to face.

It was what Edward himself had sought from his first childhood friendship with Hubert Leicester, from his first writing of the 'tune from Broadheath'. And at Christmas a year ago, British and German troops had fraternized between the trenches before resuming their deadly combat the next day.

The final scene returned to the pine forest at night, the stone now rolled away from the Star-cave. As the Sprites brightened the scene, the humans entered. The ghost of Miss Waghorn passed into the cave 'clothed in light'. At last the Star of Bethlehem rose 'where a child was born'. Orchestra and organ played 'The First Nowell' with Carillon bells descending through the entire scales which had once reflected the *Apostles* sunrise.

In just a month *The Starlight Express* had achieved the dimensions of a major work—300 pages of score containing more than an hour's music. The sheer size bore witness to the play's power over the composer's imagination. He inscribed it at the end:

Fine Edward Elgar Dec 1915 ae 15.

Fifteen was just the age of the twentieth century—the century which had followed directly on his *Dream of Gerontius*. Now, far within the fantasy of hope in the play, the faery grace of his music for *The Starlight Express* chimed another elegy.

While Edward's every energy was bent on finishing his score for rehearsals Landon Ronald had been negotiating on his behalf with the music publishers. Novellos were chary of the whole venture. Elkin agreed to a contract, but its terms were far from what Edward's music had commanded before the war—a sum down of 25 guineas and a small royalty on the sale of any songs or piano arrangements the publisher might wish to print. There was no prospect of printing the full score, at least until a stage success was assured to rival Barrie's *Peter Pan*.

On 6 December the two singers chosen by Lena Ashwell came to rehearse at Severn House. The soprano, Clytie Hine, offered '*very* delightful singing'. The baritone, Charles Mott, brought something more than that. He had been heard at Covent Garden early in 1914 by Edward, who recommended him enthusiastically for the Angel of the Agony in *Gerontius*.[39] Through rehearsals and the subsequent run Mott became a valued friend—'sympathetic' as the play's author Algernon Blackwood was sympathetic to Edward. And the young conductor at the Kingsway Theatre, Julius Harrison, was a Worcestershire man—a former composition student of Bantock's who had listened enthralled to Edward's lectures at Birmingham ten years ago. So far all promised well.

Then Blackwood wrote that Lena Ashwell had asked him to stay away from rehearsals for a week, as he was inexperienced in theatrical ways. He shared his misgivings with Edward over the choice of designer—Harry Wilson, President of the Arts and Crafts Society:

I hear that Mr. Wilson, the artist, has designed the Sprites in the spirit of Greek fantasy—Lamplighter a quasi-Mercury, Gardener as Priapus, or someone else, and Sweep possibly as Pluto. It is a false and ghastly idea. There is nothing pagan in our little Childhood Play. It is an alien symbolism altogether. It robs our dear Sprites of all their significance as homely childhood Figures. Don't you think so too? If our Play means anything at all, it means God—not the gods.[40]

When at last on Christmas Eve they saw the play in late rehearsal, it was worse than Blackwood had imagined:

---

[39] 16 Mar. 1914 to Ivor Atkins.
[40] Novello archives.

Xmas Day, 1915.

My dear Sir Edward,

I know what you're feeling. Probably you can guess what I'm feeling. Can we do anything? I am so inexperienced in these things. I have, of course, the right of veto. That means getting a new artist, postponement of opening, heavy loss of money to Miss Ashwell, and so forth. You know better than I do what a sweeping veto would involve. That our really big chance should be ruined by her strange belief in a mediocre artist is cruel. This murder of my simple little Play (qua words) I can stand, for the fate of my books has accustomed me to it; but this suburban, Arts & Crafts pretentious rubbish stitched on to your music is really too painful . . .

I . . . shall come up to the theatre on Monday—by an effort of will.　Yrs A.B.[41]

Edward agreed. He wrote to Troyte Griffith, who knew the designer:

Your friend has entirely misused any chance this play had of success—he's an ignorant silly crank with no knowledge of the stage at all & has overloaded the place with a lot of unsuitable rubbish & has apparently never read the play. He ought to be put in a Home!'[42]

To crown all, two days before the opening a taxi in which Alice was riding through streets darkened against night-time air raids met with an accident. She was brought home with concussion and confined to bed for ten days. Edward did not go to the opening night on 29 December, although it had been arranged for him to conduct the first performance.

The critics next day registered dissatisfaction with the play. Behind all their comments stood the silent contrast of *Peter Pan*—the annual Christmas staple since its production in 1904. Measured against Barrie's juxtaposing of childhood vision with fierce childhood cruelty, *The Starlight Express* was sentimental. The *Pall Mall* critic wrote:

Where it should be just humbly telling the audience a story, it is preachy and pretentious. It pretends to be meant for children, but is canvassing all the while for grown-up sentiment.

The Observer questioned the wisdom of dramatizing such a story at all:

There are times when the theatre is detestable . . . It is detestable because it stretches out its great paws to grasp exquisite little things, and crushes them . . . because the more it tries suggestion the more hopelessly it proves itself limited to the statement of action, action, action. Here is 'The Starlight Express', a charming fancy—a dainty dream between sleeping and waking. The theatre grabs at it; and the Starlight Express itself—that 'train of thought'—becomes a set of wobbly black curtains with spangles on them. The Pleiades come down from the sky, and are a bunch of ballet-girls. The delicate fancy becomes a heavy sermon . . .

The house response was described for the American *Christian Science Monitor*:

The audience, when the writer attended, was evidently for the most part made up of

---

[41] Ibid.
[42] Quoted in Young, *Elgar, O.M.*, pp. 182–3.

those whom Matthew Arnold described as Barbarians, Philistines and Populace. They deserved even ruder titles. Music or no music, whenever the curtain fell, they began to make a complicated verbal noise. Shutting their ears they refused to listen to one of the master minds of modern music.

For the music itself the reviews had nothing but praise. The critic of *The Times* recognized power below the conscious level of listening:

Apart from the charming tunes and the gleams of orchestral colour which one carries away in mind, a great deal which one does not carry away—which one hardly distinguishes—plays its part very subtly and yet very simply. Whosoever is 'wumbled' let him listen to Sir Edward Elgar.

Gradually Edward forgot the scenery and costumes which had sought to show a childhood so distant from his own. While the play ran through January 1916 Alice noted: 'Starlight Express every day.   E. there very often.' But the handwriting was on the wall. On 29 January—just a month after its opening—*The Starlight Express* had its final performance in the presence of the Elgars, Blackwood, and the others who had created it. Next day:

January 30 1916. E. felt the ceasing of the Starlight very much, so did Mr. Blackwood—indeed all the nice people concerned.

There remained one chance for the music, and Edward took it. The Gramophone Company had just offered to renew his recording contract on generous terms—'Flat payment of £100 per annum. £21.0.0 per session in addition.'[43] The contract arrived, but Edward held it up until the Company agreed to record no fewer than eight records of *Starlight Express* music. Charles Mott would repeat the Organ Grinder songs, but in the soprano part Edward requested Agnes Nicholls—for whom he had written 'The Sun Goeth Down' in *The Kingdom*. With the Kingsway Theatre orchestra augmented to about thirty players, the records were made at Hayes on 18 February. Blackwood was in attendance and Edward on the qui vive in the small wood-panelled recording room. One pressman reported: 'Sir Edward, in his high conducting seat, was in bantering mood. He gravely informed the band after they had done an excerpt from Act II that "it looked very well on the wax."

When the records were ready, they were introduced with a Press Luncheon at the Savoy, where author and composer were the Company's guests. It marked the beginning of Edward's serious interest in the gramophone. Through these records *The Starlight Express* music, which would otherwise have been interred with the costumes and sets in the Kingsway Theatre basement, was to travel far and wide. Later in the war a letter arrived at Severn House from an officer serving at the Front:

Though unknown, I feel I must write to you tonight. We possess a fairly good

[43] See Moore, *Elgar on Record*, (OUP, 1974), p. 13.

gramophone in our mess, and I have bought your record 'Starlight Express' "Hearts must be star-shiny dressed" [sic]. It is being played for the twelfth time over. The Gramophone was Anathema to me before this War because it was abused so much. But all is changed now, and it is the only means of bringing back to us the days that are gone, and helping one through the Ivory gate that leads to fairy land or Heaven, whatever one likes to call it. And it is a curious thing, even those who only go for Rag-time Revues, all care for your music. Our lives are spent in drunken orgies and parachute descents to escape shelling or Bosch aeroplanes. In fact, the whole thing is unreal, and music is all that we have to help us carry on.[44]

* * *

For Edward himself *The Starlight Express* opened another gate. The re-animating of his childhood landscape and childhood drama seemed to clear away at last his private difficulties over producing *The Spirit of England*. An occasion was offered by Clara Butt, who wanted to organize an entire week of *Gerontius* performances in May 1916 to benefit the Red Cross. Her reasons were widely publicised: 'Clara Butt believes that there is a new attitude towards death. She holds that many people that had no faith before the war are now hungering to believe that beyond the grave there is life.' To preface every *Gerontius* performance through the Red Cross week, Edward agreed to conduct 'To Women' and 'For the Fallen'. In 'The Fourth of August' he was still uncertain how to set the stanza that showed the enemy.

The Rootham setting of 'For the Fallen' now had a good start, and friends assured Edward that the other composer could not possibly object. They were wrong. Spurred on (it was said) by the Cambridge protégés of Stanford, Rootham raised a howl of protest after the die was cast.

The fuss made Edward intensely miserable. Following another short northern tour conducting the London Symphony Orchestra, he sickened with influenza. Alice followed, and unwillingly went to bed. Severn House was arctic because the fuel allowance was insufficient to run its big central heating system. When he was able, Edward made plans to go to Pollie in Worcestershire. But then his old ear trouble started again.

April 8.   E. ready to start for Stoke. Said he felt giddy & was not sure he wd. go.   A. in bed & afraid of the cold house (no coke) for him so did not persuade him to give it up, hoping change wd. do good.   A. dismayed to hear by telephone from Capt. Dillon that E. was ill in the train & Capt. D. took him out at Oxford & motored him to the Acland Nursing Home—A. told Sir Maurice [Abbott Anderson] who telephoned to the Home & gave directions.   A. telephoned to C. & she went off to Oxford & sent a comforting account. A. so mis.

April 9.   A. heard good account D.G.   C. stayed with Dora Townshend [Mrs. R. B. T.] & went over very often to see E, who said he was quite comfortable . . .

---

[44] 5 Oct. 1917 from J. Lawrence Fry (HWRO 705:445:6367).

April 10.   Thought E. wd. return but it was put off till next day.   A. down for a little while. Ordered car for E. to come home in next day.

April 11.   E. & C. arrived all safely about 6—E. a good deal better.   A. so thankful to see him safes. Very disappointing to have this touch of former illness but hope & trust it was only caused by influenza & wd. pass.

Three days later they had a visit from the young composer Arthur Bliss, now serving with the Army:

Through mutual friends, I visited the Elgars when they were at Severn House in Hampstead. I remember very well going up the hill there, entering this large house, and Lady Elgar being very kind, and hearing Elgar upstairs playing some phrase over and over again on the piano.[45] (Afterwards I discovered that this phrase came from an 'occasional piece', Le drapeau Belge . . .)[46]

He soon stopped and came down, and there he was—the first time I met him—aloof, shy, speaking to me in his soft Worcestershire voice, and taking me in with his rather curious blinking eyes . . .[45]

The Drapeau belge recitation was finished, and a piano version published by Elkin with an English translation by Lord Curzon of Kedleston. But there was no performance: the London Symphony could not find room in their schedules. Public attention had now turned from Belgium southward to the terrible battle of Verdun. There were 50,000 total casualties in February, 120,000 in March, 80,000 in April, and no perceptible progress either way.

At home, Clara Butt's Gerontius concerts were coming on at the Queen's Hall. As the chorus was to be Embleton's Leeds Choral Union, preparatory performances were planned for the north. Thus Leeds was to have the actual premières of 'To Women' and 'For the Fallen'. Far from well, Edward journeyed up to conduct them.

The critics recognized a remarkable work. 'Never since "Gerontius"', wrote one, 'has Elgar given us music that carries so unmistakable a ring of sincerity throughout the whole course.'[47] The reviews of a second performance at Bradford and the sold-out London week beginning on 8 May echoed the judgement. 'Here in truth is the very voice of England,' Ernest Newman had written on first seeing the printed music, 'moved to the centre of her being in this War as she has probably never been moved before in all her history.'[48] After hearing the music Newman expanded his view in an article 'On Funeral and Other Music':

With the exception of Elgar, none of our composers, so far as I know, has produced any

[45] 'The Fifteenth Variation', BBC broadcast, 1957.
[46] As I Remember (Faber & Faber, 1970), p. 23, where the recollection is ascribed to 1917, the year in which Le drapeau belge was first performed with orchestra. There is no recorded composition date known to me, but the piano score was published in 1916. Lady Elgar's diary makes no mention of any visit from Bliss in 1917, but does describe his visit in April 1916. As the Elkin contract covering Le drapeau belge is dated January 1916, the evidence suggests placing Bliss's recollection here.
[47] The Sheffield Daily, 4 May 1916.   [48] The Musical Times, 1 May 1916.

music, inspired by the War, that expresses anything of what the nation feels in these dark days . . . Only out of an old and a proud civilisation could such music as this come in the midst of war. It is a miracle that it should have come at all, for Europe is too shaken just now to sing . . . The artist in [Elgar] gives him the power, denied to the rest of us, of quintessentiallising his emotions, of extracting from the crude human stuff of them the basic, durable substance that is art.[49]

Alice wrote of the opening London concert:

May 8 1916. E. keeping fairly better. Enormous crowd at Queen's Hall, people not able to get seats—Performance wonderful. E. conducting most masterly. New works made profound impression, A. Nicholls' 'We will remember them' never to be forgotten. The semi-chorus perfect in Gerontius, the opening Kyrie breathed out pp. in perfect time . . .

Henry Coward, now the regular conductor of the Leeds Choral Union, was jubilant:

The record success of the series smashed the hoary fiction that oratorio is exhausted with one performance. The audiences were astounding in quality and quantity. King George, Queen Mary, and Queen Alexandra attended, and the Royal box was always a picture.[50]

The gross receipts for the week of performances amounted to £3700, of which a net sum of £2700 was handed to the Red Cross. The editor of *The Music Student*, Percy Scholes, planned an entire 'Elgar Number' of his journal with interviews, photographs, and critical studies of all the major works. He summarized:

If this country had a 'Musician Laureate' it would be to Elgar that the laurel would be offered. For he, of all our musicians, is the one to whom we turn in times of national feeling to provide us with the musical expression for which our spirits crave.[51]

The war deepened. The continuing battle at Verdun cost 110,000 casualties in May, and the death-rate in June would be higher. The British prepared a big offensive on the Somme. They went over the top to be scythed down by German artillery: on the first day alone the British lost 60,000. It dwarfed everything in history.

In July Edward returned to Stoke, trying again to find his health. When Alice Stuart-Wortley came to Stratford for a weekend, he met her at Droitwich and drove her through the forests to visit the enchanted ground. Later he wrote:

I am glad you 'feel' Stoke—that is a place where I see & *hear* (yes!) you. A. has not been there since *1888* & does not care to go & no one of my friends has ever been but you. No one has seen my fields & my 'common' or my trees—only the Windflower and I

[49] *The New Witness*, 11 May 1916.
[50] *Reminiscences*, p. 294.
[51] *The Music Student*, Aug. 1916, p. 358.

found her namesakes growing there—aborigines I'm sure—real pure sweet forest folk. Bless you.[52]

At Severn House his own Alice was in bed with a chill. Since the taxi accident in December her health had been fragile, and she was not far from seventy. When she had recovered enough to travel, Edward met her from Stoke and took her to spend three weeks of August in the Lakes.

August 16.   Fine—Went up the Lake in 'Raven' . . . A storm came on & E. & A. sat at end of bench & a man who had been standing by talking to another suddenly said, putting his face close to theirs, 'You are lùvers still, like me & my wife.'   E. said in a sweet way 'I hope so.'—it was quite sincere & very touching.

He wrote to Alice Stuart-Wortley:

Alice is much rested & this has done her any amount of good—she thinks it is good for me but I stay on for her sake. I do not know when we shall arrive in town.[53]

August 18.   Left Ullswater with regret. Very rainy. Kind Mrs. Speer wrapped A. up beautifully. E. & A. on box seat & *lovely* drive over Kirkstone Pass—did enjoy it. E. walked up Pass . . . Lunched at Ambleside very late & *so* hungry! then started to walk. E. had a nice sleep on bench, full length, covered up & umbrella over him! Then began to walk & torrents came down. Coach took us up at Rydal . . .

Next day Edward felt giddy, and he began a throat infection. Alice took alarm and got him quickly back to Severn House. Sir Maurice Abbott Anderson tried tonics, painting, and finally electric cautery before the throat seemed to respond.

They were just able to accept an invitation from the Berkeleys to stay at Spetchley Park in the Worcestershire countryside. Here again, in the scenes of childhood summers now half a century distant, he felt better. He fished the lake through fine September weather. There he was found during a long afternoon by Ernest Newman, who recalled it forty years later.

Happening to be in Worcester one summer day I met Mr. Robert Berkeley on his way to a council meeting; he told me that Elgar was staying at the Hall and suggested that I should run down there. As no vehicle was available, some races being held in the neighbourhood, I decided to walk the few miles. About half-way on the route I left the main road and made a short cut diagonally across the meadows. As I was nearing the Hall I was hailed by someone from the lake. It was Elgar. He brought the punt to the bank and conveyed me to the middle of the lake, where I passed with him one of the golden summer afternoons of my life, talking about this and that. He had felt, he told me, an irresistible urge to be alone, and had excused himself on the plausible ground that he wanted to do some composing.

It was a small revelation of an escapist mood that came to bear a symbolical aspect for me in later years. For Elgar, for all his comprehensive interest in life, was at heart, I have always thought, an 'escapist'.[54]

On the last day of the visit the Berkeleys, who were Catholic, asked Edward to write in their vocal score of *Gerontius*. He turned to the passage about 'The summer wind among the lofty pines' and inscribed it, 'In Spetchley Park, 1869.'

Back in London there were renewed air raids, Zeppelins shot down, trainloads of wounded from the Front arriving daily at Charing Cross. The new Allied offensive on the Somme was making minute gains: the total casualties for September approached half a million.

Every shred of normal life was seized on and cherished:

October 8 1916 . . . Mr. & Mrs. Elkin to tea & Mr. Blackwood who met E. coming up from Finchley Road with a toad in his pocket. E. had bought it of some boys for 2d. He did not think it was happy with them. He put it in the garden & calls it Algernon as he met A. B. at the time—He puts his head out of the window & says 'Do you think he will come out if I make a noise like a worm?'—Algernon invisible.

The story found its way into several papers under the rubric 'Toad and Verklärung'.

Edward got through several conducting engagements, including a Philharmonic performance of the Second Symphony on 27 November. The standard of playing was far from pre-war days, the reception distant. In the sparse audience—sparse on account of air raids, it was said—was the younger composer Frank Bridge, who wrote afterwards to Edward Speyer:

Poor Elgar–The orchestra positively scrambled through the Symphony as they had never before seen it. If it had been a work of mine I should have gone home & vowed 'never again without three *long* rehearsals'. In fact, as I had never heard the work before I felt quite sick about it. Performances of this kind do far more harm than good.[55]

The audiences continued sparse. At one Elgar concert someone asked where were Edward's friends. Thomas Beecham, remembering the pre-war reputation of Edward's music among German musicians, replied wickedly: 'They've all been interned.'

With the onset of winter his ear trouble grew worse. In December he wrote to Alice Stuart-Wortley: 'I am very unwell indeed and do not know what to make of it—I suppose it is the old thing but I can make no plans at all.'[56] When he learned that Charles Stuart-Wortley was to be raised to the Peerage in the New Year's Honours, his letter of congratulation turned to the unhappy contrast of himself:

Wedy [20 December 1916]

My dear Windflower:
I am out of bed for the first time since Saturday & I use the first minute to send you love & congratulation on the event,—I gave you a coronet long ago—the best I had but you may have forgotten it—now you will have a real one, bless you!

[54] *The Sunday Times*, 6 Nov. 1955. [55] 30 Nov. 1916. [56] 11 Dec. 1916.

I cannot tell you how glad I am—I expected it—wanted it for you long ago & now it's come I feel afraid of you & wonder in a vague sort of way what will be the difference? But you are still the Windflower I think & hope.

I may try to get out today but I am very shaky still.

Everything pleasant & promising in my life is dead—I have the happiness of my friends to console me as I had fifty years ago. I feel that life has gone back so far when I was alone & there was no one to stand between me & disaster—health or finance—now that has come back & I feel more alone & the prey of circumstances than ever before.

Bless you,
Your
EE

I wonder what the new name will be—this may be the last time I address you in the old, familiar way.[57]

The new name was 'Stuart of Wortley', and of course it made no difference. But 1916 had been the worst year of his life so far, and the future in every prospect was dark. Creatively the entire year had produced almost nothing—the brief and simple *Drapeau belge* (so far only performed in a recital by Carlo Liten) and a song for the 'Fight for the Right' movement requested by Gervase Elwes. The recrudescence of the 'Menière's Disease' suggested that the end of any musical life might not be far off.

On 1 January 1917 *The Musical Times* reviewed a new *History of Music* by Stanford and his pupil Cecil Forsyth that showed astonishing hostility. Elgar, it was said in the brief entry about him, 'reaped where others had sowed'. And the bitterness of his success to the academic establishment was revealed in naked enmity:

. . . Cut off from his contemporaries by the circumstances of his religion and his want of regular academic training, he was lucky enough to enter the field and find the preliminary ploughing already done.

The Musical Times made short work of this, and friends such as Nicholas Kilburn leapt to Edward's defence against 'vile innuendos': 'veiled lies', he wrote to Alice on 9 January, 'are now alas! the weapons of men who, forsooth, pose as real British gentlemen!'[58] Edward's music had overtaken Stanford's in every way. Yet it began to look as if Stanford through his teaching might have the last laugh: almost all the younger composers, from Vaughan Williams downward, seemed to be Stanford pupils. Rootham was one. The Cambridge vendetta (despite the friendship of Dr. A. H. Mann, the organist of King's, and of Arthur Benson) appeared more implacable than ever.

Early in January 1917 his throat was burned again with electric cautery. It enabled him to get out of the house a little on fine days and to attend a play or two. When Mary Anderson de Navarro came out of retirement to appear in a

[57] HWRO 705:445:7689.
[58] Letter mounted in the book of Newspaper Cuttings 1914–19, between pp. 42 and 43 (Elgar Birthplace).

War Charities week at the Coliseum in late January, Edward visited her drawing-room at the Savoy and played 'his lovely fragment of Piano Concerto over & over again' [59]

—as if the presence of another beautiful woman might galvanize the spirit to continue. But the phrase remained in detachment: he could not enter on the stresses and risks of a new major work.

There came a chance of something smaller. A matinée revue to be called *Chelsea on Tip-Toe* was planned to aid Lena Ashwell's 'Concerts at the Front'. Alice Stuart of Wortley was a member of the executive committee; Robin Legge was on the music committee; Lalla Vandervelde and Philip Burne-Jones were involved. They wanted music from Edward for a short ballet. And the author of their scenario touched a memory from the Malvern years at Craeg Lea.

In those days she had been Ina Pelly, the daughter of Canon Pelly, vicar of Malvern. She had fallen in love with young Alban Claughton, the son of another Worcester Canon whom Edward knew well. But the affair was ill-fated, and the girl was so hurt that she had run away to London and the stage, to the scandal of her family. She was a self-trained dancer, and her cleverness in the pre-Diaghilev days had attracted attention. Her style, recalled by her friend Penelope Spencer, was in the school of Maud Allan:

Sylvan flirtations with a certain amount of running about, arms held up and peepings under them, shaking of the extended finger, and so on. Not a good dancer by present professional standards, but she had charming movements. A highly intelligent, very talented woman, well-known and liked. She was ambitious, more artistically than socially, but that too.[60]

In the wake of success, she had accepted a proposal of marriage from Christopher Lowther, whose father was Speaker of the House of Commons. Thereafter she established a big social life. She arranged ballets for the endless charitable matinées of wartime, often dancing in them herself.

February 7 1917. E. in bed & his room all day but raser better—just came down in fur coat to see I. Lowther [who] came to lay the Chelsea Ballet idea before E.

She had based her scenario on a lady's fan designed in sanguine by the artist Charles Conder. The fan was owned by another member of the Chelsea Revue committee, Mrs John Lane (the wife of the publisher). It showed a forest with a long prospect opening through the centre toward distant clouds. On one side

---

[59] Quoted in a letter of 8 Feb. 1917 to Alice Stuart of Wortley (HWRO 705:445:7804).
[60] Conversation with the writer, 1973. Penelope Spencer assisted Ina Lowther (by then Lady George Cholmondely) in the newly-formed ballet classes at the Royal College of Music in 1922.

Pan piped to Echo; on the other side young couples in Louis XV costume wandered in Watteau attitudes. From this Ina Lowther had made a scenario that might project her own experience. Her glade was dominated by 'a somewhat disfigured statue of Eros'. The character she designed for herself, Echo, was to tempt Pan and the mortal lover alike.

Here were things to touch Edward in a war-torn world. It was only a short step to the musical sketchbooks of earlier years, which yielded ideas in plenty. *The Sanguine Fan* score opened with an '18th century theme' (as the diary described it)—a courtly minuet elaborately descending:

That pale sunlight was shadowed with something more passionate

before the curtain rose on Pan piping (clarinet), followed by Echo (flute). Then Pan fell asleep to a cello variant that was *Nobilmente* in all but name:

To a double-dotted variant of the passionate theme the young couples entered: their amours developed other variants to entwine with the oft-returning '18th century theme' before they left the scene.

In the second section Echo's Dance owed something to the *Airs de Ballet* Edward used to write in the 1880s, full of scampers and sly pausings. Pan's heavier energy generated the score's first *fortissimo*; but their passion was framed in a comfortable F major, returning *grandioso*.

In the final section the mortals reappeared. Echo flirted with the leading young man to a reprise of her dance. Then Pan came, *fortissimo* again, to strike the young man dead and carry off Echo. But in Edward's music this was accomplished gently enough, as the '18th century theme' (unheard since the opening section) returned to combine with Pan's piping at the end.

A quarter of an hour's continuous music evolved so quickly that in less than a fortnight he was scoring it. It was his largest work since *The Starlight Express*, and again it showed escape. He wrote to the Windflower on 5 March:

Oh! this weather! & I was dreaming yesterday of woods & fields &, perhaps, a little drive round Harrogate—or a little play journey to Fountains or some lovely remembrance of long ago idylls, & now deep snow.

Well, I have put it all in my music & also much more that has never happened.[61]

The Sanguine Fan had been written for the single performance. For the Chelsea Matinée Ina Lowther had assembled a caste of principal dancers more eminent and more amateur than herself—Fay Compton and Ernest Thesiger as the leading young couple, and Gerald Du Maurier as Pan. Du Maurier proved a delightful sunny character who dared to tease Edward on his conducting: 'said E. wd. become absorbed in his music & keep him standing on one leg while he dwelt on a note.'[62] Costumes and scenery were in the hands of the eminent theatrical designer George Sheringham.

They rehearsed for three afternoons, and then came the performance on 20 March. 'Ballet perfectly beautiful,' Alice wrote. 'Ina Lowther *very* good—just filling the part with gaiety & spontaneity.' One of the shepherds' roles was danced by Mrs Lowther's niece Winifred Lawford, whose memory was to preserve vignettes of the performance through nearly sixty years:

Ina Lowther designed her own costume. It was typical of her not to have a finished headdress, but she had wild grass and buttercups (showing the wildness of love, Pan, etc.) and a Pipes of Pan to mime playing upon. She circled round and round the young man with her pipes, eyeing him seductively.

At the end Du Maurier (Pan) looked superb, utterly sardonic and unyielding as he half carried Echo off, his arm round her shoulder, looking back on the dead young man and desolate girl.[63]

It had been so good that a second performance of *The Sanguine Fan* was arranged for another variety matinée in May. For it Edward added a Shepherd's Dance—to a fresh disarming variant of the '18th century theme'. Winifred Lawford remembered: 'At the beginning of our dance I had a moment of uncertainty, and Sir Edward looked straight at me with just the look of encouragement I needed.'[64] But the occasion was damping, as Alice wrote:

May 22 1917.   E. & A. to Palace Theatre for Conder Ballet—Most things dull, a *very* heavy dull audience—Why shd. Philanthropy make ladies so oppressively stupid? . . . E. rather depressed, such a dull house to hear that perfect work.

There was to be a single gramophone record of extracts. Elkin published only Echo's Dance in a piano reduction. Otherwise *The Sanguine Fan* remained unheard for more than half a century.

Autumn and winter had brought the loss of more friends—at the end of November 1916 the genial president of the old U. B. Quiet Club, Archibald

[61] HWRO 705:445:7678.
[62] Diary, 9 Mar. 1917.
[63] Conversation with the writer, 1974. In the programme she used the name 'Patricia Law'.
[64] Ibid.

Ramsden, with whom Edward had shared the excursion to Bayreuth in 1902 just before starting work on *The Apostles*; in December Hans Richter, separated from his English friends by the war; in January 1917 Lady Maud Warrender's husband, one of the Admirals on the Mediterranean cruise of 1905; in February Edward's Uncle Henry, the last link with his parents' generation in Worcester. But the death which hit him hardest was that of George Sinclair, the Hereford Cathedral organist, victim of a heart attack at the age of fifty-four. It broke a chain of Three Choirs memories going back to the production of *Froissart* at Worcester in 1890, when Sinclair had brought his eager young assistant Ivor Atkins to hear the new work.

February 12 1917. Thinking much of our friend George Sinclair. The funeral this day—E. keeps G.'s last letter in front of him on writing table & thinks much about him . . .'

Now Ivor Atkins, at Worcester, asked Edward to conduct 'For the Fallen' at a War Memorial Service in the Cathedral there on 15 March. Three Choirs Festivals had been suspended since 1914, and for many of the choristers this was a first experience of singing under Sir Edward Elgar. One of them was Atkins's twelve-year-old son Wulstan, who remembered:

Keyed up as we all were, the performance in the evening—in a Cathedral which was packed as I have never before or since seen it—was the most moving and thrilling experience that I have ever had. War conditions precluded the use of an orchestra, and my father had to fill in as much of the orchestral colour as possible from the organ. A special platform was erected over the choir steps for the sadly depleted Worcester Festival Choral Society . . . The King's School O.T.C. supplied the drummers. Although it is over fifty years ago . . . I can still feel the shivers which passed through me as the drums rolled, perhaps the contrast to organ instead of full orchestra making them even more impressive . . .[65]

Next morning the Elgars prepared to return to London:

March 16 1917. E. & A. left at 9 train . . . Startling news on opening paper at station: Revolution in Russia & Czar deposed . . .

The major forces on the Eastern Front after two and a half years of war had exhausted themselves. The armies of both Russia and the Austro-Hungarian Empire were in total disarray. And the fear often expressed by diplomats in the old days—that a world war would topple the thrones of Europe—had now its first fulfilment.

In the West the Germans had declared unrestricted submarine warfare. During January they sank 181 ships of all nations; in February 259; in March 325, including three United States merchant vessels. This was too much for the Americans, who at last declared war on Germany. It was only in the nick of time, for the European Allies had by now exhausted every credit for eventual repayment of gigantic war loans from the United States.

[65] MS reminiscences.

On the Western Front in the spring on 1917, the Germans made heavier use of their latest horror-weapon, the 'flame-thrower'. French generals hurled their troops again and again at the Germans with no result but staggering bloodshed, until the French army was in nearly the state of the Russians. When they were pitchforked into another new offensive on the Aisne, half the French soldiers mutinied. Through a simultaneous British offensive at Arras, the killing and maiming were more frightful than ever. The Germans remained calm behind the immense fortification of their Hindenburg Line.

So the hatred hardened. When Henry Embleton again pleaded for the remaining part in *The Spirit of England*, Edward made up his mind to set Binyon's stanza that showed the enemy of the country:

> She fights the fraud that feeds desire on
> Lies, in a lust to enslave or kill,
> The barren creed of blood and iron,
> Vampire of Europe's wasted will.

He set those lines to the Demons' music from *Gerontius*, and sent his explanation to Ernest Newman:

Two years ago I held over that section hoping that some trace of manly spirit would shew itself in the direction of German affairs: that hope is gone forever & the Hun is branded as less than a beast for very many generations: so I wd. not invent anything low & bestial enough to illustrate the one stanza; the Cardinal [Newman] invented (*invented* as far as I know) the particular hell in Gerontius where the great intellects gibber & snarl *knowing they have fallen*:

This is exactly the case with the Germans now;—the music was to hand & I have sparingly used it. A lunatic asylum is, after the first shock, not entirely sad; so few of the patients are aware of the strangeness of their situation; most of them are placid & foolishly calm; but the horror of the fallen intellect—*knowing* what it once was & *knowing* what it has become—is beyond words frightful . . . And this ends, as far as I can see, my contribution to war music.[66]

Yet there was still a little war music to appear. April 1917 saw the orchestral première of *Le drapeau belge* at long last, in a birthday concert for the Belgian King Albert. And another project long deferred reached completion. In January 1916 there had come a request from Lord Charles Beresford to set some verses by Kipling which had appeared in a recent pamphlet entitled *The Fringes of the Fleet*. In prose and poetry Kipling drew breezy pictures of life aboard the small commercial vessels now mounted with guns for minesweeping, the submarines, and patrol boats in outlying waters:

They talk about men in the Army who will never willingly go back to civil life. What of the fishermen who have tasted something sharper than salt water—and what of the young third and fourth mates who have held independent commands for nine months past? One of them said to me quite irrelevantly: 'I used to be the animal that got up the trunks for the women on baggage-days in the old Bodiam Castle,' and he mimicked their

[66] 17 June 1917.

requests for 'the large brown box', or 'the black dress basket', as a freed soul might scoff
at his old life in the flesh.

> And I'm sorry for Fritz when they all come
> A-rovin', a-rovin', a-roarin' and a-rovin',
> Round the North Sea rovin',
> The Lord knows where!

To set the verses, Edward sketched some hearty tunes for baritone and men's
chorus. But then Kipling had objected to his verses being turned to musical
entertainment. Since he had written them, his only son had been reported
missing in action. Edward dropped the project—to the intense disappointment
of friends who felt that pairing the author of 'Recessional' with the composer of
Land of Hope and Glory was too good a chance to miss. They set about
persuading Kipling to change his mind, and a year later they seemed to be
successful.

Now Edward finished *The Fringes of the Fleet* settings in what he described to
Ernest Newman as 'a broad saltwater style'.[67] They were more topical than ever
on account of the submarine warfare. The songs began in jauntiness and
darkened. 'The Lowestoft Boat' told the story of a converted herring boat
complete with unlikely crew recruited anyhow and mounted with guns: it
opened on the old descending fourths. The second song, 'Fate's Discourtesy',
reversed the fourths upward—but its sequence still descended in steps after the
pattern of the 'tune from Broadheath':

> Then welcome Fate's discourtesy
> Whereby it is made clear
>   How in all time of our distress
>   And our deliverance too,
>   The game is more than the player of the game,
>   And the ship is more than the crew!

The third song, 'Submarines', set a slow vocal line in the minor against dark
orchestral shapes:

> The ships destroy us above
>   And ensnare us beneath.
> We arise, we lie down, and we move
>   In the belly of Death.

The final 'Sweepers' just lightened the atmosphere as enemy mines were
harmlessly exploded in fairways close to home shores:

> Boom after boom, and the golf-hut shaking
> And the jackdaws wild with fright!

When the completed score was shown to Oswald Stoll at the Coliseum, he
signed up *The Fringes of the Fleet* for a fortnight's run under Edward's baton, to

---

[67] Ibid.

take its place as part of a wartime variety bill in June 1917. Charles Mott, the Organ Grinder of *The Starlight Express*, would sing the leading part.

The coming of another spring found Edward longing for the country again. It was just a year since the journey which had ended disastrously at Oxford with the recrudescence of ear illness. Now again he had ominous headaches; but at last he went to Stoke. During his absence this time, Alice set out to look for some isolated house in the country toward the south coast—in easy reach of London, but where Edward might find the remoteness he seemed more and more acutely to need from the horrors of the world in their ageing lifetime. Almost immediately she found one:

May 2 . . . A. & C. to Fittleworth to see a cottage—a 2-seater met them but after arriving at Inn preferred to walk—*Lovely* place, sat in lovely wood & heard a nightingale, turtoo doves, & many other dicksies [birds] & saw lizards & heard Cuckoo first time. Also saw swallows—lovely *hot hot* day. A. much perplexed as cottage is so very cottagy but large studio & *lovely* view & woods—dear place. Finally took it for June. Lovely walk thro' woods & by primroses to Station . . .

As *The Fringes of the Fleet* was coming on at the Coliseum in early June, they went down to Sussex in late May. The cottage was called Brinkwells, and from there on the 26th Alice wrote to Lady Stuart of Wortley:

My dear Alice
I know you will be thinking of us & I want you to have a few lines before you start on Monday. I am delighted to tell you that Edward's first exclamation was 'It is too lovely for words' & he was quite pleased with the house & has loved every minute since we came—So you may think how relieved & pleased I felt. I am in the garden & before my eyes lies a wonderful deep wood & low hills beyond & then the Downs, larks are singing as there are some fields as well, & a nightingale is heard sometimes, & in the evening the nightjars go whirring around on the fringe of the wood—It is a most extraordinarily lovely spot. Endless walks & paths in the woods—There is also a Carpenter's bench & tools &c & E. has already made me 2 rustic footstools . . .

Yr af[fectiona]t[e]
C. Alice Elgar[68]

Edward exclaimed to Frank Schuster: 'It is divine: simple thatched cottage & a (soiled) studio with wonderful view: large garden *unweeded*, a task for 40 men.'[69] They had a fortnight to garden and explore the woods and fields. Edward's sixtieth birthday was spent at Brinkwells, but then he had to go back to rehearse *The Fringes of the Fleet*.

The stage setting placed Mott and the other three singers in fishing boots and sou'westers, sitting round a table outside an idyllic English country inn to sing their songs of coastal defence. Edward himself went down to Harwich to secure

---

[68] Elgar Birthplace.
[69] 7 June 1917 (HWRO 705:445:6664).

the right costumes, and then rehearsed his men at the London office of the publisher Enoch (who had published his other Coliseum work, *The Crown of India*, in the days of peace). Alice noted: 'Crowd in street begins to hum & whistle the tunes.' On Saturday 9 June they rehearsed with the Coliseum orchestra of twenty-five—mostly girls and women, as the men had joined up. The availability of Mott himself was only arranged by Lord Charles Beresford, to whom Edward's music was dedicated. On Sunday the four baritones lunched at Severn House and went through the songs once more, ready for the première on the following day.

It was to open the fortnight's run of two performances a day in a long and varied bill—George Graves in 'What a Lady', Charles Cochran's production of 'Hello! Morton', Florence Smithson, and many others.

June 11 1917.   Rehearsal at Coliseum: had car & took down E.'s luggage. A. with E. to 1st performance. *Very* good & very enthusiastic receptn. Back to Severn House & E. rested & then again in car, & A. to evening performance. *Very* good & most exciting & much enthusiasm. Maud [Warrender] & many Admirals there . . .

When Edward came up on stage to bow with the singers, he cut a strange figure. *The Daily Chronicle* reported:

A man is generally made to look his best on the stage, but it is not so with Sir Edward Elgar at a matinée. You feel that with his morning coat and brown boots he is going to crown himself with a silk hat! He is a handsome man with a good colour, but the limelight makes him pallid of hue and patriarchal of appearance. In the conductor's chair he is a prince . . .[70]

The critics wrote quite seriously about the latest Elgar work:

The sardonic humour of the lines [in 'The Lowestoft Boat'] is faithfully reflected in the music, and the grimness of sentiment is thoroughly realised by the exponents. The second song, 'Fate's Discourtesy', is less obvious, and I thought it was taken a little too slowly, but I know Sir Edward attributes the greatest importance to every word being clearly heard, and this is best assured by the tempo being on the slow side. Number three, 'Submarines', introduces the elements of secrecy and stealth little connected with the sea until exploited by the Huns . . .[71]

The youngest of the four singers, Frederick Henry, wrote a lengthy 'Log' of *The Fringes of the Fleet*, with the adventures of himself, Mott, Frederick Stewart, Harry Barratt, and Sir Edward all retailed in sub-Kipling doggerel:

> The house was full and cordial, 'twas
>   A huge success instanter,
> The critics all remained to praise
>   Save one, who tried to banter.
>
> . . . .

[70] 30 June 1917.   [71] *The Referee*, 17 June 1917.

So afternoons and evenings passed
   With song and pipe and beer,
And crowded houses every time
   Would clamorously cheer.

We'd take our curtains modestly
   Three times, or sometimes four,
Then Charlie Mott would go in front
   And cause another roar.

No matter though how late the hour,
   The house would not go bedward
Until before the 'rag' appeared
   Our skipper, great Sir Edward.

'Tis then that we're most proud to shine
   In his reflected glory,
And ladies in the band express
   Their pleasure '*con amore*'.

With lapses of his merry men
   He's always kind and gentle,
Should Harry Barratt fluff his words,
   Or such-like errors mental.

'The golf-house' may be 'wild with fright',
   'The jackdaws shaking' too,
And 'wave and wind' change places,
   He will still forgive the crew.

Or should a singer lose his voice
   He's amiably lenient,
And tells him just to 'look the part
   And chip in when convenient.'

. . . .

And all the time he's *one of us*,
   To work with him's a treat,
So God bless the splendid skipper
   Of the 'Fringes of the Fleet'.

It was a big success. The Coliseum run was extended fortnight after fortnight—and Mott's leave along with it. In the theatre foyer, the manuscript full score was displayed for sale to the highest bidder. On 4 July the whole cycle was recorded for the gramophone by the four singers and orchestra under Edward's direction. He added a new song for the baritones alone: it was a setting of Sir Gilbert Parker's 'Inside the Bar', a sailor's yearning for safe harbour and an inn and his girl.

The Coliseum manager Arthur Croxton expansively met the press:

It is only a few days ago that Mr. Balfour confessed to beginning his real life in the year 1914. To most of us has come the same experience . . . As the poets of to-day have come from the trenches, may we not expect that the British musical future is being born in the fields of France? But whilst it is going through the pangs of birth, we can cast an eye around and see, within our own horizon, the man whose work is approaching that line of country where the 'Flammenwerfer' of musical genius is being made. I refer to Sir Edward Elgar.

. . . To me the great hope of Edward Elgar standing out as a master-mind in musical composition after the war is shown by his wonderful setting of Rudyard Kipling's songs of Sea Warfare. In these songs you get the real right magic of British seafaring spirit, of the open air, of the sea. The music smells of salt-water, and you feel that here at last is work which to its hearers gives added confidence that from the Great War Edward Elgar will obtain impressions to which his musical genius will give magnificent utterance.[72]

This was very well: only it ignored the fact that Edward's popular audience was not his chosen audience. And his Victorian background of social aspiration was bound to bring disillusion when the levelling effects of the long war effort should begin to appear in the life of the nation.

The Fringes of the Fleet continued at the Coliseum—two performances a day, week in, week out, to the end of July. Mott was at last called up—to be replaced by George Parker for a provincial tour of the songs. At the end of the Coliseum run Edward found himself exhausted. Yet every public and private reason called him to conduct The Fringes through the provinces as well. After a few days snatched at Stoke, he plunged into a week's run at Manchester. From there on 17 August Frederick Henry sent an account to Alice: 'I am glad to report that in spite of depressing weather here, Sir Edward still keeps bright and cheery and when in our dressing room gives us vast entertainment with his flow of wit and anecdote.'[73]

From Manchester they went for a week to Leicester. After that Alice took him for a few days to Brinkwells—before several more weeks of Fringes performances booked for Chiswick, Chatham, and the Coliseum again. 'Here's success!' Edward wrote to Frank Schuster.[74] But Kipling was threatening to withdraw consent to the performances continuing.

The situation reacted immediately on Edward's health, as he wrote to Alice Stuart of Wortley on 18 September after the week at Chiswick:

I have been very unwell all the week & sadly worried about everything. I 'got thro'' the Chiswick week but had to be taken down & brought back every evening—I mean convoyed as I was so giddy. This morning I feel better for the first time—Next week Chatham.

I fear the songs are doomed by R. K. he is perfectly stupid in his attitude.

I recd. the Ballads with great joy—but have been unable to read them—or anything for a week—head quite gone giddy alas!

Yes: everything good & nice & clean & fresh & sweet is far away—never to return.[75]

[72] 'Britain's Musical Genius', The Sunday Evening Telegram, 22 July 1917.
[73] HWRO 705:445:3922.
[74] 24 Aug. 1917 (HWRO 705:445:7051).
[75] HWRO 705:445:7679.

Without a satisfactory agreement from Kipling, they went ahead with Chatham. It proved a dreadful experience, for the German Zeppelins were raiding naval installations up and down the coast.

September 24 . . . E. to Theatre, almost immediately lights went out, & a raid—A. sat in Hall & watched for E.—feared he wd. be quite exhausted & at last was able to send him sandwiches—he returned & we dined—*no* performance cd. take place, Raid over, E. returned . . .

She wrote to Alice Stuart of Wortley:

At the Picture Palace wh. E. has to pass on his way, they were sitting there & singing Land of Hope & Glory in all the firing . . .[76]

September 25 . . . E. to Theatre in evening, again lights out, he returned while firing was still going on—we dined, cannonading going on furiously—very darksome only a few candles . . .

The following two days were free from raids, and the two scheduled performances each evening were given. Only one performance was possible on 28 September and next day they left.

Severn House was a relief, though he wanted to escape to Stoke before the *Fringes* revival at the Coliseum. But when it was time to leave on Monday 1 October, he was too exhausted and had to wait until the following day:

October 2\ A tangledy day—E. did not go by morning train. Settled to do so at 1.40—He & A. in car at door when *bang bang* again, so cd. not start. Ordered car for 4.10—but it did not come—so he had to give up going & was muss disappointed . . . Mansfield Evans came to stay the night . . .

Mansfield Evans was an Army Captain with whom Carice had 'a knight-and-lady relationship', as it was recalled by her friend Winifred Davidson. 'The Elgars were not at all easy as regarded young men.' Or young ladies either. Winifred was a granddaughter of Mrs Joshua, and therefore had approval at Severn House. But visits there were still likely to prove a trial, as she recalled them sixty years later:

Lady Elgar was my idea of an excessively difficult character. The atmosphere was formidable. Her standards were rigidly Victorian: there was this atmosphere of having to conform to an ideal which you knew you could never live up to. She was sweet, and welcomed young people in her way. But the flowers of conversation often withered.
    Lady E. to Winifred: 'And have you read any interesting books lately?'
    Winifred: 'Yes' (thinking rapidly and blankly of the last thing read) '—Stephen Graham's *With the Russian Pilgrims to Jerusalem*.'
    Lady E.: '*That* miserable little man!'
    One day Carice—who was by habit excessively punctual—turned up for a walk very

---

[76] 27 Sept. 1917 (Elgar Birthplace).

late. I saw she was furious, but said nothing. It turned out she had to go to Harrod's at the last moment to change her father's pyjamas. When I said I might have joined her there, Carice indicated that it was not permitted to admit an outsider to the secrets of the great man's private garments.[77]

Another friend of Carice's small circle had been introduced by Eleanor Berkeley, the daughter of the family at Spetchley Park. This was Lady Petre—a war widow, though she was younger than Carice. She recalled vividly her first visit to Severn House on a Sunday afternoon, when the big music room was resounding to the voices of 'eminent men' whom Alice invited week by week to amuse and assure Edward:

The whole place was full of interesting people. Eleanor Berkeley quickly mixed with the guests, leaving me to sit down to a good talk with Carice on a sofa. As I came to know her, I found her rather miserable at this time, asking 'Who am I amid all this throng?' Once she said: 'Kitty, I'm *sick* of eminent men!'[78]

All this was carefully hidden from her parents. But the effects of her mother's rigidity were painfully apparent in the daughter's shyness. One day, after much persuasion, Carice agreed to try her singing voice for Winifred Davidson, who recalled:

At that moment my mother came suddenly into the room, and Carice went down on all fours and dived under the table—though she joined in laughing about it afterwards.

She was like Georgina Podsnap in Dickens. I have actually heard her say under her breath at the approach of someone, 'Oh, go away! go away!'

And when Mansfield Evans asked if she had missed him Carice, ever anxious to escape from the centre of the picture, nervously jerked out: 'Oh, not in the least, thank you.'[79]

On this occasion the young man had a warmer reception from Carice's parents. Alice continued her diary entry for 2 October 1917:

Capt. M. *extremely* nice. E. seemed to enjoy the evening & pweaked [spoke] & told stories & we laughed very muss. Quiet evg—the first for some time.

October 3. Left C. & Capt. M. at front door, they looked so nice.   E. really left by 9.50 train. A. wis him to Paddn. he wished A. coming too—A. returned [to Severn House] by train & had nice talk with Captn.M. before he left.

In Worcestershire Edward found the end of the harvest and no men to gather it in:

Stoke

                                                                Thursday [4 October 1917]
My dear Windflower:

I am safely here in pouring rain which I don't mind: I had a good rest in the train & wonderful! began smoking a pipe again which is a mercy: so soothing to the nerves . . .

[77] Conversation with the writer, 1975.   [78] Conversation with the writer, 1976.
[79] Conversation with the writer, 1975.

It is sad to see the tons of fruit wasting, alas!
I was going to save the best apples but now comes the rain.
I see someone—amongst the trees & reach out but——nothing.

Bless you
Yr
EE.[80]

He got through the Coliseum week of *Fringes* in early October, through performances of the complete *Spirit of England* and *Gerontius* at Huddersfield and Leeds—where Embleton and the Committee 'spoke seriously to him that he must go on & they want the 3rd part Apostles.'[81] On 3 November he and Alice climbed thankfully into a first-class compartment of the London train: 'Nice rest, & alone to Nottingham. Then young officer got in, he & 50 more had been telegraphed for that morning to return—he had only arrived on leave previous evening . . .' It was the end of another disastrous British offensive. Haig, ignorning predictions of autumn rain, had insisted on attack in late summer. Now, three months later, they had reconquered five miles of mud to enter the ruins of Passchendaele—at a cost of 250,000 British casualties.

Back at Severn House 'E. looked tired & dreadfully worried over the war news'. He went to Stoke for another few days before the London première of the completed *Spirit of England* on 24 November. And he got through a final week of *The Fringes of the Fleet* at the Coliseum in December—'the *funeral* this week', he wrote to Alice Stuart of Wortley,[82] as Kipling had succeeded in stopping further performances. Edward was ill again, and every night at the Coliseum was touch and go: it was, he wrote to Frank Schuster, 'the old trouble'.[83]

*   *   *

When the curtain came down on *The Fringes of the Fleet* for the last time, Edward subsided into bed. Sir Maurice Abbott Anderson prescribed new remedies. They did no good. On 15 December 1917 Edward had to cancel a date with the Hallé Orchestra—in all his years of illness only the second time he had ever cancelled a conducting engagement. Sir Maurice gave another prescription. After Christmas he brought in a specialist—the senior internist at Guy's Hospital—who gave further assurance that there was no internal trouble. But at the turn of the year Alice wrote, 'These weeks are so sad & anxious to see him still ill.'

In January 1918 Sir Maurice ordered X-Rays. They revealed nothing but a 'dropped stomach'—for which electrical treatments and a belt were arranged. He seemed a little better, and was able to leave the house again. But he could not concentrate on music. On 10 February he wrote to Troyte Griffith:

---

[80] HWRO 705:445:4124.
[81] Diary 30 Oct. 1917.
[82] 27 Nov. 1917 (HWRO 705:445:4187).
[83] 25 Nov. 1917 (HWRO 705:445:7017).

I am ashamed to tell you the amount I have read during this inevitable leisure: leisure most undesired by me. But I have come across scarcely anything new worthy of note.[84]

A week later he was able to receive a visit from a young conductor who was hiring the London Symphony Orchestra for a series of concerts to survey modern British music, and wanted to go through *In the South*. It was Frank Schuster's young friend Adrian Boult.

February 17 . . . Lalla [Vandervelde] & Mr. Boult to tea. Quite a nice quiet man. E. went through 'In the South' with him—He seemed really to understand. Raid in evening—Lasted rather a long time.

February 18. E. & A. to Queen's Hall at 10.30—to hear Mr. Boult rehearse 'In the South'. Too few strings but reading of it really good. Feared audience wd. be almost none on account of raids, & so it was . . .

The Germans were intensifying their raids with the new Gotha bombers. There was widespread loss of life and property in London. National food rationing was instituted.

When Edward had another relapse, Sir Maurice could only call in another specialist. This time a different trouble was diagnosed.

March 6. E. & A. to Mr. Tilley, met Sir Maurice just arriving. A. waited in much anxiety. Sir Maurice said tonsil was not right, matter was forced out—if taste &c was not better, tonsils wd have to be taken out—horrid idea—He was *very* kind & nice—E. felt better for the moment though throat was rather hurt.

Over the next days they debated it. Carice wrote: 'It worried Mother considerably that he should have such an operation at his age . . .'[85]

They had already planned an afternoon party in the music room at Severn House for the following Sunday: W. H. Reed and his string quartet were engaged to play to a big group of their friends. And it was managed. The quartet was a great success:

March 10 . . . They played beautifully & sd. it was such a pleasure to play in this room—Everyone loved it—the rooms looked beautiful . . .

March 11. Lovely day like early summer—E. vesy badsley. A. to Sir Maurice to talk about E. Very kind & nice. Urged removal of tonsils—Pray God it may cure him. Horrid to think of E. being touched—E. telephoned to Sir Maurice that he wd. go through the operation.

March 14. E. feeling so ill, glad to go & try something—A. wis him in car about 5.30—The Home seemed nice & a cheerful room. Sir Maurice came & E. was very dood—Dreadful to leave him.

March 15. A long day of suspense as the op[eratio]n was not till 3 P.M. . . . A. spent an anxious horrible 40 mins. then Sir Maurice & Tilley came & told her all was well. Sir M. showed her the worst tonsil—all over abcess matter & a black stone, pea size, in it. A. not let go up so went home, returned at 7. Found E. in great pain not knowing how to

[84] *Letters* p. 237.
[85] Quoted in Young, *Alice Elgar*, p. 180.

bear it, agonising to A. They gave him an injection & in 10 mins he was sleeping peacefully. A. had to leave him—

She went every day, until in a week he was ready to come home.

March 22.   Preparing to leave Dorset Sqre [nursing home]. Mr. Tilley very pleased with the throat—A. & C. lunched at Canuto's & C. helped arrange all. Car was late & E. was very worried waiting, the Home had upset his nerves & all he had suffered—He was *so* peased to be home & loved everything . . .

   Recovery was slow and there were set-backs. But gradually it began to seem that the 'Menière's Disease' which had hung over all the years in London might have been chronic tonsilitis in fearful combination with worry and profound unhappiness about himself and the changing of his world.[86]

* * *

There was one unmistakable sign of relief. 'At Severn House the night I returned' Edward wrote his first music in nine months.[87] It was an easy 9/8—upward moving through one long bar, downward moving through the next:

It led him back to a key he had used near the beginning of his creative life, E minor.

March 23.   E. so peased at being home. Throat feeling better & his colour muss better . . . Very hot & lovely, E. in garden, cleared out fountain . . .

March 24 . . . Wonderful day, hot as summer. Sheen of green bushes everywhere . . .

March 25. E. began a delightful Quartett. A remote lovely 1st subject. May he soon finish it—Wrote all day . . .

   The new music pursued the memory of what had been heard in the music

[86] This diagnosis was suggested by Dr John Barretto in 1980, after a study of all the written evidence and consultation with surgeons whose experience reaches back to the 1920s. Tonsillectomy was then a favoured response to a wide range of complaints. But it is clear that after this operation Elgar did experience relief which proved to be more or less permanent. Menière's Disease could not be cured by tonsillectomy, or by any other means known then. And Menière's Disease, over the 22 years of life remaining to Elgar after the first recorded giddiness in 1912, would certainly have produced marked deafness: yet surviving friends recall no such affliction through his last years.
[87] W. H. Reed (*Elgar as I knew Him*, p. 64, and *Elgar*, p. 123) suggested that 'the night I returned' was in the autumn of 1918. But Elgar copied out the entire theme, extended to 50 bars, as 'a thank offering to a good Windflower' in Mar. 1918 (MS at the Elgar Birthplace).

room at Severn House just before the operation—the sound of a string quartet. He had sketched quartets several times in the past—in 1907, and many fragments through the 1880s. Now, after the nullity of illness and early convalescence, he sought the intimate reflection of chamber music with a new impulse.

The Quartet shared the E minor tonality of the 9/8 idea, but not its prominence of melody. The Quartet opened with two short figures—a question hesitating upward

its answer descending in fourths:

The figure emerging underneath the 'answer'

only turned the dialogue upside down—to set rising fourth against descending step.

The diction of this music followed the Quartet first movements of Beethoven. A hundred years later, living through the depths of a war which was destroying his past, Edward's Quartet shredded melody with semiquaver and pause and syncopation, with small interval mocking tonality and large interval bridging emptied emotion.

The second subject found an easier rhythm in which to develop the primary juxtaposition of intervals:

It managed to sing its song for a moment in a settled F major. When the primary semiquavers tried to interpose, the second subject overbore them *fortissimo*. The wide intervals stretched to octaves, whose juxtaposition with half-steps made the melody grotesque.

Through these rent textures, formal development was seized by the primary eccentric rhythms. The strife drove up an astonishing climax:

It was a twisted echo of Violin Concerto fantasy

Recapitulation moved to another echo of the Second Symphony's dying 'Delight':

In a coda, the descending fourths would sound one *nobilmente* as if to recall their predecessors in Edward's music of the past. Then they evaporated suddenly in a last whisper of the primary question. It was a new and barren harmonic world for him.

March 27. E. feeling so unwell—Terribly disappointing after all he has gone through . . .

March 29 . . . E. much the same—
Most anxious for news—Terrible fighting & having to give ground—Was there ever such a Good Friday so much in harmony with sacrifice & suffering? . . .

The Germans were launching the most bitter offensives of the entire war on the Western Front. In the East, the Russian communist government had signed a peace with Germany, releasing all the Eastern German forces for action in the West. It was a last hope for Germany. In March 1918 American reinforcements to the Allies were just beginning to be felt: more would be arriving every month, and by summer it might be too late. For this last chance of victory on the Western Front the Germans fought like tigers, pushing the Allies back towards the Marne.

Gradually Edward could venture out. Permission was granted by the Petrol Control for Sir Maurice to motor him down to The Hut on 12 April, and Alice followed by a slow train. The Hut was full of young wartime guests including Robert Nichols. Edward wrote to Sidney Colvin:

I cannot do any real work with the awful shadow over us, but I have some Italian paper & the pen *will* go on 'tic-tac-toe'-ing (as dear R. L. S[tevenson] said) for the sheer love of feeling the touch of the pen on the surface of the sundried, handmade lovely papyraceous stuff; it hath an edge (which I call crispisulcant) which helps to inspire; but the result is nothing worth.[88]

[88] 17 Apr. 1918 (HWRO 705:445:3466).

The clear necessity, when it could be managed, was to get him right away for a long stay at Brinkwells. Carice wrote:

Mother was meanwhile very busy arranging for the move to Brinkwells for recuperation. She had to think what would be wanted in the country, as there was a scarcity of furniture and comforts there, and also the business of shutting up Severn House safely. Father's only contribution to all this was choosing tools which he would need for the woodwork he did and the repairs and small improvements he made.

The great removal day was May 2nd [1918], and they went by train and were met by Mr. Aylwin with his pony cart and luggage cart and a warm welcome. I had joined them at Victoria, and Father and I walked up. We had a warm welcome too from Mrs. Hewitt, who 'did' for us.[89]

Here were picturesque rustic characters of the Sussex countryside such as he had met at Hookland. Mr Aylwin was the patriarchal, grandfatherly local farmer: he gave a quiet welcome to the distinguished people who travelled up in his pony cart from the station to Brinkwells. At the other extreme, the handyman Mark Holden reacted to Edward's notions of woodmanship (as W. H. Reed remembered) ' . . . with what I should describe as a pained expression if so impassive a face could be said to have any expression. Mark was a character after Sir Edward's own heart. His monosyllabic replies were quoted again and again.'[90]

These people were as the Sprites in *The Starlight Express*: they touched Edward's nostalgia. At Brinkwells he did his best to turn himself into one of them. He took to making things again—not music now, but rough footstools and penholders and doorstops from wood cut down and picked up round the place. He had the boots and sou'westers from *The Fringes of the Fleet* sent down, and put them on to tramp the woods in wet weather.

Alice, in her seventieth year, had begun to suffer from colds. Edward got Mark to put up a rough wooden shelter which would allow her to enjoy the sunlit garden protected from every draught. Another task was to place the old sundial, given by Frank Schuster in Plas Gwyn days—in the garden at Brinkwells:

May 12 1918. Sunny & rain storm . . . E. fixed the sundial & it shadowed the hour.

Time's passage was in his mind again as he wrote to Carice next day about arranging trusteeship for Alice's money 'in case anything shd happen to me or your Mother'.[91] Yet he was feeling better than he had for years. On sunny days, Alice noted, 'Air vibrating with the song of birds', and Edward noted in his own diary: 'All this time chaffinches (many) singing "Three cheers!"' The peace of Brinkwells was broken only once then. Edward noted it: 'Clock weight fell in the night 3.20!'

[89] Quoted in Young, *Alice Elgar*, p. 180.
[90] *Elgar as I Knew Him*, p. 61.
[91] 13 May 1918 (*Letters*, p. 240).

On that day, Charles Mott died of wounds in France. Almost with the news came Mott's last letter, written hours before the battle in which he was fatally wounded:

On passing over a shelled area I could not help deploring the waste of power, which, if directed in another channel might preserve life instead of shatter it. There is something still very much wrong with a world that still sanctions war & something wrong with our *practise* of the various forms of religion too. One can only hope & pray that this war may wake the whole world up with a great start and preachers & teachers rise up to the *priceless* occasion which this poor ravished bleeding world offers. It is groaning for the truth—so simple too. . . .

There is one thing that 'puts the wind up me' very badly & that is of my being wiped out & thus miss the dear harmonies of your wonderful works.

(He had written out in the pencilled letter the 'Windflower' theme from the Violin Concerto and the 'Little Bells' figure representing sleep in *The Starlight Express*.)

But I have a supreme confidence in my destiny & feel that I have some useful work to do in the world before I am called away.

Meanwhile the roar of the guns thrills me somehow, & I only dread my comrades coming to grief & seeing them wounded. I pray they may all get through safely.[92]

Mott was one of 300,000 British casulaties as the Germans pushed toward the Marne. In the next fortnight the war came closer still to Brinkwells. Edward wrote on 30 May: 'Bad war news this & succeeding days—Incessant gun fire (distant cannon).' Coming right across the English Channel, it must have made an eerie sound in the Sussex woods of early summer.

He fulfilled a request from the Ministry of Food to set a poem by Kipling about England's lifeline—her 'Big Steamers':

> Send out your big warships to watch your big waters,
> That no one may stop us from bringing your food.
>
> For the bread that you eat and the biscuits you nibble,
>   The sweets that you suck and the joints that you carve,
> They are brought to you daily by all us Big Steamers—
>   And if anyone hinders our coming you'll starve!

Alice was surprised at Edward's compliance: 'Very magnanimous to set anything more of R. Kipling's but he said, "Anything for the cause." '[93]

There was also good news. Percy Hull, Sinclair's assistant at Hereford in the old days and for long interned in Germany, was safely on his way home to succeed his master in the Cathedral that meant so much to them both. Yet when a parcel of delicacies arrived for Edward's sixty-first birthday from Mrs Joshua, his letter of thanks found his thoughts haunting the past once more:

I think over all the great days with Richter while I am working with the hoe or plane &

[92] 11 May 1918 (HWRO 705:445:6371).
[93] Diary, 14 June 1918.

little of the actual manual labour & I regret that such hours can never come again; your letters tell me that you also live on the past glorious art life sometimes; but, happily for you, you have so many interests in the present.[94]

They explored farther into the surrounding countryside—alone, together, with visitors. Alice Stuart of Wortley came, Lalla Vandervelde, Muriel Foster, Sidney Colvin (spending the summer with his wife nearby), Carice in snatched weekends and holidays from the Censorship, and Algernon Blackwood. They went as far as two old lodges near the Bedham woods with the sinister names of Gog and Magog. They fished Bognor Pond. They found the 'Octopus' beech tree in Flexham Park—a place which made a deep impression on all who went there. W. H. Reed recalled it:

A favourite short walk from the house up through the woods brought one clean out of the everyday world to a region [close to] Flexham Park, which might have been the Wolf's Glen in *Der Freischütz*. The strangeness of the place was created by a group of dead trees which, apparently struck by lightning, had very gnarled and twisted branches stretching out in an eerie manner as if beckoning one to come nearer.[95]

But all the life of the countryside through that first full summer at Brinkwells was dominated by the approaching harvest:

June 8 . . . Lovely day. Mr. Aylwin's field cut for hay—lovely to go into . . .

Edward added in his own diary: 'Clover field cut by machine—in centre one mouse.'

July 11. Wild storms after a night of almost gale & wind—Bright intervals. Glad wheat & barley do not seem to be laid . . .

July 29. Hot & fine, Mr. Aylwin had lovely corn field scythed round ready . . .

On the last evening of July Edward noted in his own diary: 'Began to use reading lamp after dinner.'

August 1. Hot—up early, put shelf in Garden House—some rain in evening. Aylwin began to cut the wheat—only the margin . . .

August 2. Thundery & rainy—longing for fine weather for the *lovely* harvest fields. Better [war] news—D.G.

As they waited on the harvest through a week of rain, news of the changing war filtered through the little post office at Fittleworth. In Russia the revolutionaries had murdered the Tsar and all his family. But on the Western Front a big Allied advance was under way. The tide was turning with the arrival of a million American soldiers and enormous supplies. Soissons was retaken, and the Germans retreated back behind their Hindenburg Line.

August 9. Great cavalry charge—Good news Western Front—such a mercy. Lovely day—Wheatfield being cut. E. not so vesy well. A. to Petworth, Mr. Aylwin driving

[94] 3 June 1918 (writer's collection).   [95] *Elgar as I Knew Him*, p. 63.

her—lovely drive: inquired about oil stove & did many useful shoppings. In wheatfield after tea E. worked hard 'stooking'. A. helped too with a good many but found the sheaves heavy. Marvellous atmosphere, & views . . .

August 10.   Lovely hot day—E. & A. watched beautiful wheatfield next garden being finished—all cut . . .

Edward added: 'Wheat finished; barley begun . . . ordered large oil stove.'

August 11.   Glorious sunny day—Splendid for harvest . . . Such good news still—Germans *must* begin to know something.

August 12.   Harvest going on—Cloudless weather.

August 13.   Perfectly beautiful day, sunny & hot—Lovely for Harvest. Mr. Aylwin's barley field being cleared . . .

August 14.   Gloriously fine—rather cool wind—Mr. Aylwin's wheatfield being rapidly cleared . . .

Gradually the suggestion of harvesting—of preparing for the cold and darkness to come after the last, best fruit of summer—had begun to touch the deepest impulse of Edward's sixty-one-year-old life. In July he had started to orchestrate the 9/8 invention written the night he returned to Severn House: its use was still undefined. He experienced a little eye trouble when a chip of wood flew up from his chopping one day. But then he worked at clearing overgrowth before the Garden Room studio window to open out a prospect to be described later by W. H. Reed: 'From this studio there is a view of a hill sloping down to more thickly wooded country; beyond this the river Arun, and, in the distance, the heights of the South Downs are visible.'[96] A window-blind came from Alice Stuart of Wortley to control the light at this studio window and so obtain the best and longest view.

At the beginning of August the Windflower came again—almost *en route* for Tintagel, with its associations of Violin Concerto and Second Symphony. After she left, he painted the studio and put in a new floor. Then he 'electrified' his own Alice by demanding a piano for the studio. It was to be the old Steinway upright given to him by Professor Sanford at Plas Gwyn, long in store at Ramsden's. On 15 August word came that it was 'on rail', and four days later Alice noted:

August 19.   Much excitement—the piano arrived, in Mr. Aylwin's waggon. He came & his son & grandson & 2 or 3 other men, & Mark, & got it through garden & the workshop to nice position in studio—It sounded so well—not the worse for journey. May it result in booful new works.

Next morning Edward noted laconically, 'Wrote some music'. Except for the little Food song, the 9/8 idea and the Quartet sketch of the spring, it was his first music for more than a year—his first serious work since the winter of

---

[96] Quoted in Maine, *Life*, p. 210.

1915–16—his first large project for abstract music since the Violin Concerto and Second Symphony. Like the Quartet, it was to be chamber music. It did not so openly pursue the Quartet's questioning. Instead, memories of Tintagel seemed to echo through it: for this was a Violin Sonata.

Within four days he had the basis of an opening *Allegro*. The music soon found its way to the E minor of the 9/8 melody and the Quartet—again through two contrasted opening figures:

These were followed instantly by longer variants of each:

The figures were explored, contrasted, inverted, and counterpointed between violin and piano through the exposition with the concentrated thoroughness of Beethoven.

The second subject inverted the opening to echo again the Second Symphony's dying 'Delight':

This lyric generated slow violin arpeggios which ruminated again over the Violin Concerto figure twisted in the Quartet development. Now at length it reached

but this only yielded to the primary variant in a figure dwelt on through wandering chromatic keys. Then the entire sequence of events was given binary repetition (recalling the big binary form of the Second Symphony *Larghetto*), before a coda sought to dismiss all quiet, uneasy reflection with a return to the opening rhetoric.

August 21.   Misty early & then *very* hot & sunny—full moon. 2nd crop (as good as the 1st) of clover cut. E. put clock gut in & made old clock go . . .

August 24 . . . Squirrels ravaging nut tree. E. busy with music . . . M.S. paper from Novello.

Later that day Alice wrote: 'Mr. Aylwin's clover field finished—lovely scent. E. writing wonderful new music, different from anything of his. A. called it wood magic. So elusive & delicate.'

It was an opening for the Sonata's slow movement *Romance*. The violin's hesitant chromatics traced another outline of the *Gerontius* 'Judgement':

Spectral figure and arpeggiated chord passed from one instrument to the other—recalling for W. H. Reed, when he heard the music, the Aeolian harp in the study window at Plas Gwyn.[97]

On 26 August a telegram came with news that Alice Stuart of Wortley had broken her leg very painfully at Tintagel. Edward's sympathy found instant expression[98] in a long-lined melody:

Like the 9/8 idea for orchestra, this melody extended to more than 50 bars. It became the entire central episode of the Sonata *Romance*.

Next day, 27 August, Landon Ronald came down to hear the Sonata thus far and the 9/8 orchestral idea. Alice noted: 'He heard the new music after lunch & loved the mysterious Orch. piece & wants it dreadfully, & much liked the Sonata. He had a walk with E. & they talked *music* . . .' Through the following morning, after Ronald left, Edward gleaned (with Mr. Aylwin's permission) in the wheat field. And the day after that Billy Reed came with his violin to try the Sonata. He was met at the station by Mr Aylwin with pony and trap:

We jogged along through some wonderfully wooded country, along a road which twisted and turned continually, until at last we came to about half a mile of straight road rising up a fairly steep hill, with chestnut plantations on either side.

[97] *The Daily Telegraph*, 11 Mar. 1919.
[98] 11 Sept. 1918 to Alice Stuart of Wortley (Elgar Birthplace; transcript HWRO 705:445:7696).

At the top of the hill, looming on the sky-line, was what at first sight I took to be a statue; but as we drew nearer I saw it was a tall woodman leaning a little forward upon an axe with a very long handle. The picture was perfect and the pose magnificent. It was Sir Edward himself, who had come to the top of the hill to meet me, and placed himself there leaning on his axe and fitting in exactly with the surroundings. He did these things without knowing it, by pure instinct.

We had to send Mr. Aylwin on to the house with the trap containing my belongings; for Sir Edward could not wait another moment to introduce me to the very heart of these woods, and to tell me all about the woodcraft which he had been learning from the woodmen who earned their livelihood here. Chemistry, physics, billiards, and music were abandoned and forgotten: nothing remained but an ardent woodman-cooper.

. . . As we walked up to the house, he told me which part of the wood went with it, and how he had the right to cut any wood there except forest-trees, which, I gathered from him, included oaks, beeches, and elms. I was glad to hear it; for he had another axe ready for me, and apparently expected me to handle it there and then as expertly as my violin-bow . . .

At my first visit the Violin Sonata was well advanced. All the first movement was written, half the second—he finished this [with a recapitulation of the 'wood magic', *con sordino*] while I was there—and the opening section of the Finale. We used to play up to the blank page and then he would say, 'And then what?'—and we would go out to explore the wood or fish in the River Arun.[99]

The Sonata Finale began by recalling the first movement:

It extended to some length before receiving answer in a vigorous plunge down the triad:

The second subject began by reducing the primary up-and-down to

[99] *Elgar as I Knew Him*, pp. 56–8, 62–3.

—which then developed to melody:

But again, as in the first movement, second-subject melody yielded to short figuration repeated through wandering keys.

A false recapitulation of primary material did not reach the big second-subject melody. The true recapitulation did—thus making another appearance of binary form.

September 6 1918. Misty morning . . . E. worked all the early morning . . . & really all day. The Sonata was vibrating through his very being. He wrote & offered dedication to Mrs. Joshua . . .

So this first large-scale music of Edward's revival in late middle age was associated with a woman old enough to be his mother. And then perhaps Marie Joshua's German origin might seek to placate the German sources in his own style.

Mrs Joshua was 'overwhelmed by the honour', but did not think she ought to accept. She was ill in bed just then, but would write in a few days to explain more fully. On 14 September came the news that she had suddenly died. That day, as he looked out of the studio window at pouring rain, Edward turned his Finale back to the long nostalgic melody at the centre of the central *Romance*. Now it made, Alice wrote, 'a wonderful soft lament—something like the ending of slow movement of 2nd Symphony.'

A coda sent the Finale primary theme to the minor—where its chastened gravity suddenly recalled the penultimate E flat minor music in the First Symphony Finale before the last return of the 'great beautiful tune'. That Symphony was the music which had brought Marie Joshua's friendship. The friendship was memorialized now, with the assent of Mrs Joshua's daughter, in the simplest epitaph above the Sonata title:

M.J.—1918.

*   *   *

On the day he finished the Sonata, 15 September 1918, Edward began a Piano Quintet. The presence of the piano again at the centre of a major work could hint at a desire to reach some understanding at last with the piano-tuner's son. And this music also pursued his mother's inspiration in nature: 'wonderful weird beginning,' Alice noted, 'same atmosphere as "Owls"—evidently reminiscence of sinister trees & impression of Flexham Park.'

The dead trees on high ground above Brinkwells had inspired a legend—possibly embellished by Algernon Blackwood when he visited them:

'Upon the plateau, it is said, was once a settlement of Spanish monks, who, while carrying out some impious rites, were struck dead; and the trees are their dead forms.'[100] Edward, an enthusiast for supernatural stories,[101] turned the hint to his renascent music.

September 16 . . . E. wrote more of the wonderful Quintet—Flexham Park—sad 'dispossessed' trees & their dance & unstilled regret for their evil fate—or rather course wh. brought it on—Lytton 'Strange Story' seemed to sound through it too . . .

The novels of Bulwer-Lytton had arrived at Brinkwells a fortnight earlier by Edward's request. *A Strange Story* brings to its English village setting motives of witchcraft, alchemy, and Eastern secrets for renewing life and youth. All this is opposed, in the novel, to the natural warmth of man and woman. Those themes could not fail to strike both the readers at Brinkwells—Alice silently sacrificing months of comfort at Severn House,[102] Edward seeking to renew creative impulse at this primitive place in the woods of autumn.

The Quintet opened with the piano turning the shape of *Gerontius* 'Judgement' to the *Salve Regina*, while the strings muttered *pianissimo* chromatics of opposition:

The strings' chromatic rise inverted to a falling sigh; through it the notes of a triad reached up toward an octave but fell short:

[100] Maine, *Works*, p. 268 footnote. History records a monastic settlement in the area, and obscure trouble connected with it, but nothing of Spanish origin. The suggestion of embellishment by Blackwood was made by Michael Pope.

[101] Among his favourite writers in this genre were Lord Dunsany and M. R. James (Wulstan Atkins in conversation with the writer, 1980).

[102] Lady Elgar's diary entry for 14 July 1918 reads in part: ' . . . Most unpleasant day—Mrs. Hewitt away . . . E. & A. did dinner—Very trying here with the hateful back premises—quite different if at home.'

These were the elements of slow introduction, as the motto theme of the First Symphony had preceded its *Allegro*; and these themes of Quintet introduction were to recur through the later music in similar ways. But the contrast of mood to the 'great beautiful tune' was total. That antediluvian 'massive hope in the future' stood up now in black tree skeletons—whose very nearness to the cottage could give the name Brinkwells sinister significance.

The Quintet's pursuing *Allegro* changed the introduction's counterpoint to a 6/8 primary subject whose vigour sought to banish every ghost:

Yet its climax found a variant of the *Salve Regina*:

Its resolution was aggressive.

A pause; a sigh from the introduction; and a second subject appeared in a little melodic figure developed from the introduction's opening counterpoint. It sang an intimate, persuading song that might be vaguely Spanish:

Like the small second subject of the Violin Sonata first movement, this expanded in wider and warmer ways. But warmth faded, and the gaunt introduction-motto returned.

Development spun the *Salve Regina* through ambiguous piano textures of four notes against three. The strings cut across it when a stern fugato: ' . . . it was meant to be square at that point,' Edward wrote to Ernest Newman, '& goes wild again—as man does . . .'[103] The end of it was a cut-and-thrust dialogue of ferocity new to Edward's music, rising ultimately to recapitulation.

There the quarrel fomented shorter and shorter bursts—until the disintegrating ensemble was seized by the melodic second subject in a huge *fortissimo*—an astonishing transformation of its former winsome persuasion. Yet again the melodic impulse faded through extending repetition. A coda closed with the bare piano notes of the *Salve Regina* sounding as at first against the muttered opposition of the strings. Revising the Quintet first movement, he wrote to Ernest Newman, ' . . . it's ghostly stuff.'[104]

He had worked at the Quintet through equinoctial storms—which on 22 September blew down and broke Alice's garden shelter. Afterwards the sun came again, but with it the winds of autumn.

September 27 1918 . . . Bright sunny day but keen wind. E. writing wonderful new music—real wood sounds & another lament wh. shd. be in a War Symphony.

But the war was nearly over. That day the ring of Germany's alliances began to break when Bulgaria asked the Allies for peace. Along the Western Front the Allies started a huge final attack synchronizing British, American, and French forces from north, south, and west. The last phase was at hand.

It was beginning to be cold at Brinkwells, and Alice wanted to get back to Severn House. Instantly Edward's illness returned. He wrote to Alice Stuart of Wortley on 1 October: 'I have had—at the thought of town life—a recurrence of the old feelings & have been just as limp as before the nursing home episode.'[105] The illness lasted through a week of autumn gales, with better and better news from the war fronts.

October 7. 'Rather finer—E. very hard at work . . . He lay on the sofa before dinner & A. read poetry things to him. Suddenly at dinner he said 'I feel all right again'—& seemed so—D.G.

But the music which called him now was not only a slow movement for the 'ghostly' Quintet. With the end of the war in sight, his thoughts went back through the recent recurrence of illness and recovery to the convalescence in March, and the String Quartet begun then:

October 8. Fine & sunny, cold wind—E. possessed with his wonderful music, 2nd movement of 4tet—Varied by excursions to wood—Much excited over news. Germans ask for terms.

---

[103] 5 Jan. 1919.   [104] Ibid.
[105] HWRO 705:445:7700.

Then Alice's entry for that day ended with a note on the eve of her own seventieth birthday:

Last day of birthday year—full of serious thoughts & thankfulness for year with beloved.

October 9 . . . Gale & rain again—E. repeats 'Storms are sweeping sea & land' [from Alice's 'In Haven']—E. *possessed* with his lovely new music—the 4tet—writing the 2nd movement, so gracious & loveable—A.'s birfday . . .

Alice was not well. Through the Brinkwells autumn she had been more and more subject to colds and coughs. And a wen had appeared on her forehead, about which they would clearly have to consult Sir Maurice Abbott-Anderson. On 11 October they went to London. Alice took the chance to reopen Severn House for a large Sunday afternoon party on the 14th, when Edward and Billy Reed played the Violin Sonata twice—once before tea and again afterwards— for a big group of admiring friends. Yet that day brought news of further deaths. Hubert Parry had died a few days earlier; now they lost their friend William McNaught of Novellos; and the same day Ernest Newman's young wife died after a long and painful illness. Edward wrote to Newman, offering him the dedication of the Quintet 'if it is ever finished'.[106]

He attended Parry's memorial service at St. Paul's. Then, while Alice went to consult about her wen, she packed him off to stay with Frank Schuster for a few days. Society at The Hut was young and diverting as always. This time it included Lalla Vandervelde fresh from Paris with the latest prospects for peace, Aldous Huxley, Adrian Boult, and the pianist Irene Scharrer. She so delighted Edward that he promised her a first performance of his long-delayed piano concerto.

He returned to London for two charity concerts. On 22 October he conducted *The Fringes of the Fleet* at a matinée for Lord Charles Beresford's fund to aid Royal Navy prisoners of war on their return to civil life. On the 28th he took part in a concert of humorous music got up by Landon Ronald to aid the Red Cross. For Richard Blagrove's *Toy Symphony*, he played cymbals in an orchestra which included Irene Scharrer, Myra Hess, and Muriel Foster (nightingales), Albani, Ada Crossley, and Carrie Tubb (cuckoos), Sir Frederick Bridge and Sir Frederic Cowen (rattles), Benno Moiseiwitsch (triangle), and Mark Hambourg (castanets), while Ronald conducted with a 9-foot baton. In the course of the afternoon George Robey auctioned a number of MS songs, including Edward's *Inside the Bar*.

But Sir Maurice had taken a serious view of Alice's wen. The day after the Red Cross affair she went into a nursing home to have it removed. Edward stayed with her, and seemed more ill over it than Alice herself. It was serious enough to delay their return to Brinkwells, as he wrote to Alice Stuart of Wortley on 7 November:

Alice's operation was much more of an event than we anticipated & than she knows

106 21 Oct. 1918.

even now—there is a large wound. The doctors refused to let us go & all plans had to be altered by telegraph . . . My writing new stuff has been held up by the confusion & I am in despair at ever overtaking it. Our coming away—which was necessary—has been a *tragedy* for my music: alas![107]

Now they planned to go down to Brinkwells on 11 November.

It was the day of the Armistice. The events which had followed on the German request for peace had brought Germany herself to the brink of total collapse. This had a meaning for the rest of the civilized world which would be spelt out afterwards by Winston Churchill:

When the great organizations of this world are strained beyond breaking point, their structure often collapses at all points simultaneously. There is nothing on which policy, however wise, can build; no foothold can be found for virtue or valour, no authority or impetus for a rescuing genius. The mighty framework of German Imperial Power, which a few days before had overshadowed the nations, shivered suddenly into a thousand individually disintegrating fragments. All her Allies whom she had so long sustained, fell down broken and ruined, begging separately for peace. The faithful armies were beaten at the front and demoralized from the rear. The proud, efficient Navy mutinied. Revolution exploded in the most disciplined and docile of States. The Supreme War Lord fled.

Such a spectacle appals mankind; and a knell rang in the ear of the victors, even in their hour of triumph.[108]

November 11, 1918. E. & A. heard Armistice was signed. Muriel [Foster] telephoned. E. put up our Flag, it looked gorgeous—Crowds out & all rejoicing. D.G. for preservation & Victory.

E. & A. to Brinkwells—Lalla came to train at Victoria—C. tried too but just missed & went to Coliseum where 'Land of Hope & Glory' was sung twice—the 2nd time the words of refrain were thrown on the screen & people stood & joined in . . .

From Brinkwells Edward wrote to the Windflower:

We arrived all safely & found everything ready. It is cold but *vividly* bright weather & the woods divine: there are still leaves & the colours ravishing. Music does not go on yet: my poor dear A. has a cold & keeps her room—I doubt if she will be able to stay here but we shall see.

We have had the threshing machine & the drone of the humming 'sorts well with my humour'. I have been cutting down some more of the wood but I spare one spot! . . .[109]

November 13. Wrote music & tried to recover the threads—(*broken*)—worked in wood . . .

Gradually he wove the broken threads into a slow movement for the Quartet. It was marked *Piacevole* ('peacefully'), and its *poco andante* began with a song-like melody:

[107] HWRO 705:445:7127.
[108] *The World Crisis 1916–1918*, Part II (Thornton Butterworth, 1927), p. 540.
[109] 14 Nov. 1918 (HWRO 705:445:7116).

Answering phrases climbed toward a momentary triumph—only to fall back in gentle negation:

The primary theme expanded in thirds to make a figure haunted by keening wistfulness:

The answer this time turned the primary answering three-note figure upwards, but without any warmth of hope:

The primary song returned surmounted with another echo of the Second Symphony dying 'Delight':

The wistful figure found its upturned answer drooping over cello *pizzicati*. When the primary song returned again, it was a muted *pianissimo*, its answer broken. At the very end it glimmered out of shape through some softest distance. The *Piacevole* was described by Alice as 'captured sunshine'. But what it really captured was the sunshine's evanescence. Listening to a gramophone record of it years later, Troyte Griffith said to Edward:

'Surely that is as fine as a movement by Beethoven.'

He said quite simply, 'Yes it is, and there is something in it that has never been done before.'

I said, 'What is it?

He answered, 'Nothing you would understand, merely an arrangement of notes.'[110]

November 26 1918. Great cleaning of rooms. Finished & copied *Piacevole*—lovely day.

[110] MS reminiscences (Elgar Birthplace).

He polished the Quartet and the Quintet first movement, ready to try them when Billy Reed arrived in a few days' time with his violin. One practical preparation for the visit was observed by the handyman Mark:

Sir Edward, finding that there was no music-stand in the studio, went down to the tool-shed and knocked up a very creditable one. Standing back and surveying his handiwork with his head a little on one side, he became aware that Mark had entered and was standing looking at it, too.

'You see I have made a music-stand, Mark,' said Sir Edward. 'Mr. Reed is coming, and will want something to put his music on. I am afraid it is rather rough; but then, you see, I am not very handy with tools.'

'No,' said Mark dully, and walked away.[111]

December 3. E. writing. Mr. Reed came . . . after 2—E. & he played & played, he was much excited, sd. the 4tet was the most advanced ever written & was *amazed* at Quintet—He was so eager & keen & deeply impressed—Left at 7.30. very dark—[112]

Alice's cough had worsened until it kept her in bed at Brinkwells for days on end. So they must go to London for the doctor again. Edward wrote to Frank Schuster:

. . . it means another interruption & the future is *dark* as A. poor dear is not well &, of course, is bored to death here while I am in the seventh heaven of delight: so we may only return here to clear up—we shall see. But it means that if I have to live again at Hampstead composition is 'off'—not the house or the place but London—*telephones* etc *all* day *and* night drive me mad![113]

Yet London held something for Edward just then—the first hearing of *Falstaff* since the initial performances of 1913–14. Landon Ronald was opening the Philharmonic season with it—only because young Adrian Boult had threatened to do it in his own Philharmonic concert later in the year.[114] Nobody in London really knew the work, and its announcement now had the air of a recovery from oblivion.

December 5. E. & A. to Queen's Hall . . . Splendid performance—wonderful insight of Lan—Great reception & he & Lan recalled about 6 times, E. dragged Lan on—Wonderful evening.

The Times critic H. C. Colles wrote of

. . . the way the orchestra talks throughout, just as if it were a human being instead of a number of skilfully used pieces of wood and catgut . . . Elgar has more power than anyone of getting a strong, wholesome diatonic feeling into the most heart-rending chromatics . . . it is the lift of a mind that can think on a big scale, but loves to play with children far more.[115]

---

[111] W. H. Reed, *Elgar as I Knew Him*, pp. 61–2.
[112] This entry appears in Lady Elgar's diary on 2 Dec. But often she made up her diary several days in arrears, and thus got a day out. Elgar's own diary shows Reed's visit on the 3rd.
[113] Tuesd[ay 3 Dec. 1918] (HWRO 705:445:7028).
[114] See *Music and Friends: Letters to Adrian Boult* (Hamish Hamilton, 1979), p. 37.
[115] 6 Dec. 1918.

Next day Alice consulted Sir Maurice about her cough: he 'found nothing the matter with lungs'. So they went back once again to Brinkwells for Edward to pursue his chamber music.

December 7.   Most lovely day, sunny & warm—Settling down again . E. gradually finding his broken threads . . .

December 8.   Misty day—quite warm.   E. writing. In aftn. to P.O. Little Bognor— Ferns lovely deep yellow colour—E. writing last movement of Quartet—Very impassioned & carrying one along at a terrific rate.

Its *Allegro molto* opened a witches' descent:

To it came three answers. The first seemed to confirm the tonality in syncopating semiquavers:

The second set an explosive rhythmic sequence rattling through empty spaces between the notes of a far-fetched triad:

The third did just manage a melodic figure of two bars, but syncopated through tritone intervals:

The three answers showed their tonality in progressive decay.

A second subject shrank the Violin Sonata's

to:

At length the second half of the phrase produced a sequence of melodic promise:

Once upon a time such an idea would have led Edward's music to a noble climax: now it merely beat its sequences against unyielding semiquavers. A furious development and recapitulation allowed no progress toward such a goal. The coda just looked at a possibility

—but it was swept into the witches' fire that crackled *sempre cresc. e stringendo* to a final cynical burst of E major. It was as if this music faced a time when all must go faster—when every cinder of melodic expression from the old world would be consumed in the energy which had fired it, now the war was over.

As the Quartet went toward its headlong end in December 1918, they had word that the long-closed Severn House had been burgled. Alice struggled up to London alone to reckon the loss of silver, Edward's suits, and the contents of his cellar including a dozen bottles of Scotch whisky. The news reached the Royal Academy of Music, where Sir Alexander Mackenzie was principal:

Someone at the Academy opened the paper on the morning when the list of depredations at Elgar's house appeared and read out the news. 'Fancy Elgar having a dozen bottles of whisky in his cellar anyway,' said the reader.

'Oh, we all knew he was a Hoarder of Merit,' said Mackenzie in his most Scottish accent.[116]

Alice returned to Brinkwells. But the deepening cold in the primitive little house, and the servant problem allied to it, were now beyond her strength. Before they packed for the final evacuation, on 24 December the Quartet was finished. Keeping an old promise from the turn of the century, he dedicated it to the Brodsky Quartet. Adolf Brodsky was now nearly seventy, and his elderly colleagues could not look forward to many performances of their music.

Back in Severn House at the New Year 1919, there was the Quintet to complete. He had been maturing ideas for it side by side with finishing the Quartet at Brinkwells. The Quintet's destinies would now be worked out amid the spaces of the great music room. But he marked the first movement with the name of woods near Brinkwells, 'Bedham, 1918'.

[116] W. H. Reed, *Elgar*, p. 123.

A trial of the new works before friends was arranged with Reed's string quartet for 7 January. Alice, far from well, still unpacking and coping with multitudes of details needed to reopen the big house, noted:

A. seemed always to be having to do impossibilities—she said so on telephone to Frank—he said 'but you always achieve them.'[117]

January 7 1919. A. arranging for afternoon. Mr. Reed came early. E. & he to Stewart's to lunch. Then nice old Harrod waiter [who had served them when the *Fringes of the Fleet* men lunched at Severn House] came. Quartet came first & rehearsed. Then Colvins, Alice S. of W., Fortescues, Evelyn Horridge, Lan[don Ronald] & his brother, Frank. Wonderful music: 4tet—1st movement beautiful especially—the Piacevole like captured sunshine—Sonata so fine & beautiful with the wonderful 2nd movement. E. played it so wonderfully—everyone delighted & astonished. Then crowning all, the 1st movement of Quintet, so marvellous—A most thoroughly enjoyable aftn. & everything went so smoothly & comfortably. D.G. for such wonderful gifts given to E.

Over the next days he worked at the second movement *Adagio*. A seamless melody turned its back on the astringencies of the Quartet to recover the vanished world of the First Symphony *Adagio*:

The reply

found again the very contour which had fulfilled *A Child Asleep* and *They are at Rest* in the old days before the war. Now it soared to a climax in which dying 'Delight' was wondrously renewed:

When the song was ended, the piano struck a question:

That might just recall the triadic figure of counterpoint against the Quintet's

[117] Diary, 4 Jan. 1919.

sighing introduction. It was followed instantly by a distinct echo of the sigh itself through the strings:

So began a second subject of small presences, moving softly to question the long lines of the primary melody.

Development held another *fugato*, a slow exploration of the primary melody. Yet its climax raised the ghost again:

Recapitulation returned to the primary melody and secondary questions. Again and again the melody was questioned: even after its last warm descent to tonic resolution, the secondary spirits followed softly to question the end.

January 14.   E. writing all day & beginning to look very tired . . .

January 15.   E. very absorbed & not out . . .

January 16.   E. raser porsley & not out . . . went on saying 'this is no home for me!' . . .

Three days later he escaped from Severn House to meet Reed's men at the house of Lord Charles Beresford. There they tried the Quartet again, the Sonata, and the two finished Quintet movements. Again all the friends were enthusiastic.

So he settled to the last movement. Its beginning—barely separated from the final question of the *Adagio*—was a return of the sighing figure from the first-movement introduction. But soon, in a new *Allegro*, the sighing figure was forged to a strong primary theme *con dignita, cantabile*:

Extending variants did not challenge the tonality: their chromatics only made a slight winter haze before the sun of A major. Short figures made the syncopated second subject, but (unlike both the previous movements' second subjects) the short figures now combined to long-lined expression:

Development extracted a bare F sharp minor triad from the primary theme:

As the melodic presence ebbed, the temperature dropped. Here especially the writing was spun out of idiosyncrasies in Edward's piano-playing. These were to be characterized by Bernard Shaw in a letter written immediately after hearing the music:

There are some piano embroideries on a pedal point that didn't sound like a piano or anything else in the world, but quite beautiful, and I have my doubts whether any regular shop pianist will produce them: they require a touch which is peculiar to yourself, and which struck me the first time I ever heard you larking about with a piano.[118]

To the eerie beckoning music of the first movement, the *Salve Regina* sounded again. And then, as from some far distance, came the winsome voice of the earlier *Allegro* second subject:

It was the development-in-reverse which had appeared at the centre of the Violin Concerto Finale.

Recapitulation healed every regret. In a radiant coda primary and secondary themes were set side by side, like the two 'Enigma' figures in the coda achieved twenty years ago. It was, Edward wrote to Sidney Colvin, an 'apotheosis'.[119] No such affirmation had closed any work of his since the First Symphony. For then, as now, a war of the spirit had been won. Then the enemy had been the quasi-military ghosts of ambition. Now they were the disintegrating self-doubts which had been faced through the wood magic of countryside peace. Thus the end of the Quintet, though finished at Severn House, was inscribed as the other chamber music with the name of Brinkwells.

---

[118] 8 Mar. 1919 (Quoted in McVeagh, *Edward Elgar: his life and music*, p. 178).
[119] 12 Jan. 1919 (HWRO 705:445:3480).

Yet such an apotheosis of melody in 1919 could leave its composer more isolated than ever. When the younger survivors returned from the war, their first resolve would be to destroy every shred of the old world—the world which created the war that had maimed and coarsened their lives. Amongst the friends invited to Severn House on 7 March 1919 to hear Reed and his colleagues play the Quartet, and with Edward the Sonata and completed Quintet, was the twenty-seven-year-old Arthur Bliss. The ex-Captain was asked to turn Edward's pages through the Sonata:

But all I can recall now was a certain embarrassment as to what I ought to say as the sonata ended. Was my disappointmnent due to the far from brilliant performance or to the belief that its musical substance had little in common with the genius of his earlier masterpieces? I hope I sat quiet, as if absorbed.[120]

Among the guests at Severn House that afternoon were Mr and Mrs Bernard Shaw. Shaw had been introduced to Edward a few days earlier at a luncheon party given by Lalla Vandervelde. The party had also included the designer Roger Fry, a dozen years younger than Edward and Shaw, representing a newer generation of taste. Fry designed home furnishings (including some for Lalla) in a deliberate primitivism which rejected every shred of Edwardian sophisitication: he was virtually excluded from the conversation that day, as Edward recognized in Shaw a kindred spirit. Shaw recalled:

Elgar, who had enjoyed my musical criticisms when he was a student [sic] and remembered all my silly jokes, talked music so voluminously that Roger had nothing to do but eat his lunch in silence.

At last we stopped to breathe and eat something ourselves; and Roger, feeling that our hostess expected him to contribute something, began in his beautiful voice (his and Forbes-Robertson's were the only voices one could listen to for their own sakes) 'After all, there is only one art: all the arts are the same.'

I heard no more; for my attention was taken by a growl from the other side of the table. It was Elgar, with his fangs bared and all his hackles bristling, in an appalling rage. 'Music', he spluttered, 'is written on the skies for you to note down. And you compare that to a DAMNED imitation.'[121]

Was it the presence of the Fry furniture in Lalla's flat, bound to strike Edward as a betrayal of all he ever learnt and stood for? Or was it some perceived implication in Fry's remark that would limit music to the 'imitation' that involved a programme?

When Reed and Anthony Bernard played the Violin Sonata in a semi-public

---

[120] *As I Remember*, p. 24. Although Bliss recalled only the Sonata, the date is fixed by his memory among the guests of the Bernard Shaws, paying their first visit to Severn House that afternoon.

[121] 10 May 1940 to Virginia Woolf (quoted in Leonard Woolf, *Beginning Again*, OUP, 1980, pp. 89–90). Virginia Woolf's biography of Roger Fry gives the date of Shaw's meeting Elgar as 1917, but I can find no record of their meeting before Mar. 1919.

performance for the British Music Society on 13 March, the composer was absent. The full public première was to be given by Reed with Landon Ronald a week later.

March 21 1919. E. to rehearsal of Sonata—returned very depressed not liking his music—At last was persuaded to go to Concert,   A. ordered car, a bitterly cold night, & the performance was beautiful & its reception overwhelming—Lan played the 2nd movement perfectly exquisitely & all beautifully –& so did Reed as if he were inspired. The audience roared & shouted when E. at last came on . . .

The audience, according to *The Daily Chronicle*, 'numbered many well-known musicians, particularly violinists.'[122] But the hall was far from full. One critic wrote:

I could hardly restrain a cry of indignation upon entering the Aeolian Hall. Imagine the announcement of a new sonata by Ravel in Paris, by Richard Strauss in Berlin, or in times gone by, by Brahms in Vienna. Why, weeks beforehand it would have been impossible to find a single seat, and here was a new violin sonata by Sir Edward Elgar, one of the great of all times and perhaps the greatest living composer, at the piano one of our best conductors, the leader of one of our orchestras as the solo violinist, and, for shame! a half empty hall.

. . . Like Brahms in the later part of his career, Sir Edward aims at ever-increasing directness, terseness and simplicity of expression. There are in the sonata no complications of rhythm or harmony, no thematic singularities, it is not exceedingly difficult to play, it seems like a protest against the far-fetched devices of the ultra-moderns—it seems to say: See what can be done yet with the old forms, the old methods of composing, the old scales . . .[123]

The judgement was amplified by Ernest Newman after the Sonata's first performance in Birmingham a fortnight later:

Elgar's style has become one of extraordinary slenderness so far as the mere notes are concerned. It may be that he has deliberately rarified the tissue of his music for the occasion; or it may be that his style in general is becoming simpler in maturity, as that of Hugo Wolf did in his last songs. As with Wolf, the simplification is in the texture only; every superfluous line has been eliminated from the design, every superfluous note from the harmony; but the music carries a surprising weight of thought and feeling. It is all calculated, too, with extraordinary certainty for each instrument: many a passage that looks a little unimpressive on paper turns out to be singularly impressive in performance.[124]

Edward read this with gratitude, and wrote to Newman immediately: 'Thanks for the beautiful notice of the Sonata—you are right about the difference of look & sound in the music.'[125] But again at Birmingham the hall had been partly empty.

[122] 23 Mar. 1919.
[123] L. Dunton Green, 'Music of the Week', *The Arts Gazette*, 29 Mar. 1919.
[124] *The Birmingham Daily Post*, 8 Apr. 1919.
[125] 9 Apr. 1919.

On 26 April 'a very distinguished gathering' was invited to Frank Schuster's house in Old Queen Street to hear the Quartet, the Sonata *Romance*, and the Quintet played by Albert Sammons, Reed, Raymond Jeremy, the distinguished young cellist Felix Salmond, with William Murdoch at the piano and Adrian Boult to turn his pages. One or two pressmen were present, including a hesitant *Pall Mall* critic:

There is no doubt that a beautiful music-room, filled with an audience of the very elect, is the ideal place in which to hear chamber music. The only danger is that the charm of the surroundings may take the sharp edge off the critical faculty . . .[126]

Four weeks later, on 21 May, the public had the chance to judge for themselves when the same performers gave all three works at the Wigmore Hall. Critical judgement was summed up by H. C. Colles, writing in *The Times* under the rubric 'Music of Yesterday and To-day':

An immediate effect of listening to Sir Edward Elgar's opp. 82, 83, and 84 in succession is to give one a new sympathy with the modern revolt against beauty of line and colour. A stab of crude ugliness would be a relief from that overwhelming sense of beauty . . . It is not really ugliness, and still less vulgarity, that one craves as an antidote to the Elgarian type of beauty. It is the contrast of a more virile mind, something less purely visionary and more touched by hardness . . .

The fertility of his musical invention in the new chamber works is amazing. But as one listened, various phases of his former work seemed to be hovering in the air and to bring back to memory, often delightfully and quite spontaneously, moments spent in the same musical company. Was there not something of 'Nimrod' in the sustained loftiness of the slow movement of the quintet, and who could forget the violin concerto in the finale of the same work, when the music dropped into a reverie over a theme which had been prominent in the first movement?

Elgar's music is always autobiographical; but the life is not completed; it is the present which one looks for most eagerly in his latest work, and not the past. What has he to say now, and have the years stamped their meaning on him in any profound way ? It was the failure to find this through the greater part of the new works which made one impatient before the end of Wednesday's performance.

Yet mercifully it is not altogether absent. The first movement of the piano quintet has a breadth of view, one might almost say a manliness of expression, which has never appeared so clearly in anything he has written before . . . It was this movement, rather than the two which followed it, which seemed to raise the quintet to a higher plane than either of the preceding works; and more than all else convinced us that Elgar is still a force among the many currents of the musical tide.[127]

But it was the Quintet first movement which questioned and doubted, the other movements which had presented Edward's answers.

*   *   *

Then the whole experience of the chamber music brought a culminating insight. From the probing Quartet begun in March 1918 to the 'apotheosis' of

[126] 28 Apr. 1919.   [127] 24 May 1919.

the Quintet Finale a year later, one question had dominated: whether melody could hold the centre of Edward's music in the future as it had done in the past. The Quintet had suggested finally that melody must hold a central place if the music was to continue to express what was central in Edward himself. But this answer carried with it the threat of growing creative isolation for the future.

He turned back to the idea written the night he returned to Severn House from the nursing home—the long sequences of 9/8 moving slowly downward to E minor. This he resolved now to make the basis of a concerto—but such a concerto of isolation, loneliness, farewell even, as had never yet been written.

As he thought over the question of a solo instrument, he might have remembered the request of long ago from Brodsky's cellist Carl Fuchs. The violoncello—lower than the violin and darker, but holding the same *cantilena*—became the solo voice for this Concerto.

The difficulty of writing a concerto for the cello is to allow its tenor voice to be heard through the orchestra. Edward's orchestra remained roughly the same as for the Violin Concerto. But now he would score much at the top and bottom extremities, leaving an empty middle space through which the lonely solo spirit was to wander. Occasional ensemble climaxes would give broad perspective. Then the orchestra must fall silent altogether as the cello explored a world of loneliness. So the traditional rhetoric of the concerto would turn to soliloquy.

The solo presence was established at the outset of the new Concerto in four almost unaccompanied chords *nobilmente*:

Their sequence looked harmonically forward to the 9/8 sequences to come. To

an upturning solo question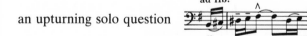

came the 9/8 in a single voice through all the violas, reaching down at last to the orchestra's cellos:

The solo cello took it up in dialogue with the orchestra—a constant slow dotted rhythm whose melody moved up and down through unfolding question and unresolving reply, their units in turn shaping larger cycles of echo and repeating echo to fill the prospect with its single rumination. Was it the rhythm of hills and answering valleys, the curving of a river through fields and orchards planted and harvested through hours, years, lives? It was no one of these, and so it might be all.

The dotted rhythm yielded to relative animation—against which the solo cello sounded a counterpointing echo from Edward's earlier music:

The counterpoint extended downward to recall again the *Gerontius* 'Judgement', the Second Symphony's dying 'Delight'. When the 9/8 returned, the counterpoint sounded above it. At last the music sank in a hushed joining of solo and orchestral voices.

Without any definite break, the opening solo chords returned *pizzicato* to begin a second movement. This was a 4/4 *Allegro molto*, juxtaposing extremities from Edward's style. On the one hand, nervous semiquavers echoed the Quartet and *Introduction and Allegro*:

On the other, a big *cantabile* recalled 'Nimrod' and 'E.D.U.', and in the far distance the shape of the 'tune from Broadheath':

But this melody was only a fragment. Repeated in the full orchestra, it was still unable to stay longer than an instant the demon energy of semiquavers. When the two commingled, the fragment of noble melody was twisted—until at the end blind impulse rushed softly to meaningless resolution.

A third movement *Adagio* in B flat major was all restrospective melody. A

rising figure

might have been developing the upturning solo question at the beginning of this Concerto; but it echoed more clearly the Violin Concerto's 'Windflower'. A longer figure looked in another way at dying 'Delight':

The climax of this long melody returned to the first movement's downward

counterpoint  :

only now the notes sounded a forlorn echo of *The Music Makers* 'You shall teach us your song's new numbers'. Beyond this the melody did not develop, but sang its song until the opening 'Windflower' wraith returned to take the *Adagio* to a soft half-close.

Out of the *Adagio*'s B flat tonality came a Finale *Allegro* beginning in B flat minor:

Here was a primary theme drawn from the Concerto's opening solo chords. The solo cello recalled its first rising question. And the primary Finale theme replied *risoluto* in the tonic E minor. The orchestra extended the downward

phrase

to recall again the opening movement's descending counterpoint. A second subject *allargando* caught echoes of the opening movement's second subject.

Through the development came a new descent which echoed the sighing figure from the Quintet:

Under a growing burden of descent the music struggled upward through chromatic ghosts of the primary theme. Then the solo cello acquiesced, joining *nobilmente* with all the cellos of the orchestra in recapitulation. Primary subject and descending answer, secondary essay and hopeful sequence all turned through smaller compass. The primary subject returned once more for what should be the coda.

Suddenly rhythmic articulation fell away to reveal the sighing figure alone, pursuing a climax outside the sonata form. Climbing higher, it turned *con passione* to an echo of dying 'Delight' that touched the memory of Wagner's *Liebestod*:

This went through sequence after slowing sequence—to find at last a memory of the vanished *Adagio*. So the ultimate discovery went into the past. The solo chords of the Concerto's beginning met the Finale primary at its end. The music was finished: it was half the length of the Violin Concerto.

Written through May and June 1919 between Severn House and Brinkwells, the Concerto was still marked at the end with the name of Brinkwells. The well this time had given back its reflection of nostalgia utterly complete. He dedicated the Concerto to the most aged of his close friends—Sidney Colvin who was seventy-four, his wife Frances who was eighty. She alone of all Edward's intimates now survived at an age just old enough to have been his mother.

At Brinkwells, side by side with woodworking, he revised the Quintet for publication—'painted' it, in his own word borrowed from the homely occupations around him. He invoked the editorial help of Harold Brooke at Novellos. Brooke was a considerable pianist with professional training. They corresponded particularly over two figured passages, which Edward named in tribute to his rustic setting 'Tadpole' and 'Straddlebug'. The *arpeggio*-sign he dubbed a 'Wireworm'. By late July the Quintet was almost ready, and he was 'painting' the Cello Concerto.

Already he had tried the Concerto at various stages with Felix Salmond, the cellist who had taken part in the chamber music premières. On the last day of July Salmond arrived at Brinkwells with his cello—to be met by the redoubtable handyman:

Mark went to the gate to help Felix in with his luggage, which included the monstrous black coffin of the 'cello. Then Mark gazed for a moment at Felix, and silently withdrew.

Later, Sir Edward said to him, 'Mark, that gentleman who came to-day with the big case is a very famous musician, a great 'cellist, a very important person, you know.'

'Well, I suppose', said Mark, to console him, 'it takes some of all sorts to make up a world.'[128]

After tea they went through the Concerto, and again after dinner. Next morning:

August 1 1919 . . . Soft wind & lovely sky & scenery . . . After breakfast played the Concerto again with Felix S.—such a delightful visitor . . . E. & Mr. Felix went down to the river to fish, no success, F. enraptured with scenery.   A. to P.O., *very* hot, & A. lagged a good deal. More music in evening . . .

Then Edward offered Salmond the Concerto première.

August 2.   A little more music & walking around—F. Salmond left after lunch & seemed so happy here—*thrilled* with the thought of playing the Concerto for the 1st time & wildly excited about it, did not sleep all night thinking about it . . .

That day Alice went down with a chill. She had to keep to the house for several days while Edward finished 'painting' the Concerto. By 8 August she was well enough to take the precious score down to Fittleworth Post Office to go to Novellos with the last proofs of the Quintet. The intensely hot weather brought the harvest, and soon Mr Aylwin's corn was cut. It was just a year since Edward had asked for the piano to be sent down to Brinkwells. Now Alice wrote to Lady Stuart of Wortley: 'The Music Room is so silent without the Concerto, gone to be printed.'[129]

Edward felt the emptiness. He wrote to the Windflower: 'I want to finish or rather commence the Piano concerto which *must* be windflowerish so I hope you will come but somehow I know you will not.'[130] She did come for two days; but then he only played for her (as his own Alice noted) '2nd Symphony & other wonderful things'.[131] A month later he would write even from Brinkwells: ' . . . I want to get away—strange: but the Studio is sad sad & I feel I have destroyed the best thing I ever wrote & that it had to be so. I am not well & worried in many ways.'[132]

Sadness met him at every turn. At Severn House they had the 'shock' of ' . . . finding dear wavy trees cut down [by the neighbour] & pit dug just under Music room window—garage being made.'[133] They consulted solicitors, but there was nothing to be done. It added discontent to the growing burden of trying to run the big house on incomes which the war had reduced in every direction. On 2 September Alice reluctantly went to estate agents: '. . . to Hampton's & said we might sell the beloved house.'

But if Severn House had been difficult to sell before the Elgars bought it, it was thrice difficult now. Its grandeur was redolent of a past which had vanished

---

[128] W. H. Reed, *Elgar as I Knew Him*, p. 62.
[129] 14 Aug. 1919 (HWRO 705:445:7915).
[130] Sunday [3 Aug. 1919] (HWRO 705:445:7703).
[131] Diary, 20 Aug. 1919.
[132] Monday [22 Sep. 1919] (HWRO 705:445:4085).
[133] Diary, 29 Aug. 1919.

with the war. The longest purse might soon exhaust itself in running such a house. Its operation depended on a good domestic staff: but the war had opened all kinds of opportunities for women, and contentment in domestic service was growing rare. The interviewing of servants was already a subject for irony, as when *The Evening Standard* published a piece called 'The Time for Tact'. Among its recommendations: 'Should the prospective housemaid start to hum or whistle, assume that she is musical, and by degrees lead to a discussion of an Elgar Symphony and offer her the grand piano.'[134] One or two people came to look at Severn House. No one showed much interest.

On 16 September they returned to Brinkwells for a last long look at the woods through the mists and waning warmth.

October 9. In wood most of the day, so perfectly lovely & warm sun. Lady Bartelot [owner of the Stopham Estate] came up in afternoon, only just returned. After tea we walked back with her to end nearly of fields—It was wonderful to see, golden & crimson sky & pearl Downs & brilliant moon . . .

October 10. Large parcel of proofs for E. Cello Concerto sounded great noble music—E. worked all day & after tea we walked down & he posted them . . . Colours in wood & road so beautiful —& sunset & moonrise . . .

October 11. E. long day in wood, clearing, Mark helping all day. Fires of rubbish, great flames shooting up . . .

The Violoncello Concerto was to have its première at a concert to open the first post-war season of the London Symphony Orchestra. They had a new conductor in the thirty-seven-year-old Albert Coates, very much on his mettle to show what he could do. In this opening concert he was to conduct everything but the new Concerto, and he was busy with the Orchestra when Edward arrived to rehearse:

October 26 . . . E. & A. & C. to rehearsal at Mortimer Hall—horrid place—the *new* work, Cello Concerto, never seen by Orchestra—Rehearsal supposed to be at 11.30. *After* 12.30—A. absolutely furious—E. extraordinarily calm—Poor Felix Salmond in a state of suspense & nerves—Wretched hurried rehearsal—An insult to E. from that brutal, selfish, ill-mannered bounder A. Coates.

October 27. E. & A. & C. to Queen's Hall for rehearsal at 12.30 or rather before—absolutely inadequate at that—That *brute*, Coates went on rehearsing 'Waldweben' [from Wagner's *Siegrfried*]. Sec[retary of the Orchestra] remonstrated, no use. At last just before one, he stopped & the men like Angels stayed till 1.30 [half an hour beyond their time]. A. wanted E. to withdraw, but he did not for Felix S.'s sake—Indifferent performance of course in consequence. E. had a tremendous reception & ovation.

But once again the hall was not full for an Elgar première.

Ernest Newman (now based in London writing for *The Observer*) was

[134] 29 Apr. 1919.

outraged and puzzled: the Coates part of the concert had been splendid, whereas in the Elgar 'never, in all probability, has so great an orchestra made so lamentable a public exhibition of itself.' Newman referred to the subtle balances needed for a cello concerto:

This scale of colour it has obviously been Elgar's preoccupation to achieve. Some of the colour is meant to be no more than a vague wash against which the solo 'cello defines itself. On Monday the orchestra was often virtually inaudible, and when just audible was merely a muddle. No one seemed to have any idea of what it was the composer wanted.
    The work itself is lovely stuff, very simple—that pregnant simplicity that has come upon Elgar's music in the last couple of years—but with a profound wisdom and beauty underlying its simplicity . . . the realisation in tone of a fine spirit's lifelong wistful brooding upon the loveliness of earth.[135]

After one hearing in such a performance, few discerned so much.

* * *

Old enthusiams awaited Edward in the old places. In Worcester for a week at the beginning of November, he listened to Ivor Atkins's plan for restarting the Three Choirs Festival in 1920, took the chair at a meeting to reconvene the Worcestershire Orchestral and Ladies Choral Society with Isabel Fitton as secretary, and attended a Peace Banquet at the Guildhall with Hubert Leicester. Hubert's son Philip, returned from the war and now happily married, asked Edward to dine at his own home in Britannia Square—nearly opposite the house where the piano-tuner's son had attended Miss Walsh's school nearly sixty years earlier. Philip found the same spirit in his father's old friend:

He looks older. His hair & moustache are now more white than grey. His face is paler. Otherwise he seems unchanged. The same low voice, rapid earnest speech, keen sense of humour, & quick movement.
    . . . He talked of the recent burglary at his house at Hampstead & his police court experiences connected with it. Said that he usually sympathised with the criminals—'they are the only poets in these days.'—Young man on trial for seducing a girl of 17 who ran away with him from an unhappy home, got 12 months, & girl wept in court.   E. said we were a curious nation. Our classical system of education caused hundreds of respectable gentlemen to spend their lives teaching thousands of school boys the pagan poets' theories of *love*—they wrote nothing about marriage—'and when two young people have the courage to defy convention & follow the poets to find perfect bliss, you send the boy to gaol & the girl home to her parents.' He spoke jestingly but I could see he was half in earnest.[136]

Two nights later, on the eve of his departure to London, he dined with the elder Leicesters. Philip and his wife were present again:

[135] 2 Nov. 1919.
[136] MS note of 4 Nov. 1919.

During dinner he quoted Belloc's lines about

> 'The men that live in West England
> They see the Severn strong,
> A-rolling on rough water brown
> Light aspen leaves along.
> They have the secret of the Rocks,
> And the oldest kind of song.'

and his eyes sparkled when he came to 'the oldest kind of song' as if he felt the appropriateness in his own case. He said he did not know Belloc personally. I quoted his verse ending

> 'There's nothing worth the wear of winning
> But laughter and the love of friends'
> which he liked very much . . .[137]

He returned to London to cross the Channel for some concerts arranged by Lalla Vandervelde in Amsterdam and Brussels. Alice was ill again, and spent almost the whole fortnight of Edward's absence in bed. She crept out for his return on 27 November, but during the next month was hardly able to do more than get to the dining-room for one meal a day. Sir Maurice could find nothing wrong beyond her persistent cough.

So Carice pursued a long-laid plan for a post-war holiday on her own—away from her parents and Severn House and great men and music and London and the Censorship—away from the memory of her young Army Captain killed late in the war. She chose Mürren in Switzerland, full of winter sports, dances, young people—and went to stay for several weeks.

Just before Christmas Edward conducted a gramophone recording of the Cello Concerto with the young Beatrice Harrison. She was one of four sisters, all deeply musical, whom their mother had aspired to turn into a string quartet of international soloists. Mrs Harrison had succeeded so far as to see her two eldest daughters, May and Beatrice, become first-class virtuosi of violin and cello respectively. They had made a friendship with Delius which resulted in his Double Concerto. Now the whole Harrison family descended on Severn House with their enthusiasm. And the Concerto's first recording, despite cuts and retakes, was a success. Alice had helped to edit orchestral parts for the recording, but was not up to attending the session on 22 December.

At the end of the year came a tribute from Bernard Shaw. He had been asked to write an essay about Elgar for the first number of a new quarterly, *Music and Letters*. His piece began by protesting the impossibility of defining Edward's place in history, and then going on to define it:

Either it is one so high that only time and posterity can confer it, or else he is one of the Seven Humbugs of Christendom. Contemporary judgements are sound enough on Second Bests; but when it comes to Bests, they acclaim ephemerals as immortals, and simultaneously denounce immortals as pestilent charlatans.

[137] Ibid.

Elgar has not left us any room to hedge. From the beginning, quite naturally and as a matter of course, he has played the great game and professed the Best. He has taken up the work of a great man so spontaneously that it is impossible to believe that he ever gave any consideration to the enormity of the assumption, or was even conscious of it. But there it is, unmistakable. To the north countryman who, on hearing of Wordsworth's death, said, 'I suppose his son will carry on the business' it would be plain today that Elgar is carrying on Beethoven's business. The names are up on the shop front for everyone to read. ELGAR late BEETHOVEN & CO. . . .

This, it will be seen, is a very different challenge from that of, say, Debussy and Stravinsky. You can rave about Stravinsky without the slightest risk of being classed as a lunatic by the next generation. Without really compromising yourself, you can declare the Après Midi d'un Faune the most delightful and enchanting orchestral piece ever written. But if you say that Elgar's Cockaigne overture combines every classic quality of a concert overture with every lyric and dramatic quality of the overture to Die Meistersinger you are either uttering a platitude as safe as a compliment to Handel on the majesty of the Hallelujah Chorus or else damning yourself to all critical posterity by a *gaffe* that will make your grandson blush for you.

Personally, I am prepared to take the risk. What do I care about my grandson? give me Cockaigne . . .

If I were king, or a Minister of Fine Arts, I would give Elgar an annuity of a thousand a year on condition that he produced a symphony every eighteen months . . . Neither I nor any living man can judge with certainty whether these odds and ends which I have been able to relate about Elgar are the stigmata of what we call immortality. But they look to me very like it . . .[138]

Shaw's article drew warm praise from Alice, who had always been suspicious of his atheism. He responded as generously to her:

10 Adelphi Terrace. W.C.2.   3rd Jan. 1920
My dear Lady Elgar
I am very glad you liked the article. These 'appreciations', as they call them, are at best pardonable impertinences; and after committing them one can never feel quite sure whether the victim's wife—to say nothing of himself—will ever speak to one again. The only excuse for them is that musical fools, like other fools, have to be taught manners; and as a great composer cannot very well condescend to such lessons in his own interest, someone has to do it for him. Besides, it comforts the people who know better to be told so occasionally: otherwise his admirers would feel almost as lonely as he does at the start.

Tell Sir Edward that he has done me a good turn; for the Americans have paid me quite handsomely for the article. Though why *I* should be paid for his genius, God alone knows!

Sincerely   G. Bernard Shaw[139]

Otherwise the beginning of another decade in the twentieth century brought no joy anywhere but in the past. On 11 January Edward finished a Preface to a

---

[138] *Music and Letters* vol 1 no. 1 (Jan. 1920 pp 7–11).
[139] HWRO 705:684:5968.

book on *Musical Notation* by Novello's editor H. Elliott Button—with another attempt to link the future with old times:

Long years of experience have not dulled the feeling that the shaping of a new work is like the coming of a child, and something akin to awe attends its birth. Printing, and all that belongs to it, has had a fascination for me since I was first permitted to pull a lever in Leicester's office in Worcester fifty years ago; with the same thrill I touched a similar machine in the Musée Plantin in Antwerp a month back. Some artists affect to despise a knowledge of the means, mechanical or otherwise, by which their ideas become realised; give me those who know the depths as well as the heights—those who can look 'unappalled upon the spears of kings, and undisdaining, upon the reeds of the river.'[140]

He wrote no new music, but only abridged 'For the Fallen' to be sung at the dedication of Lutyens's new Cenotaph in Whitehall.

On 22 January Alice took a fresh cold. She had not been out of the house since November, and W. H. Reed remembered:

She who had always been so full of vitality and energy was now often listless. She would creep up close to the fire and look so fragile that I began to feel anxious about her. She would brighten up for a little while; but every time I saw her she seemed to be getting smaller . . .[141]

Rosa Burley was 'shocked to find Alice shrunken and terribly depressed'.[142] In February she seemed well enough to venture out, and on the 24th she went with Edward to Hayes for the next gramophone session. The recording manager Fred Gaisberg wrote:

I observed how especially tender and solicitous Sir Edward was for his wife, and he seemed so very happy to have her with him . . . Motherly kindness radiated from her, and it was easy to see how much Sir Edward Elgar owed to her good advice and solicitous care.[143]

Those days were brightened for her by honours accorded to Edward— membership in the Accademia at Florence and the Institut of France; election at home to the Literary Society (whose president was Sidney Colvin), the Council of the Royal College of Music, and Chairmanship of the London committee for starting a festival and school at Glastonbury around the work of a younger composer whose genius seemed clear to almost everyone in England then, Rutland Boughton. Edward appealed for support for the Glastonbury scheme in terms reminiscent of his first lecture at Birmingham fifteen years earlier, 'A Future for English Music': 'We may ultimately, through such experiments as are being made at Glastonbury, find a means to bring a conscious artistic life to the countryside itself.'[144] Before the war several craft

[140] Preface to *Musical Notation* by H. Elliot Button, also printed in *The Musical Times*, 1 Aug. 1920.
[141] *Elgar as I Knew Him*, pp. 66–7.
[142] *Edward Elgar: the record of a friendship*, p. 201.
[143] *The Voice*, May 1920, p. 2.
[144] Quoted in *The Manchester Guardian*, 18 Aug. 1920.

communities had been set up in the Cotswolds on William Morris principles. One or two of these were still in being: so the hope seemed not yet
forlorn.

Severn House saw younger musicians in March. John Ireland brought
several of his scores for Edward's advice. And Adrian Boult was preparing the
Second Symphony to conduct at Queen's Hall. Alice wrote:

March 16 1920 . . . Wonderful performance of the Symphony. From the beginning it
seemed to absolutely penetrate the audience's mind & heart. After 1st movement great
applause & *shouts*, rarely heard till end, & great applause all through. Adrian was
wonderful—At end frantic enthusiasm & they dragged out E. who looked very
overcome, hand in *hand* with Adrian at least 3 times—E. was so happy & pleased.

So at last the Second Symphony began to find the understanding which had
been accorded instantly to the First—almost a dozen years ago in the
antediluvian world.

But Alice was spending much time in bed. She had been there when Carice
arrived home from Switzerland on 13 March. Carice had a precious secret
which she had no intention of divulging just then: at Mürren she had met an
Englishman, several years older than herself, who was interested in her. But
when she entered her mother's bedroom, Alice looked straight at her: 'Well,
Carice, and what have you tell me?' Remembering afterwards this meeting
with her mother, the daughter said, 'She could read me like a book.'[145]

Alice had managed to attend the Boult concert, and she went next day to
Harley Street to Sir Maurice 'who gave her new meddies [medicines] &
relieved her mind of some anxiety'. She was with Edward for a concert he
conducted at Woking on the 20th. Then he had to go to Leeds to conduct
Gerontius and *The Apostles* for Embleton. The journey was too great for Alice,
but one afternoon during Edward's absence she and Carice went to a
performance of the chamber music organized by Sammons and Reed.

When Edward returned on 25 March he found her 'very unwell in bed'. He
himself now kept the diary:

March 26 . . . Alice very ill—retaining nothing. Dr. Rose (Sir M's locum tenens)
came.

March 27 . . . A. better—less pain during day. E. to Savile [Club] lunch (only went to
soothe A.) Home to tea. A. not so well.

March 28 . . . A. *very unwell*. Dr. Rose came. A. awake during most of the night. alas!

A spell of warm spring weather brought no relief. The nights grew worse. On
5 April Edward noted 'Two sleeping draughts (ineffectual).'

April 6. My darling—in great distress—cd. not understand her words—very, very
painful. Dr. Rose called early—nurse arrived midday—Dr. Lakin (specialist) with Dr.
Rose—Bad report.

[145] Recalled by Mrs Philip Leicester in conversation with the writer, 1979.

It was cancer, less common then and less readily recognized than it might have been in later years—and it was in the last stage.

April 7. My darling sinking. Father Valentine [from the Catholic Church in Hampstead] gave extreme unction. Sir Maurice called at 12.30. Sinking all day & died in my arms at 6.10 pm.

The Wanderer

The date of Alice's death, 7 April 1920, was inscribed by Edward beside four lines copied from Swinburne:

> Let us go hence, my songs; she will not hear.
> Let us go hence together without fear;
> Keep silence now, for singing-time is over,
> And over all old things and all things dear.[1]

Before their marriage, his music had found no way to create a large design. The marriage had inaugurated his artistic achievement because Alice had given the constant assurance he lacked in himself. Swinburne's lines made the epigraph to a left-over life.

The most sensitive of his friends were the first to measure the shock. Ernest Newman wrote to the young second wife he had recently married:

I am sad today over the death of Lady Elgar. I am very fond of Edward, and I know that, whatever people may say, to a man of his fine and sensitive nature, the severance of a long tie like this must inevitably mean much bitterness and suffering, much dwelling in the past and self-reproach. We always seem heavy debtors to the dead: we feel they have not had their chance and that life has given us an unfair advantage over them.[2]

One friend tried to reap that advantage. Rosa Burley wrote:

So stunned was he by the blow, so withdrawn into himself, that no one at Severn House dared to approach him even when the undertaker had to be interviewed. I was in the house at the time and, realizing that something had to be done, I went into the study and told him as gently as I could that he really must pull himself together.[3]

But the message Miss Burley told herself she wanted to deliver was garbled, whether in speaking or hearing. Carice heard Miss Burley say on that occasion that she herself now stood as Carice's mother and would take over the running of the Elgar household for the future.[4] Whatever she said, at this moment it was

---

[1] Elgar Birthplace.
[2] Quoted in Vera Newman, *Ernest Newman: a memoir by his wife* (Putnam, 1963), p. 18.
[3] *Edward Elgar: the record of a friendship*, p. 202.
[4] Carice's impression is testified to by three surviving friends—Wulstan Atkins, Sybil Russell, Percy Young. She never discussed the matter with me: possibly the difference of nearly half a century in our ages precluded it. But through sixteen years of friendship, the supreme quality Carice always showed me was an unswerving loyalty to every friend's memory. Only once our

sufficient to end a friendship almost as old as Edward's marriage. He never saw her again.

The funeral was on 10 April at Little Malvern. Alice had many years earlier expressed the wish to be buried in the cemetery of St Wulstan's Church, on the side of the Malvern Hills between Craeg Lea and her childhood home at Redmarley. W. H. Reed remembered:

Frank Schuster and Carice begged me to bring my colleagues to play the slow movement from the String Quartet. Lady Elgar loved it; and they thought it would comfort Sir Edward a little.

I hurriedly arranged this: and Sammons, Tertis, Salmond and I went to Malvern and played in the little gallery at the west end of the Church. It was very sad to see Sir Edward there with bowed head, leaning on Carice's arm.[5]

From rooms he had taken with Carice overlooking the churchyard, Edward wrote to Frank Schuster:

This is the quietest place ever known & we are the only people here—we can see the little grave in the distance & nothing cd be sweeter & lovelier—only birds singing & all *remote* peace brought closely to us.

I cannot thank you for the quartet—it was exactly right & just what she wd. have loved . . .

The place she chose long years ago is too sweet—the blossoms are white all round it & the illimitable plain, with all the hills & churches in the distance which were hers from childhood, looks just the same—inscrutable & unchanging. If it had to be—it could not be better.[6]

When he asked Troyte Griffith to design the gravestone, he suggested: 'Could the motto "Fortiter et fide" go in: it suited dear A. so well.'[7]

From Malvern Wells he went to his sister Pollie. There his constant companion, as before his marriage, was a family dog. But there he was reminded that part of Alice's money now reverted to her family; and the bitterness foreseen by Ernest Newman began to emerge.

The Elms, Stoke Prior, Bromsgrove.
Friday [17 April 1920]

My dearest Frank:

Thank you for your letter. I am doing my best here—the weather of course prevents my having any walks but Juno seems to know that something is wrong & never leaves me—such wonderful things are dogs. Here my dear A. never came so I can bear the sight of the roads & fields . . .

You write to me of my music etc like the good fellow you are—but I am plunged in the midst of ancient hate & prejudice—poor dear A's settlements & her *awful aunts* who wd. allow nothing to descend to any offspring of *mine*—I had forgotten all the petty

discussion turned on Miss Burley (of whose manuscript I knew nothing then; the book was not published in Carice's lifetime): she talked of her old teacher in a level and friendly way.

[5] *Elgar as I Knew Him*, p. 67.

[6] Monday (12 Apr. 1920) (HWRO 705:445:6977).

[7] 17 June 1920 (*Letters*, p. 265).

bitterness but I feel just now rather evil that a noble (& almost brilliant) woman like my Carice should be penalised by a wretched lot of old incompetents simply because I was—well—I.

Don't talk to me of achievements. I drank spider juice for my mother's sake, went thro' penurious times to buy my dear wife a car for her old age—I failed of course—& now I am going to fail in settling the third woman for whom I ever entertained real love—so you see I must revert to my old plan as soon as convenient. But Juno's nose presses against me & says a walk is nec[essar]y

Love
E. E.[8]

He put off the return to Severn House as long as he could. Finally on the last day of April he went back to face the valuations for probate and a mountain of condolences. Near the top of the pile was one from Walford Davies. Edward replied:

Severn House, Hampstead, N.W.

May 1 1920

My very dear H. W. D.

I am just back to a cold & empty house—alas!—& find your sympathetic note—I can only say thank you for it.

All I have done was owing to her and I am at present a sad & broken man—just stunned.

My daughter does everything & is wonderful about the hideous ghoulish business which civilization makes necessary. Death we know & expect & try to bear like men, but I cry out 'Leave me with my dead'—& creatures have to come in & count over her pretty valueless little rings & the most private things to see what they are worth—it drives me to distraction, & I am no fit company for human beings

Bless you and thank you.

yr old friend,
E. E.[9]

On 6 May he invited the Bohemian Quartet to rehearse in the music room for their forthcoming performance of the Quartet, as he wrote to Ernest Newman, 'to break the awfulness of the change . . . but—without the hostess, My God.'[10] That day he wrote to Ivor Atkins, who had asked for some new music for the reviving Three Choirs Festival:

It will not be possible for me to write anything new—you cannot fathom the loneliness & desolation of my life I fear . . . I am sorry, sorry to be so unhelpful. *We* had been looking forward to the dear old festival & suddenly the whole thing is hurled away from me.

. . . To avoid smashing the whole affair I am going for Embleton to Newcastle on Saturday.

The booking to conduct *The Apostles* with Embleton's Choral Union at Newcastle had been made months earlier.

May 7. C. & E. to Newcastle 10 o'c—arrd. 3.58. Stayed at Station Hotel. To Nest

⁸ HWRO 705:445:6986.　　　⁹ HWRO 705:445:2212.　　　¹⁰ 24 May 1920.

House [Henry Embleton's home] with Embleton—a lone old house on the river bank in the midst of old & worked out mines.

May 8.   At Newcastle. Apostles rehearsal in Cathedral at 3.0, performance at 6.30, over at nine—rested. (Our wedding day 1889)

May 9.   C. & E. left Newcastle at 2.47. Home at 10. Long & hot journey. The empty house.

At the end of the month he fulfilled another long-standing engagement at Cardiff to conduct two London Symphony Orchestra concerts built around the Second Symphony. The local press reported:

Yesterday we had the spectacle of Sir Edward Elgar, the greatest figure in British music, and a composer of world-fame, conducting one of his great symphonies performed by a famous orchestra to an audience of a handful of persons lost in an array of empty chairs.

    . . . The first performance of a series of new songs by Sir Edward Elgar to be sung by Norman Allin was announced in the programme, but they were withdrawn . . .[11]

Visiting Troyte Griffith in Malvern on the way home, he had a new attack of giddiness. Carice found a doctor, and they moved him to rooms in a private hotel. From there she sent bulletins to Lady Stuart—the first on her father's sixty-third birthday, 2 June: 'I do not yet know about returning or any future plans yet—it's all very trying & difficult. It is so distressing this giddiness again—& depresses him so—& everything seems useless & hopeless.'[12]

Home at last a week later, Lalla Vandervelde was arranging concerts for him to conduct in Brussels and Prague. Passports and visas were secured, and then Edward decided he could not face it. On 21 June Henry Embleton called at Severn House; he spent several hours trying to persuade Edward to write Part III of The Apostles. A few days later Embleton's cheque arrived for £500—a 'loan' against completing the work. Edward did not return the money.

He secured another summer's lease of Brinkwells, and returned there in the last week of June with Carice. She wrote to Lady Stuart: 'Of course it was dreadful just the first day—but I think the air, & the quantity of lovely milk, butter, eggs & fruit all are doing Father good—& he seems better . . . He is quite busy with his wood & little repairs . . .'[13] Mark was there, ever ready with the retort simple. When Edward apologized for the number of circulars filling a wastepaper basket that had to be emptied, Mark replied solemnly, 'They must send 'em somewhere.'[14]

Edward wrote to the Windflower:

All goes on the same; Mark and the routine of the farm. The weather has been terribly depressing but we have not really felt it—I am 'numb'. I could have borne the many memories but they have cut down the woods so much & made a road which alters the look of the place but of course dear A. made it & it is full of remembrances—too too sad

[11] The South Wales News, 28 May 1920.
[12] HWRO 705:445:7905.
[13] 25 June 1920 (HWRO 705:445:7906).
[14] Diary, 7 July 1920. Also reported in Reed, Elgar as I Knew Him, p. 62.

for words. One odd thing I must tell you: in the shelter which Mark built for her & which I do not allow anyone to use, touch or look at, a Wren has built a nest just where A's head used to touch the roof-twigs; Muriel [Foster] always called her the Little Wren. . . . Music I loathe—I did get out some paper—but it's all dead.[15]

When another harvest was gathered in, he wrote to the Windflower: 'the fields are as bare as my mind & soul.'[16]

For the revived Worcester Festival in September, he lodged alone in his Uncle Henry's old rooms at 9, College Precincts, at the east end of the Cathedral between College Yard and the Edgar Tower. There was the old *Sursum Corda* at the Opening Service, *The Music Makers*, 'For the Fallen', and *Gerontius* (still with ecclesiastical deletions in the text). After each performance he went back to his rooms and saw nobody but Billy Reed, who wrote:

When we went for walks along the banks of the Severn he became very reminiscent, telling me all about his early married life, and when Carice was born, and everything that had happened year by year up to the time of Lady Elgar's death. In the evenings, so that he should not feel dull, I always went to his room, when I came out of the cathedral, to play cribbage . . .

After the Festival he was very quiet, and seemed quite disinclined to write any music. I tried once or twice to lead him on by asking him about the third part of the Trilogy . . . He had a cupboard full of sketch-work for it; and once or twice I succeeded in getting him to play me some of it, in the hope that I should set him on fire once more and get him to complete the trilogy. But I could not stay with him all the time; and, the moment I went away, it all went back into the cupboard and nothing was done.[17]

On 21 September he recopied some fragments from the old unfinished Trio of 1886 into a newer sketchbook, but it went no farther. A week later he marked Frank Schuster's score of the *Variations* with crosses beside the initials whose subjects had died. The first cross was set beside 'C. A. E.', the last beside 'E. D. U.'[18]

In October 1920 he conducted his own works in Brussels and Amsterdam. The scores and parts were flown over by aeroplane—'probably the first time', observed *The Musical Times*, 'Mercury has thus come to the aid of Apollo.' But when he went to Birmingham to conduct an opening concert with the new City Orchestra, his mind slipped back to the old orchestral days under Stockley, as he confided to Robert Buckley: 'At the rehearsal this afternoon, there was only one vacant seat in the orchestra. It was the one in which I used to sit. I almost expected to see myself come on with the fiddle!'[19]

At the end of November the young violinist Jascha Heifetz was to play the Concerto at a Philharmonic Concert conducted by Albert Coates. Edward invited Heifetz to tea at Severn House, reminisced over all his own early

---

[15] 18 July 1920 (HWRO 705:445:7686).   [16] 15 Aug. 1920 (HWRO 705:445:4216).
[17] *Elgar as I Knew Him*, pp. 68–9.   [18] In possession of Sir Adrian Boult.
[19] *The Birmingham Post*, 11 Nov. 1920.

fiddle-playing, and ended by writing out at Heifetz's request an 'Exercise for the 3rd finger' he had devised long ago in the Pollitzer days.[20] Three nights later he attended Heifetz's performance of the Concerto, and was summoned to the platform to share the applause; but then there were virtuoso encores. He wrote to the Windflower:

Yes, it was a tremendous display—not exactly our own Concerto: as to the noise afterwards—none of it was for me or my music—the people simply wanted Heifetz to play some of his small things with piano—the latter instrument being dextrously provided by his agent. I shd not have 'gone up' but I was called—and went, much against my will.

I have not forgotten the days at H'ford when 'it' was made—but there is nothing in it all somehow & I am sad.[21]

For Christmas he and Carice went to Pollie at Stoke in a car he had recently bought, driven by a hired chauffeur: 'I do not drive the machine *m[eae] p[ersonae]*', he wrote to Ivor Atkins; 'I ride the whirlwind and direct the storm.'[22] But how little joy the post-war world could hold was forced on his attention when they stopped for luncheon in Stratford at the Shakespeare Hotel, where he had shared remembered meals with the Windflower. He wrote to her:

Alas, alas! we can never go again: it seems unbelievable—the dear old Hotel is in new hands & much enlarged;—*smart* (!) waitresses with *very short skirts*, the dining room *very* large (The next house has been taken in & a covered 'awning' put up to the Town Hall)—(a piano in it) partly used for *dancing*,—they were jangling rag-time at *two o'c:*, a three-weeks' carnival (sort of thing) dances every night for three weeks & a *jazz band* specially imported . . . Every room booked for this venture—*Birmingham* etc fast people. Thus goes another (I think the last) of my peaceful, poetic old haunts. I know well that this sort of thing has always been but *not* at S-on-A.[23]

*   *   *

Severn House was still for sale. The atmosphere there was increasingly morbid for Carice, who at the age of thirty was trying to combine the roles of hostess-without-status and factotum. But she had not forgotten Samuel Blake, the Englishman she had met in Switzerland. She turned to her own friends for guidance, consulting lengthily and repeatedly. Winifred Davidson was not convinced that Blake, a farmer in Surrey, was in any way the right match for Carice. But Kitty Petre (lately remarried as Lady Rasch) 'pushed it as hard as I could, feeling it might be her only chance.'[24] When Blake himself hesitated, Kitty chivvied him. At last on 23 March, during one of Edward's absences at Stoke, the farmer proposed, and Carice accepted him subject to her father's

[20] Reproduced in *The Daily Telegraph*, 12 Dec. 1920.
[21] 27 Nov. 1920 (HWRO 705:445:7858).   [22] 21 Dec. 1920.
[23] 23 Dec. 1920 (HWRO 705:445:7706).   [24] Conversation with the writer, 1976.

consent. She waited four days after Edward's return before telling him, but he assented immediately and generously.

Perhaps this reminder of other lives and problems a year after Alice's death sent him back a little toward music. In April 1921 he was playing Bach fugues—not only from *The Well-Tempered Clavier* but the big organ fugues he had studied forty years earlier at St. George's Church: 'Now that my poor wife has gone I can't be original, and so I depend on people like John Sebastian for a source of inspiration.'[25] He decided to orchestrate the C minor organ fugue, as he wrote to Ivor Atkins, ' . . . in *modern* way—largest orchestra . . . So many arrgts have been made of Bach on the "pretty" scale & I wanted to shew how gorgeous & great & brilliant he would have made himself sound if he had our means.' Yet the sound of the organ was in his mind as well. He wrote to Ernest Newman: 'You will see that I have kept it quite solid (diapasony) at first;—later you hear the sesquialteras & other trimming stops reverberating & the resultant vibrating shimmering sort of organ sound—I *think*.'[27] By 25 April the arrangement was finished. In May he copied it out and sent it to Novellos, who reluctantly agreed to pay Edward a hundred guineas for the copyright.

At the same time, he suggested that the publishers might care to buy the copyrights of many of his smaller works which had ceased to earn significant royalties. The company secretary Henry Clayton replied with an account covering the last five years of sales—which altogether totalled barely £250:

Of course the valuation to-day of these Royalties is a pure speculation—or a gamble—as owing to the varying fashions, tastes, & conditions, no one can say what their future is to be. But three of us have carefully gone into the matter, & we have formulated ideas as to what we can & ought to pay for the surrender of the Royalties in question, & without further explanations that sum is £500 for the lot!!

Please consider the matter, & if you would like any items extracted from either list, let me know what they are.

I must admit that I got a bit of a shock when I saw the figures relating to 'The Variations' Pf.Solo £3.9.2 & Pf. duet £2.8.0 or £5.17.2 for both—in Five years!![28]

Severn House, Hampstead, N.W.

June 12.1921

My dear Clayton,

Thank you for the 'five year' table of the royalties on the small things; it is, of course, sorry reading, but I have never permitted myself any illusions and have none now.

. . . I wish the firm could see its way to buy me out entirely; I never really belonged to the musical world,—I detest my slightest necessary connection with it & should be glad to have done with it and get back to my (deceased) dogs & horses!

Do not trouble to write: I will come down one day this week, it will save much trouble.

Yours sincerely,
Edward Elgar[29]

Thus a long list of works became the publishers' property. They included most

---

[25] Quoted in Eugene Goossens, *Overture and Beginners* (Methuen, 1951).
[26] 5 June 1921.   [27] 26 Oct. 1921.   [28] Novellos archives.   [29] Ibid.

of the part-songs and anthems Novellos had published, many arrangements of major works for piano solo and duet, and even such orchestral pieces as the Bassoon Romance and the Coronation March of 1911. On 29 June Edward returned the signed indenture: 'I hope we have both made good bargains,— sentimentally I am glad to be "out of it".'[30]

As Carice was to be married after Christmas, they decided to sell Severn House at auction in the autumn. He started sorting through accumulations of the past. Carice wrote in her diary:

July 2 1921 . . . Father went through all his sketches, M.S.S. etc. sad work. Destroyed much & got all in order. At it all day . . .

Between conducting engagements and country visits, he spent time with microscopes, wrote long letters to *The Times Literary Supplement* on Shakespeare and other subjects, visited endless theatres and cinemas. On 20 July he opened a big new His Master's Voice shop in Oxford Street. Next day he and Carice went to Brinkwells: it was the final summer there, as the owners wanted to return to it themselves.

After the Hereford Festival in September, the auction notices were fixed to Severn House. Edward had decided to move into a service flat, and Carice found a good one for him on the ground floor at No. 37, St. James's Place, close to the many clubs of which he was now a member. There were two rooms, tiny kitchen, and bathroom. Much of the Severn House furniture was sent to auction. Carice supervised the moving of her father's remaining things to the flat, saw him into it, and returned to the immense task of clearing Severn House for auction on 8 November. But when the auction came it brought no acceptable bid, and the house was bought in at £6500.

The Bach Fugue arrangement had its première under the young conductor and composer Eugene Goossens and was encored. In December it fell to Edward to invite Goossens and Arthur Bliss to write works for the Gloucester Festival in 1922. John Ireland came to go over another new score of his own. But it was a lonely Christmas that year in the flat, as Carice was staying with Alice's relations in South London to prepare for her marriage. The small ceremony took place at St. James's Church, Spanish Place, on 16 January 1922. A few friends attended.

A week later Edward greeted Richard Strauss on his first post-war visit to London with a luncheon to meet the younger British composers. Ireland, Goossens, Bliss, Rutland Boughton, Adrian Boult, Arnold Bax, and Norman O'Neill were among the guests, with Bernard Shaw as virtuoso commentator. Edward discussed the orchestrating of Bach's organ works with Strauss, who favoured a more restrained approach than Edward's C minor Fugue. Edward challenged him to add his own orchestration of the Fugue's preceding Fantasia, but Strauss dodged the invitation. In the end Edward orchestrated the Fantasia himself.

[30] Ibid.

Otherwise his only musical activity was conducting his own works. Henry Embleton sponsored one performance after another of *The Apostles* in the hope of tempting the completion. But for Edward these occasions seemed only to raise the wind of neglect. After the first at Leeds on 29 March 1922, the host's attention seemed elsewhere. The second, at the Queen's Hall in June, drew a half-empty house. Bernard Shaw wrote to *The Daily News*:

The Apostles is one of the glories of British music . . . It places Britain once more definitely in the first European rank, after two centuries of leather and prunella.

It would be an exaggeration to say that I was the only person present, like Ludwig of Bavaria at Wagner's premieres. My wife was there. Other couples were visible at intervals. One of the couples consisted of the Princess Mary and Viscount Lascelles, who just saved the situation as far as the credit of the Crown is concerned, as it very deeply is. I distinctly saw six people in the stalls, probably with complimentary tickets.

He apologized

. . . for London society, and for all the other recreants of England's culture, who will, I fear, not have the grace to apologise for themselves . . . And, finally, I apologise to posterity for living in a country where the capacity and tastes of schoolboys and sporting costermongers are the measure of metropolitan culture.—Disgustedly yours,

G. Bernard Shaw.[31]

After a third Embleton *Apostles* at Canterbury, Edward received a single letter of appreciation—from the Windflower. He replied:

I have *seen* no one, *no one* has written or taken the *slightest* notice & I have read nothing & seen no papers: truly I *am* a lonely person if I liked to think so;—but my *'friends'*!!! where, oh, where! are they? Silence profound. I was talking to Claude [Phillips] last week & he said 'Don't be a Timon.' I am not—but do these horrible frauds expect me to continuously 'attend' them when they happen to want me?[32]

At the Gloucester Festival in September 1922, the invitations to Bliss and Goossens resulted in works which Edward did not like. He himself conducted the first performance of the completed Bach Fantasia and Fugue and the first serial performance of *The Apostles* and *The Kingdom* in many years. After *The Kingdom* they unveiled a memorial tablet in the Cathedral to Parry. This brought a large contingent of composers including Bantock, Edward German, and Stanford. The Cathedral organist Herbert Brewer made Edward and Stanford shake hands, but at that late moment in their lives it could mean little.

Conducting *The Apostles* and *The Kingdom* side by side again raised ideas of Part III. Again through the autumn he looked over sketches. But the contrast of what already existed was too much for new and lonely effort. Remembering the still earlier experience of *King Olaf*, he wrote to Ivor Atkins:

It always sweeps me off my feet: it seems strange that the strong (it is *that*) characteristic stuff shd. have been conceived & written (by a poor wretch teaching all day) with a

[31] 9 June 1922.
[32] 14 June 1922 (HWRO 705:445:7843).

splitting headache after dinner at odd, sustained moments—but the spirit & will was there in spite of the malevolence of the Creator of all things . . . But thro' it all shines the radiant mind & soul of my dearest departed one: she travelled to London (I was grinding at the High School) & became bound for one hundred pounds so that my work might be printed—bless her! You, who like some of my work, must thank *her* for all of it, not me. *I* shd. have destroyed it all & joined Job's wife in the congenial task of cursing God.[33]

* * *

In January 1923 came a definite and defined request. Laurence Binyon's play *Arthur* was to be produced at the Old Vic in March. Would Edward write the music? The story of King Arthur, like the Saga of King Olaf, showed heroism that was doomed. Nothing could have spoken more directly to Edward in 1923. He wrote to Binyon:

I *want* to do it but since my dear wife's death I have *done nothing* & fear my music has vanished. I am going to my daughter's tomorrow & shall be quiet & things arranged for me as of old: my wife loved your things & it may be that I can furnish (quite inadequate) music for 'Arthur'—Can you give me three days more to 'try'?[34]

The task was to provide introductions and entr'actes to link the nine scenes of the play, and general motives under speech and action. These were to be scored for a theatre orchestra so tiny that a piano must supply 'continuo'.

He went on 2 February to stay with Carice and her husband at their farm near Guildford, to see whether the countryside in winter might even now give back some hint. 'I am . . . struggling hard,' he wrote on 7 February. Through the next fortnight the struggle continued. Thematic ideas came from old notes in his sketchbooks, to be developed with the same careful economy that would govern their orchestration.

The result was half an hour's music. It opened with a figure moving slowly up the scale through modal harmonies to E minor. Then followed two basic motives. The first was a new echo of dying 'Delight'. It stood for the shadowy barge which must bear the final burden of heroism through misty waters:

It made a basis for rich subtle variation and development. The other leading motive showed King Arthur's Fellowship:

[33] 30 Dec. 1922.   [34] 31 Jan. 1923.

But the hearty rhythm moved downward in sequences. It was entwined with the 'fateful' music of the barge through the first two Scenes. Scene III brought a longing variant of the Fate music for the doomed Elaine.

The biggest set piece was the introduction to a scene of the Knights' Banquet. New motives were exposed. First came a festive but short-breathed

Allegro maestoso

whose strains would be heard later in hollow circumstances. A *Maestoso e grazioso* introduced Guinevere with all the springtime freshness of Edward's early inspiration:

Yet this feminine evocation distantly echoed the descent of fate and death from the music's beginning. The banquet itself developed a nervous shape from the String Quartet:

The Guinevere music introduced a scene in the Queen's tower at night. The following battle was fought to the short percussive rhythms of Falstaff's scarecrow army. Then the last scene of all was fulfilled in a Gregorian figure which reduced all the foregoing themes to utter simplicity:

When the slow chant repeated, a single bell tolled the tonic note, as years earlier it had tolled at the death of King Olaf. The 'Fellowship' figure appeared in ghostly echo. Edward wrote to Binyon: 'I shd have liked Arthur & *all* his train to march mistily past . . .'[35] The bell tolled once more, and the violins rose slowly over the modal harmonies which had opened all this music, to a final E minor.

By the end of February 1923 the score was complete. The band parts were copied at Edward's expense, as his contribution to a production which was reported to have had a cash outlay of only £15 including wood for a huge round table and the hire of armour. Nine performances were scheduled. He rehearsed the 'orchestra' in preparation for conducting the first night. 'Fancy

<hr />

[35] 18 Mar. 1923.

the greatest composer in England conducting ten musicians in the Waterloo Road!' exclaimed *The Sunday Times*.[36]

Sir Edward Elgar's patience all through the long and trying rehearsals of 'Arthur', the poetic drama by Laurence Binyon, was monumental. He composed the music to please his friend the author, and he sat about for hours at the Old Vic, last Friday, smoking his pipe, just as Barrie does during rehearsals. Dégas missed a wonderful picture as England's greatest composer, his face half-lit by the light in his pipe, waited at the conductor's desk while the stage hands struggled with the scanty properties, and Lilian Baylis sat in the box admiring him.

Elgar passed a minute of the time by paying a compliment to Jane Bacon, who will play Elaine. 'Had I seen you in that dress before,' he said, looking at the white robe she wears in her death scene, 'I should have written more beautiful music. I think I'll take it away and re-write it.'[37]

Arthur opened on 12 March. Ferruccio Bonavia wrote in *The Musical Times*:

In the tender phrases which characterise the unfortunate Elaine, as in the musing phrases which prepare us for the clash of arms, the Elgarian idiom is evident even though there is not the faintest likeness between this and any other music of his. [Here is] yet another proof of the manifold quality of Elgar's genius, which can adapt itself to the most varied situations without ever losing its typical accent . . .[38]

*   *   *

In the midst of writing the *Arthur* music, he had arranged the six months' lease of a black-and-white house in the Worcestershire countryside at Kempsey, just south of Worcester. Napleton Grange stood right away from the tiny village, but close to Kempsey Common. After the final performance of *Arthur* on 31 March, he moved down there (still keeping on the flat in St. James's Place). His household at Napleton consisted of one or other of his Bromsgrove nieces as 'matron', together with cook, parlourmaid, and gardener. He wrote to the Windflower on 19 April:

The common, five minutes away, is lovely & I walk about a great deal—lonely & thinking things out: no music yet. There is a fair Bechstein grand in the hall & we have huge peat fires etc. We are thinking if it will be possible to have visitors or whether they wd. find it too dull—I fear they would. The plum blossom is nearly over & the apple coming on—a cuckoo appeared this morning. I hope you are well—this seems far away—far away.[39]

His first composition at Napleton was literary. He wrote a letter to *The Telegraph*, published anonymously on 28 April with the title 'The Vernal Anemones: a beautiful native', and sent the cutting to his own Windflower.

Music followed in small essays, much of it answering requests. There was a piece to inaugurate a new War Memorial Carillon at Loughborough, and a slow

[36] 4 Mar. 1923.   [37] *The Sunday Times*, 11 Mar. 1923.
[38] 1 Apr. 1923.   [39] HWRO 705:445:7085.

keyboard fugue in the style of his earliest efforts in music. Then he orchestrated the Overture to Handel's Second Chandos Anthem. He wrote to Novello's music editor John E. West: 'I have known the overture from the old two stave organ arrangement since I was a little boy and always wanted it to be heard in a large form—the weighty structure is (to me) so grand—epic.'[40]

To please Ivor Atkins he added orchestral accompaniment to Wesley's anthem *Let Us Lift Up Our Heart* and (at Atkins's special request) Battishill's *O Lord, Look Down From Heaven*. Both were to be done at the Worcester Festival in September, when the Handel Overture orchestration would also have its première.

Then Robin Legge requested two male voice part-songs for the DeReszke Singers, a group of American pupils of Jean DeReszke. For the two part-songs Edward sought words to reflect his own isolation. He found the basis for one, he said, in an old Restoration anthology of *Wit and Drollery* (1661). He made an adaptation, and called it *The Wanderer*:

> I wander through the woodlands,
>   Peace to you,—day's a-dying;
>   I tune a song
>   The trees among,
> But oft-times comes a crying.
>
> I know more than Apollo;
>   For, oft when he lies sleeping,
>   I see the stars
>   At mortal wars,
> And the rounded welkin weeping.
>
> The morn's my constant mistress.
>   The lovely owl my morrow;
>   The flaming drake
>   And the night-crow make
> Me music, to my sorrow.
>
> With a heart of furious fancies,
>   Whereof I am commander,
>   With a burning spear
>   And a horse of air
> To the wilderness I wander.
>
> With a knight of ghosts and shadows
>   I summoned am to tourney:
>   Ten leagues beyond
>   The wide world's end;
> Methinks it is no journey.

A wraith of melody moving softly up and down between empty octave and

[40] 16 July 1923 (MS in possession of Raymond Monk).

skeletal harmony and a recurring drone set the five stanzas in a tiny world of variants.

The other part-song was a march with words by Edward himself.[41] He used the pseudonym 'Richard Mardon'—a surname that might suggest an ambiguous view of his own 'gift'. The marching beat was set to a vigorous soft repeated monosyllable 'Zut':

> Zut! zut! zut! zut! &c
> Come! shall we forget our old-time march-song?
> The lads sang it so,
> Long long ago . . .

It made a good square rhythm, softly sung, with each longer line lifted through a second-beat triplet. A festive *strepitoso* rose step by step toward a climax:

> Glory to them and a fame transcending
> The heroes of old in time unending.
> Hurrah!

But as the triumph touched its resolving upper octave, it fell suddenly away to *pianissimo*:

> Come! give it a lift, their old-time march-song,
> No! never forget their old-time march-tune;
> Long, long ago,
> The lads sang it so!
> Zut! zut! zut! zut! &c

till the sound was lost in nothingness. The song's subtitle was 'Remember'.

On 17 July he sent *The Wanderer* to Novellos with a note to Henry Clayton:

You wanted something primarily for competition purposes: I do not suppose you will think the piece sent with this will do,—the words are strange & weird. In any case I should like it printed at once & if the firm does not want it, it may remain my property.[42]

Clayton submitted it to the music editor West, who was unenthusiastic. So Clayton wrote a polite letter offering 25 guineas. Edward replied with the completed *Zut! zut! zut!*:

I do not think your offer is quite good enough. With this I send an[othe]r TTBB which shd. be very popular: could you give me one hundred guineas for the two? Failing that wd. you publish them for me as 'author's property' on the usual terms?[43]

West's report on the second part-song was worse:

I am sorry to say this is rather *cheap* for Elgar—*cheap* without being sufficiently *interesting*. Is it my judgment that's at fault, or is the composer *falling off* in the value of his ideas?[44]

---

[41] Extensive drafts of the poem in Elgar's hand are preserved at the Elgar Birthplace.
[42] Novello archives.   [43] 24 July 1923 (Novello archives).
[44] Novello archives.

160, Wardour Street, W.1

July 31st. 1923.

My dear Elgar,

Your two part-songs for men's voices have been very carefully considered and I am sorry to say that we cannot take them over from you on your terms, which are more than we can give for partsongs for men's voices—all 'male voice' things necessarily have a rather limited sale, which is the deciding factor in this case.

We will therefore put the two partsongs in hand and will publish them for you as your own property, as suggested in your letter of the 24th inst.

Yours faithfully,

[Henry R. Clayton][45]

Kempsey, Worcester.

Augt 11:1923

My dear Clayton:

I am too old to begin altering the method of publication I have been accustomed to &, after due thought, will not ask you to print the PtSongs as 'author's property'. Just tear up the M.S.S.—or return them to me & I can do so.

I hope you are having or about to have a good change & rest—it is lovely in the country just now

Yours sincly

Edward Elgar[46]

160, Wardour Street, W.1.

August 14th 1923

My dear Elgar,

Many thanks for your letter.

Neither I nor anyone else here would wantonly destroy an Elgar MS., so if that is to be the fate of your two partsongs for men's voices, you must apply the finishing touch yourself. I am returning them to you; not for that purpose, but in the hope that you will reconsider your judgment, and that you will at all events grant a reprieve.

To my mind it seems absurd to destroy two properties which are worth 50 Guineas to us; and, unless you do it as a protest, you had better consider whether there is any good reason for throwing away that sum.

We will gladly publish both of the Partsongs at our own expense, in the usual way, and will pay you 50 Guineas for the two—so why sacrifice everything? Your view can only be explained by a feeling of annoyance that we are either underrating your property, or are not willing to pay a proper price for it. We cannot plead guilty to either proposition.

The whole difficulty about these two compositions is, that we never do, and never can, sell partsongs for men's voices in large quantities. Send us something for mixed voices, or for women's voices, and the whole situation will be changed at once.

Yours sincerely

Henry R. Clayton[47]

In the end Edward acquiesced in the publishing of his songs at just half the value he placed on them.

---

[45] Ibid.   [46] Ibid.   [47] Ibid.

The Worcester Festival in September 1923 brought performances of the Handel Overture and the anthem orchestrations. Edward conducted the Cello Concerto (with Beatrice Harrison), *Gerontius*, *The Kingdom*, and 'For the Fallen'. The contrast of his music with the work of younger composers at the Festival seemed greater than ever, and in a letter to Alice Stuart of Wortley he added a carefully casual postscript: 'The Kingdom, Gerontius & For the Fallen are not bad: I think I deserve *my peerage* now, when these are compared with the new works!!'[48] Thus the man still showed his need for reassuring honours in the world the artist despised. But it was mostly at Three Choirs Festivals and in a few northern cities that his music was now in demand. In London the First Symphony emerged in one of Henry Wood's programmes almost as a curiosity—its first London performance in more than three years.

The six months' lease at Napleton Grange was at an end. He had felt isolated there, and returned to London and the flat. But the isolation of London was worse, and after less than a fortnight he wrote to the Windflower: 'I have been to *12* theatres since I retd: I am so desperately lonely & turn in to see anything'.[49] He booked a six-week cruise in November and December to South America and a thousand miles up the Amazon. W. H. Reed remembered:

He came back very full of his experiences; but the Amazon impressed him less than the fact that in South America, in places with quite a small population, the opera house was the handsomest and most important building in the town . . . It was after this trip that he began to talk to me of opera and the possibility of his composing one.[50]

He was asked to write vocal pieces and a March for a 'Pageant of Empire' to open the huge British Empire Exhibition at Wembley on St. George's day 1924. He wrote to Alice Stuart of Wortley on 10 January: 'I have "*composed*"! *five* things this week—one about "Shakespeare" you will love when I shew it you: slight & silly.'[51] The idea of the March stirred him more. He took a month to finish it—only to be told it would not be done at Wembley owing to the difficulty of all the contingent bands rehearsing a new piece separately. He was asked to conduct the old *Imperial March* instead, together with *Land of Hope and Glory*, Parry's *Jerusalem*, and the National Anthem.

On 30 March Edward appeared at Wembley in grey morning coat and bowler to rehearse the colossal choir—drawn from the Chapel Royal, St. George's Chapel at Windsor, Westminster Abbey, St. Paul's Cathedral, Southwark Cathedral, the London Church Choirs' Association, and many other choirs from London Churches. *The Daily Telegraph* reported:

The choir cheered the veteran conductor as he mounted the steps. The National Anthem was the first thing Sir Edward desired to test. He wanted the singers to shorten and sharpen their words, so that each word would be carried distinct and ennobling to the microphone, and thence by cable to the amplifiers, whence it is conveyed by another

---

[48] 12 Sept. 1923 (Elgar Birthplace).   [49] 10 Oct. 1923 (HWRO 705:445:7838).

[50] *Elgar as I Knew Him*, p. 88.   [51] HWRO 705:445:7842.

cable to the great sound projectors massed on a high platform and transmitted into the Stadium like an invisible sea of sound.[52]

He wrote to the Windflower:

I was standing alone (criticising) in the middle of the enormous stadium in the sun: all the ridiculous court programme, soldiers, awnings etc: 17,000 men hammering, loud speakers, amplifiers—four aeroplanes circling over etc etc—all mechanical & horrible—no soul & no romance & no imagination. Here had been played the great football match—even the turf, which is good, was not there as turf but for football—but at my feet I saw a group of real *daisies*. Something wet rolled down my cheek—& I am not ashamed of it: I had recovered my equanimity when the *aides* came to learn my views—Damn everything except the daisy—I was back in something sane, wholesome & *gentlemanly* but only for two minutes.[53]

The Opening Ceremony on 23 April was attended by the press in force:

Thrown across the greensward directly before the King were two scarlet lines of infantry of the guard of honour—the contribution of the Army. Beyond them, at the right, was the guard of honour of the Air Force, and at the left the bluejackets and marines. Between them were drawn up the massed bands of the Brigade of Guards, with the pipers of the Scots Guards and Irish Guards.

Far in the distance was a tall crimson pulpit set close against the opposite end of the arena where Sir Edward Elgar directed the massed choirs. They towered above him in the form of a gigantic white 'T', with the band of the Royal Military School of Music a mere splash of scarlet in the centre.

. . . Despite the dull sky and more than a hint of rain, the audience flowed in steadily for an hour before the beginning of the musical programme at 10.30. There were volleys of wild cheering as the massed bands marched and counter-marched to such stirring airs as 'On to Victory' and 'Voice of the Guns' and 'The Contemptibles'.

. . . The Lord's Prayer was recited reverently by the multitude. Sir Edward Elgar, a lonely figure in black poised in his lofty pulpit, raised his baton. The massed choirs above him sang 'Jerusalem'.

The King stepped forward to a low pedestal in front of the throne, upon which rested a gold model of the Globe in an open casket. He placed his finger upon a button. The trumpeters answered him from the arena; a field gun spoke outside. Instantly the flags of the Empire were unfurled from their tall staffs around the top of the Stadium, and from the buildings through the park. The exhibition was open![54]

The question of honours in the official world came up again with the death of Sir Walter Parratt on 27 March. For many years Parratt had held the Court office of 'Master of the King's Music'. Now the office was vacant, as Edward told Alice Stuart of Wortley:

I wrote to Stamfordham urging that the Master of the King's Music shd. be retained—its suppression wd. have a very bad effect abroad—where the effacement of the last shred of connection of the Court with the Art wd. not be understood. It was not S's dept so it

---

[52] 31 Mar. 1924.   [53] 16 Apr. 1924 (Elgar Birthplace).
[54] Sir Percival Phillips in *The Daily Mail*, 24 Apr. 1924.

was turned over to F. Ponsonby. He wrote to me that it was one of the offices which it was (long ago) proposed (scheduled) to cease. I wrote again offering myself (honorary)—*anything* rather than that it shd be publicly announced that the old office was abolished. No reply. Colebrooke wrote to the Ld. Chamberlain but as far as I can make out the three (?) depts. simply quarrel over these things: no grit, no imagination—no *music*: no nothing except boxing, football & racing . . . But everything seems so hopelessly & irredeemably *vulgar* at Court. I was at the processional rehearsal all the morning—quite simple but it takes time. If you like to write to Vi[s]c[oun]t Fitzalan do—but I fear the matter of the 'Master' is dead. As to any peerage I fear it is hopeless but it wd. please me.[55]

The question of the peerage remained unanswered, but on 28 April a letter came offering Edward the Mastership of the King's Music.

The announcement was greeted with universal approbation. But it focused attention again on the fact that he had completed no major works for several years. *The Daily Sketch* asked:

What is Elgar doing?—When is Sir Edward Elgar going to write another concerto, or another symphony—or any other important work? . . . He goes to revues and to Bernard Shaw plays. But he doesn't write music. He is our greatest composer and we want him to.

Every encouragment seemed to bring its own defeat. Henry Embleton arranged to take the Leeds Choral Union with the London Symphony Orchestra to France for performances of *Gerontius* in Paris and Dieppe. Edward conducted, the Lord Mayor of Leeds travelled with the Choral Union, there were official speeches and bilingual banquets. But when Edward discovered that Novellos had sent no vocal scores or texts for sale at the performance, he felt outraged. He wrote a bitter note observing that the publishers were apparently 'not disposed' to help his music. It drew a stinging reply from Augustus Littleton, now head of the firm:

160, Wardour Street W.1.

July 21.24

Dear Sir Edward Elgar

Your letter of the 18th has been laid before me. I am quite unable to understand the insinuation contained therein that Novellos were not disposed to help Mr. Embleton's Concert in Paris. No suggestion that Mr. Embleton was in need of assistance has ever reached me and I can only conclude that you must be in possession of facts which have been withheld from me or that you have discussed the matter with Mr. Embleton and that he has made some complaint to you. It is quite impossible for me to allow your letter to pass without further and most careful enquiry: so I hope you will send me all the explanation you possibly can.

Yours faithfully
Augustus Littleton[56]

[55] 16 Apr. 1924.   [56] Novello archives.

37, St. James's Place. S.W.1.

July 23.1924

Dear Mr. Augustus Littleton:

There was no question of assisting Mr. Embleton but Mr. Embleton's Concert. I did not see a single copy of the French edn of Gerontius or the libretto—these were published years ago and I thought the publishers shd see that the things are on sale—that is all.

Yours truly

Edward Elgar[57]

Littleton defended the firm by saying that they had tried to work through French publishers:

One of these publishers acknowledged our letter, the other two were silent. I consider that every possible effort was made to see that the things were on sale, but in the face of the evident wane of interest on the part of the French public in British music our efforts did not have the desired result.[58]

It was the final straw that broke the back of a publishing association which had dominated Edward's career for more than thirty years. Novellos had been involved with most of his great successes, and in these later days the memories must have embarrassed both sides. Their correspondence trickled on over the odd small work or arrangement. But even Henry Clayton now cloaked himself in a 'Dear Sir' formality to head letters that finished with the anonymous 'Novello & Co.' signature.

*   *   *

When professional life disappointed, the landscapes of home invited him again. In July 1924 he was in Worcester to speak at a Guildhall ceremony to give the Freedom of the City to the elderly Town Clerk:

Some of us have had to leave the old city, and, for reasons which it is difficult to explain, eke out a miserable existence in other parts of the Empire. I can only speak for myself, and my thoughts are always with the old place, its old institutions, its poetry and romance . . . We people who career wildly about (and don't like it) look with envy at the quiet and unostentatious way in which things are done here . . .[59]

He arranged another lease of Napleton Grange from October 1924. Again the countryside of home seemed full of promise. He decided on a long stay, ordered new Napleton stationery, and wrote to the Windflower: 'I am going here tomorrow & want to write oh! such a lot.'[60]

The intention of a long stay emerged also in the acquisition of dogs: 'Marco is the loveliest spaniel I have ever seen—quite a silly baby & cries for nothing; he loves riding in the car. Mina—the little cairn—is a love & so sharp.'[61] Another

[57] Ibid.   [58] 26 July 1924 (Novello archives).
[59] *Berrow's Worcester Journal*, 14 July 1924.   [60] 19 Oct. 1924 (HWRO 705:445:7844).
[61] 17 Dec. 1924 to Alice Stuart of Wortley (HWRO 705:445:7064).

diversion was the motor car. Having bought a car for his sister Pollie's family, he bought a new one to drive himself. But after a slight accident he gave up driving and employed as his chauffeur and valet Richard Mountford, a Kempsey man. Dick Mountford's wife Fanny came to Napleton Grange as cook, her sister Nellie as parlour maid. His secretary was Mary Clifford, an attractive and sophisticated niece of one of the old Severn House servants.

At Christmas 1924 he received a holiday wish from Frank Schuster, who wrote:

For my part I find the older one gets the less one wants—and looking forward into the year *1925*, the only thing I can think of which I eagerly desire is the *conclusion* of your trilogy—that would be the most precious gift I can conceive!—but am I at all likely to experience it?[62]

The same question haunted Billy Reed's mind during his visits to Napleton:

I went down there continually with my violin . . . I tried very hard in these days to induce him to work at Part III of the Trilogy; but he did not show any enthusiasm, and always said, 'Oh, no one wants any more of that nowadays'; but he would nevertheless sit down at the piano and play portions of it; and the old light would come into his eyes as he worked himself up and began grunting away to himself, his hands meanwhile flying about the piano. He never could sustain the mood, however, [and] could not face the drudgery of putting it on paper. The mainspring was broken somehow.

I say somehow, because even now it is not certain whether he was disabled by the beginnings of his physical breakdown or by loss of faith in any real necessity for any more oratorio. He never talked about his religion; but he was obviously more sceptical generally as a widower than he had been during Lady Elgar's lifetime . . .[63]

The summer of 1925 at Napleton produced two further part-songs. Both celebrated the same subject. Alexander Smith's poem 'The Herald' showed Death attending an old king at the end of successful battle. Edward's setting for men's voices pursued almost tuneless, almost keyless harmonies through sharply contrasted tempi and dynamics. The other, Walter de la Mare's 'The Prince of Sleep', breathed a dusky atmosphere that Edward could meet on his own doorstep any summer evening at Napleton:

> Dark in his pools clear visions lurk,
> And rosy, as with morning buds,
> Along his dales of broom and birk
> Dreams haunt his solitary woods.

In October his eldest sister Lucy died in a little house he had quietly bought for her and her husband, Charley Pipe. It was the first family break in their generation since the passing of little Jo close to sixty years before. His brother Frank was now more or less a confirmed invalid. And just before Christmas 1925 Edward himself had to undergo an operation for haemorrhoids. Young

[62] 23 Dec. 1924 (HWRO 705:445:6894).
[63] *Elgar as I Knew Him*, pp. 72, 74–5.

Dr Moore Ede (the son of the Dean of Worcester) thought seriously enough of the matter to schedule the surgery for Christmas Eve.

Afterwards Edward insisted on seeing a consulting physician in Birmingham, Dr Arthur Thomson. Dr Thomson found his new patient

. . . a neurotic, who most of all wanted reassurance. Again and again he would come in depressed, as if all useful life was over; and after reassurance, he would brighten up perfectly. One day I looked right to the back of his eyes with the opthalmoscope; there was nothing much wrong with them . . .

At our first meeting he delighted me by beginning: 'You know, Dr Thomson, I don't want any doctor's nonsense. I know all about doctors, because I started life in a lunatic asylum.' Then he recalled that the superintendant at Powick in the '70's had been considered a crank because of his notion that music might soothe patients: it was only obvious that concerts could amuse them and keep them out of mischief.

At the third or fourth meeting, I asked him how he spent his day.

'I get up at 7.30 or 8, have breakfast, and then address the serious business of the day.'

I asked him whether that was composing.

'No: making up my betting slips.'

Another time I asked him what set his music off.

'It is infinitely various. It may be the cry of a child: then I *see* a bar of music, and can write it down. It may be pain. Or it may be the sound of wind and waves, or the murmuring of a stream.'

He told me there was only one thing he really loved in life. That was the Golden Valley of the River Teme, especially a place 500 to 800 yards below the Knightsford Bridge down the right bank, near the Ankerdine Hills. He used to sit on the banks, and said he composed much of *Gerontius* there.[64]

The Worcester Festival programme for 1926 announced another serial performance of *The Apostles* and *The Kingdom*. Again the suggestion was before him. Visiting Napleton on 23 March Carice found 'Father full of writing Apostles III'. But the Worcester programme also contained an extract from Wagner's *Parsifal*, and this elicited from one of the Cathedral canons a public protest that Wagner was a 'sensualist'. Edward replied with heat:

The Canon quotes 'His emotions and spiritual experiences were those of the ordinary sensual man.' But 'Aren't we all?' If the Canon really believes that such emotions in early life debar a man from taking part in the services of the church in riper years he should at once resign his canonry and any other spiritual offices he is paid to hold.[65]

Canon Lacey replied:

Sir Edward Elgar misses the point. The writers whom I quoted were not criticising Wagner's life or character, but his art, in which they found sensuality of pietism matching the sensuality of his erotics. It was this that attracted my attention, for in my work as a priest I have had acquaintance with both kinds of sensuality, and I know what kind is the more dangerous.[66]

[64] Conversation with the writer, 1976.
[65] *Worcester Daily Times*, 17 Mar. 1926.
[66] Ibid., 18 Mar. 1926.

It made an unlooked for discouragement to the increasingly fragile pursuit of his own 'festival drama'. A thematic idea occurring to him in August was put down not for 'The Last Judgement' but once again for 'Callicles': yet neither the distant voice of Arnold's youthful singer nor the older figure of Empedocles approaching the pit tempted that music farther. When Ivor Atkins proposed the music for Binyon's *Arthur* in his draft programme for the 1926 Worcester Festival, Edward deleted it: 'I . . . must return your pretty proofing notes with dismal marks and remarks. The day (week, month, year, era, age, eon) is past for ever . . .'[67]

A full two years of country life in Worcestershire had done no more than the London flat and London Clubs to revive major composing. As autumn moved to the winter approaching his seventieth birthday year, the memory of past achievements brought the question of a peerage again to the front of his mind: perhaps such a signal recognition would fire him where nothing else had served. He had hoped for support from Lord Stuart of Wortley, but Lord Stuart had died in the spring. Edward wrote to Lady Stuart on 26 November 1926: 'I should like to know if "it" is entirely dead—my birthday (70) next June is to be "recognised" by concerts in the musical world & I do not want this or these & shall *squash* them unless the other thing turns up.'[68]

The tributes began with the opening of the birthday year 1927, and he did not squash them. In January he conducted the Hallé Orchestra in a performance of the Violin Concerto which the seventy-five-year-old Adolf Brodsky emerged from retirement to play. In February he conducted the Royal Choral Society in a performance of *Gerontius* which filled the Albert Hall. The Gramophone Company ran land lines to their recording rooms, and half the performance was preserved on discs by a new electric process which made it possible to capture virtually any sound from virtually any distance: Edward himself hailed it as 'the greatest discovery made up to this time in the history of the gramophone'. In April he conducted new records of the Second Symphony for publication on his birthday. Ernest Newman greeted it in a pamphlet entitled 'Elgar and the Gramophone':

While Elgar's name is known and honoured wherever music is cultivated to-day, it cannot yet be said that his later music is as well known as we who love it and believe in it could wish it to be . . . The greater Elgar has been too long in entering into his own. I hope, and believe, that these 'His Master's Voice' records of his splendid second symphony will do more to bring him into his kingdom than the all too few performances of the work in the concert room have yet done.

The Company gave him a life contract 'on such lines as may be agreeable to Sir Edward', with an annual retainer of £500. On the seventieth birthday night, 2 June 1927, Edward conducted a huge concert of his works with the BBC Chorus and Orchestra for national broadcast. At the conclusion he stepped to the microphone (as *The Daily Mirror* reported): 'When in the proper B.B.C.

[67] 8 June 1926.   [68] HWRO 705:445:7760.

manner he said, "Good *night*, everybody. *Good* night, Marco," he knew there would be one listener at his home at Kempsey, Worcester, who would be pleased to hear his voice.'

Late in the birthday month Frank Schuster organized his own tribute—a performance of the chamber music with Sammons, Reed, Tertis, Salmond, and William Murdoch to take place at The Hut. The beloved home had now passed from his occupancy to that of a New Zealand friend who had named it The Long White Cloud: but Frank had borrowed it back for the festive day. Edward, looking across the fissure of years that separated the present from a past in which such things were understood, tried hard to avoid the occasion. But not to hurt Frankie, he agreed at last to come.

Many old friends and a few younger gathered on the rainy Sunday afternoon. A printed programme bore the legend 'Homage to Elgar'. Among the younger contingent of Schuster's friends were Siegfried Sassoon, William Walton, Constant Lambert, Sacheverell and Osbert Sitwell. Osbert later turned his recollection to a celebrated set piece for his autobiography—a survey of the gulf between the post-war young and their elders:

I seem to recall that we saw from the edge of the river, on a smooth green lawn opposite, above an embankment, and through an hallucinatory mist born of the rain that had now ceased, the plump wraith of Sir Edward Elgar, who with his grey moustache, grey hair, grey top hat and frock-coat looked every inch a personification of Colonel Bogey, walking with Frank Schuster.

. . . The music-room was so crowded that, with Arnold Bennett, we sat just outside the doors in the open air . . . From where I sat I could watch Elgar, enthroned at the side, near the front. And I noticed, too, several figures well known in the world of English music, but in the main the audience was drawn from the famous composer's passionately devout but to me anonymous partisans here gathered for the last time. It is true that these surviving early adherents of Elgar's genius seemed to be endowed with an unusual longevity, but even allowing for this, it was plain, looking round, that in the ordinary course of nature their lives must be drawing to an end. One could almost hear, through the music, the whirr of the wings of the Angel of Death: he hovered very surely in the air that day, among the floccose herds of good-time Edwardian ghosts, with trousers thus beautifully pressed and suits of the best material, carrying panama hats or glossy bowlers, or decked and loaded with fur and feather . . . Most of them knew, I apprehend, as they listened so intently to the prosperous music of the Master, and looked forward to tea and hot buttered scones (for it was rather cold, as well as being damp), and to all kinds of little sandwiches and cakes, that this would prove their last outing of this sort. The glossy motors waited outside to carry them home . . . Some of the motors were large and glossy as a hearse.[69]

The gramophone was present again at the Hereford Festival in September 1927, when a mobile recording van was stationed at the west end of the

---

[69] *Laughter in the Next Room* (Macmillan, 1949), pp. 196–7.

Cathedral. Records were taken from Edward's performances of *Gerontius* and *The Music Makers*, and of a tiny *Civic Fanfare* he had devised at the request of the Cathedral organist Percy Hull to precede the National Anthem. It was almost his only instrumental composition in his years at Napleton—aside from a pastiche of former successes for a film entitled *Land of Hope and Glory*.

The lease at Napleton Grange could no longer be renewed. In October 1927 he took a six-month lease of Battenhall Manor, a big black-and-white house within the city of Worcester, whose owners were wintering in Egypt. He wrote to the Windflower from Battenhall on 14 December. 'I wish I could have given you a Christmas here—it is a real Yule-loggy house & you wd have met, in spirit, Oliver Cromwell, Charles I & II & a lot of agreeable restoration ghosts.'[70]

Two days after Christmas a new ghost was added with the sudden death of Frank Schuster. One clause of his will read:

To my friend, Sir Edward Elgar O.M. who has saved my country from the reproach of having produced no composer worthy to rank with the Great Masters, the sum of £7,000.[71]

The New Year's Honours for 1928 contained the result of the peerage effort: it was only another order of knighthood—Knight Commander of the Royal Victorian Order. Bernard Shaw sent his own breezy view of the matter: 'Why don't they make us duty-free instead of giving us O.M.s and the like long after we have conferred them on ourselves?'[72]

In the spring of 1928 the owners of Battenhall Manor returned, and Edward had to make another move. When he saw a house for rent on the river near Stratford-upon-Avon, he took it. It was called Tiddington House. Carice came up from Sussex to move him in, as her mother had supervised their moves in the old days—while Edward escaped to spend the days of transition with Pollie's family (now living at Bromsgrove, as the old house at Stoke had been sold over their heads).

The river at Tiddington opened a new range of diversions. Billy Reed recalled:

The fishing-rods were brought out and set up all ready for the correct fishing mood to come upon him: he also bought a very smart rowing-boat . . .

Then there was the food. Someone had sent him a case of fizzy Spanish wine, like champagne. It was a great success. We had to have a bottle up after bonfiring or boating; then we would get into the car and go over to Leamington, where 'the best sausages in England' were on sale in a certain shop he had discovered and thenceforth visited in person once a week.

On returning from one of these expeditions we had to pass some pleasure gardens where a band was performing. We slowed down to catch a strain of what they were

---

[70] Elgar Birthplace.
[71] Quoted in Adela Schuster's letter of 10 Jan. 1928 to Elgar (HWRO 705:445:6655).
[72] 4 Apr. 1928 (HWRO 705:445:2225).

playing. It was the *Meistersinger* overture; so we stopped the car and went in to listen. Fancy our feelings when the overture suddenly changed into the Jewel Song from Gounod's *Faust*! This too became bedevilled, and melted into Grieg's Anitra's dance from *Peer Gynt*, which in its turn melted into 'variety', presently solidified into the *Tannhäuser* march, and wound up with *Chant sans paroles* by Tschaikowsky.

Elgar was furious, spluttering, as we fled back to the car, 'Don't they want to hear any piece played through properly; or can no one nowadays listen to more than a few bars of anything without getting bored?' He was really angry about it, and said it would not matter so much if they made their potpourris from the jazz tunes and the lighter music, but to drag the classics into such company and make them ridiculous was to corrupt the taste of the young and degrade the world's musical heritage.

Unfortunately this sort of thing accentuated his disinclination to compose . . .[73]

In July he had a weekend visit from the Gramophone Company's artist manager Trevor Osmond Williams. Witty, elegant, every inch the sophisticated man-about-town of forty, Osmond Williams was the image of everything Edward might have dreamt of in a worldly life. And Edward's company gave the younger man undisguised pleasure: in London they had shared evenings at the Russian ballet, and one night Williams dared to take him to Jerome Kern's triumphant American musical *Show Boat*.

Osmond Williams had been trying to get Kreisler to record the Violin Concerto. Kreisler dodged and delayed: though he was still giving an occasional performance of the work in London, it was said by some of the Gramophone men that the great violinist was no longer young and was afraid of the microphone's immortality in this most daunting of the big concertos. But now it seemed that Kreisler would make the records if Edward conducted the sessions in Berlin. He and Osmond Williams celebrated with some 1848 brandy before the younger man dashed back to London in his Wolseley 6.

The American copyright of *Gerontius* fell due for renewal that summer, and Novellos saw to it for him. But then he wrote to wonder whether it would ever be worth anyone's while to print new copies there: 'I presume the work is worthless, or as nearly valueless to me in the U.S.A. as it is in England.'[74] When old Henry Embleton wrote again about completing the *Apostles* trilogy, plaintively suggesting the forthcoming retirement of the Leeds Choral Union's great conductor Henry Coward as a suitable occasion, Edward gently told him it was unlikely to emerge so soon.

He was writing a little music at Tiddington. After the Gloucester Festival of 1928 he finished a small part-song. Anticipating the Christmas season, it was a setting of Ben Jonson's verses beginning 'I sing the birth was born tonight'. The Jonson words were shaped in long unaccompanied single-voice lines, interspersed with slow modal Alleluias for the remaining three voices. The setting was slight but immaculate, with not a note too many.

Then the actor-manager Gerald Lawrence exerted his formidable charm to

[73] *Elgar as I Knew Him*, pp. 81–2.   [74] 14 Dec. 1928 (Novello archives).

persuade Edward to write the music for a new play he was about to produce about Beau Brummel. The play was described in *The Birmingham Post*:

Mr. Bertram P. Matthews, who has written this romance for Mr. Lawrence, has idealised the character of Beau Brummel, and depicted him as a gentleman who is ready to sacrifice his life, his career, his friendship with the Prince Regent, to save a woman's honour.[75]

It was an idealization like Edward's *Falstaff*, and it touched him. He reached back into old sketchbooks. There he found a grave and stately Minuet subject, which he made the unifying theme of a score that was to be his largest work since *Arthur* nearly six years earlier. Yet even with his own presence in the Theatre Royal, Birmingham to conduct the première on 5 November 1928, the play attracted little notice. Both play and music sought to recall an *ancien régime* more remote from the brave new world of the 1920s than King Arthur himself.

A rare performance of *Falstaff* brought a letter of appreciation from Arthur Bliss. Edward's answer revealed another part of his outlook on the age into which he had survived:

My dear Bliss,

You letter gave me great pleasure and satisfaction and I am obliged to you for writing it.

I do not refer to the concert now but to what you say about our friendship: this I valued and shall value again if you will allow me to do so.

Frankly, I was greatly disappointed with the way you progressed from years ago. There was so much 'press' of a type I dislike and newspaper nonsense. I can easily believe you were responsible for little or none of this but it rankled a great deal because I had great hopes for you: I had affection. It will seem vulgar to you if I add that commercially you have (I believe or was led to believe) no concern with the success of your works—an unfortunate side of art which we penniless people have always with us, and try to ignore. I hoped you were going to give us something very great in quite modern music, the progress of which is very dear to me; and then you seemed to become a mere 'paragraphist'. I am probably wrong and trust I was.

Now I have written at greater length than you will like but my reason (not excuse!) must be that you are one of the very few artists in whom I took an interest, to use the word again, affectionate interest.

Believe me, my dear Bliss,

Yours most sincerely
Edward Elgar[76]

The result was that Bliss dedicated a new choral work, *Pastorale*, to Edward—who heard it subsequently in a wireless broadcast of imperfect reception: 'But I could judge that your work is on a *large* and *fine* scale, and I like it *exceedingly*. The Pan sections suited me best . . .'[77]

[75] 6 Nov. 1928.
[76] 8 Nov. 1928 (quoted in Bliss, *As I Remember*, p. 94).
[77] 9 May 1929 (Ibid., p. 95).

In response to persistent request from Ivor Atkins for a new work, he proposed making his own choral settings of two Shelley poems. 'The Daemon of the World' traces an immortal journey like that of Gerontius, but with Demon instead of Angel guide—toward a place from which the distant world can be seen in the throes of an utterly secular Last Judgment. 'Adonais', that sustained lament for pure genius abused by the callous world, turns at last upon the singer himself:

> The breath whose might I have invoked in song
> Descends on me; my spirit's bark is driven,
> Far from the shore, far from the trembling throng
> Whose sails were never to the tempest given;
> The massy earth and spherèd skies are riven!
> I am borne darkly, fearfully, afar . . .

Did Edward seriously contemplate making Worcester Cathedral resound to such words in 1929? He asked Atkins to approach the old Dean Moore Ede (the father of the doctor), who duly produced the expected refusal: 'Can Sir Edward not find some poem suitable? He found the right thing in Gerontius . . .'[78] In view of Worcester's long record of changing Newman's text for their performances, the Dean's question might be read with an irony to amuse Angel or Devil. Edward replied to Atkins: 'Here is the Dean's letter . . . I fear I cannot turn on another subject so easily as it seems to the D.'[79]

The position of his achieved music in the later 1920s was more paradoxical than ever. On the one hand he was coming to be viewed as a figure of classical importance and remoteness from modernity. When he conducted *The Kingdom* in London in March 1929, Colles wrote in *The Times* of

. . . Elgar, whose music, in spite of the meticulous markings in his scores, is written in his head, and only there. Such things as the pauses and accents, directions for *rubato*, . . . and such indeterminate suggestions of mood as his favourite *Nobilmente*, acquire their authoritative interpretation only from him. He knows where to throw the emphasis in each phrase, so as to give it eloquence . . . His mind, especially in the oratorios, moves in a region for which notation offers no precise record. Here is a chance for the Gramophone Society. Posterity should be told exactly how Sir Edward Elgar meant his music to sound.[80]

But on the other hand, the concert-goers of 1929 could be breath-takingly callous. When Edward conducted *Gerontius* with the Hallé, Neville Cardus wrote in *The Manchester Guardian*:

A certain section of the audience on Saturday committed the barbarity of walking out of the hall and passing along the base of the platform even while Sir Edward was conducting the ineffably lovely strains that herald the close of 'Gerontius'. The time of evening was then not later than half-past nine. Behaviour of this sort is an insult to

---

[78] 3 Dec. 1928 to Ivor Atkins.
[79] 6 Dec. 1928.   [80] 4 Mar. 1929.

genius, and in a land which properly valued the things of the spirit would be regarded as a punishable offence.[81]

When Edward gave the presidential address to a Musicians' Club dinner later in the year, he observed:

The character of English people musically is extremely bad; they do not care for music in the least. They follow a man for a time and then drop him.[82]

It was not only in England. A German paper of March 1929 carried a reminiscent article by the ageing Fritz Volbach, recalling Elgar's first fame in Germany at the turn of the century. But since the war, said Volbach, he was almost forgotten—for political reasons perhaps, but also because his music was of no interest to the young atonalists who had the ear of fashionable criticism.[83]

Through it all one friend, eminent before all the rest, was pugnaciously loyal. It was Bernard Shaw. They were opposite characters in many ways, as Edward had written once to Sidney Colvin: 'GBS's politics are, to me, appalling, but he is the kindest-hearted, gentlest man I have met outside the charmed circle which includes you—to young people he is kind.'[84] Each was stimulated by the contrast of the other, and by the kindness of the other for his art. As Edward aged, he valued the iconoclasm in Shaw. As Shaw aged, he valued the mythic quality in Edward's art and presence. At the beginning of 1929 Shaw sent news of his new play *The Apple Cart* together with a challenge:

Lazy! I've not only begun a new play but finished it. Not since The Messiah has a work hurled itself on paper more precipitously.

But after St. Joan it will outrage the world as a hideous anti-climax. It is a scandalous Aristophanic burlesque of democratic politics, with a brief but shocking sex interlude.

It is to figure at the Bayreuth Festival which Barry Jackson is projecting at Malvern for the next August.

Your turn now. Clap it with a symphony.

G.B.S.[85]

In August he wrote again from Lawnside, Malvern to ask Edward to open the Festival exhibition:

Barry Jackson has taken this house, which you know well, for entertaining, and has reserved the front row in the theatre for distinguished guests, which really means you. In his first enthusiasm he was bent on getting from you an overture for The Apple Cart; but on obtaining from Boult a rough estimate of the cost of an Elgar orchestra, and letting his imagination play on the composer's fee, he went mournfully to his accountants, who informed him decisively that he could not afford a band at all and would be lucky if he came out of the affair with a shirt on his back.

My own view was that six bars of yours would extinguish (or upset) the A.C. and turn the Shaw festival into an Elgar one; but that it would be a jolly good thing so. I

[81] 28 Jan. 1929.
[82] *The Daily Telegraph*, 13 July 1929.
[83] *Am Weg der Zeit*, 20 Mar. 1929.
[84] 13 Dec. 1921 (HWRO 705:445:3527).
[85] 2 Jan. 1929 (HWRO 705:684:BA5968).

demanded overtures to Caesar, to Methuselah (five preludes), and a symphonic poem to Heartbreak House, which is by far the most musical work of the lot.

Barry had to give up, crushed; and there won't be any music except chin music . . .[86]

At the Festival on 17 August Edward paid his tribute:

Bernard Shaw knows more about music than I do. (Laughter.) We won't grumble at that. He was a musical critic and a good one in those dull days when the two Universities and the Colleges of Music used to do nothing but sit around and accuse one another of the cardinal virtues.[87]

Shaw responded:

Sir Edward has been alluded to to-day as the greatest English composer. As a matter of fact, he is one of the greatest composers in the world.

I don't believe England is proud of it, and that is the disgusting part about it. Although I am rather a conceited man and I feel I could carry my head high compared with any other artist in England, I am quite sincerely and genuinely humble in the presence of Sir Edward Elgar. I recognise a greater art than my own, and a greater man than I can hope to be.[88]

At the Worcester Festival in September he produced nothing more than an orchestral accompaniment for Purcell's motet *Jehovah Quam Multi Sunt Hostes*. On the Thursday evening the Second Symphony shared a programme with Vaughan Williams's *Sancta Civitas*, written to some of the St. Augustine texts which Edward had long ago considered for 'The Last Judgment'. Vaughan Williams recalled: 'He came to hear a performance of my *Sancta Civitas*, and gave it generous praise. And he told me that he had once thought of setting the words himself. "But I shall never do so now." '[89]

*   *   *

For the Festival week he had taken a house in Worcester which he had known all his life. Both house and garden commanded an uninterrupted view to the Cathedral and the Malvern Hills from its position on Rainbow Hill (formerly known as Red House Hill). He had written of it years earlier to Ivor Atkins:

The Red House exists & is now called Marl bank—S. T. Dutton lived there; it is an old house (200 years) & now much built in . . . If the old Red House (which has oak floors etc) had been more free from surrounding new buildings I wd. have ended my days in it—it *is* for sale.[90]

That was in 1917. In 1929 Marl Bank was for sale again. The dogs seemed happy during their week there, and one morning Harold Brooke (a director of Novellos who had maintained closer friendship) came up with a moving-picture

---

[86] 12 Aug. 1929 (HWRO 705:445:2228, quoted in *Letters*, pp. 327–8).
[87] *Worcester Daily Times*, 19 Aug. 1929.   [88] Ibid.
[89] 'The Fifteenth Variation', BBC broadcast of 1957.   [90] 17 Apr. 1917.

camera to film their antics together with Edward's encouragement of them. Before the Festival was over, Edward made up his mind to buy the house. He could move in near the end of the year.

A happy reference touching the house on Rainbow Hill came in verses by the Tudor poet Gascoigne entitled 'Goode Morrowe'. Edward set Gascoigne's words to one of his old hymn-tunes to meet a request from Walford Davies for a carol to celebrate the recovery of King George V from serious illness that autumn:

> The Rainbowe bending in the skye,
> Bedeckte with sundrye hewes,
> Is like the seate of God on hye,
> And seemes to tell these newes:
> That as thereby he promisèd,
> To drowne the world no more,
> So by the bloud which Christ hath shead,
> He will our helth restore.

Edward conducted the performance at Windsor on 9 December, and it was broadcast.

Even as this newest old piece was being sung, officialdom was seeking to change its terms—in the shape of a new Music Copyright Bill, under which a composer could not charge more than 2d. for the perpetual right of public performance of a work. Edward wrote to *The Evening Standard*:

The most serious blow ever aimed at the unfortunate art of English music is proposed to be dealt by some of the very persons to whom creative artists might not unreasonably have looked for sympathy and assistance.

A situation such as that now existing could not have arisen in any artistically civilised community.[91]

Bernard Shaw joined in the protest, and the cartoonist David Low depicted the two of them playing and singing to a street harmonium with *Gerontius* on its desk and 'All my own composition' chalked across its back: from the pavement an over-fed 'Labour Concert Favourite' tosses them tuppence while strolling arm-in-arm with a kilted and Tam O'Shantered gillie carrying a case labelled 'Cheap Music for the Caledonian Choral Society' and asking, 'For why did ye no mak' it a penny?'

In the bad economic climate of late 1929 Edward gave up his flat in St. James's, vacated Tiddington, and took possession of the house which had caused him to rejoin the ranks of home-owners at the age of seventy-two. His Christmas card from Marl Bank that year bore a quotation from Walt Whitman:

I think I could turn and live with animals, they are so placid and self-contain'd;
They do not sweat and whine about their condition;

[91] 17 Dec. 1929.

They do not lie awake in the dark and weep for their sins;
They do not make me sick discussing their duty to God;
Not one is dissatisfied—not one demented with the mania of owning things;
Not one kneels to another, nor to his kind that lived thousands of years ago;
Not one is respectable or industrious over the whole earth.

Near the turn of the year another letter arrived from Henry Embleton. It had been written by an amanuensis, for Embleton was burdened with the infirmities of age. He was still hoping for the completion of the trilogy:

I should very much like to see you if it were possible during the coming Season—We shall only I think give three concerts but wish one of these could include your third work. I do not know whether you still keep on your flat at St. James Place but as I am sometimes in London would like to pop in and have a cup of 'wishy washy' tea—(as you call it) with you.[92]

It was too late. On 7 February 1930 Embleton died. Within a week his executors sent a polite demand for the return of the £500 loan made in 1921 against the completion of Part III. Edward protested that his friend had later made it a gift. But there was nothing in writing; and it transpired that Embleton had practically bankrupted himself in the cause of choral music—much of it Edward's. The executors needed the £500 to cover their obligations. There was nothing for it but to find the money.

Already home ownership was proving expensive. Winter gales and rain brought down a large section of garden wall nearest the road, and he was in despair. He asked Novellos to return all his manuscript full scores which they had kept since the departure from Severn House. When the manuscripts arrived, he let it be known to such people as Fred Gaisberg of the Gramophone Company that they were for sale. No bids were forthcoming to match his own evaluation of them.

In February 1930, while Carice was with him again, Granville Bantock motored over with his daughter Myrrha to see them. She recalled:

We sat together before a cheerful fire in their drawing-room, which was furnished in typical English country-house style, with large comfortable chairs and settees in soft floral colours. The two old friends chatted of the past, and while his daughter entertained me and showed me the garden Elgar took my father into his study, where he confided how lonely he felt without the wife and companion who had lived only to encourage, inspire and serve him.[93]

For years Edward had been in correspondence with an editor of brass band music, Herbert Whiteley. Whiteley's great ambition was to get a new Elgar piece for the Brass Band Competition Festival held every year at the Crystal Palace. Now he begged for a work to celebrate their twenty-fifth Festival in September 1930. He offered £100 plus half the gramophone royalties, or £150

[92] 27 Dec. 1929 (HWRO 7053:445:3158).
[93] *Granville Bantock: a personal portrait*, pp. 156–7.

exclusive, or asked Edward to name terms.[94] It was a noticeable offer in comparatively hard times. And though a work of twelve to fifteen minutes was wanted, the drudgery of scoring would be spared: for Henry Geehl, with a knowledge of brass-band writing, could be asked to devise the instrumentation from Edward's short score.

March 1930 found him recopying such old fragments as the Minuet from the G minor 'Mozart' Symphony of 1878 and various Wind quintet pieces. He made a Suite—Introduction, Toccata, Fugue (a slow piece written out at Napleton in 1923), Minuet, and Coda returning to the Introduction music. The work was to be called *Severn Suite*—another celebration of his return to the home country.

Through the first half of April 1930 he sent in the short score by instalments, and met several times over the scoring with Geehl, who recalled:

During the time I was arranging Elgar's *Severn Suite* I was in continuous consultation with the composer, who provided me with a very sketchy piano part with figured bass and a kind of skeleton orchestral score, mostly in two or three parts, with an indication of the sort of counterpoint he desired me to add; the rest of the score he left to my discretion.

Elgar was not an easy man to work with. He had many preconceived ideas on brass treatment—usually unworkable—which he tried very hard to get me to adopt, and it took a great deal of argument on my part to convince him that his ideas were just not possible. I remember particularly a 'bad' afternoon when I endeavoured to persuade him to omit the mutes in the Minuet, well knowing that the sound would be entirely different from what he imagined. But all to no purpose! So the somewhat banal sound of the muted trombones will be handed down to posterity! I did, however, get my own ideas adopted in several instances, but these were always conceded rather grudgingly.[95]

Geehl himself was not an easy personality. He had been chosen for the instrumentation because he was quick at his work. But the score he produced was hastily done and in many respects careless.[96]

When the *Severn Suite* was finished, however, Edward gave it an opus number, 87—his first use of an opus number since the Bach Fantasia and Fugue transcription of 1921–2. He dedicated the new work to Bernard Shaw, who responded:

My dear Elgar

Naturally I shall be enormously honored: it will secure my immortality when all my plays are dead and damned and forgotten. I am really not worthy of a symphony; but a Serenade, say:—A Serenade for Brass Band to the Author of Captain Brassbound's Conversion—that would be about my size. . . .

your hugely flattered and touched

G. B. S.[97]

[94] HWRO 705:445:2084–5.
[95] Quoted in Young, *Elgar, O.M.* (2nd edn), p. 259.
[96] In possession of R. Smith and Company.
[97] 25 May 1930 (HWRO 705:445:2230).

And the *Severn Suite* proved the beginning of inspiration. When Percy Hull
asked for a new work for the Hereford Festival in September, Edward
suggested that 'if chances fall luckily I might "revise" an early *orchl.*
"work"(?)'[98] But he knew that Percy cast longing eyes at the *Pomp and
Circumstance* Marches dedicated to his fellow organists Atkins and Sinclair. So
he worked up an idea which had come one day in June 1929 as he was driving
through the countryside with Dick Mountford. On that drive he had suddenly
asked for some paper, and all that Dick could produce was an Ordnance Survey
Map of Worcestershire. In the margin of this the idea was promptly pencilled:

Though it seemed new, the germ of it went back fifty years to a sketch from his
days of writing music for the Powick Asylum. Now it made a vigorous opening
for *Pomp and Circumstance* No. 5. He set it off with a Trio melody as fine as
anything in the earlier Marches:

He orchestrated it brilliantly—his first original work for large orchestra since
the Cello Concerto of 1919. In early May he sent it to Booseys to add to Op. 39.
    Leslie Boosey grudgingly agreed to pay £75 and a royalty for the new March.
But it was a burden, as he wrote to Edward:

You admitted on your own account that things are not what they were in the days when
you wrote 'Pomp' No 1 . . . Actually 'Pomp' No 5 will be costing us £25 more than any of
the others did, added to which is the immense increase in the cost of engraving and
printing and the very serious falling off in the sale of the sheet copy.[100]

There were fewer amateur musicians of all kinds now, as radio broadcasting
and the cinema drew people away from the trouble of practising music for
themselves.
    Pomp and Circumstance No. 5 was the last Elgar work to be published by
Booseys. In June 1930 Edward signed a contract with another publisher, Keith
Prowse. They would pay a retaining fee of £250 *per annum* for three years. In
return, 'The Composer will in each year during the period of this Agreement
compose and submit to the Publishers manuscripts of not less than three songs

[98] 1 Mar. 1930 (*Letters*, pp. 301–2).
[99] 27 June 1929 (Elgar Birthplace).
[100] 7 May 1930 (Boosey and Hawkes archives).

or pieces.' There would be royalties payable in addition. So a steady trickle of 'new' Elgar works was once more to appear.

But health was again causing concern. In the spring of 1929 he had had a severe bout of bronchitis, and after that his ear had given intermittent pain. In August 1930 a new trouble developed suddenly. It was diagnosed as severe lumbago or sciatica. It prevented his attendance at the London rehearsals for the Hereford Festival, and at the Hereford Opening Service Percy Hull had to conduct the *Introduction and Allegro* in his place. He was just able to manage *The Apostles* on the Tuesday. Billy Reed recalled: 'He had to be helped to the conductor's rostrum, where he *sat* to conduct through the performance and then was helped down again.'[101]

As the Festival went on the sciatica improved, and in the following week he was able to get to London for gramophone sessions which included the new *Pomp and Circumstance* No.5—rehearsed and recorded two days before Henry Wood conducted the first public performance in a sold-out Queen's Hall. Bonavia wrote in *The Daily Telegraph*: 'No recent announcement can have given greater pleasure to a vast number of people than the promise of a new Military March by the Master of the King's Music.'[102] But the Master of the King's Music was not at the Queen's Hall to receive the 'rapturous' applause, for the sciatica had returned.

It kept him from the Crystal Palace Brass Band Competition featuring the *Severn Suite* on 27 September, of which G. B. S. sent a full and diverting report. Edward 'just managed to crawl to London'[103] to conduct the Second Symphony with the new BBC Symphony Orchestra for a broadcast on 2 October. But five days later he had to cancel a series of gramophone sessions booked for recording the First Symphony. It was not until late November that he was able to carry out the recording.

At all these gramophone sessions he missed the presence of Osmond Williams, who had died suddenly in July. Edward wrote to the Gramophone Company chairman Alfred Clark at the beginning of a new year:

On looking over the year just closed, I find the greatest (and bitterest) sensation was the death of Trevor Osmond Williams; his going has left a blank in my life and a pain which nothing can soften. He had become my greatest friend. At my age old acquaintances fail and depart with appalling rapidity and young friends are not easy to find. Trevor was always ready to look after me—a dinner, lunch, theatre—anything pleasant and helpful. You knew his charm and most delightful company and can realise what a loss his death made.[104]

Through the bouts of sciatica and enforced absences, another project had been forming. Its suggestion had begun with the manager of His Master's Voice English Branch, W. L. Streeton, who recalled:

[101] *Elgar as I Knew Him*, p. 105.
[102] 22 Sept. 1930.
[103] 7 Oct. 1930 to Alice Stuart of Wortley (HWRO 705:445:7795).
[104] 13 Jan. 1931 (Quoted in *Elgar on Record*, pp. 124–5).

In casual conversation with Sir Edward on one of the studio premises one morning, he mentioned that he had recently come across a relic of his early youth—a trunk or box containing various articles, some books, and some of his youthful compositions. I asked him whether any of the latter were likely to be useful.

He said: 'I don't think so. After all, they were very early efforts.'

I replied: 'Sir Edward, I appreciate that the themes might not be of the type you would use in a large-scale work today. But is it not possible that they might be suitable for "boiling up" into a little suite or something of that kind?'

He said: 'That's a possible idea. Perhaps they might. I will think that over.'[105]

Streeton had taken the idea to the publishing manager of Keith Prowse, and had suggested that the recent birth of the Princess Margaret Rose to the Duke and Duchess of York might make a 'subject'. It was a deft proposal for the Master of the King's Music with a trunk of youthful themes. Edward assembled and shaped the themes into a *Nursery Suite*, for which he wrote a programme note.

This miniature Suite was suggested by the charming group—H. R. H. The Duchess of York and the two Princesses, Elizabeth and Margaret Rose.

I. AUBADE (*Awake*). The first movement . . . should call up memories of happy and peaceful awakings; the music flows in a serene way; a fragment of a hymn tune ('Hear Thy children, gentle Jesus'—written for little children when the composer was a youth) is introduced; the movement proceeds to develop the opening theme, the hymn is repeated more loudly and dies away to a peaceful close: the day has begun.

II. THE SERIOUS DOLL. A sedate semi-serious solo for flute . . .

III. BUSY-NESS! Here busy fingers fly and there is a suggestion of tireless energy . . .

IV. THE SAD DOLL. A suggestion of a pathetic tired little puppet.

V. THE WAGGON PASSES. This explains itself: a remote rumbling is heard in the distance increasing in volume as the waggon approaches; the waggoners' song or whistle accompanies the jar and crash of the heavy horses and wheels, dying away to a thread of sound as remote as the beginning.

VI. THE MERRY DOLL. A vivacious person is here represented . . .

VII. DREAMING. Intended to represent the soft and tender childish slumbers; this leads into:

VIII. ENVOY. The solo violin plays a cadenza which, following an old and august example, introduces fragments of the preceding numbers, viz: 'The serious doll', 'The merry doll', and 'Dreaming'. A reference, somewhat more extended, to the first movement, brings the Suite to a peaceful and happy end.[106]

The old and august example could come from the heritage of Violin Concerto cadenzas from Beethoven onward.

But the transition to the 'Envoy' gave trouble. When the sciatic pain came back sharply and persistently, he sent for Dr Moore Ede, who gave him an injection. Edward jumped violently: 'When you gave me that damned prod, it came to me!' Ordering paper and pencil from the desk, he made the connecting

---

[105] *Elgar on Record*, p. 117.
[106] Quoted in *Elgar on Record*, pp. 137-8.

join he wanted between the last two movements.[107] The doctor's needle had given the little shock his brain needed then.

By December 1930 the work was finished and scored. The experience of writing again for orchestra had brought fresher music than the *Severn Suite*. The *Nursery Suite* was in every way the equal of the *Wand of Youth* Suites. Only its perspectives were a little longer.

The new Suite made one of the three works to send to Keith Prowse for the year. He added a Scottish song, *It isnae me*, written for the young soprano Joan Elwes whose voice he had liked at Three Choirs Festivals. And he revised the old Sonatina written in 1889 for May Grafton's childhood piano lessons at Stoke.

So closed a year which had produced as much music (though most of it was revival) as the entire decade of the 1920s. It had all come out of Marl Bank, overlooking the childhood city, the Cathedral, and the river. In September 1930 Edward wrote a Foreword for a book of local history, *Forgotten Worcester*, by Hubert Leicester:

It is pleasant to date these lines from an eminence distantly overlooking the way to school; our walk was always to the brightly-lit west . . . Descending the steps, past the door behind which the figure of the mythical salmon is incised, we embarked; at our backs 'the unthrift sun shot vital gold', filling Payne's Meadows with glory and illuminating for two small boys a world to conquer and to love. In our old age, with our undimmed affection, the sun still seems to show us a golden 'beyond'.

Near the turn of the year Edward's old violin pupil Frank Webb brought his son and daughter-in-law to Marl Bank for an evening. The son, Alan, was intensely interested in Edward, and had seen him often from a distance. He wrote a record of this evening immediately afterwards.

A very ordinary housemaid (not even in black: and one had rather expected a butler for the Master of the King's Music) opens the door to us, and we are ushered into a very ordinary hall. It is long and narrow, with a good, shallow flight of stairs leading straight up out of it.

The maid goes to announce us, and my straining ears catch the sound of his voice—jovial, and so English. He is talking to, or about 'Marco', his favourite dog, somewhere in a room at the end of the passage. 'Marco' must have his bone . . . And then Miss Grafton, his niece, who is keeping house for him, comes down the hall to greet us, and he is following, with the familiar slow, hunch-shouldered stride . . .

The forehead, with the silky white hair growing far back over it, is veined and intensely sensitive in appearance. The nose, mouth, and chin are so heavy as to be almost coarse, and yet the general effect is one of outstanding distinction and refinement. It is the abrupt downward break in the curve of the nose which seems to give the whole face its character. The eyes are not big, and of no particular striking colour, and yet they are astonishingly expressive, and light up now with a merry twinkle, now with a beautiful happiness at the sound of well-loved music. And they look at you directly. He blinks slightly as he talks, and moves his head from side to side. Sometimes

---

[107] Conversation with Mrs Moore Ede, 1975.

he chuckles rather thickly behind his moustache when talking. Although not absent-minded in the accepted sense of the term, he gives the impression of being intensely occupied with his own thoughts and feelings, and his voice seems to come from some inner life of thought. Nevertheless, the general impression one gets is that he enjoys his fellow-men thoroughly, and feels that the world is full of wonderful things to think and do.[108]

He gave the young Webbs an evening of gramophone records, with comments and stories that would remain vivid for the rest of their lives.

The new year 1931 opened with an article on 'Elgar and the Public' by the young song composer C. W. Orr.

The position of Elgar in England has been something of a paradox till within quite recent years. He has been made a knight and remains a very great composer. He has been given the Order of Merit by the King, and the cold shoulder by the public. He is one of the most distinguished of living Englishmen, yet there are plenty of third-rate politicians whose names are more familiar to the man in the street. And he is, with the exception of Delius, the only British composer with a high reputation on the Continent; yet at home he is politely disregarded by the older and patronized by the younger generation of musicians. That he has had to wait till his seventieth year to receive anything like widespread appreciation in his own country is a curious sidelight on English musical history . . .

Orr recalled the effect of Richard Strauss's turn-of-the-century toast to the first English progressivist'

. . . in the academic dovecotes over here, where Elgar was by no means *persona grata*, originality of any sort being held suspect at our schools and colleges in those days . . .

And he concluded:

. . . it is becoming more and more evident that he, who has never worried about 'nationalism', is the most *national* of all our composers.[109]

Orr's article drew affirming letters from E. J. Moeran and others.

It was all too much for Stanford's pupil and successor as Professor of Music at Cambridge, Edward J. Dent. Several times in the past he had published dismissive remarks about Elgar. Asked to write about 'Modern English Music' for a German *Handbuch der Musikgeschichte*, Dent dealt with Elgar summarily:

He was a violin player by profession, and studied the works of Liszt, which were abhorrent to conservative academic musicians. He was, moreover, a Catholic, and more or less a self-taught man, who possessed little of the literary culture of Parry and Stanford. . . .

For English ears Elgar's music is too emotional and not quite free from vulgarity . . .[110]

---

[108] MS dated 13 Jan. 1931 (Alan Webb).   [109] *The Musical Times*, 1 Jan. 1931.
[110] Translated from the German in Maine, *Works*, pp. 277–8.

Professor Dent could not have done Edward's music a better turn had he tried. His remarks, published in Germany, quickly found their way back to England. And when they did, they touched a nerve of growing public apprehension that justice had not been done to Elgar and his music for a long time. A letter of protest was organized by Delius's pupil Philip Heseltine and sent to all the leading newspapers of England and Germany:

We, the undersigned, wish to record an emphatic protest against the unjust and inadequate treatment of Sir Edward Elgar by Professor Dent . . . At the present time the works of Elgar, so far from being distasteful to English ears, are held in the highest honour by the majority of English musicians and the musical public in general.

Professor Dent's failure to appreciate Elgar's music is no doubt temperamental; but it does not justify him in grossly misrepresenting the position which Sir Edward and his music enjoy in the esteem of his fellow-countrymen.

The letter was signed by Heseltine, Emile Cammaerts, John Goss, Harvey Grace, Leslie Heward, Beatrice Harrison, Hamilton Harty, John Ireland, Augustus John, Robert Lorenz, E. J. Moeran, André Mangeot, Philip Page, Landon Ronald, Albert Sammons, Richard Terry, William Walton, and Bernard Shaw. A young writer, Basil Maine, came forward with the idea of writing a two-volume study of Edward's life and works. In the Birthday Honours in June 1931 Edward was created Baronet. The choice of his official style was unhesitating: he chose to become 'First Baronet of Broadheath'.

Before the new honour was announced, the *Nursery Suite* was recorded at Kingsway Hall in advance of its public première. Herbert Hughes attended for *The Daily Telegraph*:

In the artists' room—the rehearsal has not yet begun—I meet Sir Edward Elgar, looking a good ten years younger than the 74 he reaches next week. He is standing by a table on which some music is lying. On his left is 'Willie' Reed, who has led many a Three Choirs meeting, trying over, very gingerly, a cadenza that occurs towards the end of the new suite. On his right is Gordon Walker, one of the finest flute players in the world, trying over an important solo part in 'The Serious Doll'—the second movement. Sir Edward marks time with one hand, indicating the rhythm as the player reads the exquisite page for the first time.[111]

They were able to finish recording all but the last side. Hughes concluded:

As I came away it seemed to me that the composer was no more delighted than those fine players who realised that they had a new masterpiece under their fingers. The composer may call this nursery music; but those of us who have ears know well that this score, written for the Royal Duchess and her children, is the sublimation of eternal youth. There is a philosophy, a metaphysic in this music that comes from one of the subtlest intellects of our time. These moving contrasts of sadness and gaiety have only been expressed in this music because they have been deeply felt. Like Verdi in his day, our Elgar appears to grow younger and more masterful as the years pass.[112]

[111] 25 May 1931.   [112] 25 May 1931.

To finish the *Nursery Suite* recording, they scheduled a make-up session on 4 June. On that occasion the Duke and Duchess of York and Princess Elizabeth honoured the new baronet with their presence. Edward had also invited a number of friends, and *The Daily Sketch* reported:

It looked a bit of an ordeal for the orchestra, especially as they were under the immediate eye also of Sir Landon Ronald, Norman Forbes (Sir Johnston Forbes-Robertson's brother) and Cedric Hardwicke, but the amusing passages descriptive of 'The Wagon Passing' [sic] set everyone laughing, and when the Duchess asked for a repetition it went better than before. Presently it was humorously suggested that Mr. Shaw should take a hand with the baton, only to bring the rejoinder that in that case the orchestra would be conducting themselves.[113]

In the Worcestershire countryside, boyish things animated Edward himself. At Perryfield, the home of his sister Pollie and her family, Edward hatched a plot with Pollie's twelve-year-old grandson Martin Grafton—to peashoot Pollie's grocer. Martin remembered:

Mr. Weston (who was a corpulent gentleman, given to wearing a vast watch-chain across his ample person, together with a bowler hat) owned the local grocers in Bromsgrove. Since he lived not far away from my Grandmother's house, it had become his long-established custom to walk round at precisely four o'clock on Sunday afternoons, and collect a cup of tea together with the grocery order for next week. He was a pompous man.

The house had the front door in the centre, with the bathroom window above. The plan was that we would open the bathroom window at the bottom and station ourselves behind it, where we had a good view of the sweep of the drive down which Mr. Weston would proceed on leaving the house after tea.

He duly arrived, took his grocery order and his tea, and left. Uncle Edward, crouching at the window, took careful aim at Mr. Weston's receding bowler hat.

'Now,' he whispered, 'Fire!'

True to his word, he deposited a salvo of dried peas in Mr. Weston's general direction. They fell short.

'Damn,' said Uncle Edward. 'Load and fire again.' He discharged a veritable shower, which completely swamped my poor contribution.

However, these efforts to discompose Mr. Weston were of no avail. For Uncle Edward had made one serious error in his otherwise excellently thought out campaign. He had not appreciated that by the time Mr. Weston came into our sights from the bottom of the bathroom window, he would be so far down the drive as to be out of range of even the mightiest puff! And so Mr. Weston escaped unscathed.

'Well, at least it was a jolly good try,' said Uncle Edward. 'Now we'll go and tell your Grandmother!'

Now this was something for which I had not bargained, for she was a formidable lady. Uncle Edward, divining my misgivings, however, said, 'Dont worry—I shall take full responsibility.' So downstairs we went.

'Polly,' said Uncle Edward, 'we have just made a very respectable attempt to shoot up that pompous Weston man!'

---

[113] 5 June 1931.

'*Edward*,' said my Grandmother, clearly not pleased, 'what *have* you done?'

'Oh, nothing really,' he said, 'We only used peas, and unfortunately we missed him.'

'You ought to be ashamed of yourself,' said my Grandmother, 'leading Martin astray like that!'

And as she said it, I became aware of a twinkle in her eye, while at the same time Uncle Edward aimed a pronounced wink in my direction. And the incident was closed.[114]

The sciatica had seemed to abate. But to it had been added another complaint—persistent nettlerash. In March 1931 it had been necessary to send for the doctor at five o'clock one morning to give an injection of morphia. After the *Nursery Suite* recording and the attempt to peashoot Mr Weston, Edward endured misery for much of the summer. Bernard Shaw wrote to encourage him:

It is an extraordinarily puzzling, capricious, and unexplained condition; but your vital powers are still obviously very vigorous.

Damn the thing, anyhow. If only one's friends would curse these aberrations instead of praying for one![115]

In September he got through the Gloucester Festival, and his performance of the *Nursery Suite* at the Wednesday Evening concert was broadcast. In October he conducted another broadcast performance of the Second Symphony for a crowded house at the Queen's Hall—'one of his now rare public appearances in London', wrote Richard Capell, who commented on the new fashion of refraining from applause between movements.[116] *The Daily Telegraph* noted: 'The enthusiasm at the close was overwhelming, and Sir Edward had to return several times to the platform in response to the demonstrations.'[117]

On 12 November he opened the new His Master's Voice studios in Abbey Road, St. John's Wood before a distinguished audience with the first complete recording of *Falstaff*. As a preliminary, he was to conduct *Land of Hope and Glory*. The recording manager Fred Gaisberg wrote:

Pathé Pictures Ltd. had been invited to make a 'short' on this occasion for their 'News Reel Service'. In this, Sir Edward enters the studio, gives his hat and coat to his valet, Dick, and mounts the rostrum where he is greeted by the London Symphony Orchestra with a cheer and 'Good morning.' He responds with:

'Good morning, gentlemen. Very light programme this morning! Please play this as if you had never heard it before.'[118]

The Gramophone Company house journal *The Voice* reported:

In the audience were many of the most distinguished men and women in the fields of music, literature and art. Two old friends of Sir Edward's—Sir Landon Ronald and Bernard Shaw—sat together following the score of *Falstaff*. Sir Walford Davies, Sir

---

[114] Letter to the writer, 1981.   [115] 2 July 1931 (HWRO 705:445:2234).
[116] 2 Oct. 1931.   [117] 2 Oct. 1931.   [118] Quoted in *Elgar on Record*, pp. 147–8.

Barry Jackson, and Cedric Hardwicke walked round together admiring the wonders of the new studios.[119]

*   *   *

Five days earlier a new friendship had suddenly opened. In Croydon to rehearse *Gerontius*, his eye caught that of a young woman at the second desk of violins in the orchestra assembled from London Symphony and local players. Her name was Vera Hockman. She was devoted to his music, and the rehearsal was a great moment for her. She recalled:

I hardly seemed to have to look at the music, my heart and soul went out to him because his way was not to command the orchestra but to implore them to give all the fire and energy and poetry that was in them. You could feel the love and veneration like great clouds of incense enveloping him. Then somehow he became aware that there was one whose only desire was to play as he would have it played. I distinctly remember catching his eye, small and bright with the serene expression of a God who looks upon his creation and finds it good.

The first time this happened it might have been accidental, but as the rehearsal went on, it occurred again and again until it became a part of that sublime dream through which I was living that whenever the first violins had to play a passionate or appealing or 'molto risoluto' passage he would look at me beseechingly and yet with such an expression of happy confidence in his eyes as if to say 'I know *you* understand and will do it for me.' It was a smile that can rarely be seen on this earth . . .

Gerontius over—and I remember in the melting theme of the Angel's Farewell he said imploringly, 'I should like to hear a still more beautiful tone from the strings, *please*'; and then we played it again much more tenderly—he left the orchestra in his dignified manner having thanked the choir for their beautiful singing—'nobilmente' always.

A few minutes later I was talking to a small group of players—raving about him—when someone touched me on the arm, saying, 'Sir Edward Elgar has asked to be introduced to you.'

'Good God, you're joking,' I gasped, and then followed her with trembling limbs to where E. E. was standing in the hall talking to Billy Reed: 'Sir Edward has asked for you,' said W.H.R.

'I noticed your playing, and wanted to be introduced to you,' said he.

'Oh, but this is the most wonderful experience of my life to play for you.'

'I could see by your face that you understood my music,' he said in that same quiet but distinct way, the words rather detached . . .[120]

On 10 November came a second rehearsal at Croydon and the performance:

Various members of the L.S.O. had by this time observed him gazing in my direction rather frequently. I must reluctantly admit that they did not think it an unprecedented occurrence as he was renownedly susceptible, but I do think it was the first time that

[119] Jan. 1932.

[120] This and following quotations by Vera Hockman are from 'The Story of November 7 1931' (TS at the Elgar Birthplace).

anyone actually playing in the orchestra had found such favour in his sight. 'For goodness' sake don't look at him too much to-night or he will lose his place,' said Wynn Reeves.

After the performance there was a party—at which Edward monopolized her to the exclusion of other guests.

On 19 November he rang up and asked her to lunch in London at the Langham Hotel (again his London base since giving up the flat). When she arrived he asked the waiter for two 'Manhattans'. (She recalled: 'I was amazed to hear him ordering such new-fangled yankee drinks.') Then he rose to go and order their meal, saying:

'. . . *Please* don't vanish while I am gone. I am so afraid I shall come back and find it is all a dream.'

During those few moments I waited for him to return I was almost in tears, so overwhelmed by his humility. To think that just because he wore that disguise of old age—and how swiftly that youthful and exuberant soul broke through the disguise—he should have to feel humble before an inferior being forty years his junior . . .

When he came back we spoke about English poetry. His knowledge of English literature was stupendous. He never forgot anything he had read and loved. 'I can't help it, I can't forget anything I'm interested in . . .'

They found they shared tastes in music and poetry.

There seemed so much to be done together that a whole lifetime would not be long enough even if he were young:

'But I am so old, you know, and the time is so short.'

'Apropos of that,' I said, 'supposing you ever found yourself alone and free I would gladly come to Worcester just for one hour or two—distance is nothing to me.'

'I will remember—perhaps telegraph when such a time comes.'

On December 2nd came the following telegram:—

'Could you come to-morrow meet train Worcester 12.43
Reply paid Edward Elgar'

Vera replied that she would like to go through the Violin Sonata with him.

He met her at the station with Dick Mountford in the car. They went to Marl Bank and met Marco and Mina.

We lunched alone together with one dog on either side of him, each in his (or her) chair, begging for his (or her) dinner. He explained how Marl Bank had been quite a vast estate in former times, and showed old engravings of the huge lands, and other engravings—one of Bridgnorth in Shropshire I specially liked. Most of the conversation was interspersed with dog talks, because Marco had to speak before he was given tit-bits, and Mina had to conduct. There were moments when Marco would not speak in a dignified enough manner, but rather too vociferously, and other times when Mina was conducting in 6/8 instead of 3/4 in a bar. So quiet and unfussy he was with his servants, yet he was so restless and impatient, and ceaselessly active; such a gorgeous medley of Michelangelesque grand faults and virtues.

. . . He had forgotten how much he liked the Sonata, which henceforth he was to call 'my Sonata
'our   „   .'

'Isn't that a gorgeous bit—let's do it again.'

The ceiling of his drawing room was so low, the piano was so flat, but after the first few moments of embarrassment 'it went'. Of course he corrected various points, and speeded it up a good deal. 'But on the whole you understand it because you understand me and always have.' He was delirious with joy over the Romance: 'Oh, this a *lovely* passage—I nearly always cry when I hear it, but I am not lonely to-day—we are together—I am so happy.'

They met in London on 7 December—

. . . the first 'mensiversary' of our meeting. From that day the seventh of every month was to be a festival because he knew he would not live long enough to celebrate the years.

It was then that he said, 'With you that most perfect three-fold relationship is possible—so rare on earth: guardian, child, lover.'

'And friend as well,' I added.

'Yes, friend too,' he agreed, 'and that is why I am going to give you a little book—Longfellow's *Hyperion*—which for many years belonged to my mother; since then it has gone with me everywhere. I want you to have it because you are my mother, my child, my lover and my friend.'

So the generations seemed to melt and mingle, and perhaps for the first time since Alice's death he was aware of a woman who might stand between him and the world.

He spoke so naturally of his 'dear little wife' and of what a devoted and wonderful woman she had been. But she had not understood Carice their daughter—'Carice, who is so clever but is alas buried alive in a Sussex village where there is no scope for her brains and energy; but one can do nothing for her.' He spoke of his three nieces—'who flit in and out of Marl Bank, and who for some unknown reason are most proud of their uncle and think that everyone who comes to see me and every letter I receive is of the utmost importance. There are no secrets from them and my secretary, you understand?'

He took her that night to a Beecham concert which included the First Symphony, and next day they went to Wimbledon to see Frank Schuster's old sister Adela. Afterwards he sent her all the sketches he could find remaining from the Violin Sonata.

On 7 January 1932—the second 'mensiversary'—Bernard Shaw was writing from a ship taking him on a long voyage: 'Why don't you make the B.B.C. order a new symphony? It can afford it.'[121] For years Shaw had been after Edward to add to the list of his Symphonies. It was quite possible that he had spoken privately to Landon Ronald (who was a member of the BBC Music

[121] *Letters*, p. 334.

Advisory Committee) when they sat together over the *Falstaff* score during the recording in November.

However it happened, there was soon a proposal for Edward to orchestrate the Funeral March from Chopin's B flat minor Sonata as a vehicle for the BBC Symphony Orchestra's virtual gramophone début. Boult and Fred Gaisberg each pointed to the other as the proposal's originator, and no mention was made of Shaw. The Gramophone Company offered an arranger's fee of £75, and by the end of March 1932 the score was finished. When it was conducted for records by Adrian Boult, Edward came to the studio. The *Telegraph* critic Bonavia attended the session. He asked if the remaining movements of the Chopin Sonata could be orchestrated to make in effect a symphony—only to receive Edward's answer that 'he would rather write a symphony of his own.'[122] Bonavia promptly published this exchange.

It must have reached the eye of G. B. S.   For he sent Edward a postcard suggestion spun delicately out of the fact that several months earlier he had quietly lent Edward £1000 'to get him out of money troubles':[123]

Why not a Financial Symphony?   Allegro: Impending Disaster.   Lento mesto: Stony Broke.   Scherzo: Light Heart and Empty Pocket.   All° con brio: Clouds Clearing.[124]

Edward sent Shaw's postcard to Fred Gaisberg with a query:

Perhaps H.M.V. would like to commission (say £5,000) for such a symphony as *G.B.S.* suggests: the p.c. is worth more than my music![125]

Gaisberg prized the postcard and deftly replied:

I hope you do not mind if I circulate it among some of our people who would enjoy the idea of commissioning a 'Financial Symphony'.[126]

In this year of Edward's seventy-fifth birthday, life was more reminiscent than ever. Basil Maine was busy with his big biography. Edward alternately encouraged and discouraged it—as if to acknowledge the compliment while fearing the result of an attempt to read his creative life. He had invited Maine to Marl Bank for a weekend in March, and spent the Sunday morning driving the younger man over the old places, with the dogs as companions. The early spring weather itself seemed to illumine their pursuit, as Maine recalled:

The country was covered with a light mist which was invisible until one looked towards a distant point. The day was still enough to allow the warmth of the sun to touch the earth . . . We had come by way of Broadheath, for Sir Edward had wished to show me the cottage in which he was born . . .

---

[122] 31 May 1932.

[123] 28 May 1948 to Rutland Boughton (quoted in Michael Hurd, *Immortal Hour*, Routledge and Kegan Paul, 1962, p. 120). Shaw said that the money was never returned. But Carice Elgar Blake wrote to recall that the debt was paid out of her father's estate in 1934, only to be returned by Shaw as a present to herself.

[124] 29 June 1932 (quoted in *Elgar on Record*, p. 171).

[125] 1 July 1932 (ibid.).   [126] 4 July 1932 (ibid., p. 172).

'Slow down a little, Dick,' Sir Edward called to the driver as we were passing, and we looked at the little house in silence. I shall not easily forget that suspended moment . . .

We followed a bend in the road and found the village green hidden there and a shop or two. The driver began to accelerate. The dogs looked out like sentinels. Over there, Sir Edward was saying, was Birchwood, where 'Gerontius' was written. He was pointing to a far-off hill on the left. The hill itself was visible, but its outline was in a haze.[127]

A month later the garden wall which had collapsed at Marl Bank was replaced in the happiest way. Billy Reed remembered:

One day when I arrived at Marl Bank I was rushed off to see what they were doing at Worcester, widening the bridge over his beloved Severn: the old familiar bridge he had known all his life. I was taken there so often that I guessed he had something in his head about it. At last it came out. He could not bear to part with the old iron balustrades—or whatever they are called—that were being removed; so he bought two lengths of them and had them brought up on lorries to Marl Bank and set up there on a concrete bed. I thought they looked rather crazy in the garden, but took care not to say so; for he was so delighted when they were set up that we had to go out repeatedly to study them from all points of view and discuss what colour they should be painted. He decided to keep them as like their old selves as possible. I think he used to go out and imagine that the Severn was flowing under them as of old.[128]

He scored the *Severn Suite* for orchestra, and gave the movements names of local association: Introduction (Worcester Castle), Toccata (Tournament), Fugue (Cathedral), Minuet (Commandery). He recorded the new version with the London Symphony Orchestra on 14 April 1932, and sent the test pressings to Shaw, who responded: 'What a transfiguration! Nobody will ever believe that it began as a cornet corobbery. It's extraordinarily beautiful.'[129]

Instead of Shaw's symphony, however, he fulfilled the year's obligation to his publishers Keith Prowse with three simple unison or two-part songs for children, to answer the request of a young Worcester schoolmaster, Stephen Moore. The words of all three, by the Victorian poet Charles Mackay,[130] sang the symbols of childhood and age as surely as the *Nursery Suite* had sung them:

> O streamlet swiftly flowing
> Down through the cornfields going,
>   Stay thy course with me.
> For us the skylarks sing,
> For us awakes the Spring;
>   There's time to spare,
>   The earth is fair,
>   Why hurry to the sea?

[127] *Life*, pp. 1–2.
[128] *Elgar as I Knew Him*, pp. 102–3. Elgar and Hubert Leicester had been negotiating for the bridge railings for the better part of a year.
[129] 11 July 1932 (HWRO 705:445:2236).
[130] Keith Prowse wrote on 1 July 1932 about permission for use of Mackay's words.

His seventy-fifty birthday on 2 June was celebrated with a dinner in London and a thicket of articles in the press. Bonavia wrote:

It is not improbable that the historian of the future commenting on English life in the early twentieth century will find that music had taken the place of poetry not only in the heart of the people but as reflecting the mood and mind of the nation . . . If this should be so, the greatest figure in the art of our time will be Edward Elgar. Of him alone it can be said with truth, as has been said of the dramatists of the English Renaissance, that he lived and worked in full sympathy with the whole people. His art does not appeal exclusively to any one section of the community: it embraces all.[131]

The testimony was made in a different way a few days later with the performance of a short Ode by Edward for the unveiling of Alfred Gilbert's memorial statue to Queen Alexandra in a recess of the wall at Marlborough House. The verses by the new Poet Laureate, John Masefield, began, 'So many true Princesses who have gone . . .' They had been sent with an official request in mid-May, and the work was quickly and impressively done. On the dedication day, 8 June 1932, *The Daily Telegraph* described

. . . carpet laid on the gravelled road, many rows of gilt chairs, and the approach on foot of the leaders of the nation . . . The Prime Minister and the members of the Cabinet arrived at well spaced intervals, and presently came the children of the Chapels Royal, in their liveries of scarlet and gold, and the choir of Westminster Abbey. Following them was Sir Edward Elgar (Master of the King's Music), a magnificent figure in his doctor's habit surmounting Court dress . . .

Sir Edward has taken up his baton, and the wistful melody of his music rises and falls as the little puffs of wind in the June sunshine carry it back and forth . . .[132]

July brought meetings with old and young. John McCormack, having nursed resentment over Edward's behaviour at the Verdi Requiem in Birmingham for twenty years, had recently sung *Is She Not Passing Fair?* in a Musicians' Benevolent Concert at which Edward conducted *Pomp and Circumstance* No. 1. Their mutual friend Mary Anderson de Navarro seized the chance to bring them together at her Cotswold home. McCormack sang parts of *Gerontius* to Edward's accompaniment with such satisfaction that Edward wrote to His Master's Voice proposing a complete recording conducted by himself with McCormack in the title role. The answer was that in the prevailing economic depression, the Company could only record McCormack in popular ballads.

In observance of Edward's birthday, however, the long-discussed recording of the Violin Concerto was about to be made—only not with Kreisler. Fred Gaisberg had sent a last appeal to Kreisler in August 1931, pointing out the approach of the seventy-fifth birthday. When that appeal also failed, Gaisberg realized he must wait no longer. Instead, he proposed an extreme opposite—to send the solo part to the fifteen-year-old Yehudi Menuhin, already established as an international virtuoso. It was a stroke of genius, greeted with enthusiasm

---

[131] 'A National Composer: the significance of Elgar'.
[132] C. B. Mortlock in *The Daily Telegraph*, 9 June 1932.

on all sides. Yehudi studied the Concerto, and on 12 July 1932 soloist and composer met for the first time. To aid in going through the Concerto's interpretation, the pianist Ivor Newton was asked to accompany. He recalled:

Menuhin and Elgar discussed the music like equals, but with great courtesy and lack of self-consciousness on the boy's part. Listening to the discussion, I could not be other than amazed at his maturity of outlook and his ability to raise points for discussion without ever sounding like anything but a master violinist discussing a work with a composer for whom he had unbounded respect . . .

We played right through the Concerto . . . Most of the time Elgar sat back in a chair with his eyes closed, listening intently, but it was easy to see the impression that Yehudi made on him.[133]

The recording, made at the Abbey Road studios on 14 and 15 July, was one of the most famous achievements of the gramophone. The dramatic meeting of youth and age created a *frisson* fully matched by the latest HMV microphone techniques. The recording has remained in the catalogues almost continuously to this day.

The first test pressings were available in August 1932. On 2 September Bernard Shaw came to Marl Bank to hear them, with his wife and Lawrence of Arabia (now hiding himself in the Air Force under the name of Shaw). The company already gathered there for the Worcester Festival included Carice, the Grafton nieces, and Vera Hockman, who wrote:

Here we all sat, spellbound at the glorious sounds, G. B. S. with bowed head, sometimes softly singing with the music; Aircraftsman Shaw serious and silent, looking straight ahead with those unforgettable blue eyes which seemed to see into the life of things. After Menuhin had lovingly lingered over the last melting phrase of the Slow movement, E. E. whispered, 'This is where two souls merge and melt into one another.'[134]

So again perhaps Vera might to sit in the place of Alice.

At last inspiration seemed to galvanize the new Symphony for which Shaw had cajoled and bullied him. It could open with a powerful counterpoint originally sketched for 'The Last Judgement':

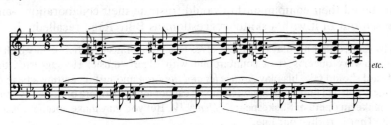

---

[133] *At the Piano* (Hamish Hamilton, 1966), pp. 185–6. After the session, Menuhin père invited Elgar to luncheon, which Elgar declined on the strength of an engagement to go to the races. In Yehudi Menuhin's memory Elgar's departure for the races came after only a few bars of rehearsal, these few bars having produced great satisfaction. There could be no doubt of the satisfaction, but Newton's full and circumstantial account seems indisputable—and in character.
[134] 'Elgar and Poetry' TS.

But the centre of the first movement would be a musical idea associated from its inception with Vera. Its first sketch was marked 'V.H.'s own theme':

Behind this music stood the motive of 'Heroic Beauty' from *King Olaf*

now the old masculine aspiration had a feminine complement. But Edward marked the new sketch ambiguously 'will never be finished?'

Other first movement material was drawn from the unpublished score of *Arthur* and the 'Callicles' sketches of 1926 and earlier. So the new symphonic opening could pursue a synthesis from the scattered music of his life since Alice's death—some final insight into whatever had beckoned his interest in the characters of Arthur, Callicles, and the Eminence which might have emerged from 'The Last Judgment'. Using his extemporizing skill at the piano, Edward showed Bernard Shaw how these disparate ideas might be pointed toward unity. Shaw was clearly impressed.

Yet Edward's powers of large-scale abstract composition were rusty with age and disuse. As in the old days before he achieved the First Symphony, the approach to large abstract composition might be easier through a work that should follow some ready-made outline of plot. If oratorio was dead, opera was still being written—notably by Richard Strauss, who was now himself not far from seventy.

Edward asked Shaw for a libretto. But G. B. S. saw that differences which stimulated their companionship would frustrate their collaboration—and got himself out of it with a skilful suggestion, as Billy Reed recalled:

Shaw replied that his plays set themselves to a verbal music of their own which would make a very queer sort of counterpoint with Elgar's music. He suggested that Sir Edward should take his play *Androcles and the Lion* and just try setting a page of it to music. 'You will find', said Shaw, 'that you cannot make an opera of it, just as you could not make an opera of Shakespeare's *Henry IV*. But you may make another *Falstaff* out of it. That is really your line.'[135]

The thought of entering on the operatic stage at this late moment of life was none the less a heady notion. Edward had spent several summer days with G. B. S. at the 1932 Malvern Festival, where Barry Jackson produced old

---

[135] *Elgar as I Knew Him*, p. 89.

English plays side by side with the moderns. And his mind went back to a scheme he had once entertained of basing an opera on one of the least-known plays of Ben Jonson, *The Devil is an Ass*.

It was a typical Jonsonian mixture of medieval morality and current vice in thickly woven plot and counter-plot. The choice of this obscure play could make a fine answer to self-important professors of music with their talk of 'literary culture'. And if Jonson's verse was not especially distinguished, it offered another chance to set what he had described to Vera as 'the best second-rate poetry', in preference to the very best poetry which '*is* music already'.[136] As Barry Jackson wrote:

> It is quite possible, indeed, that the forbidding want of ease and flow in this verse and the curious recalcitrance it shows in the matter of scanning may be a positive advantage to a musician, leaving him free to play with his own rhythmic schemes, whereas Shakespeare's wonderful lines in blank verse would continually fascinate and distract him by that soaring and surging 'verbal music of their own'.[137]

The plot of *The Devil is an Ass* revolved round an ageing protagonist, whom many schemers try to cheat of both his money and his young wife. And in the centre was a figure of special appeal to Edward—a young man who gets the heroine where he wants her and then nobly refuses to take his advantage.

It was a chance to transmute a lewd old action, as with *Falstaff* and *Beau Brummel*, into a 'warm and full thing'. He would eliminate the morality-Devil of the play title, and call his opera *The Spanish Lady* after the disguise used by the young hero to gain his advantage. He would change the old man from husband to guardian, and marry the young people happily at the end.

To do all this, he sought collaboration not with another playwright, but now with the producer and presiding genius of the Malvern Festival, Barry Jackson. Jackson described himself as

> . . . a man of the theatre who claimed to be no poet but could find out by close and continuous personal contact exactly what the composer wanted. I may at least say that I always encouraged him to air his views and to declare his wishes.[138]
> Elgar told me that he wanted to write something that was thoroughly and very typically English—'Roast beef and beer'—and could think of nothing better than one of the plays of Ben Jonson . . .[139]
> He had set his heart upon *The Devil is an Ass*, a work which always appeared to me to be quite moribund. After renewing acquaintance with this comedy, I wrote saying I felt nothing could be done with it. A fortnight later I was overwhelmed by the sensation that I might be standing between the world and a great musical work, and earnestly applied myself to the not-easy task of disentangling the imbroglio of the play in question.[140]
> Once I had allowed myself to be persuaded that behind the play's wordiness and

[136] 'Elgar and Poetry'.
[137] 'Elgar's "Spanish Lady"', in *Music and Letters*, Jan. 1943, p. 2.
[138] Ibid., p. 7.
[139] BBC broadcast, 'The Fifteenth Variation', 1957.
[140] Letter to W. H. Reed, printed in *Elgar as I Knew Him*, p. 90.

complexity of intrigue lay an excellent story for an opera that could be made clearer to an audience than the play itself could ever become even without the music, I agreed with some diffidence but also with much pleasure and enthusiasm to try my hand at a libretto.

. . . If only the bare skeleton of the tangled story of 'The Devil is an Ass' remained in my version, I did endeavour to keep to something like its spirit and certainly to the atmosphere of the period, though I also tried to leave a good many blanks in order to give the composer as much chance to enlarge to his heart's content on whatever aspect of this or that scene appealed to his imagination. For it *was* the composer's imagination, not my own, which I knew had to be stimulated and satisfied if opera was to be enriched by another great work.[141]

I delivered a rough MS. founded on the plot and text of the play, with interpolations of the poems for the more lyrical moments, and with this he appeared to be delighted. He had given me his copy, which was liberally covered with jottings and suggestions, obviously a labour of love carried on over a long period. The story was fined down to the uttermost dramatic limits in my version; but Sir Edward was determined that the work should be on a grandiose scale; for he added incidents and complications without end, always declaring that, if ever he composed an opera, it was going to be a grand opera.[142]

As with his oratorio writing in the past, Edward began with the music long before anything like a finished libretto emerged. Once again he started from his sketchbooks—now often very old sketchbooks indeed—and many of the ideas for arias, dances, and instrumental pieces were as ancient as anything in the *Severn Suite*. In fact a good deal of *The Spanish Lady* would have the atmosphere of early Elgar—almost juvenile Elgar, of the period when he had first made acquaintance with the old English dramatists. But there were also later things—other possibilities from the third oratorio and the still unpublished music for Binyon's *Arthur*, as well as excursions into apparently new thinking.

Through it all he would keep control of the developing text, just as he had done in the old collaborations with Harry Acworth long ago. But, as Barry Jackson noticed when he looked at Edward's marked copy of the play: 'The pencil notes, in fact, diminish as the play proceeds, no doubt because the task of arranging it became more and more difficult towards the end of the complicated imbroglio.'[143] Billy Reed recalled:

Elgar was very headstrong and not a little difficult when he had conceived a situation on the stage in a certain way (perhaps not very practical from the theatrical standpoint); and when Barry Jackson, with his vast stage experience, came along and put his finger unerringly on a weak spot, summer lightning would ensue. Elgar would alter an idea *here* in deference to Barry's judgment, perhaps; but he would fight for an idea *there* until he had convinced Barry that it *was* practicable.

. . . We began playing a lot of it on my violin and his piano. There was a Spanish dance, a country dance, a bolero and a saraband. Also numerous vocal portions of which I played the voice part whilst he accompanied vigorously and excitably at the piano. We could have earned our living with it at a provincial café, *al fresco*.

---

[141] 'Elgar's "Spanish Lady"', pp. 2, 9–10.   [142] Quoted in *Elgar as I Knew Him*, p. 91.
[143] 'Elgar's "Spanish Lady"', p. 3.

Then we talked it all over. It was to be Grand Opera on the biggest scale: a tremendous work, in fact. He would explain with a wealth of detail everything that was to happen on the stage at the particular bar we were trying over. He would even draw a plan of the stage, showing all the 'properties' and exactly where the characters were to stand; but if I am ever asked what it was all about I shall have to confess that I have not the faintest idea, and never had.[144]

Behind the opera project, meanwhile, there rose the dream which Shaw was keen to see and hear as reality—the Third Symphony. Through the summer of 1932 the rumours had flown thick and fast in the self-generating way of such things. One crass example actually arrived in the post at Marl Bank from the young man who edited the Gramophone Company's house journal *The Voice*:

4th August, 1932.

Dear Sir Edward,

I hope you will forgive me for writing to you on this topic, but I have heard, on what I believe to be very reliable authority, that you have practically completed a third symphony. Is there any truth in this rumour? If you could tell me, I should be delighted to make use of the information, not only in 'The Voice', but in the general press. Moreover, our mutual friend, Ernest Newman, is very anxious to know whether there is any truth in the news, which I passed on to him for what it is worth.

With kind regards,
Yours sincerely
Walter Legge.[145]

5 AUG 1932

Dear Mr. Legge:

Many thanks for your letter: there is nothing to say about the mythical Symphony for some time,—probably a long time,—possibly no time,—never.

Kind regards
Yours sincerely
Edward Elgar[146]

A month later, during the Worcester Festival, the question of the Symphony came up again. Vera Hockman was constantly with him through the Festival, and perhaps her presence inspired a brave, dismissive claim. The critic H. C. Colles wrote:

At a tea-party he spoke of a third symphony as 'written', but said that it would not be worth while to finish up the full score since no one wanted his music now. This, at a festival where 'The Dream of Gerontius', the First Symphony, 'The Music Makers' and 'For the Fallen' were all being given under his own direction, and the 'Severn Suite' was receiving its first full orchestral performance, was too obviously perverse to be passed without protest. His remarks were quoted lightly by one who heard them to another who

---

[144] *Elgar as I Knew Him*, pp. 91–2.   [145] HWRO 705:445:4505.
[146] Draft letter at the Elgar Birthplace.

had not. Next morning the *Daily Mail* came out with a demand, emphasised with large headlines, for the production of Elgar's new symphony.[147]

After the Festival was over, he seemed to give a gentle hint to Vera, who recalled:

As I was reluctantly leaving Marl Bank I was presented with a Novello's vocal and piano score of the 'Music Makers', and upon opening it I found underlined the very last line of the poem:
> A singer who sings no more.
> i.e. Edward Elgar.[148]

But the demand for the Third Symphony was now unstoppable. On the last day of September Bernard Shaw wrote to the Director General of the BBC, Sir John Reith:

May I make a suggestion?
In 1823 the London Philharmonic Society passed a resolution to offer Beethoven £50 for the MS of a symphony. He accepted, and sent the Society the MS of the Ninth Symphony. In 1827 the Society sent him £100. He was dying; and he said 'God bless the Philharmonic Society and the whole English nation.'
This is by far the most creditable incident in English history.
Now the only composer today who is comparable to Beethoven is Elgar. Everybody seems to assume either that Elgar can live on air, or that he is so rich and successful that he can afford to write symphonies and conduct festivals for nothing. As a matter of fact his financial position is a very difficult one, making it impossible for him to give time enough to such heavy jobs as the completion of a symphony; and consequently here we have the case of a British composer who has written two great symphonies, which place England at the head of the world in this top department of instrumental music, unable to complete and score a third. I know that he has the material for the first movement ready, because he has played it to me on his piano.
Well, why should not the BBC, with its millions, do for Elgar what the old Philharmonic did for Beethoven. You could bring the Third Symphony into existence and obtain the performing right for the BBC for, say, ten years, for a few thousand pounds. The kudos would be stupendous and the value for the money ample; in fact if Elgar were a good man of business instead of a great artist, who throws his commercial opportunities about *en grand seigneur,* he would open his mouth much wider.
He does not know that I am meddling in his affairs and yours in this manner; and I have not the faintest notion of what sum he would jump at; but I do know that he has still a lot of stuff in him that could be released if he could sit down to it without risking his livelihood.
Think it over when you have a spare moment.[149]

Behind the scenes at the BBC, Landon Ronald was pulling every wire he could, and it was decided that he should be the message-bearer. The

---

[147] *Grove's Dictionary of Music and Musicians,* 4th edn., Supplementary volume (Macmillan, 1945) p. 194.
[148] 'Elgar and Poetry'.
[149] Quoted in Reith, *Into the Wind* (Hodder & Stoughton, 1949), p. 163.

commissioning sum was fixed at £1000. Keith Prowse had to have knowledge of it, as they held the exclusive contract for publishing Edward's music. They suggested that the news be quietly passed to Basil Maine for his book, which was approaching completion. When the young man asked for another visit, Edward responded:

Marl Bank, Worcester.

13th Oct 1932

My dear Maine:
    I am sorry that the work is being persisted in!—I wish your publishers cd let me die in peace.
    I fear there is nothing to say in regard to the new Symphony or anything else: things take shape without my knowing it—I am only the lead pencil & cannot foresee.
    I shall be glad to see you—give me some idea,—you say after Nov. 5th. I seem to remember that is an auspicious anniversary;—let us put your MS. & mine on a good Guido Fawkes bonfire and have done with it.

Kind regards
Yrs sincy,
Edward Elgar.[150]

But the die was cast. The formal offer, involving four quarterly payments, came a month later. Edward wrote to Sir John Reith:

I have left everything in my friend Ronald's hands, and do not propose to say anything about the business side of that matter.
    I write now to thank you for your kind and generous attitude in the inception of the idea: whatever happens I shall always treasure the remembrance of your kindness and consideration.[151]

Conducting engagements were keeping him busy. October 1932 had taken him to Ireland for the first time to broadcast *Gerontius* from Belfast. A week later he was at Hanley to conduct *King Olaf* in the hall where the work had its première thirty-six years earlier: that too was broadcast. In November there was a public performance of the Violin Concerto at the Albert Hall with Yehudi Menuhin. *The Daily Mail* reported:

Once Sir Edward caught the boy's eye, at a moment when he was not playing, and Yehudi gave him a quick, shy smile . . . Side by side they carried their audience away from the world of reality on the wings of music; the boy with his life before him—the man with so much of his life behind him, a life of high endeavour and notable success.[152]

Afterwards Edward wrote to Yehudi:

At my age old friends pass away and leave the world rather empty—this is inevitable and

---

[150] Royal College of Music. Maine published a transcript with some inaccuracies in his *Twang with Our Music* (Epworth Press, 1957), p. 101.
[151] Transcript in BBC files (quoted in Reith, *Into the Wind*, pp. 163–4).
[152] 21 Nov. 1932.

has to be faced. Your friendship in any case must be—is—a remarkable thing, and it has given me a new zest in life. To hear you play the Concerto (now over twenty years old) gives me the deepest artistic satisfaction; I think you know this, as we seemed en rapport when you were playing at the vast Albert Hall. Anyway, I have never felt such a reading as you gave it, with such a thrill of expression.[153]

Another performance of the Concerto was devised by Yehudi's father for Paris in the spring. He asked Edward to go to conduct it.

Then came three BBC concerts of Edward's music planned by Adrian Boult to celebrate the seventy-fifth birthday year. The first programme included Symphony No. 1 conducted by Ronald. In the second, Edward conducted Symphony No. 2. On the night of the third concert, 14 December, the announcement of the Third Symphony commission was given out by Landon Ronald at a dinner in the Guildhall. Edward did not attend the dinner, but finished the evening quietly with Vera.

The press trumpeted: 'This is a most important event in the history of music. It establishes a unique precedent. It means in effect that the enormous listening public have become the patrons of a great musician.'[154] The post brought many letters. One of the first was from Fred Gaisberg:

15th December, 1932.

Dear Sir Edward,

No sooner do I wake up this morning than I see the bombshell which Sir Landon Ronald has thrown regarding the completion of the Third Symphony. Of course all the papers are full of it and I have had telephone calls from various people wanting to know if we are going to record it. We would certainly like to record it immediately before or after the inaugural performance by the B.B.C. Orchestra.

I see that it is scheduled for the autumn of next year and I am only just writing to tell you to keep the matter in mind for us.

I am now trying to arrange a session at which we can record the Prelude to 'The Kingdom' on 2–12" records, and 'Contrasts' . . .[155]

Marl Bank, Rainbow Hill, Worcester.

16 DEC. 1932

My dear Fred:

Thanks: I shall be glad to 'do' the Prelude etc when you are ready: as to *Sym. III*—?

Yours ever,
E.E.[156]

The approach of Christmas brought a cluster of memories. At one of the last meetings of the old Worcester Glee Club on 20 December, Edward made a little speech of gratitude to his old helper in score-and-parts reading, 'Honest John' Austin. He presented Austin with the dedication of a little *Serenade*, one of three pieces Keith Prowse were publishing for piano. The others were

---

[153] 1 Jan. 1933 (quoted in 'Elgar and the boy violinist: a batch of letters', *The Daily Telegraph*, 20 Oct 1934, p. 6.).
[154] *Irish News*, 6 Jan. 1933.   [155] Quoted in *Elgar on Record*, pp. 185–6.   [156] Ibid.

entitled *Adieu* and *Mina* (after the smallest of his dogs). He had sent them in as 'sketches', he told the publishers: 'they can be adapted to any "arrgt." you think fit.'[157]

His Christmas card that year bore an original fable. It described a group of Angels discussing The Last Judgment without enthusiasm. Lucifer recommended Shakespeare and Milton. But the 'MAKER-OF-ALL' replied by creating a puppy: 'Through the ages Man could be serenely happy with his DOG.'

Christmas dinner was at the Philip Leicesters' house, The Homestead, close to Marl Bank. There two dozen Leicesters and Graftons and friends sat down to a festive meal in the big dining-room. Afterwards Edward prepared to return home for his nap. When the Christmas crackers were opened, he took his farewell: 'Now if you'll all blow on your whistles, I'll put the sound into my new Symphony.'[158]

Acknowledging a New Year's diary for 1933 from the Windflower, he wrote: 'I wonder what I shall have to enter in it: a new symphony is promised.'[159] He had been working on and off at the opera, but the Symphony must now take precedence if it was to be ready in time. He sent for a new supply of the Italian paper on which the two previous Symphonies had been scored. Ricordi could no longer supply it; but a distinguished English printer, George W. Jones, undertook to imprint a fine rag paper with 26 staves and instrument names to Edward's requirement.[160] On 1 February Edward got out one of his old sketchbooks less written in than the others and sealed up the written pages—leaving the remaining blank pages ready to receive new ideas.

On 5 February Billy Reed brought his violin to try over as much of the Symphony as there was. They went through the first-movement sketches, including a transition figure 'which I had to play countless times in every conceivable manner.'[161] It was all still quite fragmentary.

For a second movement 'in place of Scherzo'[162] Edward wrote out again the figure he had used for the banquet in Binyon's *Arthur*:

[157] 29 Sept. 1932 (Keith Prowse archives).
[158] Recalled by one of the party present, Mrs Patricia Neal, in 1975.
[159] 6 Jan. 1933 (HWRO 705:445:7788).
[160] HWRO 705:445:2609 and 2630–8.
[161] *Elgar as I Knew Him*, p. 175.
[162] BL Add. MS 56101 fo. 29.
[163] This discussion of the Third Symphony material is indebted to Christopher Kent's *Edward Elgar: a composer at work* (thesis submitted to King's College, University of London, 1978).

Reed remembered afterwards:

He must have had the main theme for this movement (very light and rather wistful) in his mind for some years, as I have seen it scribbled in his scrap books in various forms . . . He loved this simple little theme, and we played it again and again with violin and piano.[164]

Most recently it had figured in the opera sketches. The very manuscript they played from was marked for the opera:[165] pressure to get on with the commissioned Symphony prompted him to raid the alternative project. Ideas for episodes in this movement were taken from other material in *Arthur* (notably the main 'heroic' motive), 'Callicles', and 'The Last Judgment'.

The Symphony's slow movement was to open with an idea conceived for 'The Last Judgement' and closely related to the Angel of the Agony in *Gerontius*:

Its ultimate theme was a broad *Adagio* from another old sketch:

He exhorted me to 'tear my heart out' each time we repeated it, so much was he always overcome by its emotional significance. I was never able to induce him to write down the continuation, but I was allowed to play a bar or two (looking over his shoulder) from the fragments on one or two other scraps of MSS. But I could never prevail upon him to divulge in what order they were to appear.[168]

This priming of the pump with older ideas seemed to succeed, for the Finale first subject emerged new and strong:

The Finale second subject might open with a reprise of the heroic motto from the *Arthur* music. Many possibilities for transition and subsidiary themes were sketched. But there was nothing in the shape of a conclusion. Reed recalled:

---

[164] *Elgar as I Knew Him*, p. 176.
[165] Cited in Young, *Elgar, O.M.*, p. 371 ex. 48. Young gives an exhaustive analysis of the surviving opera sketches.
[166] BL Add. MS 56101 fo. 55.   [167] Ibid., fo. 38.
[168] *Elgar as I Knew Him*, p. 177.   [169] BL Add. MS 56101 fo. 50.

[He] would leave off suddenly and abruptly when we arrived anywhere near that part, and say, 'Enough of this; let us go out and take the dogs on the Common.' Also, he would be very restless and ill at ease, and would not discuss the symphony any more, and it would be quite a while before he became calm and resumed his normal good spirits.[170]

That could be set down to no more than understandable worry over major composition after many years' quiescence and under pressure of a commission. When the first BBC quarterly cheque arrived on 25 February, he wrote to Sir John Reith:

I am hoping to begin 'scoring' the work very shortly: I am satisfied with the progress made with the *'sketch'* & I hope that the 'fabric' of the music is as good as anything I have done—but naturally there are moments when one feels uncertain: however I am doing the best I can & up to the present the symphony is the *strongest* thing I have put on paper.[171]

Time was moving inexorably. On 17 March he wrote to old Adela Schuster:

I am hoping that you have been able to enjoy the wonderful weather which has made the earth look like a promise of better things;—I fear not of better times though: I am in a maze regarding events in Germany—what are they doing? In this morning's paper it is said that the greatest conductor Bruno Walter &, stranger still, Einstein are ostracised: are we all mad? The Jews have always been my best & kindest friends—the pain of these news is unbearable & I do not know what it really means.

Please write & tell me you are well: for my part I managed to collect a very bad specimen of ordinary cold a month ago & it is still with me. I have no news but am working hard at new things which may be heard one day—if there [are] any listeners allowed to live.[172]

A month later came an enquiry from the BBC, written over the signature of Adrian Boult's assistant Owen Mase:

21st April, 1933.

Dear Sir Edward,

We have carefully considered the question of your new symphony which you are writing and the best place to put it in our big series of concerts commencing next autumn. The series opens on 18th October and we should very much like to play your symphony at that opening concert.

I wonder if you could yet tell me whether it will be finished in time? Obviously parts and score would have to be ready by the end of September at the latest so that the fullest rehearsal and preparation could be given. At the same time, you might let me know whether you would like to honour us by conducting the work yourself.

I am sorry to worry you about it so early, but you will understand that our arrangements have to be made a long way in advance.

With kind regards,
Yours sincerely,
Owen Mase.[173]

[170] *Elgar as I Knew Him*, pp. 178–9.   [171] BBC archives.
[172] *Letters*, p. 316.   [173] BBC archives.

Marl Bank, Worcester.

24 Apr. 1933

Dear Mr Mase:

Many thanks for your letter: I fear we must not announce the first performance of the Symphony until everything is printed, and, as you say, the 'material' wd. have to be ready for rehearsals in September: too much depends on the 'reading' of proofs & printing generally for me to be able to say 'yes'.

I am as forward with the work as I hoped to be &, if nothing untoward occurs, shd. be able to begin to 'feed' the publishers with M.S. shortly. In the meantime I think no announcement or reference need be made: As to conducting—we will 'wait & see'.

Kind regards,
Yours sincerely
Edward Elgar

The following night he had a sort of seizure. Carice (who happened to be at Marl Bank for a visit of several days) had to send for Dr Moore Ede three times in the course of the evening, but she noted, 'Everything passed off all right—& he was safe by 12.' It was given out as a severe chill which kept him in bed over the next day or two, but no one could say what it was.

Next morning there was another letter from Owen Mase at the BBC:

I quite understand the difficulties and dangers and certainly we must not make any announcement until we are quite sure. At the same time, such an important event in music must receive every advance notice possible, and I have been wondering whether you could tell us just a little more. Would it be possible to say now that the work would definitely be ready for our May Festival in 1934? It would be a magnificent time to produce it and it would give us plenty of time to get everything ship-shape before any announcement were made.

I know how you must dislike committing yourself, but this strikes me that it might be far enough ahead for you to say 'Yes' and everything will be clear for rehearsal and performance before May of next year.[174]

From his sick bed Edward replied:

Marl Bank, Worcester.

27 Apr. 1933

Dear Mr. Mase:

I like your idea to announce the Symphony for the May Festival of 1934: that wd. give us all full time for preparation in all departments, printing, rehearsals, etc. So if you like to proceed pray do so: you do not worry me! So if there is anything else to ask or answer be sure to write.

I had a sudden bad turn two days ago—but am getting all right again

Kind rgrds
Yrs sincy
Edward Elgar
P.T.O.

[174] Ibid.

Symphony in C minor
  I. Allegro
  II. Allegretto
  III. Adagio
  IV. Allegro
Now the trouble is that I have not decided finally the positions of II & III that is to say III might follow I [175]

On the same day he sent a further letter to Sir John Reith:

This is only to say that as far as an artist can feel satisfied with his work,—and no real artist can ever feel that,—the thing goes on happily. I have, with the invaluable aid of Mr W. H. Reed, played through portions of the MS. & he is delighted: I can say no more now but you have been so really kind in the whole matter that I felt I should like to send you this note.

    I presume it will be correct for me to send my MS—(which will be delivered in portions) *direct* to the publishers . . . [176]

That evening he was out of bed, playing 'some symphony and opera' for Carice.

    A week later he was able to go to Croydon to rehearse and conduct *The Apostles*: there he played through parts of the Symphony again for Carice and Vera. On 11 May he wrote to Keith Prowse, 'I hope to begin to send portions of the full sc. very shortly.' [177]

    He had, seemingly, arrived at a moment when the developing sketch material should make possible a final constructive effort simultaneous with the scoring. That had been the pattern of work for the Second Symphony, which he had scored from his sketches in the first two months of 1911. If there was not nearly so much short score now, his own experience was greater. He wrote out a short score of the First movement exposition, and began to orchestrate it on the new paper. This exposition was to be repeated, to make its impression doubly imposing. The development was not yet clear, but the excitement and close attention of getting to grips with the opening of the new Symphony in something like its final form should provide the impulse for that.

Basil Maine's two-volume biography appeared in May 1933 to a chorus of praise. One of the first readers was Landon Ronald, who wrote to Maine on 9 May: 'Bravo! I saw the great man on Sunday and made him sign my copy with a dedication, which he did very unwillingly! He swears he'll never read it! I don't believe it. But he spoke *so* nicely about you.' [178]

    At the end of May came the trip to Paris to conduct the Violin Concerto for Yehudi Menuhin. Fred Gaisberg was to go with him, and he suggested they fly. The aerodrome, at Croydon, was close to Billy Reed's house, to Carice, and not much farther from Vera Hockman's home Robin Hill. Carice noted:

---

[175] MS in possession Raymond Monk.   [176] MS in possession Denis Mellor.
[177] Keith Prowse archives.   [178] Quoted in *Twang with Our Music*, p. 104.

May 27 1933. Went to Croydon to be there to meet Father—Went with V[era] to Reeds about 3.30. He came about 4—Very heavy shower—Went to Robin Hill for dinner where he stayed. V. took Mrs. Reed home later. Father had had another bad turn but seemed quite all right again.

May 28. Went to church at 8. Lovely quiet am at Robin Hill—[cross-word] puzzle etc—Went to Mrs. Reed for lunch, I drove Father & Vera's car came too for luggage & saw him from aerodrome—wonderful to see it—photographers etc—& great excitement. Lovely weather . . .

Fred Gaisberg recalled the flight:

Elgar enjoyed it with just a tinge of anxiety as he would grip the rails when we struck some air pockets on his first flight. He seemed to feel like a hero and had a daring smile on his face like a pleased boy. I still possess a crossword puzzle he successfully completed on that journey. We put up at the Royal Monceau, Avenue Hoch, and celebrated our first night in Paris with a fine dinner in the bright company of Isabella Valli.[179]

Isabella was Gaisberg's sixteen-year-old niece, in Paris studying the piano with Isador Philipp.

Of the orchestra at the first Concerto rehearsal the following morning, Yehudi Menuhin's father wrote:

It was a study to watch the men. From minute to minute, as the themes were unfolding, their faces turned from superficial smiles of wickedness to admiration and appreciation, and when the glorious final cadenza was reached there broke a delirious applause and a wild ovation. It was a genuine revelation and a genuine enthusiasm.

When the rehearsal was over, Sir Edward immediately took the score, and on the front page where it reads 'Dedicated to Fritz Kreisler', he inscribed, 'and to my dear Yehudi Menuhin.' In [the opening violin entry] he discovered a theme that lends itself to be sung into the name

YE - HU - DI ME·NU·HIN

and he immediately looked it up, inscribed Yehudi's name over the notes, and then exclaimed: 'You see, Yehudi, I thought of you ten years before you were born!'[180]

The Paris days passed in a whirl of activity—rehearsals, luncheons with the Menuhins at their suburban villa, and on the afternoon of 30 May a long drive to visit Delius at Grez-sur-Loing. They had been acquainted for years, but a warm correspondence had grown up between them in the later 1920s, and when Edward proposed the visit the blind and feeble Delius responded warmly. Gaisberg accompanied Edward, and was struck by their first sight of the famous expatriate:

Delius was sitting in the middle of the room, facing the windows, very upright, with his

---

[179] Quoted in *Elgar on Record*, pp. 202–3.
[180] Cutting from an unidentified newspaper (Elgar Birthplace).

hands resting on the arms of a big rolling chair. Illuminated by the afternoon sun, his face looked long and pale and rather immobile. His eyes were closed. Mrs. Delius was sitting beside him expectantly waiting for our arrival.

Genial, resourceful Elgar quickly established a friendly, easy atmosphere and in a few minutes led off into an animated duologue that . . . reminded me somewhat of a boasting contest between two boys. Delius waved his left arm freely; his speech, halting at first, became more fluent as he warmed up to his subject and we forgot his impediment of speech.[181]

Edward himself noted:

Delius takes a keen interest in all that goes on. He is very much amongst us; little escapes his alert intelligence of what is of interest or value in contemporary art—music, painting, literature. And he is still writing.

Just now he has finished a composition for soprano, baritone and orchestra to words by Walt Whitman which he is to call 'Romance' [later entitled *Idyll*]. Delius is a man with a future . . .

I inquired what prospects there were of seeing him in London. There is nothing Delius would like better, and he is anxious to be present when his opera is performed. But the journey, the going from train to steamer and from steamer to train, is, for him, too arduous an undertaking. Having flown from Croydon to Paris, I suggested the pleasant alternative, and pointed out how after motoring to Le Bourget he could reach London by aeroplane in less than two hours.

The prospect attracted him. 'What is flying like?' he asked.

'Well,' I answered, 'to put it poetically, it is not unlike your life and my life. The rising from the ground was a little difficult; you cannot tell exactly how you are going to stand it. When once you have reached the heights it is very different. There is a delightful feeling of elation in sailing through gold and silver clouds. It is, Delius, rather like your music—a little intangible sometimes, but always very beautiful. I should have liked to stay there for ever. The descent is like our old age—peaceful, even serene.'

At the end, Delius sent for a bottle of champagne, and they drank a toast.

The time passed all too quickly, and the moment of parting arrived. We took an affectionate farewell of each other, Delius holding both my hands. I left him in the house surrounded by roses, and I left with a feeling of cheerfulness. To me he seemed like the poet who, seeing the sun again after his pilgrimage, had found complete harmony between will and desire.[182]

The performance of the Concerto in Paris the following evening was in fact the French première—a twenty-two year delay that was a monument to provincial nationalism of that time. Gaisberg wrote:

The concert was a brilliant repetition of the London performance, and the presence of the President and many Ministers of State lifted it up to an international event. The Concerto was received with enthusiasm, but one felt that it had not just made the

---

[181] Quoted in *Elgar on Record*, p. 204.
[182] 'My Visit to Delius', *The Daily Telegraph*, 1 July 1932.

impression that was its due. I fear Elgar's music will not receive a solid appreciation from the Frenchman at least in our generation.[183]

The happiest of the reviews described the Concerto as ' . . . œuvre touchante par sa sincérité et sa purité expressive, répondant à une esthétique fort en honneur au siècle passé.'[184]

Edward flew back to Croydon on 2 June, his seventy-sixth birthday, and in the Birthday Honours announced next day he was promoted to Knight Grand Cross of the Royal Victorian Order. Among the deluge of congratulations was Sir John Reith's: 'We are looking forward to next Spring.'[185] Hannen Swaffer published an 'Open Letter to Sir Edward Elgar':

You are now seventy-six. Well, Verdi went on composing opera when he was eighty. You, although not quite so old, are achieving the miracle of writing, at an advanced age, the first symphony ever written specially for the wireless. . . . When we hear your new symphony played over the wireless, we shall all feel, every one of us, that you have been writing for the ordinary people. Genius expresses itself at its highest when it includes the whole world as its audience.[186]

But the Paris trip had tired Edward, and he could settle to nothing. He wrote to Billy Reed on 28 June: 'It is sickening not to see you & I have no real reason M.S.S.lly to worry you. I hate music!'[187] He could not get to London for the GCVO investiture on 11 July because of a chill. Sir Frederick Ponsonby wrote to ask whether he could call at Buckingham Palace to collect the new order before the King's departure at the end of the month, but that too proved impossible and the decoration had to be sent.

At Marl Bank a hot summer was growing hotter. It was too hot to go to Malvern—even to hear the attractive Harriet Cohen play the Quintet with the London Symphony violinist George Stratton and his String Quartet at the Malvern Festival. But on 27 July, when the Shaws came over to tea, he and Billy Reed played 'a great deal' of the Symphony for them. After they had gone, he put the Symphony away and played opera sketches.

Three days later Basil Maine rang up from Malvern, and Edward asked him to come and hear the Symphony. It was just as in the old days, when the presence of a sympathetic listener could galvanize Edward's piano sketching toward continuous musical development. Maine wrote:

He was full of it. One evening he brought out several sheets of manuscript, sat down at the piano and, playing, reconstructed what was already existing of the work. He warned me that I should get no proper idea of its sound if I did not get past the pianoforte tone, and I recalled with what scorn he used to refer to the mere 'keyboard composer'. He played an extended melody, and, as it mounted, sang it as a string player would phrase it. 'That's for violins,' he said. 'And these harmonies, of course, are for brass. Listen. Now the theme passes to the violas.' Then, straining as if there and then to transmute the

---

[183] Quoted in *Elgar on Record*, p. 204.   [184] *Excelsior*, 7 June 1933.
[185] 3 June 1933 (HWRO 705:445:4261).   [186] *John Bull*, 17 June 1933.
[187] *Letters* p. 319.

music into orchestral tones, he played another theme. 'The fiddles will love that, eh?' he said, with the air of one appraising something he had come upon by chance and now must tend like a watchful gardener for the promise of flowering it held.

. . . What kind of music was he finding in the air which was then (as it is now) so full of harsh conflict? Would the baneful influence have wrought any fundamental change in his symphonic expression? I can only answer that certain sequences of harmony in that unfinished symphony as he played it to me that evening impressed me as being a fresh showing of his mind. He played considerably more than was actually written down and more than has since been published in the sketches.[188]

Of some symphonists, particularly of Mozart, it is true that their themes are in themselves a sign of sudden afflatus, but we have only to look back on the First Symphony, and the Second, to realise that Elgar's inspiration lies chiefly in the transmuting of his themes, and in the regions of the imagination to which they unexpectedly lead us . . . When he played parts of the [Third Symphony] to me on the piano, he relied partly on the sketches (so disjointed and disordered as to be a kind of jigsaw puzzle), partly on memory, partly I imagine on extemporisation. During the improvised (or memorised) passages, it was possible to think that one was beginning to share Elgar's vision, but the experience was so clouded and so fleeting that it could not possibly be re-captured by means of the sketches alone.[189]

The sketches provide a clue here and there as to his approach to the new work, but they give very little idea of the conception as a whole. In the process of bringing forth a new conception every creative artist waits for that final moment of crisis which determines the greatness or the ordinariness of the achievement. If the work is to be great, in that moment there comes the flash which lights up all the previous processes of thought, gives them unity, and orders their final relationship.

It is my conviction that, in this last adventure, Elgar was still waiting for that final moment. The last revealing light had not yet broken upon his mind. Or, if it had, it broke when he lacked the physical strength to set down the signs. This would explain the moody restlessness which came upon him after he had been playing some of the symphony to me, and to which Mr. Reed also refers.[190]

Before I left him on that occasion, the heat-wave had broken and his strength and enthusiasm had returned. Crossword puzzles were among his innumerable interests, and on Sunday morning he was delighted to find that he had successfully cracked one of Torquemada's hardest nuts. In his pleasure he immediately started upon another. I remember thinking that as long as the remote ingenuity of those puzzles could be maintained, the symphony would not fail to make good progress.[191]

On 17 August he went up to London to conduct the Second Symphony at a Queen's Hall Promenade Concert. The critic of *The Evening Standard* was shocked at his appearance:

He conducted from a chair, and, although he was cheerful when I went to see him after the concert . . . his voice was weak and his hands trembled. He had also lost a lot of weight. He was never a fat man, but he had an abnormally broad chest. He seemed to have shrunk to half the size. He was, as always, immaculately dressed.[192]

[188] *The Best of Me* (Hutchinson, 1937) p. 196.
[189] *Maine on Music* (John Westhouse, 1945) p. 33.
[190] *The Best of Me*, p. 198.   [191] *World-Radio*, 1934.   [192] 23 Feb. 1934.

Edward also spoke to Adrian Boult that night about Gaisberg's idea of recording the new Symphony before its première. Next day he wrote to Gaisberg:

I saw Dr. Adrian Boult last night & passed on your suggestion about recording the incipient Sym III.;—he seemed delighted at the idea & we shall hear more of it—whether you will ever hear more of Sym III or E.E. remains to be seen. I delicately put the matter.[193]

A week later Fred Gaisberg arrived for a weekend at Marl Bank. Hot weather had returned in full force, as Gaisberg recorded in his diary.

Saturday, August 26 . . . The extraordinary heat of the day seemed to affect him, so until tea we quietly chatted in the drawing room (also the music room) and library. In this room is a Keith Prowse piano—small grand always open on which Sir E. illustrates when talking on music . . .

Later, about half past five, Sir E. took me for a motor drive to the Malvern Hills. He stopped the car by a pretty little Catholic church, St. Wulstan's, standing on the shaded side of the hill and overlooking the plains of Worcestershire. In the graveyard lies . . . Lady Elgar. Today, with glorious sunshine flooding the plains below, nothing more beautiful could be imagined than the view from this grave on the side of the hill. Sir E. remarked that he had given instructions for his body to be cremated and turned to ashes, but I think it quite certain in his own mind that his remains will lie in Westminster Abbey. To my question if he attends St. Wulstan's, he replied rarely since Lady Elgar's death . . .

We returned on the side of the Malvern Hills overlooking Herefordshire. Sir E. pointed out the school where he taught and later a home where he gave his first violin lesson. During the entire ride he pointed out places associated with his various compositions—also a small shop (now a shoe shop) in the main street of Worcester where his father had his music & piano shop . . .

Sunday, August 27 . . . Tea in the Music Room—Elgar in fine humour. Started by playing me bits of his opera—a bass aria, a love duet, and other bits. He then started on his IIIrd. The opening a great broad burst *animato* gradually resolving into a fine broad melody for strings. This is fine. 2nd movement is slow & tender in true Elgar form. The 3rd movement is an ingenious Scherzo, well designed: a delicate, feathery short section of 32nds contrasted with a moderate, sober section. 4th movement is a spirited tempo with full resources, developed at some length.

The whole work strikes me as youthful and fresh—100% Elgar without a trace of decay. He makes not the smallest attempt to bring in any modernity. It is built on true classic lines and in a purely Elgar mould, as the IVth Brahms is purely Brahms. The work is complete as far as structure & design and scoring is well advanced [*sic*]. In his own mind he is enthusiastically satisfied with it and says it is his best work. He pretends he does not want to complete and surrender his baby. His secretary Miss Clifford says he has not done much recently on the Sym. and seems to prefer to work on his opera . . .

When he wrote to Edward a few days later, Gaisberg was less certain that what he had listened to was a completed Symphony:

---

[193] 18 Aug. 1933 (quoted in *Elgar on Record*, p. 210).

I feel greatly flattered at being able to hear bits of the new work. It was like a door opening and letting rays of sunshine into a dark chamber. There were sufficient jewels of melody to stud a king's crown, but the greatest satisfaction was to know that you are up to your highest form in the work you are now engaged on.[194]

Monday, August 28.   This is the 3rd oppressively hot day of brilliant sunshine—85° in the shade. Marl Bank gardens quite burnt up . . .

Sir E. & I, with Richard (Dick) his valet, took 12 o'c train for London. Lunched on train with good appetite in spite of very hot weather—stretched out for a nap after lunch. Sir E. pointed out all the beauty spots as the train rolled through Worcestershire. I wish I could remember the many jokes & stories he told me on this trip . . . He said he had been travelling on the line for 60 yrs. and everyone knew him. He seemed to have a word & smile for everyone on the line and all responded happily to his sunshine.[195]

This London visit included a recording session at the Kingsway Hall. The programme was the early *Serenade for Strings* and the *Elegy* of 1909. Gaisberg's niece Isabella was there:

Uncle had asked me to come: I think he thought I would help to keep the atmosphere genial and relaxed. He'd also invited Harriet Cohen, who was very gay—very good with him.

I came in by a side door on the right. Elgar was in the right stalls, with Uncle on his left and Harriet Cohen on his right. He was sitting with his arm around Harriet Cohen's chair, so he was facing my way. And his face lit up—really a lovely smile. He rose and spread his arms in a gesture of welcome. I had on a pale green silk dress with little flowers on it. And he said in the most affectionate way: 'Ah! Enter the Spring!'[196]

Carice and Vera were with him next day through a London rehearsal for the Hereford Festival. Then came the Festival itself. Billy Reed remembered:

He took a house for the week, with a convenient garden in front which was daily thronged with people coming to tea. They came in and out, and one hardly knew who they all were; but the Festival spirit prevailed—it was open door and hospitality throughout the week.

Sir Edward, I feel sure, enjoyed it, though he did not move about very much or seem to want to go for walks as before. He sat mostly in that pleasant garden receiving and entertaining friends when he was not busy rehearsing or conducting.[197]

Vera was there, the Shaws close by, and Carice—who was to recall in retrospect: 'He seemed so full of vitality—Yet I felt so sure all last summer that we were doing things for the last time.'[198]

* * *

---

[194] 31 Aug. 1933 (quoted in *Elgar on Record*, p. 216).
[195] Quoted in *Elgar on Record*, pp. 212–14.
[196] Conversation with the writer, 1974.
[197] *Elgar as I Knew Him*, p. 106.
[198] Letter of 16 Mar. 1934 to Clare Stuart-Wortley (Elgar Birthplace).

On 20 September Carice noted 'Father had bad turn but all right again.' Nine days later 'Father bad with sciatica.' Billy Reed learned of it:

Suddenly a letter came from Miss Clifford saying she would be glad if I could possibly come, as he had had some sort of bad attack and was staying in bed for a day or two: better, but complaining of a pain like sciatica. I thought it was probably his old enemy lumbago, and was making plans to go down and see whether I could cheer him up a little or do anything to help when I was startled to hear that he was to undergo some small operation and was going into a nursing-home for it at once.[199]

Edward himself seemed instantly to know the dimensions. He wrote to Reith:

Marl Bank, Worcester.

7th October 1933.

(Private & *confidential*)

My dear Sir John,

I have to go to a nursing home today for a sudden operation (gastric); this upsets all plans. I am extremely sorry to have to tell you that everything is held up for the present. I am not at all sure how things will turn out and have made arrangements that in case the Symphony does not materialise the sums you have paid on account shall be returned. This catastrophe came without the slightest warning as I was in the midst of scoring the work.

Perhaps it will not be necessary to refer publicly to the Symphony in any way at present; we will wait and see what happens to me.

I have written to no-one else on the subject.

Kindest regards and thanks for all you have done.

Believe me to be,
Yours very sincerely,
Edward Elgar[200]

The operation was performed next day by Moore Ede's surgeon, Norman Duggan. The result was terrible, for the intestinal obstruction proved to be inoperable cancer: the operation could provide only temporary relief. Carice conspired with the doctors to keep the truth from him, but his alert intelligence served him too well. Two days later he wrote to Hubert and Agnes Leicester: 'This is a bad affair—however we have had some good times and I love you both.'

The consulting doctor, Arthur Thomson, came down from Birmingham. Between them there were no secrets:

After all his years of worrying over imagined troubles, he displayed magnificent courage in the face of great adversity. Nothing impresses a doctor so much as that. He told me that he had no faith whatever in an afterlife: 'I believe there is nothing but complete oblivion.'

They parted in a final way.

---

[199] *Elgar as I Knew Him*, p. 109.   [200] BBC archives.

A fortnight later Dr Thomson was surprised to be sent for again:

On my arrival, Moore Ede said it was not *he* who had sent for me, but 'those people waiting downstairs.' I went down to confront a group of eminent musicians (mostly I think from the BBC) who were concerned about finishing the Third Symphony, and demanded to know whether there was nothing that could be done to relieve Elgar's pain and yet leave him a clear mind to compose.

'Well,' I said to Moore Ede privately and speculatively, 'I suppose we could cut his spinal cord across.' (In 1933 that operation was in an early experimental stage.)

The BBC men overheard this, and asked whether it should not be done. I explained that no such thing could possibly be contemplated without the patient's approval. They insisted I go up to Elgar and put this very proposal to him.

But Elgar would have none of it, and there, ill as he was, he gently brushed it all aside. He then made one of the noblest remarks I have ever heard from any patient:

'If I can't complete the Third Symphony, somebody will complete it—or write a better one—in fifty or five hundred years. Viewed from the point where I am now, on the brink of eternity, that's a mere moment in time.'[201]

The spectre of posthumous interference, once raised, would not depart. During the night of 19–20 November 1933 he had a collapse that seemed likely to prove fatal. In the morning Carice sent for Billy Reed, who went down instantly.

I arrived at Worcester in the early afternoon, and was driven at once to the nursing-home, where Sir Edward was lying still unconscious, but, as was whispered to me, showing signs of improvement.

I sat watching him for some time, noting his familiar features. They had scarcely changed during his illness: his hair was a little whiter perhaps, and his characteristic nose, with its high bridge, a trifle more prominent; but his colour was good, and he did not look very much thinner than before.

While I was letting my mind run back over the thirty or more years during which he had honoured me with his closest friendship, he suddenly opened his eyes, and, looking intently into my face for a few seconds, uttered my name, as a smile stole over his face. Closing his eyes, he seemed to lapse once more into unconsciousness, and I exchanged some whispered words with his daughter Carice and the nurse, both of whom were also watching.

Presently he stirred again, and, putting out his hand let it rest on mine, and drew me a little nearer. Silence, except for the ticking of the clock and an occasional settling down of the coal in the fireplace. Then it was evident that he was trying very hard to speak; and gradually and at long intervals the words came from him. 'I want you . . . to do something for me . . . the symphony all bits and pieces . . . no one would understand . . . no one . . . no one.' A look of great anguish came over his face as he said this, and his voice died away from exhaustion.

Leaning over him, I said, 'What can I do for you? Try to tell me. I will do anything for you; you know that.'

Again a long silence; but a more peaceful expression came back into his face, and

---

[201] Conversation with the writer, 1976.

before long he drew me down again and said, 'Don't let anyone tinker with it . . . no one could understand . . . no one must tinker with it.'

I assured him that no one would ever tamper with it in any way . . .

A little while later he said in a whisper and with great emotion, 'I think you had better burn it.'

I exchanged glances with his daughter, who was now sitting at the opposite side of the bed; and I saw that she looked, as I am sure I did, a little startled at this suggestion. Then I felt that it was only a suggestion and not really a request; so I leaned over him and said, 'I don't think it is necessary to burn it: it would be awful to do that. But Carice and I will remember that no one is to try to put it together. No one shall ever tinker with it: we promise you that.'

Hearing this, he seemed to grow more peaceful. His strugglings and efforts to speak ceased; he lay there with his eyes open, watching us . . .[202]

There was one doctor who thought he might help. This was a Dr Perkins, a friend of Walford Davies; Davies knew simply that Elgar was ill, and without knowing any details, begged Dr Perkins to see whether he could do anything. Perkins had spent some years in the United States, where he had built an electric machine to diagnose illness. The machine, he said, had curative properties as well. He arrived in Worcester and learnt the nature of the case. None the less he instructed the nurses in using the machine, and they began.

Within ten days Edward showed strong improvement. Ernest Newman came to see him, and Mary Anderson de Navarro, and Basil Maine. Maine had not seen him since the Hereford Festival.

Watching Elgar's chiselled, stone-coloured profile now, I thought I had never seen a like change so swiftly come upon a man. Yet, in spite of pain, the zestful spirit prevailed and it almost seemed that he was cheerful. When the nurse had propped up his head a little he began to tell me of a piece of country he had discovered and had shown to a few privileged friends.[203]

It was the Teme Valley. But, as he confessed a few days later to Bernard Shaw, 'I am low in mind.'[204]

Then he asked Fred Gaisberg to bring a photographer 'who could take photographs of me *in this room*.'[205] On 12 December they came:

We drove straight to the Nursing Home & found Sir E. in good spirits waiting for us. (Carice had been instructed to light a cigarette outside and enter the room smoking—Elgar enjoyed the smoke.) We set up camera—Elgar stage-managed the proceedings—he enjoyed the proceedings immensely—4 shots were made . . .[206]

One showed him with Fred and Carice sitting beside his bed, another with Carice and his secretary Mary Clifford, one holding a record Gaisberg had brought to him, one lying as if asleep . . .

[202] *Elgar as I Knew Him*, pp. 113–15.
[203] *The Best of Me*, p. 200.
[204] 6 Dec. 1933 (*Letters*, p. 321).
[205] 7 Dec. 1933 (quoted in *Elgar on Record*, p. 220).
[206] Gaisberg's diary (quoted in *Elgar on Record*, p. 220).

A gramophone had been set up for him, and Gaisberg had already sent test discs of a new Piano Quintet recording by Harriet Cohen and the Stratton Quartet. These gave great happiness. *The Sunday Times* carried a vignette:

An amusing incident happened the other day in the nursing home where Sir Edward Elgar has been a patient. He was listening to a record of his 'Cockaigne' Overture, and was using a soft needle in order not to disturb the other patients in the nursing home.

A message came from a lady in the next room asking if he would play it louder. Later there was another message, inquiring for the name of the piece as she wished to buy the record for herself.

'Whatever I may be as a composer, I think I'm a pretty good salesman,' was Sir Edward's comment.[207]

Just before Christmas he asked for a pen and music paper—to trace again the 'Angel of the Agony'-like figure from the Third Symphony for Ernest Newman. He dictated (through Mary Clifford) a letter to Newman:

With this I send the opening four bars (introductory) of the slow movement. I am fond enough to believe that the first two bars (with the F$\sharp$ in the bass) open some vast bronze doors into something strangely unfamiliar. I also have added the four final bars of this movement. I think and hope you may like the unresolved *estinto* of the viola solo.[208]

And six days later: 'I send you my stately sorrow; naturally what follows brings hope.'[209]

On 3 January 1934 they brought him home to Marl Bank, where he could be as comfortable as possible in his own room now fitted out as a hospital. Fred Gaisberg suggested a recording session which Edward could supervise by telephone. He made arrangements with the HMV London studios in Abbey Road, with the conductor Lawrance Collingwood and the London Symphony Orchestra, with Gramophone personnel to transport equipment to Marl Bank and set it up in an adjoining bedroom, with the Post Office for connecting lines.

When the day came on 22 January, Gaisberg himself arrived at Marl Bank—to find Edward in a poor state:

He began many conversations, but his voice seemed to fade and he dozed off. I was much concerned for fear he would not rouse himself in time for our recording. He recognised me but could not concentrate. I was told by Mrs. Blake that he had had frequent injections of morphia, but it was principally the toxic poisoning from the wound that was making him so drowsy.[210]

But after lunch and a long nap Edward pulled himself together. The microphone was switched on, and he addressed the Orchestra. Gaisberg wrote:

He made a most marvellous speech of thanks, telling them of his great pleasure in being able to speak to his old friends from his sick bed, and how he was thrilled with the idea of being useful again. He then went so far as to order Collingwood to wear his 'basque' (a beret worn by the Basques in France); this was a standing joke between them. Later on

[207] 17 Dec. 1933.   [208] 22 Dec. 1933.
[209] 28 Dec. 1933.   [210] *Elgar on Record*, p. 227.

he told the violins to draw out a certain *andante* passage—'ten feet long', he insisted. These were real Elgar touches that reminded one of Elgar in his prime.[211]

The repertoire that afternoon was the Triumphal March from *Caractacus*, and if they had time, the Woodland Interlude. The Orchestra played through the March, and the Gramophone man in London asked if there were any comments.

Almost before he had finished speaking, Elgar immediately commenced to criticise the tempi and make suggestions for bringing out the melody more strongly in certain passages: 'I want all the tune you can get out of the clarinets and oboes in that figure. La, la, la,' hummed Sir Edward, 'That is how I want the tune brought out.' After humming a few notes he added, 'That's a nice noise to make. My voice is like a crow's.'

No. 1 record was then made twice. Then they started on the second part of the Triumphal March. This begins with a long, sweeping melody. Elgar bent over to me and said:

'I say, Fred, isn't that a gorgeous melody? Who could have written such a beautiful melody?'

During the entire period of this recording Sir Edward was literally on the *qui vive*, much to the amazement of everybody, including Dr. Moore Ede, who from time to time gave him a little water . . .

At the end he specially asked for the Woodland Interlude from *Caractacus* to be played, and this was done. He made comments and insisted on it being done again—'I want it very much lighter and a slower tempo'—and then after this even once more because he was not quite satisfied. Certainly each time there was a great improvement on the previous performance.

This finished with farewell greetings from everybody, and at 5.15 we all bade Sir Edward good-bye.[212]

But it had been a supreme effort. When Ernest Newman asked about coming down again, Carice replied on 28 January:

This last week [he] has gone downhill a great deal—very drowsy & muddled—except that he is stronger we have gone back very much to where he was the first time you came down. Whether he can turn this corner—I wonder. If it only means a return to pain—well I hope not. How he did the recording is a perfect miracle—but he did it wonderfully & enjoyed it—& it certainly did him no harm. He tries to rouse himself to see people—but he is greatly living in the most extraordinary dreams.

That day Gaisberg came to Marl Bank with the *Caractacus* test pressings, which had been rushed through. Edward had the manuscript of *Mina* under his pillow, and was able to write a dedication to Gaisberg. Gaisberg then had *Mina* recorded by a small orchestra and the test pressing sent down: and Edward even managed to make some coherent criticism of it.

But dimensions of reality were opening more and more to the landscapes of dream. When Newman came again on 5 February, Edward 'made a single short remark about himself which I have never disclosed to anyone and have no

[211] *Music on Record*, p. 250.   [212] *Elgar on Record*, pp. 228–9.

intention of ever disclosing . . .'.[213] Carice had been out of the room at that moment, but she wrote afterwards of Newman's visit to her father: ' . . . he was very confused & they had really no proper talk.'[214]

One of the most constant visitors was Hubert Leicester's son Philip, as devoted a Roman Catholic as his parents. He was deeply shocked to hear Edward say he wished to be cremated and his ashes scattered at the confluence of Severn and Teme: 'Edward is off his head with morphia,' Philip told his wife.[215] In the end Carice persuaded her father that burial must be beside Alice at Little Malvern. The headstone designed by Troyte Griffith had space left below Alice's inscription. In a shaky hand Edward wrote:

> Add only this:
> In memory also of the above-named Edward Elgar
> Died—.   [216]

At the urgent request of Carice and the Leicesters, but against the advice of Dr Moore Ede, old Fr. Reginald Gibb came from St. George's Church to give the Catholic last rites. Afterwards Fr. Gibb was quoted in newspapers as having obtained a confession of faith. The obituary of his own career, written by the Jesuits a few years later, was circumspect:

> It ought to be recorded, for obvious reasons to all acquainted with the life of Sir Edward Elgar, that it was Fr. Gibb who went to the famous musician when he was dying and ministered to him such sacraments as he could receive.[217]

At last, on the morning of 23 February 1934, somewhere between 7.30 and 8 o'clock, came a peaceful end. 'He just slept away,' wrote Carice, who was with him, and none could tell the moment of his going.

So closed the experience opened nearly seventy-seven years earlier on a June morning across the river at Broadheath. Those years had seen change accelerate as never before in human history. His response had been to seek the illumination of time remembered. For all those of his generation and the future who would feel the insight of retrospection, he had made of that evanescence his music.

Explicit

[213] *The Sunday Times*, 13 Feb. 1955.
[214] 8 Feb. 1934 to Lady Stuart of Wortley (Elgar Birthplace).
[215] Mrs Philip Leicester in conversation with the writer, 1979.
[216] MS in the keeping of Percy Young.
[217] 'Fr. Reginald Gibb, 6 February 1865–29 November 1942', in *Our Dead: Memoirs of the English Jesuits who died between June 1909 and December 1945* (Manresa Press, 1947–8) p. 241.

Index

(Elgar's musical works are alphabeticized under the composer's name. Other composers' works are indexed only by the composer's name.)